cone shells

a synopsis of the living conidae

jerry g. walls

Frontis: *Conus barthelemyi.* From a painting by Alex Kerstitch.

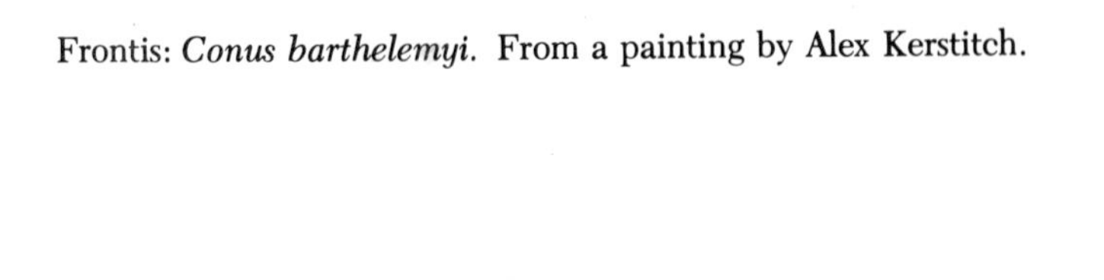

All photos by the author unless otherwise credited.

"I've never yet succeeded in producing a plot without at least six major howlers. Fortunately, nine readers out of ten get mixed up too, so it doesn't matter. The tenth writes me a letter, and I promise to make the correction in the second edition, but I never do. After all, my books are only meant for fun; it's not like a work of scholarship."—From *Gaudy Night,* by Dorothy L. Sayers, with permission of Harper & Row.

ISBN 0-87666-628-4

Distributed in the U.S.A. by T.F.H. Publications, Inc., 211 West Sylvania Avenue, P.O. Box 27, Neptune City, N.J. 07753; in England by T.F.H. (Gt. Britain) Ltd., 13 Nutley Lane, Reigate, Surrey; in Canada to the book store and library trade by Clarke, Irwin & Company, Clarwin House, 791 St. Clair Avenue West, Toronto 10, Ontario; in Canada to the pet trade by Rolf C. Hagen Ltd., 3225 Sartelon Street, Montreal 382, Quebec; in Southeast Asia by Y.W. Ong, 9 Lorong 36 Geylang Singapore 14; in Australia and the south Pacific by Pet Imports Pty. Ltd., P.O. Box 149, Brookvale 2100, N.S.W., Australia. Published by T.F.H. Publications Inc. Ltd., The British Crown Colony of Hong Kong. Printed in Singapore.

Contents

Acknowledgments

A book such as this is obviously an impossible task without the aid of many friends and acquaintances who provided specimens, photos, information, or just consolation. If I have forgotten anyone from the following list, forgive me.

Dealers are the heart of shell collecting, and they are also the heart of this book. The following dealers have loaned anything from a single specimen to thousands of dollars' worth of rarities; George Gundaker and Marvin Hume, Boardwalk Rock and Shell Shop, New Jersey; Phillip W. Clover, California; Fernando Dayrit, Philippines; Dick Kurz, Wisconsin; Wendell Lim, JW Shell Place, New Jersey; Bob and Dottie Janowsky, Mal de Mer Ent., New York; M. Marescot, Tahiti; Bob Morrison, Morrison Galleries, Florida; and Ms. I. Welch and M.E. Young, The Shell Cabinet, Virginia.

Although I have not been able to utilize scientific collections as much as I would have liked, the following people have loaned important type material or photographs in their care: W.O. Cernohorsky, Auckland Institute, New Zealand; R.N. Kilburn, Natal Museum, South Africa; Dr. T. Okutani, Japan; Ed Petuch, Florida, for an especially helpful group of material; Dr. H. Rehder, U.S.N.M., Washington; and Dr. T. Shikama, Japan.

Many collectors have loaned important material from their collection or have provided useful photos: Mr. and Mrs. S. Armington, Jr., Florida; B. Bailey, Solomons; J. Bateman, N.J.; Mr. and Mrs. W. Burgess, N.J.; P. Calhoun, Carolines; A.J. da Motta, Thailand, for a very important group of slides; M. Fainzilber, Tanzania; T.C. Good, Okinawa and Australia; N. Gray, Malawi; A. Guest, Bermuda; R. Guest, Florida; R. Jensen, Delaware; R. Jonklaas, Sri Lanka; A. Kerstitch, Arizona, for much important Panamic and Asian material; T.C. Lan, Taiwan; Dr. H.G. Lee, Florida; M. Meyer, South Africa, for specimens and literature; B. Parkinson, Papua New Guinea; R.W. Pitman, West Virginia; Mr. and Mrs. E. Skinner, North Carolina; and T. Yocius, Florida. Mr. Zengo Nagohoma, Okinawa, provided photos of *C. dusaveli.*

Finally, Dr. R. Tucker Abbott and the staff of the Delaware Museum of Natural History have been unfailingly helpful in allowing me free usage of their collections and library, without which this book could not have been completed. Dr. Herbert R. Axelrod started me on this project two years ago, and my wife, Maleta, has tolerated it for that long. . . to both my thanks.

Introduction

Of all the popular groups of snails, the cones present the greatest taxonomic problems. This statement can be made with some assurance, as most of the other popular collector's shells, such as cowries, volutes, olives, and murexes, have had some type of recent treatment usable to collectors and professionals alike. The cones consist of about 300 species plus several dozen distinctive subspecies and forms, as well as many dubious names. The last attempt to cover in a systematic fashion all the living cones was in Tryon's *Manual of Conchology*, volume 6, 1883-1884. Not only is this now an uncommon work available to few collectors, but in addition it is taxonomically unsatisfactory by modern standards.

The collecting of cones soon turns the serious hobbyist into an amateur taxonomist, whether he likes it or not. Dealers are often unable to correctly identify their offerings, although most try to do so. There are very complicated nomenclatural problems which every collector must face in order to even write a simple label. Some species come in a confusing variety of patterns, shapes, and names, and the collector must think carefully before purchasing from lists. Even then he is likely to receive a very different shell from what he thought he ordered.

This book is intended as an introduction to this complex family of snails. It is aimed primarily at the serious advanced collector who really cares about the identification of his shells. It has been possible to, for the first time, illustrate most of the valid (in my opinion) species, subspecies, and varieties of living cones in color and present detailed descriptions and synonymies of the valid taxa. Throughout this book I have tried to use the most modern and acceptable names, although this is an almost impossible task. The nomenclature of cones is so greatly confused that it will not be possible for many years to be certain of the proper name for *any* cone, nor to be exactly sure just what constitutes a valid species in the family. I am a lumper by nature and this book naturally reflects my feelings on some of the taxonomic problems in the Conidae. If I could find no characters to distinguish a supposed species or variety, I have placed it in the synonymy of another species.

Wherever possible I have obtained the original description and illustration of each name proposed and evaluated it by the rest of the literature and the specimens available to me. In a great number of cases this has revealed that the name now being used by most people for a species is not the correct one. I have taken it as my responsibility to change names wherever in my opinion the evidence leaves me no other choice. That this course results in name changes which many or most collectors and dealers will not receive with great joy is of course obvious. But someone must eventually work out the proper name for each species and correct previous errors or else the group will never be stable. The information is here—it is up to the individual whether he will profit from it or not.

A cone is of course a living animal before it becomes a dead shell in someone's collection. The next few chapters present a brief introduction to the living cone and its world, with special emphasis on features which might be useful in classification. The selected bibliography, although heavily slanted toward taxonomy, provides several titles which should serve as a starting point for the ecologically-minded.

It must be emphasized that this is a book for collectors by a collector who is perhaps a bit more interested in taxonomy than most. An attempt at a revision of the cones is many years in the future and will have to involve the fossils as well, with yet more name changes in the offing. The collector must not despair, as eventually cones will be as well known as many other groups of collectable shells. Collecting cones can only be fully enjoyed if all collectors and dealers understand each other—hopefully this book is a start in the right direction.

New Taxa List

The following three new species and one new subspecies are described and illustrated on the indicated pages (**bold** numbers indicate color photos).*

*Because of the time lag before publication of this book, these four names have been validated in *The Pariah*, No. 2, 30 Jan. 1978.

POSTERIOR

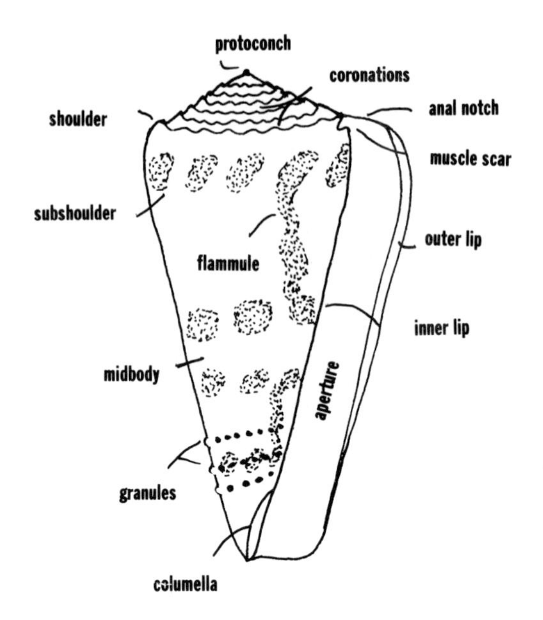

ANTERIOR

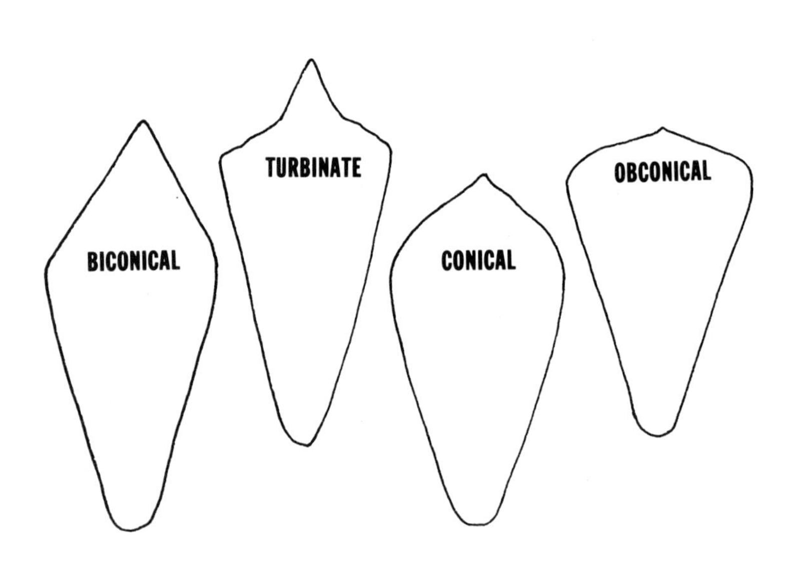

Shell Structure

In life the shell is covered by a thin layer of tissue known as the *periostracum*. In many species it is a slightly yellowish, transparent cuticle which allows the pattern of the shell to show through. This thin periostracum is typical of the tented cones with their complex patterns. At the other extreme are such species as *Conus virgo*, *C. terebra*, and *C. princeps* with a very thick, brown, opaque and often tufted covering. In most cones the periostracum is a moderately opaque or translucent brownish covering which at least partially obscures the pattern and color of the shell unless the cone is immersed in water. Commonly the periostracum is thickened or tufted on the spire and in spiral rows along the body whorl, often following the sculpture of the shell itself. Thickness of the periostracum seems to vary somewhat with age and ecology and is *not* a good taxonomic character.

For descriptive purposes the shell can be divided into two portions, the spire and the body whorl. The body whorl contains most of the body of the animal in life, including the foot and proboscis. The animal emerges through a long, usually narrow, opening, the aperture or mouth of the shell. The free edge of the mouth is the outer or labial lip, the surface of the body whorl opposing is the inner or columellar lip. The upper edge of the outer lip is separated from the body of the shell by a U-shaped *anal notch* or sinus through which the animal excretes wastes. There is usually a corresponding shallow notch on the columellar lip for the attachment of the anal muscles.

The open or anterior (basal) end of the body whorl is of course the true anterior end of the animal, as this is where the proboscis exits. Since cone shells are usually oriented with the anterior end toward the observer, this end is called the *base*. The tip of the spire is the *apex* for the same reason of conventional orientation. Ventral is the view in which the mouth and columellar lip can be seen; if the mouth is away from the observer, the dorsal surface is in view.

The outer lip of most cones is moderately strong in adults but very weak and fragile in juveniles. Often the edge of the outer lip is brightly colored compared to the rest of the shell, as it represents new growth through depositing of calcium salts and proteins. In a few species there is a thickened ridge at the middle of the outer lip, continuing into the mouth. There are no teeth or folds within the outer lip of cones (other than the continuation of spiral ridges), a point which distinguishes them from the often similar dove shells, Columbellidae. Unless broken, the outer lip is straight or slightly curved with no projections except for the anal notch posteriorly.

The inner lip is smooth over most of its surface and often a different color than the rest of the shell, commonly lacking a color pattern. At its anterior end is a very smooth, shining fold called the *columella.* The columella is relatively short in cones, less than a third the length of the inner lip. The relative shape and size of the columella is of importance in some groups of snails, but it is relatively unknown and of minor use in cone identification, although it can sometimes be used as a character to confirm the identity of confusing species.

The body whorl is said to be sculptured if it has any grooves, ridges, or other surface irregularities. Sculpture falls into several groups: 1) raised ridges, folds, or ribs; 2) depressed grooves between ridges; 3) raised granules or pustules. Sculpture is spiral if it coils around the shell, axial if it follows the long or vertical axis of the shell. Wide, heavy ridges are often called ribs, while thin, strongly raised ridges are sometimes also called ribs. If spiral and axial sculpture overlap to produce small regular squares, the surface is said to be *cancellate.* Very fine or only slightly raised ridges are often called threads or wrinkles. Scratches are fine surface irregularities which are just visible under the lens. Broad and shallow depressions are called folds or plicae. The base of virtually all cones has at least 3-4 spiral ridges.

The spire is very variable in the cones, both among species and within a single species. In few species does the spire exceed 30% of the total length of the shell, while in many species it is nearly flat or even slightly concave. Biconical shells have the spire tall relative to the body whorl; conical is used for most shells where the spire is neither exceptionally tall nor exceptionally low; obconical is used for very low or flat spires. If the spire whorls (a whorl is one complete turn of the growing shell) merge continually into one another, the spire has flat sides; if the whorls are sharply set off from each other, they are stepped or turreted. There are many intermediate conditions, such as concave sides, convex sides, etc. If the spire is flat with the earlier (apical) whorls suddenly raised into a strong cone, the shape can be called turbinate, as in *C. generalis.*

In addition to the types of sculpture found on the body whorl, the spire whorls are often characterized by distinctly raised small knobs regularly spaced along the edge of the whorls. These *nodules* may be present on all the whorls, only the early or posterior whorls, or only the later or anterior whorls. They may be present on the shoulder, which is usually the broadest part of the shell where the spire whorls merge into the body whorl. The spire whorls and shoulder may be rounded to distinctly sharp-edged (carinate).

The flattened and usually erect rather triangular knobs on the lower spire whorls and shoulder of some cones are often called coronations.

Although coronations are often useful in identification, they are also variable within some species with age or growth. The presence of the less obvious nodules on the early whorls or their absence is of more importance in determining relationship than the presence or absence of coronations.

The shell of the newly hatched cone is retained in many species into the adult stage as a set of delicate, usually shiny and translucent whorls at the very apex of the spire. These young whorls, the protoconch, are probably useful in classification, as their relative size, number, and texture vary from species group to species group. However, they are often lost well before the shell matures, whether through being physically knocked off or through gradual erosion due to sand or water chemistry. The absence of the protoconch in an adult shell should not be considered a flaw, but rather the normal effect of aging, like the loss of baby teeth in mammals.

In many cones the size of the protoconch relative to the body whorl is the only way to distinguish juveniles (with large protoconchs) from adults (with relatively small protoconchs). Any shell with a large protoconch and fewer than five whorls should be considered juvenile unless otherwise proved. Many synonyms have been caused by inability to distinguish juveniles and adults.

Color patterns in cones are very important but variable features in their identification. Colors are usually the product of depositing meta-

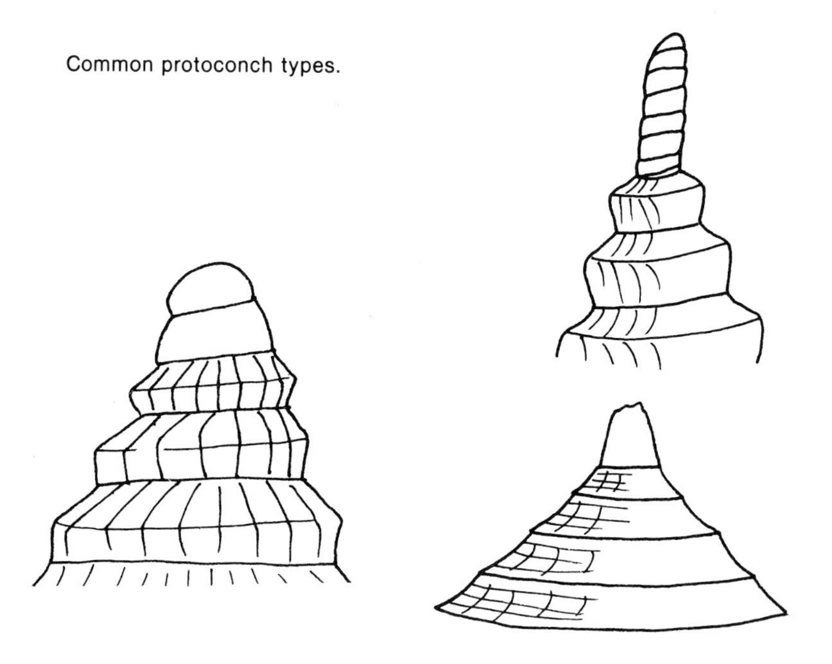

Common protoconch types.

bolic waste products, although they are not fully understood. The patterns are often more stable than the colors themselves and remain on the shell longer after death. There is no reason to go into the various patterns here, as they are amply illustrated in the following pages. Patterns can be spiral or axial or various combinations. The color of the spire is often different from that on the body whorl. In a great many species there is a white or pale spiral band near the middle of the body whorl. It is not uncommon for shells to change their pattern slightly after each period of growth, the changes corresponding with the axial growth marks or varices where the resting shell became slightly thickened at the edge of the outer lip. New deposits of color when the shell begins growing again are at first thinner and often brighter than on the old growth. Such growth marks are natural in many species of cones and not the same as axial breaks or scars.

The mouth is often colored in cones, usually some tone of bluish or pinkish. Although aperture coloration tends to fade rapidly, it is often an important differential character for closely related species. A few species have more than one color within the mouth or may even be banded.

Common operculum types.

The operculum of most cones is a small oblong brownish sliver with rough edges and the nucleus indistinct. Commonly there are one or two ridges on the back for attachment. In a few species there are heavy serrations on one margin; in a few other species the operculum is round; while in a very few species it appears to be completely absent. Although the presence of the operculum is sometimes taken as proof the cone was collected alive, the opercula of many species are virtually identical and cannot be identified from species to species.

Soft Anatomy

Cones are highly evolved predaceous snails restricted to a marine environment. Their fossil record is a long one and, although they have evolved into several distinctive types, the evolution has apparently been along the line of shell and radula modifications. What is known of cone soft anatomy (admittedly little) indicates that most species fall easily within a generalized model, the most obvious variation being in the rostrum, the verge, and the color patterns. Since most cone species are of relatively large size, the foot and proboscis—the major part of the 'body' of the snail—are easily seen in specimens prowling around the aquarium or reef.

A long muscular *foot* is the most obvious feature of the anatomy. The undersurface is broad and rather smooth, seldom conspicuously pimpled or striated as in some other snail groups. Posteriorly the foot comes to an obtuse point and bears the generally small operculum on its upper surface. The anterior or head end of the foot is truncated or broadly rounded without greatly projecting corners. Coloration of the foot varies considerably with the different species, perhaps in response to food and ecology as well as genetics. Often the sole of the foot is paler than the sides, whitish or pale tan. The rest of the foot may vary from white to deep velvety black or bright scarlet, uniform or covered with light or dark spots or blotches; in some cases the pigment is concentrated in a band near the edge of the foot (marginal), and in one case around the operculum. The cones with deep red or deep black bodies are especially attractive in the aquarium.

Almost as conspicuous as the foot is the *siphon*, an extendable rolled tube (continuous with the body covering or mantle) that directs water to and from the gill. It is on the cone's left side, is open ventrally, and protrudes from the sand or coral rubble even when the cone is buried. In most species the siphon is very active and may bear bright color bands at the tip, often red or white and contrasted with the rest of the animal's color. Black is also a common siphonal color; in many cases, however, the siphon is the same color as the foot.

In an actively extended snail the *eye stalks* are conspicuous, one on each side of the body. Generally the eye stalks are rather thick at the base, widened near the middle into the eye lobe itself, then slender and bluntly pointed at the tip. The black eye lobes are on the outside of the stalks. Possibly some of the deeper water cones are blind, though this is apparently not recorded.

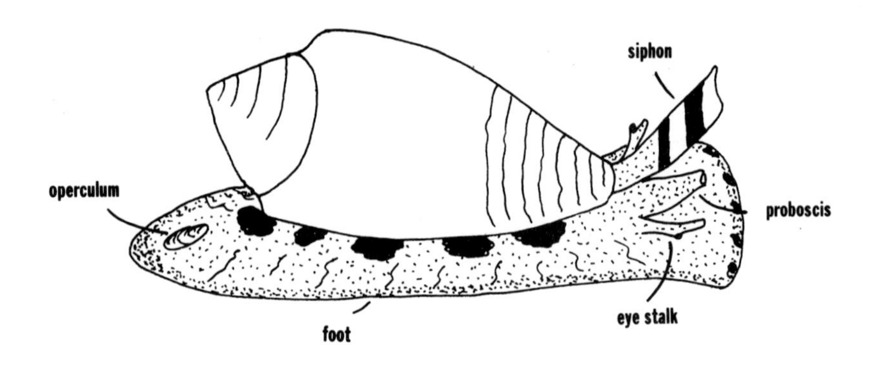
siphon
operculum
proboscis
eye stalk
foot

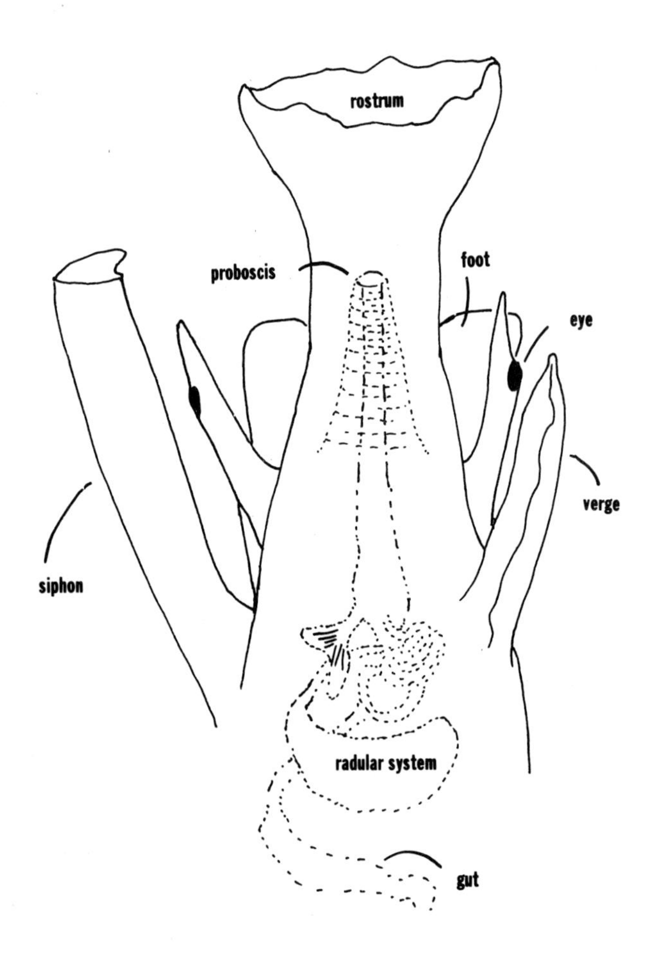
rostrum
proboscis
foot
eye
verge
siphon
radular system
gut

Projecting above the foot and anterior to the eye stalks is the *proboscis*. Actually what is normally visible is the *rostrum* or proboscis sheath which covers the very flexible and extendable proboscis when it is not in use. The rostrum may be small and have a simple margin so it is not very distinct from the true proboscis, or it may be large, hood-like, and have a margin with long fringes or knob-like protuberances. The proboscis is wrinkled and ends in the mouth opening. From it the anterior part of the pharynx can be protruded, bearing the radula tooth. The mouth opening and pharynx are both very elastic, especially in fish-eating cones, and capable of great distension to swallow large prey.

Behind the right eye stalk, more-or-less to the right of the midline, is the conspicuous *verge* or penis of male cones. This contains the sperm duct which transmits sperm to the female during mating. Little is recorded about variation in shape and size of the verge in cones, but in many other groups of snails the verge and the internal organs of the reproductive system are very important in classification. At least two types of verge are obvious in cones. One is the *tubular* type with a simple rounded opening at the extreme tip; the tube may be short or long, narrow or thick, and in life apparently may be straight or curved. This seems to be the type found in most cones. The other type is much more complicated, with a large or small rounded lobe before the tip, then a constriction to a much narrower and often curved '*hook*.' There may even be some ornamentation along the margin of this type of verge. The hook-type seems to occur in textile cones and some deep-water species. The female genital opening is a simple pore, sometimes with a small raised collar. The anus is also on the right side but more posteriorly.

At the base of the siphon, just within the opening of the shell, is the gill. In cones it consists of a single rather feathery structure lying close to the body. There is some evidence that the number and size of gill filaments vary with different species and different ecologies.

If a preserved cone is slit down the midline just behind the eye stalks, the *radular apparatus* and associated organs will be exposed. These will be covered in more detail later, and it is sufficient to state here that the radula is in a pouch opening off the pharynx and is technically a part of the digestive system. More posteriorly the digestive tract continues into the maze of small tubes and glands usually hidden deep within the whorls of the shell and seldom noticed by collectors except as 'junk' to be washed out. Also in the upper whorls of the shell are the ovaries or testes and the numerous small ducts and tubes of the reproductive system. These have not been well studied and are of little interest to collectors although they have definite possibilities in classification. The same is true of the nervous system, which is usually noticed only as a ring of large ganglia around the pharynx.

The most conspicuous muscle of the cone is the long, very strong *columellar muscle* which is attached to the internal axis of the shell and to both ends of the foot. When it contracts it pulls in the foot by folding it in

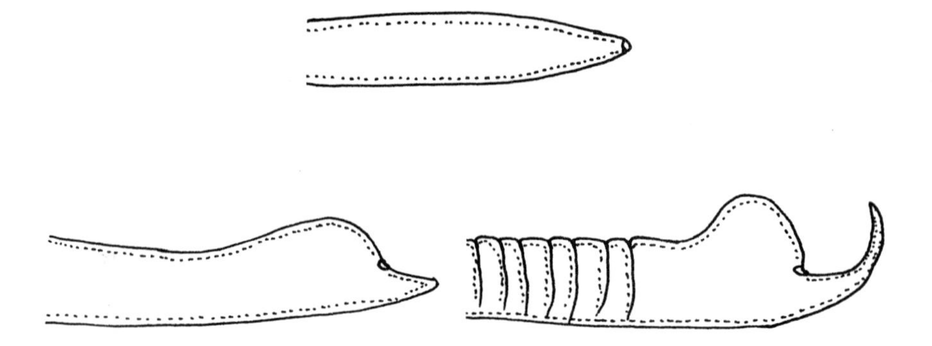

Three verge types.

the middle. In this way the posterior surface of the foot ends up closing the opening of the shell, which explains why the operculum is attached to the trailing edge of the foot.

Cones are difficult to preserve in a relaxed or extended condition even when treated with such chemicals as Nembutol or menthol crystals, but unless the cone is relaxed it is almost impossible to study the important organs deep within the shell. Persons having access to living cones should try to preserve a few specimens of each local species in a relaxed state by gradually adding small amounts of menthol crystals or Nembutol solution to the water containing the snail. Propylene phenoxetol is said to work well with cones, although it is relatively expensive. Its use is explained by Dr. J. Rosewater in *Indo-Pacific Mollusca,* 1(6):394. A good collection of properly relaxed cones preserved in alcohol or buffered formalin would be invaluable in tracing the relationships of the cones.

Radula and Feeding

The cones belong to a group of advanced snails called the Conacea or Toxoglossa ('poison tongue'). Also belonging to this group are the turrids (in a broad sense) and the terebras, both large families of small to moderate sized shells, all strictly marine. These three families share an unusual development of the radula into a device for spearing prey with a poison-bearing barb.

In most snails the radula consists of numerous rows of chitinous teeth of variable shape set in a well developed pouch or buccal mass. The tooth rows are drawn across the food in a scraping motion, each movement abrading a small amount of both food and teeth. In most snails the radula is a highly useful character in the classification of both genera and species, as it tends to be very constant in each category and varies little from individual to individual.

In the cones and their relatives the radula has lost the appearance of rows of teeth. Instead a casual inspections seems to show that the radular sac, which is inserted into the gut just behind the pharynx at the base of the proboscis sheath, contains usually two bundles of isolated, spear-like teeth, none projecting into the gut. Actually, evidence shows that in the primitive toxoglossans (some groups of turrids) the teeth are still set in rows and can be exposed within the proboscis for rasping. Gradually the teeth became more spear-like and useless for rasping and at the same time became contained within a pouch which did not permit them full access to the proboscis. Correlated with this step was the development of an efficient venom system.

Diagram of the radular system.

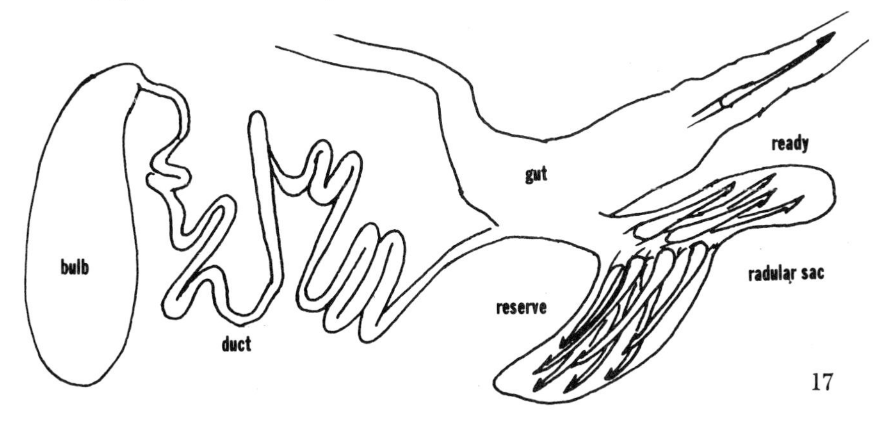

The Conidae have a radular system consisting of three basic parts: a large transverse poison bulb, the long and highly convoluted poison duct, and the radular sac containing the radular teeth. The poison bulb does not actually produce the venom, but apparently serves some function in helping force the poison through the poison duct, where it is produced, into the radular teeth. The bulb varies little in the cones and is the most conspicuous part of the system. It is generally kidney-shaped, with the venom duct coming off the more tapered right end.

The poison duct is not a true duct either, as it does not simply conduct a liquid from one point to the other. It is granular in nature and actually secretes the venom. The system by which venom moves from the duct into the teeth is not understood.

The radular sac varies in shape and size from species to species, but consists of two distinct arms, the reserve arm which apparently generates the teeth, and the ready arm which contains only a few fully developed teeth. The sac is attached to the gut near the base of the proboscis sheath and about opposite the genital opening. In some cones the sac is almost sessile on the gut, but in others it has a rather long connection.

Within the larger reserve arm of the sac can be found teeth in various stages of development, each apparently connected to the sac by a delicate sheath which stains easily. The various tooth types will be mentioned later, but all are basically similar in structure. Each tooth is a hollow tube with at least one opening near the tip, usually under a larger barb. The sides of the tube overlap slightly and have minutely serrated or irregular margins in many species. The teeth are separate but are held in groups by a heavy mucus.

By a method not fully understood, the individual, fully formed teeth are transferred from the ready arm (and the reserve arm?) of the radular sac into the gut and then to the proboscis. Somewhere along this trip they are filled with venom. In at least the fish-eating cones there is usually at least one detached, filled tooth present in the proboscis, ready for action.

The venom itself is a yellowish or whitish, rather granular and viscous liquid with an alkaline pH of 7.8-8.1. It is relatively immune to both heating and freezing, and it is still highly toxic after such treatments. The chemistry of the venom is poorly understood and very complex, so we will

Parts of a vermivorous radular tooth.

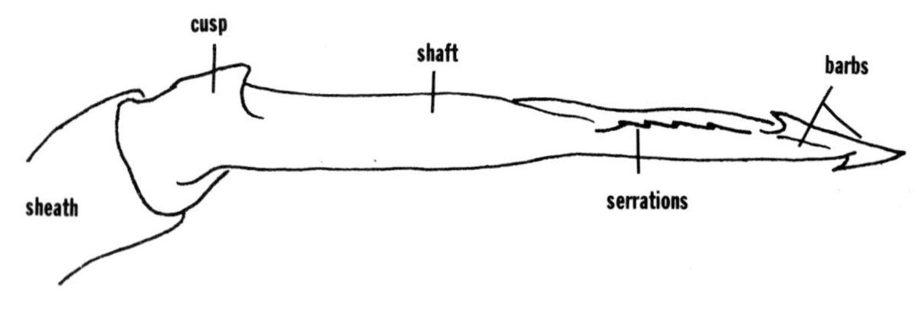

not enter into it here. Suffice it to say that it apparently affects the nervous system of the prey animal, causing muscular incoordination and eventual respiratory failure. The same type of symptoms appear in all prey, whether worm or man.

Conus radular teeth vary greatly in shape and size from species group to species group, the shape apparently reflecting adaptations for specific prey animals. Radular teeth can be very small and hard to find in such species as *Conus distans* and *C. orbignyi*, or 10-20mm long in the larger piscivores such as *Conus striatus* and *C. geographus*. Only a half dozen teeth may be present in some species or individuals, with 50-60 teeth present in other species. There is little information on individual variation in radular tooth size, shape, and number.

Radular teeth can be grouped into three principal types: 1) piscivorus; 2) molluscivorus; and 3) vermivorus. Piscivorus teeth are usually elongated, have long, curved, terminal barbs, and have a long shaft that is smooth and lacks barbs near the base. This is of course the type associated with species that prey upon fishes—and usually the most dangerous type to man. Molluscivorus teeth are characterized by being heavily serrated over most of the length of the shaft, moderately slender, and with a strong barb near the base at the end of the serrations. Cones with such teeth usually prey upon snails, including other cones. Some are very dangerous to man, but not so much as the piscivores. The vermivores are probably the most common type of feeding adaptation and tooth pattern in the cones. These teeth are usually relatively short, broad, and not strongly barbed at the tip; serrations are often strong, and there may be one or more large barbs near the middle of the shaft. The most prominent feature is a large basal cusp or barb situated on the base of the shaft and projecting forward. It has been suggested that this cusp serves to keep the tooth in the proboscis after it is injected into the worm. In this way the cone would gain extra leverage to pull worms from their burrows after being stung. Few vermivores are really dangerous to man, although the larger species can cause painfully swollen stings.

The food habits of only a few species of cones are known with any certainty, although enough are known to be able to generalize. The cones with narrow apertures and without a tented or textile pattern or spiral rows of brown dots or dashes are generally vermivores; such species include *C. arenatus*, *C. figulinus*, *C. flavidus*, *C. ebraeus*, *C. eburneus*, *C. generalis*, *C. musicus*, *C. planorbis*, *C. quercinus*, *C. virgo*, and similar species. Molluscivores have relatively narrow apertures also, but commonly have a tented pattern. Familiar examples are *C. ammiralis*, *C. marmoreus*, *C. aulicus*, *C. pennaceus* and allies, and *C. textile* and allies. The piscivores generally have wider apertures and various patterns, though often spiral rows of narrow brown dashes are present. The thin-shelled *C. geographus*, *C. tulipa*, and *C. obscurus* are all piscivores, as are *C. magus* and its relatives and *C. monachus* and relatives. *Conus striatus* is the most familiar and possibly most deadly species, but there are cer-

Conus purpurascens eating a sculpin. Photo by Dr. J. Nybakken.

LUCIDUS

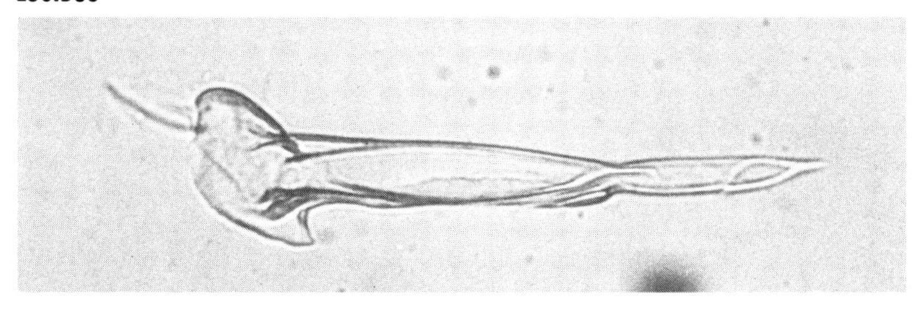

GLADIATOR

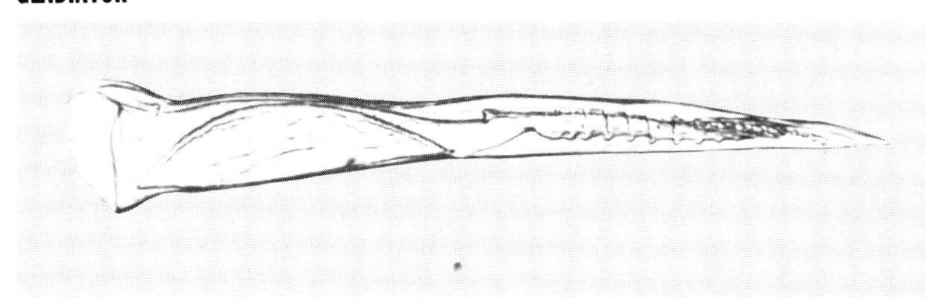

XIMENES

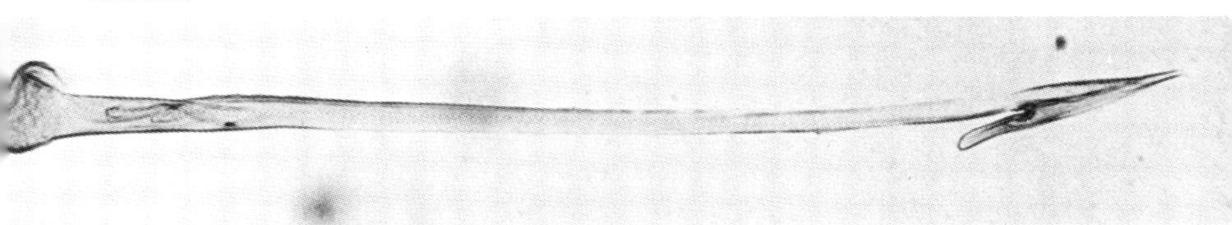

Radular teeth of three common Panamic cones. Photos by Dr. J. Nybakken.

tainly others just about as deadly. All piscivores should be considered potentially dangerous to man, regardless of size.

There is some evidence that points to the possibility that some cones may be more generalized feeders, preying on both worms and mollusks and perhaps even scavenging. Examples of such species are *C. spurius* and *C. algoensis*, neither of which has well known habits, however.

Radular teeth are held in great esteem by some taxonomists of recent years, and there has been a growing tendency to stress radular characters in classification. I feel that there is currently insufficient evidence to indicate that teeth are more trustworthy than color pattern and sculpture for indicating relationships in the cones. The teeth are such reduced and simplified structures, in any event, that they present few characters for analysis; small variations for slightly different prey species might be reflected by more obvious visual differences in the teeth. Radular teeth should be used with as much caution in identification as color pattern and sculpture, although they are sometimes useful in helping separate closely allied species which occur in different habitats and presumably have different food requirements.

Dangerous Cones

It can be truthfully said that all cones are dangerous, in much the same way that all spiders are dangerous—all have venom glands and teeth at least theoretically capable of causing pain and swelling. Fortunately most cones are small worm-eaters, so their threat to man is negligible. At worst their bites are as bad as a bee sting. The real problem species are the larger piscivores and the few molluscivores known to be harmful.

There is no longer any doubt that some of the larger fish-eaters are capable of killing adult humans. Such species as *C. geographus* and *C. striatus* are large, have wide apertures, long and highly extensible proboscises able to reach any point of their shell, and are relatively active. These two species cannot be safely picked up with bare hands under any conditions. In both species the radular teeth are often over 10mm long and easily capable of piercing thin cloth. Bites from these two species are not uncommonly fatal and at best can cause temporary paralysis of the limbs and prolonged difficulty in breathing.

Of lesser danger are the smaller piscivores which also have wide apertures. These include *C. tulipa*, *C. obscurus*, *C. magus* and allies, *C. monachus* and allies, *C. catus*, and *C. stercusmuscarum*. The relatively small size of these species and the shorter radular teeth (seldom over 5mm long) make them less of a threat than in the previous species, but they are all to be considered very dangerous. The radular teeth of *C. geographus* and allies curiously are more similar to those of molluscivores than piscivores, but the venom is just as deadly.

There is still considerable controvery over whether or not the venom of the molluscivores (mostly the tented cones) is dangerous to man. There seems to be little doubt that *Conus textile*, a very common large species in the Indo-Pacific, has caused human fatalities. Yet in laboratory tests of the venom of this species, *C. aulicus*, *C. marmoreus*, and similar species, not even mice have been killed. Fortunately these are not very aggressive species, and their characteristic brownish or reddish tented patterns are easily seen through the thin periostracum in nature. All tented cones should be considered potentially dangerous and be handled with caution, particularly specimens over 50mm long. *Conus gloriamaris* and allies are to be especially considered dangerous because of their large size.

Some of the larger vermivores, such as *C. leopardus*, *C. litteratus*, *C. quercinus*, *C. imperialis*, and *C. betulinus*, are to be considered theoretically dangerous because of their large size, although their venom is not really effective against mammals. Reactions to their stings are generally just localized swelling and inflammation. Even smaller vermivores such as *C. pulicarius* can be as painful as a bee, however.

The venom of the piscivores (and also the larger molluscivores) acts much the same way as the venom of the elapid snakes (cobras, coral snakes, sea snakes). There is only a very short time available for treatment before the bite's effects become generalized, often as little as five minutes. Thus treatment procedures are ineffective. A bite causes immediate and at first localized pain, soon followed by numbness of the bitten member. The venom gradually stops the nerve action in all the extremities, causing either numbness or tingling sensations, swelling, redness, dizziness, vomiting, and sometimes acute pain. Soon the diaphragm is affected and respiratory difficulties begin as the lungs are inactivated and unable to contract and expand as necessary for breath. This is followed by extreme dizziness, inability to focus the eyes, and difficulty in swallowing and talking. Death is usually the result of respiratory failure—the diaphragm is completely paralyzed and the lungs inactivated as a result.

As mentioned earlier, it is probably not realistic to talk about the treatment of truly dangerous cone bites. If the bite is from a small or only mildly toxic cone, the symptoms will pass in from a few hours to several days. If a dangerous bite occurs away from immediate (within a few minutes) hospitalization, probably little can be done except to keep the victim inactive and perhaps trying to apply suction at the site of the bite. The obviously most reasonable course is to avoid being bitten by any cones, whether identified or not. This means never handling cones with the bare hands or thin gloves, not carrying specimens in plastic bags or other piercable containers, not passing the bare hands through sand known to contain large cones, and being careful when turning over coral and other debris in tropical waters. Cones should be considered as dangerous as poisonous snakes and the same precautions taken in collecting them and handling them in captivity.

Finally, it should be mentioned that the thick periostracum of many cones prevents their identification to species without turning them over to check the pattern showing on the inner lip. Many books suggest that it is safe to handle cones by the shoulder, but this is certainly not so with species such as *C. geographus* and *C. striatus*, which can reach all parts of their shell. Always wear gloves of tough or thick material when handling any large cone.

Natural History

As will be readily noticed, there is very little natural history information in this book. I am a collector who lives a thousand miles from the nearest concentrations of cones (in the Caribbean), and I have not taken the time to abstract the ecological information present in the literature. Actually, such information is rather sparse and almost nonexistent except for papers by Kohn and a few other workers. Occasional notes in *Hawaiian Shell News* indicate such items as habitat and food preference for some species. Perhaps at some later date we will be able to compile the published information into a usable synopsis for the family.

However, certain generalizations are possible. Cones are certainly tropical animals, with only a small number of species extending outside the tropics, and these showing a high endemicity in such cool areas as South Africa, southern Australia, and southern Japan. Most typical cones are inhabitants of reefs in shallow or moderately shallow water, occurring on or under debris and in crevices between the coral formations. In such a habitat the vermivores may be very numerous and common, with a good diversity of species. Molluscivores are generally second in abundance, with piscivores less common or actually rare.

Few species live exposed to sunlight, seemingly being crepuscular or nocturnal, with exceptions. Some of the supposedly rare species such as *Conus barthelemyi* have been found buried in muddy sand to depths of over a meter, requiring specialized collecting techniques.

Several smaller species in all oceans are adapted to an environment of tidal sands, being buried most of the time like olives. Others can be found in distinctly muddy bottoms, presumably having adaptations of the respiratory system that allow them to survive silt accumulations on the shell and gill.

Many cones have become adapted for deep-water environments, most of them probably living on deep hard bottoms much as they would in shallow water, but others seem to be able to survive on relatively barren substrates with little hard material. Curiously, the deep-water cones can be either very thin or very thick in shell, probably depending more on which species group they are derived from than on specific adaptations to the depths.

Food specialization seems the order of things, at least among the more diverse vermivores. By having the different species adapted to diets more-or-less specialized for certain genera or families of worms, there is

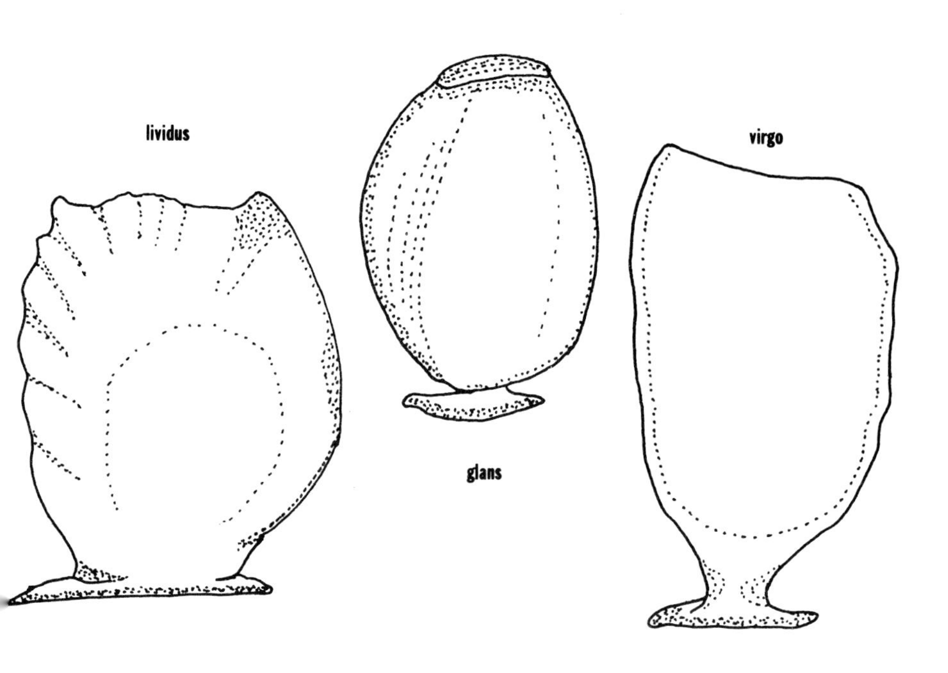

Common egg capsule types.

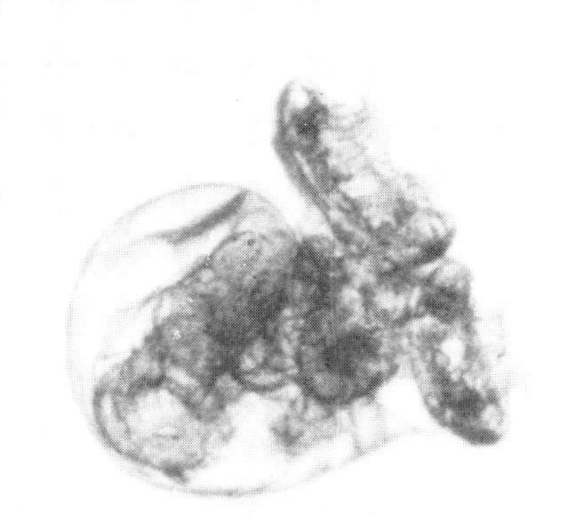

Living veliger larva of *Conus purpurascens.* Photo by Dr. J. Nybakken.

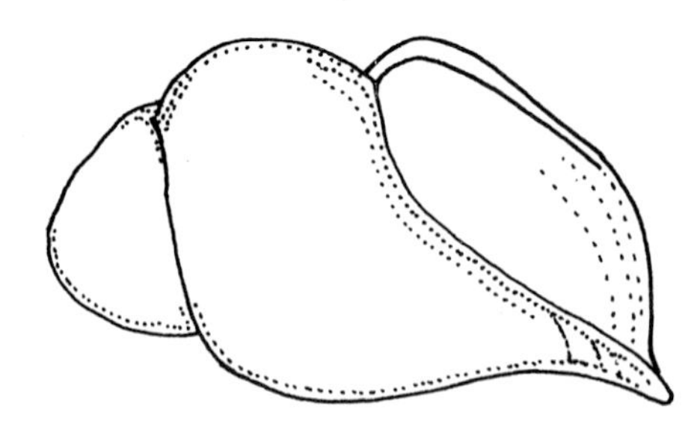

Typical veliconcha.

less competition and more species can live in the same habitat. Molluscivores seem to be less specialized, probably because there are fewer species that can possibly inhabit an area and fewer prey species in any particular area. Piscivores tend to be very wide-ranging, with little selectivity about prey; they are, as might be expected, much less common in the average habitat than the other types.

Sexes in cones are apparently always separate, though this has not been checked by dissection of many species. In some species there may be some sexual dimorphism in size or shape of shell and perhaps in color and habitat preference, but it is minor as far as known. In at least some species mating occurs only during a few short weeks and in an environment not inhabited by the species the rest of the year. Thus such species as *Conus concolor* are taken in the Solomons in shallow water of estuaries only for a few weeks before disappearing into deep water where they are seldom collected. Other species apparently breed all year in the same habitat. Information on period of breeding and habitat is very much needed.

Females usually lay several large flask-shaped or pouch-shaped egg capsules in clusters under coral slabs and rocks, with the total number of eggs varying from a few dozen to over 10,000 or more. In species with large numbers of small eggs, the larvae hatch into a veliger that is pelagic for several days or several weeks before settling and metamorphosing into a more adult-looking shell. Such species are usually widely distributed and show little variation from area to area as the veligers serve for distributing the species over the range.

Species with egg capsules containing fewer and larger eggs usually hatch directly into what is called a veliconcha, which looks like a metamorphosed shell of the more typical species. Such species have a very limited ability to expand the range as the larval stage is effectively absent and there is no free-swimming stage. This type of development results in small colonies of the species and a stronger tendency for speciation or at least variation from colony to colony. Many of the cones of Australia, South Africa, and West Africa seem to have this type of development—and corresponding variability. It also occurs in such species as *C. glans*. Obviously more work is needed on the reproduction of the cones.

It might be mentioned that older references on cone biology are very undependable sources of information because the species identifications are often incorrect. Much more work is needed on all aspects of cone natural history.

The Cone Collection

This book is for identification purposes and is already overly thick, so little can be said here about the cone collection itself. For a good review of the subject see "Cowries and the Collector" in *Cowries*, by Taylor and Walls (T.F.H. S-101). Suffice it to say, however, that the same dangers and problems exist in the purchasing and storage of cones as in cowries, although the fragile lips and spires of cones make them even more delicate.

In cone shells the lip and tip of the spire must be protected at all costs. The smallest nick in the lip usually decreases the value of any cone. Personally, I would much prefer a shell with a moderately bad growth flaw to one with the spire broken or the lip filed. (Filing means that the broken lip is filed or sandpapered until the chips disappear.) Remember, though, that many species of cones are rough shells, subject to reef breaks and encrusting growths, so they are never quite as perfect as most cowries. In some species the protoconch whorls are almost always lost before adulthood and the spire always appears incomplete. Such natural defects should not be held against a shell, providing it applies to the great majority of specimens sold. Extremely perfect specimens of all species probably exist but would fetch a heavy premium over nearly-perfect specimens.

The HSN-ISGS (Hawaiian Shell News-International Shell Grading System) standards are now in common use among the better dealers and help clearly define just what quality shell a collector is getting for his money. The grades are:

GEM — A perfect specimen with an unblemished spire, unbroken spines and lip without chips, fully adult and normally colored—a shell without a visible flaw. Well cleaned inside and out, with original natural gloss and color. Bivalves must have both valves, properly matched and unbroken. Cone lips may have minor natural roughness.

FINE — An adult shell with only minor flaws and with not more than one shallow growth mark. Must have original color and gloss. A cone lip may have one small chip, *Murex* two minor frond breaks. No repairs, such as filed lips, mended knobs or filled worm holes.

GOOD — A reasonably acceptable shell with a few defects such as growth marks, broken spines, worn spire or lip chips. Specimens may be subadult, but still must faithfully display all the characteristics of the species.

COMMERCIAL — May be obviously dead or beach collected, with chipped lips, faded color, growth faults or imperfect spires. Shells of Commercial grade are not acceptable for mail order retailing and should not be offered as collectors' specimens.

Grades may be abbreviated as G, F, Gd, and C. (Hawaiian Malacological Society.)

Although this system of grading is good in theory, it often does not apply well to cones. Many species, for example, are seldom available in other than commercial to good condition unless one is willing and able to pay gigantic premiums and wait 4-5 years for a better specimen. In several species gem condition shells are practically unknown, as there are simply very few perfect adult specimens—the common species that inhabit shallow reefs are good examples of this. In such cases dealers often "up" the grade of specimens, assuming that most collectors will understand that a gem *Conus balteatus*, for example, will certainly not be a true gem according to the HSN-ISGS.

A simplified and more practical grading system has been suggested by Bob Morrison of the Morrison Galleries and has much to recommend it. The Morrison system is as follows:

"**GEM** — A mature shell with no noticeable flaws.

FINE — A minor flaw or flaws which do not detract significantly from appearance; shell may be slightly sub-adult but this will be noted.

GOOD — A noticeable flaw or flaws, or some wear or fading.

FAIR — A shell with significant damage, and/or substantial wear."

This system is intended to be used with + (plus) signs to indicate shells grading a bit higher than the average for the grade (such as fine+) and also assumes that the dealer will mention any obvious flaws as well as good points for the specimen. Cones are better described by the Morrison system than the HSN-ISGS, which seems to be more desirable for cowries and similar shells.

A few words about cleaning cones are necessary. To be identifiable, cones must have the periostracum removed. I have found from experience that the simplest and most efficient way of removing periostracum is by simply immersing the cone in a bath of straight (undiluted) Clorox® or similar chlorine bleach. This will not hurt the shell under normal circumstances, rapidly removes the thickest periostracum, dissolves gut and foot remnants left in the shell, and softens algal and coralline crud. The only disadvantage is that the very highly polished cones (such as *C. bullatus*) must be removed and rinsed often so the gloss is not damaged (use petroleum jelly to protect the aperture if desired). Also, after working with Clorox® for awhile your fingers will become very slippery, so forceps must be used; use rubber gloves if necessary. Be sure, however, that the chlorine bleach has no softeners, brighteners, or additives added—brighteners can turn a white shell pale blue in seconds. After the periostracum is dissolved the shell is dried, coated *very* lightly with fine mineral or baby oil, then rubbed with a lintless cloth. Such specimens are attractive and useful for identification work—probably most dealers use this method or a similar one on their shells. Lye and other harsh chemicals are unnecessary, inefficient, and dangerous.

Every collector who is sincerely interested in the hobby should subscribe to one of the very few collector publications. Only two truly international publications come to mind, although some of the larger society publications are also very good. The best, in my opinion, is *Hawaiian Shell News* (Hawaiian Malacological Society, P.O. Box 10391, Honolulu, Hawaii, USA 96816), a thin monthly that is essential to the hobbyist who wants to keep up with his hobby. The articles vary from excellent to poor, but they do give a broad range of coverage and opinions, with emphasis on Indo-Pacific collectable groups. *La Conchiglia* (Via Tomacelli, 146 IV P, 00186 Rome, Italy) has an English edition that is heavily illustrated in color. However, the quality of the writing is often very low or has badly suffered in the translation, and the taxonomy follows Continental standards—many genera, many many species, and excessive interest in aberrations and freaks. Many new taxa in various groups were published in this journal over the last couple of years, fortunately none in cones, but its taxonomic standards are very low. It is certainly worth having, however. Numerous scientific journals on molluscs also exist, but these are seldom of interest to the collector; the few new species in the major collectable groups described each year are often mentioned and figured in *Hawaiian Shell News* a few months after the description appears.

Cone Identification

Cones are still the most difficult to identify group of larger shells popular with collectors. Even with a good collection of literature and a good reference collection, many specimens from even well-collected areas will defy all attempts at identification. Before attempting to identify an unknown cone, one must be familiar with the species of the world in a general way and the ocean from which the specimen comes in a more specialized fashion. One must also remember that almost all cones are very variable—an attempt at too tightly constructed a species definition will result in misidentifications and every third specimen seeming to be a new species.

First check the structure of the early whorls, using a small specimen with the least erosion if at all possible. The presence or absence of distinct nodules on the early whorls is usually very useful information, as is the type of sculpture on the spire whorls. Remember, though, that both these spire characters are sometimes variable both within different populations of the species and individually due to erosion. The same applies to coronations on the shoulder.

Obvious sculpture on the body whorl is often important, but the presence or absence of granules is highly variable and seldom a specific character. Virtually all cones have at least a few low spiral ridges at the base of the shell, but few have such obvious modifications as axial plicae, grooving only at base and shoulder and not between, distinctly cancellate sculpture, heavy spiral ridges to the shoulder, etc.

Shape and texture of a shell are often of importance, but both vary considerably in some species, probably due to ecology. The same is true of such secondary characters as periostracum and operculum, which will seldom be mentioned in the description. Spire height and shape are very variable in some species, and various proportions used in some papers have been found to vary considerably from area to area; thus meristics are not stressed in the descriptions, and I doubt the practicality of exact measurements and percentages in describing cones.

It seems certain that the most useful character for identifying cones is the color pattern. Although variable—often extremely variable—in most species, it is easy to use and constant enough in at least some features to permit recognition of a species when combined with characters of shape and sculpture. Some species have color patterns that are more constant than the shape or sculpture and more distinctive at the specific level. For this reason I have tried to avoid the use of fossil names in this book and have not checked the fossil literature when trying to determine the proper name for species. If the Recent cones were to be judged strictly on shape and sculpture, the number would be reduced by more than half of the conservative number recognized here, and I think this is simply not realistic. Color patterns, when used with caution, are very good specific characters, and they are probably better than most shape or sculpture characters.

Although many subgenera or genera have been devised for the cones, only the single genus *Conus*, with no subgenera, is recognized here. There certainly are very characteristic species groups present in *Conus*, but these commonly have very nebulous boundaries that shade into related species groups *ad infinitum*. Subgenera commonly can only be applied to adult shells, with juveniles falling into different subgenera. The classic examples of the frustrations of subgenera are the several cases where typical individuals of a species fall into one subgenus, with granulose or striate individuals from the same population falling into a different subgenus. Before subgenera can be meaningful in this genus, there must be an integration of the fossil species into the Recent species, a task which will take many years.

Because I have not been successful in applying subgeneric names to the cones and have not been entirely successful in defining species groups, the recognized species of cones are listed in alphabetical order, with the photos also arranged alphabetically. This has the obvious advantage of allowing easy reference once a name is known, but the equally obvious disadvantage of not aiding identification of unknown shells. However, this is not as great a disadvantage as might be thought, and any collector familiar with at least 50-75 of the more common species should have little problem determining most unknown specimens. Using these common species as a basis for comparison with the text and illustrations, most species should be rather rapidly identifiable with a little practice. The best bet is to scan all the plates as often as possible to gain an idea of the form and variation of the species and to read carefully through the *Comparison* section in each species discussion. Not all shells can presently be identified; many individual variations can be placed in the correct species only if series of specimens are available, and there are certainly several possibly undescribed species on the market under various names.

Each species discussion begins with a skeleton synonymy listing as a rule only original descriptions. Names are given in strict chronological order; note that I have dated all Kiener species at 1845, the first appearance of the plates, instead of 1846-1850 as is often done. Question marks before a reference indicate doubt that the name is correctly placed. Most synonymy is discussed in the text unless it is well established and familiar to most workers. The synonymy is based on the literature, examination of original descriptions and figures, examination of specimens of almost all the taxa recognized as valid, and the application of a conservative taxonomic sense to the information available. Subspecies are considered to be morphologically distinguishable populations separated by a geographic barrier of some type from other distinguishable populations, but either showing intergradation where the two types of populations come in contact or showing a morphological degree of separation that is, in my opinion, less than that of a full species. Varieties or forms are distinguishable individuals or minor populations that occur within the range of otherwise typical shells and usually intergrade fully into the normal form of the species. It should be mentioned that the current taxonomic rules do not allow formal recognition of varieties or forms, but since the names are available and often familiar to collectors I see no harm in using them in this book.

Where the synonymy comes as a shock to collectors, there is usually a mention of why the various names are placed in synonymy or why a name was changed. I have not attempted to indicate new synonymy, as the literature is too inexact and confused to be sure who first placed a name within the synonymy of another. Be sure to also check the list of doubtful names for further comments on some familiar names which could not be satisfactorily placed.

The species descriptions follow a standardized format which should allow easy reference to any particular character; the structure of the shell is described first, then the pattern, and lastly the aperture. Often the terms used in the description are not exact, and I have avoided like a plague almost all mention of proportions and other mathematical niceties—cones are simply too variable at the specific level to attempt to corset them in either exact terminology or statistical magic. Radular characters have generally been ignored as: 1) I doubt their utility in species identification; and 2) only a small percentage of the species have been figured, and these are often doubtfully identified in older literature.

Sizes given are approximate minimum and maximum adult sizes, but the current record size for many species is probably beyond the maximum size given here. The new edition of Wagner and Abbott's *Standard Catalog of Shells* (American Malacologists, Greenville, Delaware, 1977) should be consulted by size-conscious collectors.

The comparisons are usually restricted to either closely related species or to unrelated species that are commonly confused. The *Variation* section includes variation seen by me and variation reported in the literature. Subspecies and varieties are diagnosed in the variation section; it should be remembered that occasionally variants fall well outside the range of the species description given earlier in the discussion. Juveniles are mentioned where I have seen them and there are interesting differences from adults. Unfortunately juveniles of cones are very poorly known and seldom described or figured.

Under *Distribution* I have given a simple statement of range, not an exact indication of the areas occupied by the species. The literature on the distribution of cones treats in detail only a few localities and is replete

Conus moluccensis from New Britain. Notice the constricted shoulder. Photo by R. Lubbock.

A typical *Conus marmoreus* from the Coral Sea. Photo by T.E. Thompson.

with a great number of misidentifications, so it cannot be safely used in determining the exact range of a species. The same is also true of the habitat of cones, which is barely touched on here by general statements such as "found in deep water" or "common under rocks in shallow water." Although inane, these statements are about as trustworthy as any published for most species, especially when it is remembered that cones occupy different habitats at different localities and sometimes during different parts of the year.

Collectors could very profitably aid in an understanding of cones by reporting their observations on habitat, life colors, and species lists for some of the less well-known islands and mainland collecting areas. Certainly the living cone is even more poorly understood than the dead specimens.

Since I am a collector, I just could not resist a few jabs at the practices of some dealers and collectors as well as comments on some of the species commonly misidentified. Under *Notes* will be found these comments as well as other statements on general condition of the species and sometimes the market. Remember that the market fluctuates constantly and that while material is rarely available from some areas today, it might be commonly sold from there next year.

Finally, a warning. This book represents my own opinions and conclusions as to the relationships and proper nomenclature of the living cones. There are certainly errors here, but on the whole I think that it is a significant step in the understanding of the most complicated family of popular shells. Thoughts and comments would be appreciated from informed readers, although I reserve the right to be stubborn where necessary. Perhaps if more people, amateur and professional alike, would publish their findings instead of supplying an abundance of cryptic letters and reviews based on their unpublished notes and observations, the cones would not be half as confused as they presently are. During the preparation of this book I have run into several individuals who either refused to help in any way or would only supply the information that they had examined such-and-such a type specimen long ago and would get around to publishing on it later. It must be remembered that, in the sciences, unless something is published it simply does not exist; what an individual knows and keeps to himself does not help anyone.

Family Conidae

Small to large toxoglossan snails with the aperture elongated, the lips nearly parallel and open anteriorly; anal notch present at shoulder level, usually deep and distinct; columella small, less than one-third length of inner lip; sculpture variable, but at least 3-4 low oblique ridges anteriorly; body whorl not narrower posteriorly than anteriorly; spire very low to very high, adults with more than 5 whorls excluding protoconch; teeth absent from inner lip or outer lip; operculum oval or elongated, usually less than one-third length of aperture; periostracum present in living shells; radular teeth reduced, usually barbed, held in a two-armed radular sac; proboscis with or without apical fringes; siphon large, flexible.

A single living genus is recognized, with the characters of the family. The following subgenera and genera have been described but are not used at this time. References are available from the author.

Genus CONUS Linnaeus, 1758

AFRICONUS Petuch, 1975: *cuneolus* Reeve, OD (original designation) = *balteus* Wood
ASPRELLA Schaufuss, 1869: *sulcatus* Hwass, M & SD (monotypy and subsequent designation)
CHELYCONUS Moerch, 1852: *testudinarius* Hwass, SD = *ermineus* Born
CLEOBULA Iredale, 1930: *figulinus* Linnaeus, OD
CONASPRELLA Thiele, 1929: *cancellatus* Hwass, M
CONUS Linnaeus, 1758: *marmoreus* Linnaeus, SD
CORONAXIS Swainson, 1840: *bandanus* Hwass, M = *marmoreus* Linnaeus
CUCULLITES Herrmannsen, 1847: *marmoreus* Linnaeus, SD
CUCULLUS Roeding, 1798: *marmoreus* Linnaeus, SD
CYLINDER Montfort, 1810: *textile* Linnaeus, OD
CYLINDRELLA Swainson, 1840: *asper* Lamarck, M = *sulcatus* Hwass
CYLINDRUS Deshayes, 1824: emendation *Cylinder*
CYLINDRUS Fischer, 1883: emendation *Cylinder*
CYLINDRUS Herrmannsen, 1847: *marmoreus* Linnaeus, SD
DARIOCONUS Iredale, 1930: *omaria* Hwass, OD = *pennaceus* Born
DAUCICONUS Cotton, 1945: *daucus* Hwass, OD
DENDROCONUS Swainson, 1840: *betulinus* Linnaeus, SD
DYRASPIS Iredale, 1949: *pontificalis* Lamarck, OD = *dorreensis* Peron
EMBRIKENA Iredale, 1937: *pergrandis* Iredale, M & OD
ENDEMOCONUS Iredale, 1931: *howelli* Iredale, OD
FLORACONUS Iredale, 1930: *anemone* Lamarck, OD
GASTRIDIUM Modeer, 1793: *geographus* Linnaeus, SD
HERMES Montfort, 1810: *nussatella* Linnaeus, OD

Anterior end of *Conus textile* from the Great Barrier Reef, showing the eye and siphon. Photo by K. Gillett.

Conus dalli and *Conus fergusoni*, two uncommon large cones of Mexico. Photo by A. Kerstitch.

KERMASPRELLA Powell, 1958: *raoulensis* Powell, OD
KURODACONUS Shikama & Habe, 1968: *stupa* Kuroda, OD
LAUTOCONUS Monterosato, 1923: *mediterraneus* Hwass, M = *ventricosus* Gmelin
LEPORICONUS Iredale, 1930: *glans* Hwass, OD
LEPTOCONUS Swainson, 1840: *amadis* Gmelin, SD
LITHOCONUS Moerch, 1852: *millepunctatus* Lamarck, SD = *leopardus* (Roeding)
MAMICONUS Cotton & Godfrey, 1932: *superstes* Hedley, OD
NUBECULA Herrmannsen, 1846: *geographus* Linnaeus, SD
PARVICONUS Cotton & Godfrey, 1923: *rutilus* Menke, OD
PHASMOCONUS Moerch, 1852: *radiatus* Gmelin, SD
PIONOCONUS Moerch, 1852: *magus* Linnaeus, SD
PROFUNDICONUS Kuroda, 1956: *profundorum* Kuroda, OD = *smirna* Bartsch & Rehder
PUNCTICULIS Swainson, 1840: *arenatus* Hwass, M
PUNCTICULUS Coates, 1925: emendation *Puncticulis*
PYRUCONUS Olsson, 1967: *patricius* Hinds, OD
REGICONUS Iredale, 1930: *auratus* Hwass, OD
RHIZOCONUS Moerch, 1852: *miles* Linnaeus, SD
RHOMBUS Montfort, 1810: *imperialis* Linnaeus, OD. Non *Rhombus* Walbaum, 1792
ROLLUS Montfort, 1810: *geographus* Linnaeus, OD
STEPHANOCONUS Moerch, 1852: *nebulosus* Hwass, SD = *regius* Gmelin
STRIOCONUS Thiele, 1929: *striatus* Linnaeus, M
TARANTECONUS Azuma, 1972: *chiangi* Azuma, M
TEXTILIA Swainson, 1840: *bullatus* Linnaeus, SD
THELICONUS Swainson, 1840: *nussatella* Linnaeus, SD
TULIPARIA Swainson, 1840: *tulipa* Linnaeus, M. Non *Tuliparia* Blainville, 1830
TURRICONUS Shikama & Habe, 1968: *nakayasui* Shikama & Habe, OD = *excelsus* Sowerby iii
UTRICULUS Schumacher, 1817: *geographus* Linnaeus, M
VIRGICONUS Cotton, 1945: *virgo* Linnaeus, OD
VIRROCONUS Iredale, 1930: *ebraeus* Linnaeus, OD
XIMENICONUS Emerson & Old, 1962: *ximenes* Gray, OD

The following names are based on fossil species:

CONOLITHES Swainson, 1840: *antidiluvianus* Hwass, M
CONOLITHUS Herrmannsen, 1847: *antidiluvianus* Hwass, OD
CONOSPIRA Cossmann, 1896: emendation *Conospirus*
CONOSPIRUS Gregorio, 1890: *antidiluvianus* Hwass, OD
CONTRACONUS Olsson & Harbison, 1953: *adversarius tryoni* Heilprin, OD
HEMICONUS Cossmann, 1889: *stromboides* Lamarck, OD

Species Accounts

ABBAS Hwass, in Bruguiere, 1792

1792. *Conus abbas* Hwass, in Bruguiere. *Cone*, in *Ency. Method., Hist. Nat. des Vers*, 1: 750 (East Indies). Neotype selected and figured by Kohn, 1968.
1857-58. *Conus corbula* Sowerby ii. *Thesaurus Conchyliorum*, 3(*Conus*): 42, pl. 23 (209), fig. 573 (Locality unknown).
1882. *Conus textile* var. *euetrios* Sowerby iii. *Proc. Zool. Soc. (London)*, 1882: 120, pl. 5, fig. 6 (Locality not stated).
1884. *Cylindrus Gillëi* Jousseaume. *Bull. Soc. Zool. France*, 9: 188, pl. 4, figs. 1-2 (Locality unknown). Not seen.
1937. *Conus abbas* var. *grisea* Dautzenberg. *Mem. Mus. Roy. d'Hist. Nat. Belg.*, 2(18): 7 (Locality not stated). Non *C. griseus* Kiener, 1845.

Description.—Moderately light in weight, with a high gloss; elongate-cylindrical, rather ventricose; body whorl smooth except for a few weak spiral ridges at the base and nearly obsolete spiral threads to about midbody; shoulder rounded, the margin slightly set off from rest of whorl; spire moderate, the sides usually distinctly concave; apex sharp, the early whorls slightly stepped and weakly nodulose; later spire whorls with traces of weak spiral ridges and indistinct axial wrinkles. Body whorl white to bluish white, covered with a heavy and usually regular meshwork of fine chestnut reticulations; this pattern is broken into three areas (midbody, above and below midbody) by two spiral bands of small to moderately large irregular chestnut spots or blotches containing vertical blackish lines (textile lines); scattered large white tents present, sometimes abundant; spire whitish with numerous axial brown lines often connected into blotches by brown areas; apex pink, the areas between nodules often with a brown dot; base white with a few brown axial lines. Aperture slightly widened anteriorly, the margin straight; lip thin, sharp; mouth bluish white to deep blue, outer lip with external pattern showing through. Columella long and narrow, indistinctly set-off posteriorly. Length 35-77mm.

Comparison.—Undoubtedly this species is easily confused with *C. textile* and *C. canonicus*. From *C. canonicus* it usually differs in the broader, more obese form and the distinctly concave-sided spire as well as the blue or blue-white mouth as contrasted to the usually pinkish mouth of *C. canonicus*. The tenting of *C. abbas* is usually very small and regular with the textile blotches seldom connected into broad spiral bands. *C. textile* is relatively larger and has a more open pattern of larger tents, often with the textile blotches forming broad spiral bands above and below midbody; many races of *C. textile* have a strong tendency to develop an axially lineate pattern seldom found in *C. abbas;* virtually all *C. textile* have spires with nearly straight or even convex sides, and the mouth is usually pale bluish white. *C. dalli* is broader, has a lower and straight-sided spire, often three broad bands of blotches on the body whorl, and a deep violet mouth. *C. aureus* is readily distinguishable by the strong and

The sometimes similar *Conus vittatus* (above) and *Conus orion* (below) from Mexico. Photos by A. Kerstitch.

Conus brunneus (above) and *Conus bartschi* (below) differ in animal color as well as shell pattern. Photos by A. Kerstitch.

long axial black lines and the distinctly striate body whorl.

Variation.—The one feature of this species which is relatively constant is the snakeskin-like pattern of small tents. Many specimens, especially from Sri-Lanka, are deep bluish overall. Width at the shoulder varies from relatively slender to very obese, with the spire strongly concave to nearly straight. Occasional specimens have a tendency to produce broad axial dark stripes on the body whorl, obscuring the smaller tenting; such specimens vaguely resemble *C. legatus* in pattern. Rarely *C. abbas* develops nearly straight axial lines in the midbody area; a few specimens with nearly complete spiral bands of textile blotches have been seen. The extremes of variation of this species merge with some forms of *C. textile* and *C. canonicus;* this species should be identified from series only.

Distribution.—*C. abbas* is relatively uncommon throughout the Indian Ocean, including the Sri Lanka area south to Natal, South Africa, and the southern Indian Ocean islands. It also occurs in the Thailand area and even east to the Philippines. Whether specimens from the Marquesas are actually this species or simply aberrant *C. canonicus* is uncertain. Note that *C. canonicus* is generally rare in the Indian Ocean, occurring often as single individuals among dozens of *C. abbas;* similarly, *C. abbas* is uncommon in the western Pacific, usually occurring with large batches of *C. canonicus.*

C. abbas from the Marquesas, probably just a local pattern of *C. canonicus,* sold as 'panniculus'.

Synonymy.—On the basis of original descriptions and illustrations it is usually impossible to tell whether a name applies to *C. abbas, C. canonicus,* or some forms of *C. textile;* the synonyms above probably apply here. *C. textile* var. *euetrios* seems to be a fairly distinct form of *C. abbas* from the southern Indian Ocean islands with the textile blotches tending to form nearly complete spiral bands and the midbody area tending to contain nearly straight axial lines instead of very fine tents.

Notes.—Specimens of *C. abbas* from outside the Indian Ocean area and Philippines should be viewed with caution as they are probably variants of *C. canonicus.* It would probably be better to consider *C. canonicus* a subspecies of *C. abbas*, but this step should be deferred until fuller studies are made. Specimens over about 50mm are uncommon in most areas.

ABBREVIATUS Reeve, 1843

1843. *Conus abbreviatus* Reeve. *Conchologia Iconica,* 1(*Conus*): Pl. 6, sp. 86 (Wahoo, Sandwich Islands).

Description.—Thick and heavy, with a low gloss; broadly conical, the spire low and sides of the body whorl nearly straight; body whorl superficially smooth, especially in larger shells, but with 15-20 shallow, widely spaced punctate spiral grooves, strongest basally and obsolete posteriorly; shoulder heavily coronated, the coronations rather flattened; spire with straight sides, low to nearly flat, often strongly eroded; spire whorls coronate if not eroded; whorls with about 2-4 strong spiral ridges. Shell bluish gray, slightly darker basally and sometimes with a strongly contrasting paler band at the shoulder and at midbody; body whorl with about 8-15 uniform spiral rows of small squarish reddish brown spots, the interstices between spots not with white dashes; narrow brown spiral lines sometimes present; spire grayish white, sometimes with indications of brown spots between the coronations; apex white to violet-brown; base white to deep violet. Aperture narrow, the sides nearly parallel; adults sometimes with a thickened ridge internally near middle of outer lip; mouth purplish to violet-brown, with a paler or white band at midbody and posteriorly. Columella short, broad, twisted, bounded posteriorly by a strong and distinct ridge. Length 20-60mm.

Comparison.—This species is very close to *C. encaustus* and *C. miliaris* and to the allied *C. fulgetrum* and *C. tiaratus.* From typical specimens of all these it is easily distinguished by the regular arrangement of the small brown spots on the body whorl and the absence of opaque white dashes and blotches; the bluish gray color is uncommon in the other species.

Variation.—As usual with the miliaris complex of cones, there is a sizable amount of variation in color although the pattern is quite consistent. The typical bluish gray color with lighter bands at shoulder and midbody is sometimes replaced with pale brown or white; the spire may be patternless, with small spots between the coronations, or with a dark brown dash band running around each whorl. Brown specimens with a strong spire pattern closely resemble *C. encaustus* except for the absence of opaque white dashes and blotches. Adult specimens are often paler than juveniles.

Distribution.—Restricted to the entire Hawaiian Islands chain. Records of isolated individuals of *C. abbreviatus* from the Line Islands and Marshall Islands may represent either aberrant individuals of *C. miliaris* or temporary settlement by *C. abbreviatus* without forming a stable population.

Notes.—The most typical specimens of this species are juveniles or small adults; larger shells are often heavily eroded, have a reduced color pattern, and suffer from one to many heavy breaks in the body whorl. The color variants are apparently not uncommon.

The black animal of *Conus capitaneus* is in startling contrast with its often pale shell. Photo by R. Lubbock.

Left: Anterior end of *Conus geographus.* **Right:** Anterior end of *Conus papilliferus* from New South Wales. Photo by Walt Deas.

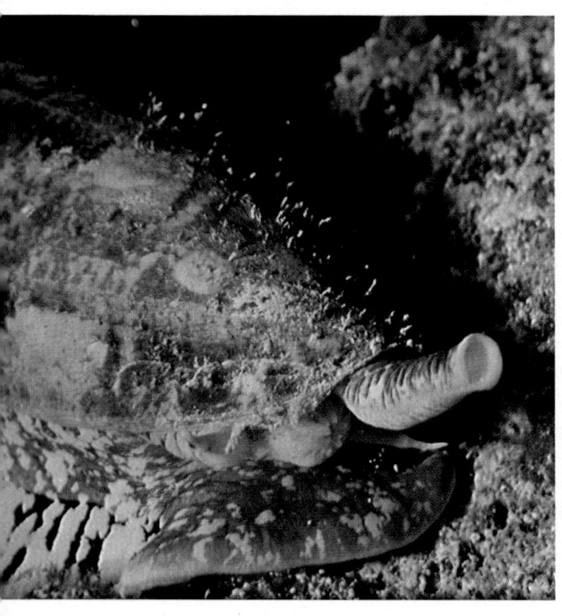

Conus striatus is one of the most deadly cones and has caused many human fatalities. Below is a close-up of the anterior end. Photos by Walt Deas on Heron Island, Queensland, Australia.

ACULEIFORMIS Reeve, 1844

1844. *Conus aculeiformis* Reeve. *Conchologia Iconica*, 1(*Conus*): Pl. 44, sp. 240b (Cagayan, Mindanao, Philippines). Lectotype here selected as the specimen figured as 240b.

1845. *Conus longurionis* Kiener. *Species gen. et icon. des coqu. viv.*, 2(*Conus*): Pl. 92, fig. 6. 1849-1850, *Ibid.*, 2: 308 (Locality unknown).

1875. *Conus gracilis* Sowerby iii. *Proc. Zool. Soc.* (*London*), 1875: 125, pl. 24, fig. 6 (Locality unknown). Non *Conus gracilis* Sowerby i, 1823.

1913. *Conus delicatus* Schepman. *Prosobranchia Siboga Exped., Monogr.* 49(5): 392, pl. 25, fig. 3 (Madura Bay, Indonesia).

1936. *Conus hopwoodi* Tomlin. *J. Conch.* (*London*), 20: 254. Nomen novum for *Conus gracilis* Sowerby iii, 1875.

Description.—Thin and fragile, with a low to high gloss; very elongate, especially anteriorly, the shoulder area broad and angular or narrower and rounded; body whorl covered with closely spaced broad flat or weakly rounded ribs separated by narrower deep grooves crossed by fine axial threads; grooves wider anteriorly, sometimes almost as wide as ribs, sometimes almost obsolete just below shoulder; shoulder broad and sharp or rounded and indistinct; spire high, sharply pointed, the sides straight to slightly convex; spire whorls sometimes slightly stepped; early spire whorls weakly nodulose or not, later spire whorls smooth; spire whorls consisting of broad anterior and posterior marginal ridges separated by a narrower off-center area containing a narrow spiral ridge and distinct axial threads; narrow ridge may be sharply set off from broad anterior ridge by a deep groove or almost fused to it; top of whorls flat, concave, or slightly convex depending on population. Body whorl and spire usually cream, grayish, or deep violet-brown; in paler color patterns the spiral ribs and spire bear scattered small brownish spots and there are usually two or three series of axially elongate brown blotches on the body whorl; with darker patterns paler spotting may show through as scattered spots or triangles. Aperture long, narrow, only slightly widened anteriorly; outer lip very thin and fragile, posteriorly sloping strongly below level of shoulder; mouth off-white marginally, becoming deep violet-brown deep within. Columella superficially obsolete. Length 25-45mm.

Comparison.—The small fragile needle-cones are greatly confused and it is currently impossible to satisfactorily distinguish the species. As treated here, *C. aculeiformis* is separable from such species as *C. insculptus*, *C. elegans*, *C. comatosa*, *C. hypochlorus*, and *C. schepmani* by the combination of a completely grooved body whorl and spire whorls consisting of two broad ridges separated by an axially striate area containing a more-or-less defined narrow spiral ridge. This type of spire sculpture is found also in *C. elegans*, which has the posterior half of the body whorl smooth, it also occurs in reduced form in *C. lemniscatus* and *C. lentiginosus*, which differ in proportions, weight, and color pattern. *C. schepmani* is very similar to *C. aculeiformis* in appearance, although it is somewhat wider at the shoulder with a more concave-sided spire; the body whorl pattern is slightly different and appears to be over an opaque white background in many specimens; the spire whorls are simply striated. *C. comatosa* in its reduced pattern form has a more distinctly carinate shoulder and spire whorls and a simple spire sculpture.

Very similar to *C. aculeiformis* (round-shoulder form) is *C. vimineus*. This species is even more fragile, has a greatly reduced or absent color pattern, body whorl with narrower rounded ribs separated by almost equally wide grooves containing strong axial threads, and spire whorls with smaller anterior and posterior ridges separated by a nearly flat area containing one or two narrow spiral ridges crossed by heavy axial threads. It is sympatric with *C. aculeiformis* and apparently distinct.

Left: Holotype of *C. delicatus* (Zool. Mus. Amsterdam) (photo Abbott).

Right: Intermediate between *aculeiformis* and *longurionis*, Philippines (Dayrit) (photo Burgess).

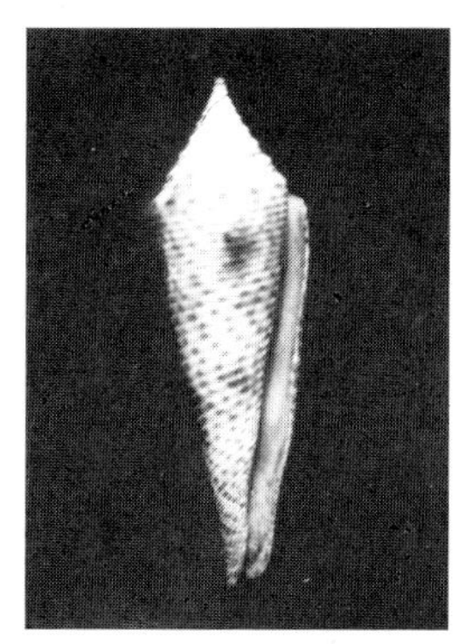

Variation.—As treated here *C. aculeiformis* is variable both in color and in width of spire. Typical *C. aculeiformis* has a broad, angled shoulder; *C. longurionis* has a rounded, indistinct shoulder. Otherwise the two taxa are very similar in spire sculpture and basic color pattern, so they are combined; intermediate specimens exist. The basic pattern of both forms is a creamy shell with two or three spiral rows of pale to reddish brown blotches and scattered smaller spots on both body whorl and spire; the spots show a strong tendency to elongate axially and fuse in certain populations, eventually obscuring the pale background and the small brown spots; at the extreme are specimens with nearly uniform blackish or dark violet shells allowing the pale background to show through only as scattered often triangular pale spots. Entirely pale or yellowish shells are uncommon, as are the entirely dark specimens.

In sculpture the species is rather uniform, although details vary from specimen to specimen. Grooves on the body whorl extend to the shoulder or sometimes the width of one or two ribs below the shoulder. The narrow spiral ridge on the spire whorls may be distinct and centered between the two broader ridges or virtually indistinguishable from the anterior ridge. In the broad-shouldered form the tops of the whorls are often deeply concave at the middle, while they may be flat in the round-shouldered form.

Turrids are closely related to cones and sometimes difficult to distinguish except by general appearance. 1. *Thatcheria mirabilis;* 2. *Gemmula unedo;* 3. *Lophiotoma leucotropis;* 4. *Brachytoma jeffreysi;* 5. *Micantapex luehdorfi;* 6. *Antiplanes contraria;* 7. *Makiyama coreanica.* From Okutani, *Shells of Japan.*

ABBAS.—Above: *Above:* Colombo Reef, Sri Lanka, 35.8 and 37.8mm. *Below:* Left: Durban, Natal, South Africa, 59.1mm (MM); Right (see text): Marquesas, French Polynesia, 48.1mm.

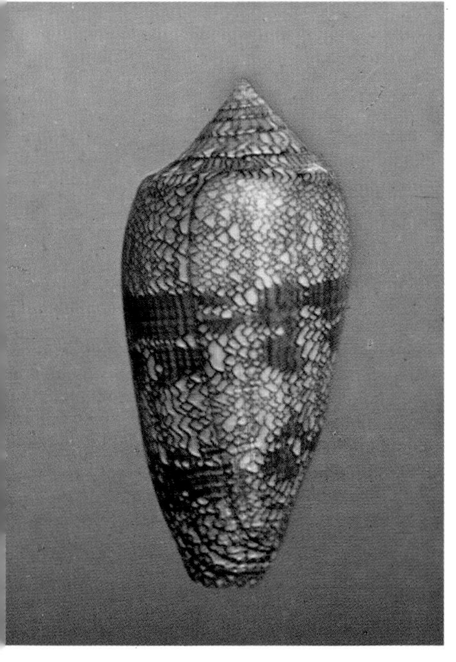

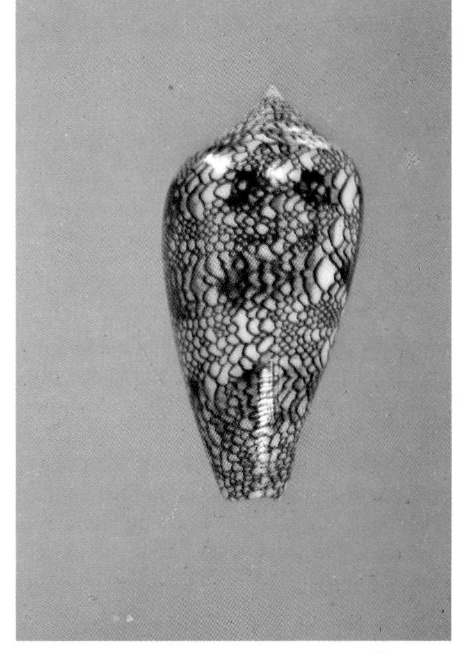

Distribution.—This appears to be a widely distributed species occurring in both the Indian Ocean and the Pacific east to at least the Solomons-Queensland area. The round-shouldered form is more common than the broad-shouldered form except in the China Sea. The species is usually trawled in over 20m.

Synonymy.—Because of the difference in shoulder shape and related spire differences, typical *C. aculeiformis* appears very distinct from *C. longurionis*, but as mentioned above there are intermediate specimens. *C. gracilis* seems to be a terminal dark specimen.

Notes.—Specimens of the round-shouldered form are sometimes sold as *C. elegans*, but that species has a smooth posterior half of the body whorl and other differences. In some respects *C. aculeiformis* is a highly specialized connecting link between the superficially very different *C. acutangulus* and *C. lentiginosus*.

ACUMINATUS Hwass, in Bruguiere, 1792

1792. *Conus acuminatus* Hwass, in Bruguiere. *Cone*, in *Ency. Method., Hist. Nat. des Vers*, 1: 688 (Amboina, Moluccas). Lectotype figure, here designated: *Tableau Ency. Method.*, Pl. 336, fig. 3.
1798. *Cucullus vicarius* Roeding. *Museum Boltenianum:* 45 (Locality not stated). Lectotype figure (Kohn, 1975): Knorr, pl. 24, fig 4. Non *Conus vicarius* Linnaeus, 1767.
1807. *Conus substitutus* Link. *Beschr. Nat.-Samml. Univ. Rostock*, 3: 101 (Locality not stated). Based on Martini, Pl. 57, fig. 638.
1833. *Conus insignis* Sowerby i, in Sowerby ii. *Conchological Illustrations:* Pt. 28, fig. 17 (Locality not stated).
1845. *Conus textilinus* Kiener. *Species gen. et icon. des coqu. viv.*, 2 (*Conus*): Pl. 103, fig. 5. 1849-1850, *Ibid.*, 2: 333 (Locality unknown).
1865. *Conus multicatenatus* Sowerby ii. *Proc. Zool. Soc.* (*London*), 1865: 519, pl. 32, figs. 10-11 (Locality unknown).
1873. *Conus Schech* Weinkauff. *Syst. Conch. Cab.*, ed. 2, 222(*Conus*): 229, pl. 37, figs. 9-10 (Massaua).
1873. *Conus cuneatus* Sowerby iii. *Proc. Zool. Soc.* (*London*), 1873: 146, pl. 15, fig. 5 (Locality unknown).
1875. *Conus Schech.* Jickell, *J.B. deutsch. Mal. Ges.*, 2: 46, pl. 1. fig. 3.
1895. *Conus Coxianus* Sowerby iii. *Proc. Malac. Soc. London*, 1: 215, pl. 13, fig. 7 (Bay of Zeyla, Somaliland).

Description.—Heavy, thick, with a good gloss; low biconic, the sides nearly straight; body whorl usually with only traces of weak spiral ridges at the base, these sometimes present to midbody or obsolete; body whorl otherwise smooth; shoulder broad, angulate, shallowly canaliculate; spire moderate, obtusely pointed, sometimes eroded, the sides concave or straight; early whorls usually eroded, but appear undulate in some specimens; whorls concave above and marked with traces of about 2-3 low spiral ridges, these most distinct on early whorls, and scattered weak curved axial threads; whorls appear distinctly stepped in most specimens. Body whorl white to cream, thickly covered with fine zigzag axial lines of dark brown to reddish brown, these overlapping to produce a very irregularly tented effect; there is a strong tendency to concentrate axial lines into axial or spiral bands with the areas between bands sparsely or not marked; often there are traces of a clear midbody band; spire largely white with regular radiating lines of dark brown thickened into blotches which often extend to shoulder; early whorls often brownish violet; base white with fine lines. Aperture moderately narrow, widest anteriorly; outer lip thick but sharp, sometimes concave at middle; mouth white toward outer margin, becoming dark violet brown deep within. Columella moderate, straight, rather short, bounded posteriorly by a low hump. Length 30-50mm.

Comparison.—In shape and texture this species is very similar to *C. erythraeensis*, which lacks any traces of a tented pattern. Superficially more similar are dark forms of *C. thalassiarchus* and *C. amadis*. *C. amadis* is thinner in texture, has heavily sculptured spire whorls, and has a white aperture; its patterns are generally (but not always) more complicated. *C. thalassiarchus* is readily recognizable by the black base, light weight, and whitish aperture. Other tented cones are usually thinner in texture, lack dark mouths, have distinctly nodulose early whorls or hard and protuberant protoconchs, and have rounded shoulders. *C. fumigatus* is usually smaller, wider at the shoulder, has flat whorl tops, and lacks tented patterns. *C. argillaceus* has a dark base, relatively longer body whorl, and three white blotch bands on the body whorl. *C. malacanus* is larger, thicker, and has an axially blotched reddish brown pattern overlaid with spiral lines and dashes.

Variation.—Although fairly constant in shape and sculpture, except for height of spire, there is great variation in the normal tented pattern. Typical specimens are completely tented from shoulder to base with a mixture of very small, small, and larger tents. There is a strong tendency to condense patterns: 1) above and below midbody so as to leave a clear midbody band; 2) into broken or complete axial bands which may be nearly black and separated by clear white areas; 3) into large blotches in spiral series above and below midbody, these sometimes united into two continuous spiral bands usually of reddish brown; 4) into a complex pattern of axial lines, dark brown blotches, and spiral dashes closely resembling some of the dark pattern of *C. amadis*. In other specimens the pattern is reduced and very openly tented or has the tents absent over large areas; at the extreme here are entirely white shells and shells with only small dark spots near midbody.

Distribution.—*C. acuminatus* is apparently confined to the Red Sea, Gulf of Aden, and adjacent areas of the northwestern Indian Ocean. Records of this species from Indonesia are abundant in early literature,

ABBREVIATUS.—Pokoi Bay, Oahu, Hawaii. *Above:* 18.0 and 19.5mm. *Below:* 17.5 and 17.7mm.

***ACULEIFORMIS.**—Above* (typical): Andaman Islands, 36.7mm. *Below* (*longurionis*): Left: Punta Engano, Cebu, P.I., 24.1mm (Dayrit); Right: Bombay, India, 30.9mm (GG).

but apparently this is in error.

Synonymy.—*C. insignis* is a broadly spirally banded form; *C. textilinus* is very openly tented; *C. multicatenatus* has large irregular blotches; *C. cuneatus* is largely white. *C. schech* looks like a miniature *C. amadis* because of the complicated color pattern, while *C. coxianus* has a reduced and largely axial pattern. It is possible that *C. coxianus* is actually a synonym of *C. fumigatus.*

Notes.—Only the normal tented patterns, including the open patterns, are usually seen. Specimens with large blotches are not uncommon, but those with complete bands are rare. Complex *amadis*-like patterns are also rare and seldom seen. Old specimens are commonly labelled "Moluccas."

ACUTANGULUS Lamarck, 1810

1810. *Conus acutangulus* Lamarck. *Ann. du Mus. Hist. Nat.* (*Paris*), 15: 286 (Mers des grandes Indes). Narrow form.
1822. *Conus acutangulus.* Lamarck, *Hist. Nat. Animaux sans Vert.*, 7: 498. Refers to Chemnitz, Pl. 182, figs. 1772-1773.
1866. *Conus turriculatus* Sowerby ii. *Thesaurus Conchyliorum*, 3(*Conus* Suppl.): 328, pl. 27 (288), figs. 643-644 (Locality unknown). Non *Conus turriculatus* Deshayes, 1865, a fossil.
1870. *Conus gemmulatus* Sowerby iii. *Proc. Zool. Soc.* (*London*), 1870: 257, pl. 22, fig. 8 (China Seas). Wide form.
1887. *Conus milesi* E.A. Smith. *J. Conchol.* (*London*), 5: 244 (Muscat, Arabia). Narrow form.

Description.—Moderately light in weight, with a dull surface; broadly to narrowly conical, the sides of the body whorl nearly straight; body whorl with broad raised ribs separated by narrow deep grooves which are shallowly punctate with axial threads; ribs of fairly uniform width, the grooves sometimes obsolete below shoulder; ribs often marked with traces of spiral and axial threads; shoulder carinate, often weakly nodulose, especially in small specimens; spire moderate to tall, about 30-40% of total length in most specimens but variable; sides straight, the whorls stepped; whorls with margins strongly nodulose on early whorls but nodules decreasing in distinctness with growth on later whorls so last whorl and shoulder often smooth; each spire whorl with a broad anterior marginal rib bearing the nodules, followed by a posterior area with two or three deep spiral grooves (rarely only one) crossed by strong, curved axial ridges usually intersected at middle by one of the spiral grooves. Body whorl white to cream, overlaid with squarish brown to reddish brown or tan spots and axial blotches which tend to form two broad solid spiral bands above and below midbody; entire shell may appear brown with only small white spots showing through or largely white with scattered brown spots; spire white, weakly to strongly streaked with brown, the nodules usually white and the interstices brown; early whorls commonly uniform white. Aperture narrow, especially so at anterior end, strongly sloping below level of shoulder posteriorly; outer lip thin and fragile; mouth pale violet to creamy. Columella indistinct. Length 15-32mm.

Comparison.—*Conus eugrammatus*, a poorly understood species, is very similar to *C. acutangulus* in appearance but readily distinguished by spire sculpture; instead of having the majority of the whorls strongly nodulose and with two or three deep spiral grooves crossing the curved axial ridges, only the first few whorls are distinctly nodulose, the later ones carinate, and the spiral grooves are represented by mere traces which do not noticeably cut the axial ridges; there are also minor differences in spire height, color pattern, and body sculpture, but these are not always reliable. *C. aculeiformis* resembles *C. acutangulus* in spire sculpture but is much thinner in texture, greatly elongated, and differs in numerous details; the same is true of such species as *C. lentiginosus*, which have a similar but reduced spire sculpture and major differences in body sculpture and pattern. *C. eucoronatus* is much larger, broader at the posterior quarter than *C. acutangulus*, and differs in body sculpture, color, and presence of several distinct spiral ridges on the spire whorls. *C. orbignyi* is more elongate, especially anteriorly, and has different body and spire sculpture; the same is also true of *C. mazei* in its various forms. *C. praecellens* is similar in general appearance and color pattern but has the spire with flatter, non-stepped sides, the whorls not nodulose and with numerous spiral ridges.

Variation.—Color pattern variation is indicated in the description. Intensity of brown pattern seems to vary somewhat with locality, as does spire height. This species seems to exist in two body whorl widths, the narrow and rather elongate form described by Lamarck and the broader form more familiar to collectors. The narrow form seems to be most common in the northern Indian Ocean, while the Pacific Ocean seems to have mostly the broad form. Since this geographic separation is not complete, subspecific recognition is apparently not required. Narrow pale specimens from the northern Indian Ocean may be difficult to place with the broad, almost all brown Solomons form, but intermediates are not uncommon. Spire sculpture appears to be very consistent, although the number of spiral grooves and development of nodules on later whorls and shoulder appear to vary individually.

Distribution.—A widely distributed species found from the eastern coast of Africa through the Indo-Pacific to northern Australia, southern Japan, the Society and Tuamotu Islands, and Hawaii. Generally uncommon or rare in sand and rubble from about 10-100m.

Synonymy.—Possibly other names apply to this species, but there is little doubt about the names listed. If the broad form is ever determined

ACULEIFORMIS.—(*longurionis*) Tolwat Beach, E New Britain, Papua New Guinea. *Above:* 30.1mm. *Below:* 27.3mm.

ACUMINATUS.—*Above:* Left: Dahlak Is., Eritrea, Ethiopia, 27.0mm (EP); Right: Taundi I., Red Sea, 32.1mm (JB). *Below:* Left: No data, 32.2mm (GG); Right: Off Saudi Arabia, Red Sea, 34.2mm (GG).

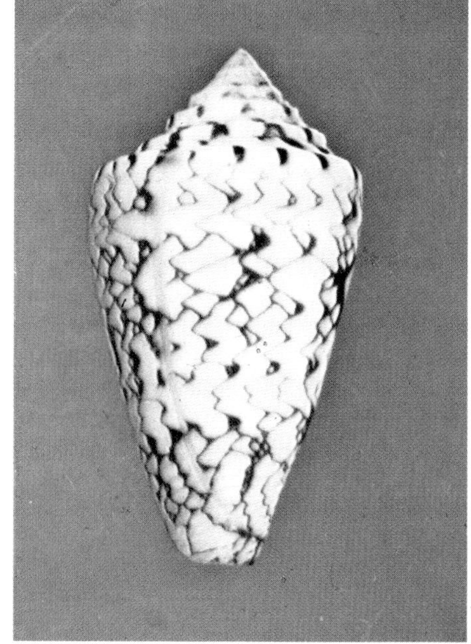

worthy of subspecific recognition, it seems that *C. gemmulatus* is available for it.

Notes.—Beware of small *C. acutangulus* from Hawaii and other localities offered as the local population of *C. eugrammatus* at very inflated prices; be sure to closely check spire sculpture. The narrow form from the Red Sea and other localities is often sold at high prices as *C. milesi*, here considered a synonym. This species is very subject to breaks and partially healed scars on the spire and body whorl and to broken lips.

ADAMSONII Broderip, 1836

1825. *Conus cingulatus* Sowerby i. *Cat. Tankerville:* xxxiv, no. 2467 (Locality not stated). Non *C. cingulatus* Lamarck, 1810.
1836. *Conus Adamsonii* Broderip. *Proc. Zool. Soc.* (*London*), 1836: 44 (Locality unknown).
1839. *Conus rhododendron* Jay. *Cat. Shells Coll. Jay*, ed. 3: 100, pl. 7, figs. 2-3 (Australasia).

Description.—A solid, bulky cone with a high gloss; subcylindrical, the sides slightly inflated below the shoulder; body whorl with 6-12 broad rounded spiral ribs anteriorly, separated by narrow, finely punctate grooves; middle of body whorl smooth, but area below shoulder with 4-8 broad rounded spiral ribs separated by narrow punctate grooves; shoulder rounded, broad; spire low, bluntly pointed, tops of whorls concave to nearly flat; spire whorls with thickened margins and fine spiral threads. Body whorl and spire pinkish white or rosy with three broad bands of small smeared brown spots in irregular fashion: an indistinct band at base, a broad band below midbody, and a narrower band below the shoulder area; at the shoulder, midbody, and base are also broad bands of pinkish to purple wavy axial blotches, the midbody band widest and most distinct; pattern becomes more diffuse with wear, the midbody band often containing pinkish brown nebulous spots; spire white with pinkish or tan streaks and blotches. Aperture widened anteriorly; outer lip thick, blubbery; mouth white to pale pink. Columella partially fused, long, thick. Length 40-65mm.

Comparison.—The shape and texture ally this species to *C. bullatus*, *C. julii*, and relatives, but it is easily distinguished by the strong spiral ribs below the shoulder and the rosy and brown color pattern. *C. boivini* has sculpture below the shoulder but differs radically in shape, texture, and pattern.

Variation.—Some specimens are nearly parallel-sided while others are strongly inflated below the shoulder. The number of spiral ribs on the body whorl varies somewhat, but they are apparently always present below the shoulder. Some specimens have spiral lines of fine white dots evenly spaced over the body whorl where spiral grooves would be expected. Details of the pattern are very variable and apparently subject to great changes with wear. The midbody band of diffuse pinkish blotches on white, bordered above and below by bands of scattered brown dots, is usually visible in all but the most badly worn specimens.

Distribution.—This is a very rare species seldom seen in good condition, so the distribution is poorly known. It is definitely recorded from the Cook, Tonga, Gilbert, and Society Islands, as well as the Solomons and the Great Barrier Reef, also extending to the New Hebrides and New Caledonia. It is subtidal in habitat and usually washed up on beaches after severe storms.

Synonymy.—The name *C. rhododendron* is more familiar to most collectors, but *C. adamsonii* definitely takes priority.

Notes.—Live-taken specimens of this shell are quite rare and even good beach specimens fetch a high price. Worn specimens seem to lose the distinction between brown and rosy in the pattern and turn purplish tan on pink. Commonly there are one or more heavy breaks on the body whorl; the thick lip breaks easily. In some areas it seems that *C. adamsonii* is not uncommon judging from the number of beach specimens, but the habitat of living specimens is hard to find.

ADVERTEX (Garrard, 1961)

1961. *Rhizoconus advertex* Garrard. *J. Malac. Soc. Australia*, No. 5: 30, pl. 1, fig. 1 (Off Moreton Is., Queensland, Australia).

Description.—Light in weight, fragile, with a high gloss; obconic, the body whorl short and with nearly straight or slightly concave sides; body whorl smooth, with a few weak spiral ridges at the base; traces of finely punctate spiral lines present over anterior half of body whorl and sometimes almost to shoulder, these lines often obsolete or very poorly indicated; numerous fine axial scratches over body whorl; shoulder broad, angled; spire nearly flat, the earlier whorls and protoconch forming a small hump at the middle of the flat plate formed by the later whorls; spire whorls flat above, with about 5-6 weak and variable spiral ridges crossed by numerous weak axial threads. Body whorl white to pale tan, variably suffused with pinkish and tan clouds; color pattern consisting of several rows of darker brown squares and rectangles in spiral lines; row below shoulder often very dark and partially fused, interrupted by continuation of white from spire; below this are 1-3 rows of small squares, then a broader belt of large squares which may be partially fused into blotches; the midbody area is usually clear or with only a few small spots; below midbody the markings become spirally elongated and line-like on a darker brown base color, with the anterior-most markings nearly con-

ACUTANGULUS.—*Above:* Left: Rove, Guadalcanal, Solomon Is., 25.0mm; Right: Eliat, Israel, 25mm (DM) (photo by Jayavad). *Below:* Left: Eliat, Israel, 28.0mm (PC); Right: Off Bombay, India, 25.8mm (DMNH).

ADAMSONII.*—Above:* New Hebrides, 45mm (photos by P. Clover). *Below:* Cook Is., 45.3mm (DMNH).

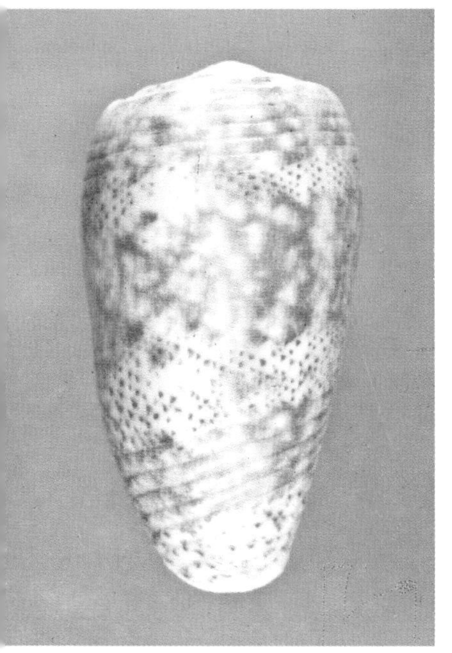

tinuous spiral lines; basal area white; spire white, with about 8-10 irregular radiating blotches of brown and scattered brown spots; early whorls white. Aperture wide, uniform; outer lip thin, fragile, sometimes weakly constricted at middle; posterior margin of lip on level with shoulder; mouth white, sometimes with pale pink suffusion. Columella with posterior half free from inner lip and forming a strong projecting cusp. Length 30-40mm.

Comparison.—*C. advertex* is very similar to *C. angasi* in all features but differs (apparently consistently) in the following features: in *C. advertex* the spire is flat and the shoulder angled, while in *C. angasi* the spire is low but convex and the shoulder rounded; additionally, *C. angasi* is more convex-sided than *C. advertex* and tends to have a less sharply defined color pattern, but this is variable. *Conus trigonus* shares with *C. advertex* and *C. angasi* the general shape and unusual columella, but has two broad bands of orange brown to dark brown above and below midbody marked with numerous narrow spiral broken brown lines; shapes similar to both *C. advertex* and *C. angasi* occur in *C. trigonus. C. sculletti* differs in much more elongate shape and distinctly banded pattern. *C. papilliferus* and *C. wallangra*, as well as *C. illawarra*, differ in color pattern.

Variation.—This is a narrowly defined species with little significant variation. In some specimens the sides are distinctly concave at the middle, in others straight. Intensity of pattern varies considerably, but as a general rule the brown squares are larger and darker just above midbody and below the shoulder; the broken lines above the base vary considerably in development and length. The background color varies from plain white, through white with pink and brown clouds, to nearly uniform very pale tan. Punctate spiral lines are better developed in small specimens and are usually visible only under the lens.

Distribution.—Restricted to deeper waters (160-300m) off southern Queensland and northern New South Wales, Australia; moderately common.

Notes.—The cusped columella is an unusual feature in cones and is shared with *C. angasi, C. trigonus*, and *C. tinianus*, all rather similar in general appearance and found in the southern oceans.

Since *C. angasi* and *C. advertex* are so similar and differ in only minor features, as well as sharing the same distribution but different depths, it would be expected that intermediates occur. Perhaps it would be best to consider *C. advertex* an ecological subspecies of *C. angasi*.

AEMULUS Reeve, 1844

1844. *Conus aemulus* Reeve. *Conchologia Iconica*, 1(*Conus*): Pl. 46, sp. 256 (Locality unknown).

1845. *Conus hybridus* Kiener. *Species gen. et icon. des coqu. viv.*, 2 (*Conus*): Pl. 83, fig. 1. 1849, *Ibid.*, 2: 256 (Locality unknown).

1845. *Conus obtusus* Kiener. *Ibid.*, 2: Pl. 109, fig. 3. 1849-1850, *Ibid.*, 2: 317 (Locality unknown).
?1853. *Conus Tamsianus* Dunker. *Index Moll. ad Guineam Inf. coll. Tams:* 28, pl. 4, figs. 22-23 (Annabon Is.). Not seen.
?1942. *Conus perryae* Clench. *Johnsonia*, 1(6): 31, pl. 15, fig. 5 (3 mi. off Little Carlos Pass, Lee Co., Florida).
?1975. *Africonus echinophilus* Petuch. *Veliger*, 18(2): 180, figs. 1-3, 7 (N'Gor, Cape Verde, Senegal).
1975. *Conus nobrei* Trovao. *Centro Port. Activ. Subaq.*, 4(1): 5 of separate, pl. 1, fig. 2 (Angola, between 12°48' and 13°51' S).

Description.—Light in weight, with a low gloss; conical, with a moderate spire and narrow body whorl with straight sides; body whorl with a few weak ridges near the base and traces of irregular spiral and axial threads; shoulder rounded; spire bluntly pointed, the early whorls eroded, sides straight; later whorls flat or slightly concave on top, with very weak spiral ridges and numerous heavy axial ribs when not badly eroded. Body whorl bluish white to pale gray with spiral lines of fine alternating white and brown spots separated by bluish background color; white spots often fused or smudged to form small triangles; interrupted axial brown bands usually present irregularly from shoulder to base, obvious only at shoulder, midbody, and base; each area may have the brown blotches fused laterally into irregular spiral bands; occasionally entire shell covered with fine meshwork of brown and white triangles and dots; basal area usually with heavy pattern of brown spiral dashes and some blotches in axial series; shoulder with bluish white elongate blotches running onto spire and separated by narrow extensions of the brown axial bands from the body whorl; spire marked as shoulder, but often losing pattern because of erosion. Aperture moderately wide basally, narrower posteriorly; outer lip thin, smoothly curved; margin of lip white followed by narrow dark brown band, then becoming deep dirty violet brown; narrow light spiral bands under shoulder and usually at middle. Columella short, narrow. Length 25-35mm.

Comparison.—*Conus aemulus* is a poorly defined species closely related to *C. ventricosus* and *C. ermineus* on one hand and to *C. bulbus*, *C. variegatus*, and *C. algoensis* on the other. It is possible that extremes of any of these species could be confused with *C. aemulus*, and it must be admitted that none of these species is identifiable with certainty from small series.

C. ventricosus is usually a darker brown shell, the spiral dashes being very fine and not easily seen, while there are at least two very broad brown bands covering most of the body whorl; the shoulder tends to be broader and more angulate than normal in *C. aemulus;* it is very possible that *C. aemulus* is an African subspecies of *C. ventricosus*, but typical *C. ventricosus*, although rare, occurs with it. *C. ermineus* is almost always larger, much broader posteriorly, and seldom has the mouth as

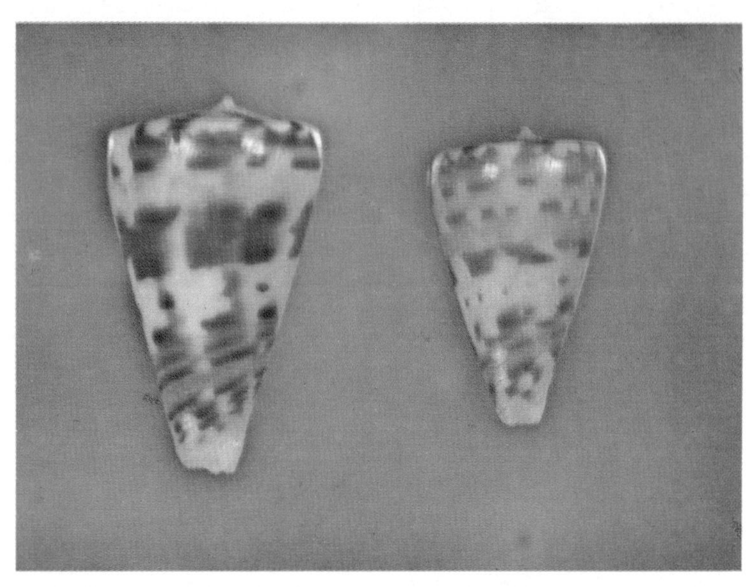

ADVERTEX.—Cape Moreton, Queensland, Australia, 26.5 and 33.2mm.

AEMULUS.—*Above:* Mussulo Bay, Angola, 31.4mm (GG). *Below:* Nova Redondo, Angola, 28.5 and 28.6mm (GG).

dark as in *C. aemulus*. Some patterns of *C. ermineus* resemble those of *C. aemulus*, but seldom are strong spiral dashes on a bluish background combined with irregular axial bands as is so common in *C. aemulus;* the two species are usually easily distinguished by the large size and broader appearance of *C. ermineus*, as well as the paler mouth.

C. bulbus and *C. variegatus* are similar to *C. aemulus* in some patterns although both are typically distinct. As a general rule, *C. bulbus* has a dark brown base, is much wider at the shoulder, and has distinct axial brown lines over the body whorl; seldom are spiral dashes present in the pattern except for some local populations in Angola; these unusual populations are wider than typical *C. aemulus*. *C. variegatus* as a rule has spiral lines or dashes over the body whorl, is thicker, wider at the shoulder, and often has a white mouth; many variants occur and some could be confused with *C. aemulus*, but shape will usually distinguish it. *C. algoensis* in some patterns has the band below the shoulder, the light spots on the shoulder, and the axial irregular bands, but these do not have the heavy spiral dashes, the bluish background, nor (usually) the dark mouth.

Variation.—Although constant in shape, pattern variation in this species is legion. At one extreme are bluish shells covered with numerous brown and white spiral dashes and occasional brown blotches; typically brown blotches develop at the shoulder and above midbody, as well as at the base, forming either broad spiral bands or connecting axially with irregular flammules; patterns with mixtures of spiral bands and axial bands are common, and sometimes the dashes become very reduced in number. At another extreme is the *hybridus* pattern, where the axial banding becomes numerous small dark brown or almost black blotches heavily infiltrated with fine brown and white spots until the blotches disappear into the fine spotted pattern, yielding a finely mottled blue shell usually with a distinct white band at shoulder and midbody.

The shell described as *Africonus echinophilus* by Petuch displays no real differences in either shape or sculpture from *C. aemulus* but does differ in the very small adult size (about 12mm) and the habitat (associated with sea urchins), as well as small differences in color patterns (grayish color with more irregular dark clouding than typical in *C. aemulus*). Animal colors also are said to differ. These small differences could easily reflect the unusual ecology of *echinophilus* and not specific differences.

Distribution.—Common in shallow water of western Africa from the Bulge to Angola.

Synonymy.—In these confused cones all synonymy is doubtful. *C. hybridus* Kiener may represent a full species, but if so it is exceedingly hard to separate from typical *aemulus* and more likely represents a terminal pattern. Whether *C. echinophilus* is a full species or just dwarf populations resulting from meager food in a very restricted and marginal habitat is very uncertain. Since I can find no differences in shell shape, sculpture, or pattern beyond the normal variation in *C. aemulus*, I am

Conus nobrei Trovao = *C. aemulus* (photo Clover).

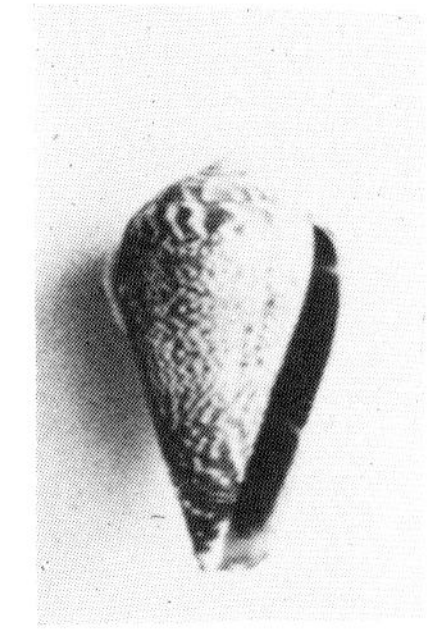

tentatively considering it a synonym. However, the differences in body color and pattern, as well as habitat, if the two are not correlated and are genetic, are constant enough to indicate full specific status to some. Since it is my personal opinion that animal colors and patterns are no less variable in some cones than shell pattern and color, I doubt the distinctness of the species.

C. perryae is possibly unidentifiable from the description, but the shape and color pattern appear to be the same as many specimens of *C. aemulus*. There has always been some doubt as to locality data on the unique shell on which the description is based.

Notes.—This shell is commonly sold under the names *C. africanus* (here considered a synonym of *C. variegatus*), *C. guineensis* (here considered a full species equivalent to *C. mozambicus*), *C. obtusus*, and *C. grayi* (here considered a synonym of *C. ermineus*). Be careful of unusual patterns sold at high prices.

ALGOENSIS Sowerby i, in Sowerby ii, 1834

1834. *Conus Algoensis* Sowerby i, in Sowerby ii. *Conchological Illustrations:* Pt. 54, fig. 66 (Locality not stated).

1845. *Conus jaspideus* Kiener. *Species gen. et icon. des coqu. viv.*, 2 (*Conus*): Pl. 55, fig. 4. 1848, *Ibid.*, 2: 218 (Locality unknown). Non *Conus jaspideus* Gmelin, 1791. [= *algoensis scitulus*]

1849. *Conus scitulus* Reeve. *Conchologia Iconica*, 1(*Conus* Suppl.): Pl. 9, sp. 283 (Locality unknown). [= *algoensis scitulus*]

1857-1858. *Conus simplex* Sowerby ii. *Thesaurus Conchyliorum*, 3 (*Conus*): 31, pl. 9 (195), fig. 199 (East Indies). [= *algoensis simplex*]

1858. *Conus Danieli* Crosse. *Rev. Mag. Zool.*, (Ser. 2), 10: 122. Nomen novum for *C. jaspideus* Kiener, 1845.

Description.—Very thin, light, and fragile, with a good gloss when fresh, but more often eroded to a dull surface; broadly to narrowly conical, the sides nearly straight or slightly convex; body whorl with

AEMULUS.—Above: Left: Luanda, Angola, 25.1mm; Right: Mussulo Bay, Angola, 26.4mm (GG). Below (*hybridus*): N'Gor, Senegal, 33.7 and 35.0mm (GG).

AEMULUS.—(*echinophilus* paratypes) N'Gor, Cape Verde, Senegal. *Above:* 10.5mm. *Below:* 9.7 to 11.4mm (all EP).

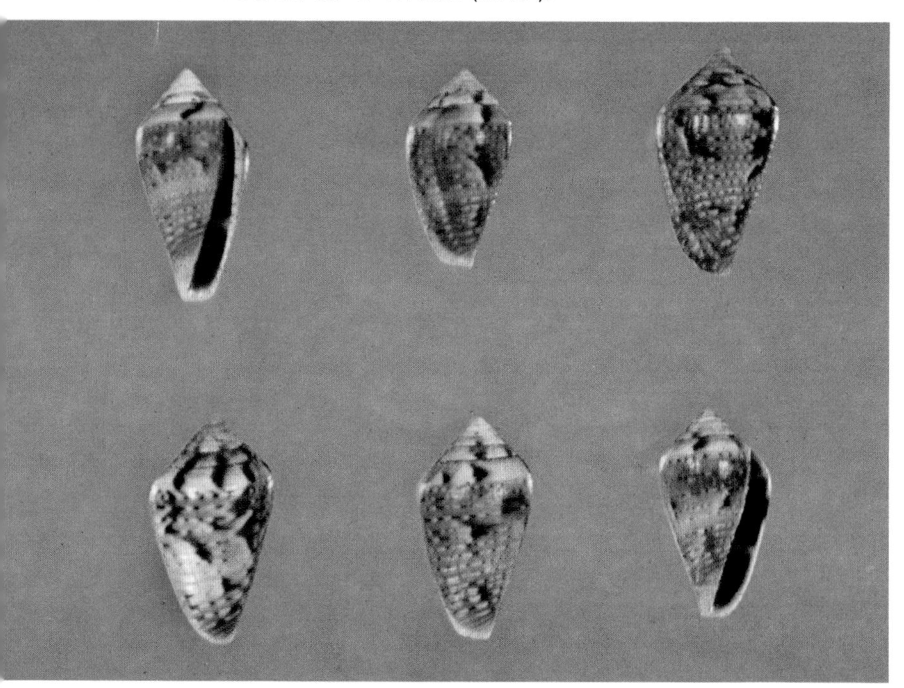

weak, closely spaced spiral ridges at base, otherwise with only traces of nearly obsolete spiral and axial threads; shoulder rounded, rarely concave on top; spire moderate, obtusely pointed, early whorls often eroded; tops of whorls slightly to distinctly concave and with traces of very fine spiral and axial ridges or threads, none very distinct, easily disappearing with erosion. Body whorl of variable color depending on subspecies, but generally white to dirty cream with a brown to orange band below the shoulder, white blotches on shoulder and spire, a broken brown band above the base, and brown stripes or spots between; body whorl may be entirely dark brown with only small white blotches at midbody; base whitish or dirty cream; spire largely white, separated into large blotches by narrow brown stripes; tip whitish. Aperture wide, uniform; outer lip thin, fragile, smoothly convex; posterior margin of lip sloping below level of shoulder in normal specimens; mouth whitish to pale violet-brown. Columella short, narrow. Length 25-70mm.

Comparison.—If not badly eroded, most patterns of *C. algoensis* are fairly easy to identify. Compared to *C. aemulus*, it is thinner, lacks the numerous brown and white dashes, and has a whitish, not bluish, background. The various colors of *C. tinianus* may sometimes resemble *C. algoensis* (especially eroded *C. a. scitulus*), but generally are distinguishable by the cusped columella of *C. tinianus*, the less sloping lip, coarser spiral sculpture on the shoulder (easily eroded, however), and lack of the broad band below the shoulder and distinct large white blotches on the shoulder and spire; bright colors usually indicate eroded *C. tinianus* and not *C. algoensis*. *C. pictus*, which is perhaps just an offshore ecotype of *C. algoensis*, differs in the spire with strongly concave whorls producing a distinctly stepped appearance and in the presence of a distinct three-banded pattern; it is otherwise extremely similar to *C. a. scitulus*. *C. guineensis* is commonly more elongate, has heavy and coarse spiral sculpture on the shoulder and spire (be careful of erosion!), usually has a lower spire, and almost never has a broad dark band below the shoulder; very dark specimens of *C. guineensis* may be almost impossible to separate from *C. a. algoensis*, but they generally have scattered white spots and blotches on the body whorl.

It is probably impossible to accurately identify at least 25% of all beach-washed cones from South Africa because of erosion and variation.

Variation.—Color pattern variation in this species is pronounced and can be used to separate the species into three subspecies having fairly well defined characters but showing a great deal of overlap. In addition, the shape varies considerably from very narrow with rounded shoulder to wider with a more angulate shoulder; the spire may be low or moderate, with concave or convex sides and spire whorls concave, weakly stepped, flat, or convex.

Conus algoensis algoensis Sowerby i, in Sowerby ii: body whorl mostly brown except for white shoulder blotches and a broken white band at midbody; base whitish, with a broad brown band above; 25-35mm long; western shore of Cape Peninsula.

C. a. scitulus Reeve: body whorl mostly whitish, with a broad and distinct shoulder band (orange in life, fading to brown); basal band weak or absent; irregular spiral bands of squarish brown spots over body whorl; occasionally an orange midbody band; 25-40mm; Hermanus to Cape Agulhas.

C. a. simplex Sowerby ii: shoulder band interrupted, connected to basal area (basal band nearly absent) by zigzag axial bands of orange or brown; spiral rows of brown spots usually absent or poorly defined; occasional specimens have shoulder band absent and are marked with broad irregular axial bands only; 25-70mm; eastern coast of Cape Peninsula.

Beach specimens of all subspecies can be found on the same beaches, so distribution can only be used with live-taken specimens.

Distribution.—Southwestern shores of South Africa, but dead shells can be found on beaches further to the east.

Notes.—Eroded specimens of this species can usually be distinguished by fine shoulder sculpture, simple columella, and color pattern including a broad band below the shoulder and white spire blotches. With further erosion the distinctive characters disappear and it cannot be completely distinguished from *C. tinianus* and *C. guineensis*. Only live-taken specimens, or beach specimens with excellent color patterns, should be purchased. Live-taken specimens of all subspecies are uncommon to rare, but some fresh *C. a. scitulus* are available.

Kilburn (1971, *Ann. Natal Mus.*, 21(1): 37-54) has covered this species in some detail and his paper should be consulted. Through an oversight Kilburn used the name *scitulus* for the species.

It should be noted that the system of three subspecies used here, although following Kilburn, is not entirely satisfactory. It would probably be best to consider the three forms, and *C. pictus*, as color variants of a single species without subspecies. It could be further 'lumped' with *C. guineensis* as the sculptural, color pattern, and shape differences given by Kilburn do not hold up well and cannot be used in many beach specimens. *C. tinianus* is usually distinguishable from the others by the columella, so it would seem to be a valid species regardless of the system used. However, Kilburn has figured the egg capsules of the three species and they seem to be very distinct; there are also minor differences in radular teeth, so Kilburn's system is followed here for the moment.

ALTISPIRATUS Sowerby iii, 1873

1848. *Conus papillaris* A. Adams and Reeve. *Zool. Voy. Samarang, Moll.*, 1: 17, pl. 5, figs. 7a-7b (Locality unknown). Non *C. papillaris* Sowerby i, in Sowerby ii, 1833.

1870. *Conus turritus* Sowerby iii. *Proc. Zool. Soc.* (*London*), 1870: 256, pl. 22, fig. 14 (Agulhas Bank, South Africa). Non *Conus turritus*

ALGOENSIS.—*Above* (*a. algoensis*): Jeffreys Bay, Cape, South Africa, 38.0mm (Meyer). *Below* (*a. scitulus*): Left: Jeffreys Bay, Cape, South Africa, 25.6mm (GG); Right: Cape Agulhas, Cape, South Africa, 17.9-23.0mm (GG).

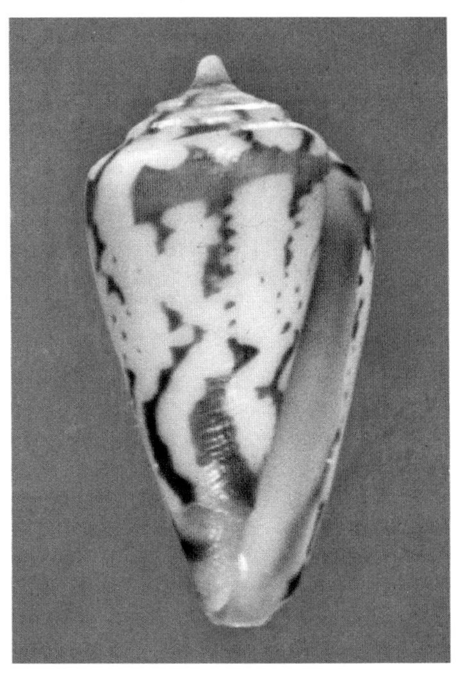

ALGOENSIS.—(*a. simplex*) *Above:* Buffelsboai, Cape, South Africa, 23.5mm. *Below:* Left: Simonstown Harbour, Cape, South Africa, 58.9mm; Right: Buffelsboai, Cape, South Africa, 21.3mm (all GG).

(Roeding, 1798).

1873. *Conus altispiratus* Sowerby iii. *Proc. Zool. Soc.* (*London*), 1873: 146, pl. 15, fig. 4 (Agulhas Bank, South Africa).

1875. *Conus gradatulus* Weinkauff. *Syst. Conch. Cab.*, ed. 2, 237 (*Conus*): 356, pl. 66, fig. 5. Nomen novum for *C. turritus* Sowerby iii, 1870.

1876. *Conus Oltmansianus* Lennep. *Cat. Alphab. Cones:* 5,9. Nomen novum for *C. turritus* Sowerby iii, 1870.

1903. *Conus patens* Sowerby iii. *Marine Invest. South Africa,* 2(3): 218, pl. 3, fig. 7 (Vasco de Gama Pk., South Africa).

Description.—Fairly thin but solid, with a high gloss in typical fresh specimens; biconic, the sides straight to convex; body whorl smooth except for a few weak ridges at the base and fine axial scratches; shoulder roundly angulate, broad, smooth or undulate; spire moderate to high, pointed in most shells, the sides straight, concave, or stepped; tops of spire whorls concave, the margins usually roundly carinate, smooth, undulate, or irregularly nodulose; spire whorls with numerous curved axial threads but no spirals. Shell glossy white, pinkish, or pale tan, with or without irregular pale reddish brown flammules or blotches arranged in about three bands below shoulder, at midbody, and above base; occasionally flammules are connected axially and extend over entire body whorl; spire same color as body whorl, unmarked or with faint scattered brownish blotches. Aperture wide, somewhat broadened anteriorly; outer lip thin, occasionally either concave or convex near middle; posterior edge of outer lip sloping below shoulder level in most specimens; mouth white to pink. Columella long, narrow, often turned abruptly to axis of body whorl. Length 50-95mm.

Comparison.—*C. altispiratus* is closely related to both *C. teramachii* and *C. sieboldii* on the basis of shape, sculpture, and color pattern. From *C. teramachii* it is distinguished by the common presence of a color pattern (absent in *teramachii*), the rounded spire carinae (sharp and distinct in *teramachii*), absence of fine regular nodules on the early whorls (although irregular and usually coarse nodules may be present), and the generally broader shape. *C. sieboldii* is a thicker shell which is generally much narrower and usually has deep spiral grooves extending to about midbody; the spire whorls have sharp erect carinae bearing distinct nodules on the early whorls. *C. smirna* is sometimes vaguely similar in shape but has a different color pattern and a tendency to nacre-over the early whorls.

Variation.—I have here combined three groups of body shapes and patterns considered by some to be full species. Actually, no two specimens are exactly alike in color, shape, and coronation. At one extreme is a broad, low-spired shell with distinctly stepped, non-coronate whorls. Most common is a broad shell with tall straight-sided spire without undulations (or with only irregular undulations on some whorls) and usually with one or two bands of brownish blotches on the body whorl. At the other extreme are relatively slender shells with very tall spires which may or may not be undulate or even strongly nodulose; the body whorl may be uniformly cream or covered with a heavy pattern of reddish brown. All possible combinations of pattern, shape, and coronation occur, even within the same locality, and I can see no reason to recognize any combination as distinct. The first available name for the complex is *C. altispiratus*, used here. There seems to be a strong tendency in slender shells to distort the shoulder into a broad flange.

Left: 35mm *C. altispiratus* (NM).

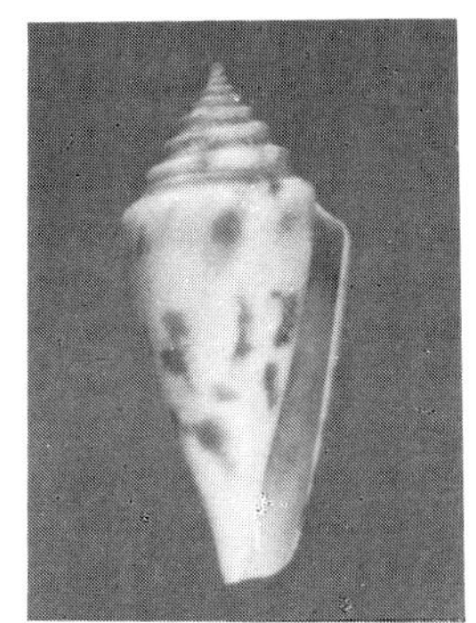

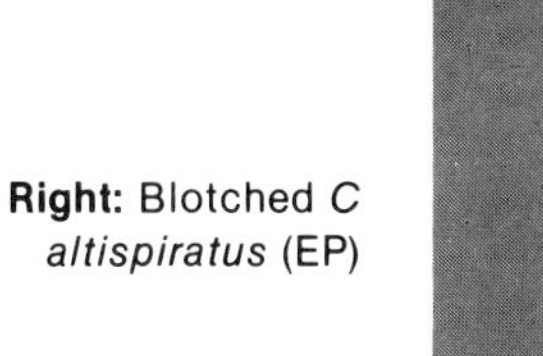

Right: Blotched *C altispiratus* (EP)

Not so simple or easily explainable is the condition of some specimens (usually the low-spired, stepped, patternless shells) with extremely adherent coarse brown periostracum. These shells usually look very rough and semi-fossil, although live-taken. The condition is not restricted to any one group of localities, but apparently most specimens from several localities are the coarse type. For the moment it seems that these could be ecological variants, but there is no real basis for this assumption.

Distribution.—Endemic to moderate (60m) to deep (700m) water off the southern African coast from Walvis Bay, South-West Africa around South Africa to southern Mozambique. Uncommon to rare in good condition. Seldom washed ashore.

Synonymy.—*C. altispiratus* is the slender, nodulose form; *C. gradatulus* is the slender, non-nodulose form. *C. patens* would be the broad form with relatively low, non-stepped spire and one or more bands of reddish-brown blotches.

Notes.—The operculum of this species is rather small and has a smooth margin; this contrasts with the serrated margin in *C. teramachii*. Collectors should not pay high prices for the coarse form as long as more typical specimens with good glossy surfaces are available. Perfect specimens of *C. altispiratus*, at least the *patens* form, are not rare, but specimens with strong nodules on the spire are rare regardless of body shape and color pattern. Specimens with strong patterns are more desirable than those with uniformly cream or pinkish color. Lips are fragile in this species and usually moderately filed.

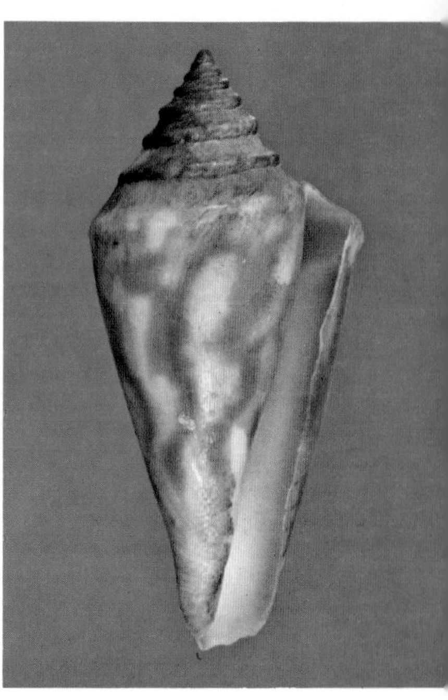

ALTISPIRATUS.—Above: 100 mi. off Cape Agulhas, South Africa, 56.1mm (NM 4479). Below: Off Luderitz, South Africa, 80.9mm (K).

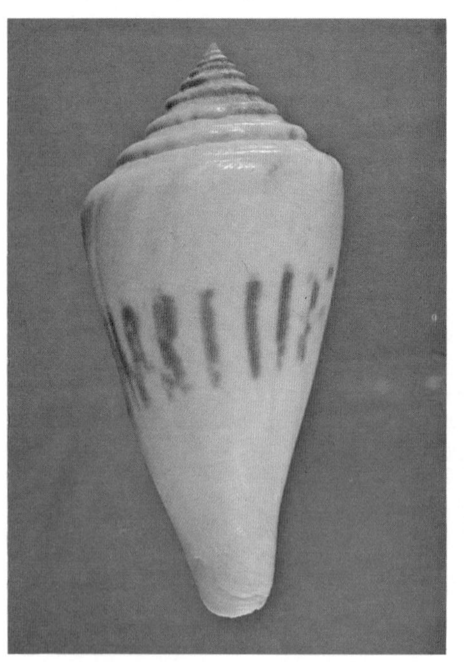

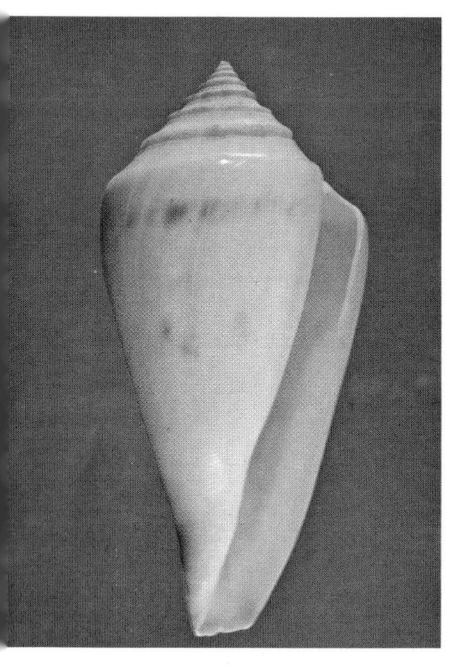

ALTISPIRATUS.—Above: Left: Off Luderitz, South Africa, 89.2mm (K); Right: 100 mi. off Cape Agulhas, South Africa, 74.7mm (NM 4479), ***Below:*** Left: Off Cape Receife, Cape, South Africa, 60.2mm (Meyer); Right: Off Walvis Bay, South-West Africa, 70.0mm (NM A2694).

AMADIS Gmelin, 1791

1791. *Conus amadis* Gmelin. *Systema Naturae per Regna Tria Naturae*, ed. 13, 1: 3388 (Locality not stated). Lectotype designated by Kohn, 1966.

1937. *Conus amadis* var. *castaneo-fasciata* Dautzenberg. *Mem. Mus. Roy. d'Hist. Nat. Belg.*, 2(18): 15 (Locality not stated). Based on Kiener, Pl. 84, fig. 2

1937. *Conus amadis* var. *aurantia* Dautzenberg. *Ibid.*, 2(18): 14 (Locality not stated). Non *C. aurantius* Hwass, 1792.

1942. *Conus subacutus* Fenaux. *Bull. l'Inst. Oceanogr. (Monaco)*, No. 814: 4, fig. 10 (Madagascar).

Description.—Light weight, with a high gloss; broadly conical, the sides straight to slightly convex; body whorl with low spiral ridges at the base separated by narrow punctate grooves; grooves usually obsolete below midbody, but continuing to shoulder in juveniles and some adults; body whorl looks superficially smooth; shoulder broad, carinate, slightly canaliculate above; spire low to high, the sides straight or concave, the whorls often slightly stepped; tops of whorls usually slightly concave to distinctly canaliculate, especially the later ones; whorls with 4-6 weak spiral ridges crossed by faint axial wrinkles to produce a faintly beaded appearance; sculpture often indistinct in large specimens; early whorls weakly nodulose, but commonly abraded. Body whorl white, overlaid with a very variable pattern of pale yellowish brown to deep chestnut, usually in one of three main patterns (but mixtures of patterns common): 1) covered with indistinct brown reticulations and blotches in no discernible order, often the brown blotches with small white triangles within; 2) covered with distinct tented pattern formed by overlapping zigzag lines, the pattern open or very tight, usually with two rows of blotches (above and below midbody) containing large white spiral dashes; 3) largely dark brown except for narrow zones at base, midbody, and shoulder consisting of large white tents, the brown bands often with scattered white tents and brown and white spiral dash lines within; base like body whorl but narrowly white at columella area; spire white, whorls with broad curved brownish blotches; amount of brown usually dependent on body whorl color and intensity, varying from nearly absent to largely all brown; margins of whorls and shoulder usually with narrow darker brown broken lines or short blotches; early whorls white. Aperture uniformly wide; outer lip strongly sloping below shoulder posteriorly, the anal notch especially broad and deep; outer lip thin, slightly convex; mouth white. Columella long, narrow, strongly indented relative to rest of inner lip. Length 50-100mm.

Comparison.—*Conus araneosus* is often similar in pattern and general shape but is distinctly coronated and thicker-shelled with a deep yellow mouth. *C. thalassiarchus* also has very similar patterns but is easily separated by the black base and the general lack of spiral sculpture on the spire whorls. *C. acuminatus* is smaller, has less spire sculpture, is generally thicker, and has a violet-brown mouth. *C. wittigi* has a less variable pattern, less stepped and carinate spire whorls, lacks nodulose early whorls, and lacks brown and white dashes in the pattern. *C. ammiralis* is usually thicker-shelled, lacks strong spire sculpture, and has a distinctly alternating brown and white spire pattern, the brown areas usually with distinct narrow darker brown axial lines. See also *C. thomae.*

Variation.—Sculpture is quite constant, but it seems to vary somewhat with size. Juveniles often have punctate grooves extending over the whole body whorl to the shoulder, while large adults have only indistinct grooving at the base; a few adult shells retain grooving over most of the body whorl, however. There also seems to be a tendency for some shells, especially deepwater populations, to greatly extend the spire and vaguely resemble *C. excelsus* in outline. Pattern is extremely variable in *C. amadis*, varying from almost all-white shells to almost all-brown ones. The three patterns mentioned in the description all appear distinct at first glance, but intermediates of all types are common. The dark brown banded shells do seem to be somewhat thinner in shell structure and a little higher spired than typical, but this is not constant. The narrow darker brown lines or blotches along the carinae of the shoulder and spire whorls appear to be distinct in all patterns, although the spire pattern usually varies in relation to body patterns, being light in light body patterns and dark with dark body patterns.

Distribution.—Moderately common in the northern Indian Ocean and extending to the Malaysian-Indonesian area. Records from the Philippines and New Caledonia, among others, are presumably based on misidentifications or inaccurate data. Most specimens seen today come from India and Ceylon.

Synonymy.—*Conus amadis* var. *castaneofasciata* is often used for the exceptionally attractive dark-banded form, which draws a premium price. The first use of this name apparently was by Dautzenberg, as Sowerby did not describe any such taxon, merely using the words as a descriptive term. Dautzenberg commonly cited as author of his new taxa the first person to adequately illustrate or describe the variant, regardless of whether or not that person supplied a name. *Aurantia* has the brown replaced with orange, a fairly common occurrence in *C. amadis.*

Notes.—Broadly dark-banded shells with high spires and a tendency to replace the brown with orange are uncommon and draw a premium price. Shells with weak and indistinctly blotched patterns are usually not as desirable as evenly tented or banded specimens. This species seldom shows breaks and flaws in better grade shells, although filed lips are usually found.

AMADIS.—Pondicherry, India. *Above:* 59.4 and 85.7mm. *Below:* 53.2-58.4mm.

AMBIGUUS.—Above (typical): Samba Bay, Angola, 41.3mm (MM). ***Below*** (*gernanti,* paratype): Off M'Bour, Petit Cote, Senegal, 57.2mm (EP).

AMBIGUUS Reeve, 1844

?1844. *Conus tabidus* Reeve. *Conchologia Iconica*, 1(*Conus*): Pl. 44, sp. 243 (Locality unknown).

1844. *Conus ambiguus* Reeve. *Ibid.*, 1: Pl. 44, sp. 244 (Locality unknown). See also Reeve, 1848, *C. I.*, 1(*Conus* Suppl.): 3-4.

1845. *Conus griseus* Kiener. *Species gen. et icon. des coqu. viv.*, 2 (*Conus*): Pl. 63, fig. 2. 1847, *Ibid.*, 2: 114 (Southern coast of Africa).

?1845. *Conus hepaticus* Kiener. *Ibid.*, 2: Pl. 97, fig. 3. 1849, *Ibid.*, 2: 227 (Locality unknown).

?1975. *Leptoconus gernanti* Petuch. *Veliger*, 18(2): 181, figs. 4-6 (15km NW of M'Bour, Petit Cote, Senegal).

?1975. *Conus amethystinus* Trovao. *Centro Port. Activ. Subaq.*, 4(2): 1 of separate, pl. 1, fig. 1; pl. 2, figs. 1-2 (Angola, between 12° 48' and 13° 51'S).

Description.—Thick, with a low gloss or dull finish; low conical, the sides nearly straight, usually convex just below shoulder and often concave at midbody; body whorl with six or more distinct spiral ridges at the base, these rapidly becoming irregular and often indistinct; rest of body whorl covered with irregularly spaced spiral ridges and threads of unequal strength, often obsolete below shoulder, and with numerous fine to coarse axial threads and flaws; shoulder usually concave above, carinate and broad to narrow and irregularly rounded; margin usually smooth or irregularly undulate; spire low, the sides straight to convex, whorls sometimes stepped; early whorls nodulose, usually prominent; later whorls usually with irregularly nodulose to weakly undulate margins, the tops concave and with 1 to 4 spiral threads of variable spacing and strength; spiral threads usually crossed by distinct curved axials to produce a very weakly cancellate appearance; whole spire shape and sculpture very variable. Body whorl white, violet, or pale tan, the base commonly faintly stained with brown; pattern absent except as indistinct to very strong tan to greenish axial blotches following flaws and rarely tending to form broad bands above and below midbody; spire colored similarly to body whorl, but the whorls usually marked with few to many narrow curved dark brown to greenish axial lines; early whorls commonly stained brownish. Aperture moderately wide, widest anteriorly; outer lip rather thin, nearly straight; mouth violet to white. Columella rather long, narrow, distinct, often indented from rest of inner lip. Length 30-60mm.

Comparison.—The combination of nodulose to undulate early spire whorls, irregularly spirally sculptured body whorl without distinct pattern in most populations, and the common presence of narrow curved dark axial lines over the later spire whorls is fairly distinctive. This species is very poorly understood. *C. venulatus* is smoother, has a tendency toward straight and carinate shoulder, and generally has a distinctive pattern. *C. mercator* is thinner in shell, has a rounded shoulder, and never has the early whorls nodulose; the nodulose early whorls also distinguish

C. ambiguus from *C. variegatus.* Other cones from other oceans of vaguely similar shape have distinct patterns or lack nodulose early whorls.

Variation.—As indicated in synonymy, I am not sure that this species as treated here is not composite. There seem to be three distinctive variants so far described, but I believe these are not well enough separated morphologically, and intermediate specimens will be found. The typical form has a carinate and concave shoulder, well marked spire pattern, and tan body whorl irregularly marked with greenish axial streaks; the spire whorls tend to have 1 or 2 fairly strong spiral ridges. A somewhat larger violet form from deep water (*C. gernanti*) is extremely similar to the typical form but has the spire whorls with only faint traces of 2 or 3 spiral threads. *C. amethystinus* is more round-shouldered, usually whitish, with pattern in broad bands if present, and seems to have heavier basal sculpture; the spire whorls may have none to several spiral threads and are generally more convex in form; the shell may be tinged with pale lilac in some forms. Juveniles seem to have more convex sides and a strongly tapered base. This species is very subject to spire irregularities and flaws.

Distribution.—Western coast of Africa from the Bulge to Angola. Found from shoreline to offshore waters, 0-100m.

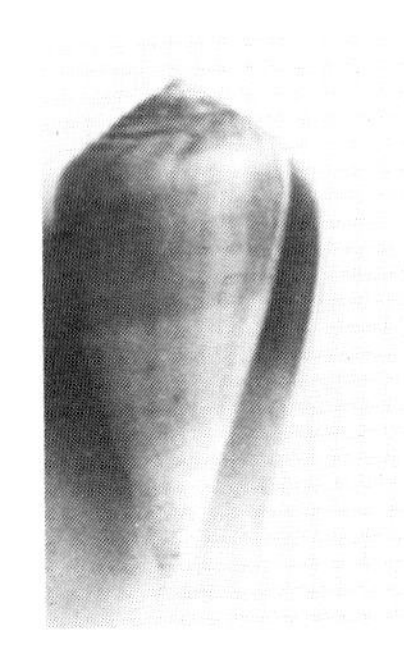

Left: *C. ambiguus* variety (photo Clover).

Right: *C. amethystinus* Trovao = *C. ambiguus* (photo Clover).

Synonymy.—*Conus tabidus* was described at the same time as *C. ambiguus*, and it might be a small specimen of the same species; the name *C. ambiguus* is retained. *C. hepaticus* has usually been associated with this species, but it seems to be lower spired, have more convex sides, and is marked with spiral ridges and brown lines in two bands; it is placed in the synonymy of *C. ambiguus* only because of tradition; some collectors have applied the name to a solid white, high-spired shell with or without coronations, probably a form of *C. marmoreus*, but this is incorrect. Reeve synonymized *C. griseus* with *C. ambiguus*, and this seems to be correct. Both *C. amethystinus* and *C. gernanti* may be distinct, especially the former, but I believe they should be considered synonyms of *C. ambiguus* until the species is better known. *C. gernanti*, at least, appears to fall easily within *C. ambiguus* except for the violet color, which may be strictly a function of the offshore habitat. *C. amethystinus* is fairly distinct.

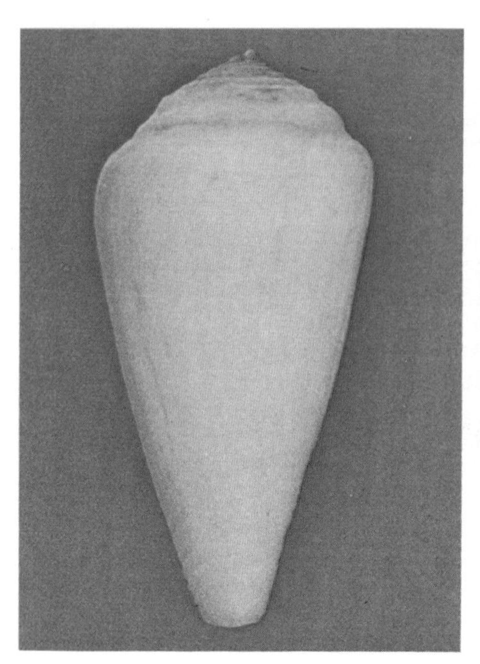

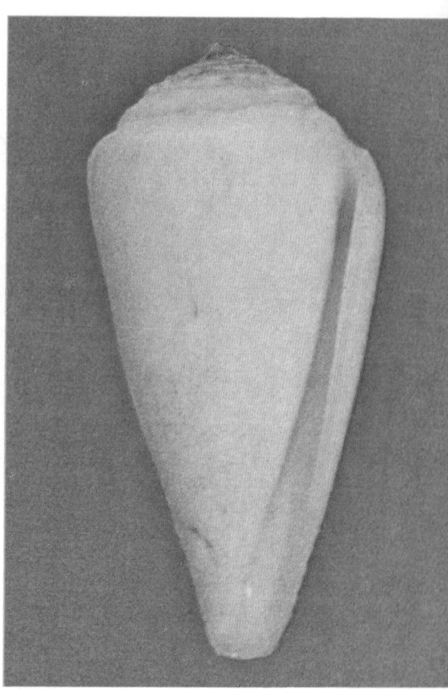

AMBIGUUS.—Above *(amethystinus)*: N'Gor, Cape Verde, Senegal, 35.6mm (EP). *Below:* Left (*amethystinus*): N'Gor, Cape Verde, Senegal, 23.7mm (EP); Right (typical): Cap des Peres, Gabon, 21.1mm (GG).

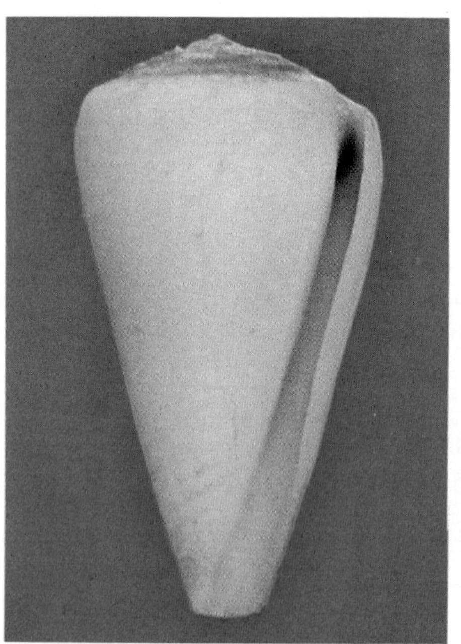

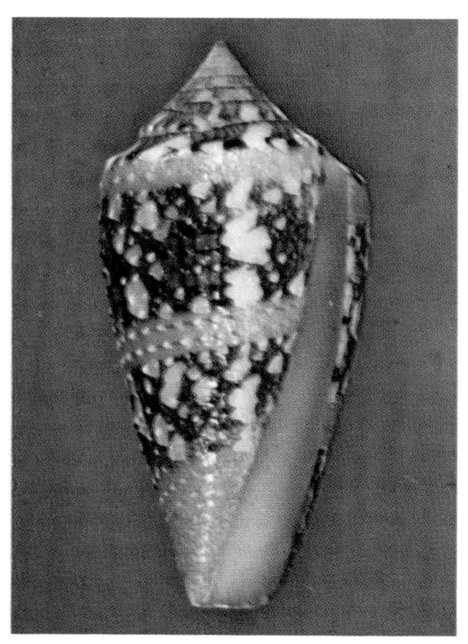

AMMIRALIS.*—Above:* Left: Off Da Nang, Viet Nam, 69.6mm (K); Right (granulose): Lungga River, Guadalcanal, Solomon Is., 37.1mm (EP). *Below:* Left: Andaman Is., 52.7mm; Right: Kwajalein Atoll, Marshall Is., 48.5mm (both MG).

Notes.—Typical broad-shouldered *C. ambiguus* from Senegal and Angola is not uncommonly seen on the market, usually with the expected irregular spires, chipped lips, and small to large breaks on the body whorl. The narrower *amethystinus* form is usually smaller, more expensive, and has eroded early whorls. Violet offshore specimens are seldom seen and command high prices; they often have the same type of flaws as shallow water specimens. Specimens of any form with distinct patterns on body whorl or spire should command a premium.

AMMIRALIS Linnaeus, 1758

1758. *Conus Ammiralis* Linnaeus. *Systema Naturae per Regna Tria Naturae,* ed. 10, 1: 713 (Locality not stated). Lectotype designated by Kohn, 1963.

1758. *Conus Ammiralis summus* Linnaeus. *Ibid.*, 1: 713 (Locality not stated).

1758. *Conus Ammiralis ordinarius* Linnaeus. *Ibid.*, 1: 714 (Locality not stated). Based on Rumphius, Pl. 34, fig. C.

1758. *Conus Ammiralis occidentalis* Linnaeus. *Ibid.*, 1: 714 (Locality not stated). Based on Rumphius, Pl. 34, fig. D.

1767. *Conus vicarius* Linnaeus. *Systema Naturae per Regna Tria Naturae*, ed. 12, 1: 1167 (Locality not stated). Holotype figure: Argenville, Pl. 15, fig. H.

1791. *Conus Ammiralis larvatus* Gmelin. *Systema Naturae per Regna Tria Naturae*, ed. 13, 1: 3378 (Locality not stated). Holotype figure: Martini, Pl. 57, fig. 635a.

1792. *Conus Ammiralis archithalassus* Hwass, in Bruguiere. *Cone,* in *Ency. Method., Hist. Nat. des Vers,* 1: 659 (Locality not stated). Lectotype (here designated): specimen in Hwass collection, 42x21mm, incorrectly considered by Kohn, 1968 to be the holotype. Granulose.

1792. *Conus Ammiralis extraordinarius* Hwass, in Bruguiere. *Ibid.*, 1: 659 (Locality not stated). Lectotype figure (here selected): *Tableau Ency. Method.*, Pl. 328, fig. 9.

1792. *Conus Ammiralis palinurus* Hwass, in Bruguiere. *Ibid.*, 1: 659 (Locality not stated).

1792. *Conus Ammiralis personatus* Hwass, in Bruguiere. *Ibid.*, 1: 660 (Locality not stated). See *Tableau Ency. Method.*, Pl. 328, fig. 7.

1792. *Conus Ammiralis polyzonus* Hwass, in Bruguiere. *Ibid.*, 1: 659 (Locality not stated). See *Tableau Ency. Method.*, Pl. 328, fig. 8.

1798. *Cucullus granulatus* Roeding. *Museum Boltenianum:* 45 (Locality not stated). Non *C. granulatus* Linnaeus, 1758.

1798. *Cucullus princeps* Roeding. *Ibid.:* 45 (Locality not stated). Lectotype figure (Kohn, 1975): Knorr, Pl. 8, fig. 2. Non *C. princeps* Linnaeus, 1758.

1798. *Cucullus summus* Roeding. *Ibid.:* 45 (Locality not stated). Lectotype figure (Kohn, 1975): Chemnitz, Pl. 141, fig. 1307. Non *C. summus* Linnaeus, 1758.
1798. *Cucullus torquatus* Roeding. *Ibid.:* 45 (Locality not stated). Lectotype figure (Kohn, 1975): Favanne & Favanne, Pl. 17, fig. 1/4. Non *C. torquatus* (Roeding, 1798, p. 38).
1798. *Cucullus equestris* Roeding. *Ibid.:* 46 (Locality not stated). Non *C. equestris* (Roeding, 1798, p. 38). See discussion in Kohn, 1975.
1810. *Conus vicarius* Lamarck. *Ann. du Mus. Hist. Nat. (Paris)*, 15: 274 (Indian Ocean ?). Non *C. vicarius* Linnaeus, 1767. Largely white.
1829. *Conus Blainvillei* Vignart. *Descr. Cone Nov. (Bibl. Francaise:* 692) (Locality not stated). See also Kiener, 2: 135. Granulose. Author also spelled Vignard and the taxon *C. blainvilli.*
1834. *Conus Ammiralis Amboinensis* Donovan. *Naturalist's Repository:* Pl. 1, fig. 1 (Amboina). See *Tableau Ency. Method.*, Pl. 328, fig. 5.
1930. *Leptoconus ammiralis temnes* Iredale. *Mem. Queensland Mus.*, 10: 80 (North-west Isle, Capricorn Group, Queensland, Australia).
1937. *Conus ammiralis* var. *australis* Dautzenberg. *Mem. Mus. Roy. d'Hist. Nat. Belg.*, 2(18): 18 (Locality not stated). See *Tableau Ency. Method.*, Pl. 328, fig. 6. Non *C. australis* Holten, 1802.
1937. *Conus ammiralis* var. *crebremaculata* Dautzenberg. *Ibid.*, 2(18): 19 (Locality not stated). Lectotype figure (here selected): Chemnitz, Pl. 141, fig. 1309.
1937. *Conus ammiralis* var. *Donovani* Dautzenberg. *Ibid.*, 2(18): 20 (Amboine). Based on Donovan, 1834, *Nat. Repos.*, Pl. 1, fig. 2.

Description.— Heavy, with a very high gloss; low conical, the sides nearly straight or slightly concave at middle; body whorl with a few weak spiral ridges at the base, otherwise smooth in normal specimens; shoulder broad, roundly angled; spire low to moderate, the sides concave to almost straight; spire pointed, the tip often weakly eroded; whorls with very weak spiral threads and axial wrinkles, often appearing smooth. Body whorl pattern very variable, but typically with two broad brown bands above and below midbody, separated and bordered by narrow parallel-edged bands of fine orange tents; base tented; whole pattern interrupted by scattered large and small white tents which may be sparse or numerous; brown bands usually with weakly defined spiral lines of long dark dashes; shoulder and spire with large brown and white blotches arranged in regular order to produce white half-moons; brown blotches usually with very fine axial dark brown lines; spire tip pinkish. Aperture moderately wide, straight; outer lip thin, sharp, slightly sloping below shoulder; mouth white. Columella narrow, relatively short. Length 45-77mm.

Comparison.— *C. ammiralis* in its various patterns is usually very distinctive and hard to confuse. *C. nobilis* has a similar spire pattern but is thinner and more elongated, usually with a lower spire and no bands of very fine orange tents in the pattern. Some color patterns of *C. amadis*

AMMIRALIS.—(coronate) *Above:* Reunion, 44mm (DM) (photos by Jayavad). *Below:* Mauritius: Left: 64.6mm; Right: 49.0mm (DMNH).

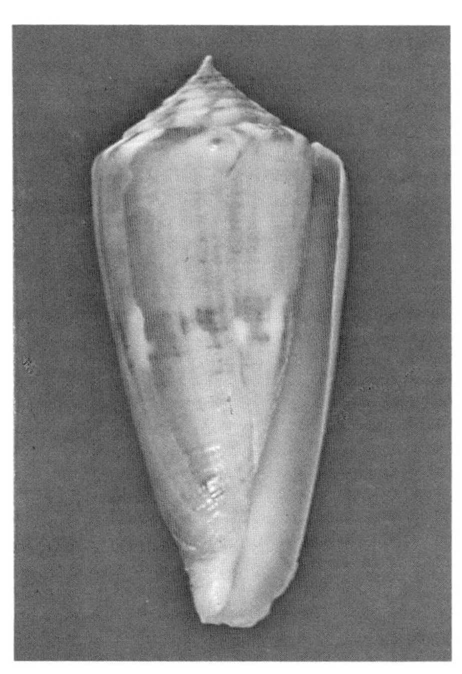

AMPHIURGUS.—80 mi. W of Boca Grande, Florida. *Above:* 42.7mm. *Below:* Left: 33.8mm; Right: 16.8 and 24.0mm.

also share the regular spire pattern and are superficially similar even to bands of very fine tents, but that species is lighter in weight, has concave spire whorls usually with distinct sculpture, and is often higher spired. *C. textile* and *C. pennaceus* are rounder shouldered and differ in many details of pattern and sculpture.

Variation.—Except for pattern, *C. ammiralis* shows relatively little variation. The spire may be low or moderate, occasionally with one or two later whorls weakly concave or stepped. Rare specimens may be covered wholly or partially with spiral rows of fine or coarse granules that commonly extend to the shoulder; such granulose specimens usually have distinctly coronate shoulders and often some coronate spire whorls. The pattern is very variable in details although there is seldom any problem recognizing its basic form. The broad brown bands may be split into several narrow bands; colors may vary from pale orange brown to almost black, the spiral brown lines nearly obsolete or very evident; white triangles may be few and scattered or large and concentrated in broad axial bands over most of the shell. Generally the midbody area has a narrow interrupted brown spiral band, but this is occasionally absent or doubled. Rare shells have the white tents greatly enlarged to occupy most of the body whorl, resulting in scattered irregular brown blotches instead of bands. All these variants appear to be purely individual and no subspecies can be recognized at the moment.

Closely related to *C. ammiralis* and perhaps belonging to the same species is a rare shell apparently restricted to the southern Indian Ocean. This shell is commonly called *C. architalassus* Solander in the belief that the name refers to a coronate form of *C. ammiralis*; however, Solander's name is a synonym of *C. cedonulli* and Hwass' later *C. archithalassus* is just the finely granulose variety of *C. ammiralis.* For these reasons the cone is presently nameless. However, it is so similar to *C. ammiralis* that the value of placing any new name on it is doubtful, so it is simply called the coronate admiral for the moment. Basically, it differs from *C. ammiralis* in being lighter in weight with distinct and strong coronations on the spire whorls; the whorls are usually distinctly concave above, perhaps in response to the coronations. The pattern is essentially that of *C. ammiralis* and shows as much variation in details, although there may be a stronger tendency to develop dark brown axial lines on the body whorl within the brown bands. The shell is rare and apparently restricted to the southern Indian Ocean islands. It should not be confused with the granulose admirals, which are also rare (but less so than the coronate admiral) and seldom have well developed coronations on the spire whorls; the body whorl in the coronate admiral is smooth.

Distribution.—Indo-West Pacific. It is apparently rare or at least uncommon in most of the Indian Ocean and only moderately common from the southern Japan-Philippines-Queensland arc. The species appears to be unrecorded from either the Hawaiian area or French Polynesia.

Synonymy.—The rather long synonymy is greatly extended by the worthless names of Roeding, most of which are probably just bad misidentifications of Linnaeus' names. The other names were given to individual variants, many of which are easily found in any large lot. Only *C. ammiralis archithalassus* Hwass (note that there is only one letter difference from *C. architalassus* Solander) for the rare granulose admiral and *C. vicarius* Lamarck (preoccupied) for the largely white admiral with scattered brown blotches are worth the collector's effort to find and remember the varietal names. *C. ammiralis temnes* is merely a barely separable deepwater variant with higher than usual spire and the normal variation from typical color pattern; it is not worthy of recognition at any level.

Notes.—Most specimens of *C. ammiralis* have one or two small axial breaks visible on the body whorl. The granulose form is rare, the whitish form rare, and the coronate form very rare. Spectacular individuals of the typical pattern always command a premium.

AMPHIURGUS Dall, 1889

1889. *Conus amphiurgus* Dall. *Bull. Mus. Comp. Zool., Harvard,* 18: 70 (27 fms near the coast of Yucatan [Cabo Catoche], Mexico). Holotype illustrated by Clench, *Johnsonia,* 1953, 2(32): 369.

1942. *Conus juliae* Clench. *Johnsonia,* 1(6): 26, pl. 12, fig. 4 (9 mi. off Fort Walton, Okaloosa Co., Florida).

1968. *Conus mayaguensis* Usticke. *Caribbean Cones from St. Croix and the Lesser Antilles:* 15, pl. 2, sp. 1003 (Pta. Guanajibos and Pta. Arenas, west coast of Puerto Rico).

Description.—Moderately heavy, with a low to high gloss; low biconic to obconic, the sides usually convex; body whorl with 6-8 low spiral ridges at the base, these rapidly becoming obsolete; rest of body whorl covered with minute spiral and axial threads, grooves, and flaws of variable size, strength, and spacing, often nearly smooth to the touch; shoulder roundly angulate, slightly concave above; spire low to moderate, sharply pointed if not eroded, the sides straight to concave; tops of whorls flat to distinctly concave, with about 3-4 irregular and sometimes indistinct spiral ridges crossed by widely spaced curved axial threads; earliest 2 or 3 whorls weakly nodulose. Body whorl usually some tone of red, pink, orange, or even yellow, rarely olive green or tan; a whitish midbody band often developed but sometimes heavily infiltrated by irregular brownish blotches; usually a row of brown squarish blotches at the posterior edge of the band; whorl covered from base to shoulder with numerous rows of small brown dots in spiral lines, these sometimes variably fused to produce irregular axial bands or large blotches; spiral bands of white blotches may be present below the shoulder and above the base; spire and shoulder

ANEMONE.—Above: Left: Torquay, Victoria, Australia, 33.0mm; Right: Rockingham, Western Australia, Australia, 29.0mm. *Below:* Left: Dampier, Western Australia, Australia, 42.7mm; Right (*compressus*): Adelaide, South Australia, Australia, 38.4mm.

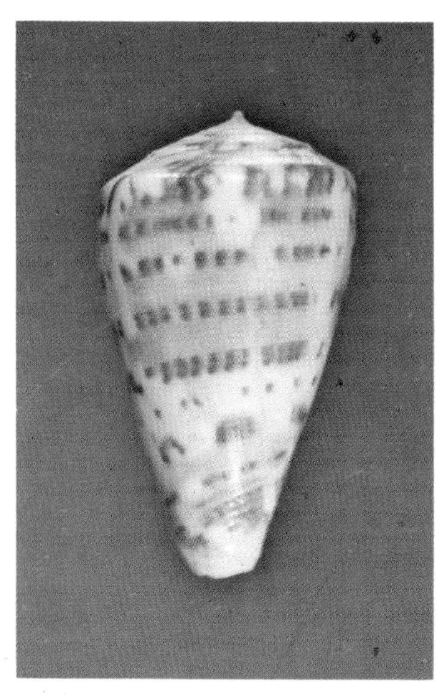

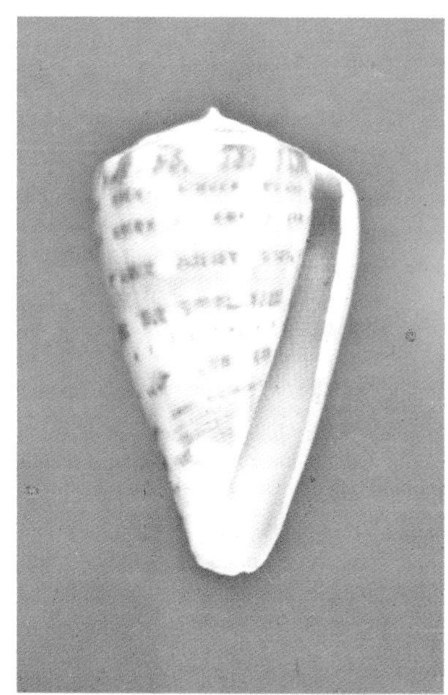

ANGASI.*—*Above: Off Southpoint, Queensland, Australia, 39.6mm (GG). *Below:* Left: Southport, Queensland, Australia, 39.1mm (DMNH); Right (very close to *advertex*): Cape Moreton, Queensland, Australia, 36mm (photo by P. Clover).

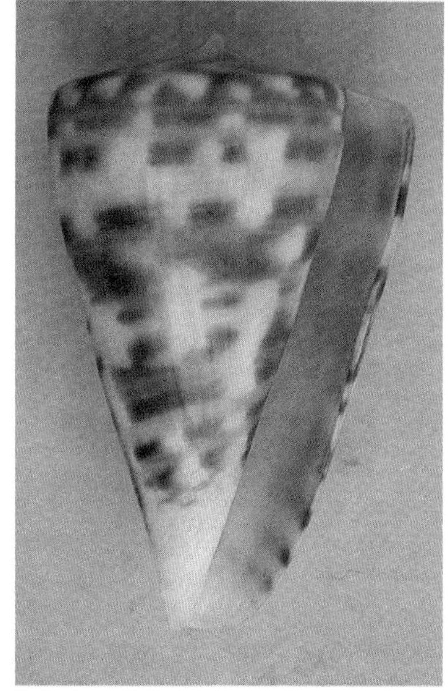

white, heavily marked with reddish to yellowish blotches about equal to white areas in size and evenly alternating with them; margin of shoulder commonly with darker brown blotches; tip of spire pale pink. Aperture moderately wide, slightly widened anteriorly; outer lip thin, posterior margin about on level with shoulder; outer lip slightly convex or straight; interior of mouth white, pinkish, or faintly violet. Columella short and largely fused to lip. Length 20-60mm.

Comparison.—In shape and general color pattern *C. amphiurgus* is very similar to the Panamic *C. vittatus. C. vittatus* is usually more cylindrical in shape with a stronger tendency to develop white blotch bands at base and below shoulder; the spire markings are almost black in typical *C. vittatus* and cover most of the spire and shoulder. Although *C. amphiurgus* is variably spirally striated, in *C. vittatus* the spiral ridges on the body whorl are fine, regular, and easily feelable with the finger-tip in all specimens, the ridges bearing the brown dots. *C. beddomei* may be similar in pattern to *C. amphiurgus* but has only the first whorl nodulose (erosion usually makes this meaningless), is usually smaller, seldom has distinct spiral rows of brown spots developed, and commonly has a deep violet aperture. *C. henoquei* has a distinctly stepped spire; *C. cumingii* differs in spire sculpture and details of color and pattern.

C. daucus is very difficult to distinguish from *C. amphiurgus* in some areas, and the two are superficially very similar in shape and pattern. Generally *C. daucus* has a sharply carinate shoulder with a very low spire with strongly concave sides. The body whorl is more elongate than *C. amphiurgus* and has a narrower base. If *C. daucus* has spiral rows of brown spots, they are commonly restricted to only three or four rows spaced over the body whorl or grouped near midbody, seldom extending from base to shoulder; some *C. amphiurgus* lack spots, however. Shape is the easiest wav of distinguishing the two species.

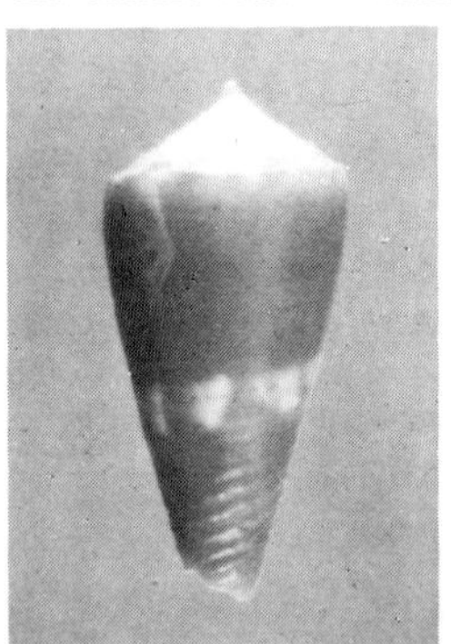

C. amphiurgus, Florida, 19mm (R. Guest).

Variation.—The most obvious variation is in body color, from yellow to bright red to greenish; reds and orange are probably most common, with yellow, greens, and tan rarest. Occasional specimens, especially small specimens and large adults, may be almost unicolor, the brown spots subdued or virtually absent. Brown spots and blotches (other than those in spiral series) may be small and restricted to the midbody area or large and developed below the shoulder and above the base; these latter areas also tend to develop variable white blotches. Occasional specimens lack the light midbody area. Occasional juveniles have undulate shoulders, while adults commonly have the spiral sculpture of the shoulder and later whorls nearly obsolete. The species may be broadly conical in shape with distinctly convex sides or relatively narrower with nearly straight sides, the spire rather high or almost flat. The regular arrangement of white and colored areas on the spire is usually consistent.

Distribution.—Uncommon on offshore reefs and other hard substrates from off the Carolinas to Florida, Yucatan, and Puerto Rico; it apparently also occurs off the Texas coast. Puerto Rico specimens are commonly dwarfed. The status of specimens from the Lesser Antilles is uncertain, as they could be *C. amphiurgus*, *C. beddomei*, or very unusual variants of *C. daucus*.

Synonymy.—Although this species is commonly called *C. juliae*, it appears that *C. amphiurgus* is more applicable. Dall's description is largely comparative with little actually said about the new species, but it is a valid description and there is a holotype preserved. The name has long been referred to *C. villepinii*, but that species has concave sides, a sharply angled shoulder region, is more attenuated, and seldom or never has a pattern similar to that shown by the holotype. The locality, 27 fathoms off Yucatan, also fits this species better than *C. villepinii*, which is usually found in much deeper water. *C. mayaguensis* differs from typical *C. amphiurgus* only in being smaller, being found in shallow water, and occurring in Puerto Rico.

Notes.—As *C. amphiurgus* is a species usually dredged or trawled from offshore reef areas, perfect specimens are seldom seen. Adults commonly have at least one large healed break on the body whorl distorting the pattern, have the lip chipped or filed, and commonly have an eroded spire. Although dealers seem to get higher prices for orange specimens with little brown pattern, the red shells with much brown seem to me to be seen less commonly on the market. Good adult shells of this beautiful species are very hard to obtain.

ANEMONE Lamarck, 1810

1810. *Conus anemone* Lamarck. *Ann. du Mus. Hist. Nat. (Paris)*, 15: 272 (New Holland). Types figured by Kiener: var. b, Pl. 46, fig. 3; var. c, Pl. 46, fig. 3a.

1833. *Conus maculosus* Sowerby i, in Sowerby ii. *Conchological Illustrations:* Pt. 24, figs. 3, 3* (Locality not stated). Non *Conus maculosus* (Roeding, 1798).

1833. *Conus maculosus* var. *pallescens* Sowerby i, in Sowerby ii. *Ibid.:* Pt. 24, fig. 3* (Locality not stated).

1854. *Conus Novae-Hollandiae* A. Adams. *Proc. Zool. Soc. (London)*,

ANTHONYI.—Cape Verde Is. *Above:* Left: 13.2mm; Right (paratype): Sao Tiago I., 12.1mm (DM). Below: 18.7mm (PC).

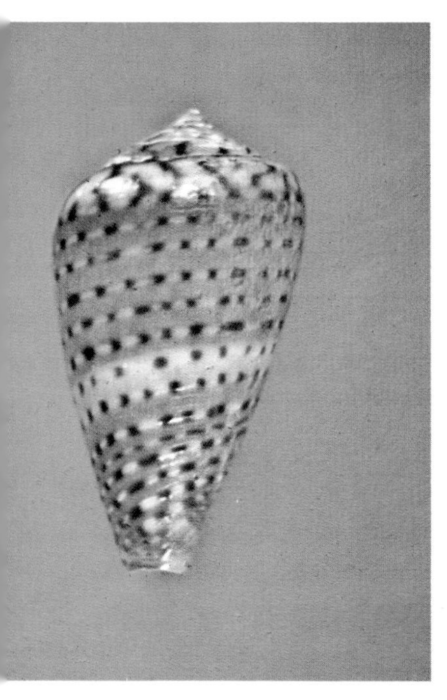

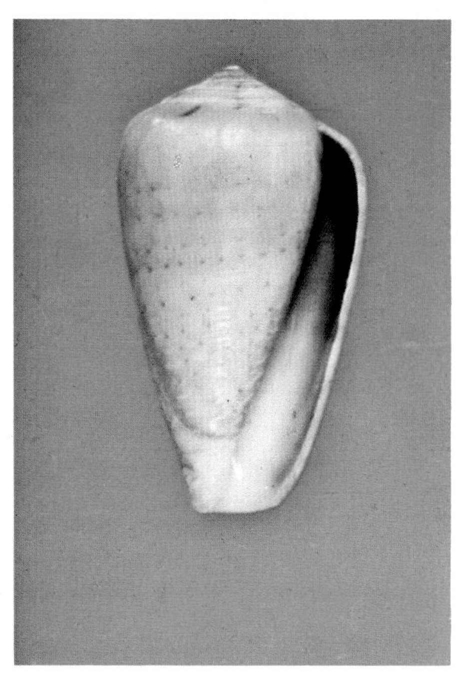

APLUSTRE.—Above: Left: Hastings Point, New South Wales, Australia, 24.4mm; Right: "Great Barrier Reef", 20.5mm (GG). *Below:* Hastings Point, New South Wales, Australia, 20.5-23.0mm.

1853: 119 (Swan River).
1857-1858. *Conus maculatus* Sowerby ii. *Thesaurus Conchyliorum*, 3 (*Conus*): 31, pl. 13 (199), fig. 296 (Capul, Philippines). Non *Conus maculatus* Bosc, 1801. This seems to be either a *lapsus calami* or replacement name for *C. maculosus* Sowerby i, 1833.
1866. *Conus compressus* Sowerby ii. *Ibid.*, 3(*Conus* Suppl): 325, pl. 25, (286), figs. 602-603 (Locality unknown). Turreted specimen.
1866. *Conus roseo-tinctus* Sowerby ii. *Ibid.*, 3(*Conus* Suppl.): 325, pl. 25 (286), fig. 604 (Locality unknown). Reddish.
1870. *Conus Rossiteri* Brazier. *Proc. Zool. Soc.* (*London*), 1870: 109 (Botany Bay, Cape Solander, New South Wales, Australia).
1877. *Conus Carmeli* Tenison-Woods. *Papers Proc. Rpt. Roy. Soc. Tasmania*, 1876: 134 (North coast of Tasmania). See May, 1921, *Checklist Moll. Tasmania:* 73.
1898. *Conus Flindersi* Brazier. *Proc. Linnean Soc. New South Wales*, 22: 780 (Flinders, Victoria, Australia). Turreted specimen.
1898. *Conus Remo* Brazier. *Proc. Linnean Soc. New South Wales*, 23: 271 (San Remo, Victoria, Australia). Type figured by Cotton, 1945, pl. 4, fig. 9. Orange blotched.
1942. *Conus incinctus* Fenaux. *Bull. l'Inst. Oceanogr.* (*Monaco*), No. 814: 3, fig. 6 (Australie).
1942. *Conus nitidissimus* Fenaux. *Bull. l'Inst. Oceanogr.* (*Monaco*), No. 814: 3, fig. 7 (Australie).
1945. *Floraconus singletoni* Cotton. *Rec. South Australian Mus.*, 8(2): 263, pl. 4, fig. 10 (Western Port, South Australia, Australia). Uniformly whitish, creamy, or pinkish.
1945. *Floraconus saundersi* Cotton. *Ibid.*, 8(2): 264, pl. 4, fig. 8 (Edithburgh, South Australia, Australia). Weak axial pattern.

Description.—Fairly light in weight, with a low gloss or dull finish; shape variable, usually low conical but occasionally pyriform or elongate, the sides usually convex and somewhat bulbous posteriorly; body whorl covered with closely spaced undulating narrow spiral ridges, these heaviest near base, often very weak at midbody, then strong again below shoulder; body whorl crossed by numerous closely spaced fine axial threads, but these do not give a cancellate appearance to the whorl; shoulder broad, roundly to sharply angled, concave to convex above; spire usually low to moderate, pointed, the sides concave or straight, occasionally weakly or strongly stepped; early whorls usually not eroded, occasionally faintly nodulose; later whorls concave, convex, or flat above, the margins straight or faintly undulate, covered with about 5-8 narrow weak often irregular spiral threads crossed by weaker spaced axial curved threads. Coloration of body whorl very variable; background color usually dirty cream, bluish gray, pale tan, or orange, overlaid by variable blotch pattern of brown, dark blue-gray, black, orange, or red; blotches may be strongly developed either axially or spirally and partially fused into large patches of color or they may be reduced to a few isolated color spots; often the blotches are heavily invaded by small spots of white or pale bluish to produce a vaguely cross-hatched appearance; commonly the blotches are best developed in spiral bands above and below midbody, leaving lighter bands above base, at midbody, and below shoulder; spire marked much as body whorl, usually light with radiating bands or blotches of same color as body whorl; apex often pinkish brown. Aperture moderately wide, especially anteriorly; outer lip thin, sharp, straight or evenly convex; mouth usually dirty violet brown or at least dirty cream, darkest anteriorly, often with a light band at midbody and below shoulder. Columella very narrow, largely fused, not easily visible. Length 30-60mm.

Comparison.—*Conus anemone* is apparently the stem or parent species of a large Australian species group. *C. papilliferus* is extremely similar in pattern in some specimens but generally is smaller, has a slightly different shape, and has a nearly smooth body whorl. *C. peronianus* is larger, usually bright pink, orange, or violet and purple, has a lower spire, and has the body whorl much more convex; it also has the spiral sculpture of the body whorl reduced or absent and the early whorls of the spire often more conspicuously nodulose. *C. klemae* is also smooth, has a higher spire with strongly concave sides, and has a distinctive color pattern involving three white bands outlined by dark blotches. *C. aplustre* is smaller, has a smooth body whorl, more convex spire whorls, and a very different color pattern. *C. clarus* is more evenly biconical, smooth, and has straighter sides to the body whorl as well as a different color pattern. *C. cocceus* is similar to some forms of *C. anemone* in being uniformly white with only a faint reddish pattern in most specimens, but the spiral ridges of the body whorl are often coarser, the shoulder is more rounded, the shape is generally more slender and evenly convex, and the aperture is white. The dingy appearance, distinct spiral ridges over the body whorl, dark aperture, and numerous fine spiral threads on the spire whorls are usually enough to distinguish *C. anemone* from any non-Australian species which might be vaguely similar.

Variation.—As indicated in the description, this is an extremely variable species, with many of the variants bearing specific names at various times in their history. Most shells have low spires which are at best weakly stepped, but the common variant *compressus* has a high, strongly stepped spire, often with nodulose margins, and the sides of the body whorl nearly straight. The color pattern of this variant assumes all the differences shown by other more typical *C. anemone* and may be heavily blotched with blue-black or nearly uniform pinkish orange. Intermediate specimens occur over a wide area, although the *compressus* form is most common in South Australia.

Specimens with bright patterns of red or orange have commonly been described as full species. Uniformly whitish shells from South Australia called *C. singletoni* appear to be local albinotic populations, and

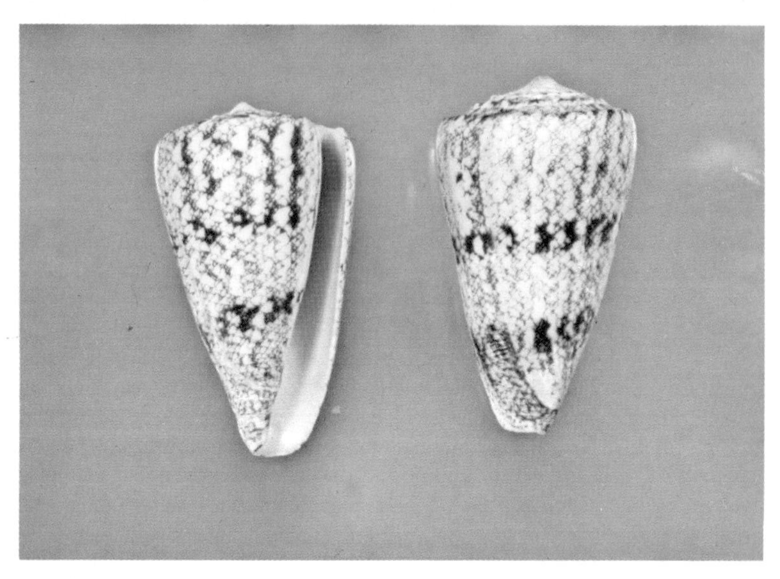

ARANEOSUS.—*Above* (*a. araneosus*): Rameswaram, India, 61.9 and 62.4mm. *Below* (*a. nicobaricus*): Sulu, P.I., 63.1 and 63.4mm.

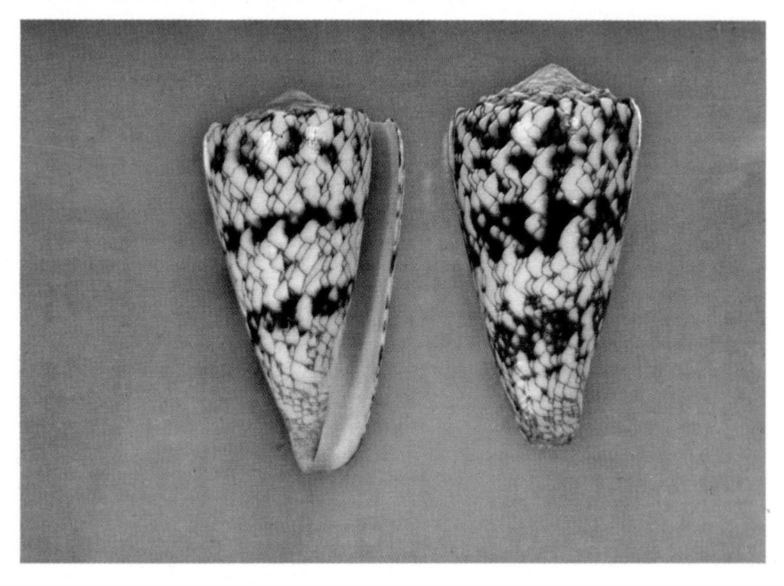

ARCHON.—Above: Left: Off Cedros I., Baja California, Mexico, 61.5mm; Right: Pedro Gonzales, Panama, 37.3mm. *Below:* Cabo San Lucas, Baja California, Mexico, 46.4-54.1mm (photo by A. Kerstitch).

the same explanation as heavily inbred local populations appears to apply to many of the other described variants.

Specimens from Western Australia north of Geraldton are somewhat more bulbous than typical *C. anemone* and have stronger and more uniform spiral ridging; these are sometimes referred to the subspecies *C. anemone novaehollandiae*, but the distinction is so vague as to be unusable on specimens of unknown locality; subspecific recognition seems undeserved.

Distribution.—Restricted to Australia. Ranges from central New South Wales around southern Australia (including Tasmania) to Western Australia at Dampier Land. Records from New Zealand, the Great Barrier Reef, and New Guinea would appear to be in error, although the presence of *C. anemone* in these areas could be explained by capture in currents; Great Barrier Reef and New Guinea records are especially possible.

Synonymy.—Because the numerous named forms all appear to be extreme variants characteristic of local populations rather than uniformly distributed species, they are not recognized as valid. Admittedly some of the variants, such as *C. compressus* and *C. singletoni*, are very distinctive in shape or color, but there are usually intermediates or the population is highly localized and inbred. These names will continue to be used by collectors as form names, I am sure. I see no reason to recognize subspecies in *C. anemone* as the variation is so great on even a local level as to encompass the entire range of differences between typical *anemone* and *novaehollandiae.* The Fenaux names are juvenile *compressus.*

Although *C. papilliferus* and *C. peronianus* are usually considered synonyms of *C. anemone*, I believe they are probably distinct (especially *C. papilliferus*) and have treated them separately.

Notes.—High-spired, brightly colored, and uniformly colored specimens are very desirable and often command very high premiums. The dingy appearance of typical *C. anemone* is often enhanced by growth flaws, unhealed breaks, encrusting organisms, and patches of persistent periostracum.

ANGASI Tryon, 1883

1877. *Conus metcalfei* Angas. *Proc. Zool. Soc.* (*London*), 1877: 173, pl. 26, fig. 13 ("Sow and Pigs" Reef, Port Jackson, New South Wales, Australia). Non *Conus metcalfii* Reeve, 1843.

1883. *Conus Angasi* Tryon. *Manual of Conchology*, 6(1): 62, pl. 19, fig. 99 (Port Jackson, Australia). Nomen novum for *Conus metcalfei* Angas, 1877; presented in form of new species description, however.

Description.—Light in weight, fragile, with a high gloss; low conical, the body whorl with convex sides; body whorl smooth except for a few weak spiral basal ridges and finely punctate spiral grooves which may extend to near shoulder in juveniles but usually end below midbody in adults; very fine axial scratches usually visible over body whorl; shoulder broad, rounded; spire low, convex; whorls slightly concave or slightly convex above, covered with about 5-6 fine spiral ridges crossed by scattered weak axial curved threads. Body whorl white, cream, or very pale tan overlaid with pinkish; pattern consisting of about 6-10 spiral rows of small brown squares of variable size, these commonly smaller and somewhat axial below the shoulder, weak or absent at midbody, and narrow and spirally elongated above the base; bands of spots often connected axially by brown blotches with irregular outlines to produce a pattern of large blotches; large brown blotches usually show spiral rows of small white spots, the background color showing through between the smaller brown spots which have been incorporated into the blotches; basal area often heavily suffused with brown, obscuring the spots; midbody area often pale and virtually unspotted; spire whitish, with 7-8 irregular radiating brownish blotches; early whorls white. Aperture wide throughout; outer lip smoothly convex, posterior edge about on level with shoulder; mouth white to pale pinkish. Columella with posterior half free from inner lip and forming a strong projecting cusp. Length 30-45mm.

Comparison.—The color pattern and partially free columella ally *C. angasi* with *C. advertex*. In actuality, the only real differences are the rounded shoulder and convex spire of *C. angasi* as opposed to the angled shoulder and flat spire of *C. advertex*. There are minor differences in body whorl shape (somewhat broader and more convex-sided in *C. angasi* than in *C. advertex*) and in color pattern (more diffuse in *C. angasi*, the basal spotting not so line-like and a stronger tendency for continuous rows of square small brown spots), but these differences are hard to evaluate. See *C. advertex* for comparisons with *C. trigonus* (flat spire or convex spire), *C. sculletti*, and *C. papilliferus*.

Variation.—There is little variation in shape in this species, but the expected color pattern variation is as in *C. advertex*. The midbody area is usually clear, the basal area and sub-shoulder area more heavily blotched. There may be several rows of distinct square brown spots with little or no nebulous clouding axially, or there may be only a few large irregular brown blotches containing small white spots. Spots in the basal area may be small, indistinct, or spirally elongated. Punctate spiral lines are most easily seen in juveniles and usually require magnification.

Distribution.—Moderately deep waters (60-150m) off southern Queensland and northern New South Wales. Not common.

Notes.—This and *C. advertex* very possibly are ecotypes of a single species, the minor differences in shape and pattern reflecting the slight differences in depth occupied by the animals. Intermediate specimens are to be expected. Most *C. angasi* I have seen have at least one heavy break on the body whorl and often small spots of erosion.

ARCUATUS.—Perlas Is., Panama, 53.6 and 56.6mm.

ARENATUS.—*Above:* Nordup, E New Britain, Papua New Guinea, 29.0 and 31.0mm. *Below:* Left: Bolo Point, Okinawa, Japan, 33.2mm; Right: Mozambique I., Mozambique, 39.1mm.

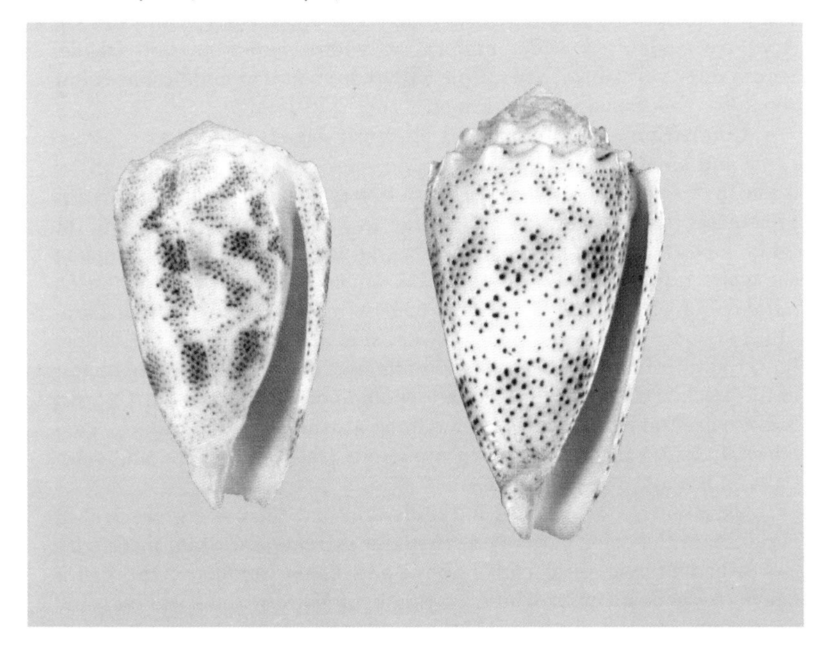

ANTHONYI (Petuch, 1975)

1975. *Africonus anthonyi* Petuch. *Veliger,* 17(3): 263, figs. 5-6 (Baia do Inferno, Sao Tiago I., Cape Verde Islands).

Description.—Very small, light in weight but solid, with a good gloss; low conical, the shoulder somewhat bulbous but the sides straight and tapering to base; body whorl with about 10-12 low wide spiral ridges at the base, then smooth or nearly so in most specimens; occasional specimens have distinct spiral ridges over most of the body whorl, these well spaced; shoulder rounded or roundly angled, slightly concave above; spire low to moderate, bluntly pointed if not eroded; sides of spire straight to concave; spire whorls with about 2-4 low weak spiral ridges crossed by scattered weak axial threads; tops of whorls flat, slightly concave, or slightly convex. Body whorl dark tan to brown, moderately or heavily covered with zigzag axial blotches and lines of cream, bluish gray, or white, producing a tented pattern in most specimens; white areas may form solid bands below the shoulder, at midbody, and above the base, or may be restricted to large blotches in these areas; usually two broad bands of brownish with irregular edges visible above and below midbody; base brown, heavily mottled with white; occasional specimens have narrow irregular brown spiral lines over the whorl, these variably developed and often corresponding with the spiral ridges which may be present; spire whorls largely white or bluish gray, heavily or weakly marked with brown to blackish radiating blotches which are best marked on shoulder and may be as wide as or narrower than white areas; early whorls brown. Aperture moderately wide, uniform in width; outer lip thin, fragile; mouth dirty violet-brown to purple with paler bands at middle and below shoulder. Columella very indistinct. Length 10-19mm.

Comparison.—*C. anthonyi* is obviously closely related to *C. mercator* and allies, differing largely in the small size and large amount of white in the color pattern. The pattern is very variable, but generally the area below the shoulder is largely white, as is the midbody area, with the white having a finely mottled or zigzag appearance; in some specimens the white is in blotches, resulting in an appearance more like that of *C. balteus,* which is a larger, thicker shell with well defined white flammules. *C. mercator* has some forms very similar in size, but these seldom have the sub-shoulder area largely white and commonly show remnants of the black or dark brown reticulations common to *C. mercator. C. taslei* has as a general rule fine axial brown lines absent in *C. anthonyi. C. ventricosus* and *C. aemulus* differ in numerous features of shape and color pattern. See also *Conus venulatus.*

Variation.—As indicated in the description, there is a great deal of variation in the color pattern. At the light extreme are white shells with two broad brown spiral bands above and below midbody, the bands heavily reticulated with white; in this light pattern the spire has only small dark blotches and lines. Intermediate specimens are rather uni-

C. anthonyi variants (photo EP).

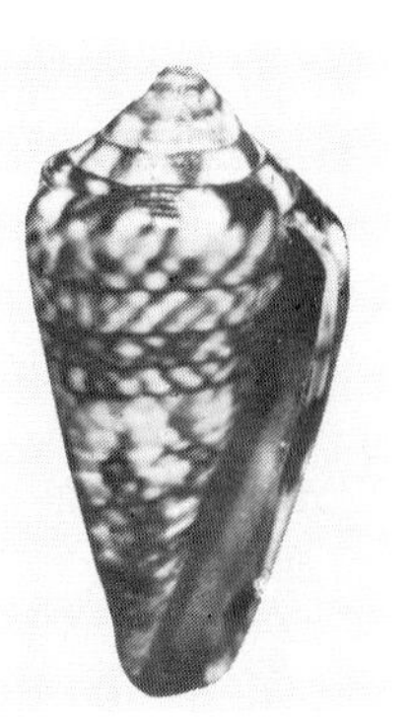

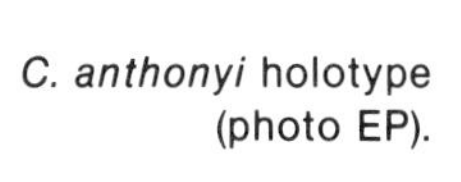

C. anthonyi holotype (photo EP).

formly mottled and reticulated with white on a brown background, no broad spiral bands being visible. At the dark extreme are specimens with the white restricted to a broad band below the shoulder and large blotches at midbody and above the base; such shells may have large dark blotches on the spire. Spire shape varies a little, from low to moderate and straight or concave. Occasional specimens have a few weak spiral ridges developed on the body whorl, these sometimes bearing fine brown lines. Rare specimens have the body whorl entirely ridged, many or most of the ridges having dark brown spiral lines. Presence or absence of ridges and spiral brown lines seems to be individually variable.

Distribution.—Seemingly restricted to the Cape Verde Islands. Uncommon in shallow rocky waters.

Synonymy.—The choice of the name *anthonyi* for this species (if it is indeed a full species) instead of the usual *C. desidiosus* is necessary because the type of *C. desidiosus* seems to be a small (23.5mm) specimen of *Conus taslei* with a high conical spire, largely brown body whorl, white midbody band and mottled white sub-shoulder band. The name cannot be applied to the small shell under discussion here.

Petuch applied the name *anthonyi* to small shells of the type usually called *desidiosus* by modern collectors, but having the body whorl covered with raised spiral ridges often bearing dark brown spiral lines; the pattern of the holotype and a paratype represent a dark specimen and an intermediate pattern, so this is variable to the same extent as the small shell more familiar to collectors. Since some otherwise typical '*desidiosus*' have spiral ridges below the shoulder and some have short dark brown spiral lines or dashes, I consider this to be an individually (or at most populationally) variable trait and use the name *anthonyi* for the small shell with variable white mottling.

It would probably be best to consider this just another variant of *C. mercator* or *C. balteus*, but it appears fairly distinct. Admittedly species in this group of shells are hard to diagnose and recognize.

Notes.—Typical specimens have at most traces of spiral ridges and brown lines; heavily ridged specimens with full spiral lines are rare. The spire is often eroded, there may be small breaks on the body whorl and

ARGILLACEUS.—Above: Yemen, 59.0mm. *Below:* Aden: Left: 49.9mm; Right: 56.4mm (both DMNH).

ARMIGER.—(*a. armiger*) *Above:* S of Pensacola, Florida, 36.3 and 40.0mm (GG). Below: Off Pensacola, Florida, 32.1mm (GG).

shoulder, and the lip is commonly rough; such minor defects should be expected in any small cone because of the size. Probably the species is more common than prices would lead one to believe, but it is easily overlooked among rocks in heavy currents of the restricted habitat.

—

APLUSTRE Reeve, 1843

1843. *Conus aplustre* Reeve. *Conchologia Iconica*, 1(*Conus*): Pl. 30, sp. 170 (Locality unknown).

?1849. *Conus lugubris* Reeve. *Ibid.*, 1(*Conus* Suppl.): Pl. 9, sp. 279 (Locality unknown).

1854. *Conus neglectus* A. Adams. *Proc. Zool. Soc.* (*London*), 1853: 117 (Locality unknown).

1870. *Conus Cooki* Brazier. *Proc. Zool. Soc.* (*London*), 1870: 109 (Botany Bay, New South Wales, Australia). Also described in *J. Conchyl.* (*Paris*), 18: 300.

1873. *Conus zealandicus* Hutton. *Cat. Mar. Moll. New Zealand:* 23 (Bay of Islands, New Zealand). See Tryon, 1884.

1887. *Conus sydneyensis* Sowerby iii. *Thesaurus Conchyliorum*, 5(*Conus* Suppl.): 260, pl. 32 (510), fig. 694 (Port Jackson, New South Wales, Australia).

Description.—Moderately heavy, with a low gloss; low conical to obconic, the sides evenly convex; body whorl with low spaced spiral ridges on basal fourth or third, then nearly smooth except for fine spiral and axial scratches; shoulder rounded, slightly concave above; spire flat to low, the early whorls heavily eroded, sides straight or slightly concave; whorls faintly stepped or with rude sutures, the tops slightly concave, with about 5-8 weak and irregular spiral ridges crossed by scattered weak axial threads; early whorls projecting when not eroded. Body whorl pale bluish white to bluish gray, covered with about equally spaced spiral bands of square brown dots separated by rectangular white dashes, these usually visible at least on basal third of body whorl and usually continuing to shoulder; brown dot bands often developed to produce short axial bands at midbody or over much of body whorl; occasionally dots largely absent over parts of shell and replaced by broad spiral bands (three or four in number) of bright pink; usually a clear midbody band; base white; shoulder with numerous narrow zigzag dark brown axial lines, these sometimes reduced; spire whorls with similar lines, but these usually obscured by erosion; tip of spire brownish. Aperture moderately broad, widened anteriorly; outer lip straight or smoothly curved; posterior edge of outer lip slightly dropped below shoulder angle; mouth pinkish violet with paler bands at middle and below shoulder. Columella short, very narrow and fused. Length 20-30mm.

Comparison.—This species has been confused with *C. papilliferus*, which may have similar but much smaller brown spots in regular spiral rows; however, that species is much thinner and lighter in weight, with a more angular shoulder and often flatter spire which has the whorls smoothly joined and flat above; in *C. papilliferus* the columella is usually long and almost free posteriorly; there is seldom a tendency toward broad pinkish bands as in *C. aplustre*, and the spots when present are not as regular or large and are not separated by the rectangular white dashes so common in *C. aplustre*. Actually, the general shape and appearance of *C. aplustre*, as well as the color pattern, remind one more of such shells as *C. tiaratus* or *C. abbreviatus*, but it is not coronated and never has a violet base; in addition, the outer lip in *C. aplustre* is thin and sharp, not thick and often ridged as in the *C. miliaris* allies. This is a very distinctive species.

Variation.—Except for spire height, the major variation is in color pattern. Specimens with complete patterns are tan on bluish gray with the spiral bands developed to the shoulder. There is a tendency to partially connect the bands to produce dark brown axial lines, especially at the midbody area and sometimes over the whole body whorl. Apparently the bluish gray background color is underlaid by three or four broad bands of pink or pinkish orange, as these commonly show through the rest of the pattern in whole or part. Extreme specimens are pale bluish gray with three broad pink bands and only faint traces of the spiral brown dots visible. The shoulder and spire are marked with a variable number of irregular narrow dark brown lines especially prominent on the shoulder and usually distinctly zigzagged; they are also present in some specimens on middle spire whorls but are usually lost through erosion.

Distribution.—Apparently restricted to New South Wales, Australia, where it is common in shallow water.

Synonymy.—*Conus lugubris* is often considered to be *C. hieroglyphus*, but the figure appears to be of *C. aplustre*, and the description could be interpreted the same way. *Conus cooki* and *C. sydneyensis* are based on specimens with well developed axial patterns.

Notes.—Because of confusion with *C. papilliferus*, this species is not commonly offered for sale. Erosion of the spire and sometimes of the body whorl appears to be common and natural, as are small worm holes. Specimens with intermediate patterns are not as attractive as either wholly spotted specimens or those with pink bands.

The relationships of this species are very uncertain. Perhaps it is actually more closely related to *C. miliaris* and allies than to the *C. anemone* stock, although some similarity can be seen with *C. rutilus* in general shape and color pattern (some *rutilus* variants).

ARTICULATUS.—Above: Left: Tolwat, E New Britain, Papua New Guinea, 19.8mm (R. Monson); Right: Mauritius, 26.4mm (DMNH). *Below:* Okinawa, Japan, 20.9mm (TCG).

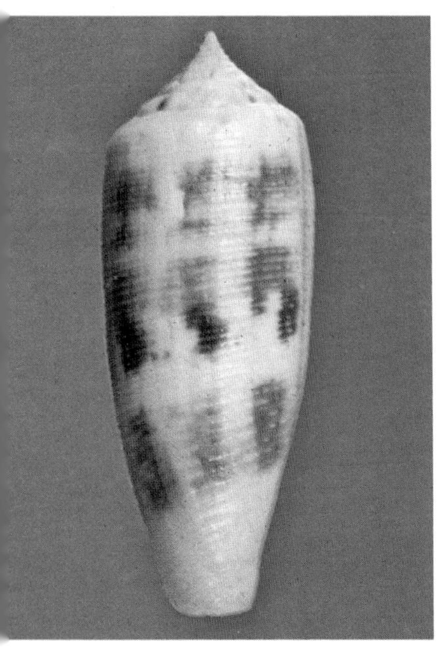

ARTOPTUS.—Above: Rockhampton, Storage Bay, Queensland, Australia, 34.8mm (R.W. Pitman). *Below:* Left: Marau Sound, Guadalcanal, Solomon Is., 41.5mm (MM): Right (*viola*): Darwin, Northern Territory, Australia, 48mm (MM).

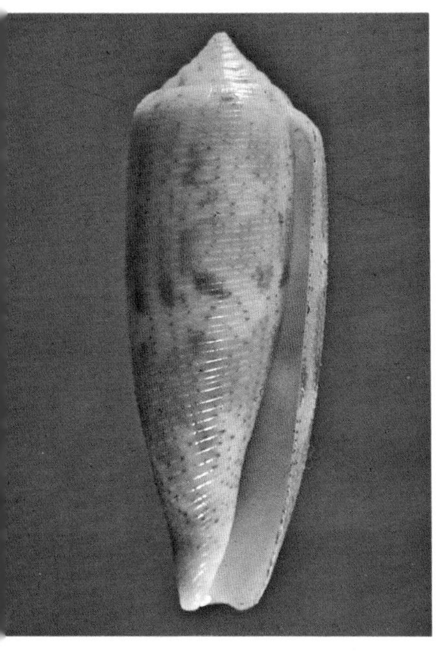

ARANEOSUS Solander, in Lightfoot, 1786

1786. *Conus araneosus* Solander, in Lightfoot. *Cat. Portland Mus.:* 76 (China; Coromandel). Lectotype figure (Kohn, 1964): Martini, Pl. 61, fig. 676.

1791. *Conus arachnoideus* Gmelin. *Systema Naturae per Regna Tria Naturae,* ed. 13, 1: 3388 (Locality not stated). Lectotype figure (Kohn, 1966): Knorr, Pl. 4, fig. 4. Misspelled *acachnoideus* in text but corrected in index.

1792. *Conus Nicobaricus* Hwass, in Bruguiere. *Cone,* in *Ency. Method., Hist. Nat. des Vers,* 1: 612 (Grandes Indes). Lectotype figure (here selected): *Tableau Ency. Method.,* Pl. 318, fig. 9; this figure was incorrectly considered the holotype by Kohn, 1968.

1811. *Conus reticulatus* Perry. *Conchology:* Pl. 24, no. 2 (Locality not stated). Non *Conus reticulatus* Born, 1778.

1838. *Conus monstrosus* Kuester. *Syst. Conch. Cab.,* ed. 2, 6(*Conus*): 77, pl. 12, figs. 5-6 (Indian Ocean). Deformed.

1857-1858. *Conus peplum* Sowerby ii. *Thesaurus Conchyliorum,* 3 (*Conus*): 3, pl. 1 (187), fig. 13; pl. 17 (203), fig. 408 (Red Sea).

Description.—Heavy, with a low to high gloss; obconic, the sides about straight until abruptly inturned at about anterior third; body whorl with a few weak spiral ridges above base, otherwise smooth except for scattered spiral and axial threads and scratches and occasional weak spiral ridges; shoulder broad, faintly canaliculate, strongly coronate, undulate, or smooth; spire very low or flat, the early whorls forming a small cone with concave sides; early whorls usually strongly eroded and with indistinct margins; later whorls concave above, the margins strongly coronate with upright coronations; tops of whorls with faint spiral scratches sometimes visible, otherwise with only scattered weak spiral threads or scratches. Body whorl white or cream, covered with fine brown axial lines which overlap to produce a finely to moderately large tented pattern over whole whorl; two well developed spiral bands of dark brown to blackish blotches above and below midbody area, commonly with a third band developed below the shoulder; blotches may be small and irregular or large and rectangular, sometimes forming nearly continuous bands; base finely reticulated brown on white; margin of shoulder and later spire whorls with black bars or blotches between the coronations or at regular intervals if coronations absent; tops of shoulder and spire whorls reticulated with fine brown lines if not eroded to dirty white; tip of spire pinkish. Aperture moderately wide, dilated anteriorly; outer lip thick but sharp, straight or slightly concave at middle; mouth with outer half white to pale violet, interiorly with a large blotch of deep yellow. Columella long, narrow, set off posteriorly by a low but usually distinct hump. Length 40-101mm.

Comparison.—*Conus amadis* and *C. acuminatus* may be similar in shape and pattern but are easily distinguished by the lack of strong upright coronations on at least the middle spire whorls. *Conus marmoreus* is very similar, and occasional specimens are difficult to distinguish. This is especially true when comparing *Conus araneosus nicobaricus* and *C. marmoreus* form *vidua*, which have very similar patterns and virtually identical shapes; the simplest way to tell the two apart is by the color of the aperture if the specimens are fairly fresh—white or pinkish in *C. marmoreus*, white or violet exteriorly with a large yellow spot deep within in *C. araneosus*. In addition, the tops of the spire whorls in *C. marmoreus* are largely black and white, while they are finely tented in *C. araneosus;* the base of many *C. marmoreus* is largely black, while it is finely tented in *C. araneosus. Conus zonatus* and *C. imperialis* might be vaguely similar in shape but are usually very distinct in pattern and color and have narrower apertures.

Variation.—As expected in tented cones, no two specimens have exactly the same pattern. In addition, *C. araneosus* has two distinctive variants which have been considered full species, perhaps justly so. At the moment I prefer to retain them as poorly understood subspecies.

C. araneosus araneosus Solander, in Lightfoot: Relatively broader, width about 57-59% of length (in adults 62-73mm long from India); gloss low, whole shell appearing coarser; body whorl usually lacking strong dark spiral blotch band below shoulder; tenting of body whorl fine, often irregular with white areas, especially near inner lip; mouth exteriorly often dark violet; coronations on shoulder usually absent or obsolete; northern Indian Ocean from Red Sea and adjacent northeastern Africa to India and Ceylon.

C. araneosus nicobaricus Hwass, in Bruguiere: Somewhat narrower, width about 52-55% of length (in adults 62-65mm long from Philippines); gloss high, the shell appearing delicately patterned; body whorl usually with well developed band of dark blotches below shoulder; tenting of body whorl very open, the tents large, continuing into mouth; mouth exteriorly white to very pale violet or occasionally pinkish; coronations on shoulder strong, erect; Philippines west to islands of northeastern Indian Ocean, probably intergrading with nominate subspecies about southern India and Ceylon.

Distribution.—Northeastern Africa across northern Indian Ocean; Philippines. Common in shallow water.

Synonymy.—Whether *C. araneosus* and *C. nicobaricus* represent distinct species, valid subspecies, or merely forms of the single species is still uncertain. I have treated them as subspecies largely because specimens from the northwestern Indian Ocean are typical *araneosus*, those from the Philippines are typical *nicobaricus*, and specimens from the eastern Bay of Bengal area (where intermediates might be expected) are not available with detailed locality data. Certainly the distinctions between the two are those that would normally be used to distinguish only forms in most species, not full species; larger collections from the entire northern Indian Ocean, with detailed locality, are needed.

ATRACTUS.—(*a. atractus*) *Above:* Off Salinopolis, Amapa, Brazil, 65.9mm (EP). *Below:* Brazil, 70mm (DM) (photos by Jayavad).

***ATRACTUS.*—**(*a. austini*) *Above:* S of Pensacola, Florida, 42.3mm. *Below:* Left: S of Pensacola, Florida, 27.0 and 27.3mm; Right: Off Dauphin I., Alabama, 42.1mm.

Notes.—*C. araneosus* is a coarse cone in the northwestern Indian Ocean and often is heavily covered with coralline growths and persistent periostracum, as well as heavy growth marks and occasional healed flaws. In the Philippines the cone is glossy and seldom has encrustations or flaws, although the thinner lip is often filed. All adults of this species I have seen have severely eroded posterior spire whorls, often extending anteriorly so that only the last two whorls and shoulder have pattern. This species has been extensively confused with *C. marmoreus* although the two are quite distinct, with occasional puzzling exceptions.

ARCHON Broderip, 1833

1833. *Conus Archon* Broderip. *Proc. Zool. Soc.* (*London*), 1833(I): 54 (Bay of Montija, Central America).

1845. *Conus granarius* Kiener. *Species gen. et icon. des coqu. viv.*, 2 (*Conus*): Pl. 98, fig. 1. 1848, *Ibid.*, 2: 215 (Locality unknown).

1845. *Conus sanguineus* Kiener. *Ibid.*, 2: Pl. 111, fig. 2. 1849-1850, *Ibid.*, 2: 356 (Locality unknown).

Description.—Heavy, solid, with a low to moderate gloss; sides almost straight; body whorl with a few low weak spiral ridges above the base, these usually soon obsolete but present in some specimens as very weak broken spiral ridges that may be irregularly present to shoulder; numerous fine and coarse axial lines, scratches, and breaks; shoulder broad, roundly angled to moderately carinate; spire moderate, sides concave; early whorls bluntly pointed but usually heavily eroded or covered with deposits; early and middle whorls with strong coronations visible in many specimens if not eroded, later whorls and shoulder smooth or at most weakly undulate; tops of whorls concave, with weak curved axial threads and often traces of 1 or 2 indistinct spiral ridges. Body whorl opaque white or cream, overlaid with large blotches of tan, chestnut, or reddish brown, the size of the blotches very variable; blotches often arranged in 2 or 3 spiral bands but just as commonly in broad axial bands over at least the posterior half of the shell; entire body whorl covered with spaced very fine spiral lines of brown and white dash and dot lines, usually not conspicuous except above the base where white dots may be large and prominent; axial growth lines usually stained with brown or yellowish to produce narrow discontinuous axial streaks over part of shell; spire white or cream with rectangular blotches of reddish brown often smaller than the white interstices; numerous very fine curved brownish axial lines may be present, especially on the shoulder and later whorls; early whorls usually stained creamy to pale tan. Aperture moderate, of about uniform width; outer lip sharp, posteriorly sloping below level of shoulder; mouth white or pale bluish white. Columella indistinct, apparently long but completely fused and turned inward. Length 35-98mm.

Comparison.—*Conus archon* is very close to *C. cedonulli* and *C. aurantius* in both shape and color pattern. It has similar spiral dot-dash lines over the body whorl and often closely resembles *C. cedonulli* in details of pattern. However, *C. archon* lacks distinct coronations on the shoulder and anterior spire whorls and has the body whorl generally opaque white with sharp-edged dark brown instead of yellowish blotches (at least most dead specimens of *C. cedonulli* are orange or yellow in color). In addition, many specimens of *C. cedonulli*. and *C. aurantius* have fine pustules in the white dots of the body whorl pattern which are typically absent in *C. archon*. *C. regius* and *C. bartschi* (as well as *C. imperialis*) may have similar shapes and patterns but generally have the shoulder strongly coronated and the spire blunt.

Variation.—In shape and sculpture this is a very constant species. Some specimens have the shoulder rounded, others have it distinctly angled. Differences in spire shape and sculpture appear to be due largely to erosion and deposits and are not real variations. In color pattern there is a tendency for at least five spiral rows of brownish blotches to be developed: two anterior to the midbody area and three posteriorly; these blotches may be variably fused into spiral or axial bands or may be reduced and absent in any row over all or part of the body whorl; occasional specimens may be nearly all white or all brown. The fine spiral lines are seldom obvious although the white dots may be part of the reason for the irregular margins of the blotches; the white dots above the base may be very large and fused into a continuous white band. Rare specimens show a great similarity to *C. cedonulli* in having the white dots raised into fine pustules over some or all of the body whorl. Yellow pigment or stains may be heavy along axial growth marks or breaks and may form a major part of the pattern in the white midbody band and on the spire.

Distribution.—Eastern Pacific from the Gulf of California south to Panama. Uncommon from below low tide to 400m.

Synonymy.—*C. sanguineus* seems to be a rather typical specimen of *C. archon*. The status of *C. granarius* is less certain, but the figures and description seem to indicate a very dark, granulose specimen of *C. archon*. It is possible that *C. granarius* represents a distinct southern Panamic population which might bridge the gap between *C. archon* and *C. cedonulli*, and I have seen photos of specimens (courtesy Ed Petuch) of *C. archon* which closely resemble *C. granarius*.

Notes.—This species is very subject to eroded spires, chipped lips, and heavy growth flaws. Most specimens have the spire whorls eroded enough to obscure the coronations and have several strong but regular growth marks; these marks and erosion are natural and do not detract from the value of the shell. Specimens with pustules are apparently very rare at the moment.

ATTENUATUS.*—Above:* Off Dania, Broward Co., Florida, 28.0mm (EP). Below: Guadaloupe, 21mm (DM) (photos by Jayavad).

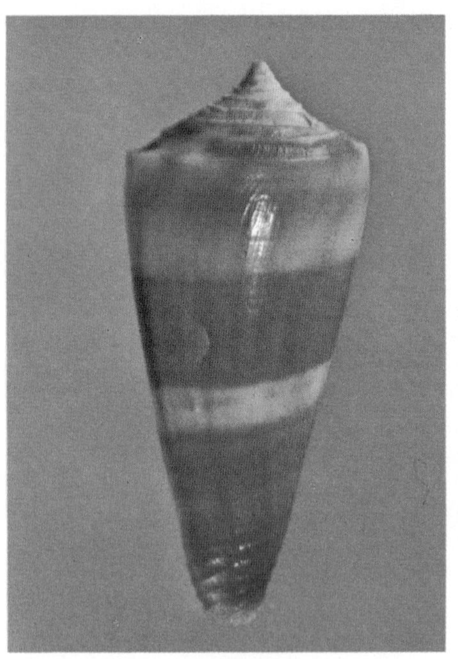

AUGUR.*—Above:* NW Madagascar, 53.6 and 56.6mm. *Below:* Left: Madagascar, 54mm; Right: Zanzibar, 58.2mm (MF).

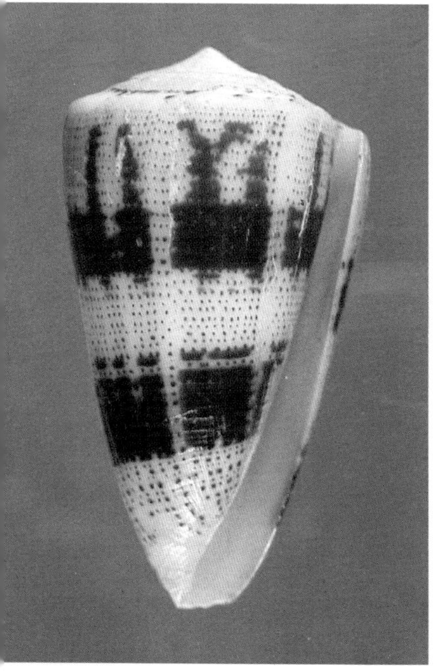

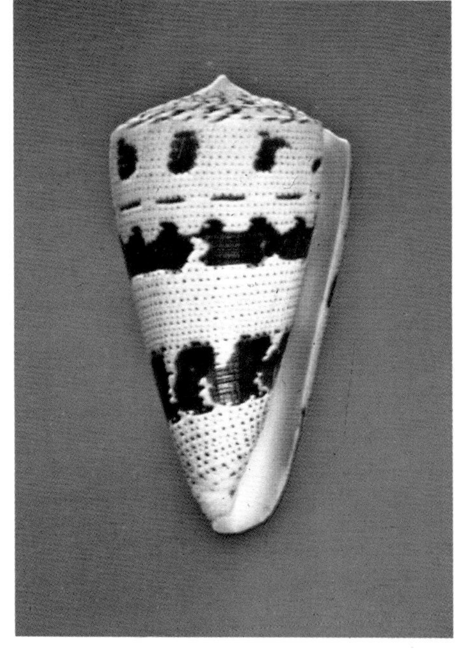

ARCUATUS Broderip and Sowerby i, 1829

1829. *Conus arcuatus* Broderip and Sowerby i. *Zool. J.*, 4: 379 (near Mazatlan).
1848. *Conus Borneensis* A. Adams and Reeve, in A. Adams. *Zool. Voy. Samarang, Mollusca*, 1: 18, pl. 5, figs. 8a-8d (Northeast coast of Borneo).
1865. *Conus Lizardensis* Crosse. *J. Conchyl. (Paris)*, 13: 305, pl. 9, fig. 5 (Lizard I., NE. Australia).

Description.—Light in weight, fragile, with a moderate gloss; sides of body whorl strongly convex posteriorly, then pinched-in and tapered to a very narrow base; body whorl with broad flat ribs from anterior third to at least midbody, often extending to shoulder; grooves between ribs wide, narrower anteriorly, containing only very weak axial threads; adult shells often smooth on posterior fourth; whole body whorl may have indistinct axial wrinkles and grooves; shoulder broad, sharply carinate; spire high or very high, pointed, the sides concave; whorls strongly stepped, carinate, concave above; earliest 3 or 4 whorls with weak but distinct nodules; tops of spire whorls with numerous fine curved axial threads and occasional traces of one or two very faint spiral ridges. Body whorl white or cream, usually with three bands of yellowish to reddish brown axial spots, one below shoulder, one above midbody, and a faint band below midbody; the bands can be relatively complete, indicated by only a few spots, or partially absent; shoulder and later spire whorls with small oval brown spots; earlier spire whorls usually patternless and pale pinkish brown. Aperture narrow, of uniform width; outer lip very thin and sharp, convex at margin, posteriorly strongly sloping below shoulder; mouth white. Columella indistinct. Length 30-53mm.

Comparison.—In shape one is instantly reminded of *C. cancellatus*, *C. armiger*, and *C. atractus*, among others. From *C. cancellatus* and *C. armiger* it is readily distinguished by the lower number of nodulose whorls (3-4 vs. 5-8) and the virtual absence of spiral sculpture on the spire; the grooves of the body whorl are also weakly or obsoletely axially threaded instead of having strong axial sculpture as in *C. cancellatus* and allies. *C. atractus* is much more similar in appearance, although *C. arcuatus* is more elongate with a higher and more stepped spire; in *C. atractus* the ribs are often narrow and rounded, not broad and flat, and there is an overall silken sheen on many specimens. *C. ione* and *C. sieboldii* differ in width at the shoulder and in many details of pattern and sculpture, although they are possibly closely related.

Variation.—Juvenile shells are uniformly ribbed from base to shoulder, but many adults have the ribs posterior to the midbody area reduced or completely absent. Grooves between the ribs are usually rather wide and very finely axially sculptured, but occasionally they are narrow and appear to be very weakly punctate. Color blotches may be strong and dark brown or indistinct and yellowish; all three major bands may be present and smaller spots may be scattered over the rest of the shell, or the pattern may be virtually absent. Usually at least the shoulder and last spire whorl have strong color blotches. Curved axial clouds of pale brown may be present.

Distribution.—Eastern Pacific from the Gulf of California to Colombia in deeper water to at least 60m. Common.

Synonymy.—*C. borneensis* is usually accepted as a synonym, but it could as easily be *C. cancellatus* or even *C. ione;* since the synonymy has been accepted so long, there is no reason to fight it. The description and figure of *C. lizardensis* apply very well to juveniles of *C. arcuatus;* Australian shells usually assigned to this name are actually *C. minnamurra.*

Notes.—Adults usually have the early spire whorls partially missing and have traces of one or more small breaks. The lip is very thin and breaks easily. Collectors do not seem to differentiate between poorly colored specimens and those with heavy dark patterns.

ARENATUS Hwass, in Bruguiere, 1792

1792. *Conus arenatus* Hwass, in Bruguiere. *Cone,* in *Ency. Method., Hist. Nat. des Vers,* 1: 621 (Philippines). Lectotype selected by Kohn, 1968.

1798. *Cucullus arenosus* Roeding. *Museum Boltenianum:* 40 (Locality not stated). Lectotype figure (Kohn, 1975): Martini, Pl. 63, fig. 696.

1822. *Conus arenatus* var. *punctisminutissimus* Lamarck. *Hist. Nat. Anim. sans Vert.,* 7: 452 (locality not stated).

1883. *Conus arenatus* var. *mesokatharos* Melvill, in Tryon. *Manual of Conchology,* 6(1): 18, pl. 27, fig. 2 (Locality not stated).

1937. *Conus arenatus* var. *aequipunctata* Dautzenberg. *Mem. Mus. Roy. d'Hist. Nat. Belg.,* 2(18): 31, pl. 1, fig. 2 (Locality unknown).

1937. *Conus arenatus* var. *undata* Dautzenberg. *Ibid.,* 2(18): 31, pl. 1, fig. 3 (Amboine). Non *Conus undatus* Kiener, 1845.

1937. *Conus arenatus* var. *granulosa* Dautzenberg. *Ibid.,* 2(18): 32, pl. 1, fig. 4 (Amboine). Non *Conus granulosa* (Roeding, 1798).

Description.—Heavy, with a low to high gloss; sides convex or nearly straight, greatest width below the shoulder; body whorl with about 6 weak spiral ridges above the base, rest of whorl smooth except for scattered spiral and axial scratches and threads; shoulder rounded, strongly coronate in unworn specimens of most populations; spire flat to low, the sides usually convex; spire whorls strongly coronated; tops of whorls concave to flat, usually with 3-7 strong undulating spiral ridges. Body whorl white to cream or very pale tan, covered with very small (about one-fourth diameter of shoulder coronations) oval brown and black spots tending to form broad wavy axial bands of tightly packed spots and occasional spiral rows of spots above and below midbody; spots always dis-

AULICUS.—*Above:* Zanzibar, 72.8 and 81.5mm (MF). *Below:* Boac, Marinduque, P.I.: Left: 102.4mm; Right: 104.8mm.

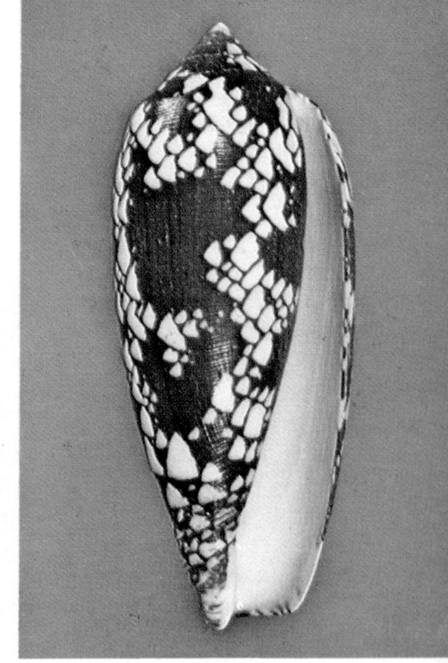

AURANTIUS.—*Above:* Off St. Barbarasbaai, Curacao, Neth. Ant., 48.4mm (photo by A. Kerstitch). *Below:* Left: Kralendijke, Bonaire, Neth. Ant., 44.5mm (R. Guest); Center: Kralendijke, Bonaire, Neth. Ant., 40mm (photo by A. Kerstitch); Right: Palm Beach, Aruba, Neth. Ant., 42.2mm (MG).

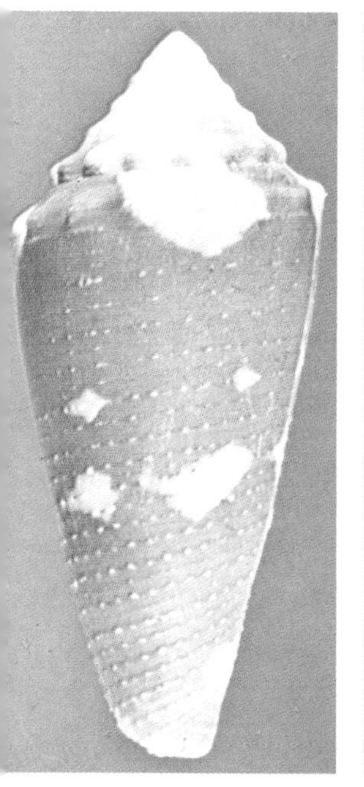

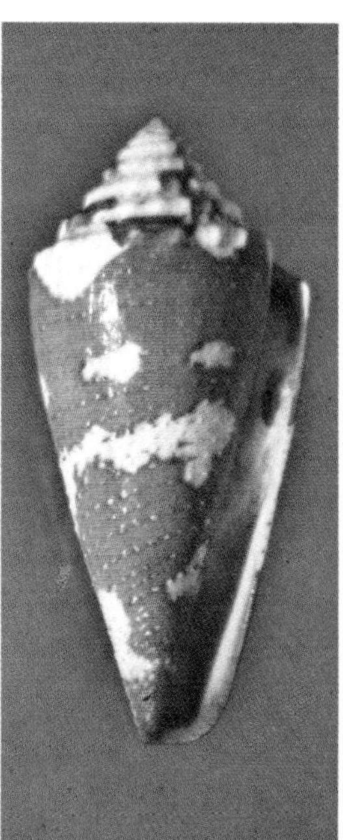

tinct, even when overlapping; in addition to the dark spots, there are numerous small spiral dashes of opaque white covering the entire body whorl but often hidden under the dark pattern and difficult to see; shoulder and later spire whorls with a few narrow dark axial stripes and some brown spots; earlier spire whorls white. Aperture narrow posteriorly, wide anteriorly; outer lip moderately thick, convex; mouth white, often with a large pinkish or orange blotch deep within. Columella wide, strong, set off posteriorly by heavy ridges. Length 30-54mm.

Comparison.—*Conus pulicarius* sometimes closely resembles *C. arenatus*, although there should never be any real difficulty distinguishing them. In *C. pulicarius* the spots are larger, at least half the diameter of the shoulder coronations and often their equal; they tend to form oval blotches above and below midbody, the blotches often suffused with pale orange and not aligned in wavy axial bands (except sometimes in *C. pulicarius vautieri*). *C. pulicarius* is also wider at the shoulder than *C. arenatus* of equal size and lacks (usually) opaque white spiral dashes. *C. zeylanicus* is more obese, has smooth or only crenulate shoulder and spire whorls, and has a pattern containing numerous fine irregular axial lines of brown on a pinkish background. *C. stercusmuscarum* and *C. eburneus* are sometimes vaguely similar but are not coronated and have numerous differences in pattern.

Variation.—Some populations of *C. arenatus* appear to be subject to great wear on the shoulder and spire coronations so they may appear almost smooth at first glance. Occasional specimens are granulose over most or all the body whorl. There are numerous differences in pattern between populations and individuals, but this is apparently not strictly geographic and the variants cannot be formally recognized. Spotting may be very strong and cover the whole body whorl in broad wavy bands, the bands may be narrow and the shell largely white, distinct spiral bands may be present, or the pattern may be confined to distinct zones above and below the unspotted midbody area (rare).

Distribution.—A common to abundant intertidal and shallow water species of the Indo-Pacific from Natal to the Marquesas and Japan, but not recorded from Hawaii.

Synonymy.—The names of Melvill and Dautzenberg are based on minor and inconsistent variations in pattern or granulose sculpture; none are worthy of recognition. The name *C.* 'armatus' Smith is sometimes applied as a subspecific name to specimens from the Red Sea and eastern Africa with high coronations and reduced spotting; actually this name is a printer's error for *C. arenatus* and was never intended as a new name by Smith—there is no reason to use this name in any literature. Dealers commonly confuse the 'armatus' phase of *C. arenatus* for *C. pulicarius* from the western Indian Ocean, although the two are easily distinguished. The African populations are as variable as any others, and no subspecies can be recognized for this area.

Notes.—Granulose specimens are not common and sometimes

deserve a premium. Wear on shoulder coronations is normal in some populations and should not detract from value. This species is not very subject to breaks and major flaws, although the lip chips easily in some specimens. Specimens with a largely spiral pattern and tendency to reduce midbody spotting are not as common as those with axial or axial and spiral patterns over most or all the shell.

ARGILLACEUS Perry, 1811

1811. *Conus argillaceus* Perry. *Conchology:* Pl. 24, fig. 6 (East Indies). Commonly misspelled 'argillaceous'.
1833. *Conus splendidulus* Sowerby i, in Sowerby ii. *Conchological Illustrations:* Pt. 37, fig. 53 (Locality not stated).
1848. *Conus luctificus* Reeve. *Conchologia Iconica,* 1(*Conus* Suppl.): Pl. 2, sp. 280 (Locality unknown).

Description.—Solid, heavy, with a good gloss; sides nearly straight or slightly convex posteriorly; body whorl with fine and irregular spiral ridges above base, these rapidly obsolete; rest of whorl covered with very fine spiral and axial threads; numerous fine axial growth lines usually present, sometimes marked with dark; shoulder broad, roundly angled, usually canaliculate or deeply concave above; spire moderate, pointed, the early whorls usually eroded or covered with deposits; sides of spire straight or slightly concave, often with last whorl abruptly and irregularly stepped relative to rest of spire; tops of whorls flat or slightly concave on early whorls, becoming concave to canaliculate on later whorls; early whorls possibly with weak nodules, remainder of whorls with traces of 2 or 3 faint spiral ridges crossed by weak curved axial threads. Body whorl dark brown to orange brown, covered with numerous fine to coarse interrupted darker brown spiral lines; usually with 3 broad bands of white (below shoulder, at midbody, and at base) broken into axial rectangles by infiltration of brown background and tendency for spiral dark lines to concentrate into axial bands; base dark brown; margin of shoulder with regular dark brown to black spots continued onto spire and alternating with white blotches; spire white, with large or small dark brown revolving blotches often forming broad lines; early whorls pale, unmarked. Aperture moderately wide, uniform or slightly widened anteriorly; outer lip straight or slightly convex, sharp; posterior margin of outer lip strongly sloping below shoulder; mouth pinkish white to bluish white, sometimes with indistinct large brown blotches. Columella long, narrow, strongly indented from rest of inner lip. Length 45-65mm.

Comparison.—*Conus argillaceus* is a very distinctive cone which has only recently become available for comparison with several other obscure northern Indian Ocean species. *C. fumigatus* is very similar in many

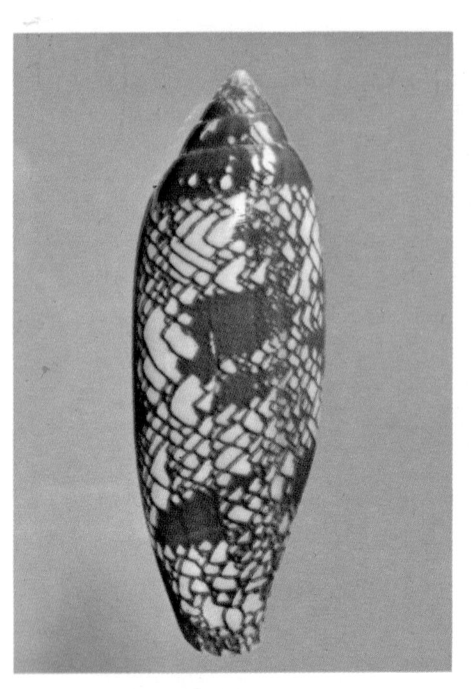

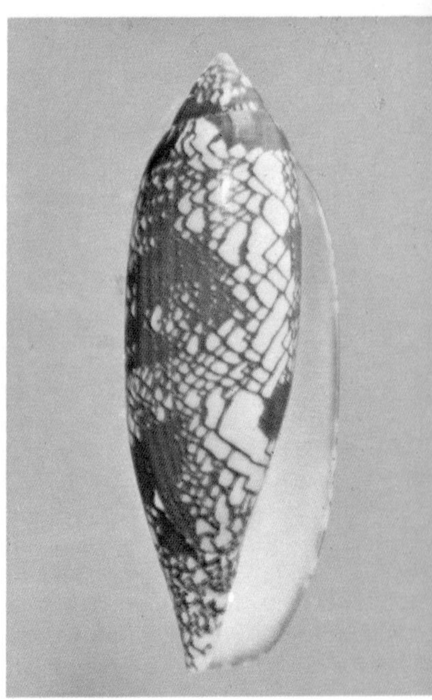

AURATUS.—Above: Tahiti, Society Is., 82mm (DM) (photo by Jayavad). *Below:* Moorea I., Society Is., 58.2mm (EP).

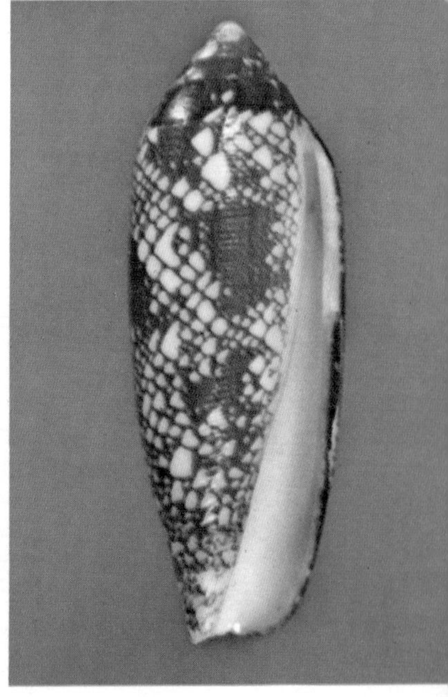

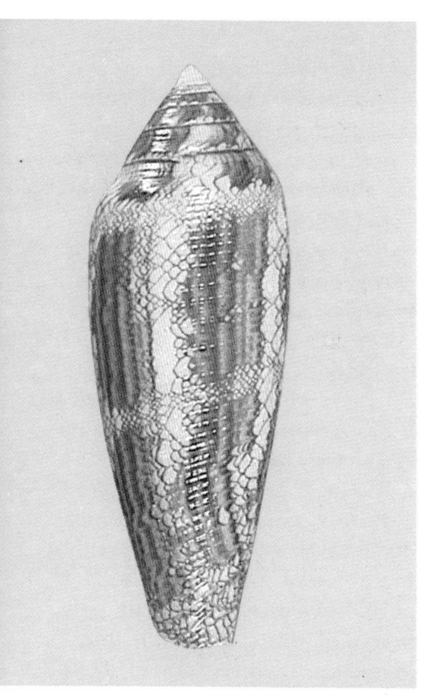

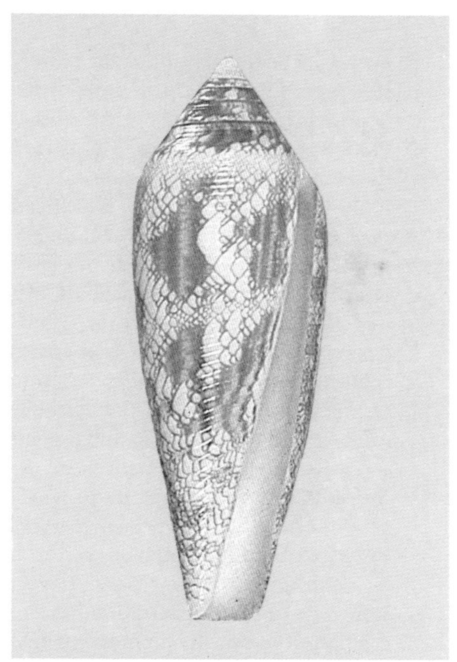

AUREUS.*—Above:* Fila I., off Efate, New Hebrides, 38.2mm (GG). Below: Left (*paulucciae*): Off St. Denis, Reunion, 62.4mm (EP); Right: Kwajalein Atoll, Marshall Is., 53mm.

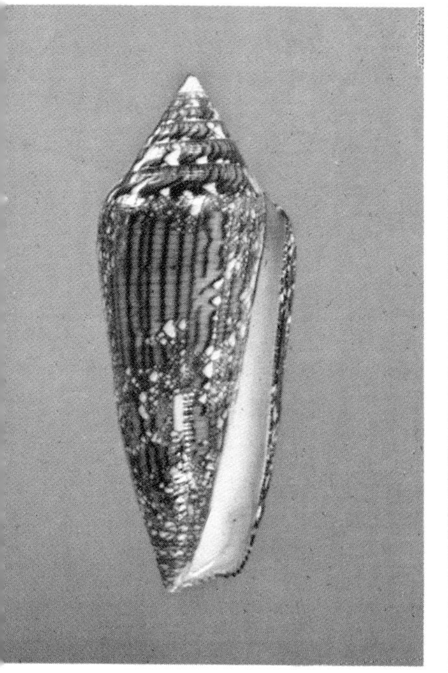

aspects of sculpture and color pattern and is probably its closest relative; however, *C. fumigatus* is a much smaller species with a short and stubby body whorl and stronger spiral and axial sculpture on the spire; the shoulder is not so deeply concave as usual in *C. argillaceus*, and the tendency to develop spiral bands of alternating white and brown rectangles on the body whorl is only weakly developed; *C. fumigatus* has a much broader range of colors and patterns than known in *C. argillaceus*. *C. malacanus* may be superficially similar but has the early whorls distinctly nodulose, later whorls distinctly cancellate in most specimens, and the last whorls and shoulder only shallowly concave and with a strongly developed carina; the shoulder area is not marked with dark brown or alternating spots as typical of *C. argillaceus*, and the base is white, not dark brown. *C. kermadecensis* lacks the spiral lines, has a normally lower spire of somewhat different form, and is generally a coarser appearing shell of different texture. *C. eximius* is more slender, thinner, has a white base, and differs in many aspects of pattern and sculpture. *C. monile* has a very narrow aperture, larger brown rectangles in spiral rows, and is generally tinted with deep salmon as well as having different spire sculpture.

Variation.—Two types of body shape and two extremes of color pattern have somewhat confused the synonymy of this species, but intermediates are not uncommon. At one extreme are relatively broad shells with slightly convex outlines and a pattern of orange brown with the spiral lines poorly developed and the white bands often nearly continuous and little interrupted; the mouth of such pale shells may be pinkish or pale bluish white. At the other extreme are relatively slender shells with very straight outlines and a body whorl largely dark brown with numerous spiral lines and few traces of the usual white bands; dark shells tend to have dark mouths. Like other related cones, there is a strong tendency to a distorted spire; spire erosion is common and may distort the appearance of the spire as well.

Distribution.—An uncommon shell found in relatively shallow water (about 5-10m) on sand channels in coral reefs. I have seen specimens only from Yemen and Aden, but presumably the species has a wider distribution in the northwestern Indian Ocean.

Synonymy.—It seems impossible to avoid the use of Perry's name for this species although the shell is much better known as *C. splendidulus*. Perry's figure also resembles some variants of *C. monile*, but the dark spots at the shoulder are plainly indicated. *C. luctificus* seems to be based on a dark slender specimen.

Notes.—The name of the species is commonly misspelled 'argillaceous', while the synonym *C. luctificus* has been misspelled 'luctifer'. Eroded spires and fine growth lines do not detract from the value of specimens. This species has only recently entered the market in large quantities and current prices are not stable. The range and maximum size are uncertain.

ARMIGER Crosse, 1858

1845. *Conus crenulatus* Kiener. *Species gen. et icon. des coqu. viv.*, 2 (*Conus*): Pl. 109, fig. 1. 1849-1850, *Ibid.*, 2: 355 (Locality unknown). Non *Conus crenulatus* Deshayes, 1835, a fossil.
1858. *Conus armiger* Crosse. *Rev. Mag. Zool.*, (Ser. 2), 10: 200. Nomen novum for *Conus crenulatus* Kiener, 1845.
1951. *Conus clarki* Rehder and Abbott. *J. Washington Acad. Sci.*, 41(1): 22, figs. 1-6 (Louisiana, Iberia Co. (*sic.*), 50 mi. SSW of Marsh Island).
1952. *Conus frisbeyae* Clench and Pulley. *Texas J. Sci.*, 1952(1): 59, Pl. A, fig. 1 (Campeche Banks, Yucatan, Mexico).
1968. *Conus bajanensis* Usticke. *Caribbean cones from St. Croix and the Lesser Antilles:* 28, pl. 4, sp. 1020 (South of Barbados).
?1968. *Conus pseudo-austini* Usticke. *Ibid.:* 29 (South of Barbados).
1973. *Conus guyanensis* Van Mol. *Zool. Meded. Rijksm. Leiden*, 46(19): 261, fig. 1; pl. 1, figs. 3-4; pl. 2, fig. 10 (Off Paramaribo, Surinam).

Description.—Light in weight but solid, with a low gloss or dull; broadly fusiform, the sides strongly convex below shoulder then greatly constricted to produce a long, narrow basal half to fourth of body whorl; body whorl covered with broad rounded or flattened ribs separated by narrow axially striate grooves; ribs sometimes less distinct near shoulder; surface of ribs with fine axial threads and usually with squarish beads or pustules on at least the posterior ribs; basal area with narrower spiral ridges; shoulder in adults weakly carinate to narrowly rounded, distinctly nodulose or undulate in most specimens; spire high, sharply pointed, the sides deeply to shallowly concave; at least first 5-6 whorls strongly nodulose, remainder nodulose or undulate at margin; tops of whorls concave, bearing about 3 sharp strong spiral ridges (weak or absent from earliest whorls) crossed by regularly spaced heavy curved axial ridges. Body whorl and spire white or cream, usually with a variable pale brown pattern; entire body whorl may be covered with distinct brown squares on the ribs, these extending as square blotches to the early spire whorls, or the body whorl may have indistinct brown clouds in a broad spiral band below the shoulder; early whorls of the spire largely brownish. Aperture narrow, of uniform width; outer lip thin, convex anteriorly, its posterior margin strongly sloping below level of shoulder; mouth white, glossy. Columella not visible externally. Length 25-40mm.

Comparison.—*C. armiger* is very similar to *C. cancellatus* in many aspects of shape and sculpture, especially when the southern subspecies of *C. armiger* is considered. In *C. cancellatus* the posterior ribs of the body whorl are rarely (if ever) strongly pustulose; the shoulder is distinctly carinate without nodules, as are the last 2 or 3 spire whorls; and the color pattern in *C. cancellatus* consists of about 3 distinct spiral bands of axially elongated pale brown rectangles which are occasionally absent in part or variably fused; the posterior portion of the body whorl has the sides paral-

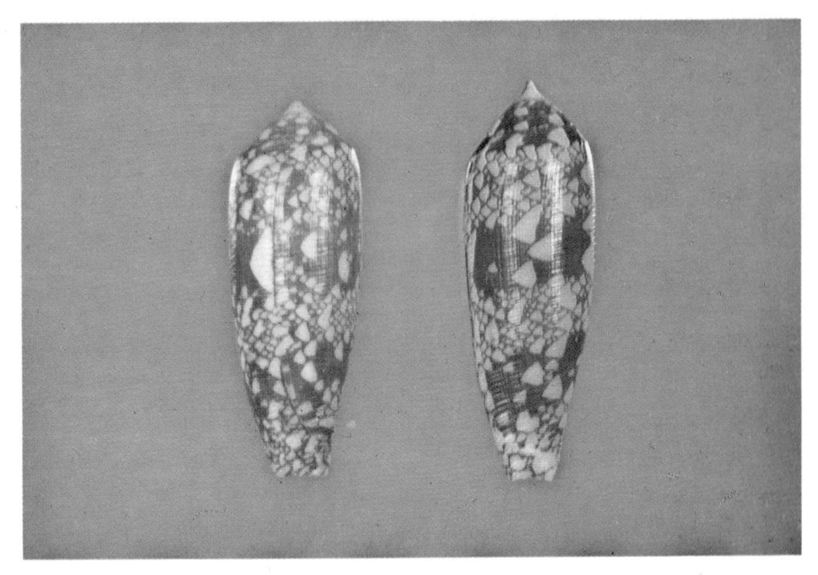

AURICOMUS.—Above: Left: Marau, Guadalcanal, Solomon Is., 35.3mm; Right: Kwajalein Atoll, Marshall Is., 37.6mm. *Below:* Sulu Sea, P.I., 41.9mm (GG).

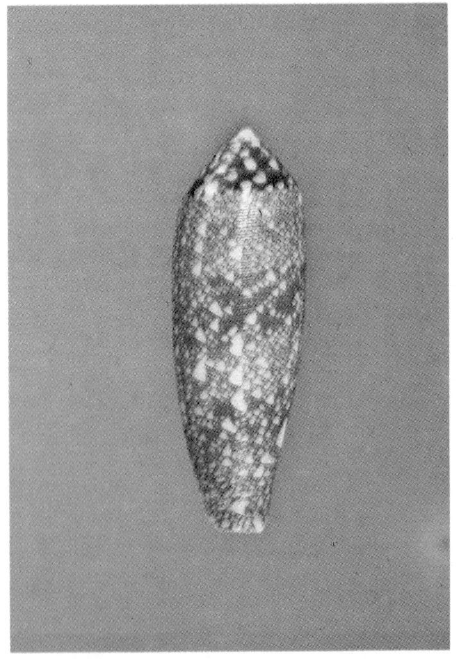

AURISIACUS.*—Above:* Banda I., Moluccas, 42.1mm (MM). *Below:* "Moluccas" (DMNH): Left: 65.0mm; Right: 49.8mm.

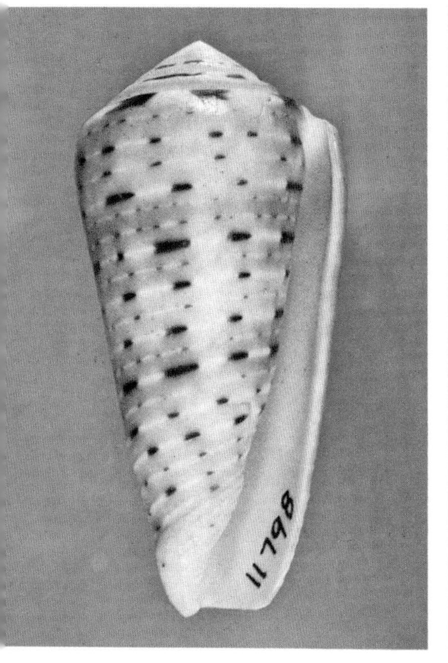

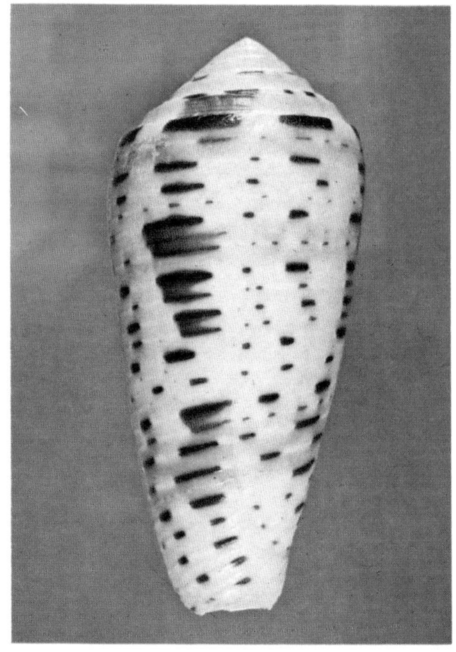

lel for a much longer distance than in *C. armiger*, where the sides are parallel for only a short distance below the shoulder. Juvenile *C. cancellatus* may very closely resemble *C. armiger bajanensis* juveniles in general shape and pattern, but the spire blotches in *C. cancellatus* are generally larger and more distinct on the late whorls.

Of Caribbean-Panamic cones, *C. armiger* has often been compared with *C. atractus, C. arcuatus,* and *C. stimpsoni*, among others, but the relationship here seems very weak. These species all have only the first few spire whorls with distinct nodules, the later whorls carinate; the spire whorls usually have only weak and often indistinct spiral ridges crossed by fine axials, never the strong and distinct ridging of *C. armiger.*

Variation.—This species is poorly known although often not uncommon, and the southern populations are especially not understood. At this time it seems best to recognize two subspecies which are geographically isolated and fairly distinct in details of sculpture, shape, and color pattern.

C. armiger armiger Crosse: Relatively slender; the shoulder nodulose as are all the spire whorls; body ribs very heavy in most specimens, usually with strong squarish beads over most of the body whorl, the interstices often marked with squarish brown spots; adult size reaching at least 40mm. Gulf of Mexico.

C. armiger bajanensis Usticke: Much broader at the shoulder and posterior part of body whorl, the spire sometimes appearing relatively lower and with somewhat straighter sides; shoulder weakly nodulose, undulate, or rarely simply carinate, the later spire whorls weakly nodulose or undulate; body ribs perhaps a bit weaker (at least in some populations), often only the posteriormost ribs with beads; body whorl with broad nebulous spiral brown band over posterior half or less, not spotted; adults apparently seldom exceeding 35mm; juveniles uniform brown or patterned like adults. Off Barbados, Colombia, and Surinam, probably southern Caribbean.

Adults of both subspecies are commonly uniform white without pattern. There are radular differences between the subspecies. In addition, specimens from off Barbados are broader than those from Colombia and Surinam and have a slightly different appearance.

C. armiger bajanensis, Colombia, 23mm (photos Kerstitch).

Distribution.—Dredged in moderately deep water (about 40-120 meters) from sand or mud bottoms in the Gulf of Mexico from southern Florida north to Alabama, Louisiana, and Texas, south at least to Yucatan; reoccurs in similarly deep water in the southern Caribbean off Colombia and in the southern Atlantic off Barbados and Surinam; there are apparently no records from the gap between the subspecies ranges. Uncommon to rare.

Synonymy.—This distinctive shell has a complex nomenclatural history which is still uncertain. Kiener described it under a preoccupied name on the basis of a single adult specimen from an unknown locality; later it was rumored that the holotype was a fossil, but this seems unlikely. Tradition assigned the species to the Pacific Ocean. In the early 1950's the same shell was rediscovered in the Gulf of Mexico and named twice in two years, the describers not comparing their taxa with extra-Atlantic or unknown locality species. In 1968 Usticke gave a fairly good description and figure of the southern form from off Barbados; his *C. pseudo-austini*, from the same locality, is unfigured and probably no type material exists—it is placed with this species for convenience. Van Mol described the Surinam population in detail as a new species and pointed out radular differences between his shells and typical *armiger* (as *C. clarki*).

Until and unless a similar cone corresponding to Kiener's description is discovered in the Pacific or Indian Ocean, there seems to be no other choice than applying the name *armiger* to the Atlantic species. It is possible that an earlier name may be found for the southern subspecies among the synonymy of other shells of somewhat similar appearance.

Notes.—Although elegant in shape, all-white adults are often very ugly and do not compare favorably with heavily marked specimens. The Gulf of Mexico populations have a strong tendency toward heavy growth marks, broken spires, badly broken lips, and tar stains. Length of the anterior canal is very variable, perhaps due to breakage during life and failure to regenerate the lost portion. Although the Gulf of Mexico subspecies is only uncommon or even locally common, the southern subspecies is seldom offered. Be careful of small *C. cancellatus* offered as *C. armiger bajanensis* at fantastic prices.

ARTICULATUS Sowerby iii, 1873

?1844. *Conus minutus* Reeve. *Conchologia Iconica*, 1(*Conus*): Pl. 47, sp. 259 (St. Vincent, West Indies). Non *Conus minutus* (Roeding, 1798).

1873. *Conus articulatus* Sowerby iii. *Proc. Zool. Soc.* (*London*), 1873: 146, pl. 15, fig. 3 (Mauritius).

1881. *Conus lombei* Sowerby iii. *Proc. Zool. Soc.* (*London*), 1881: 637, pl. 56, fig. 6 (? Mauritius).

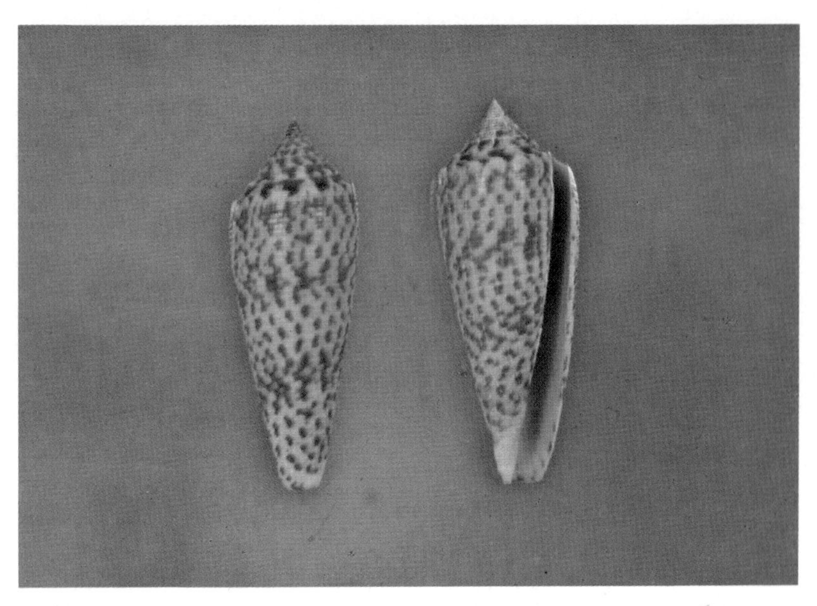

AUSTRALIS.—*Above* (*a. laterculatus*): Left: North Borneo, 47.2mm; Right: Tayabas Bay, Luzon, P.I., 49.0mm. *Below* (*a. australis*): Taiwan, 57.0 and 65.0mm.

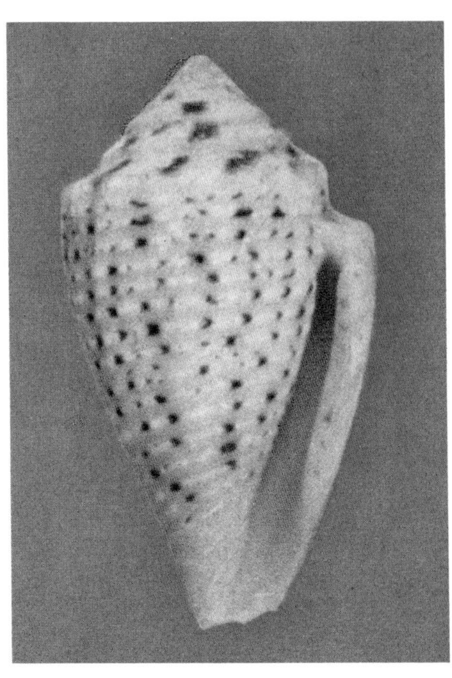

AXELRODI.—Samarai I., Papua New Guinea (paratypes) (EP). *Above:* 17.5mm. *Below:* 16.8mm.

1970. *Conus (Pionoconus) tosaensis* Shikama. *Sci. Rpts. Yokohama Nat'l. Univ.*, Sec. II(16): 25, pl. 1, figs. 22-23 (Tosashimizu, Kochi Pref., Japan).

Description.—Moderately light in weight, with a high gloss and waxy texture; low biconical, the sides nearly straight or very slightly convex; body whorl with broad flat ribs over anterior half to two-thirds, these separated by narrow deep faintly punctate grooves; rest of body whorl smooth or in small shells with faint traces of spiral ridges posteriorly; shoulder broad, sharply angled and carinate to weakly rounded in aged shells; spire low, the sides deeply concave to nearly straight, occasionally weakly stepped; early 4 or so whorls weakly nodulose, later whorls carinate or sometimes weakly undulate; tops of whorls flat to weakly concave, covered with crowded weak curved axial threads and ridges. Body whorl chestnut brown, tan, yellowish, pinkish, or pale lilac, the anterior fourth usually dark violet; body whorl covered with fine dashes in spiral lines, these often obsolete or nearly so and inconspicuous except on new growth; usually there is a midbody belt of large white ovals separated by axial bands of dashes or fused into a continuous white belt bordered above by a distinct brown belt; shoulder and area just below it white, covered with deep chestnut axial lines or blotches continued onto body whorl; spire white to bluish white or pale lilac, with chestnut lines or blotches; early whorls pinkish. Aperture narrow, slightly widened anteriorly; outer lip thin, sharp, thickening with age; posterior margin of lip on level with shoulder or slightly below it; mouth pale to deep violet within. Columella short and narrow, set off above by a ridge. Length 20-30mm.

Comparison.—This small waxy shell is very easily identified by general appearance, usual presence of fine brown dashes on the body whorl, usual presence of oval white spots near midbody, and the spire markings extending anterior to the shoulder onto the body whorl. Juveniles of *Conus muriculatus* are similar in shape and general texture and have the violet base and often a violet mouth; however, *C. muriculatus* has (normally) at least traces of spiral ridges on the spire whorls, usually has the whorl margins undulate or nodulose, and tends to have the darker brown spiral band (when present) below the pale midbody zone; the spire pattern does not extend below the shoulder. Some *C. polygrammus* may be similar to *C. articulatus* in color and shape, but this species seldom has a violet base, has distinctly undulate or coronate whorl margins, and has a reduced spire pattern. Both *C. muriculatus* and *C. polygrammus* have the basal ridges weak and often irregular, not broad ribs separated by narrow deep grooves. The small form of *C. mindanus* has similar basal sculpture but differs greatly in shape and details of pattern.

Variation.—Although apparently widely distributed, this small shell is poorly known. It is variable in color as indicated in the description and fades to very pale pink or whitish in beach specimens. The midbody white blotches may be large and obvious, fused into a broad belt, reduced, or absent. The spiral dashes may be strong over the entire shell, obvious only on new growth, visible only near the midbody belt, or apparently absent. Occasionally the violet base is indistinct. Spiral grooving may be confined to the anterior half of the body whorl or extend in weaker form to near the shoulder. Occasionally the margins of the spire whorls are weakly undulate and the whorls are weakly stepped; stepped whorls are possibly more characteristic of fully adult shells. Some specimens show traces of a single spiral ridge at the middle of the later spire whorls. Large shells may have faintly convex sides, while some smaller shells may be pinched-in near midbody.

Distribution.—Apparently this shell is not uncommon in the western Pacific from southern Japan and Okinawa south to at least New Britain.

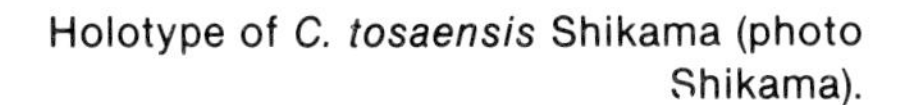

Holotype of *C. tosaensis* Shikama (photo Shikama).

Although described from Mauritius, it is rarely seen from the Indian Ocean; it definitely occurs there. Usually found dead on beaches, but occasionally taken live on sand and rubble bottoms at 30-60m.

Synonymy.—The great variability of this species has resulted in several synonyms, and perhaps more will be recognized with later work. The description of *C. articulatus* is based on a small shell with strong dashes and large oval white blotches. *C. lombei* is apparently an adult shell which has the body whorl almost uniformly colored and the spire whorls slightly stepped; I have not seen a specimen which matches the description exactly, but I have seen specimens obviously intermediate between typical *C. articulatus* and typical *C. lombei*. *C. tosaensis* is based on a large beach specimen (29.5mm) with faded pattern and slightly stepped spire whorls.

Notes.—This shell is only occasionally offered at the moment, and then only in the form of faded beach specimens. These are commonly sold as *C. semisulcatus*, a rather doubtful name which is possibly a synonym of *C. rutilus*. Live-taken shells with good color and pattern are apparently rare. Small shells have the most pleasing shape and color, adult shells apparently becoming more cumbersome and losing the bright colors.

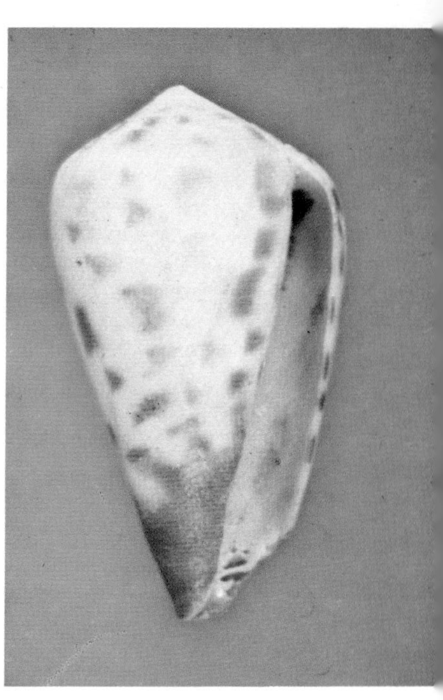

BAIRSTOWI.—Above: Hougham Park, South Africa, 33.2mm. *Below:* Left: Gonube, Cape, South Africa, 25.6mm; Right: Jeffreys Bay, Cape, South Africa, 25.0mm (GG).

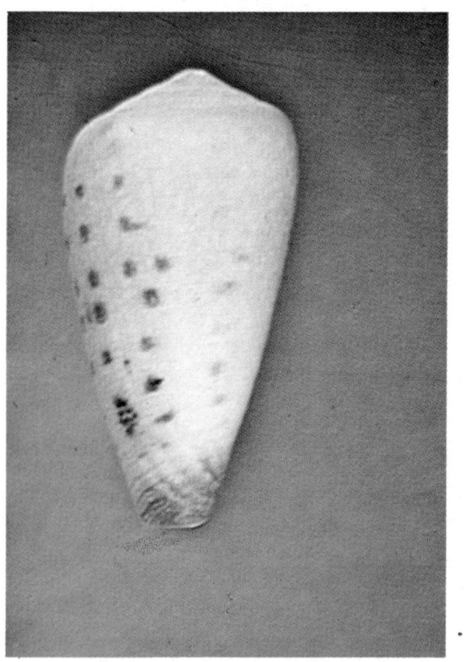

BALTEATUS.—Above: Light House Reef, Ponape, Eastern Carolines, 32.4 and 34.6mm. *Below:* Ona Village, Okinawa, Japan, 28.8 and 30.8mm.

ARTOPTUS Sowerby i, in Sowerby ii, 1833

1833. *Conus artoptus* Sowerby i, in Sowerby ii. *Conchological Illustrations:* Pt. 33, fig. 35 (South Seas).

1844. *Conus violaceus* Reeve. *Conchologia Iconica,* 1(*Conus*): Pl. 44, sp. 241 (Matnog, Is. Luzon, Philippines). Non *Conus violaceus* Gmelin, 1791. Lectotype designated and figured by Cernohorsky, 1977.

1854. *Conus spectabilis* A. Adams. *Proc. Zool. Soc.* (*London*), 1853: 117 (Australia).

1977. *Conus viola* Cernohorsky. *Nautilus,* 91(2): 72, figs. 1-3. Nomen novum for *Conus violaceus* Reeve, 1844.

Description.—Moderate in weight, with a low gloss or dull; cylindrical, the sides nearly straight and parallel until the base, occasionally slightly convex posteriorly; body whorl with weak and indistinct crowded spiral ridges at the base, these soon replaced by distinct and usually strong closely spaced spiral ridges which normally continue to the shoulder or in occasional specimens may become obsolete posterior to midbody; numerous fine axial threads present, these not normally producing granules, but giving the ridges a rough and uneven appearance; shoulder rounded or roundly angled, with about 4-6 strong narrow spiral ridges and numerous fine curved axial threads; spire moderate, pointed, the sides straight or slightly concave; whorls often weakly stepped, usually convex above; early 2 or so whorls with faint nodules; remainder of whorls with rounded margins, the tops with about 4-6 strong but narrow raised spiral ridges and weak axial threads. Body whorl white, cream, or pale violet, usually heavily covered with spiral rows of small sometimes indistinct brownish dashes or dots; usually three broad spiral bands of irregular axial clouds or blotches of yellowish tan, pale violet, or various shades of brown, one below shoulder, one above midbody, and one below midbody; usually the base, midbody area, and sub-shoulder area unmarked except for small brown dots or dashes; midbody band usually margined posteriorly by darker brown squarish blotches; spire about same color as body whorl, the tip pinkish to pale violet; usually small to large brownish spots or blotches present on shoulder and later whorls, these sometimes absent or greatly reduced. Aperture narrow posteriorly, widened anteriorly; outer lip convex, thin, sharp; mouth whitish, often with pale pink or violet tones. Columella narrow, moderately long, set off from inner lip by a weak ridge in some specimens, often curved away from axis of shell. Length 30-50mm.

Comparison.—The cylindrical shape and numerous spiral ridges on both body whorl and spire indicate a very close relationship to *C. nussatella,* which also shares some aspects of the color pattern. However, in *C. nussatella* the first 4 or 5 spire whorls are strongly and distinctly nodulose, the spiral rows of brown spots are more distinct and regular, the spots are larger and usually darker, and the axial clouds do not tend to form 3 broad spiral bands. In *C. granulatus* the body whorl tends to be more attenuated anteriorly and convex posteriorly, with the spire whorls tending to be distinctly stepped; the pattern in most *C. granulatus* is less distinct except for the dark blotches posterior to the midbody area; and the spiral ridges of the body whorl are much broader and fewer in relative number (although they are apparently formed by the fusion of 2 or more spiral ridges into a single rib-like ridge). *C. tenuistriatus* tends to be more convex posteriorly, the base is violet or dark brown, the body whorl ridges are often distinctly granulose, and the spire sculpture is weaker. *C. violaceus* is more elongate and has a distinctive pattern; the spire sculpture is greatly reduced. See also *Conus luteus.*

Variation.—Typical specimens of *C. artoptus* are rather uniform in shape and sculpture, although the spire varies somewhat in height and straightness of the sides. The major variation is one of color and pattern. Usually the 3 spiral bands on the body whorl are partially fused axially as well as spirally, but occasionally the individual blotches may be distinct all around the whorl; at the other extreme are shells with the bands very strongly developed and obscuring most of the background. As a general rule spiral rows of small (sometimes inconspicuous) brown dashes are present somewhere on the body whorl. The virtually unmarked subshoulder area, midbody band, and base are fairly constant.

Included here are shells from western Australia and the Northern Territory which, although tending to be very dark brown on violet instead of the paler patterns more typical of the species, resemble *C. artoptus* in shape and general appearance. However, these shells have the spiral sculpture of the body whorl reduced and usually obsolete above midbody, although weak spiral threads are present in place of the ridges.

Distribution.—Known from the Philippines and the New Guinea-Solomons area south to Queensland and northern Australia; it probably also occurs in the Indonesian area, but whether it occurs in the Indian Ocean other than in Western Australia is uncertain. Generally the species is found in moderately shallow water in sand and coral rubble. It is uncommon to rare and poorly known.

Synonymy.—This species is known to most collectors under the name *C. tenellus*, but this name is a synonym of *C. nimbosus; C. artoptus* seems to be the oldest available name. *C. spectabilis* has never been illustrated, but Sowerby ii and Tomlin assigned it here. The populations in northern Australia and Philippines have been called *C. violaceus* Reeve, recently renamed *C. viola.* These smooth shells fall within the range of variation of *C. artoptus.*

Notes.—Few gem condition *C. artoptus* are offered. Usually there is at least one healed scar on the body whorl and the tip of the spire is missing. The lip chips easily. Specimens with very dark or very light patterns are somewhat less attractive than those with fully developed tan or yellowish patterns. Northern Australian shells often have small worm holes.

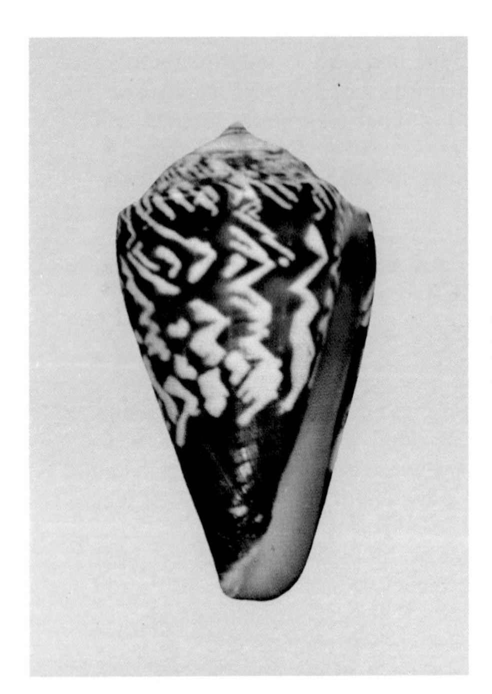

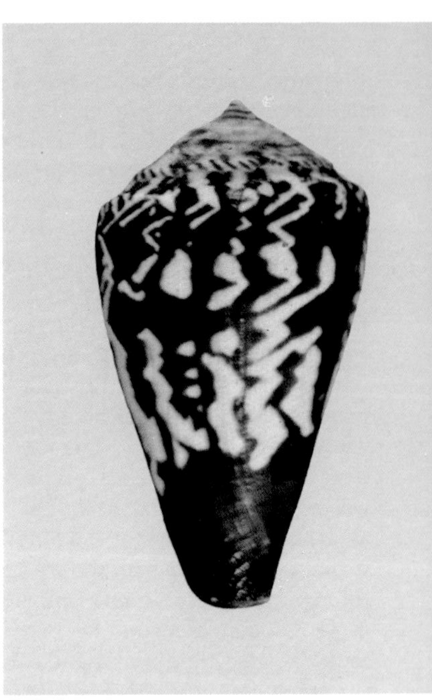

BALTEUS.—Above: Ilha do Sol, Cape Verde Is., 25.9mm. *Below:* Left: Sal I., Cape Verde Is., 25.4mm (EP); Right: Cape Verde Is., 27.8mm (MM).

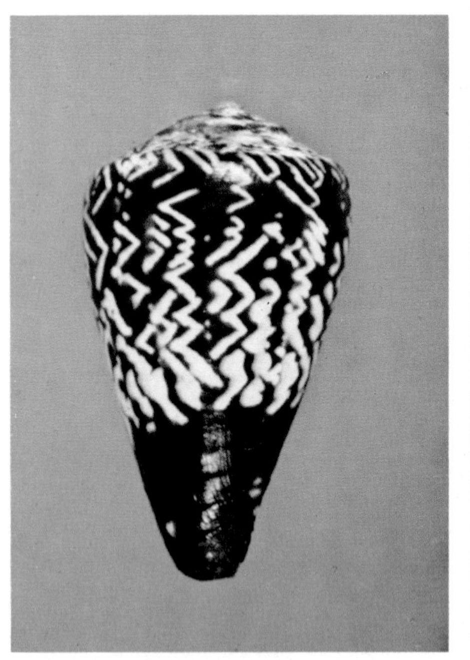

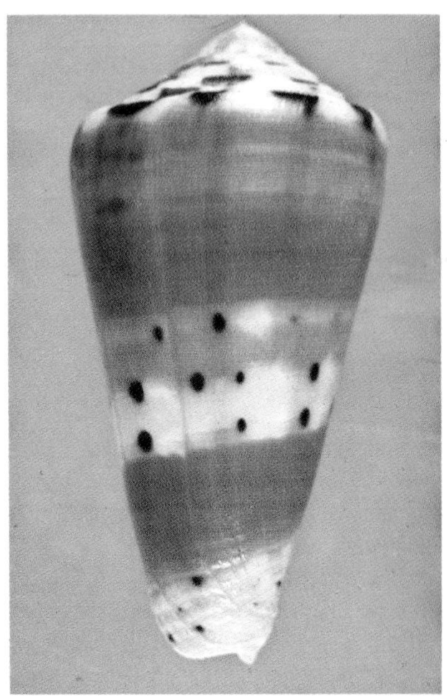

BARTHELEMYI.—Above: Reunion, 62mm. *Below:* Mauritius, 70.6-78.7mm (photo by A. Kerstitch).

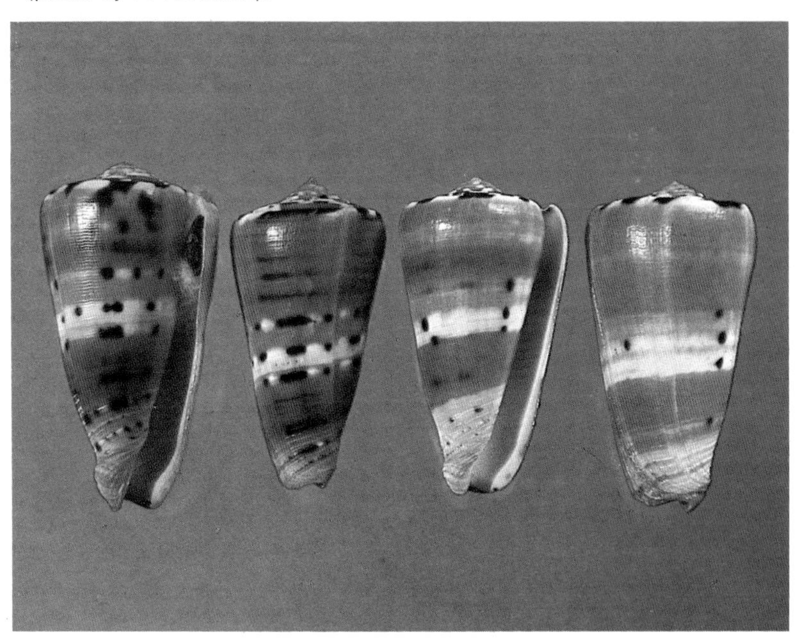

ATRACTUS Tomlin, 1937

1810. *Conus fusiformis* Lamarck. *Ann. du Mus. Hist. Nat. (Paris)*, 15: 423 (? Pacific Ocean). A syntype was figured by Kiener, Pl. 76, fig. 3. Non *Conus fusiformis* Fischer, 1807, probably a stromb.
1937. *Conus atractus* Tomlin. *Proc. Malac. Soc. London*, 22(4): 206. Nomen novum for *C. fusiformis* Lamarck, 1810.
1951. *Conus austini* Rehder and Abbott. *J. Washington Acad. Sci.*, 41: 22, fig. 7 (Off Loggerhead Key, Dry Tortugas, Florida). = *atractus austini*
1967. *Conus capricorni* Van Mol, Tursch, and Kempf. *Ann. l'Inst. Oceanogr.*, 45(16): 238, fig. 3, pl. 10, figs. 1-1a (Off Brazil, *Calypso* station 150, 30° 40' S, 49° 35.5' W, 141-135 meters). = *atractus atractus*

Description.—Fairly heavy, with a good gloss; biconical or broadly fusiform, the sides concave near middle, often nearly parallel posteriorly; body whorl covered with closely set usually low rounded spiral ridges or ribs from above base to below shoulder, these often alternating strong and weak in pairs; ridges heaviest toward base, sometimes obsolete below shoulder; numerous distinct axial grooves, threads, and sometimes plicae over body whorl, these producing a faintly cancellate appearance; shoulder broad, carinate, rather flat above; spire tall, sharply pointed, the sides weakly to strongly concave; spire whorls slightly stepped, especially the earlier whorls, the margins usually carinate; first 3-4 whorls weakly nodulose; tops of whorls flat or slightly concave, with about 3-4 distinct but low spiral ridges crossed by weak curved axial threads. Entire shell white or creamy white, sometimes with pale brown spots or blotches on body whorl either as axial clouds along growth lines or as smaller spots in two irregular rows above and below midbody; spire white, sometimes with small tan spots on shoulder and later spire whorls; tip of spire stained brown or yellow. Aperture moderate, of uniform width; outer lip straight or slightly convex, rather thin; mouth white. Columella not visible externally. Length 35-70mm.

Comparison.—The distinctive shape and weak but distinct spiral ridges and ribs over the entire body whorl in typical specimens are very distinctive, as are the weakly nodulose early spire whorls and weak but distinct spiral sculpture on the spire. *C. stimpsoni* is very similar but has somewhat straighter sides to both the body whorl and spire, very weak spiral sculpture on the later spire whorls, nearly obsolete flat spiral ribs on the body whorl, and usually a distinctive pattern of two or three broad spiral tan to salmon bands. *C. stimpsoni* and *C. atractus* are very similar and occasional specimens are hard to place. *C. arcuatus* and *C. cancellatus* are narrower and have more stepped spires, distinct color patterns, and several differences in spire sculpture. *C. armiger* is more rounded posteriorly and much more heavily sculptured.

Variation.—It would seem that two very similar but distinct subspecies can be recognized in this species.

C. atractus atractus Tomlin: More-or-less biconical, the body whorl not strongly constricted anteriorly and not strongly parallel-sided posteriorly; about 4 early whorls nodulose; white or with two irregular spiral bands of indistinct tan spots on the body whorl; juvenile with spire whorls distinctly stepped; off central and southern Brazil.

C. atractus austini Rehder and Abbott: Broadly fusiform, the body whorl strongly constricted anteriorly and distinctly parallel-sided posteriorly; first 3 whorls weakly nodulose; white or with indistinct axial clouds of tan along growth lines and with occasional pale tan spots on late spire whorls; juvenile with spire whorls not as strongly stepped; southern Florida and the Gulf of Mexico south through the Caribbean to Surinam and northern Brazil.

Shape distinguishes the two subspecies with some certainty; Kiener's figure of *C. fusiformis* matches the central Brazilian shell rather well. As a general rule, the spiral sculpture on the body whorl of *C. atractus atractus* is weaker and crossed by more axial threads than in *C. atractus austini*. The two apparently occur very close together off central Brazil, but intergrades would be hard to recognize.

Distribution.—Southern Florida south through the Gulf of Mexico and Caribbean (Antigua) to southern Brazil. Offshore bottoms, either mud or sand, in about 60-400 meters. Moderately common.

Synonymy.—Kiener's description and figure of one of Lamarck's syntypes leave little doubt that this is the shell Lamarck named *C. fusiformis*. *C. capricorni* seems to be a synonym of the typical form of the species, while *C. austini* is a distinct but very similar subspecies. The relation of *C. atractus* to *C. stimpsoni* is deserving of much more work, especially since some shells from off Yucatan combine the characters of both species; these have been called *C.* 'demarcoi', but as far as I can tell this is a nomen nudum.

Notes.—Chipped lips and large healed breaks on the body whorl are characteristic of this species, and true gems of large size are very seldom found. Juveniles with distinct patterns are sometimes very attractive for white cones. *C. atractus austini* is fairly common and not expensive, but the Brazilian *C. atractus atractus* is seldom offered and sells for a high price, perhaps undeservedly so.

ATTENUATUS Reeve, 1844

1844. *Conus attenuatus* Reeve. *Conchologia Iconica*, 1(*Conus*): Pl. 47, sp. 263 (Locality unknown).

1959. *Conus ustickei* Miller, in Usticke. *A Check List of the Marine Shells of St. Croix, U.S. Virgin Islands:* 80, pl. 1, fig. 14 (St. Croix).

Description.—Moderately light in weight, with a high gloss; low

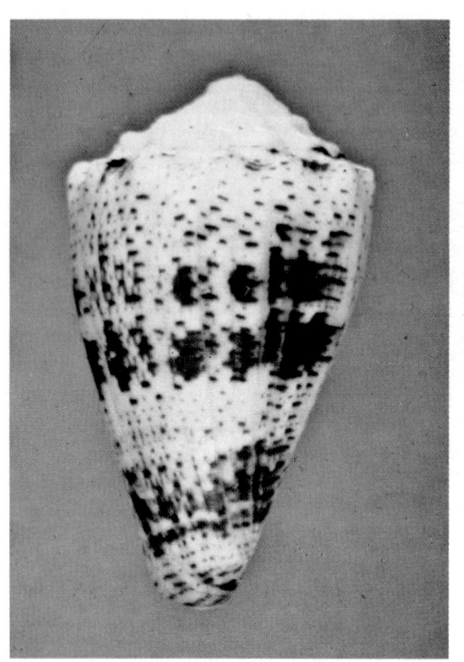

BARTSCHI.—San Pedro Martir I., Sonora, Mexico. *Above:* 37.7mm. *Below:* 29.8-34.3mm (photo by A. Kerstitch).

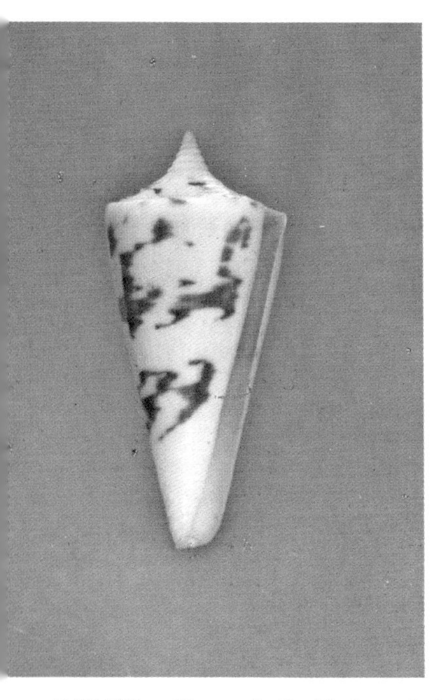

BAYANI.*—Above:* Left: Madras, India, 47.3mm; Right: India, 47.8mm (GG). *Below:* Sri Lanka, 62.8mm (T.C. Lan).

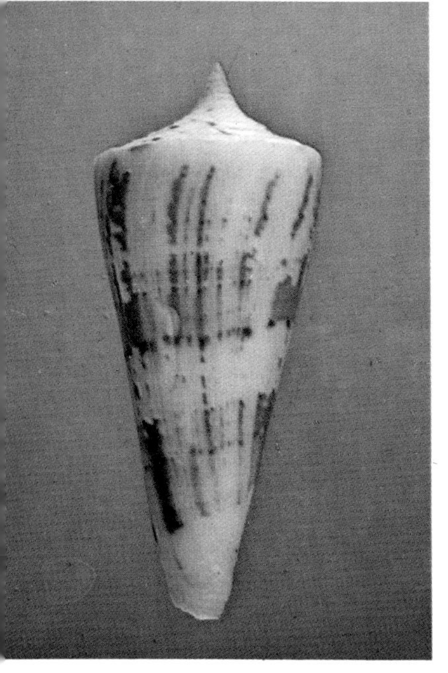

conical, body whorl very elongate, the sides straight; body whorl covered with very fine spiral threads, these most distinct near base, where about 8-10 may be quite strong; shoulder narrow, sharply carinate and smooth; spire low to moderate, sharply pointed, the protoconch prominent; early whorls commonly forming a somewhat rounded dome and weakly set off from rest of spire; sides of spire straight or slightly convex to slightly concave; tops of whorls usually distinctly concave, at least the earlier whorls with about 6 or more very fine but distinct spiral threads crossing equally fine axials. Color and pattern very variable; commonly white or pale pinkish with two very broad spiral bands of deep orange, red, or yellowish orange, the whitish background showing through as narrow to broad bands below shoulder and near midbody; base often deeper violet-pink than rest of body whorl; two other extreme patterns result in nearly white shells with two spiral bands of irregular reddish axial blotches or nearly all reddish shells with few to many irregular whitish axial blotches; spire and shoulder whitish to pinkish or occasionally pale reddish, with or without numerous squarish reddish blotches; early whorls white. Aperture narrow, uniform in width or slightly concave near middle; outer lip thin, sharp, fragile; mouth deep violet-pink. Columella not visible externally. Length 20-28mm.

Comparison.—The bright colors and distinctive shape show a close relationship with *C. daucus*. *C. attentuatus* differs from the many color variants of *C. daucus* in the relatively narrower body whorl which appears very elongate and is not widened below the shoulder; the early whorls are not nodulose in any specimen seen by me (although nodulose specimens perhaps should be expected), and the tops of the spire whorls are more deeply concave; no variant of *C. attenuatus* shows spiral rows of small brown dots on the body whorl as is so common in *C. daucus*. *C. magellanicus* is often somewhat similar in shape and even pattern to *C. attenuatus*, although seldom so elongate; in addition, *C. magellanicus* is usually more dully colored and more strongly patterned than in *C. attenuatus* and the spire whorls of *C. magellanicus* lack distinct spiral threads (occasional specimens may show traces of 1 or 2 indistinct threads).

Variation.—In shape and sculpture this is a very constant species, although the height of the spire and shape of the early whorls vary somewhat. The color and pattern are very highly variable, apparently individually. Typical specimens are of an extreme pattern where the body whorl is largely reddish brown with scattered irregular axially elongated whitish blotches; the spire is heavily spotted. Such a pattern occurs in series from Martinique, where another extreme pattern consisting of a largely white shell with axial reddish brown blotches in two spiral rows is developed. The pattern more familiar to collectors and described from St. Croix consists of largely reddish or orange shells with two narrow whitish bands below the shoulder and at midbody; width of the bands is very variable, and the shoulder band is often countershaded with reddish; in this pattern

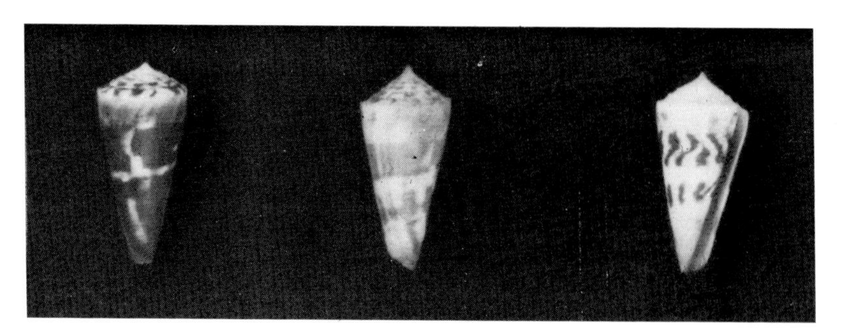

Variation in *C. attenuatus* from Martinique (photo EP).

there are seldom axial blotches, but the spire may be spotted or not. Typically the base of the shell is tinted with deeper violet-pink which is continued into the aperture.

Distribution.—Poorly known, but apparently ranging widely in the Caribbean. Occurs from off Florida and Cuba through the Virgin Islands and Guadeloupe south to at least Martinique. Taken by divers in a few meters and occasionally found on beaches.

Synonymy.—The description by Reeve of *C. attenuatus* is of a typically elongated shell with a pattern of axial white blotches on reddish brown. *C. ustickei* is very similar in shape but quite different in pattern, having two broad reddish bands with no axial breaks. However, through the courtesy of Mr. Ed Petuch, I have examined photos of series from Martinique which closely match the two types of patterns and indicate that both are within the normal range of variation of a single species. Florida and St. Croix specimens are apparently usually broad-banded, while those from the southern Antilles are more commonly axially blotched. It seems unlikely that *C. ustickei* could be a valid subspecies with a somewhat different pattern, but I have seen too few specimens to come to any conclusions. Obviously this species requires much more work. I have also seen a very similar Caribbean shell (locality unknown) with distinctly nodulose spire whorls which might fall within this species.

Notes.—This species seems to be moderately rare or at least uncommon and fetches good prices when available in any pattern. It would seem to be locally not uncommon in some localities in the southern Antilles. *C. attenuatus* is apparently not greatly subject to injury during life, although the lip is easily chipped. Beach specimens fade in intensity but retain obvious traces of their bright coloration.

AUGUR Solander, in Lightfoot, 1786

1786. *Conus augur* Solander, in Lightfoot. *Cat. Portland Mus.:* 44 (Locality not stated). Lectotype figure (Kohn, 1964): Knorr, Pl. 13, fig. 6.

BEDDOMEI.—Guarapari, Esp. Santo, Brazil, 22.2 and 23.0mm.

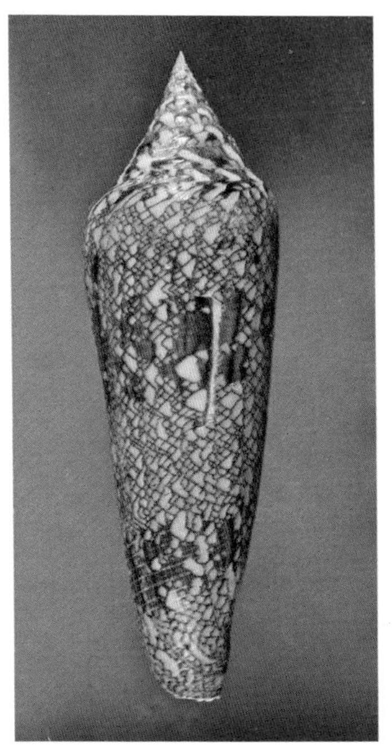
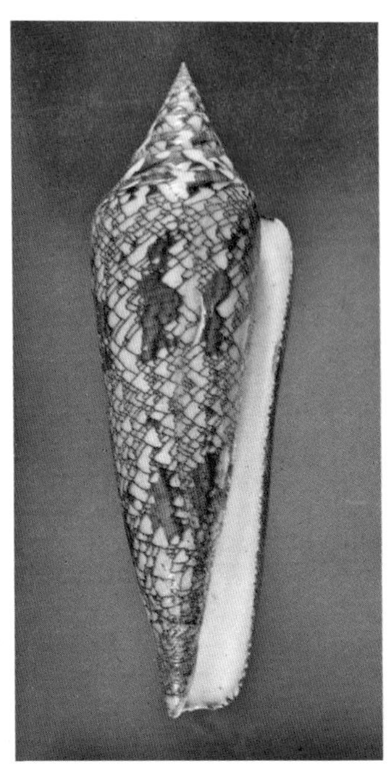

BENGALENSIS.—Above (holotype): Bay of Bengal, 96.7mm (photos by Dr. T. Okutani). *Below:* Off Burma, 93.0mm.

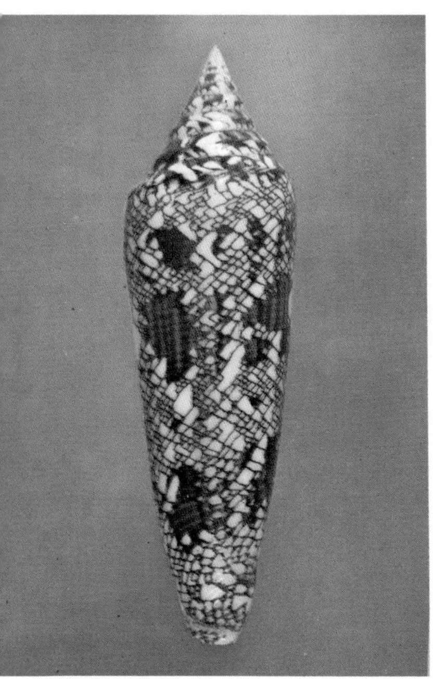
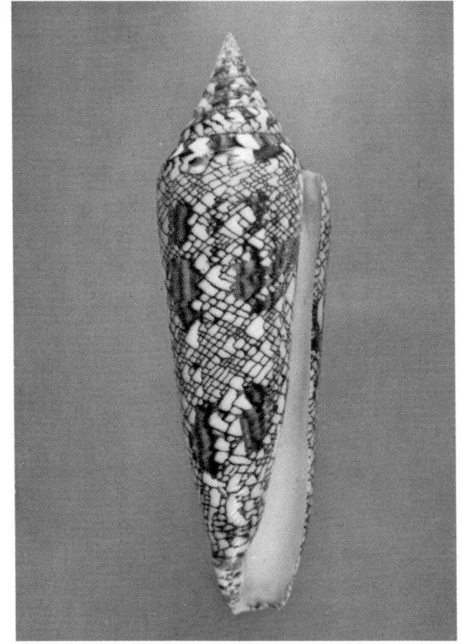

1791. *Conus punctatus* Gmelin. *Systema Naturae per Regna Tria Naturae,* ed. 13, 1: 3389 (Locality not stated). Holotype figure: Knorr, Pl. 13, fig. 6.

1798. *Cucullus pulverulentus* Roeding. *Museum Boltenianum:* 44 (Locality not stated). Lectotype figure (Kohn, 1975): Martini, Pl. 58, fig. 641.

Description.—Very heavy and coarse, with a low gloss or dull finish; obconic, the sides nearly straight or occasionally slightly concave near middle; body whorl with scattered low weak spiral ridges at base and on anterior fourth, occasionally with other spiral ridges at random over remainder of whorl; numerous fine and coarse axial threads, growth marks, and flaws over body whorl; shoulder broad, roundly angled; spire low, sometimes nearly flat, only earlier whorls at all raised in most specimens; spire whorls concave to canaliculate above, usually with traces of about two weak spiral ridges and numerous irregular and often coarse axial ridges and plicae. Body whorl creamy white to pale yellow, covered from base to shoulder with numerous (about 50) spiral rows of small to very small squarish dark brown spots; usually two broad spiral rows of darker brown squarish blotches developed above and below midbody area, these sometimes doubled or with axial extensions; base often pale salmon; shoulder and spire same color as body whorl, covered with numerous fine reddish brown curved axial lines; early whorls whitish or strongly eroded. Aperture moderately narrow, slightly wider anteriorly and sometimes constricted at middle; outer lip rather thick; mouth pale pink or white. Columella wide, rather short, with a heavy ridge above. Length 40-80mm.

Comparison.—The large size, great weight, and pattern of small brown spots make this a very distinctive shell. *Conus leopardus* is perhaps its closest relative, but it differs greatly in the color and disposition of the spotting. *C. betulinus* and *C. suratensis* are much rounder at the shoulder, have the spots in somewhat different arrangements and sizes, and commonly have brighter orange coloration. *C. biliosus* and *C. hyaena* may be superficially similar but have at least the early whorls nodulose, generally have more rounded shoulders, and differ greatly in pattern details and usually in mouth color. *C. planorbis* and allies have the base deep violet.

Variation.—The spire varies from nearly flat with only the earliest whorls prominent to moderate with rather straight sides; there seems to be a tendency to distort the last whorl. Adults seldom show distinct spiral sculpture on the spire whorls and may have the axial sculpture on the shoulder developed as plicae. Although the entire body whorl is usually covered with small spots, large adults tend to reduce the spots in the midbody area and sometimes over all the new growth. Most variable is the development of the darker blotching. Generally there are two broad spiral bands of rather squarish dark blotches, often with a fine line of dark brown just posterior to the anterior band. Occasionally the blotches are very small or even absent, especially the posterior band. In other specimens the blotches are fused to produce continuous solid spiral bands. Commonly there are axial extensions from the posterior band, and these extensions may be partially fused to produce another narrower spiral blotch band. The spire pattern may be strongly developed or reduced to just small dots at the margins of the whorls. Younger specimens tend to have bright pink or salmon tints at the base and in the mouth.

Distribution.—*C. augur* is fairly common in shallow reef areas throughout the Indian Ocean. It now seems to come mostly from India, the East African area, and Madagascar, but probably other areas of large populations exist.

Notes.—This is a rough cone, and large specimens in decent condition are not easy to find. Usually smaller specimens (under 50mm) show a minimum of axial breaks, but even these may have small eroded areas on both the spire and body whorl. Large adults have commonly lost much of the pattern, have bad breaks on the body whorl, have suffered at least one major lip chip, and commonly have large eroded areas on the spire and near the shoulder. Specimens with reduced patterns are common, as are those with typical blotched patterns. Complete spiral bands of dark brown are not uncommon, but it is rare to have two or three complete bands on the same specimen; no patterns deserve a premium.

AULICUS Linnaeus, 1758

1758. *Conus aulicus* Linnaeus. *Systema Naturae per Regna Tria Naturae*, ed. 10, 1: 717 (Asia). Lectotype figure (Kohn, 1963): Gualtieri, Pl. 25, fig. 2.

[1810. *Conus particolor* Perry. *Arcana:* 5 (Locality not stated). Considered non-binomial.]

1834. *Conus Aulicus* var. *roseus* Sowerby i, in Sowerby ii. *Conchological Illustrations:* Pt. 55, fig. 71 (Anna I.). Non *Conus roseus* Fischer, 1807.

1900. *Conus aulicus propenudus* Melvill. *J. Conchol.* (*London*), 9(10): 310 (Locality not stated).

Description.—Moderately heavy, with a good gloss; broadly fusiform, the body whorl widest just below shoulder; body whorl with irregular grooves and spiral ridges above the base, these rapidly changing into evenly spaced fine spiral threads which continue to shoulder in most specimens and are easily felt with the finger-tip; shoulder rounded, poorly defined and sometimes slightly concave above, grading into the spire; spire rather high, sharply pointed in unworn specimens, the sides straight or slightly concave; early whorls when not eroded do not stand out sharply from the remainder of the spire whorls; tops of whorls with faint

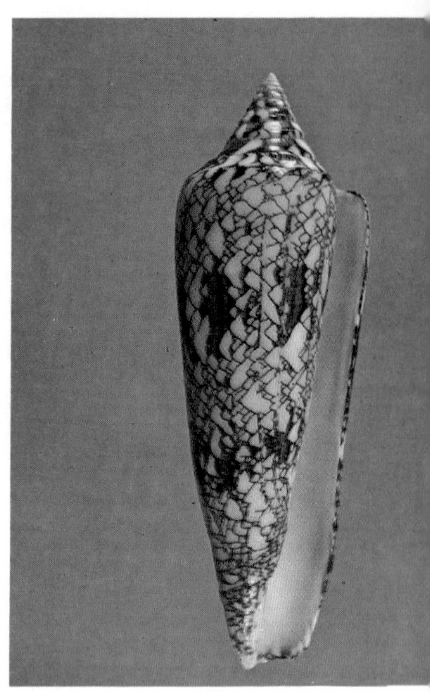

BENGALENSIS.—Above (shouldered form): Off Anakan Coast, Burma, 80.1mm (MM). *Below:* Andaman Sea off Burma, 62.0mm (GG).

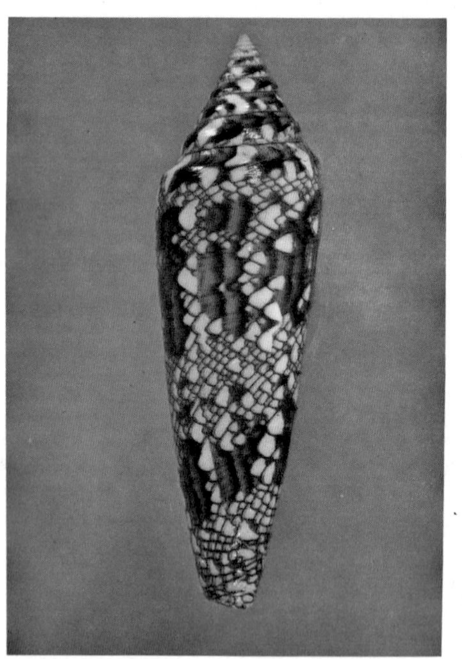
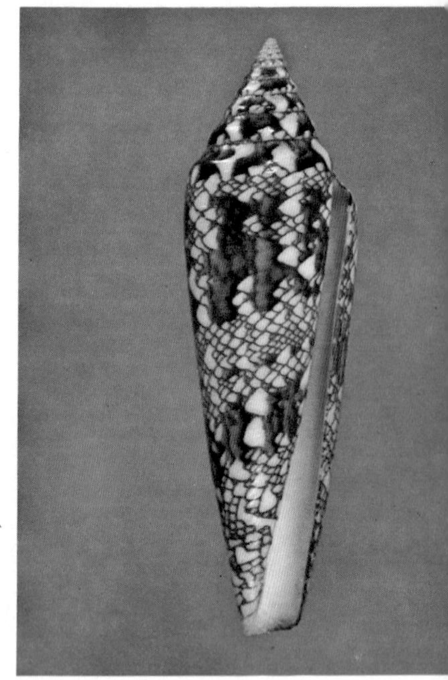

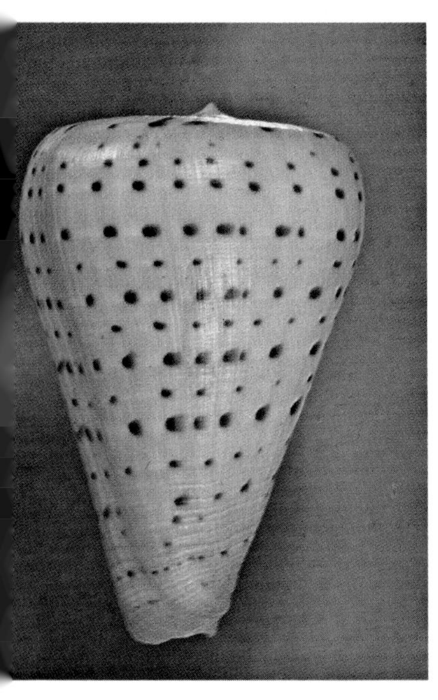
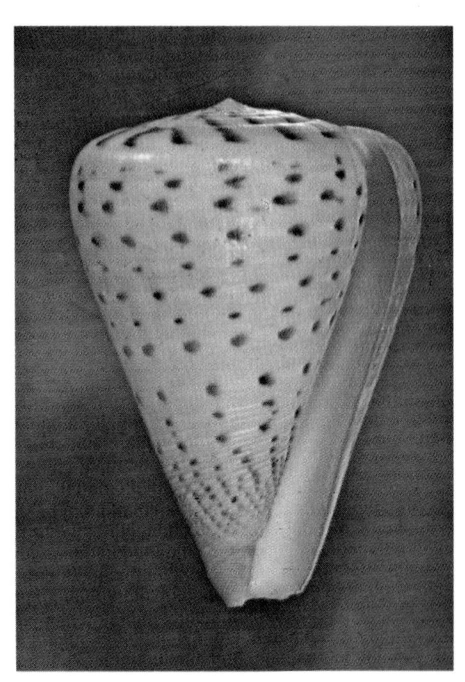

BETULINUS.—Above: Davao Bay, Mindanao, P.I.: Left: 90.7mm; Right: 87.8mm. *Below:* Left: Port Louis, Mauritius, 50.5mm; Right: Inhambane, Mozambique, 106.7mm.

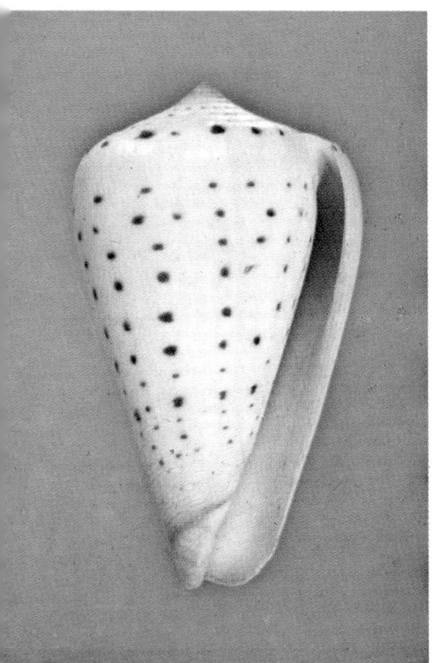
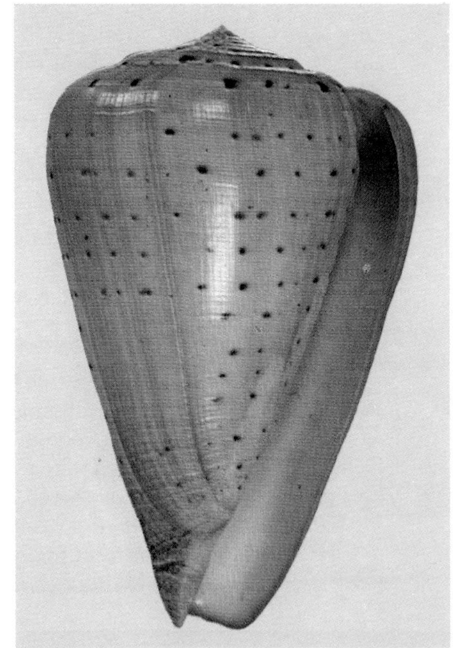

traces of 1 or 2 spiral threads and sparse axial threads, the early whorls sometimes with the margins visibly carinate, not nodulose. Body whorl white, covered with zigzag axial lines and large blotches of deep reddish brown, golden tan, or occasionally blackish brown; blotches usually axial, irregular in form, and separated by large and small white tents that tend to form 3 axial bands, one below shoulder, one at midbody, and one above the base; within the dark blotches and following the spiral ridges are fine blackish lines interrupted irregularly by very small white dots or tents, these sometimes absent; spire and shoulder same color as body whorl, largely brownish with white tents, the shoulder mostly covered with brown blotches; earliest whorls whitish. Aperture wide throughout, very wide anteriorly; outer lip thick, straight; mouth whitish to creamy yellow. Columella long, narrow, set off above by a weak ridge. Length 70-162mm.

Comparison.—*C. aulicus* stands out from most other tented cones by the combination of large size, relatively slender shape, sharply pointed spire, and distinct spiral ridging on the body whorl. From *C. textile* and *C. aureus* it differs in lacking vertical blackish 'textile lines' in the dark blotches and in not having fine nodules on the early spire whorls. From *C. magnificus* and other *pennaceus*-like cones it differs in not having the early whorls sharply set off from the other spire whorls and in the stronger and distinct spiral ridging of the body whorl. From the extremely similar *C. auratus* it is (barely) distinguishable by slight differences in shape (*C. auratus* being somewhat more evenly convex-sided), slightly weaker spiral ridging, and tendency to have larger white tenting and somewhat duller colors. Obviously *C. auratus* is very close to *C. aulicus* and possibly not distinct. Juvenile *C. aulicus* have been sold as *C. aureus*, but they are easily told by the absence of spire nodules and 'textile lines.'

Variation.—Although quite constant in shape, there is considerable variation in color and pattern. Normally the dark color is reddish brown, but blacker tones are not uncommon, sometimes over only part of a specimen; pale golden tans are also found, but these may represent fading or post-mortem changes. Although the tents are commonly white, pinkish or even orange tones are not uncommonly seen. Some specimens or parts of specimens, especially near a serious growth mark, may have the tents greatly enlarged and the blotches absent, resulting in large white areas; such shells were given the name *propenudus*, but this is obviously an individual variation.

Distribution.—Moderately common throughout the Indo-Pacific, including French Polynesia; absent from Hawaii.

Synonymy.—*Roseus* (preoccupied) was applied to pinkish specimens; *propenudus* was given to specimens with large white areas. *C. auratus* probably belongs here as well but is tentatively recognized as valid for no real reason.

Notes.—This magnificent tented cone is popular with collectors and hard to confuse with anything else. I have seen a few juveniles (which apparently are not common) sold as *C. aureus*, but this should not fool anyone. *C. aulicus* commonly has minor growth marks on the body whorl and on the base, as well as a few small lip chips. Golden tan specimens sometimes are sold as *C. auratus*, which they are not; collectors should show great caution when buying *C. auratus* not from the French Polynesia area.

AURANTIUS Hwass, in Bruguiere, 1792

1792. *Conus aurantius* Hwass, in Bruguiere. *Cone,* in *Ency. Method., Hist. Nat. des Vers,* 1: 606 (Philippines). Lectotype selected by Kohn, 1968.
1792. *Conus Cedonulli Curassaviensis* Hwass, in Bruguiere. *Ibid.*, 1: 602 (Curacao). See *Tableau Ency. Method.*, Pl. 316, fig. 4.
1792. *Conus Cedonulli Dominicanus* Hwass, in Bruguiere. *Ibid.*, 1: 603 (Dominica). Lectotype figure (Clench, 1942): Chemnitz, Pl. 141, fig. 1306.
1792. *Conus Ammiralis Grenadensis* Hwass, in Bruguiere. *Ibid.*, 1: 603 (Grenada). Holotype figured by Mermod, 1947, fig. 3a.
1792. *Conus Cedonulli Martinicanus* Hwass, in Bruguiere. *Ibid.*, 1: 603 (Dominica). See *Tableau Ency. Method.*, Pl. 316, fig. 3.
1792. *Conus Cedonulli Surinamensis* Hwass, in Bruguiere. *Ibid.*, 1: 603 (Surinam). Lectotype figure (here selected): *Tableau Ency. Method.*, Pl. 316, fig. 9. Non *Conus surinamensis* Gmelin, 1791.
1792. *Conus Cedonulli trinitarius* Hwass, in Bruguiere. *Ibid.*, 1: 603 (Trinidad). Lectotype figured (as holotype) by Kohn, 1968; here selected as lectotype.
1798. *Cucullus corona civica* Roeding. *Museum Boltenianum:* 38 (Locality not stated). Lectotype figure (Kohn, 1975): Martini, Pl. 61, fig. 678.
1968. *Conus karinae* Usticke. *Caribbean Cones from St. Croix and the Lesser Antilles:* 9, pl. 2, sp. 992 (St. Croix, to the north of Frederiksted; Aruba). Juvenile.

Description.—Moderately heavy, with a low gloss; biconical, somewhat fusiform, the body whorl rather elongate and with nearly straight sides or slightly convex; body whorl covered with broadly spaced fine spiral ridges, the anterior 10 or 12 usually most distinct and often bearing small, regularly spaced granules; ridges may be obsolete above midbody or strong and granulose to the shoulder; usually distinct axial growth marks and threads obvious; shoulder rounded, usually with distinct rather axial coronations or only vaguely undulate, slightly concave above; spire moderate, bluntly pointed, the sides straight to slightly concave or slightly convex; at least early 5-6 whorls bearing distinct coronations at margins and usually traces of 2-3 fine spiral ridges crossing weak and sometimes indistinct axials; in occasional shells the spiral ridges may be distinct and strong or may be completely obsolete on later whorls; tops of whorls slightly to strongly con-

BILIOSUS.—*Above:* Sri Lanka, 39.1mm. *Below:* Left: Sri Lanka, 40.1mm; Right: Rameswaram, India, 46.5mm.

BILIOSUS.—Above: Left: Galle, Sri Lanka, 36.3mm; Right: Wattala, Sri Lanka, 36.9mm (RJ). *Below:* Genezzano, Natal, South Africa: Left: 44.0mm; Right: 41.9mm.

cave. Body whorl white, often with pinkish or pale violet tones, variably covered with numerous spiral rows of brownish to blackish dashes and dots, these sometimes nearly obsolete and inconspicuous except near the lip; spiral ridges often bearing distinct but very small white dots covering granules, these evenly spaced and most distinct near base in many shells; variable pattern of larger blotches usually present in at least 4 rows: 1 above base, 1 above and 1 below midbody area, and 1 below shoulder area; often smaller blotches in secondary rows or blotches fused to cover most of shell except for irregular pale blotches above base, at midbody, and at shoulder; occasionally blotches very small and almost absent; pattern never greatly complex and forming 'animal figures' as in *C. cedonulli;* blotches may be margined in all or part with a narrow blackish line; blotches very variable in color, usually some shade of orange or brown, but occasionally black or yellow; spire whitish, sparsely marked with large blotches similar to those of body and small axial lines at the sutures; early whorls white, usually eroded. Aperture moderately wide, slightly widened anteriorly; outer lip straight, sharp; mouth white to deep violet or even pink, color perhaps depending on environment and how long dead. Columella narrow, largely internal, not set off posteriorly. Length 30-55mm.

Comparison.—*C. aurantius* is a variable species closely related to *C. regius* on one hand and *C. cedonulli* on the other. From *C. regius*, which it often closely resembles in color and pattern, it differs in the usually more straight-sided spire, rounded shoulder, weaker and usually more rounded shoulder and spire coronations, and—most importantly—the relatively longer and more straight-sided body whorl which gives a completely different appearance to the shell. *C. regius* is usually much broader at the shoulder with a shorter, more convex body whorl; the spire is normally strongly concave-sided and bears numerous fine but distinct spiral ridges on each whorl; the fine axial color lines marking the sutures of later spire whorls in *C. aurantius* are apparently typically (? always) absent in *C. regius*. *C. cedonulli* is even more similar to *C. aurantius*, and I am not sure that extremes of the species can be separated with certainty. As a general rule, *C. cedonulli* is more inflated, has a somewhat broader shoulder and slightly concave spire, tends to be deep waxy orange or waxy yellow in long-collected specimens (perhaps deep brown, almost black in fresh living material), and has the pattern much more complex. The spiral rows of white dots are large, usually outlined with bright purplish brown, and frequently fused into broad bands; the blotches are very irregularly margined and frequently produce very complex patterns at the midbody area, the patterns often resembling 'animal figures.'

Three other species belong to the group and could be confused with *C. aurantius*. *C. scopulorum* from northeastern Brazil is very similar in appearance and is perhaps subspecifically related; it is smaller at apparent adult size, has a taller and narrower spire, more prominent shoulder, and perhaps stronger axial ridging on the spire whorls; the species is poorly known. *C. archon* of the Panamic area usually differs greatly in spire shape and lacks conspicuous coronations; it has more opaque white in the pattern and generally lacks granules. *C. varius* of the Indo-Pacific is much more angled at the shoulder, with prominent and rather pointed coronations; the pattern consists of two broad bands of dark brown axial blotches connected by a spiral band of yellowish brown in most specimens, with the midbody white area large and almost unmarked except for the numerous small rather axial brown dashes present in spiral rows; strong granules are usually present to the shoulder; and the mouth usually has at least a large blotch of yellow deep within. *C. cabritii* is also vaguely similar but is much smaller, different in pattern and color, and has somewhat different spire sculpture.

Variation.—As indicated in the description, this is a very variable species whose many different patterns have been given individual names. There is some reason to believe that individual islands of the Lesser Antilles have somewhat differentiated color patterns, but the variation within a single population may include all the major pattern types. See Van Pel, 1969 (*Hawaiian Shell News*, October: 5) for illustrations of specimens from the Netherland Antilles. Collectors commonly recognize two major variants, but these are very hard to define and probably meaningless by themselves: *aurantius*—blotches of body whorl large, fused spirally so only small white blotches are present above midbody; blotches often orange; *dominicanus*—blotches small, not greatly fused spirally, the shell largely white; blotches usually brown or black; black specimens occur in both major pattern types.

Other than pattern, there is some variation in spire shape (concave sides are rare), coronations (usually present to shoulder and rather axially elongate but rounded), and amount of granulation on the body whorl (usually at least 10-12 anterior ridges granulose, but the granules small and often not picked out by white dots). Presence or absence of a narrow purplish brown line outlining the blotches seems to be individually variable or at least variable from population to population. Specimens with very greatly irregular blotches and pale orange in color could be confused with *C. cedonulli*, but they would seldom also have large, outlined white spots associated with the granules.

Distribution.—Uncommon to rare in 5-10 or more meters on sand and rubble from St. Thomas south through the Lesser Antilles perhaps to Venezuela and Surinam (continental records perhaps doubtful). Specimens usually available to collectors only from Netherland Antilles.

Synonymy.—The many names of Hwass are placed in synonymy of *C. aurantius* after examination of the various type figures and descriptions. In figures this species can usually be distinguished from *C. cedonulli* by the less complex pattern, narrower shoulder, and straighter-sided spire. There is of course the possibility that some of the synonymy may actually apply to *C. cedonulli*. Since *C. aurantius* was described as a full species and is in current use, I prefer to continue its usage as the proper name for the species. *C. karinae* Usticke appears to be a juvenile specimen of *C. aurantius* and does not hâve any real distinguishing features.

Notes.—The exact status of this species is far from settled, as should be

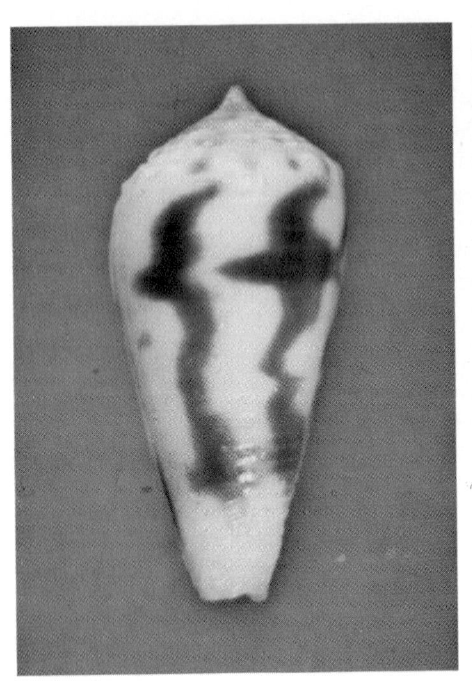

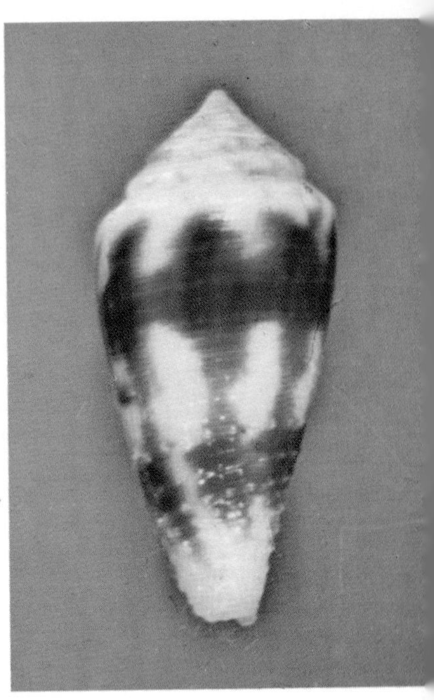

BOETICUS.—Above: Left: Kwajalein Atoll, Marshall Is., 28.3mm; Right: Orode Point, Guam, 27.8mm (TCG). *Below:* Buckner Bay, Okinawa, Japan: Left: 19.0mm (TCG): Right: 24.3mm (DMNH).

BOETICUS.—Left: Boac, Marinduque, P.I., 25.7mm; Right: Off Tosa Shimizu, Shikoku I., Japan, 28.8mm (EP).

BOIVINI.—Holotype from Kiener, Pl. 64, fig. 2, locality unknown, 62mm.

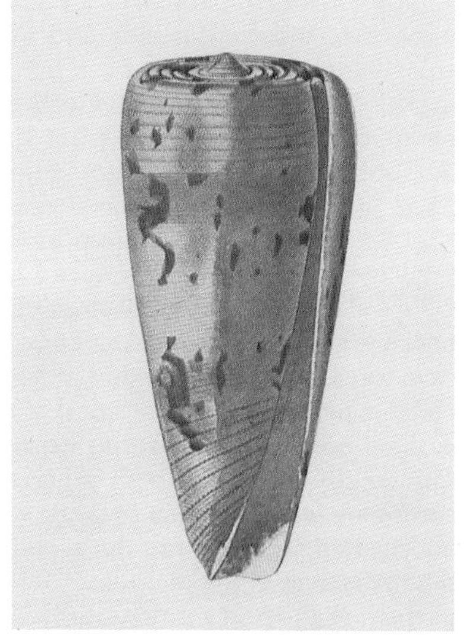

obvious from the comments above. I can see no reason to recognize more than one taxon here and believe that the various island forms are exactly that—populational and individual variants. Good specimens of this species, regardless of the name used by dealers, are rare and still expensive, although not so much as *C. cedonulli*. Gems of this species usually have eroded spire tips and as much as 1 body growth mark but are otherwise perfect.

Vink (1977, *Zool. Mededelingen*, 51(5); see also *Hawaiian Shell News*, 25(8), Aug. 1977) has recently reviewed the cones of the *C. cedonulli* complex and recognized five full species, *C. cedonulli, C. aurantius, C. insularis, C. mappa*, and the new *C. sanctaemarthae* Vink, 1977 (from Santa Marta, Colombia). *C. sanctaemarthae* is here considered to fall within the range of variation of *C. aurantius* as diagnosed earlier in this book, being a fairly well defined but local variant. The relationship of the Caribbean *C. cedonulli* and *C. aurantius* with the Panamic *C. archon* is now obviously very close, with the Colombian *sanctaemarthae* serving as a nice link with certain variants of *C. archon*. Obviously the group needs much more work.

AURATUS Hwass, in Bruguiere, 1792

1792. *Conus auratus* Hwass, in Bruguiere. *Cone*, in *Ency. Method., Hist. Nat. des Vers*, 1: 740 (Indian Ocean). Lectotype selected by Kohn, 1968.

Description.—Heavy, with a high gloss; broadly fusiform, the sides slightly convex but parallel near midbody, then distinctly curving back to the axis of the shell at base and shoulder; body whorl with irregular grooves and spiral ridges above the base, these rapidly becoming evenly spaced distinct spiral ridges which usually extend to shoulder; shoulder rounded, not distinct from spire; spire moderate, sharply pointed, the sides straight; early whorls grade into later whorls; tops of whorls with traces of faint spirals and crowded axial threads; early whorls not nodulose. Body whorl white to deep pink, covered with a network of zigzag axial lines overlapping to produce a rather finely tented pattern; lines and 2-3 spiral rows of blotches deep golden brown to deep reddish brown, the blotches above and below midbody usually rather squarish and isolated from each other and smaller series below shoulder; within the dark blotches and following the spiral ridges are fine dark and light dots and dashes, some of the light dots expanded into small tents; shoulder and spire mostly brown like body whorl, white tents usually reduced in size; early spire whorls pinkish or white. Aperture wide throughout, distinctly widened anteriorly; outer lip smoothly curved at both ends, straight or slightly convex at middle, thick; mouth whitish to pink. Columella long, narrow, slightly set off posteriorly. Length 50-110mm.

Comparison.—*C. auratus* is so similar to *C. aulicus* in all respects that it should almost certainly be considered a synonym of that species. However, the few specimens I have seen have been very consistent in shape and pattern and have a certain almost indefinable 'feel' that makes them distinct from *C. aulicus*. The pattern is easily found in many *C. aulicus*, although there does seem to be a tendency in *C. auratus* for the large blotches to be isolated and surrounded by finer tenting than usual; the spire also may average darker than usual in *C. aulicus*. The pink color of *C. auratus*, and its golden tan shades for that matter, is not constant even in the few specimens I have seen and will not separate the species from many *C. aulicus*. The regular curving of the shoulder and base seem to be fairly distinct and gives the shell an appearance different from most *C. aulicus* of similar size, which tend to have a stronger angle at the shoulder and a somewhat higher and narrower spire. The early whorls grading into the later ones and the absence of nodules will distinguish this species from *C. textile* and *C. magnificus* and allies. Small *C. auratus* strongly remind one of *C. aureus* in form and texture but lack 'textile lines.'

Variation.—Although the pink tones and fine tenting are common in this species, they are not constant; neither is the deep golden tan color. Occasionally the body blotches above midbody may be connected to those below the shoulder.

Distribution.—As treated here, this species appears to be endemic to the Society and Tuamotu Islands and perhaps other areas in French Polynesia. It is rare.

Synonymy.—Whether this is actually the *C. auratus* of Hwass is very doubtful; his taxon is in all likelihood a synonym of *C. aulicus*. The name *C. auratus* is retained only because it is currently in use for this shell and because eventually this taxon will almost certainly be shown to be an ecological variant of *C. aulicus*. Because this form sells for such a great deal more than typical *C. aulicus*, it should have a working name for collectors.

Notes.—Philippine specimens sold as this species usually are narrow, rather golden *C. aulicus*. Specimens over 85mm appear to be rare. This shell should be purchased only with extreme caution and after comparison with a good series of similar-sized *C. aulicus*. As in *C. aulicus*, this species seems to be subject to one or two short growth marks above the base and in the aperture.

AUREUS Hwass, in Bruguiere, 1792

1792. *Conus aureus* Hwass, in Bruguiere. *Cone*, in *Ency. Method., Hist. Nat. des Vers*, 1: 742 (Indian Ocean). Neotype selected by Kohn, 1968; same specimen illustrated by Kiener, Pl. 82, fig. 2a.

1810. *Conus auricomus* Lamarck. *Ann. du Mus. Hist. Nat.* (*Paris*), 15:

BOSCHI.—Mashira I., Muscat Oman, 24.2 and 26.4mm.

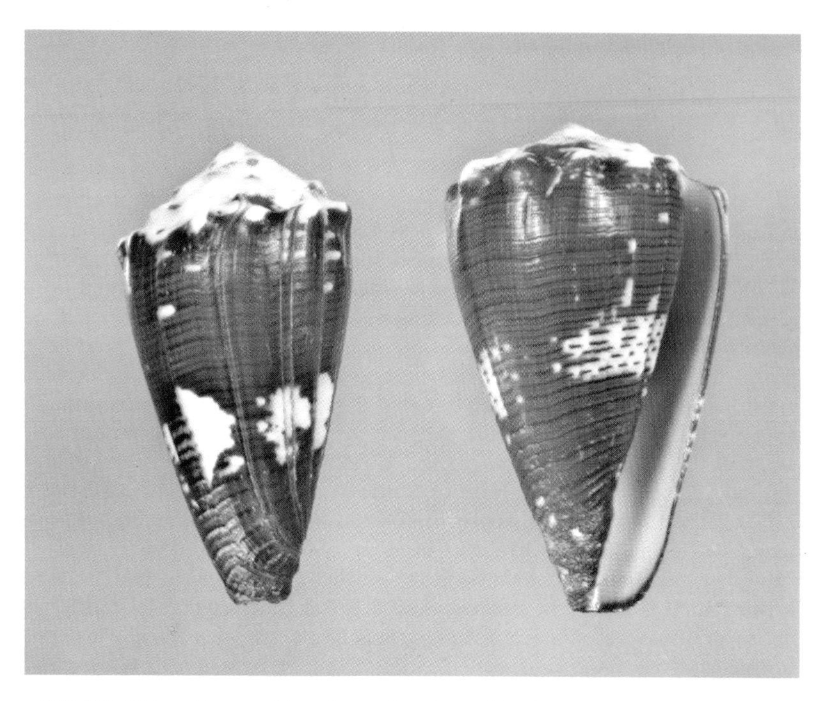

BRUNNEUS.*—Above:* Mazatlan, Sinaloa, Mexico, 43.9 and 46.0mm. *Below:* Left: Cabo San Lucas, Baja California, Mexico, 10.8-20.4mm (AK); Right: Mazatlan, Sinaloa, Mexico, 41.2mm.

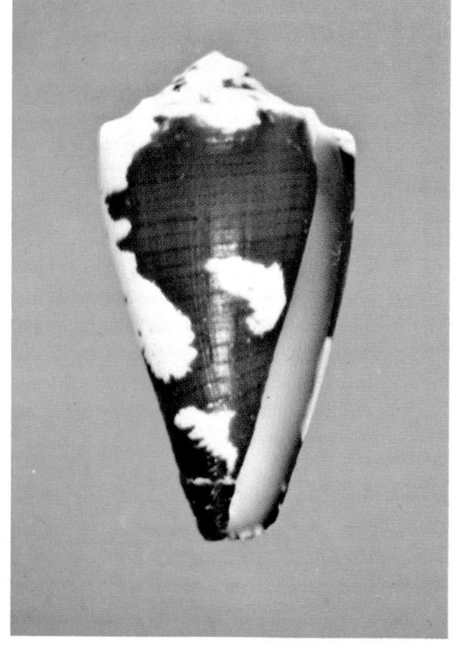

433 (Indian Ocean). Non *Conus auricomus* Hwass, 1792.
1877. *Conus paulucciae* Sowerby iii. *Proc. Zool. Soc.* (*London*), 1876: 752, pl. 75, fig. 3 (Mauritius).

Description.—Moderately heavy, with a high gloss; elongate, rather cylindrical or weakly fusiform, the sides slightly convex below shoulders; body whorl with usual oblique grooves and ridges at base, these rapidly becoming heavy, evenly spaced spiral ridges which commonly extend to shoulder and are easily felt and seen; shoulder rounded, narrow or moderately broad, blending into spire; spire moderate to quite tall, sharply pointed, the sides straight or slightly concave; tops of whorls weakly concave or nearly flat, usually with traces of 2-4 weak spiral ridges and fine axial wrinkles; earliest 4-5 whorls with low but distinct nodules and heavier spiral ridging. Body whorl orange to yellowish brown on white, the darkest pigment taking the form of elongate axial blotches separated by axial and spiral bands of usually small whitish or golden tents; the tents tend to form a narrow band at the shoulder, one at midbody (often interrupted or broken by the blotches), and one above the base, but their axial arrangement is often very irregular and the dark blotches may be continuous spirally as well; dark blotches contain numerous long black or deep brown axial 'textile lines,' these extending from the base to near the spire although usually broken in 2 or 3 places; spire and shoulder brownish or orange, marked with whitish tents like the body whorl; early whorls white to pink. Aperture narrow, somewhat wider anteriorly; outer lip slightly convex, sharp, thick; mouth white to deep pink. Columella long, moderately narrow. Length 30-65mm.

Comparison.—This striking cone is obviously close to *C. textile* and allies in having nodulose spire whorls; *C. aulicus*, *C. auratus*, and *C. magnificus* may be vaguely similar but lack the nodulose spire whorls and differ in many pattern details, especially lack of 'textile lines.' *C. textile* and *C. retifer* are much more ventricose than *C. aureus* and may have the 'textile lines' short or even absent. *C. canonicus* and *C. legatus* have the brown blotches smaller, the 'textile lines' short or absent, and the body whorl with only weak or obsolete spiral ridging above midbody; they are not nearly as elegant in shape as the average *C. aureus. C. telatus* (as currently treated) has the sides somewhat more convex, the tops of the whorls concave with prominent and often coronate margins, the body blotches often smaller with shorter 'textile lines,' and a tendency to have irregular small black spots at the shoulder; most *C. telatus* are also commonly either heavily granulose on the body whorl or virtually smooth, not distinctly and strongly striated. *C. auricomus* has a more slender shape with parallel sides, a short and rather rounded spire, and lacks distinct 'textile lines.'

Variation.—Like most other textile cones, *C. aureus* varies greatly in texture of the tenting, from very fine and uniform to irregular and large. The spire also varies considerably, as does the tone of the brown (pale golden orange to deep reddish brown). The blotches may be relatively narrow but continuous axially from the base to the shoulder, or broken into broad spiral continuous bands. A good idea of variation in this species can be obtained from Plate 24 of Veillard, 1976 (*Cones et porcelaines de l'Ile de la Reunion*, privately printed), where is it treated under the names *C. aureus*, *C. legatus*, and *C. paulucciae*.

Many specimens from the southern Indian Ocean islands of Reunion and Mauritius (and adjacent areas) tend to have relatively high spires, rather broad shoulders, very fine and dark tenting, weaker spiral ridging, and pink mouths. These have been called *C. paulucciae* and recognized either as full species or a subspecies of *C. aureus*. However, it seems that both typical, cylindrical *C. aureus* and intermediates also occur in the area, and perhaps throughout the Indian Ocean for that matter. At best this is a form of extremely elegant shape and great beauty.

Distribution.—Found through much of the Indo-West Pacific, but scarce and spotty in distribution. In the Indian Ocean virtually all records have been assigned to *C. paulucciae* regardless of the characters of the shell; the species occurs (rarely) from the Maldives to the southern islands; the recent records from Thailand are based on misidentifications of a form of *C. canonicus*, but the species certainly occurs there. In the western Pacific the species ranges from southern Japan and the Philippines through the Indonesian area south to New Caledonia; it also occurs in most of the island groups of Oceania and in French Polynesia; there appear to be no valid Hawiian records.

Synonymy.—I can see no real basis on which to recognize *Conus paulucciae* as more than an individual or perhaps localized populational variant. It would seem that large, exceptionally beautiful specimens are called *Conus paulucciae*, while smaller and more typically colored specimens are called *C. aureus* by some dealers. Extreme shells of the *paulucciae* type are quite distinct because of the exceptional colors and broad shoulders. *Conus auricomus* of Lamarck is not *Conus auricomus* of Hwass, but a separate naming.

Notes.—Adults of this beautiful species are rare and desirable. Even worn specimens with lip chips and knicked spires bring good prices and are desirable spacefillers. Juveniles (under 40mm) are less uncommon. Juvenile *C. aulicus* have been sold for this species but are easily distinguished by lacking 'textile lines' and numerous other differences. Indian Ocean specimens of the *paulucciae* form fetch high prices and deserve them.

AURICOMUS Hwass, in Bruguiere, 1792

1792. *Conus auricomus* Hwass, in Bruguiere. *Cone*, in *Ency. Method., Hist. Nat. des Vers*, 1: 742 (East Indies). Holotype illustrated by Kohn, 1968.

1845. *Conus dactylosus* Kiener. *Species gen. et icon. des coqu. viv.*, 2 (*Conus*): Pl. 97, fig. 2. 1849-1850, *Ibid.*, 2: 306 (Locality unknown).

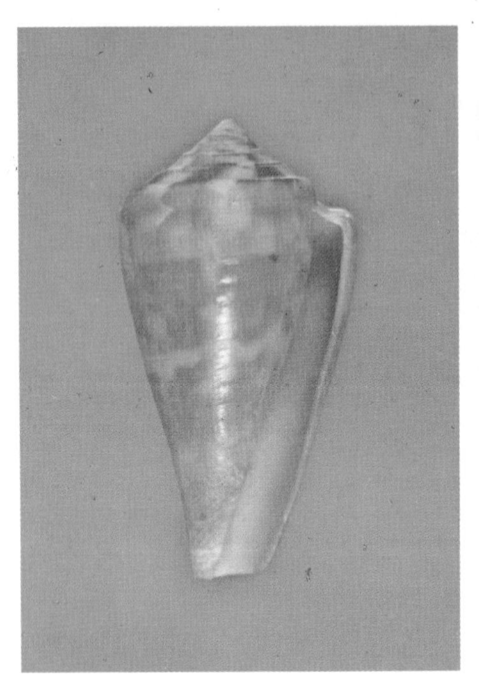
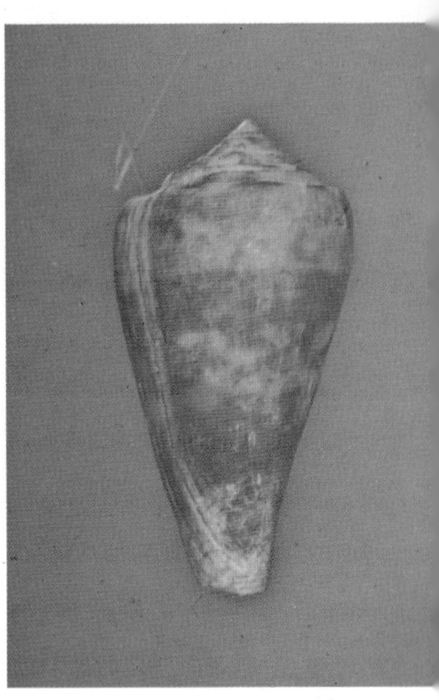

BRUUNI.—Raoul I., Kermadec Is. *Above* and spire below: 42.8mm (Auckland Inst. and Dr. W.O. Cernohorsky). *Below:* Left (holotype): 43.8mm (photo by Dr. W.O. Cernohorsky).

BULBUS.—Above: Lobita Harbor, Angola, 20.4 and 20.7mm. *Below:* Left: (top) Coarta Bay, Benguela, Angola, 25.9mm (MM), (bottom) Fernando Poo, 18.3mm (K); Right: Lobita Bay, Angola, 23.4mm (EP).

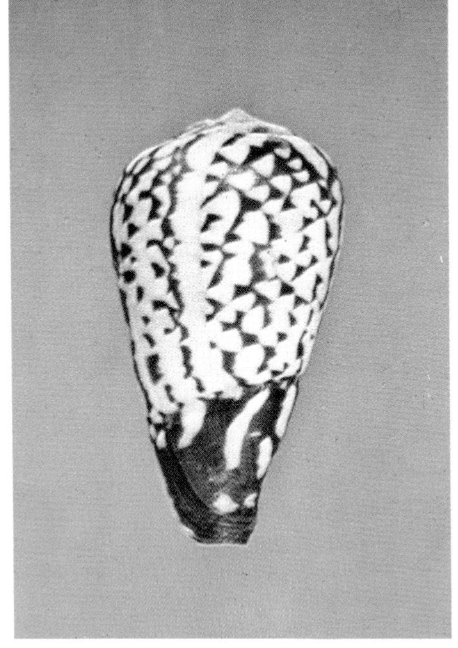

1943. *Conus debilis* Fenaux. *Bull. l'Inst. Oceanogr.* (*Monaco*), No. 834: 4, fig. 11 (New Guinea). Non *Conus debilis* Monterosato.

Description.—Light in weight, fragile, with a good gloss; elongate cylindrical, the sides nearly parallel; body whorl with weak basal grooves and ridges, these turning into weak but distinct close-set spiral ridges that continue to shoulder; shoulder rounded, not distinct from spire; spire low, rounded, the sides convex; early 3-4 whorls distinctly but finely nodulose, very small compared to later whorls; tops of whorls slightly convex, covered with 2-3 distinct spiral ridges and traces of several weaker ones plus fine axial threadlets; whorls somewhat overlapping at margins. Body whorl white to creamy, heavily covered with numerous fine tents of orange to reddish brown, these often interrupted by scattered large tents; tents tend to form three broad spiral bands (below shoulder, at midbody, and above base) separated by two broad spiral bands of brownish blotches which may be separate blotches divided by large tents or largely fused into continuous spiral brown bands; a spiral row of small blotches may be present below shoulder; dark blotches containing fine spiral dot-dash lines of dark brown and white; shoulder and spire largely brown, white tents restricted to a relatively few oval white spots scattered over spire; early whorls whitish or pinkish. Aperture narrow, distinctly widened anteriorly; outer lip thin, sharp, fragile, straight, strongly sloping below level of shoulder; mouth white to pale bluish white. Columella rather short, narrow. Length 20-46mm.

Comparison.—The combination of a cylindrical shape, low and rounded spire, tented pattern, and nodulose early whorls distinguishes this species from all others. *C. aureus* of similar size may be rather cylindrical but have a higher spire and strongly marked 'textile lines.' *C. nussatella* and *C. artoptus* are similar to *C. auricomus* in shape but very different in color pattern, lacking any indication of tents; these species also have the spiral ridges stronger and often more widely spaced. Juvenile *C. aulicus* can look amazingly similar to large *C. auricomus* but are distinguishable by the coarser tenting, taller spire with straight rather than convex sides, and the lack of nodulose early whorls; also, *C. aulicus* at all sizes is much thicker and heavier than the thin and fragile *C. auricomus.*

Variation.—This is a very consistent species which shows almost no variation in shape and general pattern. The usual variation in density of tenting occurs as well as in degree of fusion of the spiral blotches into solid bands. Young specimens are very fragile. Rare specimens have a solid band of color below the shoulder instead of the few small blotches more commonly present.

Distribution.—Widespread in the Indo-Pacific, including Hawaii and French Polynesia, but seldom common.

Synonymy.—*C. dactylosus* seems to be based on a specimen with extremely fine and uniform tenting and subdivided spiral bands; I am not familiar with other specimens matching this specimen, and it is presumably an individual variant.

Notes.—The delicate lip is commonly chipped, sometimes badly, but this detracts little from the value of the shell. Healed growth marks are also rather common on the body whorl. Very large *C. auricomus* are probably actually small *C. aulicus.*

AURISIACUS Linnaeus, 1758

1758. *Conus Aurisiacus* Linnaeus. *Systema Naturae per Regna Tria Naturae*, ed. 10, 1: 716 (Locality not stated). Lectotype figure (Kohn, 1963): Rumphius, Pl. 34, fig. A.

Description.—Moderately heavy, with a good gloss; rather broadly cylindrical, widest just below shoulder; body whorl with heavy spiral ridges from base to just below shoulder, usually about 20 principal ridges and 20-30 additional weaker ones; ridges more closely spaced and stronger anteriorly than posteriorly; shoulder roundly angled, rather broad, deeply canaliculate above; spire low, rounded if not eroded, the protoconch protuberant; tops of whorls slightly to deeply concave, especially the later whorls, with about 2-4 distinct spiral ridges. Body whorl whitish, heavily toned with pink; usually 2 broad deeper pink to pale orange-pink spiral bands above and below midbody area and occasional axial flammules of similar tones; spiral ridges bearing large and small dark brown to chestnut or blackish spots and dashes arranged in vaguely axial rows; pattern very variable in development; shoulder with large blackish spots; spire pinkish, the margins of spire whorls usually with large to small deep brown to blackish spots; early whorls whitish. Mouth moderately wide, especially anteriorly; outer lip rather thin but strong, straight; mouth white to pinkish. Columella short, curved, bounded posteriorly by a heavy rounded ridge. Length 40-70mm.

Comparison.—*C. aurisiacus* is likely to be confused only with *C. circumcisus* and perhaps *C. barthelemyi*. From *C. barthelemyi* it is easily distinguished by the delicate colors, narrower body whorl, and distinct spiral ridges absent in *C. barthelemyi. C. circumcisus* is very close to *C. aurisiacus* in some patterns and the two are perhaps impossible to distinguish with certainty in all cases. As a general rule, *C. aurisiacus* has the last 2 or 3 whorls plus the shoulder canaliculate, while in *C. circumcisus* only the shoulder is ever canaliculate, and even then it is only shallowly concave. In *C. circumcisus* the colors tend to be tan, white, and violet in various combinations, the spiral bands on the body whorl usually distinctly tan or brown, not pinkish. Many *C. circumcisus* lack distinct spiral ridging on the body whorl and do not have distinctly spotted patterns. In summary, *C. aurisiacus* is a spotted pink and white shell, while *C. circumcisus* is spotted or not and tan, often with violet tones.

Some forms of *C. magus* and of *C. pohlianus* may resemble *C. auri-*

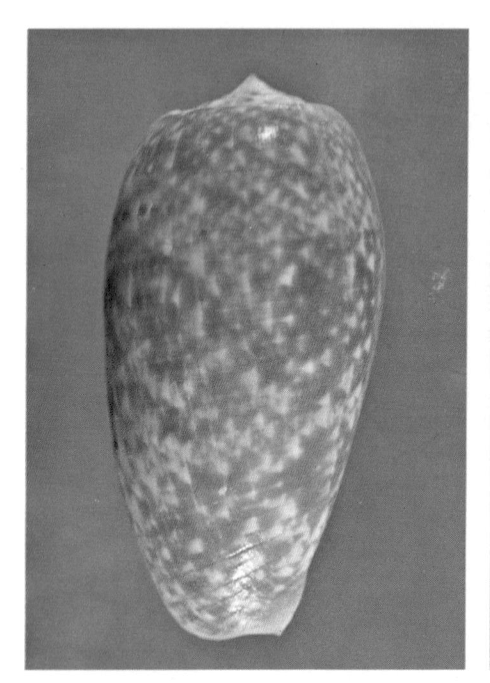

BULLATUS.—*Above:* Truk Atoll, Caroline Is., 60.9mm (MG). *Below:* Andaman Sea off Thailand, 50.9mm (photos by A. Kerstitch).

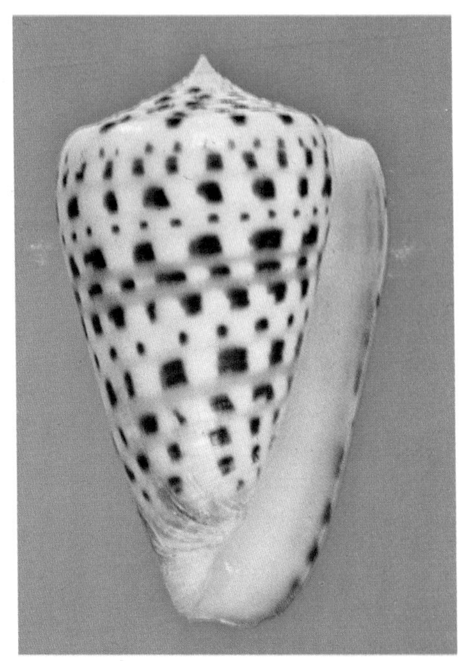

BYSSINUS.—Villa Cisneros, Spanish Sahara. *Above:* 43.7mm. *Below:* 41.3 and 41.4mm (all EP).

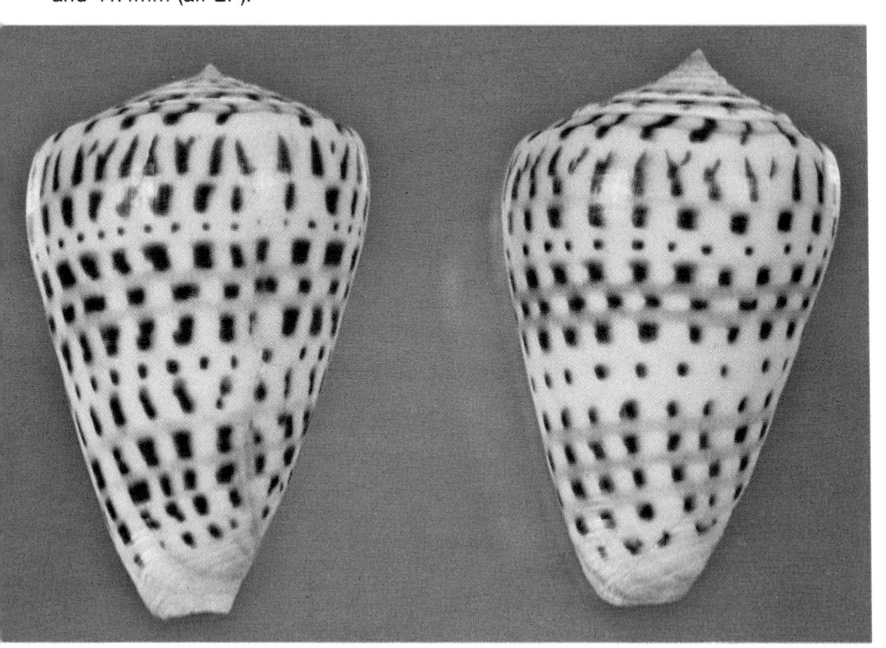

siacus, but they have much wider mouths, lack the distinct spotting, usually have different spire sculpture, and seldom or never have canaliculate spire whorls.

Variation.—There is great variation in amount of dark spotting on the body whorl and the arrangement of the spots. Usually at least a few dashes are present on some of the major ridges. The pink tones may be faint and indistinct or very strong and obvious; occasionally it is so strong that it appears orangish (fresh shells). The strong spots on the shoulder are constant. In some specimens the spire whorls and shoulder are only slightly concave, not deeply canaliculate.

Distribution.—Apparently typical *C. aurisiacus* is restricted to the Indonesian area, few recent specimens being known. Most older specimens are from the 'Moluccas,' 'Amboin,' or 'Banda I.' Recent records from the Philippines, New Guinea, and Solomons apparently refer to various forms of *C. circumcisus,* as do records from Oceania. Presumably the actual range of *C. aurisiacus* includes Java and other Indonesian islands, adjacent Malaysia, and perhaps part of western New Guinea.

Synonymy.—It might be noted in passing that for some reason early records of this species were mixed with the Caribbean *Conus daucus,* resulting in old records for Surinam for *C. aurisiacus.*

Notes.—Most available specimens are somewhat faded from age, presumably the pinks being brighter and darker (more orange) in fresh condition. Many specimens in circulation now are from old collections and bear various identifying numbers on the lip. Filed lips seem to be the norm here. Beware of *C. aurisiacus* offered at bargain prices—they will probably be *C. circumcisus* in one of its pale phases. *C. aurisiacus* of adult size in good color is rare.

If the range of *C. aurisiacus* and *C. circumcisus* were better known it might be possible to demonstrate that they were subspecies. A juvenile of *C. circumcisus* from New Guinea is practically indistinguishable from adult *C. aurisiacus* in pattern and color.

AUSTRALIS Holten, 1802

?1792. *Conus strigatus* Hwass, in Bruguiere. *Cone,* in *Ency. Method., Hist. Nat. des Vers,* 1: 735 (Grandes Indes). Holotype figure: *Tableau Ency. Method.,* Pl. 342, fig. 1.

1802. *Conus australis* Holten. *Enumeratio Systematica Conchyl. beat Chemnitzii:* 39, no. 487 (Locality not stated). Based on Chemnitz, Pl. 183, figs. 1774-1775.

1810. *Conus australis* Lamarck. *Ann. du Mus. Hist. Nat.* (*Paris*), 15: 439 (Locality not stated). Based on Chemnitz, Pl. 183, figs. 1774-1775.

1823. *Conus gracilis* Sowerby i. *Genera Recent and Fossil Shells,* 2 (Pt. 16): Pl. 267, fig. 4 (Locality not stated).

1870. *Conus laterculatus* Sowerby iii. *Proc. Zool. Soc.* (*London*), 1870:

255, pl. 22, fig. 3 (Locality unknown).
1963. *Asprella alabasteroides* Shikama. *Sci. Rpts. Yokohama Nat'l. Univ.*, Sec. II(10): 65, pl. 1, figs. 9a-9b (Japan, Kii Channel, Tatsugahama, Wakayama Pref.).

Description.—Moderately heavy, with a low gloss or dull surface; elongate, rather cylindrical, somewhat inflated below shoulder; body whorl covered with wide flattened or rounded spiral ribs, these closely spaced from base to shoulder and separated by only narrow grooves containing fine axial threads; grooves sometimes obsolete, especially posteriorly; shoulder carinate, moderately narrow, distinct from spire; spire low to rather tall, sharply pointed, the sides straight to deeply concave; early whorls not nodulose or at most weakly undulate; later whorls with carinate margins which may be slightly undulate, slightly concave above; tops of whorls with about 6 distinct spiral ridges, these sometimes crossed by weak axial threads. Body whorl creamy white to very pale straw, with three spiral rows of rather small brownish to reddish brown blotches, weakest below shoulder, strong above and below midbody; blotches somewhat axially elongate, irregular, usually partially connected with each other and base and shoulder by fine irregular axial lines; often many rows of widely spaced squarish brown dots present on the spiral ribs, probably formed by breaking up of fine axial lines; base whitish, sometimes bounded posteriorly by a row of small brown blotches; spire whitish, tessellated with brown squares; margins of spire whorls with small brown spots; early whorls whitish, sometimes obscured by tar. Aperture moderately wide; outer lip straight to broadly convex, thin, sloping below level of shoulder; mouth light to dark violet, paler exteriorly. Columella long, narrow, largely fused, bounded posteriorly by a weak ridge and somewhat indented from rest of inner lip. Length 40-100mm.

Comparison.—The elongate shape, lack of distinct nodules on the early spire whorls, heavy color pattern in most specimens, and heavy ribbing are distinctive. Most likely to be confused with *C. australis* is *C. duplicatus*, which is much more inflated below the shoulder, has a lower and more rounded-appearing spire with the early whorls produced into a much finer point, and has the spiral ridging of the body whorl almost obsolete and often appearing impressed rather than raised. In *C. duplicatus* the axial lines are fine and nearly continuous, although flexuous, from shoulder to base, with the three bands of blotches of *C. australis* visible only in some specimens. *C. mucronatus* is somewhat similar in shape and general appearance to *C. australis laterculatus* but is more attenuated anteriorly, thinner in texture, has a much more diffuse and reduced pattern, and has the body ribs more widely spaced and separated by wide grooves with strong axial threads; the mouth is also much lighter, often white. *C. aculeiformis* (broad-shouldered form) may be vaguely similar in appearance but is much more attenuate and lacks the numerous strong spiral ridges on the spire whorls.

CABRITII.—Above: Thio, New Caledonia, 17.9mm (GG). *Below:* Left: Magonta, New Caledonia, both 19.4mm (DMNH); Right: Ngo Bay, New Caledonia, 24.2mm.

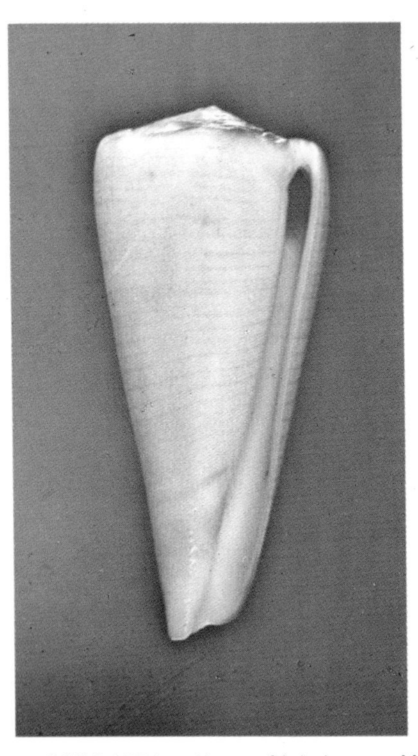

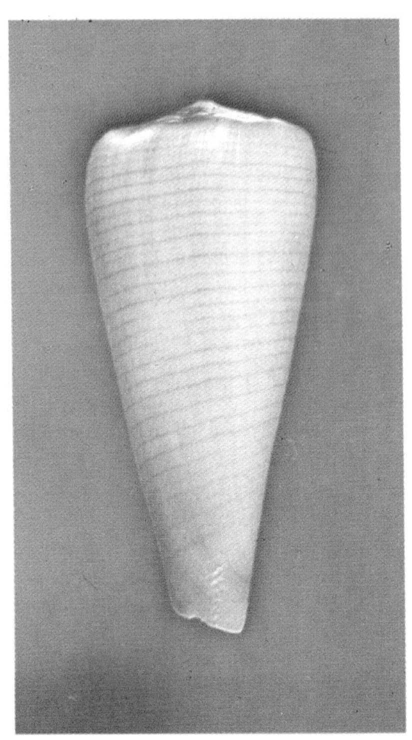

CAILLAUDI.—Above: Mahebourg, Mauritius, 47.3mm (EP). *Below:* Left: Holotype from Kiener, Pl. 55, fig. 5, locality unknown.

CALEDONICUS.—Below: Right: From Kiener, not holotype, "New Caledonia".

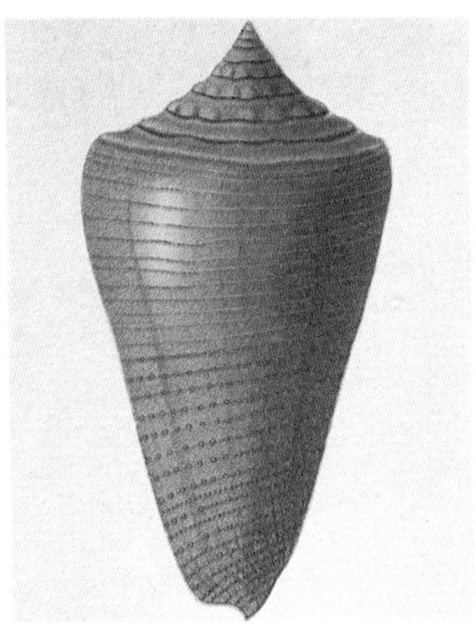

Variation.—Once two subspecies are distinguished, *C. australis* shows relatively little variation. Most of this is in degree of pattern development, especially the axial lines and small spots, which may be strong or virtually absent. There is a good amount of variation in the angulation of the shoulder and spire, especially in *C. australis australis*, with some specimens appearing more round-shouldered than others. Very large specimens of the nominate subspecies may be very obese with the pattern virtually absent or developed as broad brown bands covering most of the body whorl. Specimens are commonly granulose on some or many ribs.

C. australis australis Holten: spire tall, rather turreted, sometimes 20% or more of total length, the sides deeply concave in many specimens; spiral ribs of body whorl commonly in contact with each other, the grooves and their axial threads nearly obsolete; blotches strong, usually connected by fine irregular axial lines; sides of body whorl usually distinctly convex below the shoulder; size often exceeding 60mm; southern Japan and the China Sea.

C. australis laterculatus Sowerby iii: spire low, the sides nearly straight in most specimens, height about 15% of total length; spiral ribs of body whorl usually separated by grooves containing fine axial threads; blotches usually somewhat reduced, the connecting axial lines absent and replaced by small squarish brown spots; sides of body whorl usually nearly straight below shoulder; size seldom exceeding 60mm; Philippines south to northern Borneo.

Distribution.—Commonly dredged in about 60m from southern Japan and the Taiwan area south through the China Sea to Borneo; common in the Philippines. Probably the species ranges further south, but I am not sure if records for Queensland refer to this species or not.

Holotype of *C. alabasteroides* Shikama (photo Shikama).

Synonymy.—*Conus strigatus* Hwass may very well represent this species, but the figure is of a relatively small shell with a broad and straight-sided spire; the pattern is not typical of *C. australis*. *C. gracilis* was correctly synonymized by Sowerby i at the time of the original description. Although I know of no actual intergrades between *C. australis* and *C. laterculatus*, the non-overlapping distribution and similarity of characters certainly indicate subspecific relationship as the differences are just those of degree. *C. alabasteroides* seems to be a fairly normal *C. a. australis* with a shoulder more carinate than typical of Japanese specimens.

Notes.—Large specimens commonly have one or more growth flaws on the body whorl or spire and may be variously stained or with tar deposits. Small specimens tend to have the best patterns and most perfect spires. The violet mouth tends to fade with age.

AXELRODI Walls

Holotype: Delaware Mus. Nat. Hist.; 16.0mm; Philippines, Palawan.
Paratypes: Collections of Phillip Clover, Ed Petuch, and the author.
Diagnosis: A small thick cone covered with broad spiral ridges separated by deep, punctate grooves; 3-4 crisp spiral ridges crossed by widely spaced heavy axial threads on concave, heavily coronated spire whorls; body whorl yellowish, pinkish, tan, or gray, with many small brown dots in vaguely axial lines. 13-22mm length.

Description.—Thick, light in weight, with a low gloss; low conical, the upper sides strongly convex; body whorl with narrow spiral ridges at the base, these soon replaced by broad rounded spiral ridges extending to the shoulder and separated by deep narrow grooves with heavy axial threads; entire whorl granulose, crossed by fine axial threads and scratches; shoulder broad, sharply angulate, with many small but sharp coronations, concave above; spire moderate to tall, sharply pointed, the sides straight to slightly concave, the whorls often slightly stepped; early 2-4 whorls finely nodulose, the later whorls with strong but small erect coronations; tops of whorls concave, with 3-4 narrow but strong spiral ridges crossed by widely spaced but strong axial threads; shoulder sculpture similar, strong. Body whorl yellowish, pinkish, tan, or grayish, with many small brown dots on the spiral ridges, these arranged in vaguely axial rows; base unmarked; spire colored like body whorl, with sparse, scattered large dark brown spots; early whorls pink. Aperture moderately narrow posteriorly, much wider anteriorly; outer lip often very thick at the edge, convex, its posterior edge sloping below the level of the shoulder; mouth white, often tinted with pale yellow to pink. Columella very narrow, inconspicuous. Length 13-22mm.

Comparison.—The small size and pattern remind one of *Conus musicus* and *Conus miliaris*. From both these it differs in the taller spire with small but clean coronations, the raised spiral body ridges extending to the shoulder, and especially the strong spiral and axial sculpture on the spire whorls and shoulder. In some strongly granulose *C. musicus* (Pacific) the spots may appear to be in axial lines, but the pattern of that species usually has axial flammules in the midbody area, black spots

CALIFORNICUS.—Above: Coho Anchorage, California, 28.2 and 28.3mm. *Below:* Santa Barbara, California, 27.9 and 31.1mm.

CANCELLATUS.—Above: Off Peng Hu Is., Taiwan, 32.8 and 33.1mm. *Below:* Left: Wakayama Pref., Japan, 46.5mm; Right: Punta Engano, Cebu, P.I., 22.6mm (Dayrit).

between the coronations, and a white background. *Conus sponsalis* and its forms are more weakly sculptured on the spire, have different patterns, and have purple at the base. The relationships of this species are very uncertain, and it is difficult to place with any species known to me.

Variation.—As mentioned in the description, the overall color of the shell varies considerably, although it is usually pale yellowish or pinkish. The small brown spots are sharp and distinct in all specimens seen, with the axial alignment usually quite distinct. The spire sculpture is sharp and constant, although the spire varies considerably in height and stepping of the whorls. The coronations are small but sharp and seldom eroded in the specimens seen, although they may be weak on the shoulder. The lip is sometimes very thick.

C. axelrodi, paratype, Taiwan, 22.2mm (PC).

The largest shell seen (22.2mm) was a white specimen with reduced spotting and an apparently distorted spire, but the sculpture agreed closely with typical specimens from shallow water; this specimen was dredged in 85 fathoms off Taiwan.

Distribution.—Moderately common in shallow to deep water from Taiwan and the Philippines to the Indonesian area and New Guinea. It is perhaps more widely distributed but overlooked because of its small size and resemblance to *Conus musicus* or *C. miliaris*.

Synonymy.—This little but very distinctive cone is familiar to collectors under several names, all in my opinion incorrectly used. It is often sold as *Conus puncturatus* Hwass, but the type of that taxon appears to be a rather normal *C. sponsalis* intermediate between the typical pattern and the *nanus* form. Another popular name is *Conus papillosus* Kiener, but the description and figure of that taxon (and also *C. pustulatus* Kiener) are almost certainly granulose individuals of *Conus puncticulatus*, probably of *C. p. perplexus*. Kiener clearly did not have this cone or he would have mentioned the heavy spire sculpture; the figure seems to show a large columella like that of *C. p. perplexus*, and the entire shape and appearance of his *C. papillosus* is that of a granulose *C. puncticulatus*. *Conus anaglypticus* Crosse has also been suggested for this species, but that shell is clearly the small *duvali* form or subspecies of *C. min-*

danus, agreeing perfectly in shape, color, and sculpture with that species; again no spiral sculpture was mentioned or shown on the spire whorls, which were implied to be smooth. I can see no other choice but to describe this shell as new.

Notes.—This is a thick little cone with a very distinctive appearance and sculpture not likely to be confused with anything else. The maximum size and complete range are unknown at present.

Conus axelrodi is named as a token of appreciation for Dr. Herbert R. Axelrod, publisher and popularizer of tropical fish as a hobby, who made this book possible.

BAIRSTOWI Sowerby iii, 1889

1889. *Conus Bairstowi* Sowerby iii. *J. Conchol.* (*London*), 6: 9, pl. 1, fig. 12 (South Africa).

Description.—Moderately light in weight, with a good gloss in fresh specimens; broadly low conical, the upper sides convex; body whorl with traces of low spiral ridges on the basal third or so, these more widely spaced and strongest anteriorly; rest of body whorl smooth except for traces of fine spiral and axial threads; shoulder broad, rounded, not very distinct from spire; spire low, bluntly pointed, the sides slightly convex to slightly concave; tops of spire whorls rather convex, sometimes with faint traces of 1 or 2 weak spiral threads; spire usually heavily eroded. Body whorl white to creamy yellow, perhaps depending on degree of erosion, covered with about 6-8 spiral rows of rather irregular axially rectangular or squarish yellowish brown spots extending from above base to shoulder; entire base, both dorsally and ventrally, yellowish brown; spire and shoulder whitish, covered with small curved brownish spots if not eroded. Aperture wide, rather uniform; outer lip thick, slightly convex; mouth pale, perhaps tinted with brown in life. Columella short, indistinct. Length 25-50mm.

Comparison.—*C. bairstowi* has usually been considered a form of *C. infrenatus*, but I cannot see how erosion or normal variation of that species could result in the pattern of *C. bairstowi.* In shape the two are similar, and presumably the spire sculpture is similar as well. *C. bairstowi* has the basal sculpture more prominent than typical in *C. infrenatus*, but this is perhaps more a factor of preservation than genetics. In pattern the two are very dissimilar, *C. infrenatus* having many more spots in regular series, with a tendency to have several rows of very fine spots as well; in *C. infrenatus* the spire marks are weak and narrow, fading with even minor erosion; most importantly, the base of *C. infrenatus* is whitish with only a few scattered spots and streaks of brown or pink, never all brown as in *C. bairstowi;* with erosion, *C. infrenatus* becomes pinkish or lavender, while even eroded *C. bairstowi* are whitish or creamy yellow. The

CANONICUS.—Above: Marau Sound, Guadalcanal, Solomon Is., 44.9 and 46.5mm. *Below:* Left (see text): Marquesas, French Polynesia, 42.3mm; Right: Querimarane, E New Britain, Papua New Guinea, 34.0mm.

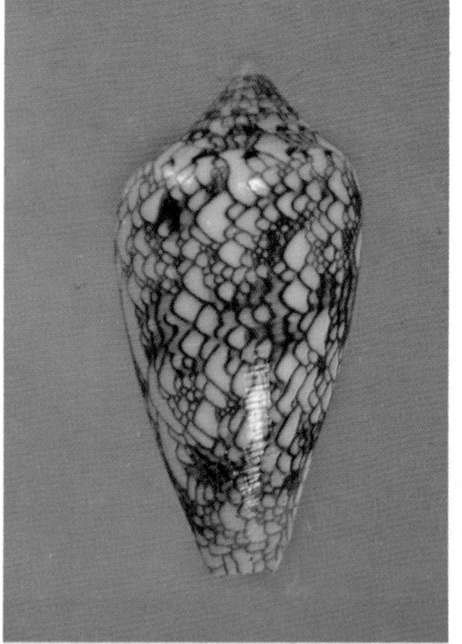

CAPITANELLUS.—Above: Pescadores Is., Taiwan, 40.1mm (T.C. Lan). *Below:* Hachijo I., Japan: Left (granulose): 19.6mm; Right: 20.8mm (photos by Dr. T. Okutani).

two species seem quite distinct even by South African cone standards, and the relationship between the two is certainly not like that between *C. natalis* and *C. gilchristi*. Eroded *C. natalis* or those without distinct body patterns are easily distinguished from *C. bairstowi* by the more cylindrical shape and the largely brown spire whorls even after considerable erosion; the base in *C. natalis* is largely white. Caribbean cones such as *C. spurius* and *C. lorenzianus* might be vaguely similar in some aspects of pattern, but they are usually different in shape, much thicker, and seldom have brown bases.

Variation.—The few specimens I have seen are quite constant in pattern although the number of spots and their exact size and shape vary considerably. Small specimens are less inflated posteriorly than larger ones and may have smaller, more squarish spots. In juveniles the spiral sculpture of the body whorl is stronger and extends to about midbody, being easily visible in the aperture. Adults may have a distinctly marked shoulder outlined by the junction of the brown spots on the spire and those below the shoulder.

Distribution.—Known to me only from beach specimens collected in Cape Province, South Africa; I have no idea of its home in life.

Synonymy.—This taxon has been commonly considered a form of *C. infrenatus*, with which I do not agree.

Notes.—This is one of the rarer beach cones of South Africa, and good specimens are not common in collections. I have not seen a truly glossy specimen with full sculpture, all specimens showing strong erosion over the body and spire. Chipped lips are also to be expected, of course. Specimens with good and rather bright patterns are available, so presumably the home of this species is not in deep water. Even badly worn specimens show traces of the brown base, a very distinctive character. Some specimens show a tendency to almost fuse the spots axially, so striped specimens should be looked for.

BALTEATUS Sowerby i, in Sowerby ii, 1833

1833. *Conus balteatus* Sowerby i, in Sowerby ii. *Conchological Illustrations:* Pl. 37, fig. 58 (Locality not stated).

1845. *Conus Cibielii* Kiener. *Species gen. et icon. des coqu. viv.*, 2 (*Conus*): Pl. 107, fig. 2. 1849, *Ibid.*, 2: 242 (Locality unknown).

1848. *Conus pigmentatus* A. Adams and Reeve. *Zool. Voy. Samarang, Moll.*, 1: 18, pl. 5, figs. 11a-11b (Locality unknown).

1933. *Conus pigmentatus concolor* Barros e Cunha. *Mem. Estud. Mus. Zool. Univ. Coimbra*, (Ser. 1), 71: 52 (Mares orientais, exact locality unknown). Non *Conus concolor* Sowerby i, 1834.

Description.—Heavy, with a low gloss; obconic, the upper sides slightly convex or nearly straight, somewhat pinched-in toward base; body whorl with heavy oblique spiral ridges above base, these continued as narrow low spiral ridges almost to shoulder, the ridges close-set anteriorly, becoming more widely spaced and low posteriorly; ridges often obsolete for a variable distance before the shoulder; shoulder strongly angled, heavily coronate with rather sharp, erect coronations; spire low, often almost flat, bluntly pointed, the early whorls often eroded; spire whorls coronate, tops slightly concave or flat, with about 3-8 fine but distinct rather undulate spiral ridges. Body whorl off-white to pale violet, usually covered with broad spiral bands of dark brown, the bands usually broadest on anterior third or more of shell, more posteriorly often narrow and allowing pale background to show through in rather even bands; subshoulder area usually without much brown pigment; usually several broad and rather weak axial bands of brown present, sometimes extending to shoulder area; spiral ridges often slightly darker brown than rest of pattern, especially contrasted in regions of new growth; ridges usually randomly marked with small white flecks in no particular sequence, these most common anteriorly, sometimes almost absent; base only slightly darker than rest of shell, brownish with pale violet tones; shoulder and spire uniformly off-white with pale violet tones, usually without brown blotches or flammules; margins of whorls usually with very fine brown line visible, occasionally with small brown spots between coronations, sometimes the shoulder pale brown over new growth; tip of spire pinkish. Aperture moderately narrow, the sides about parallel; outer lip thick posteriorly, nearly straight or slightly concave at middle; mouth pale to dark violet, usually with broad paler bands at shoulder and middle. Columella long and narrow, but twisted so largely internal. Length 25-45mm.

Comparison.—Many patterns of *C. balteatus* strongly resemble patterns of *C. rattus*, a species with which it shares the numerous spiral ridges; however, *C. rattus* has a heavily marked spire and is not coronate (some specimens tend to have slightly undulate whorls, however), as well as having a tendency to form large white blotches on the body whorl as well as the small white flecks. *C. cernicus* is very similar to *C. balteatus* but probably distinct. It is somewhat more inflated posteriorly, has a more roundly angled shoulder with lower and sometimes indistinct coronations, and has the spire whorls heavily blotched and spotted with dark brown; the base is commonly more heavily speckled with white than in *C. balteatus*, as is the entire shell; there is apparently no difference in number of coronations on the shoulder, *C. balteatus* varying from 10-16, with a small number of *C. cernicus* ranging from 10-14.

Most commonly confused with *C. balteatus* are *C. lividus* and *C. sanguinolentus*, which are somewhat similar in shape and pattern and have violet mouths. However, both have the body whorl nearly smooth except for typically granulose basal ridges which may extend to midbody but are widely spaced; there are no white flecks on the body whorl, which is commonly a very different shade of brown than typical of *C. balteatus*;

CAPITANEUS.—Above: Boac, Marinduque, P.I., 30.2-55.5mm. *Below:* Guam, 34.5 and 39.2mm (TCG).

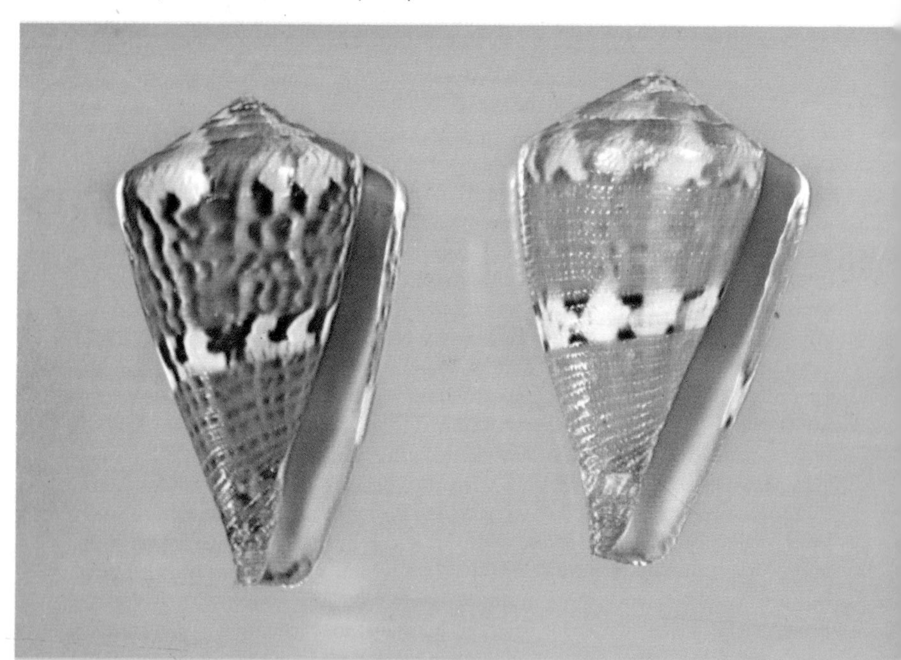

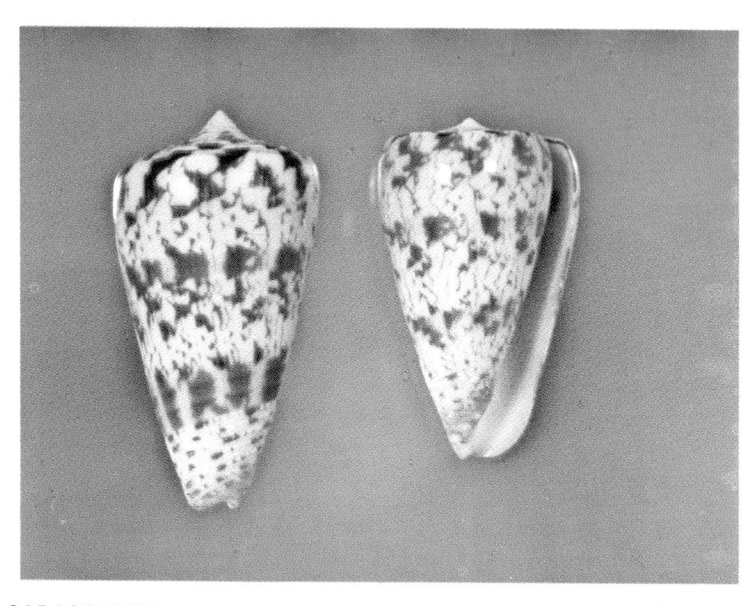

CARACTERISTICUS.—Above: Manila Bay, Luzon, P.I., 44.0 and 52.2mm.
Below: Colombo Reef, Sri Lanka, 47.5 and 50.4mm.

C. lividus usually has a very sharply defined narrow whitish midbody band, and in both species the base is usually deep violet and somewhat contrasted with the remainder of the body color. *C. parvulus* and *C. biliosus* generally have numerous brown dots or flecks over the body whorl and have at least small brown spots on the spire, as well as differing greatly in general pattern and appearance in most populations. *C. miliaris* and allies may be vaguely similar in shape (especially the strong coronations), but they are generally wider relative to length and differ in numerous aspects of pattern and sculpture, especially in generally having strongly punctate lines on the body whorl.

Variation.—In shape the sides vary from nearly straight to distinctly convex, with the spire flat to low and with straight to concave sides. Some specimens have the entire body whorl brown, while in others the dark base and one brown band above midbody are almost the only markings; most specimens show distinct broad pale brown axial bands, probably following old growth flaws; the sub-shoulder area is commonly unmarked with brown or just has extensions of the axial bands. White flecks may be heavy or very sparse; they are most likely to be found at the anterior fourth of the body whorl in typical specimens. Occasional specimens are much more strongly toned with violet than others. Only a few specimens of this species with distinct brown spotting on the spire have been seen, although the shoulder is commonly suffused with brown over major break points showing new growth, and the margins of the whorls are finely marked with pale brown.

Distribution.—Broadly distributed in the Indo-Pacific from the eastern African coast to at least the Carolines, southern Japan, and the Solomons. There are apparently no records for Hawaii or French Polynesia, and many Indian Ocean records probably refer to *C. cernicus*. This species is apparently not as common as indicated in the literature, occurring in small and spotty colonies in relatively shallow water.

Synonymy.—*C. cibielii* looks very much like a typical *C. balteatus* in shape but differs in some aspects of pattern, particularly in having the spire whorls completely brown. It might be an unusual pattern variant of any of several species, but is assigned here for temporary convenience. *C. pigmentatus* seems to be a fairly typical specimen and not referrable to *C. biliosus* as has been suggested by some authors. *C. pigmentatus concolor* is known to me only from the unillustrated description, but it suggests a uniformly violet shell very similar to *C. balteatus* in shape and sculpture; some *C. balteatus* do have strong violet tones, but I have not seen any uniformly colored specimens.

Notes.—This is a very difficult species to obtain in any condition, as shells sold under this name are commonly misidentified *C. lividus*, *C. sanguinolentus*, or even *C. moreleti* or *C. cernicus*. It is a very rough cone subject to large and jagged breaks and irregularities which mar the appearance. Apparently in life specimens are covered with a thick layer of calcium deposits which envelop them like a cocoon.

BALTEUS Wood, 1828

1828. *Conus Balteus* Wood. *Index Test., Suppl.*: 8, pl. 3, fig. 5 (Locality unknown).
1843. *Conus cuneolus* Reeve. *Conchologia Iconica*, 1(*Conus*): Pl. 37, sp. 205 (Locality unknown).
1933. *Conus lugubris fuscus* Barros e Cunha. *Mem. Estud. Mus. Zool. Univ. Coimbra*, (Ser. 1), 71: 194 (Locality not stated). Non *Conus mediterraneus* var. *fusca* Bucquoy, Dautzenberg, and Dollfuss, 1882.

Description.—Moderately heavy, with a good gloss; obconic, the sides nearly straight, almost parallel just below shoulder; body whorl with a few faint spiral ridges and grooves above the base, otherwise smooth except for weak axial and spiral threads scattered over rest of whorl; shoulder broad, strongly angled, concave above; spire low, sometimes almost flat, the sides straight or convex, the whorls somewhat stepped; tops of whorls slightly concave, later whorls (when not eroded) with 2-3 heavy flat spiral ridges crossed by fine axial threads; early whorls usually eroded, earliest whorls usually prominently produced. Body whorl whitish, heavily covered with 3-4 bands of deep brown to blackish brown, one at shoulder, one above midbody, sometimes one below midbody, and a broad one at the base; these bands variously crossed by narrow or broad irregular axial bands and flammules to produce a cross-hatched pattern; the white background may show through as squarish spots most prominent near midbody or as fine zigzag marks over most of the shell; pattern very variable; base all brown; shoulder marked with fine brown and white axial lines, as is the spire; eroded spire whorls whitish, outlined with brown; tip pale violet. Aperture moderately narrow, widened anteriorly; outer lip thin, sharp, straight, sloping posteriorly below level of shoulder; mouth whitish, with pale pinkish-violet tones. Columella narrow, short, sometimes bounded posteriorly by a ridge. Length 24-34mm.

Comparison.—This species of the *mercator* series is obviously very close to two other Cape Verde endemics, *C. anthonyi* and *C. venulatus*. From *C. anthonyi* it is distinguished by the larger adult size (doubtful), the strongly angled shoulder, the brown base with few or no white markings, the generally thicker and broader shell, the whitish mouth, and the more sharply defined white axial markings. From *C. venulatus* it is perhaps impossible to always distinguish, although the average *C. venulatus* pattern is easily distinguished; as a rule, the spire whorls of *C. venulatus* are without spiral ridging or with only traces of 1-2 weak ridges; the spire is often higher and sharply pointed even in adults, the sides concave; *C. venulatus* is commonly larger than *C. balteus* and has the sides of the body whorl somewhat more convex, although this is not a very useful character. In pattern *C. venulatus* is commonly almost white or at least pale brown and there are distinct traces of numerous spiral brown lines or fine zones over the body whorl; the spire markings are commonly weak or

CARDINALIS.—Above: Left: Gonave I., Haiti, 28.3mm (EP); Right: Palm Cay, Belize, 28.9mm (photo by A. Kerstitch). *Below:* Left: Gonave I., Haiti, 28.3mm (EP); Right: Curacao, Neth. Ant., 28.5mm (JB).

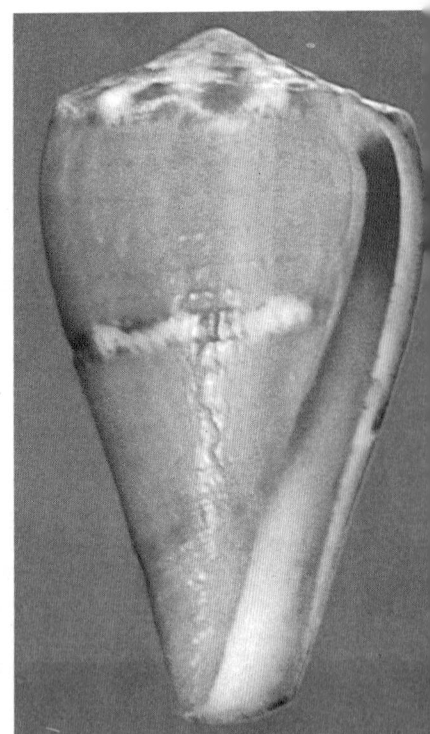

CATUS.—Above: Left: Java, Indonesia, 29mm; Right: Marquesas, French Polynesia, 32.1mm. *Below:* Left and Center: Port Vila, New Hebrides, 39.0 and 39.3mm; Right: Zanzibar, 41.3mm (MF).

absent; the dark phase of *C. venulatus* is almost black with 'tattered' white blotches at midbody and shoulder, a pattern hard to duplicate in *C. balteus.* These differences are not very convincing for three apparently sympatric species, and it is probable that larger series will reveal complete intergradation among the three 'species.'

Variation.—As treated here, this is a species which is quite constant in shape and size but very variable in pattern. The pattern currently most available to collectors is quite different from those illustrated by either Wood or Reeve. In this common form the shell appears largely brown with the posterior ⅔ of the body whorl heavily axially streaked with zigzagged narrow and wide white stripes which are commonly continuous from above the base to the shoulder; also commonly the white streaks are broken into several bands by larger brown blotches in 2-3 spiral rows, resulting in large rectangular axial blotches of white about midbody; the shoulder area is well marked by especially narrow brown and white marks giving a cross-hatched pattern. It takes little variation from this pattern to produce a more typical pattern consisting of about 3-4 broad broken brown spiral bands separated by narrow white bands often broken into a cross-hatched pattern by fine axial brown zigzags; further reduction of pattern results in a brownish shell with large white spots in about 3 spiral bands or in a shell completely covered by a finely cross-hatched brown and white pattern.

Variation in *C. balteus* (photos EP).

Distribution.—Apparently endemic to the Cape Verde Islands, where small colonies are found on several islands in rather shallow water. Very uncommon, as are all the small cones of this island group.

Synonymy.—Wood's figure of *C. balteus* shows a nearly flat-spired shell which has broad, angled shoulders; the shell is largely brown except for white at the shoulder and in several rather squarish spots at midbody. This could fit either *C. venulatus* (dark form) or a dark *C. cuneolus*-type shell. Since base and sub-shoulder area are largely dark in the figure and there is much white at midbody, it matches the *cuneolus*-type patterns somewhat better; Tryon and other early authors considered *balteus* a synonym of the later *C. cuneolus.* I would have no strong objections to returning to the use of *C. cuneolus* for this species and referring *C. balteus* to the synonymy of *C. venulatus.* The Barros de Cunha name fits here as well as anywhere else and is probably unrecognizable from the unillustrated description.

Notes.—The spire is almost always strongly eroded in this species, but otherwise fine specimens are available. Unfortunately the most common pattern for sale at the time of writing is not very typical of the species, although it fits well into the variation series. This shell, like most Cape Verde species, commands a high price for a small shell. This is apparently due more to the difficult collecting conditions than to actual rarity of the shells once colonies are found. This species seems to always have fine healed axial growth marks which also define abrupt pattern variations.

BARTHELEMYI Bernardi, 1861

1861. *Conus Barthelemyi* Bernardi. *J. Conchyl.* (*Paris*), 9: 285 (Chagos Archip.). 1862, *Ibid.*, 10: 46, pl. 1, fig. 12 (Islands of the Chagos Archipeligo or Diego Garcia).

1973. *Conus* (*Dendroconus*) *gauguini* Richard and Salvat. *Cahiers du Pacifique*, No. 17: 25, figs. 1-3 (Iles Marquises).

Description.—Heavy, with a good gloss; widely low conical, the sides straight or slightly concave at the middle; body whorl with crowded low ridges in the basal area, these usually obsolete over rest of whorl except as very fine spiral wrinkles; axial wrinkles and growth marks present, sometimes dense; shoulder broad, carinate, deeply canaliculate; spire low or moderate, rounded, with straight sides (occasionally concave); protoconch prominent; early whorls of spire forming a rounded dome, the later whorls canaliculate; 4-5 strong spiral ridges on each whorl, often more on shoulder. Body whorl pale to deep reddish or orange-brown, the color in the form of two or three darker and lighter broad spiral bands with parallel edges; a broad central band of pinkish white to pale tan present, usually containing at least a few large black spots alternating with white dashes in spiral rows; base usually same pale color as midbody band and with some black spots; a few other vague spiral white lines interrupted with black spots, the lines widely spaced, are commonly present over body whorl; shoulder pinkish white with large radiating black blotches continuing onto pinkish spire, turning into brownish lines on early whorls; coloration very variable. Aperture moderately wide, the sides parallel, sometimes with outer lip slightly concave at middle; interior white to pale bluish white. Columella elongate, rather wide, with a heavy twisted ridge above. Length 40-90mm.

Comparison.—This heavy and brightly colored shell is often con-

CEDONULLI.—St. Vincent. *Above:* 50mm (photos by P. Clover). *Below:* 48.4mm (GG).

CENTURIO.—Above: Off Georgetown, Guyana, 39.4mm (GG). *Below:* Gulf of Venezuela, off Falcon State, 49.6mm (R.W. Pitman).

fused with *Conus gubernator* and *Conus terminus*, although it is probably most closely related to *C. aurisiacus* and *C. circumcisus*. From the former two taxa it differs most constantly in the domed shape of the early spire whorls and the mammillate protoconch; in *C. gubernator* and relatives the early whorls are gradately conical and the protoconch, when preserved, merges into the spire. In coloration and often shape the approach between *Conus gubernator* and *Conus terminus* with *C. barthelemyi* is often extremely close, as are some variants of *C. striatus*. The spire difference seems to be constant, however. *C. aurisiacus* is a rather narrower shell covered (when not badly beach worn) with heavy widely spaced spiral ridges; the coloration is pale pink with hints of two slightly darker pink spiral bands, and the numerous black markings are commonly elongated dashes on the spiral ridges. *C. circumcisus*, in the smooth form, is much narrower at the shoulder; the shoulder and later spire whorls are seldom more than shallowly concave, not canaliculate.

Variation.—*C. barthelemyi* is one of the most variable of the larger cones, the described color pattern being the most common and valued by collectors. The pinkish to rosy background color can be confined to the base, midbody, and shoulder or may cover the entire shell except for two narrow dark bands. The dark color may be orange, deep rose, tan, or various shades of reddish brown and may occasionally obscure all the light background. Large and small black spots may be prominently scattered over the entire body whorl or restricted to the base and/or midbody area, or they may be entirely absent; rare specimens have axial or spiral black markings. The fine white spiral dashes are seldom obvious. The spire may be nearly flat or relatively high.

Distribution.—Known at present only from the islands of the western and southern Indian Ocean: Reunion, Mauritius, Comores, Chagos, and some smaller islands. Records from the Maldives are apparently *Conus gubernator*, but one or two definite specimens of *C. barthelemyi* appear to have been collected there.

Synonymy.—The rarity and distinctiveness of this species have prevented new names. *C. gauguini* appears to agree with the unspotted form of *C. barthelemyi* in every respect except its larger size (87mm, while few *barthelemyi* exceed 75mm) and locality (supposedly the Marquesas). The lack of black spots and the moderately high spire are common variants of typical *C. barthelemyi* and do not separate *C. gauguini* at the specific level, and it has been rumored (perhaps wrongly) that the locality is in error. Until more details are available, it seems best to synonymize *C. gauguini* with *C. barthelemyi*. If the locality should prove to be correct, the relationship with the smooth form of *C. circumcisus* should be carefully checked, although the color of the type is brighter than any *C. circumcisus* I have seen.

Notes.—Highly marked specimens are considered more collectable than weakly spotted specimens, as are those with orange to reddish backgrounds instead of tans. Heavy growth marks often occur near the outer lip and are considered minimal flaws by most collectors. The spire is almost always intact. Specimens with long black dashes like *C. striatus* or large areas of violet on the body whorl are apparently rare. The species is not uncommon in collections in recent years and is not really rare where it occurs.

BARTSCHI Hanna and Strong, 1949

1949. *Conus bartschi* Hanna and Strong. *Proc. California Acad. Sci.*, (Ser. 5), 26(9): 271, pl. 5, fig. 5 (Off Cape San Lucas, Baja California).

1955. *Conus andrangae* Schwengel. *Nautilus*, 69(1): 14, pl. 2, figs. 8-11 (Bahia El Coco, Costa Rica).

Description.—Solid, moderately heavy, with a low gloss or dull finish; low conical, the upper sides straight or distinctly convex, then rather pinched-in to the base; body whorl with a few weak basal ridges and numerous fine and heavy axial lines, threads, and flaws; sometimes traces of fine spiral threads over rest of body whorl; shoulder broad, sharply to roundly angled, strongly coronate with rather sharp projections; spire low to moderate, blunt, the early whorls strongly eroded; sides of spire usually straight or deeply concave, sometimes slightly convex; tops of whorls concave, the margins strongly coronate, tops without distinct spiral ridges but with fine axial threads; occasional specimens show traces of 1-2 faint spiral ridges on later whorls. Body whorl creamy white to pale straw, covered from base to shoulder with numerous fine spiral lines of short brown dots, dashes, and axial points, the strength of the markings variable; generally this is overlaid with a broad band of dark brown blotches above the base and another above midbody, the blotches often connected to form continuous or nearly continuous spiral bands; commonly these bands become jagged with axial elongations which may extend to cover most of the shell, resulting in a brown shell with relatively small white blotches; dark brown areas have distinct spiral lines of darker brown dashes within; shoulder and spire whitish, heavily eroded; shoulder and sometimes last 1-2 spire whorls with a few scattered dark brown blotches and fine brown points. Aperture moderately narrow, uniform in width; outer lip thin, sharp, slightly convex; mouth mostly glossy white, occasionally with a small yellowish blotch deep within. Columella very narrow and short, largely internal. Length 30-50mm.

Comparison.—*C. bartschi* obviously is closely related to *C. brunneus* and *C. regius*, although there is a superficial pattern similarity with *C. imperialis* as well. From *C. imperialis* it is easily distinguished by the broader form and more convex sides as well as the distinctly shorter body whorl; *C. imperialis* usually has the anterior end of the shell, including the mouth, heavily toned with violet-black, generally completely lacking

CERNICUS.—Above: Rodrigues I., 26.6 and 26.9mm. *Below:* Zanzibar, 36.0mm (MF).

CERVUS.—Moluccas. *Above:* 95 and 113mm (BMNH) (photo by P. Clover). *Below:* Left: From Kiener, Pl. 75, fig. 1, 94mm; Right: 103.5mm (Zool. Mus. Amsterdam) (photo by Dr. R.T. Abbott).

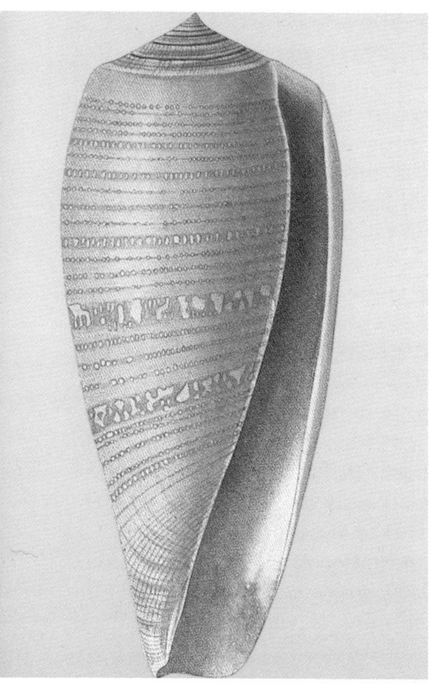

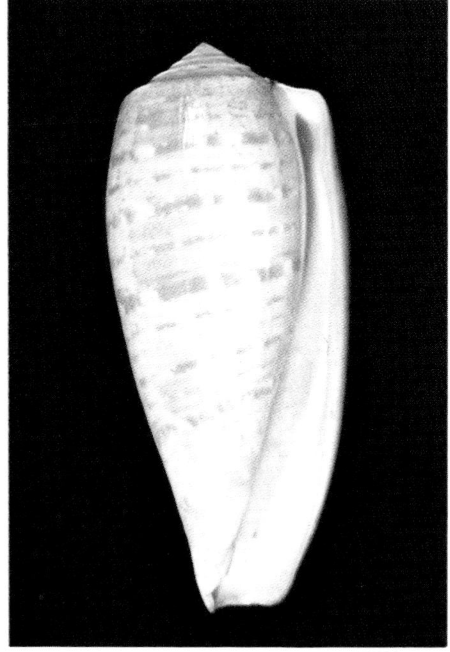

in *C. bartschi.* The sympatric *C. brunneus* is often confused with *C. bartschi,* but that species generally has strong spiral ridging on the spire whorls, a largely deep brown body whorl with only small and irregular white patches, and the spiral lines of dashes stronger and longer, without many finer points confusing the pattern; the body whorl of *C. brunneus* is relatively more slender, the shell appearing longer at similar widths than *C. bartschi;* in *C. brunneus* the mouth is deep grayish yellow throughout in fresh specimens, becoming white only in long-dead specimens subject to fading. In addition, the base of *C. brunneus* is generally darker brownish black than the rest of the whorl, while it is largely pale in *C. bartschi.* The closest relative of *C. bartschi* appears to be the Atlantic *C. regius,* which is often very similar in general shape and pattern. In *C. regius* the spire whorls are usually strongly spirally striate, the early whorls not strongly eroded, and the body whorl commonly with strong spiral ridges to near the shoulder; these ridges are sometimes finely granulose, a feature apparently seldom found in *C. bartschi.* In *C. regius* the spiral lines of brown points and dashes are seldom strongly developed, although there is a strong tendency to have distinct round white spots in rows; the pattern in *C. regius* of darker brown blotches is often stronger in development than in *C. bartschi,* forming broad axial bands which cover most of the body whorl, the spire also being heavily marked. As a general rule *C. regius* is larger than *C. bartschi,* has a more deeply concave spire which is less eroded, is a paler brown in tone, and has the basal sculpture much heavier; there should be no problem distinguishing the two species.

Variation.—The sides vary from straight to distinctly convex, with the spire convex to deeply concave; the spire whorls may be stepped or not. Most specimens have the tops of the whorls smooth except for fine axial threads, but some show traces of a few spiral ridges on the uneroded whorls. The background of the body whorl is always strongly marked with assorted sizes and shapes of small brownish points, but the darker blotching may be absent or cover most of the shell; as a rule at least a basal and post-midbody band are at least indicated by thickening of the dashes. Generally the mouth is white with no other color, but some specimens show a small but distinct dirty yellowish blotch deep within the aperture (perhaps a result of improper cleaning rather than natural coloration). The base is usually mostly white except for the brown points, but some specimens show blackish lines and streaks in this area. Rare specimens have pustulose ridges.

Distribution.—Apparently spottily distributed in the Gulf of California area south to Costa Rica. Records are from off Cape San Lucas in 20-25 fms., Isla San Pedro Martir, Sonora, in 30-40 ft., and from the beach at Bahia El Coco, Costa Rica. Presumably the shell is more widely distributed in the Panamic area but it is rare or at least uncommon.

Synonymy.—The well-illustrated description of *C. andrangae* leaves little doubt that this form is the same as *C. bartschi,* although the holotype is a specimen with a greatly distorted spire. It seems likely that there should be an earlier name for this species lost somewhere in the numerous synonyms and misunderstood names of Reeve and the Sowerbys.

Notes.—This is a rough cone which always has strong growth marks on the body whorl and heavily eroded spire. Small growths and deposits, with an occasional small hole in the spire, should be expected even in the best specimens. The lip is thin for a cumbersome shell and apparently chips easily. The shells currently available are mostly from Sonora and are generally lightly patterned.

BAYANI Jousseaume, 1872

1872. *Conus Bayani* Jousseaume. *Rev. Mag. Zool.*, (Ser. 2), 23: 200, pl. 18, fig. 1 (Bourbon).

Description.—Moderately light in weight, with a good gloss in fresh specimens; elongate, turbinate, the sides of the body whorl nearly straight and tapering to a narrow base; body whorl with about 12-15 moderately strong oblique ridges at the base, these separated by shallow grooves which are finely punctate; rest of body whorl smooth and satiny except for very fine spiral and axial threads and occasionally axial growth marks; shoulder broad, carinate; spire moderately tall, extremely deeply concave, the early whorls forming a projection on a nearly flat plate; spire sharply pointed, the early 5-8 or more whorls finely nodulose in uneroded specimens; tops of whorls flat or slightly concave, covered with very fine axial threads and occasional traces of 1-2 obsolete spiral threads on later whorls. Body whorl white to pinkish or cream, variously covered with deep yellowish brown blotches and streaks; these tend to form two broad broken spiral bands above and below midbody, surrounding a clear midbody band; midbody band sometimes distinctly marked posteriorly with large or small dark brown squarish spots; commonly few to many fine or broad axial streaks and flammules, often angular, following growth marks; in specimens with fuller patterns the dark bands may have fine spiral lines of darker brown visible within; base white, not toned with blackish or violet to my knowledge; spire largely white, the later whorls with scattered blotches of brown, these often sparse. Aperture very narrow, straight; outer lip straight, sharp, thin and fragile; mouth white, glossy, usually with violet tones. Columella very short and narrow. Length 40-70mm.

Comparison.—This 'species' has been unnecessarily confused with many unrelated cones, especially *C. recluzianus*, *C. urashimanus*, and *C. tribblei*. These species are all easily distinguished by the distinctly undulate spire whorls and/or distinct spiral sculpture on the spire whorls, as well as major differences in weight, shape, and pattern. *C. eximius* is broader and has distinct spiral sculpture on the spire whorls, which are

CHALDEUS.—Above: Boac, Marinduque, P.I., 26.0 and 26.5mm. *Below:* Fananii I., Truk Is., 22.4-22.8mm.

CHIANGI.—Above: Taiwan Straits, 17mm (DM) (photo by Jayavad). ***Below:*** Hachijo I., Kurose Bank, Japan, 14.8 and 17.0mm (photo by Dr. T. Okutani).

not nodulose. *C. voluminalis* is thinner in texture, waxy, has a flatter spire with heavy pattern, and has a body whorl pattern usually showing strong spiral lines of brown dashes and a spire pattern extending onto the body whorl below the shoulder. The closest relative of *C. bayani* is *Conus generalis maldivus*, of which I strongly suspect it is just a deep-water form; certainly the two are in complete agreement in shape and sculpture, and specimens of *C. bayani* with relatively heavy patterns closely resemble *C. g. maldivus* with reduced patterns. However, *C. bayani* commonly has a nearly uneroded spire showing distinct nodules (though very small), while in *C. g. maldivus* the spire is commonly eroded to a flat plate; *C. bayani* of moderate size are glossy, while *C. g. maldivus* of even moderate size are dull and chalky. Lastly, and most importantly, the base of *C. g. maldivus* is deeply toned with blackish violet which extends to the anterior margin of the aperture; in *C. bayani* the base is white with no traces of blackish; this is probably the only character that will consistently separate the two 'species.'

Variation.—Although constant in shape, the extent of pattern is very variable from specimen to specimen. As indicated in the description, there are generally traces of a blotch band above and below midbody and of squarish spots posterior to the midbody band; development of the bands and of the axial streaks and flammules is very variable, from heavy with jagged margins to absent. The spire pattern is apparently always weak.

Distribution.—Moderately common in 40 or more meters from the northern Indian Ocean south to Reunion. Only recently have specimens become available, these mostly from off Madras, India and the Bay of Bengal.

Synonymy.—Almost certainly this should be considered a synonym of *C. generalis maldivus*, as the only real differences are pattern reduction and texture, both subject to great ecological variation. However, the absence of a dark base in any specimen I have seen is a good enough reason to recognize *C. bayani* as valid for the moment until complete intermediates are found.

Notes.—This is a rather fragile dredged cone, so perfect specimens are not common. Generally the tip of the spire is knicked and the lip is chipped. There is a tendency to healed breaks and strong growth flaws on the body whorl. At best the pattern of this species is asymmetrical and dull, but smaller specimens have the higher gloss and darkest colors.

BEDDOMEI Sowerby iii, 1901

1901. *Conus beddomei* Sowerby iii. *J. Malac.*, 8: 101, pl. 9, fig. 1 ("West Indies").

1942. *Conus brasiliensis* Clench. *Johnsonia*, 1(6): 24, pl. 12, fig. 2 (Victoria, Brasil).

Description.—Moderately light in weight, with a good gloss; broadly conical, the sides straight or slightly convex; body whorl with about 6-12 low ridges above the base, otherwise smooth except for fine spiral and axial threads and scratches; shoulder broad, roundly angled, concave above in most specimens; spire low to moderate, bluntly pointed, the early whorls commonly eroded; in uneroded specimens the first whorl is sometimes faintly nodulose, but this is an almost impossible character to use; tops of whorls slightly to distinctly concave, the sides of the spire straight to distinctly concave or weakly convex; whorls with about 3-4 distinct low spiral ridges crossed by numerous fine axials. Coloration and pattern very variable; commonly it is white to pinkish on the body whorl, this crossed by 2 broad spiral bands of yellowish brown, tan, pinkish tan, or red, these often connected by axial flammules and streaks; the bands may be continuous or of just rounded dark blotches, the axial connections strong or absent; commonly the resultant pattern is of a pinkish midbody band irregularly crossed by zigzag lines and flammules and a band of rounded or rectangular pinkish spots below the shoulder; many specimens have fine spiral lines of brownish dots obscured in the background and visible under magnification, occasionally these quite prominent; shoulder and body whorl just below shoulder with strong dark reddish brown axial lines sometimes strongly angled at the sutures; spire whitish to pinkish, marked like shoulder if not eroded; early whorls pinkish. Aperture fairly wide, somewhat widened anteriorly; outer lip straight and sharp, fragile, its posterior margin only slightly sloping below the level of the shoulder; mouth deep pink to dark lavender. Columella not distinct. Length 20-32mm.

Comparison.—This little but usually brightly colored cone should only be confused with *C. amphiurgus*, with which it is admittedly extremely similar. As a general rule, the spiral rows of small brown spots are not conspicuous in *C. beddomei*, although occasional specimens are heavily spotted. The shoulder is more angulate than most *C. amphiurgus*, with the spire whorls more stepped in many specimens; a large adult *C. beddomei* with eroded spire is about the size of most small *C. amphiurgus* with clean and complete spires. The strong tendency of the pattern in *C. beddomei* to form two or sometimes three spiral bands of oval or rectangular pale spots is seldom pronounced in *C. amphiurgus*. The protoconch and early whorl characters given by Van Mol and Tursch to separate these two species (*amphiurgus:* protoconch 3 whorls, first 3 early whorls axially nodulose; *beddomei:* 1¾ whorls, first early whorl sometimes axially nodulose; *Nautilus*, 1968, 82(1):5-6) are useless on typical specimens, although specimens about 20-30mm long with 2-3 distinctly nodulose early whorls are almost certainly *C. amphiurgus*. *C. daucus* is much flatter-spired in most specimens, with a broader shoulder and very different color patterns.

Variation.—As indicated in the description, this is a highly variable species. The spire varies in shape from convex to concave, although the

CINEREUS.—Above: Boac, Marinduque, P.I., 45.4 and 46.7mm. *Below* (series transitional to *gubba*): Sulu, P.I., 32.5-41.7mm (AK).

CINGULATUS.—*Above:* Cartagena, Colombia, 41.5mm (EP). *Below:* Left: Off mouth of Rio Magdalena, Colombia, 29.1mm (DM); Right (*castaneus*): Off Cartagena, Colombia, 35.5mm (EP).

whorls are commonly slightly stepped. The cross-banded pattern is probably most common, but some shells have only large isolated blotches of dark. Fine spiral rows of spots may be absent, obscured in the background, or conspicuous. The spire pattern of alternating dark brown to red flammules and pinkish white areas is constant and extends to the sub-shoulder area. The mouth varies from bright dark pink to brownish violet. Some specimens give the impression of being covered with bright reddish zigzagged axial flammules on pink. The strong spiral ridges of the spire whorls are constant.

Distribution.—Found in moderately shallow waters (3-40m) off the coast of central Brazil. There are also isolated records of this species from Puerto Rico, where it is said to be sympatric with both *C. amphiurgus* and *C. daucus.* Although these Puerto Rico records are presumably correct, one cannot help but wonder if perhaps *C. amphiurgus* is especially variable in that area and subject to confusion with true *C. beddomei.*

Synonymy.—This uncommon species has generally been known as *C. brasiliensis* Clench, although it seems that the figure and description of *C. beddomei* are clear enough to leave no doubt as to the species described. The assignment of *C. beddomei* to the synonymy of *C. 'jaspideus'* (a very complex species by previous interpretations) is impossible because of the pattern, color, and presence of strong spiral sculpture on the spire whorls, all clearly indicated by Sowerby.

Notes.—This usually brightly colored shell also occurs in tones of tan and white which are seldom sold to collectors. The lip is very thin and fragile, and the spire is almost always strongly eroded, as may be parts of the body whorl. Small healed breaks on the body whorl are not uncommon. Patterns with strong spiral rows of spots and reduced patterns are not commonly seen and are not especially attractive. I am not sure that it would be possible to make a valid separation of Puerto Rican specimens of this and *C. amphiurgus* except by size and relative erosion of the spire—the two species are sometimes extremely similar in color and pattern, and the minor differences in spire sculpture and radula are probably worthless at a practical level.

BENGALENSIS (Okutani, 1968)

1968. *Darioconus bengalensis* Okutani. *Venus*, 26(3/4): 66, pl. 7, fig. 2 (Bay of Bengal).

Description.—Moderately heavy, with a high gloss; elongate cylindrical, strongly biconic; body whorl smooth except for a few weak spiral ridges near the base and traces of fine axial and spiral threads and scratches; shoulder narrow, rounded, only slightly angled, slightly concave above and merging into the spire; spire very high, sharply pointed, the sides straight or slightly concave; whorls slightly stepped, with a weak and rounded carina near the middle; early 5 or so whorls finely nodulose and with distinct axial ridges; later whorls slightly concave above, with traces of fine spiral ridges and axial threads. Body whorl white, heavily covered with a dense network of small and medium reddish brown tents; two rows of dark reddish brown blotches, one above the midbody area and one below; the blotches are axially oval and do not tend to form complete spiral bands; blotches contain weakly defined 'textile lines' and traces of fine spiral dashes; a third row of smaller blotches usually developed below the shoulder, sometimes absent; shoulder with similar reddish brown blotches alternating with white blotches, the anterior margin of shoulder often distinctly spotted with blackish; spire with similar brown and white blotches and narrow axial lines of brown connecting the blotches from whorl to whorl; tip of spire whitish to pale pinkish. Aperture narrow, parallel, but a bit wider anteriorly; anal notch very deep; outer lip sharp, straight or shallowly concave at middle, posteriorly strongly sloping below the level of the shoulder; mouth glossy white. Columella long and narrow, sometimes bounded posteriorly by a ridge. Length 60-121mm.

Comparison.—The only other tented cones of similar size and shape are *C. gloriamaris* and *C. milneedwardsi*. *C. gloriamaris* is much broader at the shoulder with straighter sides which do not appear concave at the middle as in many *C. bengalensis*. In *C. gloriamaris* the spire is relatively broader and lower and is marked with finer tents instead of large blotches; the body whorl is more heavily patterned with very fine tents often deeply toned with yellowish or bluish. In *C. gloriamaris* the blotches on the body whorl are often larger than those of *C. bengalensis* and occasionally come close to forming spiral and axial bands. *C. milneedwardsi* is very similar, especially in some forms from the China Sea, but the spire is more strongly stepped with the carinae stronger, the whorls often more deeply concave above; the spire pattern in *C. milneedwardsi* is even more blotchy than typical of *C. bengalensis*. The body whorl pattern consists of very coarse tents and small blotches barely showing 'textile lines.' *C. milneedwardsi* is sometimes even more slender than *C. bengalensis* with more concave sides. *C. aureus*, especially the *paulucciae* form, has a much lower spire, is sometimes more broadly shouldered, and has the body whorl with more heavily developed spiral ridges reaching to near the shoulder. In *C. aureus* the 'textile lines' are strongly developed and sometimes run with few interruptions from shoulder to base; the tents are commonly very fine and greatly restricted in extent. *C. excelsus* has a taller spire which is very strongly stepped; the body whorl has many deep spiral grooves over its entire extent.

Variation.—*C. bengalensis* is fairly constant in both shape and pattern for a tented cone. The blotches are rather variable in shape and size, with the row below the shoulder usually small or even absent. The spire is apparently not distinctly tented but always blotched. Tenting is generally fine, with few large tents present in the pattern. Occasional specimens are

CIRCUMCISUS.—Above: Between Mactan and Bohol Is., P.I., 64.5mm (GG). *Below:* Left: Between Mactan and Bohol Is., P.I., 60.7mm (GG); Right: data unknown, 60mm (photo by Dr. H.G. Lee).

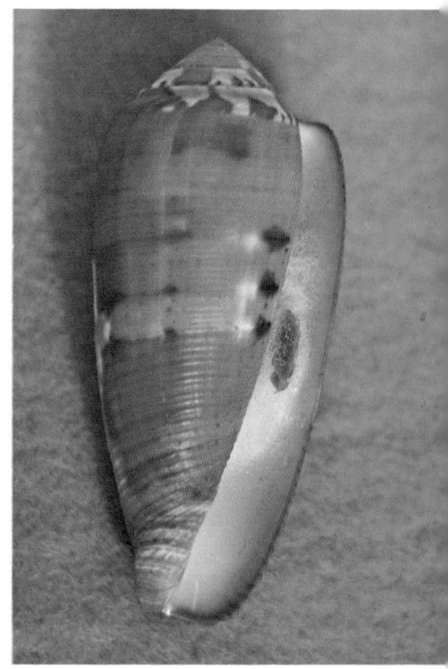

CLARUS.—Above: Left: Encounter Bay, South Australia, Australia, 30.0mm; Right: Port Lincoln, South Australia, Australia, 21.3mm. ***Below:*** Off Middleton Beach, Albany, Western Australia, Australia, 33.4mm (GG).

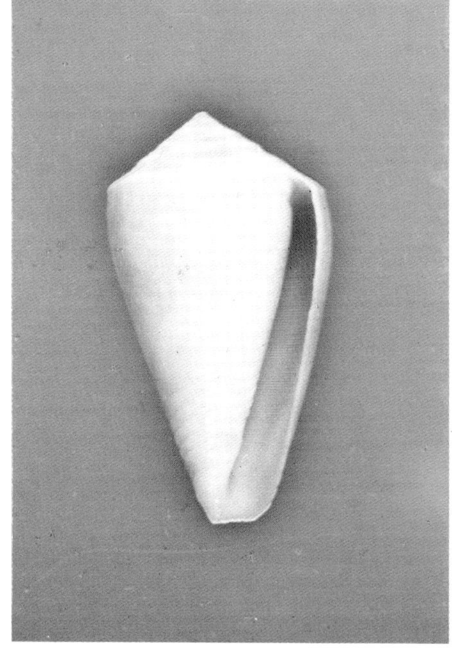

distinctly shouldered with concave spires, while very weakly patterned pale specimens are also known; seldom is the shell strongly tinted with yellow or blue as in *C. gloriamaris*.

Distribution.—Apparently restricted to the Bay of Bengal in relatively deep water, where it is moderately uncommon. Most specimens now come from off the Thai coast or off Burma.

Notes.—Perfect specimens of *C. bengalensis* are expensive but not hard to obtain. Few specimens are completely flawless, having small healed scars on the spire and often near the aperture. The lip chips easily at both ends, and the protoconch is seldom intact. Long unhealed breaks on the body whorl or spire detract greatly from the value.

The seemingly isolated distribution of this species is interesting because it suggests the pattern of *C. nobilis*, where the typical form is found in the Philippines and in the Andaman Sea. *C. gloriamaris* is relatively common in the Philippines, and certainly *C. bengalensis* is very similar to it. *C. milneedwardsi* occurs from India south and apparently does not overlap the range of *C. bengalensis* (the exact status of China Sea *milneedwardsi* is uncertain).

An early specimen of *C. bengalensis* seems to have fetched the highest price ever paid for a cone at a public sale, over $3,000. This is more than 10 times the current value of the shell, a good indication that rare shells are not always good investments.

BETULINUS Linnaeus, 1758

1758. *Conus betulinus* Linnaeus. *Systema Naturae per Regna Tria Naturae*, ed. 10, 1: 715 (Locality not stated). Holotype figured by Kohn, 1963.

1798. *Cucullus lacteus* Roeding. *Museum Boltenianum:* 41 (Locality not stated). Based on Martini, Pl. 61, fig. 673.

?1811. *Conus coralloides* Perry. *Conchology:* Pl. 25, fig. 6 (Sumatra and Ceylon).

1906. *Conus betulinus* var. *immaculata* Dautzenberg. *J. Conchyl. (Paris)*, 54: 27 (Ambodifoutra, Madagascar).

1937. *Conus betulinus* var. *alternans* Dautzenberg. *Mem. Mus. Roy. d'Hist. Nat. Belg.*, 2(18): 48 (Locality not stated).

1937. *Conus betulinus* var. *tabulata* Dautzenberg. *Ibid.*, 2(18): 48 (Locality not stated). See *Tableau Ency. Method.*, Pl. 333, fig. 5.

1937. *Conus betulinus* var. *plurizonata* Dautzenberg. *Ibid.*, 2(18): 49 (Locality not stated). Based on Chemnitz, Pl. 142, fig. 1321 (in part).

1937. *Conus betulinus* var. *scripta* Dautzenberg. *Ibid.*, 2(18): 49 (Locality not stated). Based on *Tableau Ency. Method.*, Pl. 333, fig. 2. Non *Conus scriptus* Sowerby iii, 1857-1858.

1937. *Conus betulinus* var. *paucimaculata* Dautzenberg. *Ibid.*, 2(8): 50 (Locality not stated). Lectotype figure (here selected): Martini, Pl. 61, fig. 673.

Description.—Very heavy, cumbersome; gloss low or dull surface; broadly pyriform, the sides very convex; body whorl with about a dozen rounded and closely spaced basal spiral ridges, remainder of whorl smooth except for fine spiral and axial scratches and occasional random spiral ridges; often with heavy axial growth marks; shoulder very broad, rounded, not distinct from spire; spire very low, nearly flat except for the first few whorls forming a small projecting pointed cone; tops of whorls flat or slightly convex, unmarked except for numerous fine axial threads and occasional fine traces of 1-2 irregular spiral threads. Body whorl uniformly creamy white, yellowish, to deep orange, sometimes more strongly tinted at the base and shoulder; covered with few to many spiral rows of small to large blackish spots, these usually regularly spaced; spots may be large squares or rounded dots, sometimes with secondary rows of finer brown dots between; sometimes black squares alternate with white dashes in continuous lines; commonly the spiral rows are closely spaced anteriorly and more widely spaced posteriorly; in some populations the spots are sparse and appear scattered at random over the shell; spire and shoulder about same color as body whorl or paler, with many narrow axial blackish brown blotches smaller than the pale interstices or with small blackish spots near the anterior margins (commonly early whorls with spots, later whorls with blotches); earliest whorls tan, unmarked. Aperture wide, nearly uniform; outer lip thick, sharp, straight; mouth white, sometimes with exterior margin pale orange. Columella thick, rather narrow, long, bounded posteriorly by a narrow sharp ridge. Length 75-135mm.

Comparison.—Under normal conditions *C. betulinus* is likely only to be confused with the similar *C. suratensis*. *C. suratensis* is somewhat lighter in weight than *C. betulinus* of similar size and more slender in build; the spots are reduced to narrow dashes arranged in irregular axial bands and generally numerous; the spots of *C. suratensis* are not markedly smaller anteriorly than at midbody as typical of most forms of *C. betulinus*. In addition, *C. suratensis* tends to have a whiter body than most *C. betulinus*, with the orange tones restricted to the base and shoulder. *C. glaucus* is smaller, has fine dashes instead of spots, and has a strong blue-gray color to the body whorl and mouth. *C. pulcher* is thin, fragile, has a more angled spire and longer body whorl, and generally a different pattern of spots. *C. byssinus* is narrower, white, tends to have fine orange bands in the pattern, and has a higher spire with more concave sides; it is of course much smaller at adult sizes. *C. genuanus* is narrower, with a more convex spire, tends to be bluish, has a very distinct pattern, and commonly has the spire blotches large and bright reddish brown, contrasting to most of the other markings. Some difficulty can be found separating *C. betulinus* from certain freak specimens of *C. figulinus* collected in the Madagascar Channel. These *C. figulinus* are relatively pale and have the brown lines thick, obscured in the background, and broken into irregular dashes; the spire is largely brown, however, a situation seldom found in *C. betulinus* but typical of *C. figulinus*.

CLERII.—(typical) *Above:* Bom Abrigo I., Sao Paulo coast, Brazil, 39.6mm (GG). *Below:* Left: Off Est. Rio de Janeiro, Brazil, 49.5mm; Right: Santana I., Rio de Janeiro, Brazil, 40.1mm (GG).

CLERII.*—(*clenchi*) *Above: Off Est. Rio de Janeiro, Brazil, 50.9mm (MM). *Below:* Left: Off Amazon River, Amapa, Brazil, 39.8mm; Right: Off Rio de Janeiro, Brazil, 55.3mm (both EP).

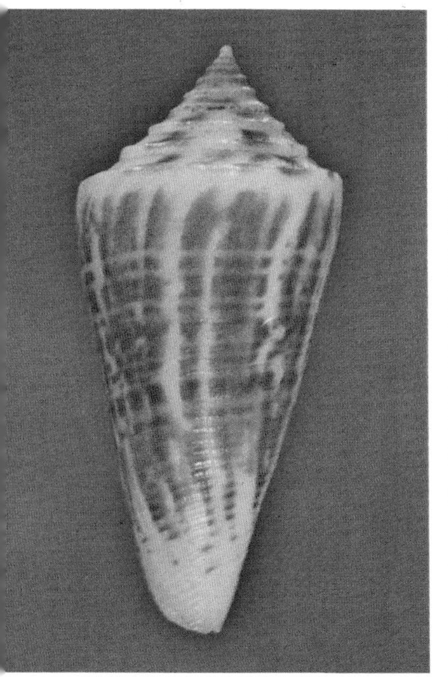
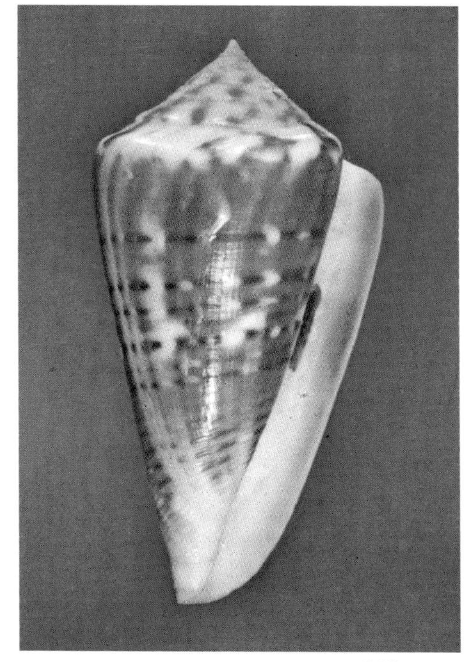

Variation.—Small specimens are relatively narrower than adults and have the basal ridging extending to midbody; juveniles are commonly whitish instead of orange or yellow. The extent of spotting is highly variable in this species. In the Indian Ocean populations are common in which the spots of the body whorl are greatly reduced or even completely absent; in such specimens the spire pattern consists exclusively of small dots at the margins of the whorls. Other Indian Ocean specimens have numerous rows of rather small and oval or irregular spots, with the shoulder and late spire whorls blotched, the early whorls finely spotted. In the Pacific Ocean similar patterns occur, but the tendency is toward larger, more squarish spots. At the extreme development of this form are speoimens with large black squares alternating with large white squares in regular series, the spire having large black blotches. At first I thought the Indian Ocean form with sparse spotting might be subspecifically distinct from the square-spotted Pacific form, but populations from both oceans have considerable variation in pattern, with moderately spotted patterns found commonly in both areas. Perhaps there is a strong cline from the Indian Ocean to the western Pacific, but many more lots with good locality data will have to be examined. The sparsely spotted Indian Ocean form was the basis of Linnaeus' description.

Distribution.—Common in shallow water from the eastern African coast throughout the Indian Ocean into the western Pacific from the southern Japanese islands and Philippines south at least to Queensland and some western islands of Oceania. Apparently absent from Hawaii and rare in French Polynesia.

Synonymy.—The holotype of *C. betulinus* is a sparsely spotted specimen. Several of the Dautzenberg names apply to the square-spotted form, but all possible pattern intermediates were also named by Dautzenberg based largely on early illustrations; none are apparently worthy of recognition.

Notes.—Even large specimens of *C. betulinus* seldom show more than one major growth mark, although the lip becomes cumbersome and subject to chipping. Pacific specimens with large black and white square spots and orange background color are especially attractive, but even some sparsely spotted Indian Ocean shells have bright orange background colors. Dealers normally do not differentiate among the various patterns.

BILIOSUS (Roeding, 1798)

1792. *Conus punctatus* Hwass, in Bruguiere. *Cone,* in *Ency. Method., Hist. Nat. des Vers,* 1: 628 (Africa). Non *Conus punctatus* Gmelin, 1791. Lectotype here designated as the specimen incorrectly called the holotype by Kohn, 1968.

1798. *Cucullus biliosus* Roedtng. *Museum Boltenianum:* 39 (Locality not stated). Lectotype figure (Kohn, 1975): Martini, Pl. 59, fig. 656.

1817. *Conus piperatus* Dillwyn. *Descr. Catalogue Recent Shells*, 1: 401 (East Indian Ocean). Lectotype figure: Chemnitz, Pl. 139, fig. 1294.

1866. *Conus concinnus* Sowerby ii. *Thesaurus Conchyliorum*, 3(*Conus* Suppl.): 329, pl. 28(289), fig. 646 (Locality unknown). Non *Conus concinnus* J. de C. Sowerby, 1821, a fossil.

1874. *Conus sapphirostoma* Weinkauff. *JB. deutsch. Mal. Ges.*, 1: 268. Nomen novum for *Conus concinnus* Sowerby ii, 1866. Commonly misspelled *Conus saphyrostoma*.

Description.—Moderately heavy, with a low gloss or dull finish; low conical, the sides nearly straight or somewhat inflated posteriorly; body whorl elongate, covered with numerous low and undulating spiral ridges from base to shoulder, these sometimes weaker near midbody and below shoulder, heaviest near base; numerous axial and spiral growth threads, lines, and flaws often present; shoulder wide, roundly angled, weakly coronated or undulate, usually slightly concave above; spire low, bluntly pointed, the sides straight to slightly concave or convex; spire whorls distinctly coronate, the early whorls commonly eroded; tops of whorls slightly concave, usually with 5-6 distinct spiral ridges which may be nearly obsolete on anterior whorls. Body whorl usually dirty creamy tan to light brown, often with strong bluish gray tones, covered with numerous spiral rows of small brownish spots following the spiral ridges from base to shoulder, the spots often tending to produce irregular axial bands; usually a lighter midbody band is visible; base orange-tan to violet-tan, not much darker than rest of whorl; heavy black axial flammules may be present over the body whorl; spire pale tan to dirty cream, usually heavily marked with irregular curved blotches or stripes of dark brown to black; early whorls sometimes pinkish. Aperture moderately wide, uniform; outer lip nearly straight or slightly convex, moderately thick but sharp; mouth deep violet, the margin of outer lip with a brownish violet or yellowish stripe. Columella narrow, indistinct. Length 35-55mm.

Comparison.—*C. biliosus* is most likely to be confused with *C. hyaena* and *C. parvulus*. In *C. parvulus* the body whorl is relatively shorter and wider, the spire whorls are more strongly coronate and have only 2-4 spiral ridges, the brown flecks tend to be sparse and form distinct axial lines or bands, the base is darker and strongly contrasted with the rest of the body whorl, and the spire is almost unmarked except for small dark brown spots often present between the coronations. *C. hyaena* is usually distinctly wider below the shoulder than in *C. biliosus*, with the shoulder straight or only weakly undulate, the body whorl is nearly smooth except for the basal ridges and some heavy axial lines, the spire has a relatively longer and sharply pointed tip, the later spire whorls are not distinctly coronated, small brown flecks on the body whorl are usually absent but distinct rows of long dashes are usually prominent at least near midbody, the base is often paler than the rest of the body whorl, and the mouth is clear bluish white with only pale violet tones; *C. hyaena* tends to be an

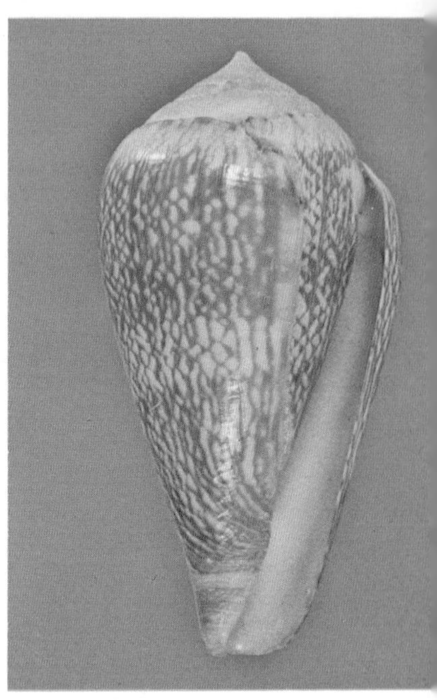

CLOVERI.—Above: N'Gor, Cape Verde, Senegal, 37.8mm (EP). *Below:* Senegal, 22mm (DM) (photos by Jayavad).

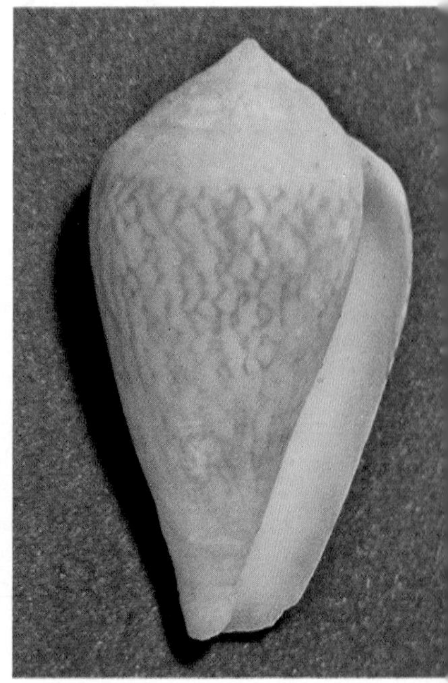

COCCEUS.—Above: Left: (top) Canal Rocks, Western Australia, Australia, 18.7mm; (bottom) Cavatomp Bay, Western Australia, Australia, 19.1mm; Right: Albany, Western Australia, Australia, 30.8mm. ***Below:*** Torbay East, nr. Albany, Western Australia, Australia, 39.4mm (K).

overall lighter color than *C. biliosus* and is more indistinctly patterned. *C. mus* and *C. gladiator* may be vaguely similar in appearance to *C. biliosus*, but they both differ in shape and details of color pattern. *C. lohri* is smooth and bright yellow or orange.

Variation.—Although this species varies somewhat in width at the shoulder and height of spire, it is fairly constant in shape. The major differences between populations and individuals are those of color and pattern. General tones vary from pale straw-tan to deep bluish gray, with the rows of spots weak or strong and crowded; axial blackish flammules may be completely absent or strong enough to obscure most of the pattern. The base is usually some shade of dark tan, generally matching fairly closely the shade of the body whorl or slightly darker. Generally the dark markings on the spire are heavy and distinct, but they are sometimes nearly obsolete. The dark mouth is quite constant.

In populations from Natal and Mozambique, which may be isolated from the northern Indian Ocean form, the body whorl and spire are somewhat more convex, with the shell approximating *C. hyaena* in shape. The base is usually blackish brown, darker than the rest of the body whorl, which is marked with numerous irregular axial flammules as well as small brown dots in spiral rows. The body whorl and spire sculpture are similar to typical *C. biliosus*, but the mouth apparently is paler and more uniformly bluish. This form perhaps deserves a distinct name.

Distribution.—Found commonly in shallow water on sandy bottoms in the India-Sri Lanka area; apparently restricted to the northern Indian Ocean except for (? isolated) populations in Natal and Mozambique. Specimens labelled as from Madagascar and Mauritius have also been seen, but I am not sure the species actually occurs there.

Synonymy.—The nomenclature of this species seems accursed. It was called *C. punctatus* for many years, but that name is preoccupied. Then *C. piperatus* Dillwyn came into use. It now seems that the earliest name is *C. biliosus*. See Walls, *Hawaiian Shell News*, 1976, 24(5):9, for a discussion of the problem. *C. concinnus* appears very similar to pale tan *C. biliosus* in pattern and shape, even to having the yellow margin of the mouth; Sowerby described it as smooth, which is not typical of the species, but individual variants commonly have the spiral ridging reduced or obsolete at midbody and below the shoulder; Sowerby's type appears to be a juvenile shell, which may also explain the absence of distinct ridging.

Notes.—At its best this is certainly not a pretty cone. The pattern is often very irregularly developed over the body whorl, and there are always strong axial flaws or growth marks. These flaws appear typical of the species. The early spire whorls are usually at least partially eroded. Large shells apparently lose part of the pattern and reduce the coronations on the shoulder. The lip is commonly chipped, but it should never be filed; filing would affect the narrow yellowish stripe which seems characteristic of the species. The species is apparently quite common in localized populations, although the Natal-Mozambique form is uncommon.

BOETICUS Reeve, 1844

?1834. *Conus pauperculus* Sowerby i, in Sowerby ii. *Conchological Illustrations:* Pt. 56-57, fig. 78 (Locality not stated).

1844. *Conus boeticus* Reeve. *Conchologia Iconica*, 1(*Conus*): Pl. 42, sp. 226 (Philippine Islands).

1848. *Conus Ruppellii* Reeve. *Ibid.*, 1(*Conus* Suppl.): Pl. 2, sp. 273 (Red Sea).

1848. *Conus cerinus* Reeve. *Ibid.*, 1(*Conus* Suppl.): Pl. 3, sp. 283 (Island of Mindanao). Distorted spire.

1849. *Conus lachrymosus* Reeve. *Ibid.*, 1(*Conus* Suppl.): Pl. 6, sp. 258 (Locality unknown).

1849. *Conus rivularis* Reeve. *Ibid.*, 1(*Conus* Suppl.): Pl. 6, sp. 261 (Moluccas).

?1849. *Conus fucatus* Reeve. *Ibid.*, 1(*Conus* Suppl.): Pl. 7, sp. 271 (Philippine Islands).

1865. *Conus Moussoni* Crosse. *J. Conchyl.* (*Paris*), 13: 299, pl. 10, fig. 3 (Seychelles).

1877. *Conus superscriptus* Sowerby iii. *Proc. Zool. Soc.* (*London*), 1876: 753, pl. 75, fig. 6 (Madagascar).

1882. *Conus dianthus* Sowerby iii. *Ibid.*, 1882: 118, pl. 5, fig. 4 (Locality unknown). Albinistic.

1887. *Conus Fultoni* Sowerby iii. *Thesaurus Conchyliorum*, 5(*Conus* Suppl.): 273, pl. 36 (512*), fig. 758 (Singapore).

1913. *Conus meleus* Sowerby iii. *Ann. Mag. Nat. Hist.*, (Ser. 8), 11: 558, pl. 9, fig. 3 (Kii, Japan).

?1913. *Conus optimus* Sowerby iii. *Ibid.*, (Ser. 8), 12: 235, pl. 3, fig. 7 (New Caledonia).

Description.—Moderately heavy, with a good gloss; low conical, the sides generally straight and the body whorl long or short; body whorl with heavy, widely spaced rounded spiral ridges above the base, these commonly strongly granulose; ridges may continue to midbody or occasionally to shoulder, but commonly become obsolete above basal third of whorl; usually posterior half or more of body whorl appears smooth and polished, but fine axial scratches are present; shoulder broad to relatively narrow, roundly to rather sharply angled, distinctly coronate, undulate, or occasionally apparently smooth; spire low to moderate, the sides straight or slightly concave, bluntly pointed; whorls all nodulose or distinctly coronate, usually most conspicuous on the earlier whorls; whorls slightly concave above, with about four strong spiral ridges crossed by low axial ridges to produce a cancellate or faintly punctate effect. Color pattern and colors very variable, generally falling into one of three common patterns: 1) body whorl creamy white, with two spiral bands of golden to orange-brown irregular blotches above and below midbody; blotches usually connected by a narrow spiral band of the same color and with

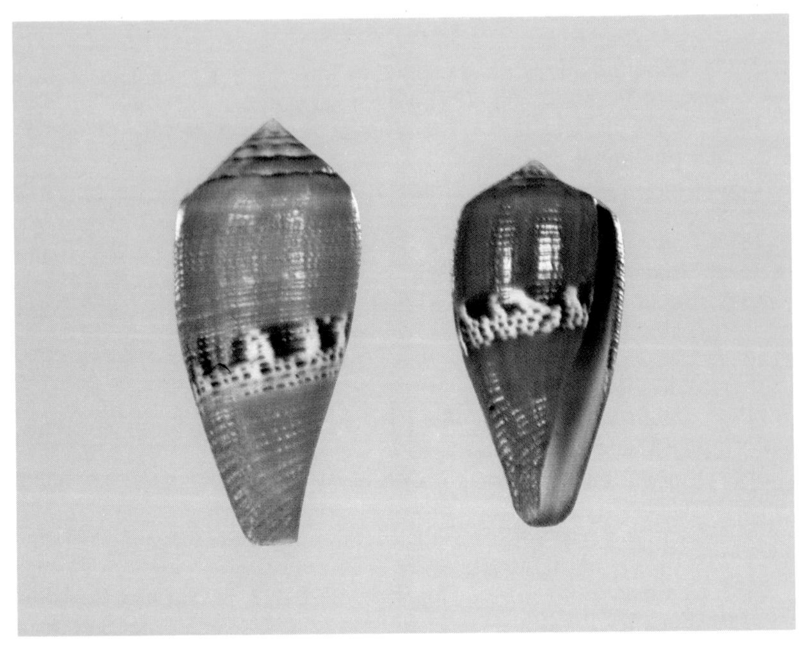

COCCINEUS.—*Above:* Left: Honiara, Guadalcanal, Solomon Is., 44.0mm; Right: Samar, P.I., 38.0mm. *Below:* Left: Samar, P.I., 53.4mm (GG); Right: P.I., 38.6mm (W.E. Burgess).

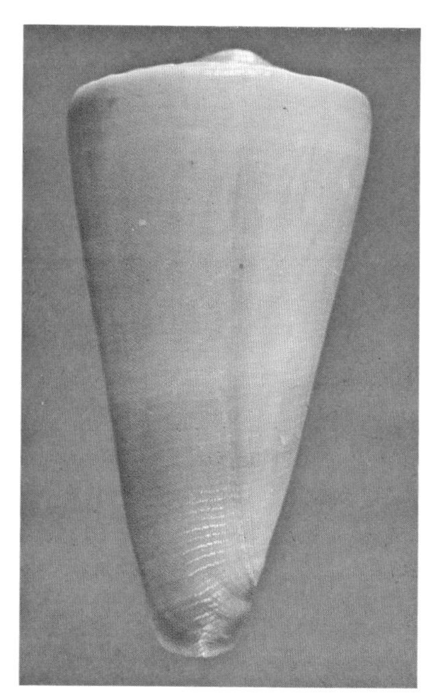
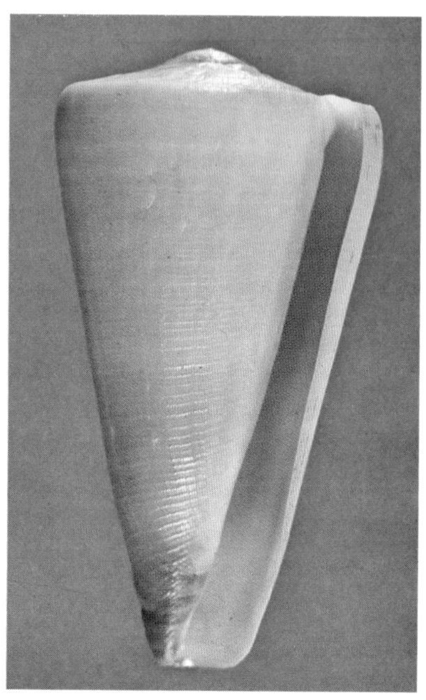

COELINAE.—(*c. coelinae*) ***Above:*** New Caledonia, 52.0mm (MM). ***Below:*** Kwajalein Atoll, Marshall Is., 59.4mm.

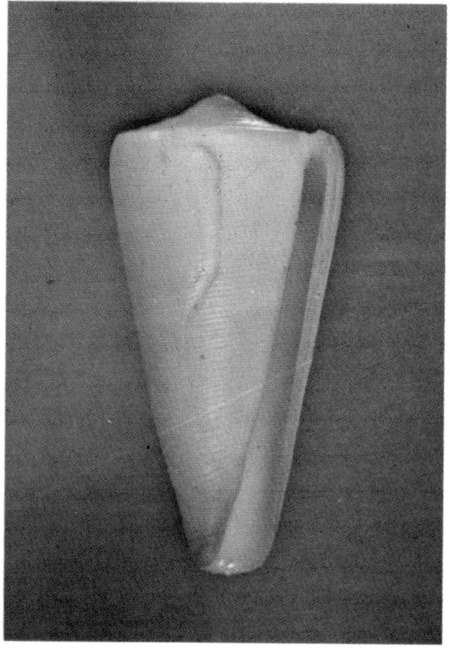

axial extensions which may extend in irregular fashion from base to shoulder; midbody area almost unmarked; base white; several spiral rows of small squarish brown spots usually visible in background but may need magnification; spire whitish, sparsely marked with orange-brown spots between coronations, the spots sometimes extending into short lines or streaks; early whorls pinkish to white; this form commonly has a narrow and relatively long body whorl; 2) body whorl creamy white to pale bluish, largely obscured by two broad, somewhat broken bands of pale tan to dark brown formed by fusion of axial flammules; midbody area largely pale, crossed by narrow irregular brownish axial flammules; whole shell marked with widely spaced spiral rows of long and usually distinct dark brown dashes; base pale; shoulder and spire marked with curved brownish blotches on whitish; in this form the body whorl is relatively long and narrow, the spiral ridging is generally reduced to a few non-granulose ridges above the base, and the shoulder is rather sharply angled and commonly only weakly undulate; 3) somewhat similar to the second pattern, but the broad brownish bands are replaced by deep brownish black flammules which usually fuse into two broad blackish bands above and below midbody; this results in a strongly contrasting white base, a white midbody area broken by irregular flammules, and a strongly contrasting series of white rectangles separated by curved black blotches below the shoulder; within the blackish areas are visible spiral rows of dark brown and white dashes, these often obscured by the dark background color; spire white, heavily marked with dark brownish black, resulting in the coronations being picked-out in white on a dark background; this form commonly has the body whorl relatively broad below the shoulder and the shoulder itself rather narrowly angled; the coronations are usually strongly developed and very obvious because of the contrasting color pattern. Aperture moderately narrow, uniform in width; outer lip thin or thick, usually sharp; mouth glossy white in the paler forms, deeper bluish white with a dark brown band at the margin in the darkest forms. Columella not visible externally. Length 20-38mm.

Comparison.—This very variable and apparently common shell has been little understood, as seems obvious from the synonymy. From *C. balteatus* and *C. cernicus*, which it sometimes resembles, it differs in having the base white, lacking white flecks, and having a series of distinct spiral rows of brown spots or dashes; the mouth is seldom violet except in very dark specimens; few *C. balteatus* or *C. cernicus* show a tendency toward widely spaced granulose basal ridges. *C. biliosus* has a dark mouth and dark base regardless of color pattern and has spiral rows of smaller spots which are very irregular in shape and much more crowded; in *C. biliosus* the spire whorls do not appear strongly cancellate. *C. proximus* and *C. moluccensis* may be vaguely similar in shape but have erect, strong, and usually pointed coronations, the outer lip is sometimes concave at the middle, and the grooves between spiral cords on the body whorl are usually axially ridged or punctate; there should be no real confusion with

C. boeticus, although the patterns of pale *C. boeticus* and some *C. moluccensis* are similar; both *C. proximus* and *C. moluccensis* have long, narrow columellas usually set off by a strong curved ridge. The same differences apply to *Conus marielae*. *C. sazanka* usually has weak body sculpture, less variegated patterns, and a more regularly biconical shape.

Variation.—Most of the common variant patterns are described in the description. There are all types of intermediates between patterns and between various body whorl shapes and shoulder angles. Generally the coronations are well marked and easily visible, but those of the shoulder are commonly very weak or obsolete. It is very hard to relate extremes of pattern without seeing series containing intermediates. The black form and pattern two are often very distinctive.

Distribution.—Apparently *C. boeticus* is very widely distributed in the Indo-Pacific and often common in shallow water over sandy bottoms with coral rubble. It is known from the Red Sea through the Indian Ocean islands to Taiwan, the Philippines, and southern Japan; it is also commonly collected in the New Guinea-Solomons area and ranges to at least the Marshall Islands and Fiji in the central and southern Pacific. It probably occurs in French Polynesia, but I know of no actual records.

Synonymy.—Most of the type figures of the synonyms listed can be placed with some confidence in the first two major patterns, but some are rather doubtful. *C. pauperculus* in particular is a troublesome name which may apply to a pattern two specimen; if so, it would become the proper name of the species; I hesitate to take this step because the figure could also represent a juvenile *C. biliosus*. Taken separately, the synonyms break into the following patterns: **Pattern 1:** *boeticus*, *cerinus* (distorted spire), *rivularis*, *moussoni* (pale), *dianthus* (albinistic), *meleus* (pale), and perhaps *optimus* (very dark with a vaguely tented pattern); **Pattern 2:** *ruppellii* (pale with strong spirally lineate pattern), *lachrymosus*, *superscriptus*, *fultoni*, and perhaps *fucatus* (body whorl all brown); **Pattern 3:** none as such, although *fultoni*, *fucatus*, and perhaps *optimus* tend in this direction. It is likely that other names will eventually be shown to be synonyms of *C. boeticus*. Patterns 1, 2, and 3 all broadly intergrade and seem to be found in several areas with no obvious geographical restriction, so they are at best forms, perhaps ecological in nature. Pattern 2 is sometimes found adjacent to patterns 1 and 3 in nature.

Notes.—Pattern 3 specimens currently come from the Philippines, where they are sometimes called *ruppellii*. Pattern 1 comes in typical form from the Solomons-New Guinea area and from the Philippines; those from the Philippines, when small, are often called *C. sphacelatus*, a dubious name which probably is not related to *C. boeticus*. Pattern 2 appears to be least commonly offered at the moment, with occasional specimens coming from southern Japan and Okinawa. Of course all three patterns can be found occasionally offered from other localities.

Since this is a small cone, it is seldom heavily flawed. Some popula-

COELINAE.—*Above* (*c. spiceri*): Pearl Habor, Oahu, Hawaii, 120mm (MG). *Below* (*c. berdulinus*): Punta Engano, Cebu, P.I., 72.6mm (Dayrit).

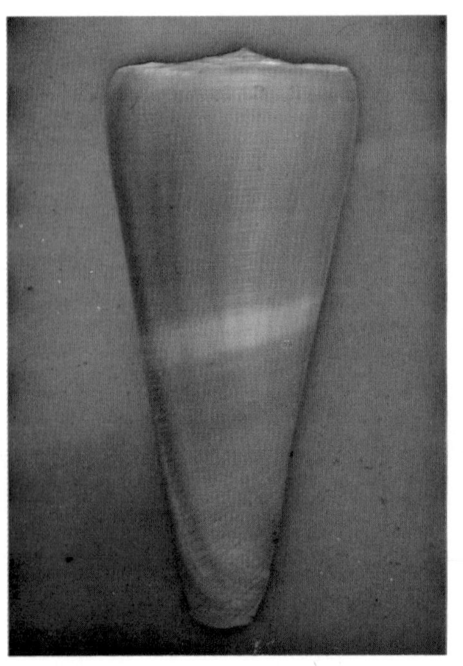
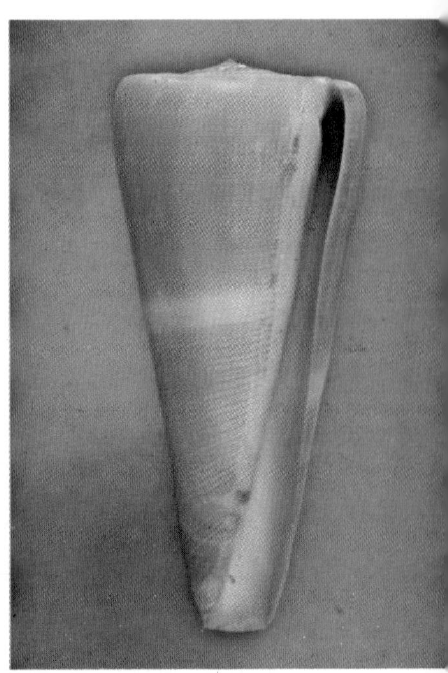

COLUBRINUS.—Above: Kwajalein Atoll, Marshall Is., 59.2mm (MG). *Below:* Off Lungga River, Guadalcanal, Solomon Is., 49.2mm (EP).

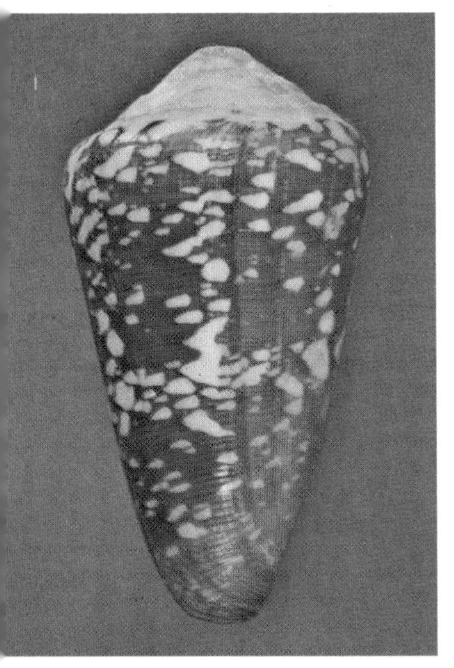

tions of pattern 1 may have heavy healed growth scars on the dorsum and commonly have chipped lips. Patterns 2 and 3 seem to come from environments which result in fewer breaks. Patterns 1 and 3 are about equally valued, while pattern 2 is uncommon to rare in excellent condition and good colors.

BOIVINI Kiener, 1845

1845. *Conus Boivini* Kiener. *Species gen. et icon. des coqu. viv.*, 2 (*Conus*): Pl. 64, fig. 2. 1849-1850, *Ibid.*, 2: 282 (Locality unknown).

Description.—Heavy, with a good gloss; cylindrical, the sides straight below shoulder; body whorl with about 8 deeply punctate spiral grooves on the basal third, widely separated by flat ribs; midbody area smooth except for fine spiral threads; posterior third of whorl with about 8 shallow grooves separated by wide ribs; shoulder angulate, concave above; spire very low, deeply concave so the early whorls form a small cone at the middle of a sunken plate; spire bluntly pointed, not projecting much beyond level of shoulder. Body whorl pinkish white, with a broad spiral band of pale reddish brown above the basal sculpture and just below the sub-shoulder sculpture; these bands and their vicinity are overlaid with irregular axial dark reddish brown blotches and spots which barely penetrate the midbody area; spire pinkish, with a few dark brown spots. Aperture narrow, uniform in width but slightly wider anteriorly; outer lip straight; mouth pinkish white. Columella rather long and narrow. Length 62mm.

Comparison.—In pattern this species resembles some *C. terminus*, as does the shape of the body whorl and the general appearance of the spire (if mentally depressed). However, the sculpture of deep punctate grooves at the base and similar but shallow grooves below the shoulder is unique in a shell of this shape. *C. adamsonii* has similar sculpture but is very different in shape and pattern and probably not closely related. *Conus spectrum* and allies are thinner in texture but may have a very similar spire, although the body whorl is almost always much wider and the pattern does not closely approach that of the type of *C. boivini; C. boivini* is much larger than *C. spectrum* and of course has a very different sculpture.

Variation.—The description is of the unique holotype following fairly closely Kiener's description. A similar specimen, probably belonging to this species, is illustrated in Marsh and Rippingale, 1974 edition, pl. 16, fig. 11. This shell, called *C. conspersus* by the authors, is much wider at the shoulder, lacks the spiral bands of reddish brown, and has axial reddish brown flammules or blotches in three series instead: small blotches at the shoulder, long flammules at midbody, and blotches above the basal sculpture. The grooves below the shoulder are apparently only 4 in number, but the midbody area appears smooth. This specimen apparently belongs to *C. boivini* and may perhaps be more typical of the species than the larger type specimen. .

Distribution.—The type came from an unknown locality. The Marsh and Rippingale specimen tentatively assigned here came from several fathoms off Swain's Reef, 150 miles off central Queensland, Australia. The species is apparently rare or being confused with another species; I know of only the type and the Marsh and Rippingale specimen.

Notes.—Certainly the figure in Kiener is of a distinctive and very interesting large cone. If the species is rediscovered and found to conform to Kiener's description, it will probably fetch a high price.

BOSCHI Clover, 1972

1792. *Conus* (*Chelyconus*) *boschi* Clover. *Venus*, 31(3): 117, figs. 1-2 (Museera Is. to Muscat Oman in South East Arabia).

Description.—Light in weight but solid, with a low gloss; low conical, the sides convex; body whorl smooth except for low spiral ridges near base, these sometimes rather strong; rest of body whorl with traces of spiral threads and some axial ones; shoulder rounded, not very distinct from spire; spire low, the sides usually convex; protoconch large, mamillate, retained by adult shells; tops of spire whorls slightly convex or flat, with weak axial threads and sometimes traces of very weak spiral ridges; whorls weakly stepped in some specimens. Body whorl creamy white, sometimes with a bluish tone, variably covered with very thin spiral brownish lines that are widely spaced, about 15-20 on the whorl; short axial brownish lines run across the spiral lines in a very irregular fashion to produce a highly variable brick-wall effect which may cover most of the body whorl or be restricted to the base and shoulder area; just above midbody is a spiral band of rather squarish irregular deep reddish brown blotches, large to small in size; another band of smaller spots and blotches may be developed above the base; shoulder and spire whitish, with narrow dark reddish brown axial lines or squarish blotches separated by larger white areas; early whorls unmarked, sometimes eroded. Aperture moderately narrow, somewhat widened anteriorly; outer lip straight, rather thick but sharp; mouth dark violet deep within, paler exteriorly. Columella narrow, largely internal. Length 20-30mm.

Comparison.—Although this very distinctive species is morphologically perhaps closest to *C. minnamurra*, there is little superficial resemblance to that species. The pattern is very distinctive, being approached only by *C. lucidus*, *C. puncticulatus* (rare patterns), *C. kimioi*, and *C. hirasei*. *C. hirasei* is larger, has large blackish spots at the shoulder,

COMATOSA.—Off Tosa, Shikoku, Japan. *Above:* 53.1mm. *Below:* 35.6mm (both EP).

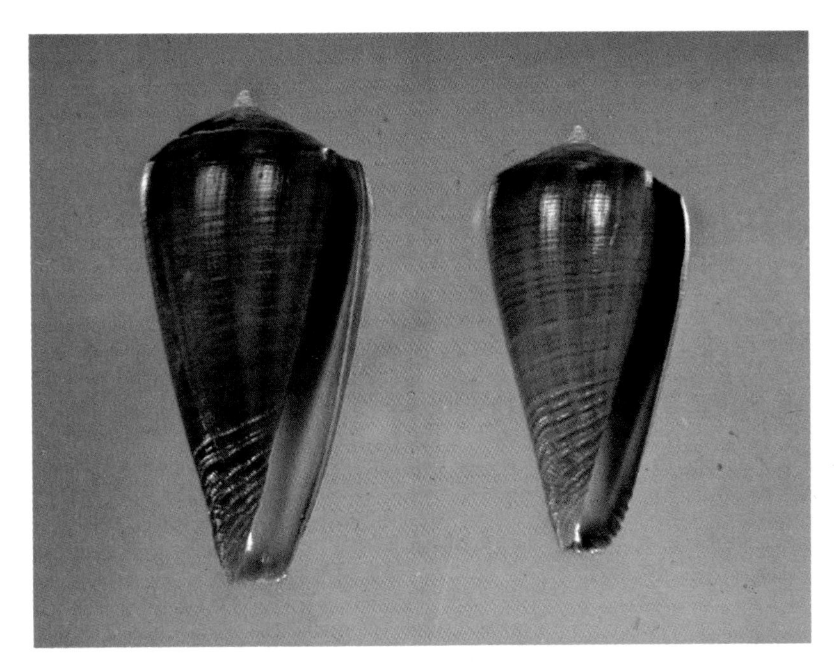

CONCOLOR.—Above: Off Lungga River, Guadalcanal, Solomon Is., 34.3 and 39.2mm. *Below:* Left: Coron, Palawan, P.I., 37.9mm (Dayrit); Right: Russell I., Solomon Is., 21.1 and 34.3mm (MG and BB).

and typically lacks axial lines. *C. kimioi* is of more similar size but is very fragile, more angled at the shoulder, attenuate anteriorly, and brown in tone instead of white; it is possible that *C. kimioi* is the juvenile of *C. hirasei.* Rare patterns of *C. puncticulatus* may produce a finely lined pattern with some short axials similar to those of *C. boschi*, but that species is usually bluish gray in general tone and lacks the large mamillate protoconch. *C. lucidus* is very similar in pattern and on first inspection it and *C. boschi* may appear identical. However, *C. lucidus* lacks the large mamillate protoconch and is probably not really closely related; it also has a taller, pointed spire with concave side, has stronger basal sculpture, and usually has the aperture pale bluish white to pale violet instead of deep violet.

Variation.—The spire may be almost flat or quite high and stepped. The pattern is variable in details, but usually the spiral lines are well developed and at least a few axials are present near the base; the blotches may be large and well developed in two bands or just small irregular spots above midbody. In shape the body whorl is quite constant.

Distribution.—Known mostly from Al Masirah Island off the coast of Oman, in the Arabian Sea, where it is not uncommon. Lately specimens are said to have been collected in deep water off the western Indian coast.

Notes.—This is a very distinctive little shell. Healed axial growth marks may be present but few other major flaws. Slight erosion of the spire does not detract from the appearance of this species. In dead specimens the deep violet of the mouth apparently fades rapidly.

BRUNNEUS Wood, 1828

1828. *Conus brunneus* Wood. *Index Test., Suppl.:* 8, pl. 3, fig. 1 (Locality unknown).

Description.—Heavy, with a low gloss; obconic, the sides slightly convex posteriorly and gently sloping to the base; body whorl with a few heavy raised ridges near the base, these continuing in weaker form to midbody or higher; numerous weak axial growth lines and occasional growth marks present; shoulder broad, angulate, heavily coronate, the coronations broad and sharp; spire low to nearly flat, the sides concave or straight; bluntly pointed, the early whorls often eroded; spire whorls heavily coronated, the tops with about 5-6 distinct spiral ridges (when not eroded). Body whorl generally dark brown on a white background; white shows through in irregular patches, especially at midbody and near the shoulder; occasional specimens all brown, others with body whorl largely white; numerous closely spaced spiral lines of darker brown or black over body whorl, these usually broken into long dashes, most prominent in the white blotches; base usually dark violet brown, somewhat darker than rest of body whorl; spire and shoulder white, with large brown blotches between coronations, the blotches often visible only on shoulder because of erosion; early whorls pinkish white. Aperture moderately wide, uniform in width; outer lip straight, sharp; mouth waxy yellowish gray, darkest deep within, exterior pattern showing through edge of lip. Columella narrow and short, often not visible externally. Length 35-71mm.

Comparison.—Although the pattern may vaguely resemble *C. imperialis*, *C. brunneus* is a much shorter and relatively wider shell which will not be confused with that species. *C. regius* is also very similar, but it has a white mouth and differs markedly in color pattern. The closest approach to *C. brunneus* is found in *C. diadema*, which is very similar in shape and general appearance. However, *C. diadema* has a waxy feel, is uniformly olive to dark brown with narrow paler bands at midbody and the shoulder, seldom or never with white blotches or spiral lines of dashes, and has the aperture pale violet with a large deep violet blotch posteriorly. In *C. diadema* the spire whorls are usually more distinctly stepped than in *C. brunneus*. *C. bartschi*, commonly considered a form of *C. brunneus*, is quite distinct in the relatively broader and shorter body whorl, the white mouth with only a touch of yellow in some specimens, and the numerous spiral rows of small brown flecks instead of large distinct dashes; it also lacks spiral ridges on the spire whorls in typical specimens, *C. cabritii* is a much smaller and narrower species with punctate spiral grooves over the body whorl.

Variation.—Other than such common variation as height of spire, the major individual variation in *C. brunneus* is in pattern. Specimens in a population may be solid brown, have only small irregular white blotches at midbody, have scattered larger white blotches, or be largely white with less than half the body whorl brown. Usually the spire whorls are mostly white with only brown spiral lines showing, and even these may be eroded; the shoulder is often mostly dark brown. In almost all specimens the spiral rows of dashes are prominent wherever there are white blotches. The basal ridges may be weakly granulose.

The juveniles (under 25mm) of *C. brunneus* are very distinctive. They are white with relatively narrow axial stripes of dark brown from the base to the shoulder; fine spiral dashes are visible in the background; the spire is sharply pointed with deeply concave sides. By a length of about 25mm the brown stripes have become wavy and begun to overlap in places, resulting in a shell which is largely brown. My thanks to Alex Kerstitch for juveniles.

Distribution.—Common from intertidal areas to moderately deep water throughout the eastern Pacific from the Gulf of California and Baja south to Ecuador, including many of the offshore islands.

Notes.—This is a common species which shows no variation of special value to collectors. Most specimens have one or many heavy axial growth marks, but these are to be expected in a rough cone and do not detract from values. Erosion of the spire and occasional small coralline growths are also to be expected. Dark and light specimens are not especially valued over moderately blotched patterns.

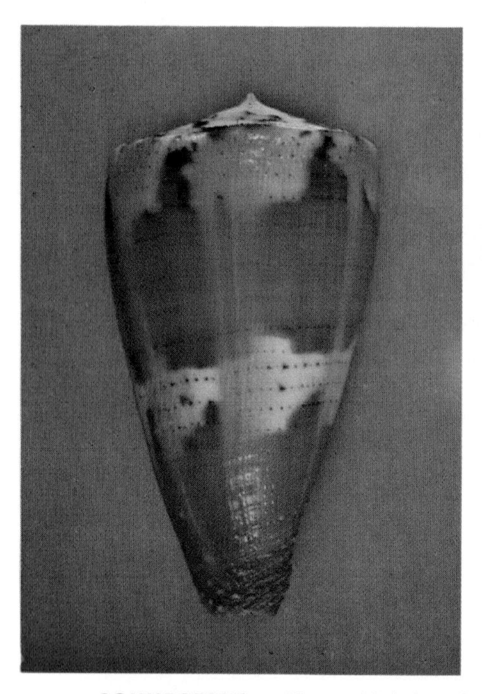

CONNECTENS.—Above: Makaha, Oahu, Hawaii, 45.3mm (MG). ***Below:*** Left: Zamboanga, P.I., 41.3mm; Right: P.I., 55.0mm (both DMNH).

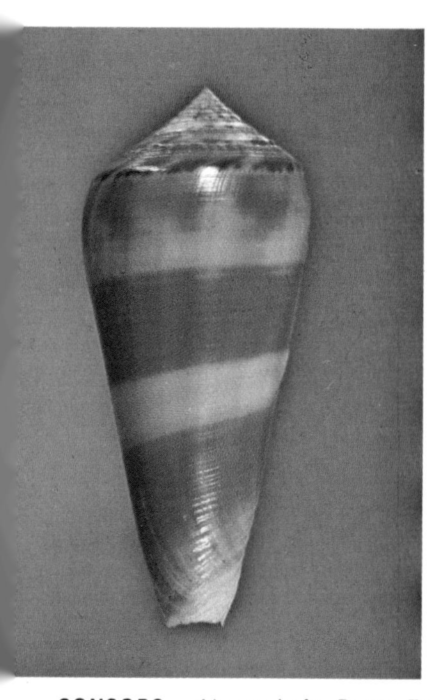

CONSORS.—Above: Left: Punta Engano, Cebu, P.I., 66.2mm (Dayrit); Right: Andaman Sea, 70.1mm. *Below:* South China Sea, 85.8mm (GG).

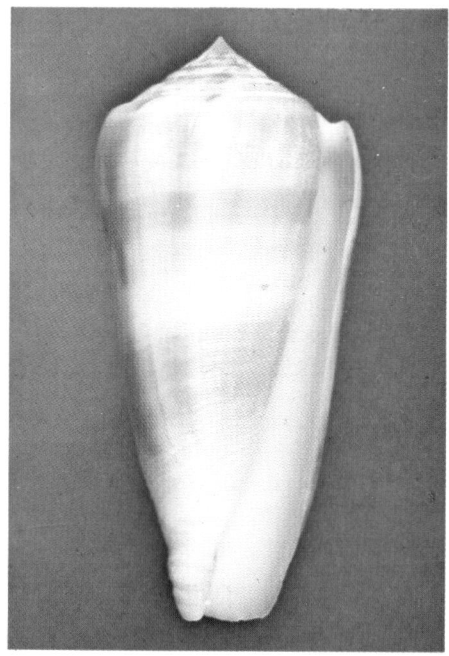

BRUUNI Powell, 1958

1958. *Conus* (*Dauciconus*) *bruuni* Powell. *Rec. Auckland Inst. Mus.*, 5 (1/2): 84, pl. 10, fig. 3 (Off Raoul Island, Kermadecs). See also Cernohorsky, *Hawaiian Shell News*, 1976, 24(9): 3-4.

Description.—Moderately heavy, with a low gloss or dull finish; low conical, the upper sides distinctly convex and then tapered to the base; body whorl with about 10-12 basal ribs, these commonly granulose; rest of body whorl with traces of low spiral ridges and axial scratches; shoulder roundly angled, rather broad, concave above in adults; spire moderate, sharply pointed, the sides straight or slightly concave; early spire whorls nodulose, then becoming undulate, later spire whorls and shoulder with straight margins; tops of whorls with about 6-8 spiral ridges and weak axial threads. Body whorl pinkish violet, with three broad bands (at shoulder, above and below midbody) of reddish brown to orange-brown; midbody area with a narrow pale band; base pale pinkish violet; shoulder and spire like body whorl, pinkish violet marked with irregular large blotches of reddish brown. Aperture moderately wide, uniform in width; outer lip thin, sharp, sloping distinctly below level of shoulder posteriorly; mouth pale pinkish violet. Columella narrow, indistinct. Length 30-44mm.

Comparison.—*C. bruuni* is extremely similar to pale *Conus kinoshitai* in shape, color, pattern, and sculpture. In addition to the presence of heavily corded juveniles apparently not known in *C. kinoshitai*, it differs in having a broader shoulder which is concave above in definite adults (a doubtful character), having granulose basal ridges in rather large specimens, having a more uniformly banded pattern with few blotches on the body whorl, and lacking the numerous fine spiral dashes commonly found in *kinoshitai* (these may be weakly developed on the basal ridges in *C. bruuni* and completely absent in *C. kinoshitai*). *C. magus* differs in lacking the pinkish violet coloration of the body whorl and aperture. *C. pohlianus* under 35mm may be tan toned with violet, but they are nearly smooth, unlike the heavily striated juveniles of *C. bruuni*.

Variation.—Few adult or near adult specimens of this species are known, but Cernohorsky has described and illustrated the variation from juvenile to adult. Juveniles about 25mm in length or less are covered with heavy nodulose spiral cords, have the spire whorls strongly nodulose, and are yellowish tan with a few brown blotches; the grooves between spiral ridges have fine axial threads. With growth the later spire whorls become less distinctly nodulose, then undulate, the shoulder in adults being straight or nearly so. The body sculpture becomes weaker, the axial threads largely disappearing and only the basal ribs being distinctly granulose; full adults (perhaps actually senile specimens) have no basal spiral ridges with granules. With growth the body whorl becomes relatively broader with more convex sides and the shoulder broadens and becomes concave above. The pattern intensifies with growth, the background becoming more distinctly pinkish violet. Three broad brownish bands, rather blotchy in appearance, develop, with the midbody band pale and demarked above and below by small squarish brown blotches; the blotches largely disappear in full adults, leaving only broad brown bands with somewhat irregular margins. The basal ribs of subadults may have fine dark brown spiral dashes developed.

Distribution.—Known only from moderately deep water (60-85m) in the Kermadec Islands north of New Zealand. Rare.

Notes.—This species apparently seldom or never enters private collections at the moment. However, it is so extremely close to some *C. kinoshitai* in all aspects that there must be strong doubt as to its specific status. I have seen *C. kinoshitai* with patterns very similar to sub-adult *C. bruuni*. It is presumed, perhaps wrongly, that juvenile *C. kinoshitai* are not heavily spirally corded and pustulose, about the only real difference between the two species. *C. kinoshitai* occurs in deep water south to at least the Solomons, so it is not impossible that its range extends somewhat closer to the Kermadecs than now believed. Because it might be impossible to distinguish all adults or sub-adults of *C. bruuni* from the relatively more common *C. kinoshitai*, caution is suggested in purchasing *C. bruuni* when it enters the market. Sub-adults are more attractive than full adults, which may actually be senile specimens.

BULBUS Reeve, 1843

1843. *Conus bulbus* Reeve. *Conchologia Iconica*, 1(*Conus*): Pl. 30, sp. 169 (Cabenda, west coast of Africa).

1845. *Conus zebroides* Kiener. *Species gen. et icon. des coqu. viv.*, 2 (*Conus*): Pl. 105, fig. 5. 1849, *Ibid.*, 2: 257 (Locality unknown).

1957. *Conus angolensis* Paes da Franca. *Trab. Miss. Biol. Marit.*, No. 13: 80, pl. 1, figs. 7-8, pl. 2.

1975. *Conus musivus* Trovao. *Centro Port. Activ. Subaq.*, 4(2): 3 of separate, pl. 1, fig. 2, pl. 2, figs. 3-4,6 (Angola, 12°32'E, 13°26'S). Non *Conus musivum* Sowerby i, in Sowerby ii, 1833 (probably not exact homonym).

1975. *Conus naranjus* Trovao. *Ibid.*, 4(2): 4 of separate, pl. 1, fig. 3, pl. 2, figs. 5,8 (Angola, 12°40'E, 12°22'S).

Description.—Light in weight with a low gloss; pyriform, the sides convex below the shoulder, the body whorl rather elongate; body whorl smooth except for a few widely spaced shallow grooves above the base and weak axial and spiral threads and scratches over the rest of the whorl; shoulder rounded, not distinct from spire, narrow to broad; spire low, usually bluntly pointed if not eroded, the sides straight to convex; whorls with weak but crowded spiral and axial threads. Body whorl creamy

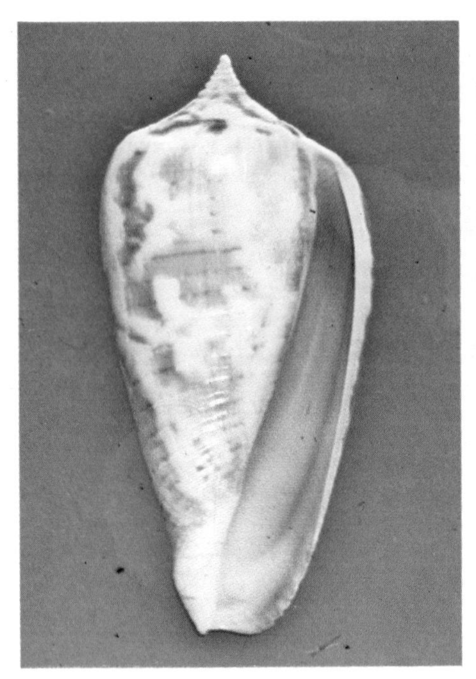
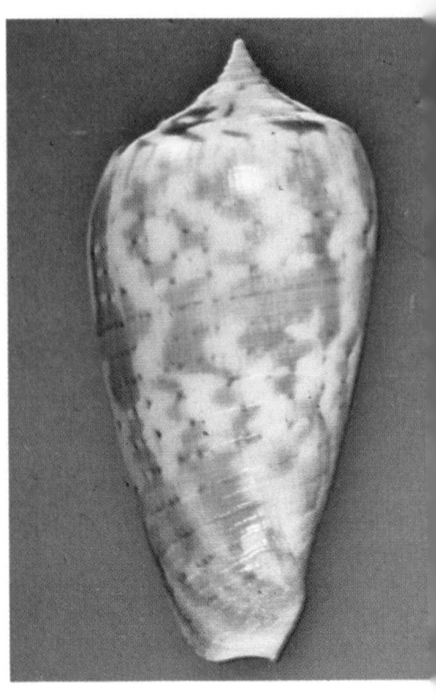

CONSPERSUS.—Above: Palawan, P.I., 37.5mm (GG). *Below:* Sulu Sea, P.I., 36.1-39.8mm (GG).

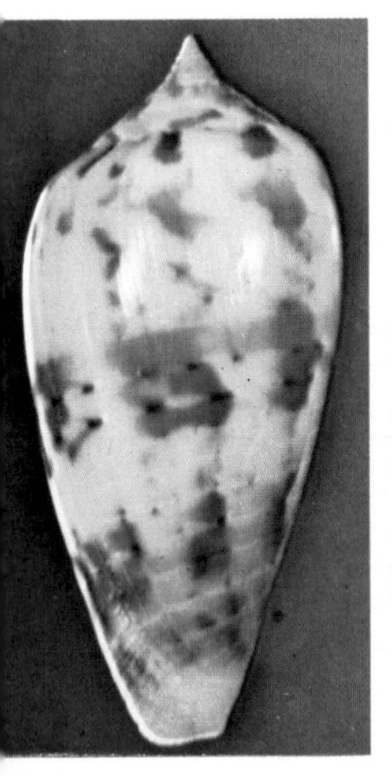

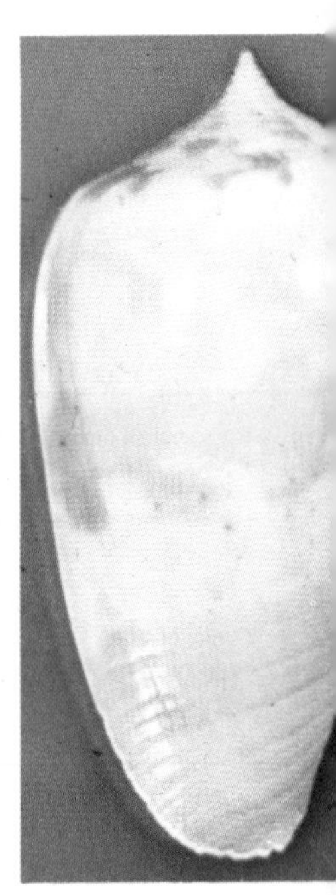

CORONATUS.—Above: Moorea, Society Is., 21.3-24.1mm. ***Below:*** Left: Mackay, Queensland, Australia, 34.6mm; Right: Samar, P.I., 26.0 and 27.3mm.

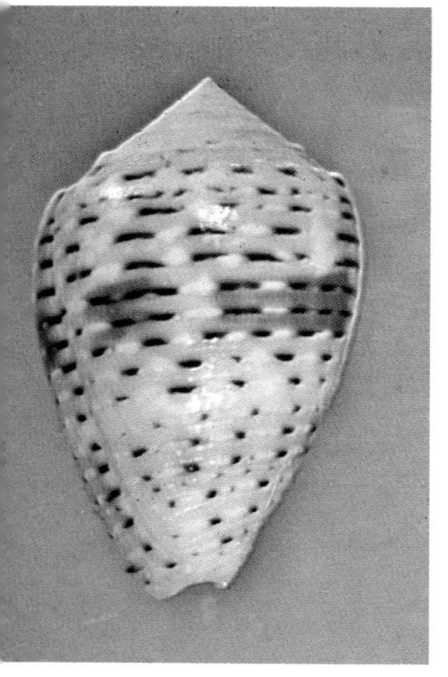

white, sometimes with pale tan or bluish gray tones, covered with narrow brown axial lines; the lines may be strong and nearly straight or very fine and branched, sometimes broken at midbody and above base or below shoulder; occasionally the lines branch and may even produce a pattern vaguely resembling a tented cone; base and shoulder usually with large dark brown blotches or even complete brown bands; spire whitish, with large to small brown lines and blotches. Aperture moderately narrow, slightly widened anteriorly; outer lip thin, fragile, sloping below level of shoulder posteriorly; mouth pinkish white to creamy white, usually with a large violet blotch posteriorly. Columella narrow, short. Length 20-45mm.

Comparison.—Typical patterns of this species are quite distinct for a western African cone and could only be confused with *C. algoensis*, *C. aemulus*, and *C. variegatus*. *C. algoensis scitulus* and *C. a. simplex* have axial lines, but these are usually bright orange brown on glossy white; there are usually several spiral rows of squarish spots on the body whorl, and the shoulder is marked with alternating brown and white areas in a very regular pattern. *C. aemulus* is often larger, has a higher spire, is covered with numerous spiral rows of brownish dashes and squares on a bluish gray background, and has a deep violet-brown aperture. *C. variegatus* as a general rule is broader, thicker, and has the highly variable pattern arranged in spiral rows or bands instead of axial ones; the mouth is usually glossy white. Admittedly some patterns of *C. bulbus* overlap some of *C. variegatus*, and the two may hybridize in some areas; however, almost all *C. bulbus* are rather elongate, narrow shells with axial patterns, while *C. variegatus* are usually shorter, broader shells with spiral patterns.

Variation.—The development of the axial lines is highly variable and perhaps somewhat constant with small inbred populations, several of which have been given specific names. Branching of the lines is common,

Left to right: *C. angolensis* da Franca, *C. musivus* Trovao, and *C. naranjus* Trovao, all = *C. bulbus* (photos Clover).

as is a tendency to break into shorter lines. The dark brown blotching at the base and shoulder is usually well developed. The spire is usually very low but may occasionally be moderately tall. Smaller specimens tend to have broader shoulders than larger specimens.

If desired by collectors, four forms of this species can be recognized, none very distinct and all easily joined with the other forms through intermediates. Typical *C. bulbus* has rather straight thin lines over the body whorl and seldom exceeds 25mm in length. The form *zebroides* is larger, narrower, often with a higher spire, and tends to have fewer and broader lines over the body whorl. In *naranjus* the lines are extremely fine and broken into several spiral rows; the shell is exceptionally broad, and the brown blotching at the shoulder and base is reduced. *Musivus* is a very attractive form with the lines irregularly branched and overlapping to produce a tented appearance over at least part of the shell. Doubtless the relentless describer (of which there appears to be no shortage) will be able to find dozens of other local variants and individuals worthy of nomenclatural immortality.

Distribution.—West Africa south of the Bulge to Angola. Common in intertidal and shallow water. Most specimens currently available come from Angola.

Synonymy.—Reeve's figures of *C. bulbus* are very good and leave no doubt as to what he had. The other names are all obviously close to Reeve's *bulbus* and differ in minor details of pattern and shape which are easy to find in large series from several localities. The problem seems to be that this species and *C. variegatus* form small populations with little dispersal ability; every bay may have its own distinctive color form. These minor forms are certainly not worthy of formal recognition at any level, unless the collector prefers to use them as pattern names.

Notes.—Like most West African cones, the spire is usually eroded, there are commonly small basal growth flaws, and the lips chip easily. Collectors should beware of minor pattern variants offered under new names at very high prices; apparently at least a dozen other such variants of *C. bulbus* and *C. variegatus* have been given manuscript names and will eventually find their way to the market.

BULLATUS Linnaeus, 1758

1758. *Conus bullatus* Linnaeus. *Systema Naturae per Regna Tria Naturae*, ed. 10, 1: 717 (Locality not stated). Neotype selected by Kohn, 1963.

1791. *Conus nubecula* Gmelin. *Systema Naturae per Regna Tria Naturae*, ed. 12, 1: 3396 (Locality not stated). Lectotype figure (Kohn, 1966): Seba, Pl. 42, fig. 14.

1798. *Cucullus parvus* Roeding. *Museum Boltenianum*: 46 (Locality not stated). See Kohn, 1975.

CUMINGII.—Off Mataniko River, Guadalcanal, Solomon Is. *Above:* 30.8 and 33.0mm. *Below:* 21.0-25.4mm.

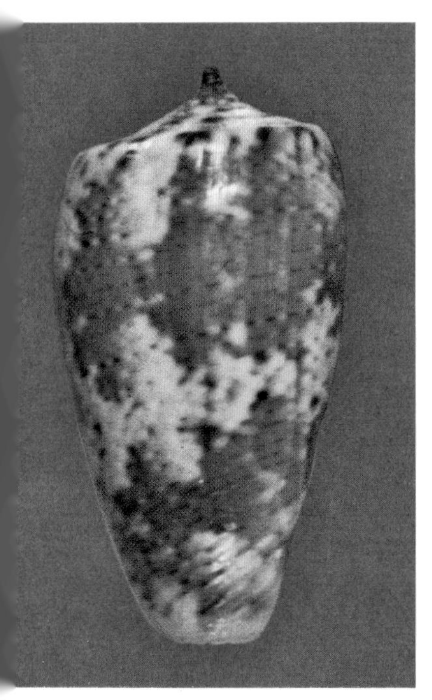
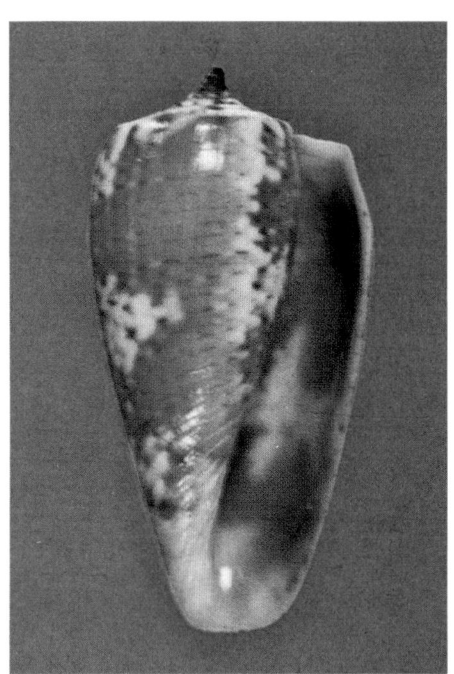

CUVIERI.—Above: Dahlak, Eritrea, Ethiopia, 26.8mm (EP). ***Below:*** Left: French Somaliland, 40mm (DM) (photo by Jayavad); Right: Dahlak, Eritrea, Ethiopia, 40.5mm (EP).

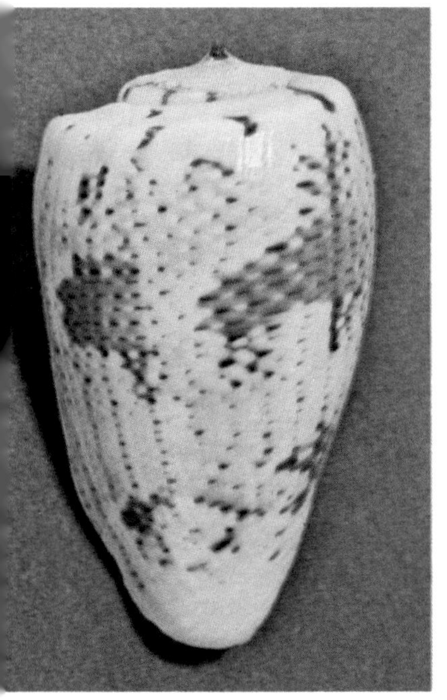

1798. *Cucullus laganum* Roeding. *Ibid.:* 51 (Locality not stated). Lectotype figure (Kohn, 1975): Knorr, Pl. 11, fig. 4.
1937. *Conus bullatus* var. *articulata* Dautzenberg. *Mem. Mus. Roy. d'Hist. Nat. Belg.*, 2(18): 55 (Locality not stated). *Tableau Ency. Method.*, Pl. 339, fig. 6 cited. Non *Conus articulatus* Sowerby iii, 1873.

Description.—Thick, solid, but light in weight, with a very high gloss for a cone; broadly or narrowly ovate, the sides convex at posterior third, occasionally less convex and more cylindrical; body whorl with about 8-12 shallow widely spaced spiral grooves at base, these sometimes punctate; occasionally traces of grooves may be seen to midbody or even shoulder, but body whorl is generally very smooth and featureless; spire low, the sides straight, bluntly pointed, the whorls weakly stepped; early 2-3 whorls weakly nodulose, later whorls undulate or with straight margins; tops of whorls slightly concave to canaliculate, with 1-2 low spiral ridges and fine axial threads. Body whorl generally pinkish brown to pinkish white, irregularly covered with lighter and darker somewhat triangular suffused spots of white and pink to bright orange or red; usually 2-3 broad spiral bands of deep orange to red blotches at shoulder and above and below midbody; often numerous spiral rows of brownish to black dashes and dots, especially strong at margins of blotch bands; often there are short axial flammules of opaque white present over the shell; base pale salmon; spire orangish to pinkish tan, with paler and darker orange blotches; tip whitish. Aperture relatively narrow posteriorly, becoming very wide anteriorly; outer lip thick but often with a sharp edge, posteriorly sometimes raised above angle of shoulder, strongly convex and often sharply angled at posterior third; mouth pale salmon-pink to deep orange, usually paler exteriorly. Columella long, narrow, set off by a broad rounded ridge; columellar area often damaged and foreshortened. Length 40-71mm.

Comparison.—*C. bullatus* is such an unusual cone that it can only be confused with two or three related species which are all much rarer and less well known. Its closest relative is *Conus cervus*, a very large species (over 80mm) which is more attenuated anteriorly and has the spiral rows of dark dashes more obvious and regular; it has been suggested that *C. cervus* is a giant *C. bullatus*, and there is certainly nothing in the morphology or color pattern to dispute this. *C. cervus* is so rare that its fresh colors are unknown, although it has been suggested it is red or orange like *C. bullatus*. What a small *C. cervus* would look like is anyone's guess. *C. julii* is very similar to *C. bullatus* in general appearance and size, but it seems to have a much higher spire in average specimens, the tops of the whorls not as concave as typical of *C. bullatus*. The outer lip is more strongly angled at the posterior third than in typical *C. bullatus*, but this is variable. The pattern of *C. julii* consists of two blotch bands above and below midbody with another indicated at the shoulder; instead of being solid, they are composed of short curved reddish brown axial lines which consolidate into blotches or bands; the same axial lines are also scattered in varying concentrations over the rest of the body whorl; reddish axial lines are apparently absent in *C. bullatus* and certainly never developed to the extent found in even juvenile *C. julii*. *C. adamsonii* is similar in shape and texture but has heavy grooving at the base and shoulder and a very different pattern. *C. floccatus* has a different spire, has a strongly spotted and axially lineate pattern, and runs to violet and yellow with black colors.

Variation.—*C. bullatus* varies in degree of distinctness of the pattern from just smudged orange and white spots to heavily banded with bright orange-red and salmon. There may be numerous dashes of deep reddish brown to black, or they may be entirely absent. Much opaque white is visible in some specimens, but in others it is present only as small spots overlaid by orange or salmon. The spire is always low, but the width of the body whorl varies from rather slender with nearly straight sides to very broad and ventricose. The mouth may be pale pinkish or salmon to deep bloody orange-red. Specimens from the western Indian Ocean tend to have the darker mouth colors, but this is not constant. Specimens with heavy black spotting are not commonly seen.

Distribution.—Widespread in the Indo-Pacific, including Hawaii and French Polynesia, but never common. This is considered a rare and very desirable shell, although its abundance on the market certainly indicates it is far from rare in the proper habitat and in some areas.

Synonymy.—Only the varietal name *articulata* has ever been used much by collectors or scientists; this refers to specimens with strong black dash and dot patterns.

Notes.—As with other reddish cones, the colors fade rapidly after death if exposed to light. Perfect gems of this species are very uncommon, as there is usually at least one minor flaw on the body whorl and the lip is chipped. Dead specimens commonly have one side with bright colors, the other faded. Many otherwise very attractive specimens have the area of new growth near the lip devoid of color except for the brown dashes. Specimens with heavy black dashes are most uncommon; this is a highly desirable shell.

BYSSINUS (Roeding, 1798)

1798. *Cucullus byssinus* Roeding. *Museum Boltenianum:* 41 (Locality not stated). Lectotype figure (Kohn, 1975): Martini, Pl. 60, fig. 669.

Description.—Heavy, thick, with a good gloss; obconic, the sides strongly convex below the shoulder; body whorl with a few weak spiral ridges at the base, these soon obsolete; rest of body whorl smooth except for numerous fine axial and spiral scratches and occasional threads;

CYANOSTOMA.—South of Keppel I., Queensland, Australia (K). *Above:* 19.1mm. *Below:* 20.5mm.

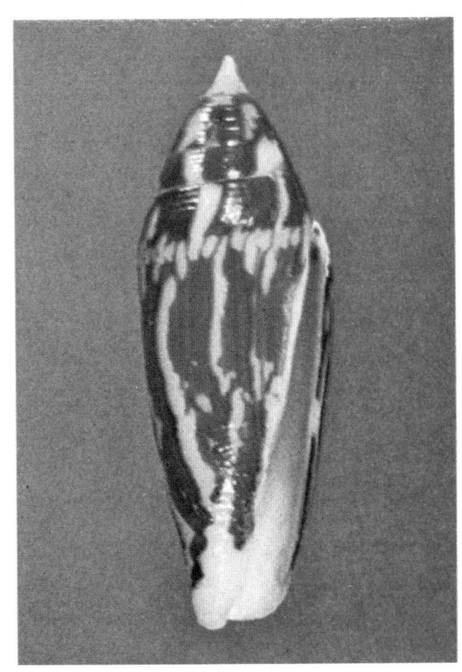

CYLINDRACEUS.—Above: Rabaul Market, Papua New Guinea, 25.3mm. *Below:* Kwajalein Atoll, Marshall Is.: Left: 17.0mm; Right: 27.0mm.

shoulder wide, rounded; spire low, sharply pointed, the sides concave to straight; tops of whorls concave, the margins raised, tops with numerous fine axial threads and no spirals; early whorls not strongly eroded. Body whorl white, the base faintly tinged with salmon; whorl with about 10-15 spiral rows of squarish blackish spots, these variable in size and development, but usually largest in several rows above and below midbody; midbody row usually small; spots below shoulder axially fused into short lines; between the rows of spots above and below midbody are usually groups of 2-3 distinct deep salmon bands, these usually without small dark dashes within; there may be another salmon band just below the shoulder; spire white, covered with narrow and sharply defined blackish axial blotches, these in a regular and usually continuous arrangement; early whorls unmarked. Mouth wide, uniform in width; outer lip moderately thick, sharp, convex, posteriorly about on level with shoulder; mouth glossy white, the exterior pattern showing through at the margin. Columella heavy, rather short, set off above by a heavy, ragged ridge. Length 30-60mm.

Comparison.—This species is extremely close to *C. pulcher* and has often been confused with juveniles of that species. *C. byssinus* is smaller at adult size (usually under 50mm), which can be recognized by the thick, heavy shell; *C. pulcher* of this length are very thin and fragile, with an extremely thin lip. *C. byssinus* of adult size have sharply defined black spots with bright salmon bands generally not carrying small dashes within; in *C. pulcher* of 30-70mm the spots begin to get smudgy and diffuse, with the color banding weakening and often disappearing before 60mm; the color bands, when distinct, usually have distinct small dashes within. *C. byssinus* has a more rounded shoulder and relatively higher, sharper spire than *C. pulcher*, which commonly has a nearly flat spire. The spire markings of *C. byssinus* are distinct blackish blotches, very distinct from the more diffuse brownish or yellowish brown stripes and blotches of *C. pulcher*. It would seem that *C. byssinus* is a distinct species, although obviously close to *C. pulcher*. Such species as *C. eburneus*, which may have a similar shape and pattern, usually have distinct spire sculpture and should prove to be no problem in identification.

Variation.—The few specimens of this species I have seen are quite constant in shape and pattern. There are minor differences in the height of the spire and in the strength and exact arrangement of the spots and color bands on the body whorl, but this is very minor.

Distribution.—Western Africa from Spanish Sahara south to Senegal. It is apparently sympatric with *C. pulcher* over most of this range. Moderately common in shallow water.

Synonymy.—This species has been generally called *C. papillonaceus* Hwass for many years, but that name almost certainly applies to a juvenile or sub-adult of *C. pulcher* and must be dropped. Fortunately the lectotype figure of *C. byssinus* designated by Kohn appears to represent the present species because of its relatively small size (about 60mm, large for the species but small for *C. pulcher*), well defined spots and color bands on the body whorl, and distinct narrow blotches on the spire. Kohn considered *byssinus* to be a synonym of *C. pulcher*, but he apparently did not realize that a small *pulcher*-like species existed in western Africa. Certainly the type figure of *C. byssinus* is the oldest available one which can be placed with the small species with any certainty, so that name is adopted for the *Conus papillonaceus* of authors, non *C. papillonaceus* Hwass.

Notes.—Many of the specimens sold under this name are actually juvenile *C. pulcher;* such shells are extremely thin and fragile and differ from *C. byssinus* as mentioned in the comparison. This is a pretty little species which does not seem as subject to heavy breaks and areas of bad pattern as *C. pulcher.* Adults of this species have a fairly high gloss, contrasting with the dull finish of most small *C. pulcher.* Average adult *C. byssinus* are about 35-50mm long and quite heavy for their small size.

CABRITII Bernardi, 1858

?1810. *Conus exiguus* Lamarck. *Ann. du Mus. Hist. Nat.* (*Paris*), 15: 39 (Mers de l'Asie). Holotype figured by Kiener, Pl. 11, fig. 1.

1858. *Conus Cabritii* Bernardi. *J. Conchyl.* (*Paris*), 7: 377, pl. 13, fig. 2 (New Caledonia).

1872. *Conus Vayssetianus* Crosse. *J. Conchyl.* (*Paris*), 10: 154 (New Caledonia). *Ibid.*, 10: 349, pl. 16, fig. 1.

1880. *Conus taylorianus* E.A. Smith. *Proc. Zool. Soc.* (*London*), 1880: 480, pl. 48, fig. 3 (? Australia).

Description.—Moderately light in weight, with a low to high gloss; low conical, the sides nearly straight or distinctly concave just below level of shoulder; body whorl with numerous closely spaced spiral ridges on the basal fourth to half, these sometimes with large granules; grooves separating ridges deeply punctate; rest of body whorl usually with punctate grooves and traces of weak spiral ridges, occasionally heavily ridged throughout; shoulder angulate, with heavy rather pointed coronations, sometimes with a distinct shallow channel just anterior to the coronations; spire moderate to low, bluntly pointed, the whorls weakly stepped, the sides straight or slightly convex; whorls strongly coronate, with 1-2 distinct spiral ridges and traces of other spiral ridges and fine axials. Body whorl bluish white to white, heavily overlaid with broad deep brown to blackish brown axial bands which are commonly connected by broad spiral bands above and below midbody to result in a largely brown shell with irregular whitish blotches at the shoulder, midbody, and above the base; occasionally the brown bands are poorly developed and more irregular, leaving more of the whitish background visible; basal granules commonly picked out with white; base paler violet brown than rest of

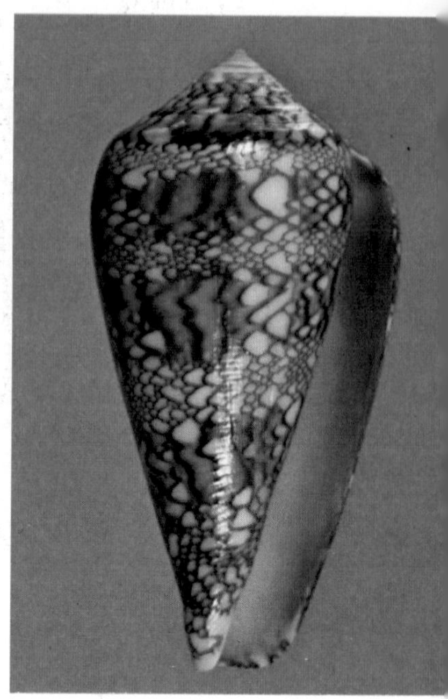

DALLI.—Above: Off Puerto Penasco, Sonora, Mexico, 43.4mm (GG). *Below:* Bahia Santa Maria, Cabo San Lucas, Baja California, Mexico, 47.1-55.2mm (photo by A. Kerstitch).

DAUCUS.—Above: Left: Puerto Maria, Curacao, Neth. Ant., 22.2 and 24.6mm; Right: South Bimini I., Bahamas, 32.1mm (photo by A. Kerstitch). *Below:* Off Ft. Lauderdale, Florida, 33.8mm (DM).

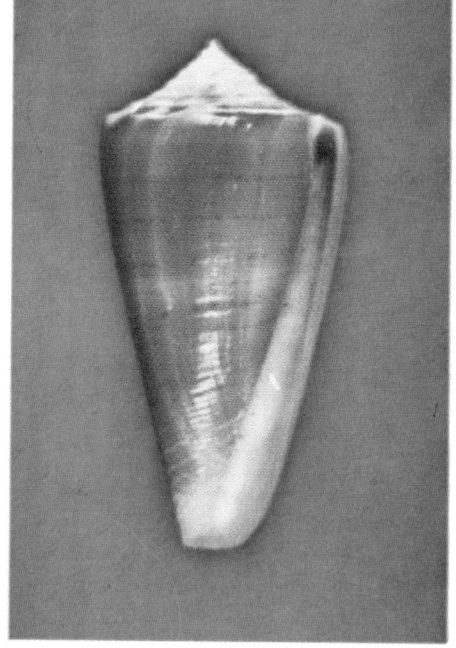

body whorl; spire whitish, with sparse to numerous large brownish blotches; early whorls whitish, usually eroded. Aperture narrow, slightly widened anteriorly; outer lip straight, thin; mouth brownish violet, usually with a whiter blotch at the middle and pale at the margin. Columella very narrow, largely internal. Length 15-30mm.

Comparison.—*C. cabritii* is hard to confuse with any other species because of the heavy coronations, distinctive pattern, dark mouth, and usual presence of granulose spiral ridges separated by punctate grooves over at least part of the body whorl. *Conus varius* may be vaguely similar in shape and is coronated, but it is covered with numerous spiral rows of small dots or dashes and commonly has granulose ridges extending to the shoulder, the grooves not deeply punctate. Juvenile *C. brunneus* have a similar pattern, although the spiral bands are not developed, but they also have dashes in a spiral pattern and lack the strong grooves and ridges typical of *C. cabritii. C. aurantius* is somewhat similar in shape but lacks the punctate grooves, has a pale mouth, and has numerous dashes in spiral rows.

Variation.—This small shell is quite constant in shape, although some specimens have more convex sides with a shallow channel just below the shoulder. The spire is low to moderate and usually weakly stepped. The major variation in sculpture is in how strongly developed the granulose ridges are; typically at least some of the strong ridges of the basal third or half the shell are granulose, but all may be smooth; in other specimens the granulose ridges extend to the shoulder. The punctations at the base are commonly very large and pit-like, especially near new growth, while posteriorly on the body whorl they may be visible only under magnification. Spire sculpture is generally weak. The basic color pattern consists of several broad axial bands crossed by two broad spiral bands, but the axial bands may be irregular enough to produce very large white blotches, and the spiral bands are commonly badly broken; in the basal area the spiral band may be broken into several narrow spiral lines by the ridges being picked out in white. Usually at least part of the body whorl shows the white blotches at shoulder, midbody, and base typical of the species.

Distribution.—Apparently restricted to New Caledonia, where it is uncommon in about 10m.

Synonymy.—I can see no differences between *Conus cabritii* and *C. vayssetianus* as described and figured in the original descriptions. Both are from New Caledonia, are about the same size, have about the same color pattern and color, are the same shape, and have basically the same deeply punctate sculpture. *C. taylorianus* also agrees in all aspects of pattern, size, and sculpture with *C. cabritii.*

The very poorly known species *C. exiguus* is referred here with a query. The holotype was illustrated by Kiener, but according to Mermod the figures leave something to be desired. Mermod described Lamarck's holotype as follows: 18.5 x 9.75mm; white, with large blotches of brown covering most of the surface; there are about 26 spiral ridges; the spire is coronated, brown, pointed, the mouth marked with brown. This description and the figure of Kiener resemble *C. cabritii*, but there is no mention of strong punctate grooves typical of that species, the spire is more sharply pointed than typical, and the coronations on the spire appear more crowded and rounded like those of *C. moreleti*. Certainly it would seem that *C. exiguus* is very similar to *C. cabritii*, and further examination of the type may result in *exiguus* replacing *cabritii*.

Notes.—This shell commonly shows erosion on the spire and posterior part of the body whorl, has a chipped lip, and often has heavy growth marks. Dealers show little interest in different patterns and sculpture, as the species is uncommon and difficult to obtain in truly gem condition regardless of pattern.

CAILLAUDII Kiener, 1845

1845. *Conus Caillaudii* Kiener. *Species gen. et icon. des coqu. viv.*, 2 (*Conus*): Pl. 55, fig. 5. 1849-1850, *Ibid.*, 2: 285 (Locality unknown).

Description.—Moderately light in weight, with a high gloss; elongate cylindrical, narrow, sides somewhat concave at midbody; body whorl with about 10-15 distinct spiral ridges above the base, the fine spiral ridging sometimes continued to shoulder, undulating; shoulder narrow, roundly angled, irregularly undulate; spire low, rather flat, the early whorls forming a small projecting cone when not eroded; early whorls nodulose, later ones broadly undulate, irregular; tops of whorls with faint traces of 2-3 spiral ridges and scattered axial threads. Body whorl deep yellow-orange, covered with about 30-40 very fine somewhat undulating spiral lines of deep reddish brown, these widely spaced and following the spiral ridges (if present); base white, contrasted to rest of body whorl; shoulder and spire heavily marked with large reddish brown blotches and paler brown blotches; entire shell has a very waxy appearance, especially the spire. Aperture narrow, slightly widened anteriorly; outer lip thin, posteriorly about on level with shoulder; mouth white or pale pink, glossy. Columella internal. Length 45-50mm.

Comparison.—This beautifully colored and elegantly shaped cone is difficult to compare with any other. In many respects it is very similar to *C. typhon*, having the same basic pattern of fine spiral brownish lines, a heavily marked spire, and a pale base. However, it is a deep golden yellow or orange shell, not white or pale tan, is much narrower at the shoulder and appears very elongate, and has the later spire whorls undulate and very irregular instead of straight-margined as in *C. typhon*. Kiener's type of *C. caillaudii* was spirally ridged to the shoulder, but the specimen I have seen had only weak basal ridges as in *C. typhon*. The differences in shape, color, and spire will easily distinguish the two species.

DELESSERTII.—Above: Off Egmont Key, Florida, 48.1 and 57.1mm. *Below:* Left: N of Loggerhead Light, Tortugas, Florida, 61.3mm; Right: Off St. Augustine, Florida, 14.1-18.3mm (GG).

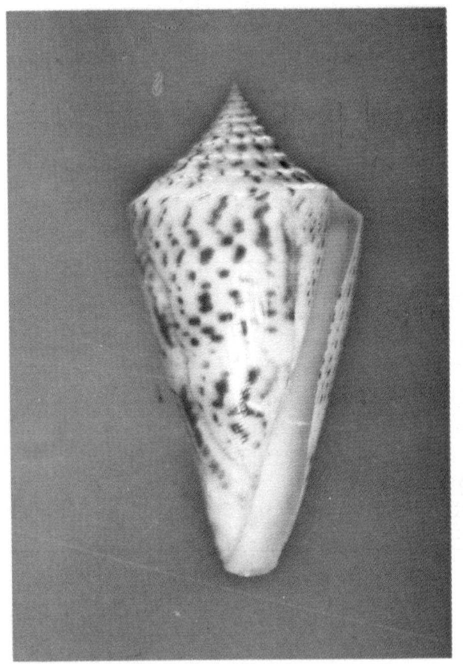

DIADEMA.—*Above:* Left: Guaymas, Sonora, Mexico, 37.7mm; Right: Cabo San Lucas, Baja California, Mexico, 35.3mm. *Below:* Cabo San Lucas, Baja California, Mexico, 29.2mm.

C. caledonicus is apparently similar in color and pattern to *C. caillaudii*, but that species is very broad at the shoulder with a higher spire with large nodules; the base in *C. caledonicus* has granulose ridges.

Variation.—This is apparently a rare species and little has been written about it. The shape, color, and pattern are constant, as is the irregularly undulate spire. With wear the color fades to deep yellow instead of bright orange. The sides of the body whorl may be straight or slightly concave. The major variation apparently occurs in the presence or absence of weak spiral ridging above the base; Kiener's type had fine ridging corresponding with the reddish brown lines to near the shoulder; the specimen examined had only weak basal ridging, but it was slightly worn, although still very waxy in appearance.

Distribution.—A rare species seldom taken. Known from the southern Indian Ocean islands, especially Mauritius. Perhaps more widely distributed; a specimen from Natal, although assigned to *C. typhon*, shows a distinct approach to *C. caillaudii* in general appearance.

Notes.—This rare species is acceptable to collectors in almost any condition. The bright colors apparently fade somewhat after death, but this is still a brightly colored cone when slightly worn.

CALEDONICUS Hwass, in Bruguiere, 1792

1792. *Conus Caledonicus* Hwass, in Bruguiere. *Cone*, in *Ency. Method., Hist. Nat. des Vers*, 1: 634 (New Caledonia). Holotype figured by Kohn, 1966.

Description.—Moderately heavy, with a good gloss; widely biconical, the upper sides convex then somewhat concave at the middle of the sides; body whorl covered with numerous very fine thread-like ridges, otherwise smooth except for occasional axial growth lines and plicae; basal ridges granulose, occasionally the granules extending to midbody; shoulder broad, roundly angled, concave above; spire moderately high, the sides concave, bluntly pointed; early spire whorls strongly nodulose, later ones more undulate, last spire whorl concave above, others rather flat. Body whorl deep yellowish orange, uniform, covered with fine reddish brown spiral lines from base to shoulder, these presumably following the spiral threads; spire toned with deep orange, unblotched. Aperture moderately narrow posteriorly, widened anteriorly; outer lip straight, posteriorly about on level with shoulder; mouth glossy white. Columella apparently short. Length 55-58mm.

Comparison.—The bright orange coloration is shared with few other species, especially when combined with fine spiral reddish brown lines. *C. caillaudii* is similar in pattern but very narrow and elongate, with a low spire which is darker than the body whorl. Such species as *C. planorbis* and allies may have vaguely similar color patterns, but they lack nodulose spire whorls and always differ in many aspects of pattern and shape. *C. cedonulli* to my knowledge always has a very different pattern and is more narrowed at the shoulder with a very different aspect. It is hard to imagine the existence of a bright orange *C. zonatus*, which in addition is narrower and has a different spire shape. *C. polygrammus* is narrower when adult, tan instead of orange, has dotted spiral lines, has the mouth commonly violet tinged, and has a variegated spire; the nodules of the spire whorls are also usually less conspicuous than those of *C. caledonicus.* It would appear that *C. caledonicus* is a valid but very rare and poorly known species; another possibility is that it is an extreme variant of another species, perhaps *Conus regius.* Some yellow-orange *C. regius* have distinct spiral brown lines and are shaped and sculptured very much like *C. caledonicus;* I have not seen a *C. regius* yet that exactly matched the figures of *C. caledonicus*, however.

Variation.—Based on the two specimens figured by Kiener, Reeve, and Kohn, there seems to be considerable differences in the degree of concavity of the sides of the spire, from nearly straight to deeply concave. The point may be rather sharp or definitely blunt. The sides of the body whorl may vary from nearly straight to distinctly concave compared to the convex sub-shoulder region. Apparently the granulose spiral ridges may vary from just a few at the base to numerous ridges extending to mid-body. Kiener mentions that weak axial plicae may be present on the body whorl. The strongly nodulose spire whorls appear to be constant, as do the distinctive color and pattern.

Distribution.—The two specimens known to the early authors were supposed to have been collected in New Caledonia, but this has often been doubted. I am not aware of recent collections of this species, but because of the confusion of names (see below) with *C. marmoreus* fresh specimens could have been lost in a nomenclatural tangle.

Synonymy.—For some reason the name *caledonicus* has come to be generally associated with the white variant of *C. marmoreus* from New Caledonia. Admittedly this variant sometimes has two broad bands of pale salmon on white, but it bears very little resemblance to Hwass' *caledonicus* other than supposed locality. The proper name for the white *marmoreus* variety is *C. marmoreus* var. *suffusus.*

Notes.—Presuming this species is distinct and not some type of fantastic freak, it is certainly rare. Apparently the same two specimens have been the only ones in collections since the mid-1700's, and I have not heard of any other specimens being collected. The locality New Caledonia is subject to doubt as are most other early locality data. When this species is rediscovered, if it has not already been, it will rank with the choice species of the genus.

DISTANS.—*Above:* Left: Samar, P.I., 61.7mm; Right: Marau, Guadalcanal, Solomon Is., 62.6mm. *Below* (types of *chinoi*): Ogokuda, Shiono-misaki, Wakayama Pref., Japan, 22.6 and 32.0mm (photo by Dr. T. Shikama).

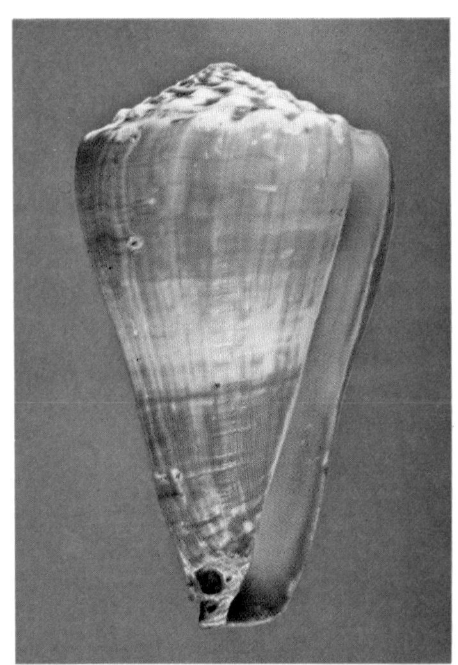

DISTANS.—Above: New Britain, Papua New Guinea, 73.0mm. ***Below:*** Samar, P.I.: Left: 85.0mm; Right: 83.1mm.

CALIFORNICUS Reeve, 1844

1844. *Conus Californicus* Reeve. *Conchologia Iconica*, 1(*Conus*): Pl. 42, sp. 224 (California).

1853. *Conus ravus* A.A. Gould. *Boston J. Nat. Hist.*, 6: 386, pl. 14, fig. 21 (Santa Barbara, California).

?1854. *Conus dealbatus* A. Adams. *Proc. Zool. Soc.* (*London*), 1853: 117 (Locality unknown).

Description.—Moderately light in weight, with a dull finish; broadly to narrowly ovate, the upper sides distinctly convex; body whorl with numerous closely spaced undulate spiral ridges on the basal third, these followed by similar but weaker ridges which extend at least to midbody and commonly to just below the shoulder; shoulder area usually smooth; numerous fine axial ridges, coarse growth marks, and healed scars present; shoulder broadly rounded, not distinct from spire; spire moderate, the sides convex, bluntly pointed, the early whorls usually badly eroded; whorls weakly stepped, the sutures irregular posteriorly, sometimes set off from rest of whorl by a distinct channel; tops of whorls convex, with traces of fine spiral and axial threads, usually indistinct and very irregular. Body whorl dirty bluish white with strong tan overtones; basal third of shell commonly darker tan than rest of whorl, while the shoulder area is often very pale tan with the bluish white showing through; the spiral ridges are commonly marked with fine brown lines, occasionally crossed by weak axial lines to produce a faintly reticulated pattern; posterior margins of shoulder and spire whorls often with a distinct dark brown line; spire dirty bluish white, weakly to heavily blotched with tan. Aperture wide, especially anteriorly; outer lip very thin and fragile, straight or slightly convex, posteriorly strongly sloping below level of shoulder; mouth dirty tan, paler exteriorly, usually with a large pale violet-tan blotch deep within. Columella narrow, fairly long, set off posteriorly by a low rounded ridge. Length 25-49mm.

Comparison.—I know of no other species which could be easily confused with *C. californicus.* The ovate shape is shared with such species as *C. tenuistriatus* and *C. scabriusculus*, but these are thick, solid species with stronger and often granulose spiral ridges and violet mouths; the spire sculpture is usually heavier, and there is little resemblance in color or pattern. *Conus mercator* and other West African species may be very similar in shape and even color, but they lack the distinct spiral ridging of the body whorl. *C. ermineus* and allies also lack the spiral ridging.

Variation.—The body whorl varies in width from relatively narrow to relatively wide, with the spire low to moderately high; much of this variation is apparently associated with localized populations. The color and pattern vary little except for intensity of the tan on the body whorl. Immature specimens sometimes have a very faintly reticulated pattern of pale brown axial and spiral lines, but such specimens are apparently not common.

Distribution.—Restricted to shallow to moderately deep water from the Farallon Islands off San Francisco, California south along Baja California to Cape San Lucas and into the Gulf of California. It is common in California but becomes less so in southern Baja, while it is rare in the La Paz area.

Synonymy.—*C. ravus* is almost certainly this species, but *C. dealbatus* is at best doubtfully referred here from tradition. The type of that name is a bleached white shell with heavy basal ridges and a concave-sided spire which is quite pointed; probably it would best be considered unidentifiable.

Notes.—This is a very rough cone with many breaks and eroded spire whorls. The color pattern is nothing to admire at its best. Specimens over 40mm are not commonly seen.

CANCELLATUS Hwass, in Bruguiere, 1792

1792. *Conus cancellatus* Hwass, in Bruguiere. *Cone,* in *Ency. Method., Hist. Nat. des Vers,* 1: 712 (Hawaii). Holotype figured by Kohn, 1966.
1845. *Conus pagodus* Kiener. *Species gen. et icon. des coqu. viv.,* 2(*Conus*): Pl. 70, fig. 4. 1849-1850, *Ibid.,* 2: 310 (Tahiti).

Description.—Light in weight, glossy; biconical, the sides almost parallel posteriorly then strongly pinched-in toward the base; anterior third of body whorl elongated; body whorl with broad rounded to sometimes flattened strong ribs separated by narrower, punctate grooves, the ribs overlaid with obvious axial growth lines; shoulder carinate, broad, slightly concave; spire very high and sharply pointed, the sides strongly concave; early whorls strongly nodulose, fifth and later whorls becoming crenulate then carinate; whorls slightly stepped, their tops with about four distinct spiral ridges crossed by heavy, spaced curved axial ridges to produce a cancellate appearance. Body whorl white to pale cream, usually with three spiral zones of pale brown variably developed: a row of axial oblique flammules on upper third of shell below shoulder, a narrow band above midbody, usually of partially fused blotches, and occasionally a weaker third band below midbody; spire and shoulder heavily spotted and blotched with light brown; early whorls light brown. Aperture fairly wide; outer lip thin, sharp, nearly straight posteriorly then pinched-in toward base; mouth white. Columella not visible. Length 30-50mm.

Comparison.—The distinctive body shape will instantly distinguish *C. cancellatus* from such superficially similar species as *C. otohimeae*, *C. eugrammatus*, and *C. memiae*, from all of which it differs in many details of sculpture and color pattern. Closest to it in appearance is the eastern Pacific *C. arcuatus*, which is similar in shape and often pattern but has greatly reduced spire sculpture, only the first 3-4 whorls being weakly nodulose and the spiral ridges of the whorls absent or indistinct.

DORREENSIS.—Above: Bumbary, Western Australia, Australia, 34.2 and 35.0mm. *Below:* Exmouth, Western Australia, Australia, both 24.7mm.

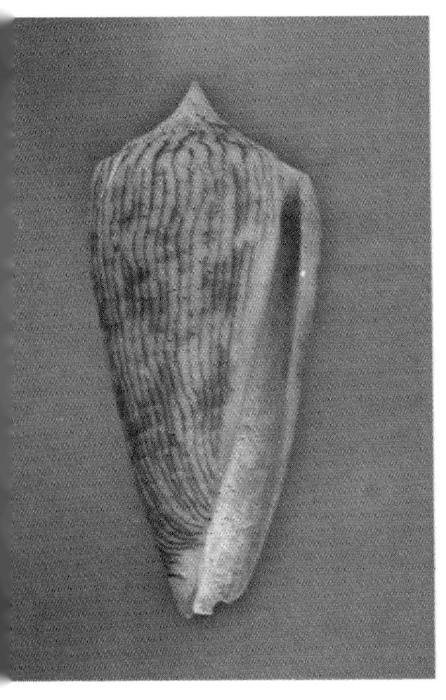

DUPLICATUS.—Above: Left: NE of Taiwan, 61.4mm; Right: 5 mi. NW of Ogasawara Is., Japan, 50.9mm (EP). *Below:* Left (holotype of *armadillo*): Taiwan, 72mm (photo by Dr. T. Shikama); Right: Russell I., Solomons, 49.1mm (MG and BB).

C. armiger is extremely similar in some variants, especially *C. armiger bajanensis*. However, *C. armiger* usually is much more elongated anteriorly, has the posterior sides more convex, generally has all the spire whorls and the shoulder at least crenulate and often strongly nodulose, has the posterior ribs of the body whorl granulose, and lacks the color pattern of *C. cancellatus*. *C. atractus* has the spire sculpture reduced and very weak, as is the body sculpture, and lacks the color pattern.

Variation.—The degree of cancellation of the body whorl varies from individual to individual since it is due largely to the development of growth lines; spire height varies little, however. Color pattern variants are common, usually exhibiting loss of the anteriormost band and reduction of the subshoulder band; occasional individuals are entirely white, while others may have all three brown zones strongly developed and partially fused into a single broad band. Occasional specimens have a more inflated body whorl posteriorly than typical, accentuating the pinched-in base.

Distribution.—Common in moderately deep water from southern Japan and Taiwan south at least to the Philippines.

Synonymy.—Except for *C. pagodus*, based on an exceptionally inflated variant, this shell has caused no great taxonomic problems.

Notes.—A common shell and typical of many heavily sculptured deeper water cones. Heavily marked specimens are especially attractive but do not command a premium. *C. cancellatus* is subject to the usual minor breaks on the body whorl, and the thin lip chips easily; tar is often found on the spire. Juveniles of *C. cancellatus* have been offered as *C. armiger bajanensis*, and I am not certain that it is possible to distinguish in all cases juveniles of these two taxa.

CANONICUS Hwass, in Bruguiere, 1792

1792. *Conus canonicus* Hwass, in Bruguiere. *Cone*, in *Ency. Method., Hist. Nat. des Vers*, 1: 749 (East Indies). Neotype selected and figured by Kohn, 1966.

1810. *Conus pyramidalis* Lamarck. *Ann. du Mus. Hist. Nat. (Paris)*, 15: 438 (Tropic Zone). Lectotype figure (here selected): *Tableau Ency. Method.*, Pl. 347, fig. 5.

1857-1858. *Conus scriptus* Sowerby ii. *Thesaurus Conchyliorum*, 3 (*Conus*): 41, pl. 23(209), fig. 563 (Locality unknown).

1857-1858. *Conus tigrinus* Sowerby ii. *Ibid.*, 3: 41, pl. 23(209), fig. 569 (Madagascar).

1864. *Conus rubescens* Bonnet. *Rev. Mag. Zool.*, (Ser. 2), 16: 282, pl. 22, fig. 6 (Anam I.). Non *Conus rubescens* Schroeter, 1803.

1866. *Conus condensus* Sowerby ii. *Thesaurus Conchyliorum*, 3(*Conus* Suppl.): 326, pl. 26(287), fig. 622 (Sandwich Islands).

Description.—Moderately heavy, with a high gloss; elongate cylindrical, the sides slightly convex; body whorl with a few weak ridges near the base, otherwise nearly smooth except for weak spiral threads and occasional fine ridges; shoulder rounded, narrow, not very distinct from spire, slightly concave above; spire fairly high, sharply pointed, the sides nearly straight; early 2-3 whorls with weak nodules, the interstices usually without brown spots; early whorls with distinct axial sculpture, other whorls and shoulder with a few weak spiral ridges and axial lines indicated, none very obvious; tip of spire pink. Body whorl white to pale pink, covered with large and small tents of reddish brown lines; two spiral bands of dark brown, sometimes nearly black, blotches commonly present, one above and one below midbody; these blotches are dark enough to obscure the axial dark 'textile lines' they contain; spire and shoulder with spaced axial brown lines and a few white tents, no brown blotches normally developed. Aperture moderately wide anteriorly, narrower posteriorly; outer lip nearly straight, often thickened but sharp; mouth whitish to very pale blue, but with a large pinkish blotch deep inside. Columella long, narrow, indistinctly set off posteriorly. Length 35-60mm.

Comparison.—*C. canonicus* is extremely close to *C. abbas* in all respects, and it is possible that these two taxa represent only one real species. As a general rule, *C. abbas* has a white or bluish mouth lacking a distinct pinkish blotch; the tenting is much finer and more regular than that of *C. canonicus*, which commonly is very irregular and often discontinuous. In *C. canonicus* the 'textile lines' are obscured within relatively small dark blotches rather than distinct as in *C. abbas*. *C. abbas* also tends to be somewhat broader and have a more concave-sided spire than typical *C. canonicus*. These distinctions break down locally in the Marquesas, where typical *C. canonicus* intergrade imperceptibly into typical-appearing *C. abbas*; a similar situation probably occurs locally in the Philippines and in some areas of the northern Indian Ocean (in the Indian Ocean the more common *abbas* grades into the rarer *canonicus*).

C. textile is of course also similar to *C. canonicus*, but it is almost always more ventricose and has distinct 'textile lines' in the pattern; many rather common *textile* patterns are apparently absent in *C. canonicus*.

C. dalli also falls into the group of species related to *C. abbas* and *C. canonicus*, and again it is doubtful if it can be separated with certainty from the other two. As a general rule, the mouth in *C. dalli* is bright, deep violet, much darker than in the other species; however, this violet can fade to bluish white or (rarely) even pinkish. In pattern *C. dalli* is closer to that of *C. abbas* than *C. canonicus*, and the spire is generally lower and more convex-sided than in either species. *C. legatus* is very similar in several respects to juvenile *C. canonicus* and doubtless derived from it; the deep pink of *legatus* and reduced tenting distinguish it from all *C. canonicus* I have seen, although admittedly some *C. canonicus* approach it in general appearance.

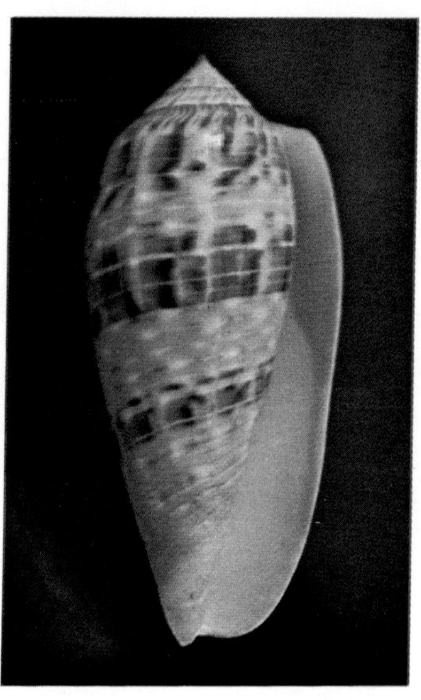

DUSAVELI.—Above: Okinawa, 68mm (Zengo Nagohoma) (photos by T.C. Good). *Below:* P.I., 79mm (photo by P. Clover).

EBRAEUS.—Above: Boac, Marinduque, P.I., 25.2 and 30.6mm. ***Below:*** Nonga, E New Britain, Papua New Guinea, 22.9 and 26.4mm.

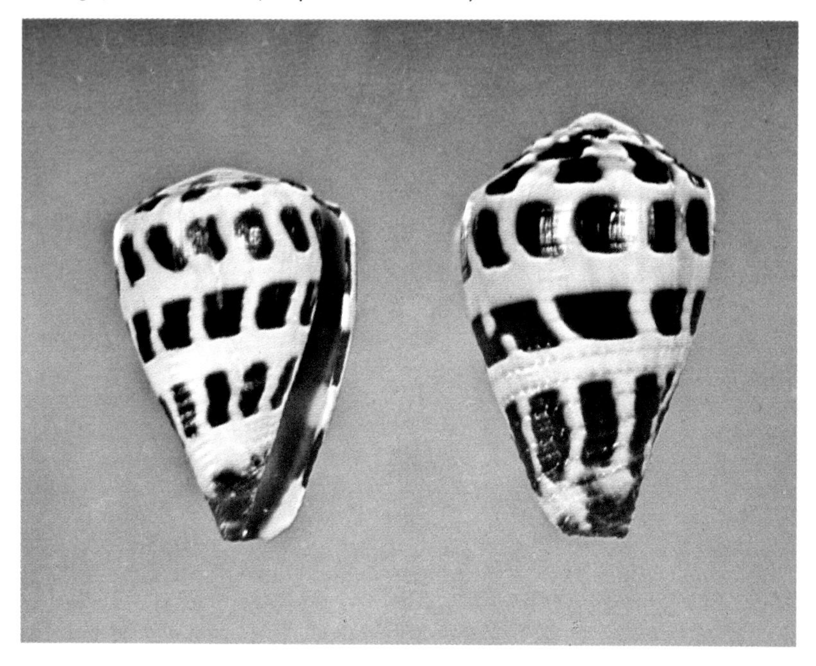

C. canonicus var. *pyramidalis* (DM) (photo Jayavad).

Such tented species as *C. aureus* and *C. magnificus* or *C. pennaceus* may vaguely resemble *C. canonicus; C. aureus* has the 'textile lines' very prominent and the spiral ridges on the body whorl are distinct and feelable; in *C. magnificus* and *C. pennaceus* there are no nodules on the early spire whorls, no traces of 'textile lines' can be found within the dark blotches, and there are fine spiral dashed lines of brown and white over the body whorl.

Variation.—In addition to the usual variation in height of spire, convexity of body whorl, and density of tenting common in all tented cones, the tendency of this species to vary toward *C. abbas*-like patterns has been mentioned above. Occasional specimens have heavy axial bands of deep brown like those found in *C. legatus.* The pink blotch in the mouth is usually obvious and occasional specimens indeed have the entire mouth bright pink; some specimens in large series have bluish white mouths without any obvious pink, however. Although generally very smooth-bodied, a few specimens with distinct spiral ridges to about midbody have been seen. The spiral zones of dark blotches may be greatly reduced or absent in many specimens, but sometimes they are broad and nearly continuous bands of brownish with a few 'textile lines' showing through.

Distribution.—Found throughout the Indo-Pacific from the eastern African coast and northern Indian Ocean (rare or uncommon) through the western and central Pacific (common) to the Society Islands and Marquesas; apparently absent from Hawaii or overlooked there.

Synonymy.—The listed synonyms seem to apply here, although admittedly some may properly belong with *C. abbas* or *C. textile*—or even *C. pennaceus.* In many cases the quality of the original figure and description is poor, and even an examination of existing type specimens may not answer all the questions, as many names are based on obviously variant specimens with uncommon patterns or shapes. The inclusion of *C. pyramidalis* here seems fairly certain, especially since abnormally elongated, high-spired *C. canonicus* very much like the type figure have been found in the Bay of Bengal and off southern Japan; these apparently represent abnormal individuals or at best local populations and are not worthy of nomenclatural recognition in this complex group of species.

Conus condensus may belong in the synonymy of *C. victoriae nodulosus*, but this is almost impossible to determine from the description.

Notes.—Typical specimens of *C. canonicus* are common in the Philippines, New Guinea, Solomons, and adjacent areas, as well as east to the Society Islands. Specimens of *C. abbas* from the Philippines and Marquesas (those from the Marquesas are commonly sold as *C. panniculus*) are uncommon to rare, but their status is uncertain. This species seldom has obvious flaws, although in some areas the pattern can be extremely irregular and reduced. The high-spired '*pyramidalis*' form is rare, as is to be expected of what is probably an individual variant.

CAPITANELLUS Fulton, 1938

1938. *Conus capitanellus* Fulton. *Proc. Malac. Soc. London*, 23(1): 55, pl. 3, figs. 1-1a (Kii, Japan).

Description.—Moderately light in weight, with a good gloss; somewhat obconical, the upper sides nearly parallel or convex, wider than shoulder in some specimens, then strongly tapering to a rather narrow base; body whorl smooth except for about 6-8 low and weak ridges at the base and scattered axial and spiral threads and lines; shoulder broad, roundly angled; spire low, pointed, the sides straight or nearly so; tops of whorls flat, with about 3-4 spiral ridges plus indistinct finer spiral and axial threads; first 3-4 whorls with small nodules, the later whorls progressively more weakly undulate or irregular, the shoulder straight. Body whorl white or cream, with two broad bands of reddish brown blotches which are axially elongated and often fused to produce continuous spiral bands; regularly spaced axial flammules commonly extend from the margins of both bands, invading the white midbody area and sometimes fusing into continuous axial lines; spire whitish, with numerous radiating brownish blotches continuous with the axial flammules of the body; early whorls pale brown. Aperture moderately wide, of uniform width; outer lip thin, slightly concave near the middle; mouth whitish. Columella short, narrow, sometimes weakly set off posteriorly. Length 25-41mm.

Comparison.—This apparently rare or at least uncommon species is very distinct in shape and pattern and difficult to relate to other species. In some ways it is like *C. kinoshitai*, but of course the differences in shape and color pattern should prevent any confusion between these two species. There is a strong and probably superficial similarity between *C. capitanellus* and such Caribbean species as *C. centurio* and *C. clerii*, but these species have weaker spire sculpture and numerous obvious differences in pattern and shape. *C. capitaneus* is only vaguely like *C. capitanellus*, having much heavier spiral ridges on the body whorl in specimens of similar size, these usually deeply punctate; of course the color and pattern differ greatly.

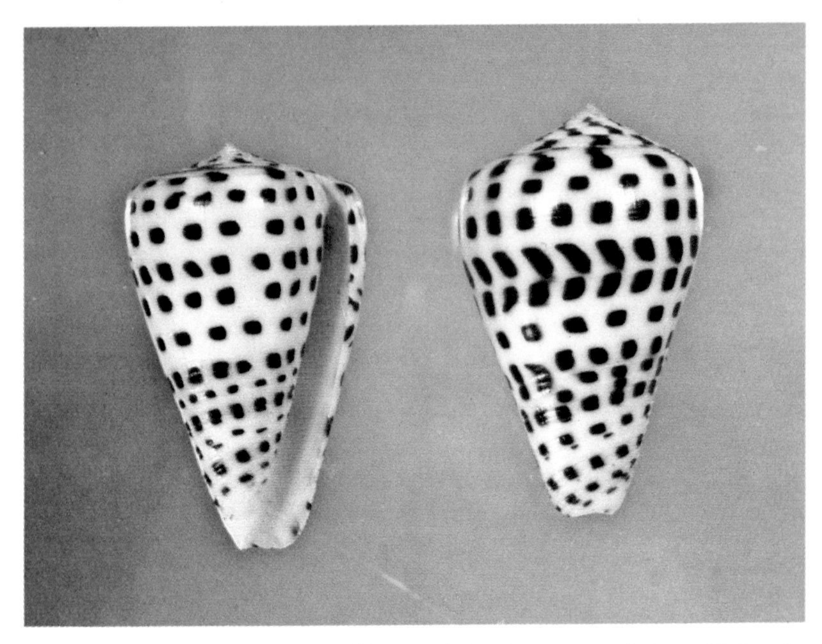

EBURNEUS.—Above: Nonga, E New Britain, Papua New Guinea, 33.6 and 34.1mm. *Below:* Left: (top) Heron I., Queensland, Australia, 21.9mm; (bottom) Manus I., Papua New Guinea, 25.4mm; Right: Bolo Point, Okinawa, Japan, 43.1mm.

EBURNEUS.—Above (*polyglottus*): Tabaco Bay, P.I., 29.8 and 31.7mm. Below: Left (*crassus*): Nananu-i-ra I., Fiji Is., 40.1mm (EP); Right (freak): P.I., 38.9mm (W.E. Burgess).

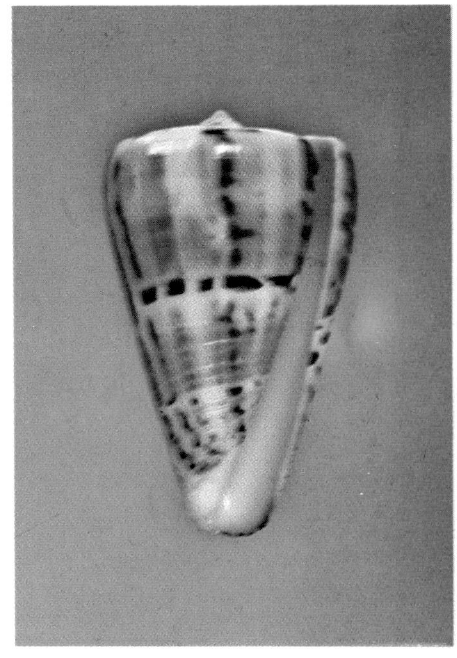

C. *capitanellus*, Japan, 27.3mm (DMNH).

Variation.—Apparently the posterior part of the body whorl becomes a bit more convex with growth, small specimens having nearly straight sides. The reddish brown blotches may be heavily fused into broad bands or partially separate, with the axial flammules short or long; the appearance of large white axial rectangles below the shoulder is apparently constant. Rare specimens are covered over the body whorl with spiral rows of weak granules.

Distribution.—Taken by trawling in deep water from southern Japan to off Taiwan; poorly known.

Notes.—This species is seldom offered at this time. Large specimens with bright reddish markings are quite attractive. Axial growth flaws are common, as are minor lip chips.

CAPITANEUS Linnaeus, 1758

1758. *Conus capitaneus* Linnaeus. *Systema Naturae per Regna Tria Naturae*, ed. 10, 1: 713 (Asia). Lectotype selected and figured by Kohn, 1963.

?1792. *Conus classiarius* Hwass, in Bruguiere. *Cone*, in *Ency. Method., Hist. Nat. des Vers*, 1: 705 (Indian Ocean). See Kohn, 1968, for a discussion of the status of this name.

1801. *Conus ferrugineus* Bosc. *Hist. Nat. Coqu.*, 5: 128 (Guinea). Based on Martini, Pl. 59, figs. 600, 602. Non *Conus ferrugineus* Hwass, 1792.

1858. *Conus Ceciliae* Crosse. *J. Conchyl. (Paris)*, 7: 381, pl. 14, fig. 5 (Locality unknown).

Description.—Light in weight, with a low gloss; obconic, very broad across the shoulder; body whorl with large rounded ribs near the base, these crossed by smaller and closely spaced axial ridges; area between base and midbody, and sometimes up to shoulder in small specimens, with spiral rows of closely spaced deep punctations, these becoming partially obsolete with growth but occasionally still obvious in adults; shoulder roundly angled, wide, merging into spire; spire low to flat, the early whorls usually eroded; spire and shoulder with 3-5 (usually 4) evenly spaced spiral rows of large to small punctations separated by flat ridges. Body whorl various shades of light tan, occasionally yellowish or olive, sometimes dark brown; base violet brown, with numerous crowded small brown spots within the cancellate area; a white midbody band bordered above and below by irregular, generally rectangular, black spots which are commonly connected across midbody area by fine dark lines; shoulder with a broad white band containing regularly spaced black blotches sometimes irregular and bifurcate in shape and continuing onto spire; upper half of shell developing numerous rows of small brown spots with growth, these often fusing to form narrow brown axial lines; similar rows of brown spots also sometimes developed below midbody but there seldom fusing into axial lines; spire whitish with broad reddish brown blotches, continuations of shoulder markings, alternating with similar white blotches; tip of spire yellow, the yellow sometimes diffuse and present over most of spire. Aperture wide, straight; outer lip thin, sharp, slightly convex; interior of mouth deep violet with broad white bands under shoulder and at midbody, sometimes entire mouth bluish white. Columella moderately long, narrow, twisted, set off posteriorly by a weak hump. Length 40-70mm.

Comparison.—From *C. vexillum*, which sometimes has a similar pattern, *C. capitaneus* differs by the distinct white midbody band, whitish shoulder band, and distinct spotting above and below midbody (all poorly developed in most *C. vexillum*); juvenile *C. vexillum* have fewer and less distinct spots above and below midbody and fewer spire blotches; average adults of *C. vexillum* are much larger than adult *C. capitaneus* and have a white, not violet, aperture. *C. trigonus* may have a somewhat similar pattern, but it has more strongly developed spiral brown lines, a white mouth, and a columella with a projecting posterior margin. Some patterns of *C. fumigatus* and *C. excavatus* resemble *C. capitaneus*, but these species lack basal punctations and have simple, not punctate, spire sculpture; they also lack the small brown flecks near the base typical of *C. capitaneus* and allies.

The only species with which typical adult *C. capitaneus* are likely to be confused is *C. mustelinus;* both species are very variable in pattern and color and vary greatly in height of spire and width at shoulder, although usually *C. mustelinus* adults have a more elongate shell which is narrower at the shoulder and have a higher, more convex-sided spire. In *C. mustelinus* the midbody band is bordered by discrete, usually oval, black spots which are almost never connected across the midbody area; there are seldom more than two or three spiral rows of large black spots between the midbody area and the shoulder, and usually only one or two below—these spots show little tendency to fuse axially. Many *C. mustelinus* have few or no brown flecks in the basal area. As a general rule, young *C. mus-*

ELEGANS.—Above: Off Chinde, Mozambique, 27.0mm (NM G5044). *Below:* Karachi, Pakistan, 19.3 and 21.8mm (DMNH).

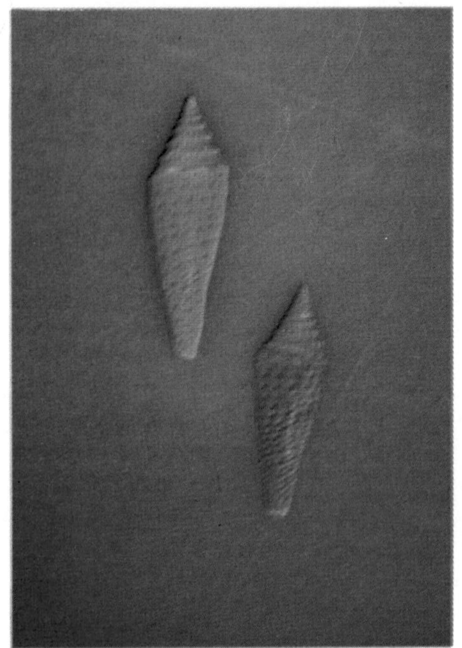

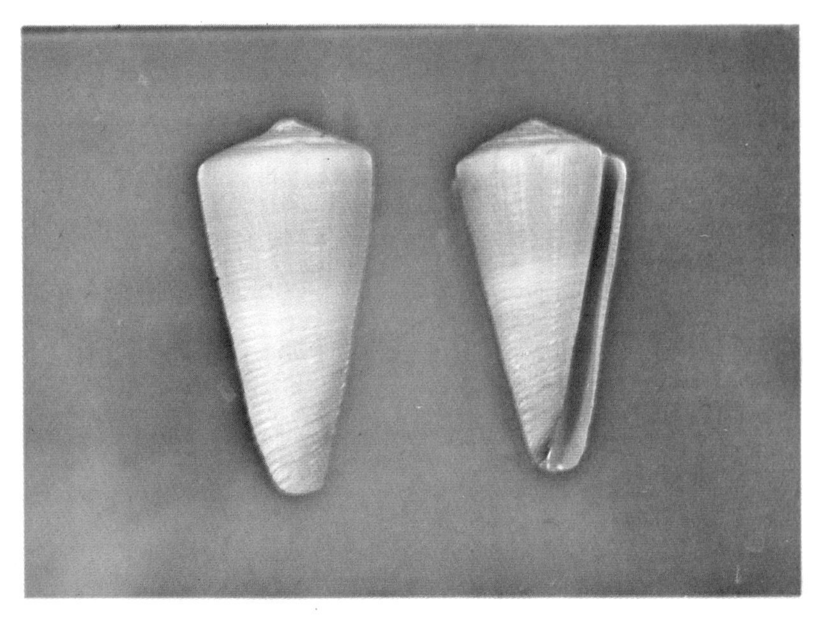

EMACIATUS.—Above: Boac, Marinduque, P.I., 42.1 and 45.0mm. ***Below:*** Left: Honiara, Guadalcanal, Solomon Is., 33.6mm; Right: Praslin, Seychelle Is., 45.6mm.

telinus are distinctly waxy green, changing to mustard yellow with growth, seldom the tans of *C. capitaneus*.

Variation.—As mentioned earlier, there is considerable variation in height of spire, development of dark spots and axial lines, mouth color, and overall color of the body whorl. As a general rule, most *capitaneus* have a dingy appearance due to the small spots and axial lines obscuring the background color. Juveniles have deeply punctate grooves alternating with heavy ridges from the base to the shoulder; with growth, the sculpture becomes obsolete above midbody and often is absent except for the cancellate basal sculpture; however, it is not uncommon for fully adult but small specimens to have some ridges and punctations to the midbody area and occasionally to the shoulder.

Distribution.—Most of the Indo-Pacific, from eastern Africa to Hawaii (rare) and the Fiji area, but apparently not recorded from French Polynesia.

Synonymy.—*C. classiarius* is probably unrecognizable; the name has commonly been applied to the cones here called *C. fumigatus* and *C. excavatus*, but I cannot agree with this. *C. ceciliae* represents relatively large specimens of *C. capitaneus* which have retained the heavily sculptured juvenile characters and are often brightly colored; I have seen a pair of Okinawan specimens collected under the same coral head, one the *ceciliae* form and the other typical *capitaneus*, and other specimens with intermediate characters, so I can see no reason to retain the name *ceciliae* at any level.

Notes.—This common and rather attractive species commonly has chipped lips and some body flaws, as well as encrusting growths on the spire whorls. The early spire whorls are usually eroded. Juveniles are not uncommon and make an interesting contrast to the adults in both sculpture and pattern. Heavily sculptured specimens over 35mm (the *ceciliae* form) are not common and probably worth a premium.

CARACTERISTICUS Fischer, 1807

1807. *Conus caracteristicus* Fischer. *Mus. Demidoff*, 3: 139 (Locality not stated). Based on Chemnitz, Pl. 182, figs. 1760-1761.

1810. *Conus quaestor* Lamarck. *Ann. du Mus. Hist. Nat.* (*Paris*), 15: 281 (American Ocean?). See Kiener, Pl. 42, figs. 1-1b.

1810. *Conus muscosus* Lamarck. *Ibid.*, 15: 281 (Locality unknown).

1817. *Conus characteristicus* Dillwyn. *Descr. Catalogue Recent Shells*, 1: 367 (West Indian Seas on coasts of Island of St. Bartholomew). Holotype figures: Chemnitz, Pl. 182, figs. 1760-1761.

1845. *Conus Paulina* Kiener. *Species gen. et icon. des coqu. viv.*, 2 (*Conus*): Pl. 108, fig. 3. 1849-1850, *Ibid.*, 2(*Conus*): 314 (Locality unknown).

1874. *Conus Masoni* Nevill and Nevill. *J. Asiatic Soc. Beng.*, 43(pt. 2)1:

22 (Andaman Islands).

1877. *Conus brevis* E.A. Smith. *Ann. Mag. Zool.*, (Ser. 4), 19: 222 (Locality not known). Non *Conus brevis* J. de C. Sowerby, 1840, a fossil.

Description.—Heavy, with a good gloss; sides straight to slightly convex, obconical to weakly turbinate; body whorl with about six heavy spiral ridges anteriorly, separated by wide grooves; remainder of body whorl smooth except for a few weak spiral ridges extending to near midbody; shoulder broad, roundly angled; spire nearly flat, the early whorls forming a small bluntly pointed cone; early whorls distinctly nodulose or undulate, usually eroded; spire whorls and shoulder divided on top into three more-or-less distinct regions: a large, high anterior ridge, a narrow ridge at the middle, and a low but fairly wide posterior ridge; each of these ridges may bear smaller and less distinct spiral ridges and a few weak axial threads. Body whorl glossy white or cream, with three irregular spiral bands of deep reddish brown blotches at shoulder and above and below midbody; two anterior bands may be accentuated by suffused tan between the darker blotches and may appear as continuous broad dark spiral bands; spaces between rows of blotches usually with thin, often angular pale brown axial lines forming a very broken, weakly netted pattern; basal ridges with brown dashes; shoulder with alternating deep reddish brown and white blotches, the same pattern continued onto spire; shoulder blotches often connected obliquely to those below shoulder; tip of spire white. Aperture moderately wide, uniform in width; outer lip fairly thin, slightly convex; mouth deep yellow within. Columella short, wide, bounded posteriorly by a heavy twisted ridge. Length 40-67mm.

Comparison.—Although several other cones have similar but not identical color patterns, the depressed spire, nodulose early whorls, smooth shoulder, and fairly distinctive spire sculpture set *C. caracteristicus* apart. *C. araneosus* and some patterns of *C. spurius* and *C. pulcher* are probably most similar superficially, but the former has the later whorls and shoulder distinctly coronate and the latter two lack coronations and nodules entirely, as well as differing in many aspects of color and pattern.

Possibly the closest relative of *C. caracteristicus* is *C. zeylanicus*, which is very similar in shape and weight and general color and pattern; both have similar columellas, although the spire sculpture of *C. zeylanicus* is weaker than *C. caracteristicus; C. zeylanicus* has the whorls all distinctly coronate (if not eroded) and the mouth deep pink, as well as being generally suffused with pinkish over the entire shell.

Variation.—There is considerable variation in height and shape of the spire and distinctiveness of the spire sculpture, as well as convexity of the sides of the body whorl. The dark blotches vary from distinctly reddish to deep chestnut or yellowish tan and from regularly rectangular to vaguely triangular; the blotches may be widely separated or completely

ENCAUSTUS.—Marquesas, French Polynesia. *Above:* 24.1 and 28.7mm. *Below:* 27.5 and 28.0mm (GG).

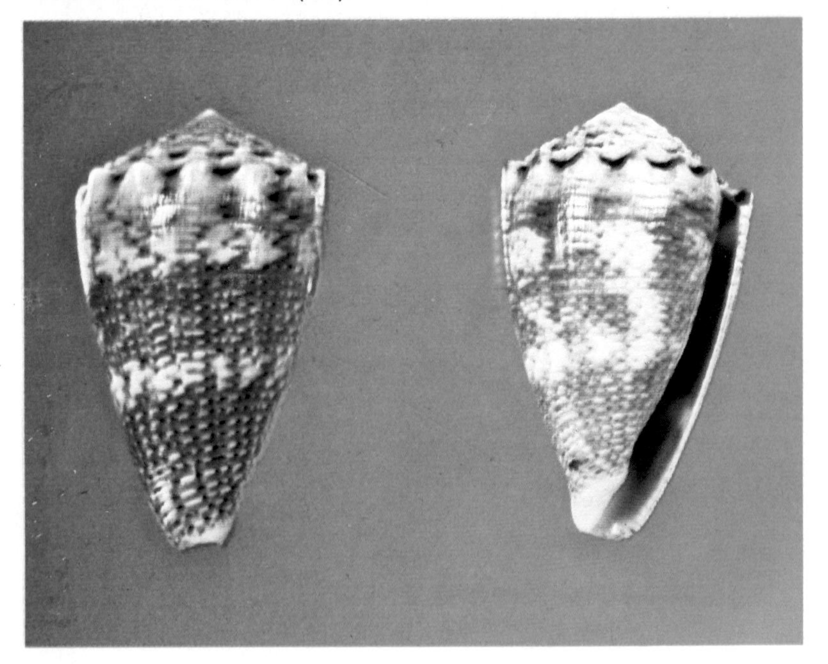

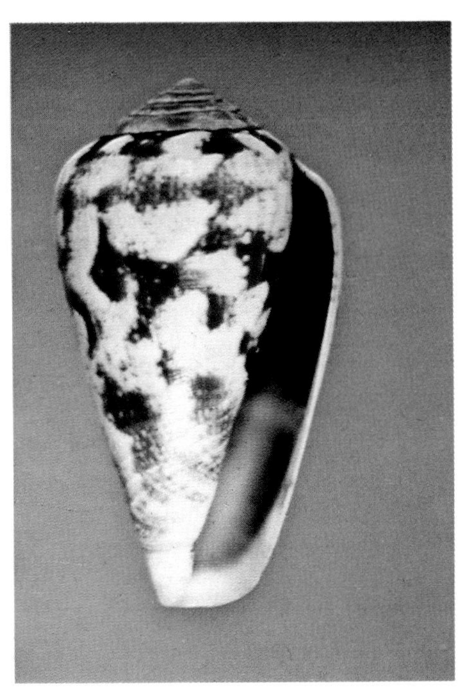

ERMINEUS.—Above: Left: St. Pierre, Martinique, 56.1mm; Right: Villa Cisneros, Spanish Sahara, 32.3mm (GG). *Below:* Left: Off Jacksonville, Florida, 62.2mm (TY); Right: Tucker's Bay, Trinidad, 53.4mm.

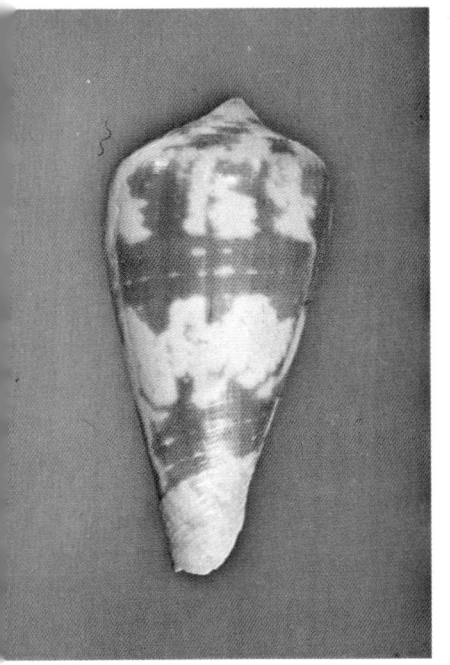

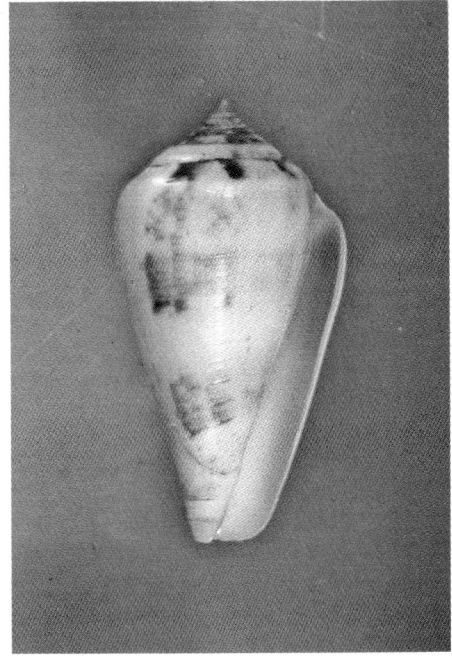

fused into solid spiral bands. The posteriormost band of blotches may be strongly developed or almost absent. Juveniles have a more nebulous pattern and more sharply angled shoulder as well as being lighter in weight.

Distribution.—Moderately common in shallow water over the Indian Ocean and east to at least the southern Japan-Philippines area; presumably it occurs south to the New Guinea-Solomons area, but I have not seen specimens.

Synonymy.—The names of Lamarck and Kiener have long been referred to this species, while *C. brevis* has usually been interpreted as a small specimen. *C. masoni* was not illustrated, but there is nothing in the description or locality to argue with its placement here. *C. characteristicus* should be treated as a new species description, not a misspelling or emendation, because Dillwyn was validating a non-Linnaean name proposed for the species by Chemnitz; the earlier description of *C. caracteristicus* by Fischer is the spelling which must be followed.

Notes.—This rather attractive shell is usually available from the northern Indian Ocean and the Philippines. Fresh specimens are often very brightly colored. The erosion of the spire whorls appears to be rather constant; the shell is not otherwise normally defective.

CARDINALIS Hwass, in Bruguiere, 1792

1792. *Conus cardinalis* Hwass, in Bruguiere. *Cone*, in *Ency. Method., Hist. Nat. des Vers*, 1: 632 (Santo Domingo, restricted by Kohn, 1966). Holotype figured by Kohn, 1966.

1833. *Conus ornatus* Sowerby i, in Sowerby ii. *Conchological Illustrations:* Pt. 29, fig. 22 (Locality not stated). Non *Conus ornatus* (Roeding, 1798). See Tomlin, 1937.

1845. *Conus cidaris* Kiener. *Species gen. et icon. des coqu. viv.*, 2(*Conus*): Pl. 63, fig. 1. 1846, *Ibid.*, 2: 57 (Locality unknown). Figure 1a, cited as a variety, may refer to *C. magellanicus.*

1848. *Conus speciosissimus* Reeve. *Conchologia Iconica*, 1(*Conus* Suppl.): Pl. 2, sp. 274 (Curacao).

1861. *Conus lubeckianus* Bernardi. *J. Conchyl. (Paris)*, 9: 169, pl. 6, figs. 7-8 (Guadeloupe).

1865. *Conus archetypus* Crosse. *Ibid.*, 13: 313, pl. 10, fig. 7 (Locality unknown).

1887. *Conus jucundus* Sowerby iii. *Thesaurus Conchyliorum*, 5(*Conus* Suppl.): 260, pl. 32(510), figs. 696-697 (Locality unknown).

1942. *Conus regius abbotti* Clench. *Johnsonia*, 1(6): 6, pl. 4, figs. 2-3 (Arthurstown, Cat Island, Bahamas).

Description.—Moderate in weight, thick, with a good gloss in fresh specimens; pyriform or low conical, the shoulder wide and tapering strongly to the base; body whorl with broad spiral ridges near the base, these sometimes strongly granulose; widely spaced granulose ridges may also be present more posteriorly, especially just below and above midbody, sometimes a few extending to shoulder; whorl otherwise smooth and polished; spire low, the sides straight or concave, the whorls sometimes weakly stepped; apex blunt; spire whorls and shoulder with large, low and sometimes indistinct coronations; tops of whorls with fine axial ridges and sometimes weak spiral ridges visible, occasionally sculpture absent. Body whorl generally some shade of red, pink, or pale red-brown, occasionally olive or dark brown; narrow or moderately broad white spiral bands may be present at base, midbody, and shoulder, those at midbody and shoulder most commonly present and most distinct; these bands are usually white blotches containing brownish spots and streaks in varying patterns, those of shoulder often very distinct and axial; occasionally body whorl with larger brown spots and axial flammules; spire with alternating blotches of white and reddish brown, all suffused with pinkish. Aperture moderately wide, of uniform width; outer lip slightly convex, fragile; mouth variable with fading after death, commonly pinkish or rosy but sometimes violet. Columella not visible externally. Length 20-45mm.

Comparison.—This poorly understood species is most likely to be confused with *C. regius*, of which it is often considered a synonym. Actually, the patterns and shape of these two species almost never appear similar, the spire in *C. regius* being much higher and with more concave sides and higher coronations. *C. regius* is usually brown and white, varying to yellow, while *C. cardinalis* is usually some shade of reddish; the mouth in *C. regius* is whitish, not pinkish. In addition, *C. regius* has strong spiral sculpture on the spire, usually absent or very weak in *C. cardinalis*. *C. vittatus* is not coronate, is regularly covered with spiral ridges, and is less granulose. See also *C. havanensis*.

Variation.—Again, variation in this species is poorly understood and it is likely that one or more of the synonyms may represent distinct species. The body whorl may be almost completely smooth or with several strongly granulose spiral ridges; the spire sculpture is similarly variable as already mentioned. Generally the shape is constant, with the spire varying from nearly flat to low. The entire body whorl may be reddish or there may be up to three well marked white bands heavily spotted and flammulated with brown; often the midbody band is distinct and contains many small brown spots; the shoulder band has distinct axial brown lines in many specimens, a very distinctive feature of the species when present. Mouth color is variable, but how much of this is due to genetic differences and how much to foods or to fading after death is unknown.

The variant described as *C. jucundus* (*abbotti* appears to be the same) is very distinct in pattern although apparently not separable from *C. cardinalis* structurally. In this variant the brown spots and streaks are strongly developed as axial flammules which extend over a large portion of the body whorl and are not restricted to the white spiral bands as in

ERYTHRAEENSIS.—Above: Massawa, Ethiopia, both 19.3mm. ***Below:*** Aden, 26.4mm.

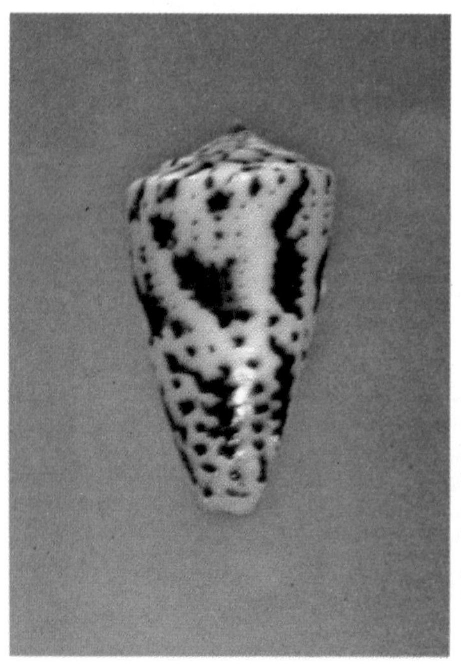

EUCORONATUS.—Above: Off Durban, Natal, South Africa, 42.8mm (Meyer). *Below:* Left: Off Mozambique, 42.0mm (MM); Right: E of Durban, Natal, South Africa, 36.2mm (NM 9491).

typical forms. Both reddish and brownish body whorl colors have been described for this variant, which is very poorly known but apparently sympatric with *C. cardinalis* typical and often larger (35-45mm).

Distribution.—Caribbean, from Florida and the Bahamas south through the Cuba-Haiti area to the Virgin Islands and perhaps some of the more southern Antilles to Curacao; also reported from the Central American coast (Honduras). Apparently rare to locally common in shallow reef areas, usually taken as beach shells.

Synonymy.—The synonymy is very uncertain because I do not understand the limits of variation in this species, or perhaps complex of species. *C. cidaris* (fig. 1) is the common well marked type, but Kiener's fig. 1a seems to be the axially striped pattern of *C. magellanicus. C. speciosissimus* definitely seems to be this species, as does *C. lubeckianus. C. archetypus* has also been referred to *C. daucus*, but the figure and description indicate *C. cardinalis* instead. *C. jucundus* is quite possibly a full species which would include the holotype of *C. abbotti* as a synonym; the illustrated paratype of *C. abbotti* appears to be a typical *C. cardinalis* with the granulose ridges demarked with distinct spiral rows of brown dots. Doubtless other names also apply to *C. cardinalis* but are not recognizable with certainty from the descriptions.

Notes.—In bright fresh condition this is at best an uncommon species. Large red specimens are rare, but smaller (20-25mm) pink specimens are more commonly available. Beach specimens often have badly broken lips and other flaws as well as faded colors. The *jucundus* patterns are apparently quite rare.

It is by no means certain that *C. cardinalis* is actually related to *C. regius*, although they are admittedly very similar superficially. The actual relationship could easily be with *C. magellanicus* and *C. daucus* and allies (including *C. amphiurgus* and *C. beddomei*) because of the color pattern and often bright colors of this complex, the generally weakly sculptured spire, and the poorly developed coronations.

CATUS Hwass, in Bruguiere, 1792

1792. *Conus catus* Hwass, in Bruguiere. *Cone*, in *Ency. Method., Hist. Nat. des Vers*, 1: 707 (Mauritius, restricted by Kohn, 1966). Lectotype selected by Kohn, 1966.

1798. *Cucullus nubilus* Roeding. *Museum Boltenianum:* 46 (Locality not stated). Based on Martini, Pl. 55, fig. 610.

1810. *Conus pusillus* Lamarck. *Ann. du Mus. Hist. Nat.* (*Paris*), 15: 39 (Guinea). See Kiener, Pl. 43, fig. 2, for a rather poor rendition of a syntype.

1833. *Conus discrepans* Sowerby i, in Sowerby ii. *Conchological Illustrations:* Pt. 29, fig. 28 (Locality not stated).

?1863. *Conus purus* Pease. *Proc. Zool. Soc.* (*London*), 1862: 279 (Niihau Is.)
?1877. *Conus reflectus* Sowerby iii. *Proc. Zool. Soc.* (*London*), 1876: 754, pl. 75, fig. 6 (Locality unknown). Name emended to *Conus reflexus* by Sowerby iii, 1887, *Thesaurus Conchyliorum*, 5(*Conus* Suppl.): 252.
1937. *Conus catus* var. *rubra-papillosa* Dautzenberg. *Mem. Mus. Roy. d'Hist. Nat. Belg.*, 2(18): 62, pl. 1, fig. 7 (Tjilaoet Eureun).
1937. *Conus catus* var. *fusco-olivaceus* Dautzenberg. *Ibid.*, 2(18): 62, pl. 1, fig. 8 (Tjilaoet Eureun).

Description.—Heavy for its size, thick, with a good gloss; broadly conical, the sides convex; body whorl with numerous rounded spiral ribs covering at least anterior half, these sometimes extending to shoulder or occasionally almost absent; ribs often with large rounded granules, especially anteriorly; grooves between ribs smooth or with weak axial threads; upper half of body whorl, especially just below the shoulder, smooth; shoulder rounded, not very distinct from spire; spire low, the sides straight or slightly convex; spire whorls marked with 3-5 low but distinct spiral ridges at middle of whorls, these sometimes crossed by distinct axial wrinkles; shoulder similarly sculptured; tip of spire usually eroded. Body whorl white to opaque white, variably covered with pale brown to reddish brown or almost black blotches, these mostly axial but often fusing to cover whole body whorl except for small to large white spots at shoulder and midbody; spiral ribs sometimes with slightly darker brown dashes, the granules usually pearly white and contrasting with the rest of the whorl; spire mostly whitish, with a few large blotches anteriorly and on the shoulder; apex pink. Aperture widened basally; outer lip convex, thick; mouth creamy white to pale bluish white. Columella short, narrow. Length 25-50mm.

Comparison.—This easily recognized species is very close to *C. monachus* and *C. nigropunctatus* but easy to recognize once one is familiar with it. It is heavier and broader than these two species and almost always thicker-shelled; the white background color usually appears distinctly opaque white and glossy. Typical specimens have numerous heavy granulose ridges which appear less ragged than in granulose *C. monachus.* Few specimens have the distinct small brown dashes so typical and strongly developed in *C. monachus* and *C. nigropunctatus* patterns. *C. regius* and allies are heavily coronated although sometimes similar in appearance. *C. magus* is usually much more elongated and thinner in texture and has numerous small brown dashes in the pattern.

Variation.—Although constant in form and texture, this species varies considerably in sculpture and color. Most specimens are heavily ribbed over about three-fourths of the body whorl, with at least some basal ribs nodulose. The area below the shoulder is usually smooth or nearly so; occasionally the ribs become obsolete at midbody, and in some

EUGRAMMATUS.—*Above:* Oahu, Hawaii, 23mm (DM) (photo by Jayavad). *Below:* Kusi, Nada-cho, Wakayama Pref., Japan, 33.0mm.

EUGRAMMATUS.—Above: Off Mie, Japan, 30.1mm (K). *Below:* Left: Okinawa, Japan, 29mm (DM) (photo by Jayavad); Right: Kii Channel off Tanabe, Wakayama, Honshu I., Japan, 26.7mm (EP).

specimens they may be almost absent. Granules of large size are commonly developed basally and may extend to the shoulder. The opaque white background color is constant, with all specimens I have seen showing at least a few small white spots at midbody or below the shoulder. Some specimens are entirely white with only small brownish blotches buried deep within the nacre, while others are entirely dark brown, almost black, except for small white flecks or spots at midbody.

Distribution.—A common and widespread shallow-water species. Found from the eastern African coast through the Indo-Pacific to Hawaii and French Polynesia.

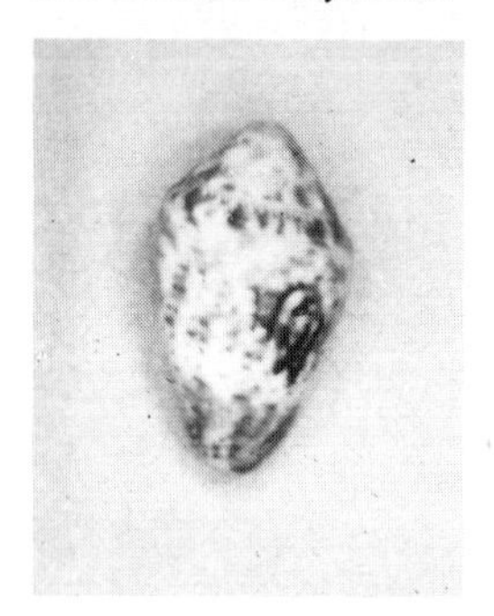

Holotype of *C. pusillus* LamarcK (M.N.H. Geneve) (photo Abbott).

Synonymy.—*C. pusillus* seems to be based on a juvenile of *C. catus*, but it could also possibly be a juvenile of some other species in the *C. monachus* complex. *C. purus* is poorly described but seems to belong here. *C. reflectus* would seem to be a *C. catus* with the pattern more regular than usual. The Dautzenberg names are simple individual variants, although the nearly black *fuscoolivaceus* seems to be the dominant form in some populations.

Notes.—This common shell has so much individual variation in color and pattern that dealers and collectors do not try to distinguish them. The very dark shells with almost no white showing (var. *fuscoolivaceus*) are sometimes segregated and sold at a small premium. The thick shell is not fragile, so flaws are uncommon except for irregular patterns deep within the nacre and eroded spires.

CEDONULLI Linnaeus, 1767

1767. *Conus Ammiralis cedo nulli* Linnaeus. *Systema Naturae per Regna Tria Naturae*, ed. 12, 1: 1167 (South Sea). Lectotype figure (here designated): Seba, 1758, Pl. 48, fig. 8. See Dunn, 1970, *Veliger*, 13(3): 290-291, for the history of this pattern.

1786. *Conus mappa* Lightfoot. *Cat. Portland Mus.*: 116 (China). Lectotype figure (Kohn, 1964): Knorr, Pl. 8, fig. 4.

1786. *Conus architalassus* Solander, in Lightfoot. *Ibid.*: 89 (Locality not

stated). Lectotype figure (Kohn, 1964): Argenville, App. Pl. 1, fig. M.
1791. *Conus Ammiralis coronatus* Gmelin. *Systema Naturae per Regna Tria Naturae*, ed. 13, 1: 3379 (Locality not stated). Lectotype figure (here selected): Martini, vignette 26, page 214, fig. 1. Non *Conus coronatus* Gmelin, 1791, p. 3389.
1791. *Conus Ammiralis Surinamensis* Gmelin. *Ibid.*, ed. 13, 1: 3380 (Locality not stated). Lectotype figure (here selected): Martini, vignette 26, page 214, fig. 5.
1792. *Conus cedonulli Caracanus* Hwass, in Bruguiere. *Cone*, in *Ency. Method., Hist. Nat. des Vers*, 1: 603 (near Caracas). Holotype figured in *Tableau Ency. Method.:* Pl. 316, fig. 6, and in Kiener, Pl. 16, fig. 1b.
1798. *Cucullus geographicus* Roeding. *Museum Boltenianum:* 39 (Locality not stated). Lectotype figure (Kohn, 1975): Martini, Pl. 62, fig. 683.
1798. *Cucullus petreus* Roeding. *Ibid.:* 39 (Locality not stated). Lectotype figure (Kohn, 1975): Martini, vignette 26, page 214, fig. 1.
1968. *Conus nulli-secundus* Usticke. *Caribbean cones from St. Croix and the Lesser Antilles:* 20, pl. 3, sp. 1010 (3 figs.) (St. Vincent). This should be considered a nomen novum for *C. cedonulli*, not a new species description.
1968. *Conus holemani* Usticke. *Ibid.:* 21, pl. 3, sp. 1011 (St. Vincent).

Description.—Moderately heavy, with a good gloss; broadly biconical, the body whorl rather long and with slightly convex sides; body whorl covered with broadly spaced fine spiral ridges, the ones at the base broader; ridges usually bearing fine granules, these sometimes not distinct except anteriorly; above midbody the ridges may become obsolete; axial growth marks often conspicuous; shoulder broadly rounded, heavily coronate, the coronations rounded; spire moderate to somewhat high, the sides straight or concave, tops of whorls slightly concave; whorls distinctly nodulose on the early whorls, becoming more roundly coronate on the later whorls; spiral sculpture of spire whorls weak and usually indistinct, sometimes 2-3 fine spiral threads present, cutting across fine and irregular axials. Body whorl white, often with violet tones, heavily covered with deep yellow or orange (perhaps dark brown in life) bands which usually leave only a little white background visible; a broad, very irregularly scalloped white midbody band present in most shells, the edges often assuming very complex patterns; numerous fine brown to purple-brown spiral lines over body whorl, these broken by large oval or rhomboidal white spots, the spots sometimes fusing to produce narrow bands; occasionally some white axial banding present; overall, the pattern appears very complex, often assuming 'animal shapes' at midbody; white spots commonly outlined with narrow purple lines; spire mostly white with a few scattered yellow blotches and fine axial brown streaks at the sutures; tip of spire white. Mouth rather wide, somewhat widened

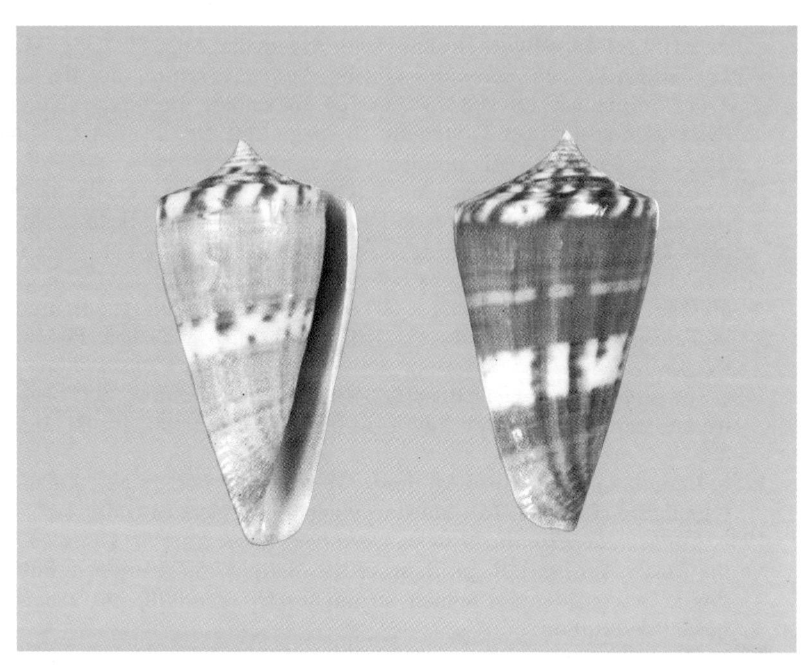

EXCAVATUS.—Red Sea. *Above:* 42.4 and 43.1mm. *Below:* 51mm (DM) (photos by Jayavad).

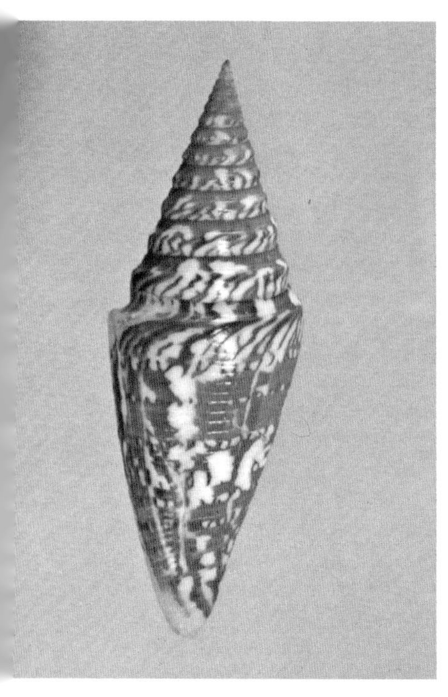

EXCELSUS.—*Above:* "Indian Ocean", 78mm (DM) (photos by Jayavad). *Below* (holotype of *Turriconus nakayasui*): Kashiwajima, near Cape Ashizuri, Kochi Pref., Shikoku, Japan, 90mm (photos by Dr. T. Shikama).

anteriorly; outer lip straight, sharp; inside of aperture white to pale bluish white. Columella narrow, largely internal. Length 40-70mm.

Comparison.—This species is extremely close to *C. aurantius* and possibly the two are not distinct at the species level. However, as treated here, *C. cedonulli* is relatively broader than *C. aurantius*, is usually bright waxy yellow or orange, has the white spots larger and more distinct, has a white midbody area but few larger white blotches, and has the margins of the white area highly complex and commonly forming 'animal figures.' The dark brown *C. cedonulli* (see *Variation*) agree with typical shells in complexity of the pattern.

C. regius is even broader, has a much more concave spire on the average, has distinct and strong spiral sculpture on the spire whorls, and lacks the fine axial brown streaks on the sutures. Some unusual *C. regius* are very similar in appearance to *C. cedonulli* but still easily distinguishable.

Variation.—As in other similar species, there is considerable variation in the height and shape of the spire and the strength of the spire sculpture. The pattern also varies considerably as would be expected, but within narrower limits than either *C. regius* or *C. aurantius*. I am not certain whether typical living *C. cedonulli* are bright waxy yellow or whether this color is the result of fading after death. The majority of specimens in modern collections come from harbor dredgings. *C. holemani*, which agrees with *C. cedonulli* in all other respects, is distinctly deep purplish brown, almost black, according to Usticke; this may perhaps represent the color of living animals of at least some populations or just a dark variant. Apparently brownish (pale) *C. cedonulli* also exist.

Distribution.—Uncommon to rare through the Antilles from the Bahamas to the Venezuelan coast, but many records are probably *C. aurantius*.

Synonymy.—As mentioned under *C. aurantius*, it is sometimes difficult to decide from the early descriptions and figures which species are represented. The names listed here seem to fit *C. cedonulli* better than *C. aurantius*, but of course some are doubtful. *C. mappa* especially is doubtful and may actually apply to *C. aurantius*, but I do not want to make such a change at this time—the name *mappa* has lately been applied to almost every possible combination of populations in this species complex and I feel the name should no longer be used at a formal level to avoid confusion.

For many years the name *C. cedonulli* was very incorrectly applied as a synonym of *C. ammiralis*, although the Seba figure on which the name was based obviously is the Caribbean cone; the resulting name changes led to great confusion about the name and status of this lovely and rare shell which once was the pride of only the largest European collections of the last century. Dunn (1970) convincingly demonstrated that the Seba figure matches at least some specimens exactly, so there can no longer be any doubt as to the status of the name. The decision to recognize two Caribbean species, *C. cedonulli* and *C. aurantius*, is less certain as there are certainly specimens which are difficult to assign to either species; but at least this system is better than the recognition of up to a half-dozen supposed species which are completely impossible to distinguish.

Notes.—Although no longer an outstanding rarity, *C. cedonulli* is still extremely difficult to obtain and fetches a good price regardless of condition. Most specimens offered are dredged dead and faded. Minor chips in the lip and growth flaws are to be expected; the spire is usually lightly eroded at the tip.

CENTURIO Born, 1780

1780. *Conus centurio* Born. *Index Rerum Naturalium Musei Caesari Vindobonensis*, 1(*Testacea*): 133 (Puerto Plata, Santo Domingo, as selected by Clench, 1942). Holotype illustrated by Kohn, 1964.

1791. *Conus tribunus* Gmelin. *Systema Naturae per Regna Tria Naturae*, ed. 13, 1: 3377 (Locality not stated). Holotype figure: Martini, Pl. 59, fig. 655.

1791. *Conus bifasciatus* Gmelin. *Ibid.*, 1: 3392 (Locality not stated). Lectotype selected by Kohn, 1966.

1946. *Conus woolseyi* M. Smith. *Nautilus*, 60(1): 1, pl. 1, fig. 5 (Off Ocho Rios, Jamaica).

(1968. *Conus centurio* form *cruzensis* Usticke. *Caribbean cones from St. Croix and the Lesser Antilles:* 12, pl. 2, sp. 997 (St. Croix, Ham Bay). Listed as *C. centurio cruzensis* on plate. This is an infra-subspecific name and does not enter priority.)

(1968. *Conus centurio* form *caribaensis* Usticke. *Ibid.:* 13, pl. 2, sp. 998 (Krause's Bay, St. Croix). Listed as *C. centurio caribaensis* on plate. This is an infrasubspecific name and does not enter priority.)

Description.—Moderately heavy, with a high gloss; low biconical, the sides nearly straight, tapering to narrow base; 4-5 low rounded spiral ridges anteriorly, followed by about 6-10 broader rounded ribs separated by finely axially threaded grooves; rest of whorl with sculpture nearly absent except for fine spiral and axial threads; shoulder broad, carinate, concave above; spire low to moderate, the sides straight or slightly concave; whorls concave above, the margins carinate; earliest two whorls weakly nodulose; tops of whorls with numerous fine curved axial threads and traces of 2-4 weak spiral threads. Body whorl white, covered with rather wide zigzag flammules of yellowish tan to deep chestnut, widely spaced; flammules tend to condense into three spiral bands of large blotches, one below shoulder, a broad and often continuous band above midbody, and a weaker band above the base; bands may be very blotchy

EXIMIUS.*—Above:* Burma, 29.8 and 30.8mm (GG). *Below:* Left: Andaman Is., 41.2mm; Right: Thailand, 36.0mm (MG).

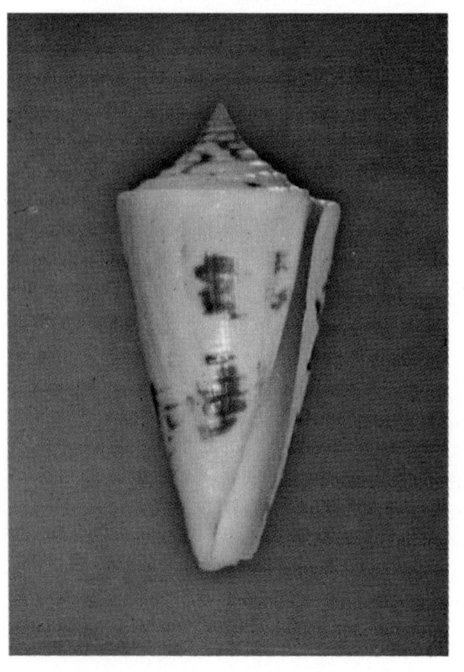

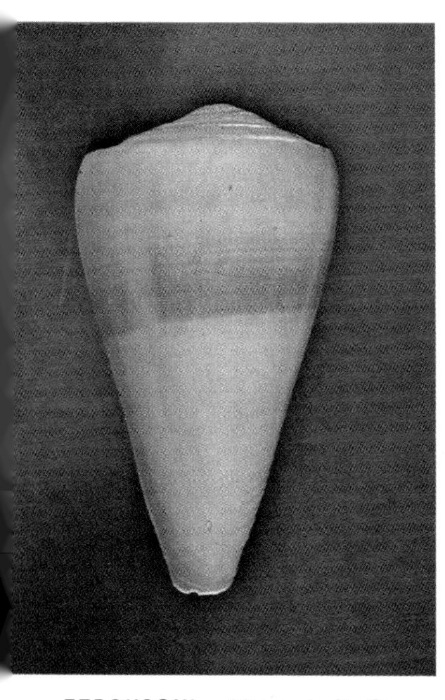

***FERGUSONI.**—Above:* Left: Guaymas, Sonora, Mexico, 65.0mm; Right: Cabo San Lucas, Baja California, Mexico, 19.9mm (photo by A. Kerstitch). *Below:* Cholla Bay, Sonora, Mexico: Left: 87.0mm; Right: 91.1mm.

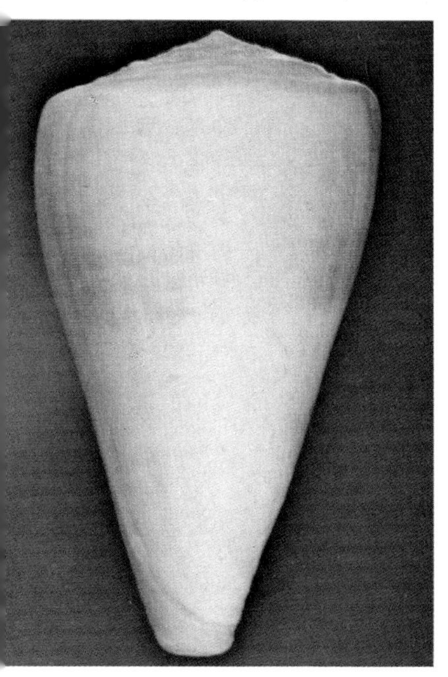

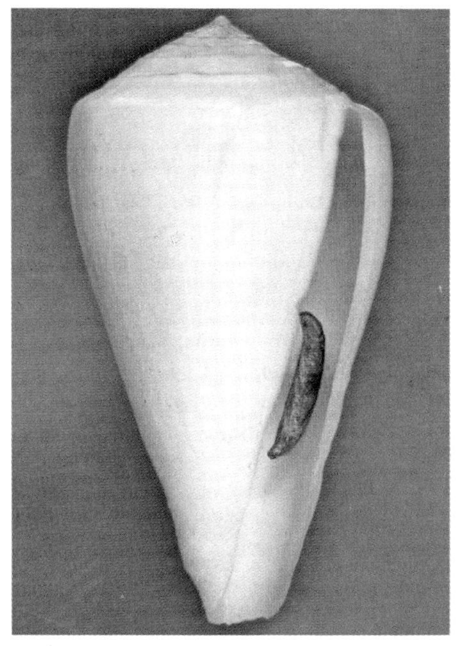

or almost continuous and with straight edges; flammules often badly broken and incomplete, sometimes forming spiral rows of short brownish dashes on the anterior fifth; base white to yellowish tan; spire white, with brown lines and scattered blotches, continuations of flammules of body whorl; tip of spire brownish or white. Aperture rather wide, uniform in width; outer lip straight, posteriorly sloping below level of shoulder; mouth white. Columella very narrow, largely internal. Length 30-65mm.

Comparison.—*C. centurio* is very closely related to *C. delessertii*, but they seem to be distinct. *C. delessertii* is relatively narrower with a somewhat higher spire, the first 4-5 spire whorls bearing strong nodules separated by small brown dots; the color pattern is more broken than in *C. centurio*, consisting in most specimens of numerous large brownish dashes resulting from the breakdown of the axial flammules (but many *C. delessertii* have distinct flammules in the pattern, while some *C. centurio* have very indistinct flammules). In *C. delessertii* the spiral brown bands are usually absent, but there are three narrow to broad bands of bright salmon in their place, these absent in *C. centurio.*

C. philippii has some patterns similar to extreme *C. centurio*, but it has a straighter sided spire with the tops of the whorls nearly flat; also, the basal sculpture is weaker or nearly absent. *C. capitanellus* may be superficially similar but differs in many aspects of shape, sculpture, and color pattern.

Variation.—The spire varies from low to moderately high and is usually sharply pointed; the whorls may be strongly stepped and concave above or only weakly concave. In some specimens the body whorl is covered with distinct zigzag axial flammules which are only indistinctly thickened in the region of the three bands; in others there are strong thickenings and the flammules break above and below the band area; other specimens are distinctly banded and have strong flammules as well, while others have strong bands and almost no flammules. There is a weak tendency to break the flammules into large spiral dashes, especially basally. The base may be tinted with tan or even salmon or may be white. Basal ridges may be weak or strong; the more posterior ribs are usually well developed and extend to about midbody.

Distribution.—Widespread in the Caribbean from the northern coast of South America along the Central American coast and the Antilles to Florida and north to Georgia. Usually dredged in moderately deep water or found dead near beach. Uncommon to rare.

Synonymy.—The synonymy of this species is rather firmly settled. The two Usticke names are described as forms and thus have no status under current rules; they were used for shells with exceptionally well defined axial flammules.

Notes.—*C. centurio* is a rare and desirable shell which is seldom found in perfect condition. It apparently leads a rough life and suffers bad growth flaws and breaks; the lip is also fragile. Specimens with one or two obvious flaws are still collectable as long as the colors are not badly faded. Live-taken specimens are rare and not often seen.

CERNICUS H. Adams, 1869

1869. *Conus* (*Coronaxis*) *cernicus* H. Adams. *Proc. Zool. Soc.* (*London*), 1869: 272, pl. 19, fig. 1 (Barkly Island, Mauritius).
1873. *Conus tenuisulcatus* Sowerby iii. *Proc. Zool. Soc.* (*London*), 1873: 145, pl. 15, fig. 2 (Mauritius). Non *C. tenuisulcatus* Sowerby iii, 1870.
1877. *Conus propinquus* E.A. Smith. *Ann. Mag. Nat. Hist.*, (Ser. 4), 19: 223. Nomen novum for *Conus tenuisulcatus* Sowerby iii, 1873.
1942. *Conus circumclausus* Fenaux. *Bull. l'Inst. Oceanogr.* (*Monaco*), No. 814: 3, fig. 8 (I. Maurice).

Description.—Moderately heavy, with a low gloss; low conical or obconical, the upper sides convex; body whorl covered with narrow, low, rounded or flattened spiral ridges, heaviest at base and sometimes obsolete below shoulder; ridges crossed by very fine irregular axial threads most distinct basally; shoulder roundly angled, broad, with low and rounded coronations; spire low, pointed, the sides straight to slightly convex; spire whorls coronate, the tops flat to slightly concave, with about 3-5 or more fine and sometimes undulate spiral ridges; ridges crossed by very fine curved axial threads. Body whorl cream, pale tan, violet, or lilac, usually at least partially covered with dark brown on basal third; a narrow or broad brown spiral band present near midbody, sometimes fused with basal brown area; area below shoulder usually pale, with little brown visible and often toned violet; base tinged with violet; whole shell covered with small to large white flecks and dashes, these sometimes forming rather regular spiral rows and alternating with small brown dashes; axially arranged tan to violet clouds often visible; shoulder and spire whitish with violet tones, heavily spotted and smeared with dark brown blotches in a faintly radiating pattern; tip of spire pink; occasionally entire shell unicolor, violet or lilac, the spire pattern indistinct but present. Aperture moderately narrow, uniform in width; outer lip straight, sharp-edged, sometimes faintly concave near middle; mouth dark violet-brown, the margin sometimes light. Columella very narrow, mostly internal. Length 20-40mm.

Comparison.—*C. cernicus* is poorly understood and seldom seen under this name; usually it is confused with *C. rattus* and *C. balteatus*. Some small *C. rattus* may be somewhat similar in pattern, although usually there are larger white blotches near the midbody area and the spire is covered with larger dark blotches than in *C. cernicus*; in addition, the spire whorls of *C. rattus* are smooth or at most undulate, not distinctly coronate. *C. balteatus* is even closer, but in that species the body whorl tends to be less inflated posteriorly, the shoulder coronations are higher and sharper, the spire is virtually unmarked, and commonly the white specks are almost absent. Small *C. cernicus* with unicolored lilac or violet shells resemble *C. pertusus* superficially, but they lack the strongly punctate spiral grooves of that species and are coronated. Some *C. tiaratus*

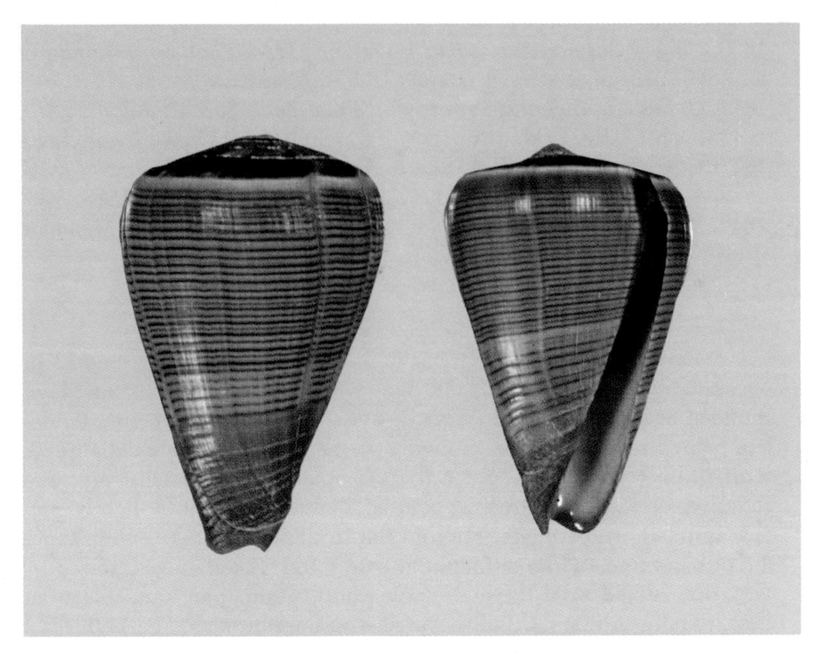

FIGULINUS.—Above: Basilan I., P.I., 53.2 and 57.6mm. *Below:* Left (*loroisii*): Andaman Sea, 73.2mm; Right: Palm Beach, Nossi Be, Madagascar, 32.5 and 34.9mm.

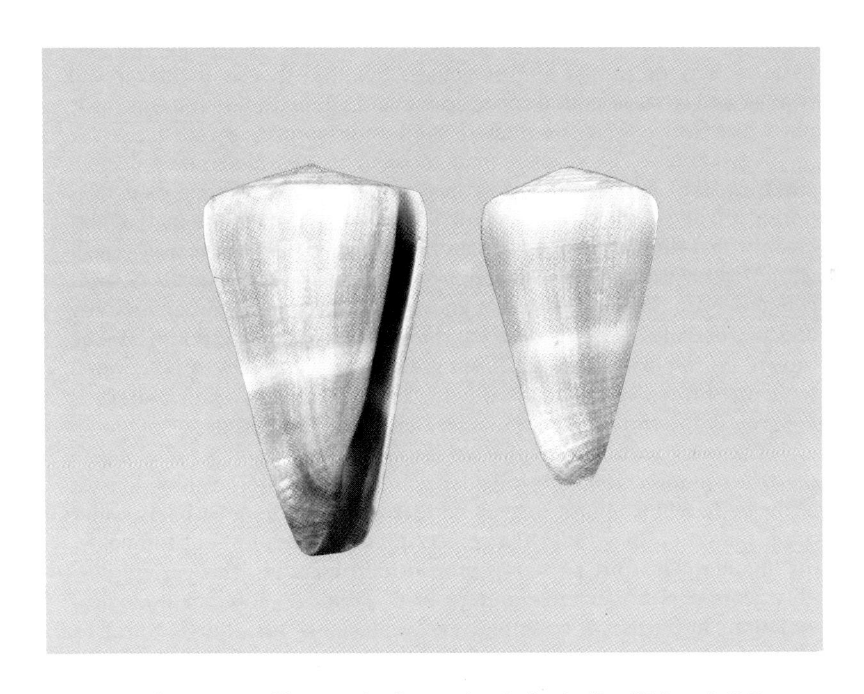

FLAVIDUS.—*Above:* Hayman I., Queensland, Australia, 37.9 and 46.7mm. *Below:* Tahiti, Society Is., 29.6 and 34.3mm.

patterns may be similar at first glance, but that species is thicker and heavier and covered with distinct brown and white dashes in spiral rows, not white flecks; there are numerous other differences of course.

Variation.—This species seems to occur in two moderately distinct variants, but I have seen too few species to exactly determine their relationship. The common type is relatively broad and short, the body whorl creamy to yellowish white or pale tan, covered with numerous small white flecks; the spire is heavily covered with brown blotches. In this common form the body whorl commonly has a broad brown band basally and another one near midbody, but the majority of the whorl may also be brown. At the extreme of this form are shells which seem to be covered with fine brown spiral lines following the spiral ridges. This pattern is apparently intermediate to the other, less common, variant.

The less common variant is relatively narrower and more elongate than the common type, with a tendency to have the anterior three-fourths of the body whorl almost unicolored lilac or tan on a violet background color, the violet (sometimes lilac) especially strong as a broad band below the shoulder; the spire pattern is present but obscured. This variant possibly represents the juvenile pattern of *C. cernicus* or minor ecological variation; in pattern it resembles the *'semivelatus'* form of *C. rattus* or even some variants of *C. pertusus.* Since there seems to be intermediates in both shape and pattern and the sculpture is the same in both types, they are tentatively considered to represent the same species.

Distribution.—Seemingly restricted to the western Indian Ocean from the Mauritius-Reunion area to the African coast. Locally common in shallow to moderately deep water. Distribution poorly known, but apparently *C. cernicus* is sympatric with both *C. rattus* and *C. balteatus.*

Synonymy.—Although the figure of *C. cernicus* presented by Adams has a distorted spire, it seems to represent this species. The next available name, should *C. cernicus* prove to be applicable elsewhere, is *C. propinquus.* Both names, as well as *C. circumclausus,* apply to the common broad variant.

Notes.—This species is almost never offered under the right name, although it sometimes appears on the market as *C. balteatus* at a low price. The brighter specimens of the slender unicolored form are sold at high prices under the name *C. propinquus,* but usually only juveniles are seen. This does not seem to be as rough a cone as *C. balteatus,* and there are seldom as many healed scars and breaks, although the pattern is often interrupted and irregular.

The similarity of some patterns of *C. cernicus* to both the closely related *C. rattus* and the less-close *C. pertusus* can be very confusing. Although closely related to *C. balteatus,* the relationship with *C. rattus* is perhaps closer, as seen in the occasional undulately-spired *rattus* and the *'semivelatus'* pattern of that species.

CERVUS Lamarck, 1822

1822. *Conus cervus* Lamarck. *Hist. Nat. des Animaux sans Vert.*, 7: 510 (Locality unknown).

Description.—Light but solid, glossy; nearly cylindrical, the sides gently convex, widest at posterior third of whorl; body whorl smooth except for several weak ridges near the base and weak axial growth lines; shoulder angulate, weakly concave above; spire low, the sides straight or slightly concave; spire whorls slightly concave above, with about 3-5 low spiral ridges crossed by fairly heavy axial ridges. Body whorl (in dead specimens) deep yellowish with pinkish tones, covered with widely spaced spiral bands of reddish brown dashes separated by white spots, dashes, or triangles; the bands are of unequal width, some wide, others fine; spire yellowish with faint white radiating blotches. Aperture narrow above, wide basally; outer lip thin, slightly convex; mouth white. Columella long, moderately broad. Length 70-120mm.

Comparison.—This large and very poorly known species is obviously close to *C. bullatus* and has often been considered a giant variant of that species, perhaps correctly. From *bullatus* it differs in the less oval shape (not constant in *C. bullatus*), perhaps the narrower aperture (variable in *C. bullatus*), the relatively thinner, lighter texture (approached by some *C. bullatus*), and the distinct spiral bands of dark and light dashes instead of nebulous spots (some *bullatus* are distinctly spotted). Few *C. bullatus* exceed 60mm, while *C. cervus* appears to commonly exceed 90mm (I know of no mention of small *C. cervus*).

C. vicweei is similar to *C. cervus* in size and shape as well as texture, but it differs in the pattern of narrow white axial zigzags in spiral bands and the darker, often brown, background color. However, very fresh *C. cervus* are apparently not reported in the literature.

Variation.—The few dead specimens of *C. cervus* reported are generally similar in shape and pattern, allowing for fading and other post-mortem changes. The spiral color bands may be broad and well marked or narrow and not very distinct. The width of the aperture and height of the spire apparently vary a bit. Fresh specimens would presumably be bright red or at least reddish brown with chestnut or blackish spotting.

Distribution.—Old specimens are poorly localized, usually with labels such as Moluccas or Amboina. Presumably this species is found in the Indonesian area. A doubtful modern record is supposed to be from the Bangka Straits, eastern Sumatra.

Synonymy.—This species has been confused with *C. bullatus* and even *C. cuvieri*. Its validity is subject to question.

Notes.—A freshly collected specimen of *C. cervus* would certainly be one of the great discoveries in cones. Almost all specimens of this species (at the time of writing) are in old museum collections, and the species has

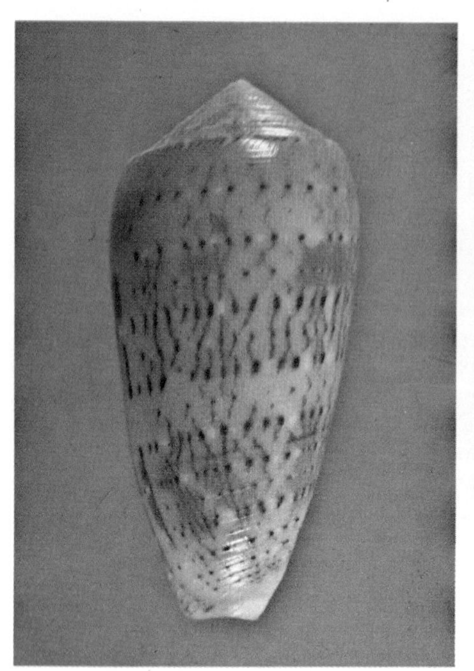

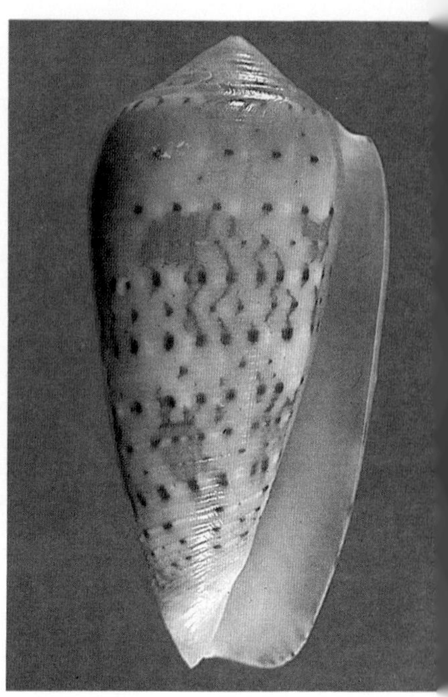

FLOCCATUS.—Kwajalein Atoll, Marshall Is. *Above:* 54.9mm (MG). ***Below:*** 42.6-52.9mm.

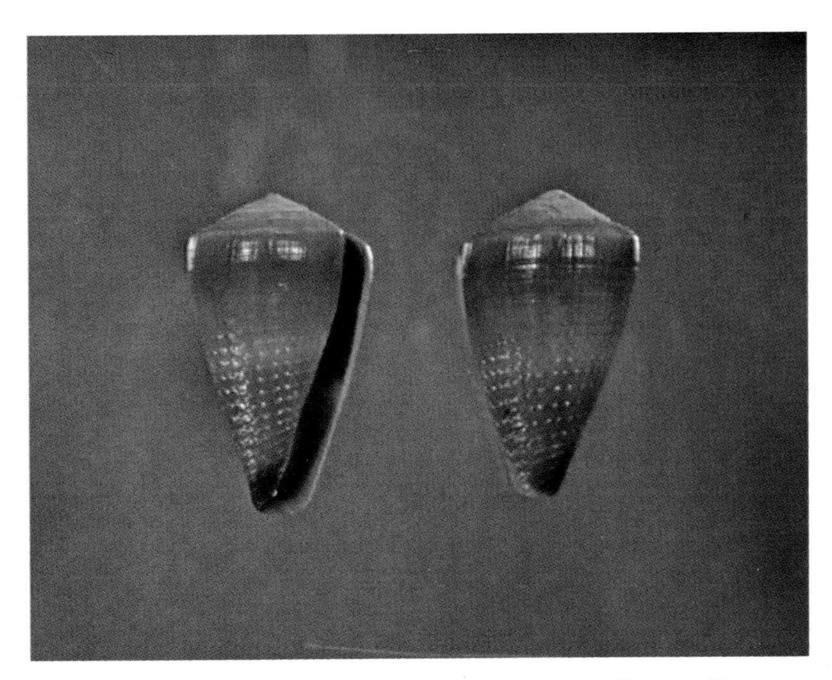

FRIGIDUS.—Honiara, Guadalcanal, Solomon Is. *Above:* 22.3 and 23.1mm. *Below:* 24.6 and 27.8mm.

apparently not been rediscovered this century. The yellow color of most specimens is probably due to fading, and it is assumed that the shell would be reddish in life. It should be looked for in dredgings from the Indonesian area in moderately deep waters. I am not at all certain that a small *C. cervus* could be distinguished from a large *C. bullatus* in a spotted pattern with any certainty, but fresh specimens of *C. cervus* may be more distinctive than dead museum specimens.

CHALDAEUS (Roeding, 1798)

1798. *Cucullus Chaldaeus* Roeding. *Museum Boltenianum:* 42 (Locality not stated). Lectotype figure (Kohn, 1975): Knorr, Pl. 4, fig. 2.

(1802. *Conus princeps* Holten. *Enumeratio Systematica Conchyl. beat Chemnitzii:* 34 (Locality not stated). Based on Chemnitz, Pl. 63, figs. 699-700. Non *C. princeps* Linnaeus.)

1810. *Conus vermiculatus* Lamarck. *Ann. du Mus. Hist. Mus. (Paris)*, 15: 34 (Tropical oceans). See *Tableau Ency. Method.*, Pl. 321, figs. 1-8.

1895. *Conus judaeus* Bergh. *Nova Acta Leop. Carol. Deutsch. Ak. Naturforscher*, 65(2): 161, pl. 4, fig. 91; pl. 6, figs. 128-131 (Philippines).

Description.—Heavy, thick, with a low gloss; broadly conical, the sides distinctly convex; body whorl with low, wide rounded ridges basally, these often extending to midbody or more posteriorly and sometimes bearing large granules; shoulder roundly angulate, coronated, but coronations often low and indistinct; spire low, the sides usually convex, bluntly pointed or eroded; spire whorls weakly coronate and with about five heavy spiral ridges on top. Body whorl white to pinkish, with two series of wavy black axial flammules, the longest ones above midbody, those below midbody about two-thirds length of posterior series; an irregular band of light ground color separates the two series of flammules, and another narrower band separates the posterior band of flammules from the heavy blotches of the shoulder; base black; shoulder with heavy black blotches separated by narrower white or pink areas; spire with irregular black blotches; tip often pinkish. Aperture narrow, somewhat widened basally; outer lip nearly straight, thick, middle often forming a thickened cord; mouth pale bluish white, sometimes with a darker blackish violet blotch posteriorly; edge of lip with continuation of external pattern. Columella short, thick, wide, bounded posteriorly by a strong ridge. Length 20-41mm.

Comparison.—Although vaguely similar and probably closely related to *C. sponsalis*, which has weaker brown to red axial flammules and a lower spire with smaller but higher shoulder coronations, *C. chaldaeus* must be closely compared only with *C. ebraeus*. The principal difference between the two species is that in *C. ebraeus* there are three distinct series of roughly rectangular axial black blotches which are usually well separated by the ground color. This usually separates the two with no difficulty, but occasional individuals of *C. ebraeus* have the spots partially fused into elongate flammules like those of *C. chaldaeus.* Then mouth color will help: nearly solid bluish white in *C. chaldaeus* versus white with three black bands (the exterior pattern showing through) in *C. ebraeus.* Although sympatric and almost certainly occasionally hybridizing, the two occupy slightly different ecological niches and appear to maintain their specific identity.

Variation.—*C. chaldaeus* shows a great tendency to fuse the axial flammules laterally to produce two nearly complete black bands above and below the light midbody stripe; all degrees of fusion are evident in even limited series. Occasional specimens have a spiral series of small but distinct rectangles at midbody, and occasionally the axial flammules may show a tendency to break into short axial rectangles. A pink background is evident in some populations. Some individuals may be almost smooth, while others within the same population may be heavily granulose over the entire body whorl. Commonly specimens are so weakly coronate that the shoulder appears smooth on casual examination.

Distribution.—An abundant sand-dweller from the eastern coast of Africa throughout the Indo-Pacific, including Hawaii and French Polynesia. Also east to the Central American coast and its offshore islands, where uncommon or rare.

Synonymy.—The listed names present no problems. *C. vermiculatus* is still sometimes used for this species, but it is certainly second in priority to *C. chaldaeus. C. judaeus* was based largely on differences of radulae.

Notes.—One of the cheapest cones, but generally not as abundant as *C. ebraeus* in any locality and often found in somewhat deeper water than that species. Specimens which must be interpreted as hybrids or at least extreme variants are not uncommon in some areas, but I see no reason to doubt the specific status of the two species. Specimens with bright pink background color sometimes draw a small premium.

CHIANGI (Azuma, 1972)

1972. *Taranteconus chiangi* Azuma. *Venus*, 31(2): 59, figs. 5-6 (South China Sea, about 200 fathoms).

Description.—Light in weight, with a good gloss; low conical, the sides straight or slightly convex posteriorly and tapering to a narrow base; body whorl covered with low, rounded spiral ridges separated by shallow, punctate grooves, the ridges somewhat weaker posteriorly; numerous heavy but straight and regular axial growth lines constant; shoulder sharply angulate, with very sharp, thin, sometimes tubular coronations resulting from the growth lines; spire low to moderate, weakly stepped, the margins all strongly coronate except for earliest 2-3, which

FULGETRUM.—Above: Batanes I., P.I., 21.2 and 26.5mm. *Below:* Amami Is., Japan, 28.1mm.

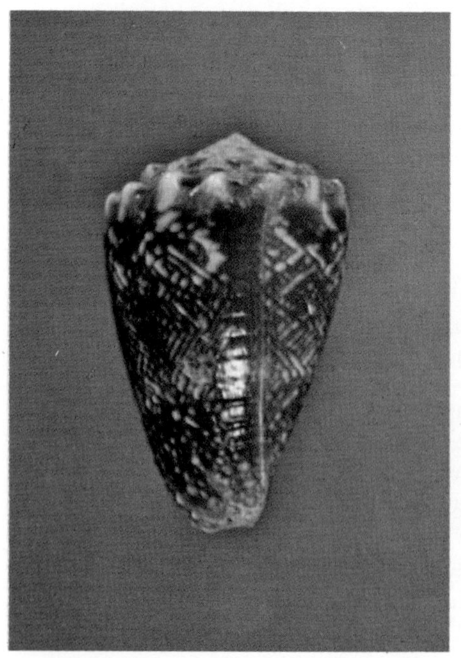

FULMEN.—*Above:* Kusui, Nada-cho, Wakayama Pref., Japan, 67.0mm. *Below:* Left (*kirai*): Kusui, Nada-Cho, Wakayama Pref., Japan, 41.3mm (GG); Right (holotype of *wistaria*): Bungo Channel, Japan, 60.5mm (photo by Dr. T. Shikama).

are strongly nodulose; tops of whorls flat to slightly concave, with traces of about three low spiral ridges and numerous widely spaced sharp raised curved axial ridges; these ridges are the results of the growth lines and form the coronations. Body whorl cream to tan, often distinctly trizonate; anterior third deep reddish brown, wide midbody band of white, and rest of body whorl paler brown; entire shell covered with spiral lines of opaque white dashes and smaller reddish brown spots, the opaque white especially heavy in the midbody band; sometimes entire shell almost unicolored tan, sometimes seemingly covered with narrow spiral brown lines; spire and shoulder tan with few to many darker brown blotches; early whorls white. Aperture narrow, uniform in width or slightly widened anteriorly; outer lip sharp, straight or slightly convex, sometimes slightly thickened; mouth creamy white. Columella narrow, short, curved anteriorly into outer lip and oblique to inner lip. Length 16-25mm.

Comparison.—This small and uncommon cone, if it really is a cone, defies comparison with other Recent species. Although the shape and pattern are not unique and somewhat resemble that of *C. otohimeae* or *C. memiae*, the strong and constant growth lines resulting in sharp, often tubular coronations on the spire whorls and shoulder are found in no other cone. Occasional injured specimens of almost any cone will develop similar growth flaws, including tubular coronations, but these are always restricted to just one or two lines, not structures that are consistently developed over the entire life of the shell.

Mention must be made of *Kenyonia pulcherrima*, a poorly known shell described by Brazier (1896, *Proc. Linnaean Soc. New South Wales*, 21(3): 346-347, New Hebrides) and now considered a turrid related to *Conopleura*. This shell is apparently very similar to *Conus chiangi* in shape and the unusual sculpture, but differs in having a mostly axial color pattern. *Conus lamellosus* Hwass, 1792 (see *Tableau Ency. Method.*, Pl. 322, fig. 5) is also certainly in this vicinity, although it has never been identified with certainty and the figure is too poor to be recognized; however, Hwass clearly described the spire sculpture and it agrees closely with *Conus chiangi*. Until these two names are clarified, I see no reason not to continue recognizing *Conus chiangi* as the proper name for this unusual species—with a reminder that it has probably been named earlier by either Hwass or Brazier.

Variation.—Spire height varies somewhat, as does the straightness of the sides; the coronations may be simply triangular plates or complete tubular pipes. Japanese specimens tend to have the tubular coronations and convex sides, with the color pattern of the body whorl much less defined than China Sea (off Taiwan) specimens. It seems possible that Japanese specimens might represent a distinct subspecies, but I have seen too few specimens from either area to be sure. Taiwanese specimens in fresh condition with full color and sculpture are amazingly bizarre and attractive shells.

Distribution.—Known from offshore coral banks in deep water from southern Japan to Taiwan and perhaps more widely distributed. Poorly known, but apparently locally common in some areas. Usually dredged dead.

Synonymy.—See *Comparison* for status of the names *Kenyonia pulcherrima* and *Conus lamellosus.* If transferred to a different genus than *Conus* in either the Conidae or Turridae (*sensu latu*), *Kenyonia* would definitely have precedence over *Taranteconus.*

Notes.—Although usually seen in dead condition when available at all, this little shell is worth adding to the collection in any condition. The columellar area is fairly distinctive and not really similar to that of any other cone known to me, so it seems likely that this species is actually a turrid in the broad sense. It emphasizes the close relationships of the cones, at least some species groups, and the turrid-like shells.

Since live-taken specimens are seldom available, dead-dredged shells must currently be accepted. These usually have the color pattern poorly defined and faded and the outer lip chalky and broken; there may be small worm tubes and boring sponge burrows as well. Coronations are commonly broken and irregular even in fresh specimens. This species is apparently not really rare, just seldom taken because of its deep habitat, so fantastically high prices do not seem deserved.

CINEREUS Hwass, in Bruguiere, 1792

?1758. *Conus rusticus* Linnaeus. *Systema Naturae per Regna Tria Naturae*, ed. 10, 1: 714 (Africa). Probably unrecognizable, but used by later authors for *cinereus* on several occasions.

1792. *Conus cinereus* Hwass, in Bruguiere. *Cone*, in *Ency. Method., Hist. Nat. des Vers,* 1: 673 (Indian Ocean). Lectotype figured by Kohn, 1966.

1810. *Conus caerulescens* Lamarck. *Ann. du Mus. Hist. Nat.* (*Paris*), 15: 423 (Seas of Moluques). See Chemnitz, Pl. 183, figs. 1776-1777. Also spelled *C. 'coerulescens'* by various authors.

1817. *Conus nisus* Dillwyn. *Descr. Catalogue Recent Shells*, 1: 388 (Eastern Seas). Holotype figures: Chemnitz, Pl. 183, figs. 1784-1785.

1833. *Conus modestus* Sowerby i, in Sowerby ii. *Conchological Illustrations:* Pt. 28, fig. 19 (Locality not stated). Yellow form.

1845. *Conus Gabrielii* Kiener. *Species gen. et icon. des coqu. viv.*, 2 (*Conus*): Pl. 74, fig. 4. 1849-1850, *Ibid.*, 2: 315 (Locality unknown). Somewhat brownish with a broken pattern.

1845. *Conus Bernardii* Kiener. *Ibid.*, 2(*Conus*): Pl. 100, fig. 2. 1848, *Ibid.*, 2: 220 (Locality unknown).

1845. *Conus Gubba* Kiener. *Ibid.*, 2(*Conus*): Pl. 104, fig. 1. 1849-1850, *Ibid.*, 2: 289 (Locality unknown).

FUMIGATUS.—*Above:* Massawa, Ethiopia, 29.0 and 31.5mm. *Below:* Red Sea, 19.6-26.7mm.

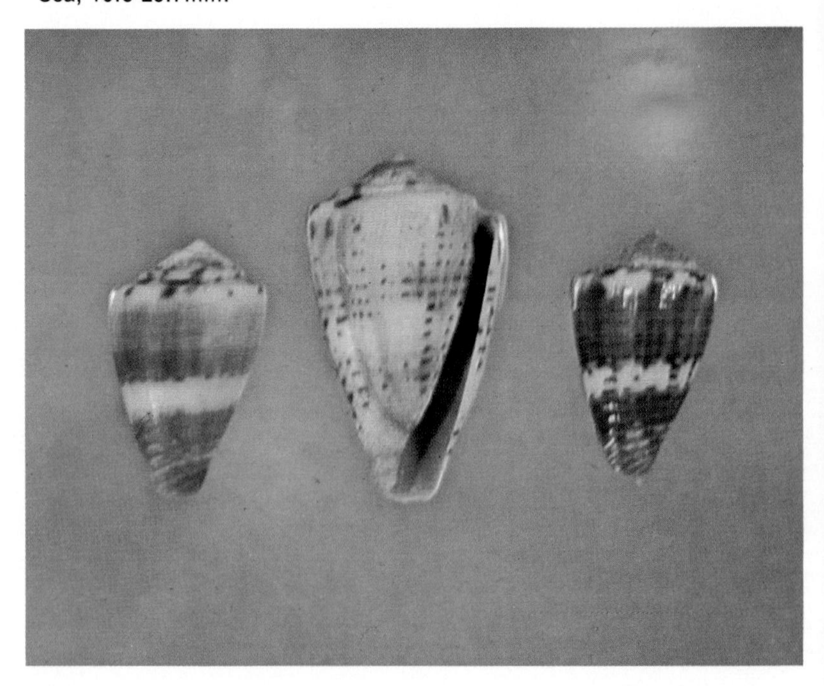

FURVUS.—Above: Left: "Bay of Bengal"(?), 47mm (MG); Right (granulose): P.I., 38.5mm. *Below:* Sulu, P.I., 31.2-41.8mm.

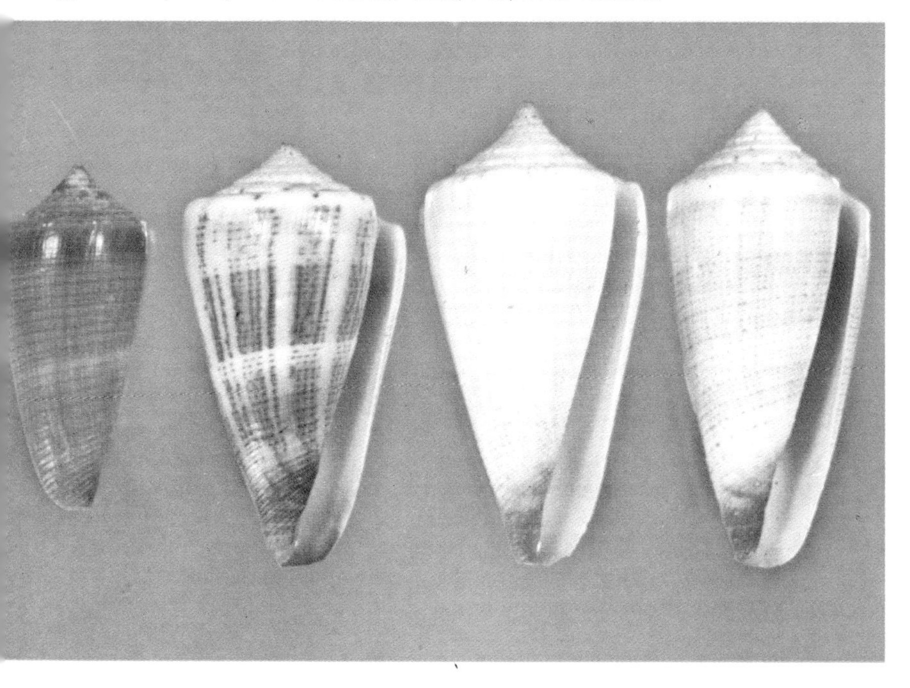

?1845. *Conus ardisiaceus* Kiener. *Ibid.*, 2(*Conus*): Pl. 108, fig. 1. 1849-1850, *Ibid.*, 2: 316 (Locality unknown).
1875. *Conus politus* Weinkauff. *Syst. Conch. Cab.*, ed. 2, 237(*Conus*): 334, pl. 62, figs. 2-3 (Locality unknown).

Description.—Light in weight with a high gloss; elongate cylindrical, the sides weakly convex and evenly curving to the base; body whorl with narrow spiral grooves over at least anterior half and sometimes all of whorl; the grooves are punctate and are separated by broad flat ribs; shoulder rounded, not very distinct from spire; spire low, sharply pointed, the sides usually deeply concave; spire whorls and shoulder usually appear smooth, but they may have 2-3 very weak spiral ridges in some specimens; juveniles have stronger spire sculpture than adults; early whorls not nodulose. Color and pattern very variable, but usually either bluish gray or yellowish with numerous small brown and white spots in widely spaced spiral rows over body whorl; usually with three spiral bands of brown to bluish very irregular and variable axial blotches, the blotches separated from each other by about their width in most specimens or sometimes partially fused laterally; shoulder with irregular brownish blotches following the axial rows of body blotches, these blotches continuing onto spire as finer lines; tip of spire grayish to whitish; in some specimens the entire pattern is greatly reduced or mostly absent, and in others the brown blotches may be greatly increased in size and contrast; base light. Aperture wide throughout, but widest basally; outer lip thin, evenly convexly curved; mouth variable, usually with a dark brownish violet band just behind outer lip edge and paler violet within; mouth sometimes all white. Columella long, slender, set off posteriorly from inner lip and distinctly indented. Length 35-65mm.

Comparison.—From *C. spectrum* and *C. conspersus*, *C. cinereus* differs in the more elongate shape, less expanded aperture which is violet and not usually white or orange, and the different shape. *C. lienardi* and *C. neptunus* have the early whorls nodulose and have broken tented patterns. The Philippines variant of *C. pilkeyi* often looks very much like a yellowish *C. cinereus*, but it has stronger basal grooves and heavy spire sculpture. *C. ochroleucus* and *C. radiatus* also have heavier sculpture and lack spiral spotting. *C. parius* lacks a dark pattern, is waxy in texture, and commonly has the tip of the spire darker than the body whorl; most *C. parius* are smaller than typical *C. cinereus*.

C. dusaveli is very similar to *C. cinereus* in some aspects of general appearance. However, in *C. dusaveli* the spire ridges are strong and about four in number, the pattern usually consists of larger square spots with a bluish tone, and the shoulder and spire are covered with narrow brown axial lines much more regular than those of typical *C. cinereus*.

Variation.—This species is highly variable, although the most common pattern, distinct spiral spotting on bluish gray with a variable number and intensity of axial blotching, is fairly constant. Even specimens with very pale or very dark backgrounds have at least some spots visible under a lens. Specimens with the brown blotches very dark and partially fused into spiral bands are not common and look quite different from more typical patterns. When the fusion is complete over the entire body whorl, a solid dark brown or even almost black specimen results, the desirable variety *gubba. C. bernardi* is a name given to a specimen of *C. cinereus* in which the brown spots have fused into regular brown spiral lines over the body whorl; this appears to be a very uncommon variant. Solid golden yellow shells are also found but are rare.

Although the pattern is variable, the shape of *C. cinereus* is quite constant. There is some variation in spire height and shape and in the strength of the spire sculpture; the spiral ridges may be distinct, but they are seldom strong and more commonly appear to be absent on the later whorls. The body grooves extend to the shoulder in juveniles and some adults.

Distribution.—Common in moderately shallow water in the western Pacific from southern Japan and the Philippines to the Solomons-Queensland area. How far west it ranges into the Indian Ocean is uncertain, but presumably it occurs on the western coast of Indonesia and the Malaysia area.

Synonymy.—*Conus rusticus* is considered unrecognizable, although it has commonly been used for this species. *C. nisus* has commonly been aligned with *C. stramineus* or *C. subulatus*, but the type figures seem to fit *C. cinereus* much better. Several other names here considered synonyms of *C. stramineus* or *C. subulatus* could also be considered under *C. cinereus*, as descriptions of these species are sometimes hard to distinguish although the shells themselves are very different. The various Kiener names all appear to be color variants of different rarity, although none is really common. *Conus ardisiaceus* has been applied as a synonym of *C. anemone* and *C. tinianus* and has been used for an apparently undescribed Arabian species, but it seems to fit *C. cinereus* best. *C. politus* is a pale *C. cinereus* with a yellowish to white background and reduced pattern; it is very uncommon. Doubtless other names apply to *C. cinereus* but are here considered unrecognizable; Dautzenberg, for instance, considered *C. bicinctus* Donovan a variant of *C. cinereus* and reported seeing specimens agreeing with Donovan's figure.

Notes.—The common bluish gray *C. cinereus* is the only color usually available, although *gubba* variants and intermediates are often seen. This species is seldom flawed and must lead a rather protected life, although the spire whorls are sometimes lightly eroded. Unusual colors and patterns fetch high prices and appear to actually be rare individuals. Beware of dark *Conus radiatus* sold as *gubba* variants at inflated prices; *C. radiatus* is a much coarser shell and lacks the small spots usually visible even in very dark *C. cinereus.*

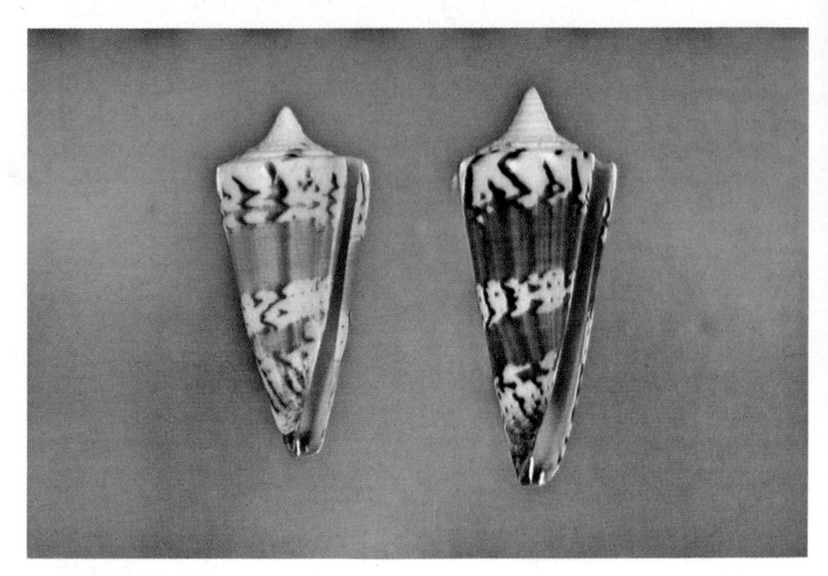

GENERALIS.—(*g. generalis*) Sulu, P.I. *Above:* 46.6 and 53.1mm. *Below:* 66.0 and 68.0mm.

GENERALIS.—(*g. maldivus*) Inhambane, Mozambique. *Above:* 39.9 and 44.0mm. *Below:* 57.7 and 59.6mm.

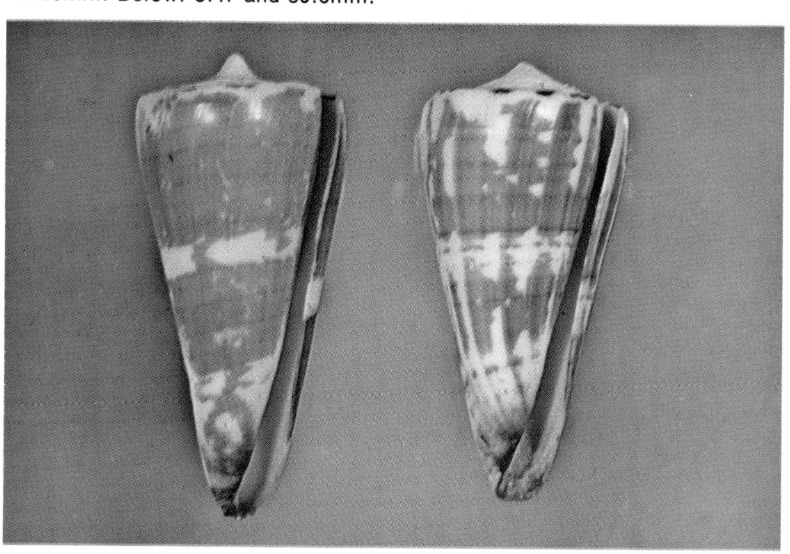

CINGULATUS Lamarck, 1810

1810. *Conus cingulatus* Lamarck. *Ann. du Mus. Hist. Nat.* (*Paris*), 15: 274 (Ocean Indien). The type is apparently lost (see Mermod, 1947), so to stabilize the name I here select as neotype the specimen in the Sollier collection (44x21mm) figured by Kiener, Pl. 93, fig. 2.
1845. *Conus undatus* Kiener. *Species gen. et icon. des coqu. viv.*, 2 (*Conus*): Pl. 94, fig. 1. 1848, *Ibid.*, 2: 210 (Indian Ocean).
1845. *Conus castaneus* Kiener. *Ibid.*, 2(*Conus*): Pl. 104, fig. 3. 1848, *Ibid.*, 2: 209 (Locality unknown).

Description.—Moderately light in weight, with a good gloss; low biconical, the sides weakly convex posteriorly and the base narrow; base with low oblique spiral ridges, these becoming heavier posteriorly and extending as flat ribs to near shoulder; ribs separated by wide, axially punctate grooves; occasionally ribs absent or reduced above midbody; shoulder angulate, carinate; spire moderately high, sharply pointed, the sides concave to straight; tops of whorls flat or slightly concave, margins carinate; spire sculpture consists of 2-3 weak spiral ridges crossed by prominent axial curved ridges. Body whorl cream to pale tan, covered with spiral lines of small brown dashes of variable size and intensity, the lines almost uniformly spaced over body whorl; occasionally the dashes are smudged and indistinct, and often they are connected into strong axial flammules; whole or part of body whorl often overlaid with dark brown axial streaks partially or entirely hiding background (if overlying brown color is strong, the pale background often shows through as small whitish tents); spire whitish, with curved brown blotches, these often fused to leave only isolated white spots. Aperture narrow, straight; outer lip thin, sharp, sloping below level of shoulder; mouth pale bluish white to white. Columella not visible externally. Length 30-50mm.

Comparison.—In form and pattern this species shows a close approach to *C. philippii*, especially the deeper water varieties; however, it lacks the variegated background colors of yellow, white, and reddish brown characteristic of this species and has heavy or at least conspicuous spiral ridges over most of the body whorl (with some exceptions). Also, the spire and shoulder in *C. cingulatus* are often largely brown with isolated white spots rather than largely whitish with small brown spots as commonly found in *C. philippii*. *C. lorenzianus* has a more rounded shoulder, lacks distinct spire sculpture, has the body sculpture generally restricted to the basal third or so, has the pattern more strongly spotted, with few and narrow axial flammules in typical patterns, and generally reminds one of *C. spurius* rather than *C. philippii*. These three species (*C. philippii*, *C. cingulatus*, and *C. lorenzianus*) have often been confused in Western Atlantic literature.

Conus virgatus of the Pacific coast is somewhat similar in shape and pattern, but it typically has no or very few spiral brown dashes and has very weak sculpture above midbody and on the spire. *C. tornatus* is much more slender and cylindrical, while *C. ximenes* also differs in shape and has much smaller dotting in the pattern.

C. eugrammatus in the broad, low-spired variant is superficially similar in shape and pattern as well as body sculpture, but it is easily distinguished by the nodulose early whorls and very strong axial ridges.

Variation.—There is some variation in width of the body whorl, but otherwise this species is quite constant in shape. The ridging of the body whorl is narrow and strong over the anterior third, becoming wider and somewhat weaker near midbody; above midbody the ridges may become narrower and even weaker, becoming obsolete below the shoulder in some specimens. Rarely, only the basal ridging is developed, the rest of the body whorl becoming smooth and glossy; this development is sometimes associated with a mostly brown pattern, resulting in a quite distinct shell.

Pattern-wise, the tendency seems to be to increase the amount of brown in the background. This is basically a creamy shell covered with numerous rows of distinct brown spiral dashes, the dashes smaller and more closely spaced anteriorly and somewhat diffuse below the shoulder. Usually at least a few irregular brown axial flammules are developed, but the dashes are easily seen. With more brown deposited, the dashes either become more indistinct or seem to fuse into continuous narrow brown spiral lines. At the rare extreme is the pattern called *C. castaneus* by Kiener, in which the entire shell is dark brown, there are fine brown spiral lines visible in the background, and a few whitish or yellowish tents and blotches break through occasionally (the whitish background). I have seen intermediates between all these patterns, but it is still hard to believe that *C. castaneus* is just an extreme development, especially as it seems that the dark pattern is also associated with reduced body sculpture.

Distribution.—As currently understood (which is very poorly), *C. cingulatus* occurs rarely in deep water and as beach specimens in the southern Caribbean near the Central American (Panama) and northern South American (Colombia, Venezuela) coasts. It seems possible that its relationship with *C. philippii* in Central America could be subspecific, but specimens are lacking.

Synonymy.—This species has been commonly, and I think incorrectly, called *C. largillierti* (=*delessertii*) and *C. lorenzianus* in the Atlantic literature, but I have no doubt that this is the species called *cingulatus* by Lamarck. To reinforce this, I have designated as neotype the specimen figured by Kiener as *C. cingulatus*, as it agrees well with Lamarck's description and Kiener's opinions on Lamarckian species are usually well founded.

Conus undatus seems to be a specimen with exceptionally open flammules and a reduced dash pattern, while *C. castaneus* is the dark form discussed under *Variation*. Both patterns seem to fall within the range of variation of *C. cingulatus*, although *castaneus* is a striking and distinctive

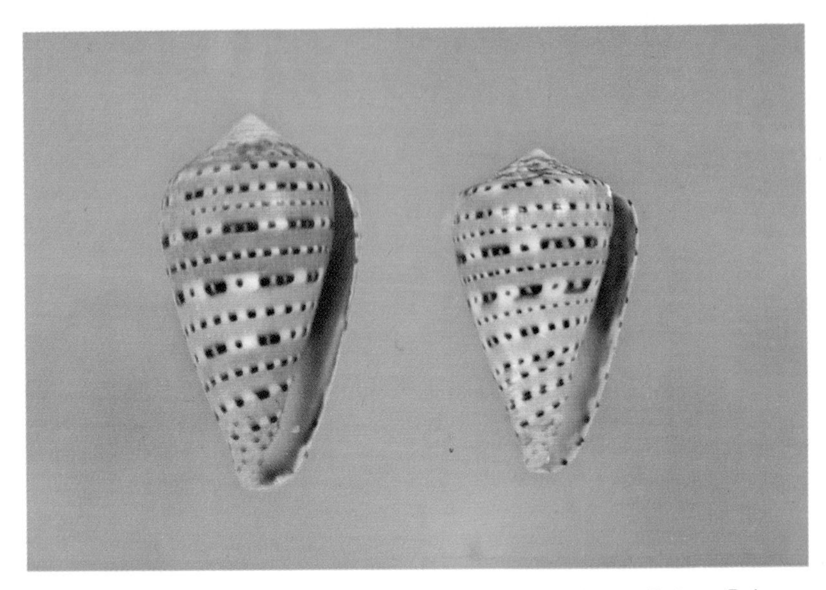

GENUANUS.—Above: Cape Verde Is., 29.5 and 34.3mm. *Below:* Dakar, Senegal, 51.2mm (GG).

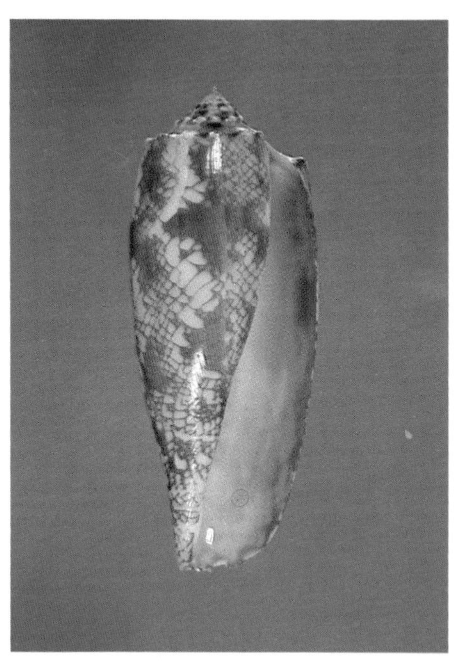

GEOGRAPHUS.—*Above:* Left: Matupit, E New Britain, Papua New Guinea, 66.4mm; Right: Kwajalein Atoll, Marshall Is., 46.5mm. *Below:* Samar, P.I.: Left: 96.4mm; Right: 96.1mm.

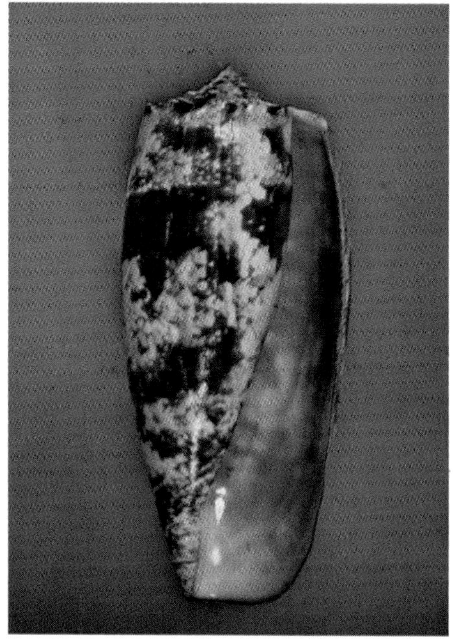

variant when the dark pattern and reduced sculpture are combined.

Notes.—Except as beach specimens, this species is not commonly available at the moment. Beach specimens have badly broken lips, reduced gloss, and sometimes eroded sculpture. Fresh specimens commonly have growth marks and chipped lips. Although probably not rare, *C. cingulatus* occurs in an area which is not heavily collected today. Good *castaneus* patterns are rare.

CIRCUMCISUS Born, 1778

1778. *Conus circumcisus* Born. *Index Rerum Naturalium Musei Caesari Vindobonensis*, 1(*Testacea*): 147 (Locality not stated). Lectotype selected and figured by Kohn, 1964.

1791. *Conus affinis* Gmelin. *Systema Naturae per Regna Tria Naturae*, ed. 13, 1: 3391 (Locality not stated). Lectotype figure (Kohn, 1966): Martini, Pl. 52, fig. 571.

1791. *Conus laevis* Gmelin. *Ibid.*, 1: 3391 (Locality not stated). Lectotype figure (Kohn, 1966): Valentyn, Pl. 8, fig. 70.

1792. *Conus dux* Hwass, in Bruguiere. *Cone*, in *Ency. Method., Hist. Nat. des Vers*, 1: 732 (East Indies). Lectotype selected and figured by Kohn, 1968 = lectotype of *C. circumcisus*.

1798. *Cucullus purpuratus* Roeding. *Museum Boltenianum:* 50 (Locality not stated). Holotype figure: Martini, Pl. 52, fig. 571.

1798. *Cucullus terebellum* Roeding. *Ibid.:* 50 (Locality not stated). Holotype figure: Martini, Pl. 52, fig. 572. Non *Conus terebellum* Linnaeus, 1758.

1881. *Conus Brazieri* Sowerby iii. *J. Conchology* (*London*), 3: 234, pl. 1, fig. 9 (Solomon Islands).

Description.—Moderately heavy, with a good gloss; cylindrical, the sides evenly convex; body whorl covered from base to shoulder with narrow closely spaced low spiral ridges, usually with every fourth or fifth ridge distinctly heavier and often granulose; numerous fine axial threads across ridges, imparting a satiny sheen to shell; shoulder rounded, narrow, not very distinct from spire, weakly to distinctly concave above; spire low, tip rounded except for projecting protoconch, the sides straight to convex, usually appearing domed; first 3-5 whorls distinctly nodulose, later whorls occasionally irregular or undulate, but usually with rounded margins; tops of whorls concave, the earlier ones nearly flat, with about three faint spiral ridges usually visible. Body whorl white to pinkish, occasionally pinkish tan or toned with violet, with two or three light brown spiral bands, one occasionally below shoulder and one above and below midbody area; the bands are usually broad and distinct, sometimes partially broken into narrower bands; spiral ridges commonly marked with variably sized brown to black or purple spots or dashes, these commonly alternating with poorly defined whitish dashes; spotting may be totally absent; weak axial flammules of brownish or violet may be present in some specimens; base of shell light; shoulder with large dark brownish to purplish axial blotches separated by pinkish white blotches, the edge of the shoulder often almost black; spire white with brownish radiating blotches; early whorls white. Aperture narrow posteriorly, much wider anteriorly; outer lip thin, sharp, evenly convex; mouth pinkish to bluish white. Columella long, narrow, sometimes set off posteriorly by a low ridge. Length 40-80mm.

Comparison.—In typical form, heavily sculptured with widely separated spiral ridges and distinct dark spots, this species is very close to *C. aurisiacus*. If it can be separated at all, it is by the tendency for *C. circumcisus* to be a brownish shell with some white and violet tones, usually with scattered dark spots, while *C. aurisiacus* is a pinkish shell with only weak brown tones and usually strongly spotted. In *C. circumcisus* the shoulder and later whorls are only shallowly concave above in most specimens, while the shoulder and last two whorls are distinctly and strongly canaliculate in *C. aurisiacus;* however, I am not at all sure that this will always separate the two species, as *C. circumcisus* is very variable in this regard. Also, *C. aurisiacus* is more distinctly shouldered than *C. circumcisus*, but this is also variable.

C. barthelemyi is strongly shouldered and in its typical patterns usually some tone of reddish seldom or never present in *C. circumcisus;* its spots, if present, are larger and more rounded than in *C. circumcisus*. If *C. gauguini* is by any chance an aberrant *C. circumcisus* population rather than just *C. barthelemyi* with poor data, then the species are even more closely related than now suspected.

C. pohlianus and various *C. magus* variants may bear a close resemblance to the smooth, unspotted variant of *C. circumcisus*, but they usually have fine brown dots in the background, more strongly nodulose early whorls picked out with small brown spots, and less concave spire whorls with heavier and more numerous spiral ridges; *C. pohlianus* and *C. magus* are usually much lighter in weight and smoother than any *C. circumcisus*.

Variation.—As is obvious from the description, this is a very variable species in both pattern and sculpture. In sculpture, it varies from being covered with numerous very weak spiral ridges which are barely feelable to having many heavy narrow rounded spiral ridges bearing strong granules. In pattern it varies from brownish with brown bands and strong violet tones to very pale tan with the bands barely defined; occasional variants are whitish with narrow brown bands and a violet band at the shoulder and sometimes at midbody. Spotting may be strong and heavy, rounded or dash-like, reduced to a few scattered small dots, or completely absent. The spire and shoulder pattern is apparently constant, however. Concavity of the shoulder varies from distinctly canaliculate to almost flat.

GILVUS.—Honiara, Guadalcanal, Solomon Is. *Above:* 22.1 and 23.7mm. *Below:* 25.6 and 27.1mm.

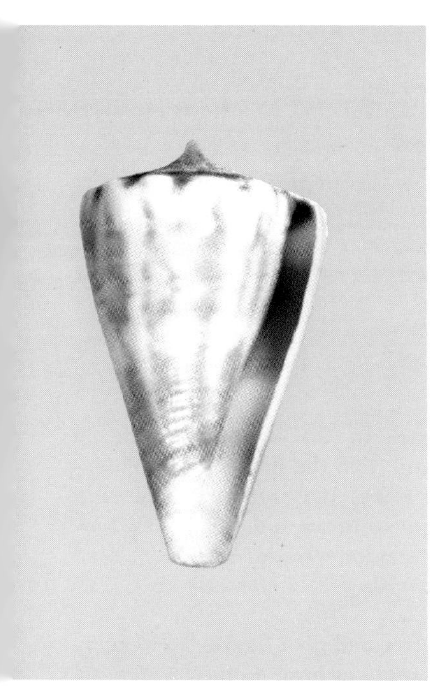

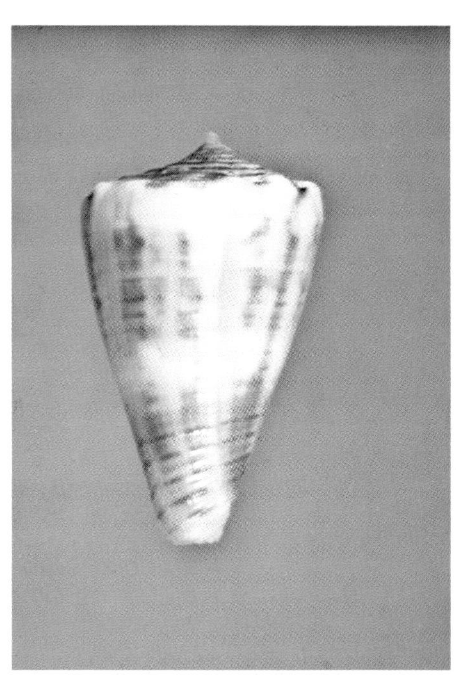

GLADIATOR.—Above: Mexico, 23.1mm (GG). *Below:* Left: Kobbe Beach, Panama, 24.0 and 25.4mm; Right: Isla Corva, Panama, 33.1mm.

Smooth, pale variant of *C. circumcisus* (DM) (photo Jayavad).

Distribution.—Apparently restricted to the western Pacific in shallow to moderately deep water. Found from the Ryukyus and Philippines south to New Guinea and the Solomons and probably to the New Hebrides. Its relationship with *C. aurisiacus* in the Indonesian and New Guinea area is uncertain, and the two may intergrade. Records of dead *C. aurisiacus* from the islands of the Hawaiian area apparently are *C. circumcisus.*

Synonymy.—The variability in pattern and poor figures led to many early synonyms which are no longer in use at any level. *Conus brazieri* is a smooth, unspotted pattern of *C. circumcisus* which is not uncommon in the Solomons and New Guinea. All possible intermediates exist with typical *circumcisus* from *brazieri*, and the name is barely worth using even as a variety.

Notes.—Gem specimens of *C. circumcisus* with heavy sculpture and bright, spotted patterns are hard to obtain and fetch a high price. An alarming number of subadult or juvenile specimens have reached the market recently, which may indicate over-collecting. Although no longer a rare shell, most specimens sold of this species have growth marks or breaks and filed lips; many are of the inferior and unattractive pale *brazieri* patterns. The *brazieri* pattern is not uncommon compared to more attractive patterns and does not deserve a premium.

I have seen dead, golden subfossil *C. circumcisus* from the Philippines being sold as *C. aurisiacus*, an error which should not fool any collector. Philippine specimens of *C. circumcisus* are usually (but not always) strongly ridged and heavily spotted, but they are quite cylindrical and not strongly shouldered like *C. aurisiacus*; the shoulder is not canaliculate.

CLARUS E.A. Smith, 1881

1881. *Conus clarus* E.A. Smith. *Ann. Mag. Nat. Hist.*, (Ser. 5), 7: 442 (West Australia).

1891. *Conus Segravei* Gatliff. *Victorian Nat.*, 7: 179, pl. (near Shoreham, Victoria, Australia).

Description.—Light in weight, with a high gloss; biconical, the upper sides straight or slightly convex; body whorl with about a dozen rounded, closely spaced spiral ridges above the base, these sometimes continued to midbody as very weak ribs; rest of body whorl smooth except for very fine spiral and axial threads; shoulder sharply angled, carinate, with a distinct marginal ridge; spire moderate to tall, sharply pointed, the sides nearly straight, the whorls distinctly stepped; spire whorls about flat above, with about 5-8 narrow, low, but distinct spiral ridges crossed by numerous fine axial threads. Body whorl pink to pinkish white, often with few or many spiral rows of fine brown dots; sometimes with several irregular axial flammules of orange-brown, these containing small triangular whitish dots; spire pinkish, with several pale to dark orange-brown radiating blotches; apex pinkish white, sometimes weakly eroded. Aperture moderately wide, uniform; outer lip very thin, sharp, strongly sloping below shoulder level, straight; mouth creamy white, turning pinkish or orange deep within. Columella not visible externally. Length 20-35mm.

Comparison.—The small size, thin structure, and biconical shape remind one at first glance of such shells as *C. mindanus*, *C. magellanicus*, and *C. jaspideus*, from all of which it differs by having distinct spire sculpture. The strong ridge at the margin of the shoulder resembles *C. jaspideus*, but also such species as *C. teramachii* and *C. ione*, although certainly *C. clarus* is not closely related to any of these.

Conus cocceus is somewhat similar to *C. clarus* in general appearance, especially pattern, but has a more rounded shoulder and is covered with numerous fine spiral and axial threads over the entire body whorl. *C. anemone* in its many forms may be similar in shape and in spire sculpture, but it is distinctly spirally striated from base to shoulder, much larger on the average, and much coarser in texture. Both these species are probably closely related to *C. clarus*, but not as closely as is *C. cyanostoma*. In *C. cyanostoma*, the shape is very similar, as are the size and the general pattern. However, in *C. cyanostoma* the shell is somewhat wider and thicker in texture, the mouth is brownish violet, the early whorls are weakly nodulose, the spiral ridges on the spire whorls are fewer and less distinct, and the color seldom is distinctly pinkish as in *C. clarus*.

Variation.—The form of *C. clarus* varies little. The color varies from pure pale pink or creamy white without obvious pattern to a mostly whitish or pinkish shell with faint traces of darker pinkish axial flammules and very fine spiral rows of brown dots. At the dark extreme are shells covered with many distinct axial flammules of bright orange-brown, each enclosing many small triangular pinkish dots (adjacent to the small brown dots, which are obscured in the flammules). Even the palest shells usually have some type of spire pattern, although it may be very faint.

GLANS.—*Above:* Coron, P.I., 28.1 and 30.0mm. *Below:* Boac, Marinduque, P.I., 28.7mm.

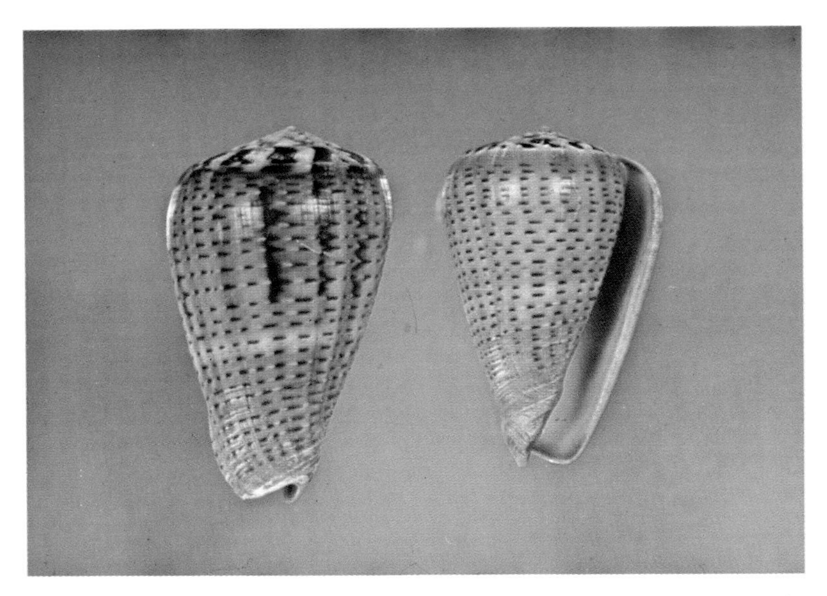

GLAUCUS.—*Above:* Lungga R., Guadalcanal, Solomon Is., 43.6 and 48.3mm. *Below:* Left: Sulu, P.I., 44.8mm; Right: Lungga R., Guadalcanal, Solomon Is., 33.1mm.

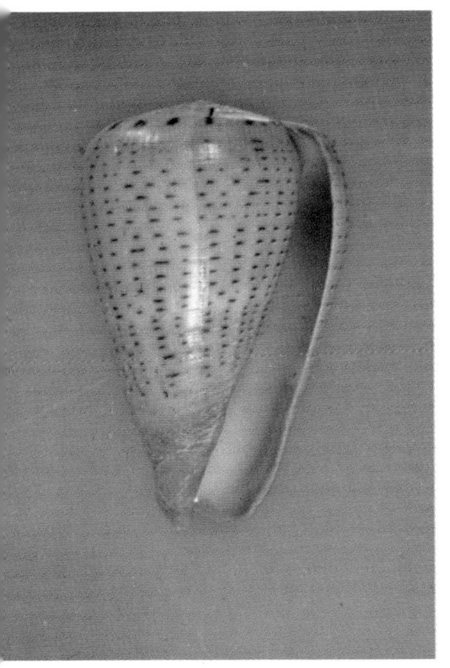

Distribution.—Confined to shallow and moderately deep waters from Victoria, Australia, westward to Western Australia. Not common.

Synonymy.—This species has long been known as *C. segravei*, with the name *C. clarus* assigned, for some unknown reason, to a shell here considered a variant of *C. nielsenae*. The original description of *C. clarus* compares the new species to *C. cyanostoma*, which is wider, describes it as pink, and comments on the heavy carinate ridge at the margin of the shoulder; the diagnosis agrees in all respects with the shell here described, and I can see no reason not to call it *C. clarus*. Although it has been hinted by various workers that *C. clarus* (*segravei*) is just a form of *C. anemone*, there seems to be no evidence to back this up; *C. cyanostoma* appears to be its closest relative, although it definitely belongs in the *C. anemone* assemblage of species.

Notes.—Although not rare and apparently locally common in some areas, *C. clarus* is seldom seen in truly gem condition. The lip is extremely thin and very fragile, while there are commonly growth flaws on the body whorl. The spire may be weakly eroded. Specimens with strong orange-brown flammules are apparently less common than paler pinkish specimens, but usually no premium is charged for bright patterns.

CLERII Reeve, 1844

1844. *Conus Clerii* Reeve. *Conchologia Iconica*, 1(*Conus*): Pl. 43, sp. 229 (Cape St. Thomas, Brazils).

1943. *Conus clenchi* Martins. *Bol. Mus. Nac. Rio de Janeiro, Zool.*, No. 12: 2 (Barra do Furado, Municipio de Campos, Rio de Janeiro, Brasil). Holotype figured by Clench, 1953 (*Johnsonia*, 2(32): Pl. 182, figs. 2-4).

1945. *Conus carcellesi* Martins. *Notas del Museo de la Plata*, 10(88): 260, figs. (Mar del Plata, Buenos Aires, Argentina).

1946. *Conus iheringi* Frenguelli. *Ibid.*, 11(Paleont.) (88): 234, pl. 1, figs. 1-8. Fossil.

Description.—Light but solid, with a good gloss; low biconical, the sides straight or slightly convex posteriorly; base narrow; body whorl with a few low spiral ridges at the base and scattered spiral and axial threads and scratches, otherwise smooth; shoulder roundly to sharply angled, broad; spire low to moderate, sharply pointed, the sides straight to slightly concave; whorls sometimes distinctly stepped, flat or weakly concave above, with traces of faint spiral ridges (2-3) and heavier curved axial ridges; first 1-4 whorls weakly nodulose, usually eroded; later whorls sometimes almost smooth and featureless. Body whorl white to pale pinkish or cream, with a variable pattern of reddish brown square spots in spiral rows and irregular long or short brownish axial flammules; occasionally most of pattern reduced and indistinct, or flammules so strong as to obscure spotting; usually the midbody area is fairly well defined and has only traces of pattern, although narrow; base white; spire white, with small to large brownish axial blotches. Aperture moderately wide, uniform in width; outer lip thin, sharp, only slightly sloping below level of shoulder, straight or slightly convex; mouth white or pale pinkish. Columella internal. Length 30-61mm.

Comparison.—*C. clerii* is very similar in appearance and sculpture to several American deep-water cones. *C. recurvus* is most distinct because of its strongly dropped outer lip and concave spire whorls, as well as differences in texture. *C. villepinii* is more distinctly nodulose on the early whorls, relatively more slender and elongated anteriorly, has more distinct spire sculpture, and has the pattern consisting mostly of two distinct irregular bands of axial reddish brown blotches above and below midbody. *C. poormani* has nodulose early whorls but also has low but distinct spiral ridges scattered over the body whorl; it has the sides distinctly parallel posteriorly just below the shoulder, which is not characteristic of *C. clerii*, and the pattern is distinctly axial, not spiral. *C. villepinii*, *C. clerii*, and *C. poormani* form a complex of very closely related species of similar ecology.

Variation.—Most pattern variation consists of differing intensities of the darker brown axial flammules and their extent. In some specimens the shell is largely white with distinct square spots and only scattered short axial blotches; in many the axial flammules are distinct and very zigzagged over most of the body whorl, forming a mostly axial pattern over the obscured spots. In others the body whorl is almost uniform brownish with the spots either not visible except near midbody or partially fused into elongate dashes. Almost solid white shells are also known. Some shells are brightly flushed with pink.

Spire shape and sculpture vary considerably and has resulted in recognition of two sympatric species differing only in minor details. In typical *C. clerii* the spire is often low, usually distinctly stepped, and the first 3-4 whorls are weakly nodulose. This type of shell also tends to have a very open body pattern. In *C. clenchi* the spire tends to be higher with straight sides, the whorls not or only indistinctly stepped, and the first whorl only nodulose; such shells also tend to have heavier patterns with more tendency to develop large areas of brown axial flammules. However, the degree of nodulation of the early whorls seems very variable, as in some other cones, and with even minor erosion such a character is worthless. The shape of the spire varies without regard to nodules on the early whorls, while intensity of pattern also appears to vary much separately. I am unable to recognize two species here, although well marked examples with taller spires that are not stepped are perhaps worth bearing the varietal name *clenchi*.

Distribution.—Found from northeastern Brazil south to northern Argentina in deeper water. Not uncommon.

Synonymy.—As mentioned under *Variation*, I can see no basis on

GLORIAMARIS.—*Above:* Bohol I., P.I., 90.0mm (MG). *Below:* Mindanao, P.I., 97.0mm (MG).

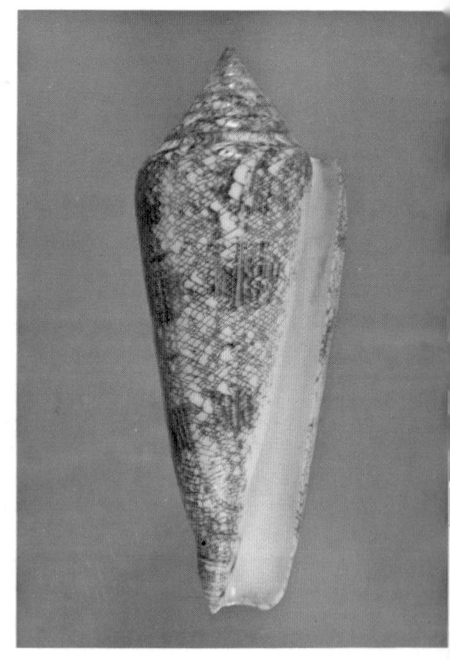

GLORIAMARIS.—Above (holotype in Zool. Mus. Amsterdam): Locality unknown, 91.2mm (photo by Dr. R.T. Abbott). ***Below:*** Guadalcanal, Solomon Is., 91mm (N. Gray).

which to recognize *C. clenchi* as a full species; it is apparently sympat ic with typical *C. clerii*. *C. carcellesi* is a *clenchi* type with numerous short axial flammules which are very irregular. *C. iheringi* is a fossil but it has also been applied, on what basis I cannot tell, to some individuals of *C. clerii*.

Notes.—Typical *C. clerii* is no longer rare and has appeared on the market in fairly large quantities. The *clenchi* variant, however, which dealers seem to distinguish more by pattern and spire shape than sculpture, is less common at the moment and fetches a heavy premium. Except for minor spire erosion, perfect specimens are common in this species.

CLOVERI Walls

Holotype: Delaware Mus. Nat. Hist.; 25.7mm; Senegal, Dakar, Harbor near Anse Bernard.

Paratypes: Collections of Phillip Clover and the author.

Diagnosis: Related to the *Conus mercator* series, especially *C. taslei*, covered with fine axial brownish lines which overlap to produce a finely tented pattern; differs from other species in having a convex and mucronate spire, pale midbody band, and broad white band on shoulder; mouth white with pale violet tone. 21-38mm length.

Description.—Moderately heavy, with a low to good gloss; low conical, the posterior sides strongly convex; body whorl with about 2-6 low spiral ridges at the base, otherwise smooth except for fine spiral threads and axial growth lines; shoulder broad, rounded, convex on top, with 2-3 low spiral ridges; spire moderately tall to tall, sharply pointed, the tip mucronate; sides of spire convex anteriorly, concave posteriorly in some specimens; early whorls often retained though eroded, sometimes first 1-2 whorls appearing very finely nodulose (because of erosion?); later whorls convex, with 2-3 low spiral ridges on top and scattered axial threads. Body whorl creamy white, covered from below shoulder to above base with wavy narrow brown axial lines that overlap occasionally to produce fine tents; base white to pale tan; a distinct broad lighter spiral midbody band with sharp edges, formed by the lines being slightly narrower and more widely spaced; a broad white band below shoulder and onto the shoulder, without distinct markings, the axial lines usually stopping abruptly below the shoulder; spire white, without pattern or with short fine brown lines at the margins of the later whorls; early whorls eroded white. Aperture moderately wide, slightly wider anteriorly; outer lip thin, sharp, evenly convex; anal notch shallow or obsolete; mouth white, with pale violet tones. Columella very narrow. Length 21-38mm.

Comparison.—This species needs to be compared only with *Conus mercator* and *C. taslei*, which it closely resembles. From most *C. mer-*

cator it is easily separated by the more distinctly mucronate spire, the tip being rather distinctly set off from the rest of the spire; it differs greatly in pattern from *C. mercator* (even when broadly defined), especially in the presence of the broad colorless band at and below the shoulder and the virtual absence of a spire pattern. *Conus taslei* is similar in shape on occasion and also has a pattern of axial brownish lines; however, the lines in *C. taslei* are typically parallel or nearly so, seldom forming a distinctly tented pattern; in *C. taslei* the lineate area is paler than the two or three broad darker spiral bands where the lines are more condensed; there may be small white spots on the body whorl; and the lineate pattern usually continued onto the spire, which may be blotched.

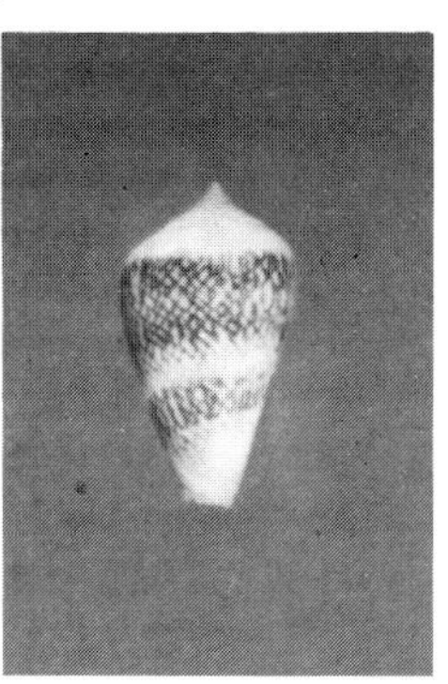

C. cloveri, paratype, Senegal, 21mm (MM).

It is of course possible that *C. cloveri* is one of the many color patterns of the complex species known as *C. mercator*, *C. taslei*, and even *C. ermineus*. However, it seems distinctive enough in shape, coarsely tented pattern, and color to be described as new; if it must be sunk when someone finally untangles the West African cones, so be it. In the meanwhile, the name serves the purpose of identifying a very distinctive shell.

Variation.—The spire varies considerably in height and development of the tip, from rather low to very high. The axial lines are usually strong and form distinct tents, the pattern being easily visible without magnification; the pale midbody band is sharp and distinct, even in specimens with pale patterns, and the shoulder band marks the abrupt end of the axial pattern. Occasional specimens are white, without pattern, but retain the distinctive shape.

Distribution.—So far known only from Senegal in shallow water; it is uncommon.

Notes.—This species has been distributed under the name of *Conus textilinus* Kiener, but that name is a synonym of *Conus acuminatus* and has nothing to do with the present species.

Perfect specimens are rare, most specimens having heavy erosion on the spire and bad scars or axial growth flaws. The animal of this species is pale, quite different from the dark animals of *Conus mercator* found sympatrically, but the value of animal coloration in cones is very uncertain.

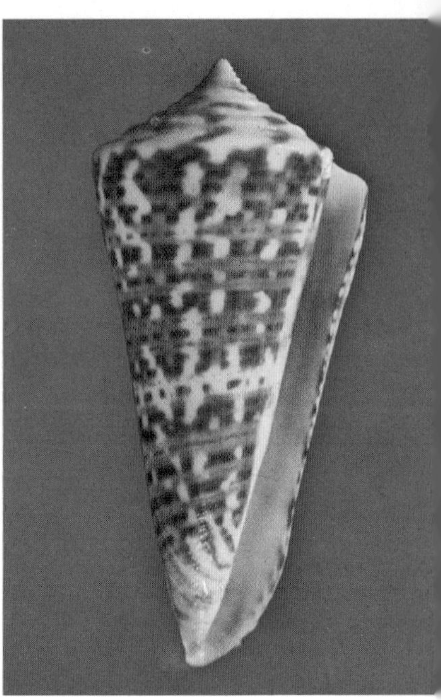

GRADATUS.*—Above:* Left: San Pedro Martir I., Sonora, Mexico, 56.2mm; Right (typical): La Paz, Baja California, Mexico, 66.2mm (photos by A. Kerstitch). *Below:* Guaymas, Sonora, Mexico: Left: 41.8mm; Right: 32.7 and 34.4mm (GG).

GRADATUS.—Above: Left: Guaymas, Sonora, Mexico, 48.4mm (photo by A. Kerstitch); Right: San Carlos, Sonora, Mexico, 23.7mm. *Below:* Left: San Pedro Martir I., Sonora, Mexico, 39.0 and 41.2mm; Right: Guaymas, Sonora, Mexico, 35.8-44.1mm (photos by A. Kerstitch).

As mentioned earlier, it is possible that *C. cloveri* will eventually be shown to fall within the range of variation of *C. mercator* or some other western African cone, but at the moment it seems quite distinct and more easily recognizable than most other taxa of that area.

Conus cloveri is named for Mr. Phillip W. Clover, dealer and collector, who collected the type series.

COCCEUS Reeve, 1844

1844. *Conus cocceus* Reeve. *Conchologia Iconica*, 1(*Conus*): Pl. 42, sp. 228 (New Holland).

1845. *Conus citrinus* Kiener. *Species gen. et icon. des coqu. viv.*, 2 (*Conus*): Pl. 59, fig. 6. 1849, *Ibid.*, 2: 248 (Ceylon). Non *Conus citrinus* Gmelin, 1791.

1845. *Conus decrepitus* Kiener. *Ibid.*, 2(*Conus*): Pl. 99, fig. 4. 1849, *Ibid.*, 2: 263 (New Holland).

1858. *Conus Kieneri* Crosse. *Rev. Mag. Zool.*, (Ser. 2), 10: 122. Nomen novum for *C. citrinus* Kiener, 1845. Non *C. kieneri* Reeve, 1849.

Description.—Light in weight, with a low gloss; low conical, rather cylindrical, the sides convex; body whorl covered from base to shoulder with low but distinct undulating spiral ridges, these somewhat weaker near midbody; numerous fine axial threads also cover body whorl, giving the shell a finely cancellate appearance; shoulder rounded, not very distinct from spire; spire moderate, the sides straight to slightly convex, bluntly pointed; spire whorls slightly convex, tops with about 3-5 distinct spiral ridges, continuous with those of body whorl, crossed by fine axial threads. Background of body whorl and spire usually pinkish white to pale yellow, this overlaid with either or both of the following patterns: 1) almost invisible spiral rows of small opaque white dots; 2) dark yellow to light carmine reticulations which tend to form two weak and irregular rows of blotches above and below midbody; these patterns may vary greatly in intensity and contrast from shell to shell; spire usually with continuation of body pattern. Aperture fairly wide, especially anteriorly; outer lip thin, straight, strongly sloping below level of shoulder; mouth pinkish to white. Columella not visible externally. Length 25-50mm.

Comparison.—*C. cocceus* is distinctive in combining a rather cylindrical shape, numerous spiral ridges, and a basically whitish color with usually indistinct pattern. *C. clarus* is distinctly biconical with a carinate shoulder and half smooth body whorl. *C. anemone* in pale phases usually has a broader shoulder and lower spire and would lack the spiral rows of small opaque white dots. *C. cocceus* is close to *C. anemone* but seems distinctive enough, especially when the red pattern is well developed.

Variation.—As indicated in the description, this species is very variable in color and pattern. Most all-white shells will, under close examination, reveal spiral rows of fine pearly spots; most white shells also have indistinct pink tones and traces of pattern under low illumination or blacklight. Juvenile shells tend to have more solid pink coloration, assuming the adult pattern with growth. The amount of yellow to reddish reticulation and blotching is highly variable from shell to shell; both colors tend to turn orange after death and exposure to sunlight.

The spiral sculpture is very fine in some specimens and appears to be easily abraded after death or under rough conditions. Axial threads on the body whorl may be strong or almost invisible. The protoconch is projecting in juveniles but commonly lost in adults.

Distribution.—Restricted to moderately shallow waters of Western Australia between Albany and Cape Naturaliste. Locally common.

Synonymy.—The Kiener names have long been considered synonyms of *C. cocceus*. Reeve's specimen was quite bad, which led to the confusion.

Notes.—Adult specimens with bright red patterns are not commonly seen, most large specimens now offered being pure white. Juveniles are more likely to have bright colors, at least pink overtones. Beach specimens seem to be especially ugly, often assuming dirty brown casts and losing much of the sculpture with relatively little wearing. The outer lip is thin and subject to chipping. In addition, *C. cocceus* seems to be subject to large breaks which leave bad scars. A large, truly gem specimen with bright colors would be a really exceptional specimen.

COCCINEUS Gmelin, 1791

1791. *Conus coccineus* Gmelin. *Systema Naturae per Regna Tria Naturae*, ed. 13, 1: 3390 (Locality not stated). Holotype figure: Knorr, Pl. 24, fig. 2. It should be pointed out that this figure is in all likelihood a specimen of *C. vittatus*; the situation should definitely be investigated.

1791. *Conus Ammiralis anglicus* Gmelin. *Ibid.*, 1: 3379 (Locality not stated). Lectotype figure (Kohn, 1966): Knorr, Pl. 24, fig. 2 (see above).

1830. *Conus Solandri* Broderip and Sowerby i. *Zool. J.*, 5: 50, suppl. pl. 40, fig. 4 (Tahiti).

Description.—Moderately light, with a low gloss; rather cylindrical, the sides nearly straight and parallel until suddenly pinched-in to base; body whorl with numerous closely spaced spiral ridges from base usually to near shoulder, these often strongly granulose over part of body whorl; occasionally spiral ridges reduced or even absent over most or all of whorl; fine axial threads sometimes present; shoulder roundly angled, undulate; spire low to moderate, bluntly pointed, the sides about straight; tops of whorls with four or five strong spiral ridges crossed by axial

GRADATUS.—(*scalaris*) *Above:* Cabo San Lucas, Baja California, Mexico, 40.1-51.3mm. *Below:* Guaymas, Sonora, Mexico, 29.8-49.1mm (all photos by A. Kerstitch).

GRANGERI.—Above: Off Lubang I., P.I., 47.2mm. *Below:* Left: Luzon, P.I., 65mm (DM) (photo by Jayavad); Right: Marinduque, P.I., 46.2mm.

threads; early whorls distinctly nodulose, later whorls becoming undulate to crenulate. Body whorl light tan, dark brown, red, or yellowish, with a single wide midbody band of white interrupted from above and below with squarish black spots and several spiral rows of small squarish black dots; spiral ridges occasionally demarked with narrow brown lines or dashes; base paler than rest of whorl; spire and shoulder dark brown to reddish, sometimes with faint yellowish tan blotches; early whorls pale. Aperture narrow above, distinctly widened anteriorly; outer lip thin, fragile, nearly straight; mouth bluish white, the exterior pattern usually showing through. Columella not visible externally. Length 30-50mm.

Comparison.—Of the several cones with a broad white central band interrupted with brown, only *C. coccineus* has a distinctly undulate or crenulate shoulder and almost unicolored spire. *C. vittatus* is very similar in shape and sometimes body sculpture, but it commonly shows three white bands and has a contrastingly patterned spire. *C. granulatus* and *C. artoptus* are narrowly cylindrical with more domed spires which have only the early whorls distinctly nodulose. *C. cumingii* is usually smooth, has a white-blotched spire, and has fine brown lines and dashes visible through the dark brown coloration. The doubtful *C. henoquei* has stepped spire whorls which are divided at the middle by a strong groove; it is more strongly angled at the shoulder.

Variation.—Although *C. coccineus* is usually distinctly spirally ridged, there does appear to be a tendency to reduce the sculpture in some specimens. Thus the ridges just below the shoulder are commonly reduced or absent, and in some specimens the whole body whorl is almost smooth except for the basal ridging. At the other extreme, some heavily sculptured specimens have the ridges with large granules for at least a few spiral rows, usually above midbody. The shoulder is variably undulate, sometimes strongly so, other times barely so.

In color there is great variation, apparently individually. The most common color appears to be light to dark brown, but this can change on a single specimen to a distinct reddish tone. Dark reds and bright scarlets are also moderately common, while yellow or pale orange seems to be uncommon. Fine brown spiral lines may be present in some individuals. If pale blotches are present on the spire, they are usually yellowish tan and not strongly contrasting with the brown of the spire whorls.

Distribution.—Moderately common in shallow water from the Ryukyus through the Philippines to New Guinea, the Solomons, and probably Queensland. Its western limit is unknown to me, but presumably it occurs throughout the Indonesian area.

Synonymy.—*C. solandri* is certainly this species. The lectotype of *C. anglicus* as selected by Kohn and the holotype of *C. coccineus* appear to be a figure of *C. vittatus* instead of the species currently known as *C. coccineus*. Since *C. coccineus* is based only on that figure, I do not see what would be gained from a complicated exchange of names, this species taking the name *solandri* and our *vittatus* becoming *coccineus*. This would seem to be a problem in complicated nomenclature which may eventually call for a ruling from the Commission. The holotype figure of *coccineus* has a strongly variegated spire and distinct brown spots on the spiral ridges of the body whorl, both characteristic of the shell here called *vittatus*. For the moment it is best to otherwise ignore this problem and continue current usage.

Notes.—Small growth flaws are common in this species, as are minor chips in the delicate lip. Bright red specimens with heavy granulations should be worth a premium over normal brown specimens, but most dealers do not distinguish the colors.

COELINAE Crosse, 1858

1858. *Conus Coelinae* Crosse. *Rev. Mag. Zool.*, (Ser. 2), 10: 117, pl. 2, fig. 1 (New Caledonia).
1943. *Conus spiceri* Bartsch and Rehder. *Proc. Biol. Soc. Washington*, 56: 87 (Sand Island, Midway Island Atoll, Hawaiian Islands).
1972. *Conus berdulinus* Veillard. *Of Sea and Shore*, 3(4): 176, figs. (Off coast of Reunion, near port of Pointe des Gialets).

Description.—Heavy, with a low gloss in clean specimens; obconical, the sides straight or somewhat concave, strongly angled to narrow base; body whorl elongate; body whorl with weak spiral threads and ridges, these strongest basally and usually merely wrinkles above midbody in most populations; numerous fine axial threads and growth lines over body whorl produce a satiny sheen; shoulder broad, sometimes concave above, its margin strongly angulate or carinate; spire nearly flat, somewhat concave on later whorls but forming a low and distinct central cone that is bluntly pointed; spire whorls separated by strong margins, covered with very weak spiral and axial threads. Body whorl solid white, waxy yellow, occasionally distinctly pinkish, orangish, or violet, the base not contrastingly colored but usually somewhat paler tan or cream; spire usually cream or white on the later whorls, the central cone often waxy yellow. Aperture narrow, uniform; outer lip thin, sharp, straight or concave at middle; mouth white to bluish white, sometimes pink or violet. Columella long and slender, not set off by a strong ridge in most specimens. Length 50-150mm.

Comparison.—There has always been a great deal of doubt that this species was really distinct from *C. virgo*. However, *C. coelinae* is a more elongated and slender cone with a sharply angled shoulder and a distinct central cone on the spire; in *C. virgo* the shape is more cumbersome and broader at the shoulder, the shoulder is roundly angled in most populations, and the spire is uniformly convex. *C. virgo* of course also has the base and adjacent lip strongly colored with bright purple or violet. *C. emaciatus* is similarly marked at the base, smaller at adult size, and

GRANULATUS.*—Above:* Plantation Key, Florida, 27.2mm (GG). *Below:* Left: Off Kralendijke, Bonaire, Neth. Ant., 40mm (photo by A. Kerstitch); Right: Agony Beach, Aruba, Neth. Ant., 38.0mm (GG).

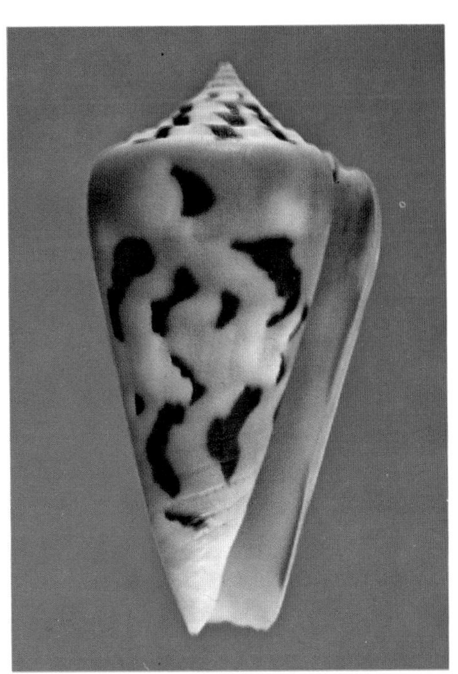

GUBERNATOR.—Maldive Is. *Above:* 48.9mm. *Below:* Left: 61.4mm (MM); Right: 32mm (photo by A. Kerstitch).

usually has heavier spiral ridging over the anterior half of the body whorl. *C. terebra* has a higher spire in most areas and a strongly rounded shoulder, as well as strong spiral sculpture on the body whorl and spire. *C. quercinus* is much broader, has distinct sculpture on the spire, and usually is covered with fine spiral brown lines.

Variation.—This poorly known species seems to be separable into three questionably distinct subspecies or at least seemingly isolated geographic types.

C. coelinae coelinae: Rather glossy white to waxy yellow, slender and strongly tapered. Adult length about 50-70mm. New Caledonia and the Marshall Islands, probably more widely distributed but confused with *C. virgo.* Uncommon.

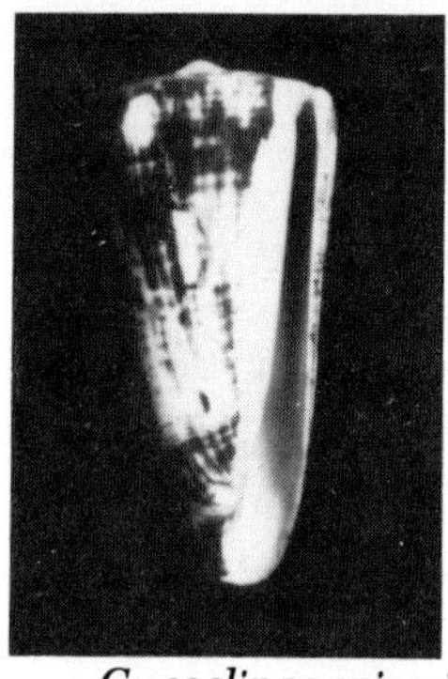

Holotype of *C. coelinae* (Inst. Roy. Sci. Nat. Belgique) (photo Abbott).

C. coelinae spiceri: Roughened with heavy spiral threads and axial lines in large adults, which also tend to be broader and less tapered to the base; color waxy yellow to white. Adult size 80-150mm. Juvenile shells like *C.c. coelinae* in shape and color. Large adults tend to distort the spire and body whorl shape. Restricted to Hawaiian Islands chain. Is it possible that this is just the result of typical *coelinae* larvae that have been swept from the Marshalls or other islands and reach gigantic, distorted sizes in Hawaii? Rare.

C. coelinae berdulinus: Very much like *C.c. coelinae,* but a little more elongated and more slender at the shoulder; colors tend to be brighter, often deep pink, orange, or violet. Adults 80-120mm. At the moment specimens come mostly from the Philippines, where it is considered rare; the taxon was described from a specimen supposedly from Reunion. The spiral ridges on the body whorl of this subspecies may be a bit stronger than in typical *C. coelinae.* Commonly known as *C. 'kintoki,'* which appears to be a nomen nudum.

Distribution.—Poorly known; at least New Caledonia, Hawaiian Islands, Philippines, and Marshall Islands; reported from Reunion. Probably widely distributed over the Indo-Pacific but confused with *C. virgo.*

Synonymy.—The three subspecies are obviously poorly defined and all could easily be encompassed within the binomial *C. coelinae;* they are retained as subspecies because of geographic separation (doubtful) and current differentials in rarity and general appearance.

Notes.—This heavy cone usually has a small break or two and some erosion on the spire. Brown stains seem to be common at the base and spire. The doubtful status of the Philippines shells make their value doubtful, although they are apparently not common and are dredged in deep water; their bright colors are certainly worth a premium over typical *C. coelinae*, but it is doubtful whether it is really worth the 20-fold difference in values.

COLUBRINUS Lamarck, 1810

1810. *Conus colubrinus* Lamarck. *Ann. du Mus. Hist. Nat.* (*Paris*), 15: 433 (Mers des grandes Indes). See Kiener, Pl. 82, fig. 3, for a representation of the holotype.

Description.—Heavy, thick, with a good gloss; low conical, cylindrical or broad, the sides about straight; body whorl covered from base to near shoulder with numerous very fine but feelable spiral ridges, these becoming weaker and less distinct above midbody and often absent below shoulder; fine axial threads and growth lines usually present; shoulder broad, rounded, not very distinct from spire; spire low to moderate, domed, bluntly pointed, the sides slightly convex to slightly concave; tops of whorls with traces of fine spiral and axial threads, not really distinct; early whorls, when not eroded, without nodules; sutures between whorls weak and indistinct. Body whorl bright orange to orange-brown, the base strongly tinted with lavender to produce a bright violet-brown tint; whole whorl covered with spiral rows of small white triangles and spots, these randomly combining to form many large to small white tents; tents may be concentrated in a midbody band or completely scattered; fine spiral lines of dark brown dashes visible in background; shoulder and spire orange with a few large white, mostly axial, blotches and tents; early whorls usually eroded white. Aperture moderately wide, almost uniform except for some widening anteriorly; outer lip straight, thick; mouth white. Columella rather short, narrow, weakly set off from inner lip. Length 40-110mm.

Comparison.—The fine brown dashes in the background and the lack of nodules on the early spire whorls ally this species with *C. pennaceus* and *C. magnificus*. In shape it is somewhat intermediate, the typical form having a rather tall spire and slender body whorl, while the broad form is much more *pennaceus*-like. The distinct spiral ridges usually extending over at least two-thirds of the body whorl will distinguish it from most *C. pennaceus*, while the violet-brown base will separate it from most specimens of *C. magnificus* (where the base is commonly whitish). The bright orange to orange-brown coloration is seldom found in either *C. pennaceus* or *C. magnificus* and I have never seen it in com-

GUINEENSIS.—*Above* (*g. informis*): Left: Jeffreys Bay, Cape, South Africa, 43.8mm; Right: Swakopmund, South-West Africa, 35.8mm. *Below* (*g. guineensis*): Left: Jeffreys Bay, Cape, South Africa, 39.7mm; Right: Cape, South Africa, 18.6-22.9mm.

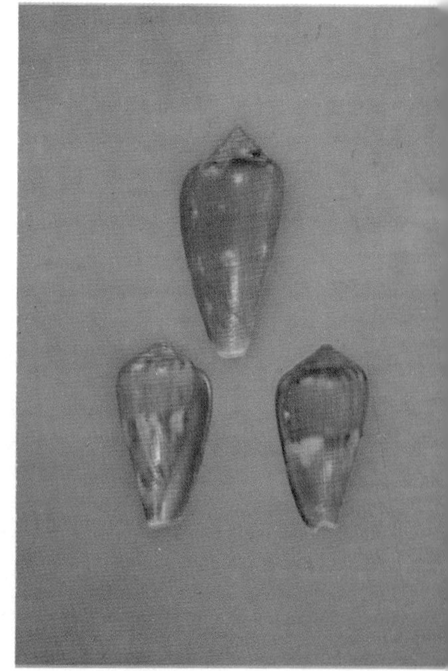

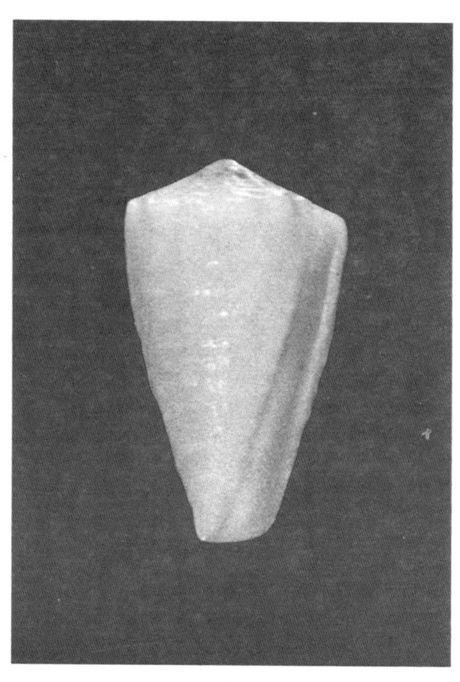

HAVANENSIS.—Greater Exuma I., Bahamas. *Above:* 18.8mm. *Below:* 14.4 and 18.7mm (all EP).

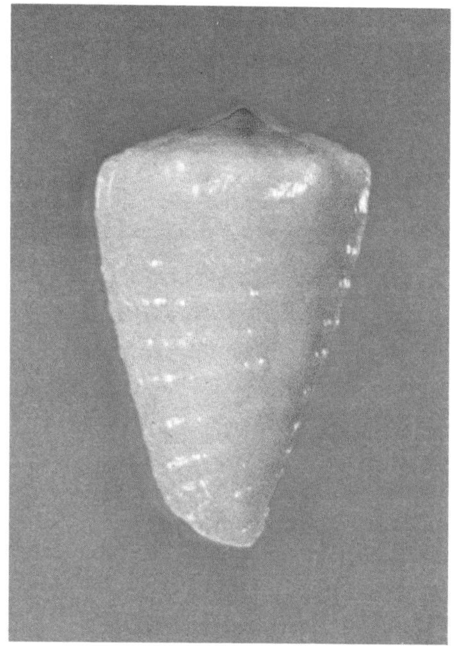

bination with the dark base. This appears to be a distinct species of the *pennaceus* complex.

Occasionally variants of *C. magus* may approach *C. colubrinus* in shape and color enough to confuse the collector. However, *C. magus* has strong spire sculpture and usually traces of small brown dots over the body whorl (not like the brown dashes of *C. colubrinus*); *C. magus* is also much lighter in weight. Some bright orange *C. ermineus* have long caused confusion, but again these have distinct spire sculpture (at least on the early whorls), usually have broader apertures, are always different in shape, and do not have the brownish tinting at the base; in addition, the white spotting in *C. ermineus* is seldom triangular and not formed from the numerous lines of white dots as in *C. colubrinus*; there are seldom traces of brown dashes in regular rows in *C. ermineus*. See also *C. nobilis lamberti*.

Variation.—Although not as variable as most tented cones (probably simply because it is so narrowly defined), this species seems to have two ecological forms. One is the typical or narrow form, usually with an elongate-cylindrical shape, the shoulder rather narrow and the spire moderately high. The other is called the broad form and has a lower spire, broad shoulder, and generally chunkier and more cumbersome body whorl. The white tenting may be sparse or dense; the basal coloration appears more-or-less constant. With extreme size the sides become concave and the white tents fuse into large axial bands.

Distribution.—A rare or at least uncommon shell over the western Pacific from the Ryukyus and Philippines south to New Caledonia and east and north at least to the Marshall Islands; also recorded from Samoa and Ceylon. Found in moderately deep water by divers.

Synonymy.—This is the species which has long been called *C. crocatus* Lamarck. However, Lamarck's description and placement of that name, as well as Kiener's figure of the type, leave little doubt that *C. crocatus* is based on a bright orange specimen of *C. ermineus*; it is described after *C. narcissus*, an admitted variant of *C. ermineus*, and I have seen orange *ermineus* which resemble closely Kiener's figure of the type. *C. colubrinus* as described by Lamarck and figured by Kiener is certainly this species and not the axially blotched variant of *C. pennaceus* for which the name *colubrinus* has been variably used.

Very large, rather deformed specimens of *C. colubrinus* from New Caledonia have been sold recently as *C. lamberti* on the basis of their orange coloration and large size. As indicated by several photos in the 1976-1977 *Hawaiian Shell News*, there is complete intergradation between typical small *C. colubrinus* and the large *'lamberti.'* As noted under *C. nobilis*, I have found the name *C. lamberti* almost impossible to place, but the description of the spire sculpture indicates a form related to *C. nobilis* and not to *C. colubrinus*. Perhaps the true *C. lamberti* is being confused with giant *C. colubrinus* in New Caledonia.

Notes.—The slender shape of the typical form makes it a more attractive shell than the broad form, but the bright orange coloration of either form is very attractive. Apparently the spire is normally eroded in this species, although pattern remains over the last 3-4 whorls in most cases. Minor growth flaws detract little from the value of this shell, which seems to be rare everywhere except New Caledonia, where it is uncommon.

COMATOSA Pilsbry, 1904

1904. *Conus dormitor* Pilsbry. *Proc. Acad. Nat. Sci.* (*Philadelphia*), 56: 6, pl. 1, figs. 9-9a (Kikai, Osumi, Japan; fossil). Non *Conus dormitor* Solander, 1766, a fossil.
1904. *Conus comatosa* Pilsbry. *Ibid.*, 56: 550. Nomen novum for *C. dormitor* Pilsbry, 1904.

Description.—Light in weight but solid, with a good gloss; elongate biconical, the sides straight and tapering to a very narrow base; body whorl covered with rather narrow flat ribs from base to shoulder, heavily developed although narrower and less distinct below shoulder; ribs separated by narrow, deeply punctate spiral grooves; shoulder narrow, carinate, with a raised marginal rim; spire tall, sharply pointed, the sides concave; tops of whorls deeply concave, bearing about 3-4 low spiral ridges crossed by heavy, widely spaced curved axial ridges; first 5-7 whorls strongly nodulose, next 1-2 undulate, last and shoulder strongly carinate with raised rims. Body whorl white to cream, covered with numerous spiral rows of distinct squarish brown spots on the spiral ribs; usually with four spiral bands of large, laterally fused brownish blotches, one at shoulder, one above and below midbody, and one at base; shoulder carina with distinct squarish brown spots alternating with white spots; spire whitish, marked with many squarish brown spots and more irregular blotches. Aperture moderately narrow, almost uniform in width; outer lip straight, fragile, strongly sloping below level of shoulder; mouth white to pinkish, the external pattern sometimes showing through. Columella long, narrow, somewhat oblique. Length 30-55mm.

Comparison.—In shape and sculpture, only *C. orbignyi* really is close. It is readily distinguishable by the nodulose shoulder lacking a raised carina and by the more convex upper sides. *C. ichinoseana* lacks spiral sculpture on the tops of the spire whorls and has a weaker body sculpture as well as a different pattern. *C. schepmani* is usually under 35mm long, has a more attenuated anterior extremity, has the tops of the spire whorls perhaps more heavily cancellate, and has a distinctive pattern of opaque white on the body whorl. *C. eugrammatus* is broader and lower-spired, not nearly as elongated, with generally narrower body ribs and weaker grooving. *C. aculeiformis* is smaller, more delicate, lacks the raised carina on the shoulder, lacks the strong pattern usually found in

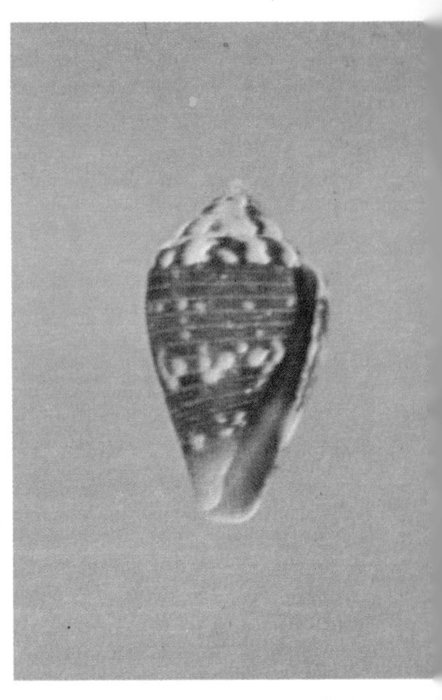

HIEROGLYPHUS.—Aruba, Neth. Ant. *Above:* Left: 16.0mm; Right: 14.6mm (PC). *Below:* 18.0mm (PC).

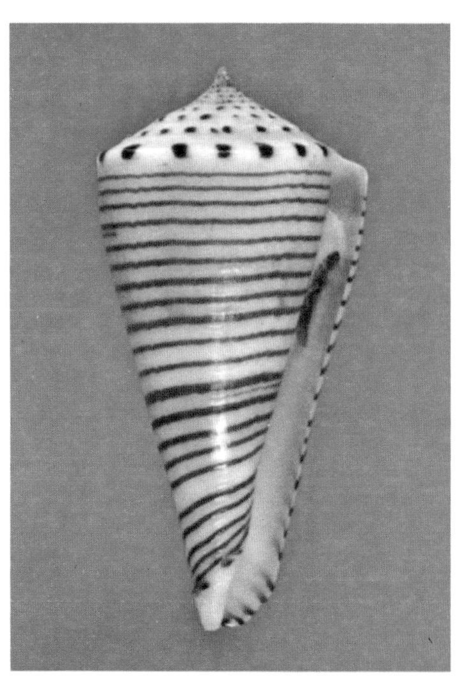

HIRASEI.—Above: Off Tosa, Shikoku, Japan, 55.8mm (EP). *Below:* Taiwan, 44.2mm (GG).

C. comatosa, and has fewer nodulose whorls. *C. hypochlorus* is deep yellow in color without a distinct pattern; only the first few whorls are nodulose. Of all these species, some patterns of *C. aculeiformis* could easily be confused with the weak pattern of *C. comatosa;* however, typical *C. aculeiformis* has the spire whorls with a modified spiral sculpture, two broad ribs separated by a groove containing a narrow ridge or two; there is no distinct shoulder carina in *C. aculeiformis*; obviously the two are closely related.

Variation.—In shape and sculpture this is a very constant species, with only a little variation in width of the ribs below the shoulder and in exactly how many whorls are nodulose and how many undulate. In pattern, however, there are two distinct types: strong and weak. The strong pattern is the most characteristic, with four spiral bands of more-or-less axial brown blotches that tend to fuse into complete or nearly complete spiral bands. The weak pattern has only the numerous squarish brown spots on the ribs, often aligned in an axial pattern, with the large blotches absent or very poorly defined. In both patterns the spire whorls and shoulder carina are distinctively patterned with alternating brown and white squares, a distinctive feature of the pattern.

Distribution.—Seemingly restricted to the China Sea in deep water, from southern Japan to the Taiwan area; probably also in the Philippines. Uncommon to rare.

Synonymy.—*C. comatosa* is of course a replacement name for the preoccupied *C. dormitor* Pilsbry; *dormitor* cannot be used for this species. The relationship of this species to *C. aculeiformis*, *C. orbignyi*, and *C. schepmani* should be carefully checked as more specimens become available; the weakly patterned form especially bears a close resemblance to *C. aculeiformis* in some patterns. *C. comatosa* was described as a fossil; I hesitate to use a fossil name based on a specimen without color pattern for a living cone because of the importance of color pattern in cone taxonomy. However, *C. comatosa* has distinctive sculpture and shape for a shell its size, and there is no other available name, so I guess it must be used.

Notes.—This deep-water species is seldom offered in either pattern; it is delicate and subject to rather bad growth flaws and chipped lips. It seems likely that the weak pattern has been confused with *C. aculeiformis* in some instances. The strong pattern is an attractive shell and should be worth a heavy premium over the weak pattern when the species becomes available in larger quantity.

CONCOLOR Sowerby i, in Sowerby ii, 1834

1834. *Conus unicolor* Sowerby i, in Sowerby ii. *Conchological Illustrations:* Pt. 54, fig. 59 (Locality not stated). Non *Conus unicolor* Sowerby i, 1833.

1834. *Conus concolor* Sowerby i, in Sowerby ii. *Ibid.*: large list. Nomen novum for *C. unicolor* Sowerby i, 1834.

Description.—Light in weight, with a low gloss; obconical, the upper sides nearly straight or distinctly convex, the base narrow; body whorl with about 6-10 strong spiral ridges at the base, otherwise smooth except for scattered spiral and axial threads; basal ridges separated by wide shallow grooves which are sometimes punctate; shoulder wide, roundly angled, with many low spiral threads crossed by scattered axial threads, nearly flat above; spire low to flat, the sides almost straight; whorls irregular at margins, sometimes appearing crenulate; whorls with about 3-6 low spiral ridges crossed by numerous axial ridges to produce a finely but weakly cancellate appearance, especially on the early whorls. Occurs in two rather distinctive patterns; one has the body whorl almost uniform deep blackish brown to olive brown, sometimes with faintly visible spiral brown dashes at growth flaws and new growth; the other pattern is pale tan to light brown with few to many spiral brown dashes visible, especially concentrated in the midbody area; spire same color as body whorl, unpatterned, the early whorls sometimes pale or pink. Aperture rather narrow, uniform in width; outer lip thin, slightly convex; mouth pale bluish white in light phase to violet with darker brownish blotch deep within in dark phase. Columella very narrow, mostly or all internal. Length 30-50mm.

Comparison.—*C. concolor*, although of simple structure and pattern, could only be confused with *C. gilvus* in the dark phase. *C. gilvus* is very similar in general appearance but readily distinguished by the presence of distinct bluish bands at the midbody and shoulder, with small brown spots visible on the shoulder and spire whorls. *C. concolor* is somewhat more slender than *C. gilvus* and perhaps a bit thinner in structure, as well as being darker in most specimens. The pale phase of *C. concolor* could perhaps be confused with a dwarf *C. quercinus*, but it is much lighter in weight, usually more slender across the shoulder, has a distinctly although weakly cancellate spire, and has a greatly reduced columella. Paler colors of *C. hyaena* are also similar in general appearance, but that species has a higher spire with a mucronate tip, usually has many spiral brown lines visible, has the brown pattern rather axially flammulate in most shells, is heavier and more cumbersome, and has the early whorls nodulose.

Variation.—The two patterns seem fairly distinct, although the dark pattern apparently can revert to a light pattern with new growth. The spire may be almost flat or moderately low, the tip either sharply pointed or blunt. Convexity of the sides varies a bit, but not enough to cause confusion. Whether the two patterns are just ecologically separated or geographic is uncertain. The dark pattern is currently the easiest to obtain.

Distribution.—Uncommon in the western Pacific from the Philippines south through New Guinea and the Solomons to the New Hebrides.

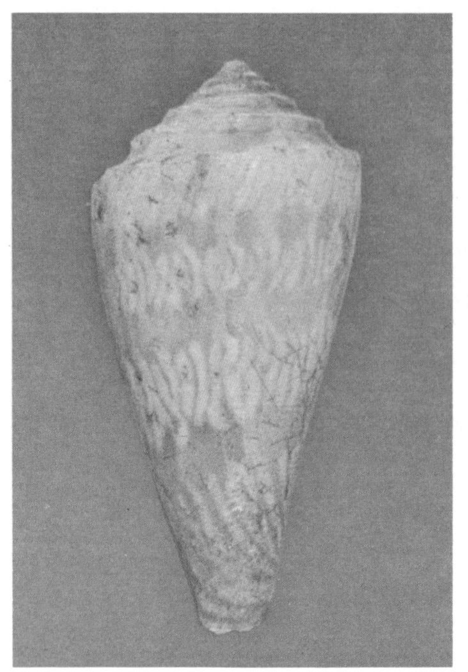

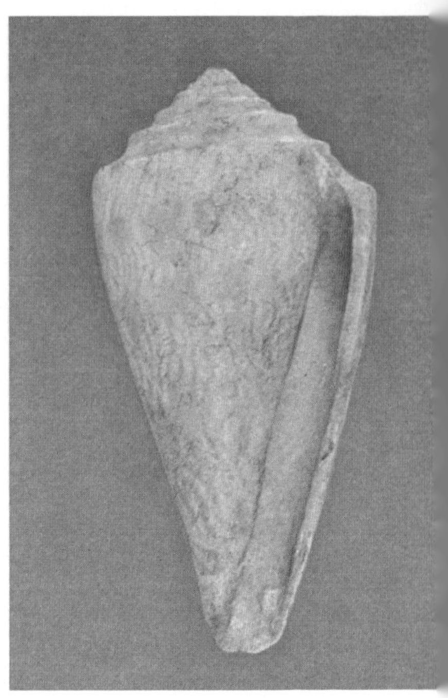

HOWELLI.—*Above:* Off Terrigal, New South Wales, Australia, 25.9mm (EP). *Below:* Moreton Bay, Queensland, Australia, 22mm (DM) (photos by Jayavad).

HYAENA.—*Above:* Bay of Bengal, 59.3mm (MG). *Below:* Left: Bombay, India, 45.2mm (RJ); Right: Off Burma, 41.2mm (GG).

Moderately shallow water.

Synonymy.—*C. unicolor* was preoccupied. The later figures of this species by Reeve look just as much like a pale *C. hyaena* as *C. concolor*, but I doubt that Reeve would have confused these two species.

Notes.—A not unattractive shell in a subdued way, it still sells for a rather high price although locally common seasonally in the Solomons. I have not seen the light phase for sale recently, but I doubt if it would deserve a premium.

C. concolor is subject to large breaks on the body whorl and jagged growth marks. The spire may be heavily eroded or covered with coralline growths that do not remove easily.

CONNECTENS A. Adams, 1855

1822. *Conus pulchellus* Swainson. *Zool. Illustrations,* 2: Pl. 114 (Amboyna). Non *Conus pulchellus* (Roeding, 1798).

1822. *Conus cinctus* Swainson. *Ibid.*, 2: Pl. 110 (Locality not stated). Non *Conus cinctus* Bosc, 1801.

1854. *Conus fasciatus* A. Adams. *Proc. Zool. Soc.* (*London*), 1853: 119 (Locality unknown). Non *Conus fasciatus* Schroeter, 1803.

1855. *Conus connectens* A. Adams. *Ibid.*, 1854: 136 (China).

1857-1858. *Conus bifasciatus* Sowerby ii. *Thesaurus Conchyliorum*, 3 (*Conus*): 23, pl. 14(200), fig. 302. Nomen novum for *Conus fasciatus* A. Adams, 1854. Non *Conus bifasciatus* Gmelin, 1791.

1929. *Conus circumactus* Iredale. *Mem. Queensland Mus.*, 9: 281. Nomen novum for *Conus cinctus* Swainson, 1822.

1943. *Conus hammatus* Bartsch and Rehder. *Proc. Biol. Soc. Washington*, 56: 86 (near Kauai Island, Hawaiian Islands).

Description.—Moderately heavy, with a high gloss; low conical, the sides nearly straight or slightly convex and gently curving to the base; body whorl with several crowded oblique ridges at the base, these followed by about 4-10 widely spaced strongly granulose spiral ridges which extend to about midbody; occasionally weaker spiral threads may be seen between the granulose ridges and extending to shoulder; numerous fine undulating axial threads present; shoulder sharply angled, broad, sometimes irregular; spire low, sometimes nearly flat, the sides straight or weakly concave, bluntly pointed; spire whorls concave above, with about three distinct spiral ridges crossed by weak curved axial threads. Body whorl pinkish white, sometimes with a violet tinge, usually with two broad scalloped bands of orange-brown to tan, the bands formed by fusion of large axial blotches; posterior band relatively narrow, anterior band broad and usually extending to base; usually several, sometimes many, spiral lines of small brown dots over body whorl; base pinkish tan to violet-tan, contrasting with anterior brown band; spire and shoulder pinkish white with a few scattered curved dark brown spots. Aperture narrow posteriorly, widened anteriorly; outer lip thin, sharp, curved; mouth with a large pinkish to violet blotch at anterior end, rest pinkish white to bluish white. Columella rather long and narrow, pinkish to violet. Length 30-65mm.

Comparison.—*C. connectens* is very close to *C. vitulinus*, *C. striatellus*, and *C. planorbis*. In general pattern it is closest to *C. striatellus* in having two broad brown bands and a brownish base, but the pink to violet staining of the base and lip is absent in *C. striatellus. C. planorbis* in some patterns is similar, but it lacks the blotch in the outer lip. *C. vitulinus* has a similar basal and lip spot, usually of a darker violet-brown and never pinkish, but it has the spire heavily covered with large curved blackish stripes; *C. vitulinus* is also usually thicker and has dark axial flammules in most patterns. None of these species has the bright pink tones characteristic of *C. connectens*.

Variation.—As would be expected, there is considerable variation in extent of fusion of the two brown body bands, from separated axial blotches, very weak posteriorly and broken anteriorly, to very broad solid bands with nearly regular edges. In one pattern the brown bands are actually pale tan heavily suffused with pink and barely distinguishable from the lighter background. The base and lip staining is constant, although variable in extent. The spiral rows of dots are usually few in number and distinct, partially fused into dashes. The spire is lightly spotted in most cases, never heavily. The amount of pink in the general coloration is variable, some specimens having only the palest tones and looking more creamy white than pinkish white; some specimens are heavily toned with violet as well.

I have not seen specimens with undulate shoulders, although they are reported in the literature. In a few specimens there were indistinct growth lines on the body whorl which became more obvious at the shoulder, giving it an irregular appearance which might be confused with undulations. The early whorls appear to be weakly nodulose in some specimens, but they are usually too eroded to tell for sure.

Distribution.—Uncommon to rare throughout the western Pacific east to at least Fiji and north to the Hawaiian Islands. Indian Ocean records are uncertain and probably apply to variants of *C. striatellus.* Found in moderately shallow water.

Synonymy.—This is one of those species whose synonymy is cursed. The first two names are based on distinctly and indistinctly banded phases, but both names were preoccupied. The first available name is *connectens*, but that is based on a large beach-worn specimen of the indistinctly banded phase. Since intermediates between the strongly banded and indistinctly banded phases have been seen, I chose to apply the name *connectens* to this species. In recent literature it has been usually called *C. circumactus* (often misspelled 'circumactis'), although *C. hammatus* is often used for Hawiian specimens. Undoubtedly all these names

HYPOCHLORUS.—Off Marinduque, P.I. *Above:* 26.7mm. *Below:* 27.9mm (both AK).

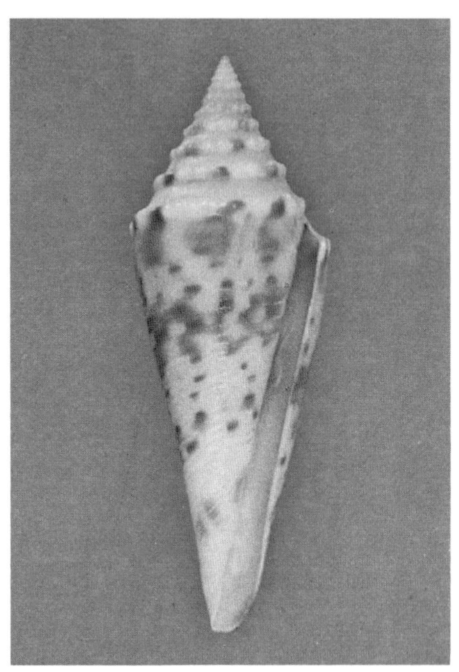

ICHINOSEANA.—Above: Off Tosa, Shikoku, Japan, 43.8mm (EP). *Below:* S of Shikoku, Japan, 46.4mm (GG).

apply to the same species in various phases and intensities of pattern.

Notes.—In bright pink patterns this is an extremely attractive cone. However, many specimens are not much more attractive than an average *C. striatellus* or *C. planorbis*, and in fact these species are sometimes incorrectly sold as *C. connectens.* These are usually minor growth flaws hidden in the body whorl under the heavy axial threads, but these are generally inconspicuous and do not detract from value or appearance. The lip is thin and easily chipped, so beware of heavy filing.

CONSORS Sowerby i, in Sowerby ii, 1833

1833. *Conus consors* Sowerby i, in Sowerby ii. *Conchological Illustrations:* Pt. 36, fig. 42 (Philippines).
1844. *Conus ustulatus* Reeve. *Conchologia Iconica*, 1(*Conus*): Pl. 44, sp. 239 (New Holland).
1854. *Conus anceps* A. Adams. *Proc. Zool. Soc.* (*London*), 1853: 119 (Moluccas).
1858. *Cònus Daullei* Crosse. *Rev. Mag. Zool.*, (Ser. 2), 10: 119, pl. 2, figs. 2-2a (Mayotte).
1882. *Conus Kobelti* Loebbecke. *Jahrb. deutsch. Malak. Ges.*, 9: 189, pl. 4, figs. 4-5 (Locality unknown).

Description.—Heavy, with a low gloss; low conical, the sides concave at the middle and more-or-less bulbous below the shoulder; body whorl with about a dozen low, rounded spiral ridges above the base, separated by shallow, weakly punctate grooves; rest of body whorl with numerous spiral and axial threads and growth lines, some of them conspicuous; shoulder broad, rounded, narrower than body whorl immediately anterior to it, shallowly concave; spire low, sharply pointed, the sides almost straight; whorls shallowly concave or flat above, with about 3-4 narrow spiral ridges crossed by fine axial threads to produce a weakly cancellate effect; first 5-6 whorls distinctly nodulose, next 2-3 with irregular sutures. Body whorl pale yellow, the base whitish; usually two or three broad, straight-edged spiral bands of yellowish tan to burnt orange, the midbody area yellowish; bands sometimes split into several narrower bands of slightly different shades; spiral rows of small brown dots usually absent; occasionally with traces of indistinct brownish axial streaks; spire yellowish to tan, at most with a few pale brown spots and streaks near sutures, the pattern indistinct; occasionally with small brown spots between nodules on early whorls. Aperture fairly wide, almost uniform; outer lip slightly concave at middle; mouth white to bluish white. Columella long, narrow, indented relative to inner lip. Length 50-80mm.

Comparison.—*C. consors* is obviously close to *C. magus* and its relatives, especially *C. pohlianus.* However, it seems to be constant enough in shape and pattern, especially virtual absence of spire pattern, to be considered a species. As a rule, *C. magus* has the early whorls less distinctly nodulose, has many blackish or brownish spots on the spire, has numerous spiral rows of small brown dots in the pattern, lacks the bulbous appearance below the shoulder (occasional exceptions), is a much rougher shell usually of different colors and not as regularly banded, and has a straight or convex outer lip. *C. pohlianus* has the same general pattern as *C. consors*, probably indicating adaptation to similar ecology, but is a thinner shell of somewhat smaller adult size; it is not bulbous below the shoulder, has convex spire sides, has distinct spotting on the early whorls, and is whitish with brownish bands, occasionally with a violet tone, never yellow and orange.

Variation.—Although the sides of the body whorl are convex below the shoulder and are wide, they are not always distinctly bulbous. There is not too much variation in pattern or color, although some specimens are much brighter than others. In a few specimens are to be seen traces of spotting on the spire whorls and fine brown dots on the body whorl, sure signs of a *magus* ancestry. In some specimens the bands are deep burnt orange, almost red, on an orange background. Axial flammules or blotches are seldom developed and never conspicuous.

I am certain that good intermediates between the *consors* pattern and more typical *magus* patterns exist, but at the moment I can see no use in adding one more name to the already long *magus* synonymy, especially as *C. consors* can be readily distinguished in almost all cases.

Distribution.—Spottily distributed in the Indo-West Pacific from the African coast and Bay of Bengal to the Philippines and Solomons; it is probably more widely distributed in deeper water.

Synonymy.—The names listed here are almost certainly pertinent to *C. consors*, except possibly for *C. ustulatus*, which is a reddish-banded shell with a few spots on the spire. *C. anceps* is a form with faint axial blotching. Probably some names listed under *C. magus* could almost as easily be placed with *C. consors* on the basis of the original description.

Notes.—Many shells sold as *C. consors* are just normal *C. magus* with unusual patterns and have nothing to do with the species as considered here. Several *C. magus* with normal patterns but distinctly bulbous areas below the shoulder (apparently abnormal) have been seen under the *consors* label as well. Although this would appear to be a rough cone, it apparently suffers few major breaks; even the spire is seldom eroded. Specimens from the Philippines appear to have especially brilliant colors.

CONSPERSUS Reeve, 1844

1844. *Conus conspersus* Reeve. *Conchologia Iconica*, 1(*Conus*): Pl. 47, sp. 262 (Locality unknown).

1845. *Conus Verreauxii* Kiener. *Species gen. et icon. des coqu. viv.*, 2

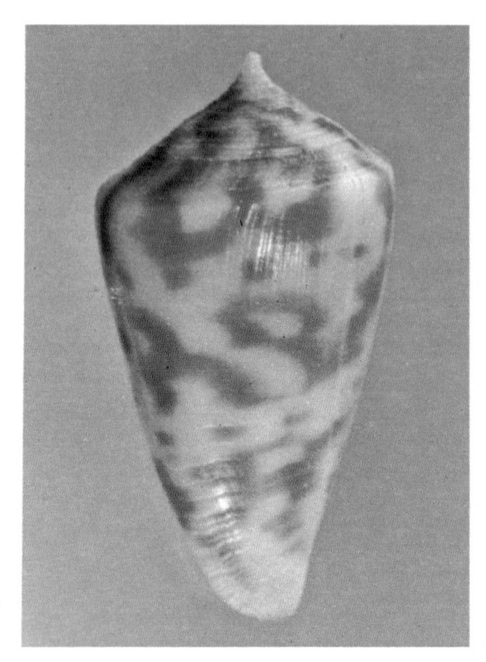
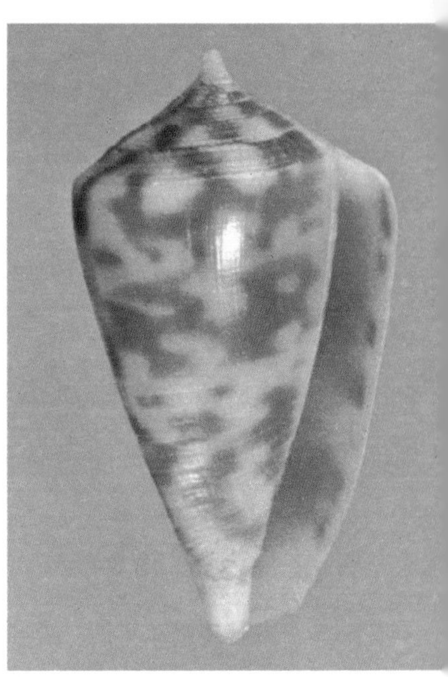

ILLAWARRA.*—Above:* Scot's Head, New South Wales, Australia, 26mm (DM) (photos by Jayavad). *Below:* Off Terrigal, New South Wales, Australia, 26.9mm (EP).

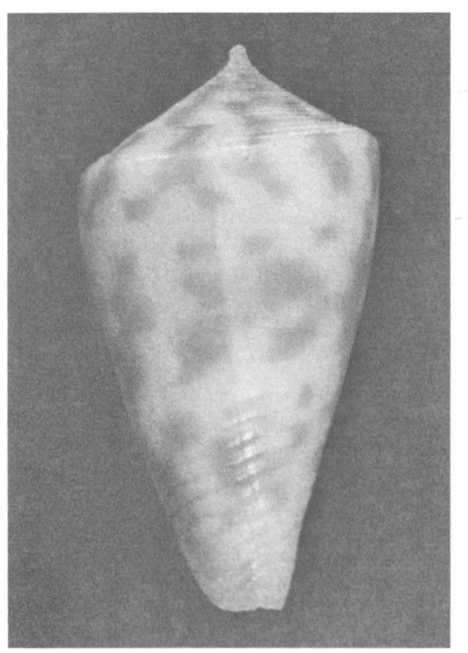
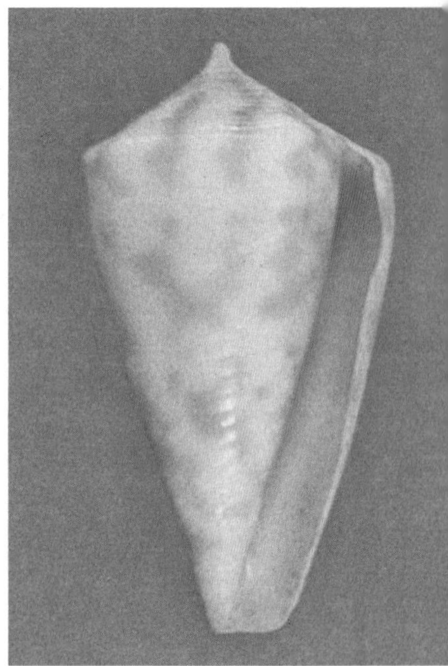

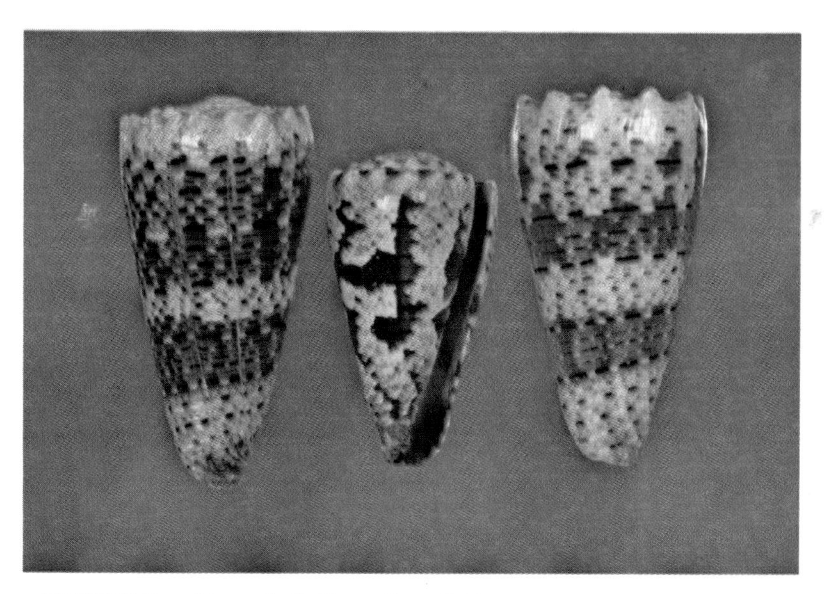

IMPERIALIS.—*Above:* Left: Kona, Hawaii, Hawaii, 50.8mm; Center: Mauritius, 41.1mm; Right: Samar, P.I., 49.9mm. *Below:* Zanzibar: Left: 66.5mm; Right: 72.8mm (both MF).

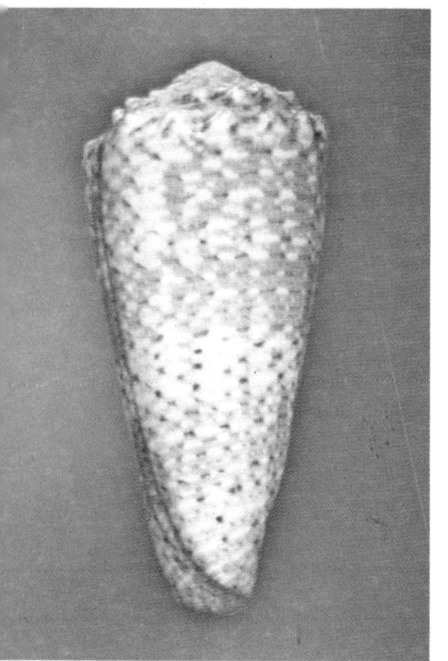

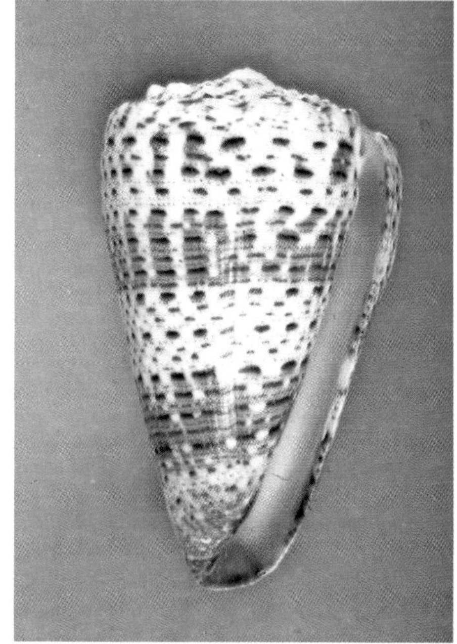

(*Conus*): Pl. 60, fig. 5. 1849, *Ibid.*, 2: 249 (Cape of Good Hope).
1864. *Conus Daphne* Boivin. *J. Conchyl.* (*Paris*), 12: 35, pl. 1, figs. 7-8 (Indian Ocean).
?1864. *Conus dolium* Boivin. *Ibid.*, 12: 38, pl. 1, figs. 3-4 (Locality unknown).

Description.—Moderately light in weight, with a good gloss; rather ovately conical, the sides convex; body whorl with closely spaced narrow spiral grooves at the base, these becoming more widely spaced higher on the body whorl and sometimes extending almost to shoulder; numerous fine axial threads over whole whorl, giving the more anterior grooves a punctate appearance; ribs between flat; shoulder rounded, broad; spire low, with a long, sharp, mucronate apex; tops of whorls flat or slightly convex, covered with about 2-3 low spiral ridges crossed by fine axial threads so faintly cancellate; early whorls not nodulose. Body whorl creamy white with pinkish or yellowish tones; usually with three irregular spiral bands of brown to orange or yellow axial blotches, these sometimes faint or laterally fused into complete bands; numerous widely spaced spiral rows of small brown dashes from base to shoulder, these often obscure and difficult to see except in the brownish blotches; fine axial clouds of yellowish tan to pinkish tan sometimes present over body whorl; occasionally entire pattern absent except for spiral rows of dashes; spire whitish with small darker brown radiating spots and blotches; early whorls sometimes pale brown. Aperture wide throughout, very wide anteriorly; outer lip thin, sharp, convex; mouth white near lip, becoming dark pink or pinkish orange within. Columella long, narrow, set off posteriorly by a strong raised ridge. Length 25-45mm.

Comparison.—Whether this is the proper name for the shell or not, *C. conspersus* as treated here is close to *C. spectrum* and allies but quite distinct. The strongly mucronate spire and pink to orange-pink mouth are found in combination in no *C. spectrum* variant known to me; although some *C. spectrum* have yellow mouths, there is no approach to the mouth of *C. conspersus*. *C. spectrum* var. *filamentosus* is more elongate and has the spiral brown lines complete, very fine, and very closely spaced. *C. subulatus subulatus* may be similar in shape, but it has a violet mouth, often is bluish, and lacks the small brown dashes in spiral lines (although the remainder of the pattern may be almost identical). *C. wittigi* differs in details of shape, has a white to very pale pinkish mouth, and is covered with a distinctive tented pattern.

Variation.—In shape and sculpture there is little variation in the samples I have seen. Occasionally the spiral grooving extends to the shoulder, but more commonly it stops just above midbody. Sometimes the spire whorls are very slightly stepped, their edges carinate. The pattern on the body whorl may be very indistinct, only traces of yellowish tan blotches or bands on pinkish white, or very heavy, dark orange-brown axial flammules crossing three spiral brownish bands. In any case the traces of spiral dash rows can usually be found with careful examination. In a minority of shells the dashes are very strong and form complete lines around the body whorl. Sometimes the midbody area is distinctly whiter than the rest of the body, sometimes it is the same intensity of pinkish and heavily marked with brown. The spire pattern is fairly constant. Mouth color varies from bright deep pink to definitely toned with orange, apparently without regard to color of the body whorl. There is little or no violet toning in the mouth as in *C. subulatus* and *C. stramineus*. It might be noted that the brown dashes seem to follow the spiral grooves rather than the ribs.

Distribution.—Apparently uncommon in moderately deep water, this species is known to me only from the Philippines and Indonesia, but presumably it is more widely distributed in the western Pacific.

Synonymy.—*Conus conspersus* seems to be the proper name for this shell, but Reeve's original specimen was far from perfect. The other names seem to fall in the range of variation of this species, being based mostly on poorly colored specimens, although in a couple of cases the bright pinkish mouth is mentioned. The name *daphne* appears to be a synonym of this species, while *C. conspersus* of dealers is here called *C. spectrum* var. *filamentosus*.

Notes.—Although its name is uncertain, this attractive little cone is quite distinct, especially in the unusual mouth color. Philippine specimens usually come from the Sulu Sea and are somewhat cheaper than Indonesian specimens, although with a somewhat weaker color pattern on the average. *C. conspersus* usually has a high gloss and is easy to find in gem condition.

CORONATUS Gmelin, 1791

1791. *Conus coronatus* Gmelin. *Systema Naturae per Regna Tria Naturae*, ed. 13, 1: 3389 (Locality not stated). Neotype designated by Kohn, 1966.

1798. *Cucullus coronalis* Roeding. *Museum Boltenianum:* 38 (Locality not stated). Based on *Conus coronatus* Gmelin.

1857-1858. *Conus Aristophanes* Sowerby ii. *Thesaurus Conchyliorum*, 3 (*Conus*): 9, pl. 4(190), figs. 81-82 (Philippines and Sandwich Islands).

1864. *Conus minimum* var. *Condoriana* Crosse and Fischer. *J. Conchyl. (Paris)*, 12: 334 (Poulo Condor).

Description.—Heavy, usually dull but sometimes with a good gloss; rather bulbous or ovate, the greatest width usually just below the shoulder, the sides convex and then pinched-in to base; body whorl with narrow spiral ridges on basal half, these sometimes broken, the ridges separated by weakly punctate grooves; ridges may become obsolete above midbody, continue to shoulder, or become broad flat ribs separated by

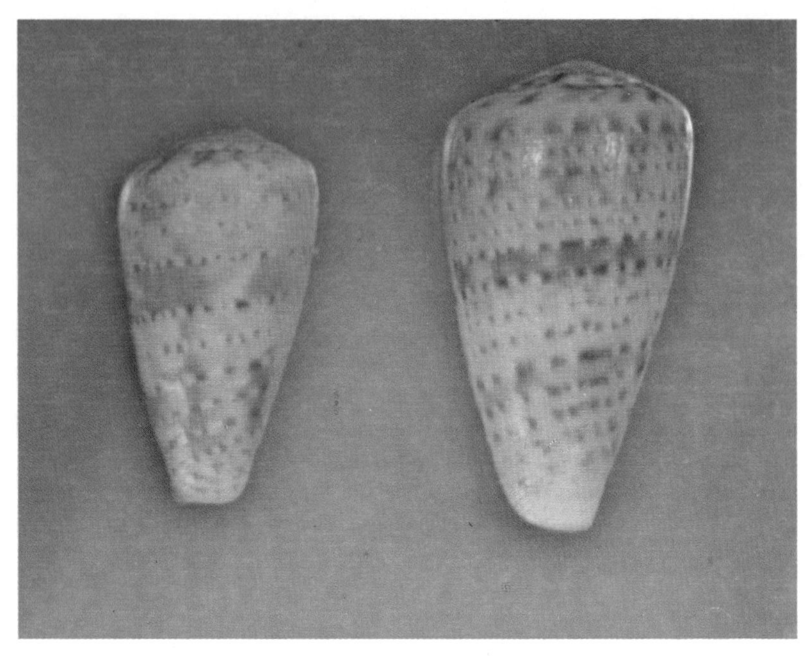

INFRENATUS.—*Above:* Left: E London, Cape, South Africa, 27.2mm; Right: Jeffreys Bay, Cape, South Africa, 34.0mm. *Below:* Jeffreys Bay, Cape, South Africa, 32.9mm (GG).

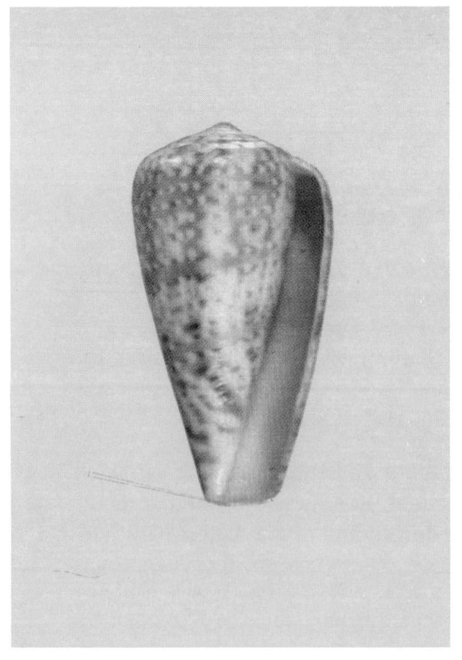

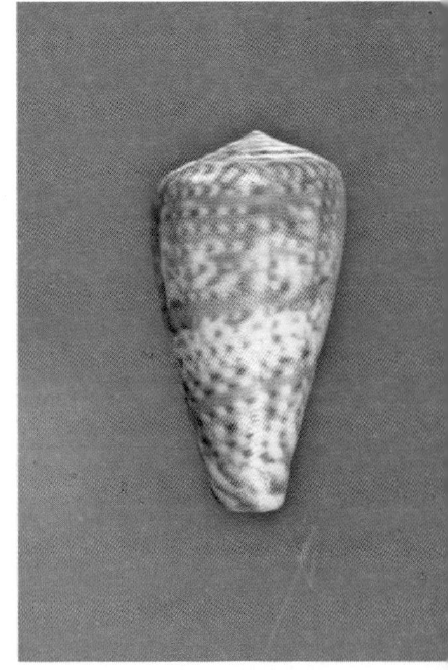

INSCRIPTUS.—*Above:* Basuk Bay, Andaman Is., 45.7 and 47.0mm (Shell Cabinet). *Below:* Left: Djibouti, 29mm (DM) (photo by Jayavad); Right: Arabian Gulf, 37.2mm.

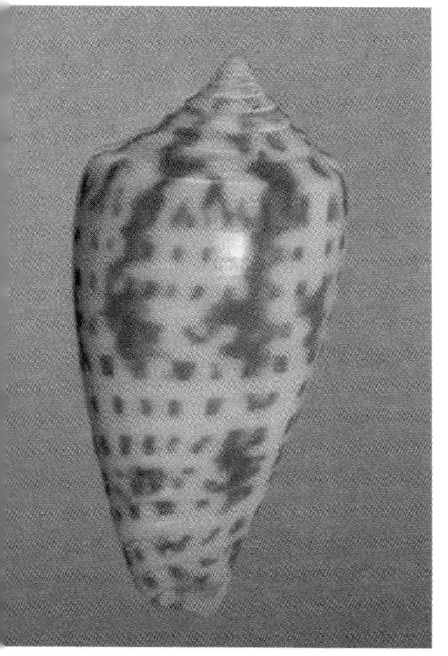

narrow grooves; basal ridges often strongly nodulose, the nodules occasionally present to shoulder; shoulder angled, weakly or strongly coronated; spire low, conical, the sides straight or convex; later whorls coronated, early whorls usually eroded; whorls with from 1-7 distinct spiral ridges, the number individually variable in large series; tip of spire usually blunt, eroded. Body whorl usually pinkish to bluish or tan, with indistinct patterns; most common is two broad bands of bluish brown irregular blotches above and below midbody, or interrupted spiral lines of brown and white dashes; both patterns or parts of these patterns may be present on one specimen, as well as random smaller blotches; base whitish; spire pinkish, bluish, or tan, much like body color, the coronations generally white with brown interstices; tip pink if not eroded. Aperture fairly wide above, flaring to very wide anteriorly; outer lip evenly convexly curved and not parallel to pinched-in base; mouth with white edge, deeper within with two broad bands of dark violet separated by a median pale band and a smaller violet band under shoulder usually separated from other bands by another white narrow band. Columella short, wide, heavily set off by a strong posterior ridge. Length 15-44mm.

Comparison.—This common species is very similar to *C. miliaris* and allies and sometimes confused even by experienced dealers. As a general rule, *C. miliaris* is less bulbous, the sides nearly straight throughout and the outer lip almost parallel to the inner; the aperture is narrower and not as flared anteriorly; most *C. miliaris* are basically tan cones with darker brown and white patterns, seldom bluish, a common color in *C. coronatus*. The spire in *C. miliaris* is lower, has straighter sides; in *C. miliaris* there are distinct spiral rows of small punctations which extend to the shoulder, while in *C. coronatus* such punctations are usually restricted to obvious grooves anteriorly. In other words, *C. coronatus* is a chubby little cone with a conical spire and wide mouth, while *C. miliaris* (and relatives) is much more 'trim' with a flatter spire and narrower mouth.

Variation.—The pattern is highly variable, as is the color, although blue is common. Some specimens are heavily mottled and blotched, others have only a few scattered brownish dashes. Spire height varies considerably.

Various authors have tried to separate this species into two, *C. coronatus* (elevated spire; sharply cut coronations; 5-7 well defined spiral ridges on the spire whorls; and strong and interrupted basal ridges) and *C. aristophanes* (spire lower; coronations nodulose, often obsolete; 1-2 broad spiral ridges on spire whorls; and basal ridges strong and continuous). However, these characters are not constant in large series obviously representing a single species and often occur in varying combinations. Thus a shell may have 3-4 spiral ridges on the spire whorls, a high spire, and indistinct coronations. I can see no reason to recognize *C. aristophanes* at even the varietal level, the separation being purely artificial.

Distribution.—Widespread and abundant in the entire Indo-Pacific, probably including Hawaii. Shallow water.

Synonymy.—See *Variation* for comments on *C. aristophanes*. This species has been commonly called *C. minimus*, based on a misidentification by Born of Linnaeus' species (*minimus* of Linnaeus, if recognizable, is probably *C. figulinus*). *Coronalis* of Roeding is strictly synonymous and cannot be used for especially violet or bluish specimens from the Red Sea as is commonly done.

Notes.—Every collection should have several good series of this species and *C. miliaris* in order to fully appreciate the distinctions between the two species. Dealers commonly use the name *aristophanes* for either very large specimens (over 30mm) or those with especially bright colors. The "subspecies" name *coronalis* is often used for the species in the Red Sea, but there is little distinction from other Indo-Pacific populations and *C. coronalis* is a strict synonym of *C. coronatus*, taking the same type. There are often small growth marks in this species, but *C. coronatus* is so cheap that no one seems to mind.

CUMINGII Reeve, 1848

1848. *Conus Cumingii* Reeve. *Conchologia Iconica*, 1(*Conus* Suppl.): Pl. 3, sp. 282 (Island of Mindanao, Philippines).

Description.—Moderately heavy, with a low gloss; low biconical, the sides almost straight; body whorl with about 12 heavy spiral ridges near the base and on the anterior fourth, the grooves between with or without faint axial threads; rest of whorl smooth except for often conspicuous spiral and axial threads and growth marks; shoulder wide, rounded, smooth; spire low, rather domed, the whorls not distinctly stepped; early whorls forming a small but sharp point; early whorls weakly nodulose or undulate, rest smooth; tops of whorls with about 6-12 weak but distinct spiral ridges crossed by closely spaced curved axial threads, all sculpture weak. Body whorl golden tan to olive on basal half, followed by a broad white midbody band; midbody band edged above by closely spaced spiral brown lines and blotches, some entering band; conspicuous axial rows of small brown spots over most of band as well; upper half of shell darker brown than base but usually still with golden or olive tones; fine spiral lines of dark brown, badly broken, may occur on any part of the whorl, especially the basal half, but also heavy near midbody band and sparse on posterior half of shell; shoulder with a distinct pattern of heavy dark brown blotches and irregularly alternating white blotches, these continued onto spire; tip of spire sometimes pinkish. Aperture relatively narrow, little widened anteriorly; outer lip thin, straight; mouth bluish white, the midbody band sometimes showing through. Columella internal. Length 20-35mm.

Comparison.—The shape and midbody band remind one of *C. coccineus*, which is usually heavily sculptured, has an undulate shoulder,

INSCULPTUS.—Above: Off Marinduque, P.I., 25.0mm (AK). ***Below:*** Left: Thailand, 25.6mm (MG); Right: Taiwan, 22.7 and 23.4mm.

IODOSTOMA.—Above: Mozambique, 38.1mm. *Below:* Left: Mozambique, 39.5mm; Right: Zanzibar, 30.6mm (MF).

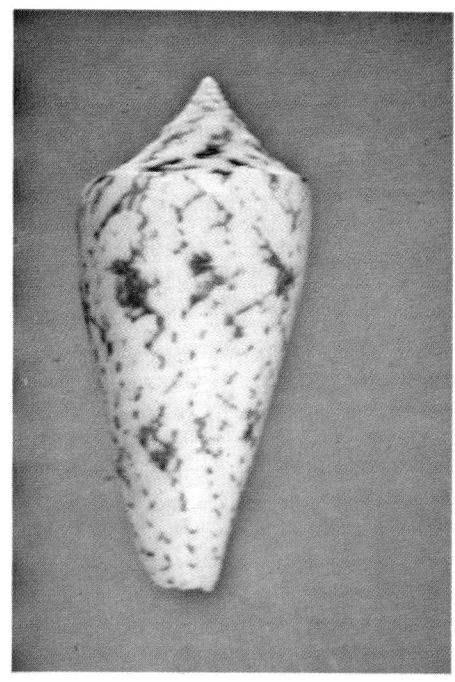

and lacks strongly contrasting blotching on the spire. *C. vittatus* and *C. amphiurgus* are also somewhat similar, but they are usually more brightly colored and lack the weakly cancellate spire sculpture. *C. sazanka* may have a somewhat similar pattern, but it is more biconical, has the spire whorls distinctly stepped and usually undulate, and lacks the numerous fine spiral ridges on the spire whorls. The greenish tones of *C. cumingii* are seldom found in other species, but not all *cumingii* are greenish.

Variation.—The intensity of general coloration varies considerably, from pale golden tan to bright olive or even dark brown. The midbody band may be clear white or only weakly marked with sparse axial spots of brown, or it may be heavily clouded and spotted with brown. The upper edge of the band is usually strongly delimited with dark brown blotches and lines, but it may also be weakly marked with a few broken lines; the basal edge of the band is almost always more weakly demarked than the posterior margin. Spiral bands of dark brown may be heavy on the base and almost continuous or may be virtually absent. Generally the dashed lines are very weak and sparse posteriorly, but occasional specimens (or parts of specimens) show heavy and continuous banding there. The spire pattern is usually strongly alternating dark and white, but sometimes it is reduced, the dark blotches becoming narrow curved lines.

Distribution.—Uncommon in the western Pacific from the Philippines to the Solomons and Queensland, Australia. Moderately shallow water. Records from Hawaii refer in all or part to *C. sazanka.*

Synonymy.—Kohn referred a large series of Hawaiian specimens to *C. cumingii*, but these specimens appear to mostly represent the poorly understood *C. sazanka. C. sazanka* is highly variable in pattern and to some extent sculpture, but it should not really be confused with *C. cumingii*. The species as treated here agrees well with Reeve's description and figure.

Notes.—This is a rather fragile cone which commonly has very bad breaks running over the body whorl. The lip is also fragile and often chipped. Specimens with contrasting colors are not very common, most specimens having dark and rather drab patterns. The shell is far from common at any locality, although it is sometimes found in fair numbers locally in the Solomons. The distribution is very uncertain, specimens of this species having recently been taken (supposedly) in the Bay of Bengal. It would seem that 'Bay of Bengal' as a locality is now in the same league as 'Taiwan' or 'Singapore' for indicating an uncertain locality.

CUVIERI Crosse, 1858

1838. *Conus cervus,* Sowerby i, in Sowerby ii. *Conchological Illustrations:* Pt. 147/148, fig. 194. Misidentification.

1843. *Conus Deshayesii* Reeve. *Conchologia Inconica,* 1(*Conus*): Pl. 5, sp. 28 (Swan River). Non *Conus deshayesii* Bellardi and Michelotti, 1840, a fossil.

1858. *Conus Cuvieri* Crosse. *Rev. Mag. Zool.*, (Ser. 2), 10: 123. Nomen novum for *C. deshayesii* Reeve, 1843.

Description.—Very thin and light but solid, with a high gloss; ovately obconical, the sides convex; body whorl with about 12 wide but low spiral ribs at the base, otherwise smooth and polished; shoulder roundly angled, sometimes weakly carinate; spire low to nearly flat, the early whorls forming a sharp point; tops of whorls flat, covered with traces of several spiral ridges and numerous crowded axial ridges. Body whorl white to creamy yellow, usually with two spiral series of large to small nebulous axial blotches variably connected laterally and axially; whole body whorl covered with spiral rows of brown and white dashes, these sometimes heavy and prominent; shoulder and margins of spire whorls usually with short axial brown lines, otherwise whitish to olive; early whorls brown. Aperture very wide, especially anteriorly; outer lip thin, sharp, convex; mouth variable, uniform pinkish or violet or with two large brownish violet blotches. Columella short, narrow. Length 30-50mm.

Comparison.—Although *C. cuvieri* has been confused with *C. cervus* and has been considered related to *C. bullatus* as well, it certainly seems much more similar to *C. tulipa* and relatives. It is extremely broad even for a cone of these two groups, the mouth being especially wide. The pattern of nebulous blotches and numerous fine lines is like that of *C. tulipa* or *C. obscurus*, but the spire whorls are lacking distinct nodules and there is little or no blue in average specimens. The texture is like a tulip cone rather than the thick but light *bullatus* type of structure or even the thinner structure of a *vicweei* or *cervus*. *Conus cervus*, *C. bullatus*, and *C. vicweei* all have very different patterns and proportions and should present no recognition problems.

Variation.—Structurally constant, there is quite a bit of variation in color and pattern. The typical pattern apparently is a whitish or yellowish shell heavily covered with more-or-less obscure brownish dashes in spiral rows and with two spiral rows of large olive to brown blotches more-or-less fused laterally and connected randomly through the midbody area. It takes little imagination to produce a largely white or yellow shell from this or one that is almost all olive brown. The spire is generally about the same color as the body whorl and weakly marked with brown spots and lines. Mouth color is light in light shells, dark in dark ones. The spiral dash rows may be inconspicuous or very heavy and a major part of the pattern.

Distribution.—Restricted to shallow water in the Red Sea and probably adjacent areas. Uncommon.

Synonymy.—The reference by Sowerby i to *C. cervus* is of course a misidentification. *C. deshayesii* is preoccupied and should not be used.

IONE.—Off Kii, Japan, 54.6 and 64.9mm (GG).

JANUS.—*Above:* Left: Zanzibar, 51.5mm (MF); Right: NW Madagascar, 51.7mm. *Below:* NW Madagascar, 48.6 and 54.4mm.

Notes.—Although *C. cuvieri* has been placed with *C. bullatus* and relatives, the texture and pattern certainly indicate relationship with the *C. geographus* complex of species. Like other species in this group, the thin body is subject to bad breaks which do not heal fully. This shell is probably not as uncommon as the high price indicates.

CYANOSTOMA A. Adams, 1854

1854. *Conus cyanostoma* A. Adams. *Proc. Zool. Soc.* (*London*), 1853: 116 (West Africa).
1875. *Conus Coxeni* Brazier. *Ibid.*, 1875: 34, pl. 4, fig. 10 (Moreton Bay, Australia).
1877. *Conus cuneiformis* E.A. Smith. *J. Conchology* (*London*), 1: 202, fig. (Locality unknown).
1892. *Conus innotabilis* E.A. Smith. *Proc. Zool. Soc.* (*London*), 1891: 487, pl. 40, fig. 1 (Rocky Point, N.S.W., and Port Stephens near Sydney, Australia).

Description.—Light in weight but solid, with a low gloss; low biconical, the sides straight or slightly convex posteriorly and tapering to a narrow base; body whorl with wide flat ribs separated by shallow, sometimes punctate grooves, these prominent over anterior half to two-thirds of body then usually obsolete; sometimes the ribs continue to just below shoulder; two or three narrow flat ridges directly below angle of shoulder, not a continuation of anterior body sculpture; shoulder wide, sharply angled or carinate; spire low to moderate, the sides flat, sharply pointed; whorls generally flat, sometimes distinctly concave and stepped, especially on earlier whorls; early whorls with faint indications of nodules, later whorls weakly carinate; later spire whorls with about four weakly defined flat spiral ridges separated by shallow weakly punctate grooves; the ridges are overlaid with faint curved axial growth lines. Body whorl usually cream or pale bluish white, overlaid with nebulous and indistinct pale brown blotches in two spiral bands and occasional spiral rows of brown spots; sometimes all color pattern absent; base whitish; spire usually with a few spots or lines of pale brown; tip often distinctly brown. Aperture narrow, uniform in width; outer lip thin, straight; mouth dirty brownish violet, at least deep within. Columella indistinct. Length 15-25mm.

Comparison.—This little shell lacks striking characters but is not difficult to recognize. The small size, biconical shape, lack of a striking pattern, broad basal ribs, and dirty mouth color are fairly unique. *C. puncticulatus* var. *columba* can be very similar superficially in every way, but the spire lacks spiral sculpture and the basal ribs are sometimes granulose. *C. clarus* is similar, but it is even more fragile, lacks nodulose early whorls, has a bright pink mouth, and has the shoulder with a definite raised rim.

Variation.—The color pattern may be totally lacking, resulting in a dirty white shell with pale violet tones, or well developed with axial dark brown blotches and lines with a few bands of spiral brown dashes or dots. The narrow ribs just below the shoulder are obvious if not obscured by the basal ribs extending to shoulder. The most striking variation is in the shape of the spire. Generally it is low or moderate and conical with straight sides, but in occasional specimens, such as the type of *C. coxeni*, it is distinctly high, stepped, and concave. However, I have seen a specimen with the early whorls distinctly stepped and concave, but the lower whorls flat-sided and conical; it would appear that this condition is caused by ecological factors.

Distribution.—Although commonly stated to be Indo-Pacific in range, I am not aware of records from other than Queensland and New South Wales, Australia. Perhaps restricted to moderately deep water off the eastern Australian coast, or at least the western Pacific.

Synonymy.—*C. coxeni* is the form with a distinctly turretted spire as discussed under *Variation.* This species seems to have caused E.A. Smith exceptional trouble, and he described poor individuals with almost no color as *C. cuneiformis* and *C. innotabilis*; these differ largely in the extent of the basal ribs. Apparently all these names fall easily within the range of variation of *C. cyanostoma.*

Notes.—This not particularly attractive little cone seldom has major body flaws, although there do seem to be more distorted spires in this species than in most other cones. Specimens usually have very little pattern and look like white shells unless examined under low illumination or with a lens.

CYLINDRACEUS Broderip and Sowerby i, 1830

1830. *Conus cylindraceus* Broderip and Sowerby i. *Zool. J.*, 5: 51, suppl. pl. 40, fig. 5 (Locality unknown).

Description.—Moderately heavy, with a high gloss; cylindrical, tapered at both ends, fusiform; body whorl with very narrow crowded spiral ridges basally, some of these weakly granulose in juveniles; rest of body whorl with crowded undulate spiral and axial threads, not conspicuous; shoulder very indistinct, merging into spire; spire high, ovate, the sutures overlapping; first 3-4 whorls with indistinct nodules, rest with traces of 1-4 very faint spiral ridges, appearing smooth. Body whorl white, almost covered with reddish brown to yellowish brown or occasionally dark brown axially rectangular blotches in two rows; blotches heavily fused both axially and spirally, leaving only narrow axial white lines and blotches; shoulder and spire heavily marked with similar

JASPIDEUS.—Above *(j. jaspideus)*: Big Pine Key, Florida, 17.5-21.0mm. *Below:* Left *(j. jaspideus)*: Bimini, Bahamas, 23.2mm; Right *(j. stearnsii)*: (top) Crystal Beach, Citrus Co., Florida, 19.3mm; (bottom) Tarpon Springs, Pinellas Co., Florida, 16.5mm.

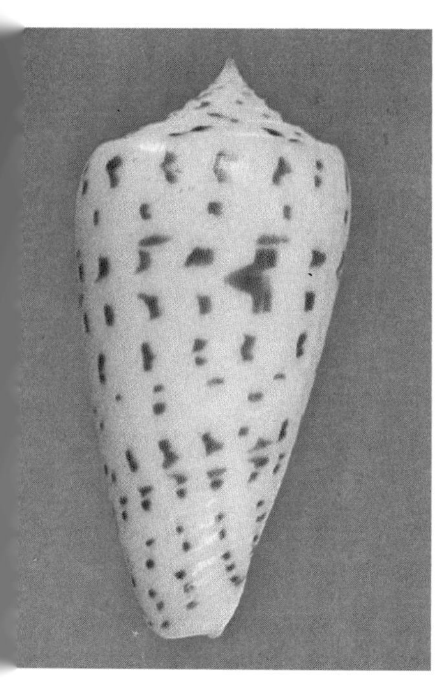

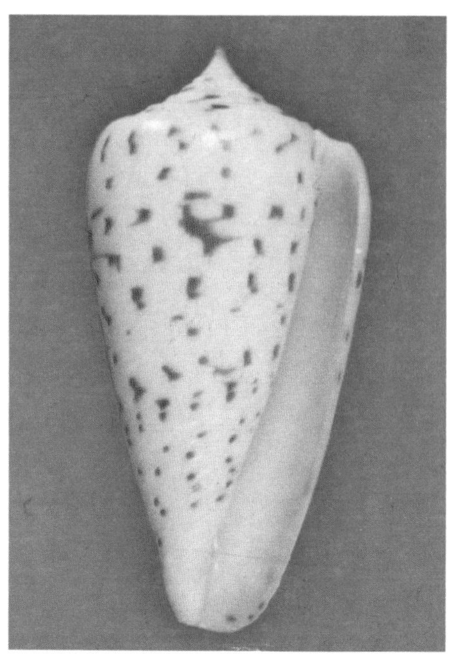

JICKELII.—Above: Hellville, Nossi Be, Madagascar, 36.2mm (EP). ***Below:*** Left: Massawa, Ethiopia, 35mm (DM) (photo by Jayavad); Right: "Indian Ocean", 34.4mm (MM).

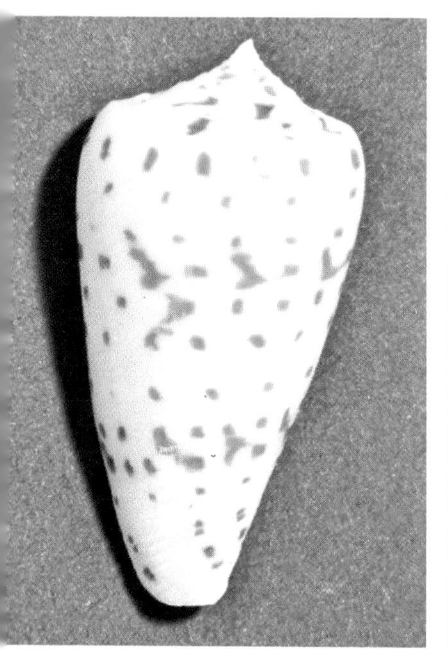

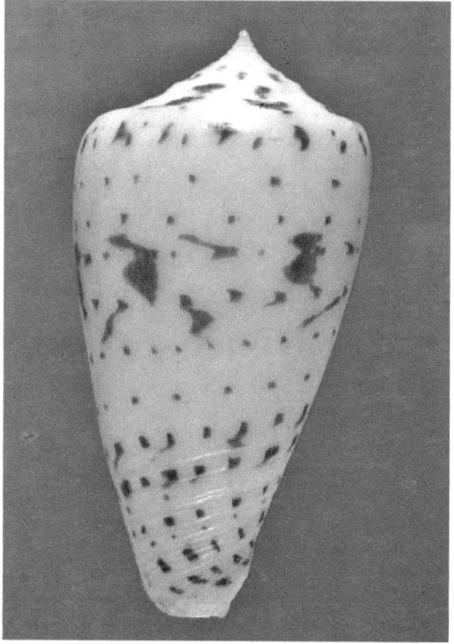

brownish blotches, the shoulder pattern separated from body pattern by a narrow, irregular spiral white line infiltrated by narrow brown lines; base white; early spire whorls whitish. Aperture narrow, especially posteriorly; outer lip straight; mouth white. Columella largely internal. Length 20-45mm.

Comparison.—The fusiform shape with overlapping spire margins is quite distinctive and only approached by a few other species. Most similar is *C. mitratus*, which is covered with numerous widely spaced rows of small granules; the spire whorls have about 4 distinct spiral ridges; and the blotches on the body whorl are not fused to the extent of *C. cylindraceus; C. mitratus* is not nearly as glossy as *C. cylindraceus* and is more inflated at the shoulder. *C. luteus* is usually wider with a lower spire and tends to have a midbody band of white blotches, as well as commonly being partially covered with small brown dashes in spiral rows; the early whorls are not distinctly nodulose.

In *C. tenuistriatus* and allies the body whorl is heavily sculptured, there are usually strong violet tones, and there are many differences in shape and pattern.

Variation.—With wear or growth the spire becomes blunter and more typically bullet-shaped than in younger shells. The basal sculpture is very weak except in occasional juveniles where some of the ridges are finely granulose. In juvenile shells the brown blotches are often very dark, almost black, and are smaller and less connected, leaving more white background visible as in *C. mitratus*. With growth the blotches become lighter, often eventually yellowish, and fuse repeatedly in every direction, leaving only small axial lines and rectangles of white. A white midbody band may be visible or not, but usually the narrow white band at the shoulder is fairly distinct.

Distribution.—Widely distributed over the Indo-Pacific from eastern Africa to Hawaii and French Polynesia, but seldom common.

Notes.—The combination of an unusual shape that is exceptionally graceful and the bright reddish brown colors makes this one of the most attractive and interesting of the cones. Apparently erosion of the early spire whorls is normal, although never excessive. The lip fractures easily and is often broken. Juveniles with very dark blotches provide a striking contrast with adults. This is not a common cone at any locality, most specimens being fresh beach or collected dead; a high gloss is retained even in dead specimens.

DALLI Stearns, 1873

1873. *Conus Dalli* Stearns. *Proc. Calif. Acad. Sci.*, 5(1): 78, pl. 1, fig. 1 (Gulf of California).

Description.—Moderately light in weight, with a high gloss; low conical, the sides convex; body whorl smooth except for a few low basal spiral ridges, these sometimes visible to about midbody; axial threads sometimes conspicuous; shoulder broad, rounded, not distinct from spire, slightly concave above; spire low to moderate, sharply pointed, the sides straight or nearly so; first 2-3 whorls with small nodules, the rest slightly concave to flat above and with traces of 2-3 fine spiral ridges and indistinct axial threads. Body whorl white, often with a pinkish or violet tone, heavily covered with small to medium open tents produced by the overlapping of zigzag axial chestnut lines; tents sometimes small and scale-like in midbody area, but often larger and of mixed sizes; usually with two nearly continuous spiral belts of chestnut blotches containing distinct axial blackish 'textile lines'; belts sometimes partially broken by larger white tents; shoulder and spire with tents and blackish axial lines as well as chestnut blotches; early whorls pinkish. Aperture moderately wide, especially anteriorly; outer lip sharp, rather thick, convex; mouth deep violet, sometimes fading to pinkish violet or even white. Columella rather short and narrow. Length 40-80mm.

Comparison.—*C. dalli* is extremely close to *C. abbas* and *C. canonicus* and, like them, could probably best be placed in a single taxon, *C. canonicus*. However, from tradition and because of its seemingly isolated geographic locale it is retained as a full species. The spire is straighter-sided than in *C. abbas*, and the body whorl is broader at the shoulder than typical examples of either of the other two species. The pattern is more heavily tented than in typical *C. canonicus*, although the tents are seldom as small and regular as those of *C. abbas*. The nearly solid belts of 'textile lines' are seldom duplicated in *C. canonicus*, but they are not uncommon in *C. abbas*. From both species *C. dalli* is readily distinguished when fresh by the deep violet mouth that is darker and more solidly colored than typical of *C. abbas* and almost never the pale pink found in *C. canonicus*. It should be noted, however, that the mouth color can fade to white or even pink in some specimens, so this cannot be used alone.

Specimens from the Marquesas here assigned to *C. abbas* (although they admittedly are probably just terminal variants of *C. canonicus*) are almost indistinguishable from some *C. dalli;* even the mouth color is similar, as is the shape and general pattern. Certainly these three taxa should eventually be combined into a single taxon in which three pattern variants, not subspecies because they overlap distributionally, would be recognized by varietal names.

Variation.—The spire is usually rather low with nearly straight sides, but sometimes it is moderately high or has distinctly convex sides. The wide shoulder is rather constant, although it grades imperceptibly into a narrower, more *C. canonicus*-like, appearance. The tenting is usually small and rather confused, appearing as though a background of fine, regular, scale-like tents had been covered with a looser network of larger and irregular tents. The two spiral bands of chestnut blotches with strong

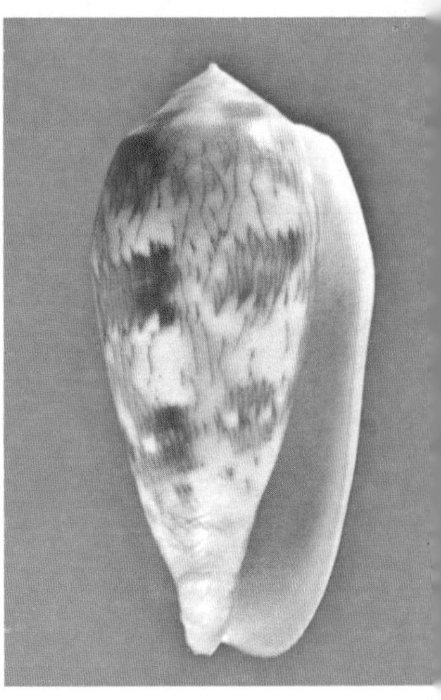

JULII.—Above: Off SW Mauritius, 52mm (S. Armington, Jr.) (photos by B. Lipe). *Below:* Mahebourg, Mauritius, 18.8mm (EP).

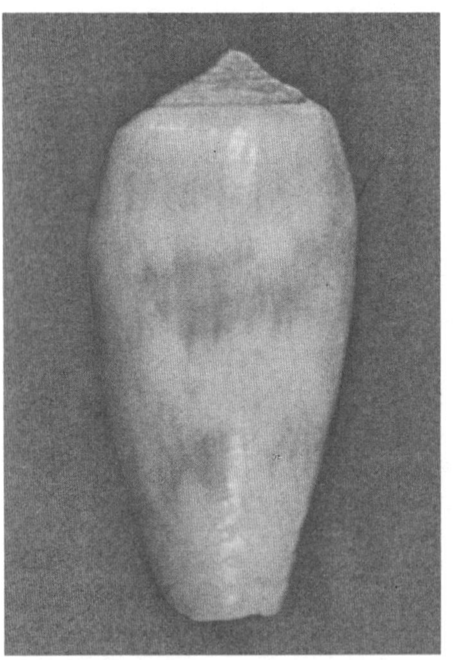
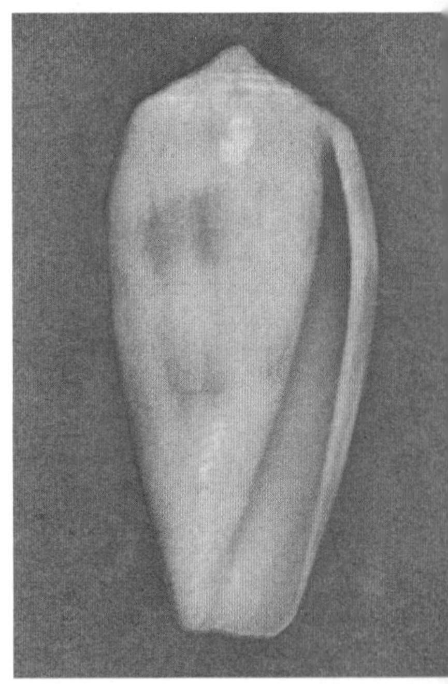

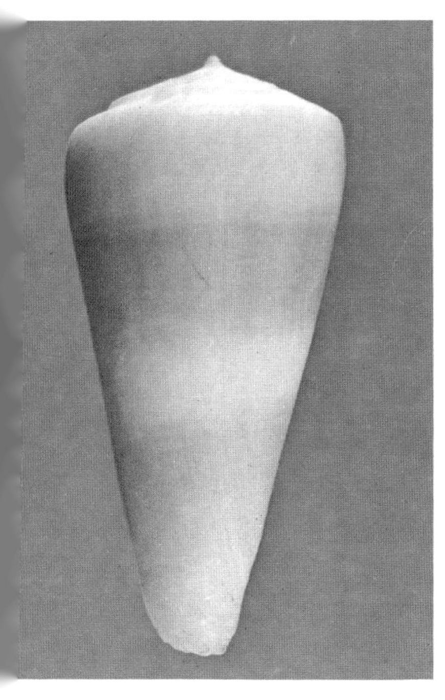
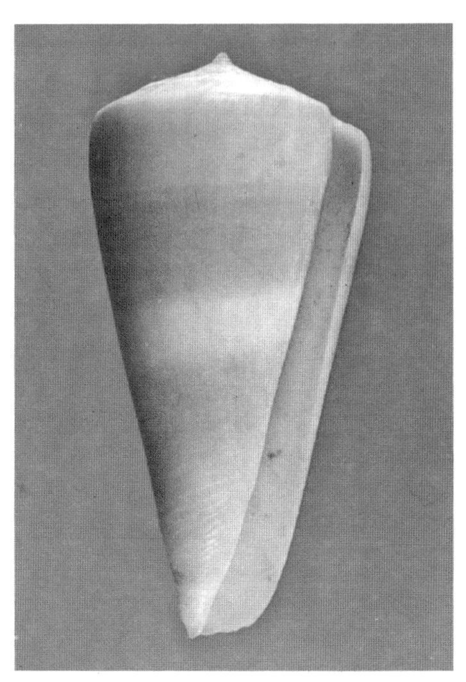

KASHIWAJIMENSIS.—Above: Taiwan Straits, 46.9mm (T.C. Lan). *Below:* Left (holotype): Kashiwajima I., SW Kochi Pref., Japan, 33mm (photo by Dr. T. Shikama); Right: Pescadores, Taiwan, 86mm (DM) (photo by Jayavad).

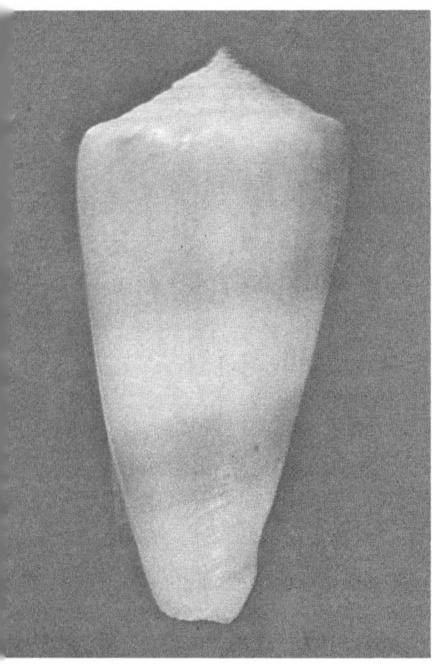
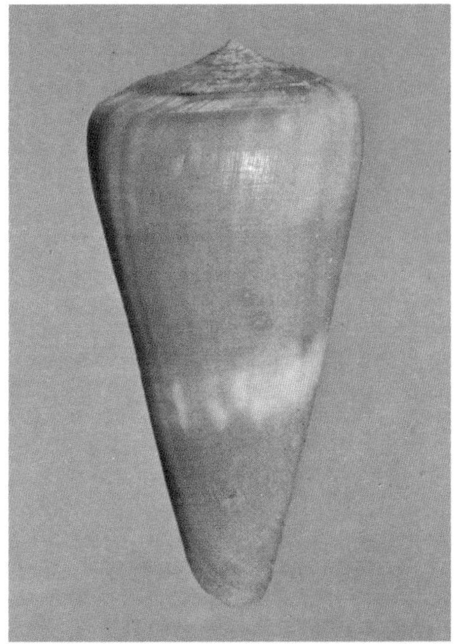

'textile lines' are sometimes broken by axial rows of small tents but are sometimes solid and continuous; a weaker band may be present below the shoulder. As mentioned, the mouth is usually deep solid violet, sometimes with a pinkish tinge, but it is subject to fading after death, at least in some populations. Some juveniles have a more open pattern like *C. canonicus.*

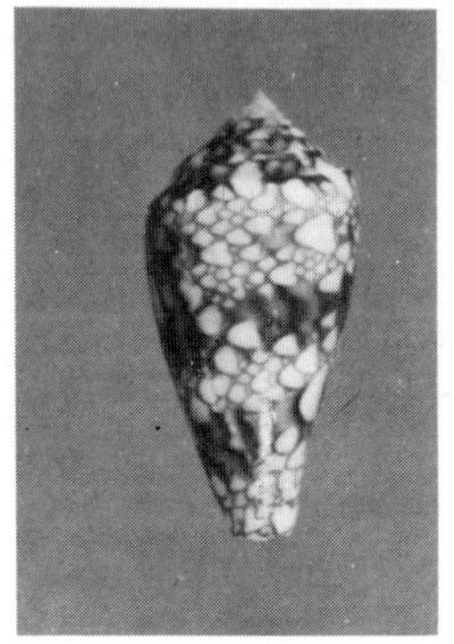

C. dalli, Galapagos, 26mm juvenile. (GG).

Distribution.—Uncommon in moderately shallow water of the eastern Pacific from southern Mexico to at least Panama and the offshore islands.

Synonymy.—The resemblance of this species to the *textile-abbas-canonicus* type of cones has long been commented upon.

Notes.—Other than chips on the lip and occasional short breaks basally, this species is not hard to find in gem condition. Perhaps keeping it out of light would prevent the mouth fading. Specimens with an especially heavy violet tone are perhaps more attractive than those with pale pinkish tones, but there is little real difference. This species is probably being over-collected in Panama, as are other uncommon but moderately expensive Panamic shells.

DAUCUS Hwass, in Bruguiere, 1792

1792. *Conus daucus* Hwass, in Bruguiere. *Cone,* in *Ency. Method., Hist. Nat. des Vers,* 1: 651 (Guadeloupe). Lectotype figure (Clench, 1942): Chemnitz, Pl. 144A, fig. L.

1798. *Cucullus cardinalis* Roeding. *Museum Boltenianum:* 42 (Locality not stated). Lectotype figure (here selected): Chemnitz, Pl. 144A, fig. L. Non *Conus cardinalis* Hwass, 1792.

1810. *Conus pastinaca* Lamarck. *Ann. du Mus. Hist. Nat.* (*Paris*), 15: 266 (Locality unknown). See Old, 1965, *Nautilus,* 79(1): 23.

1830. *Conus mamillaris* Green. *Trans. Albany Inst.,* 1: 123, pl. 3, figs. 5-6 (Florida).

1833. *Conus croceus* Sowerby i, in Sowerby ii. *Conchological Illustrations:* Pt. 29, fig. 27 (Locality not stated). Fide Tomlin, 1937.

1844. *Conus castus* Reeve. *Conchologia Iconica*, 1(*Conus*): Pl. 47, sp. 267 (Locality unknown).

1849. *Conus sanguinolentus* Reeve. *Ibid.*, 1(*Conus* Suppl.): Pl. 8, sp. 274 (Locality unknown). Non *Conus sanguinolentus* Quoy and Gaimard, 1834.

1854. *Conus comptus* A. Adams. *Proc. Zool. Soc. (London)*, 1853: 119 (Natal). Non *Conus comptus* Gould, 1853.

Description.—Moderately heavy, with a low gloss; obconical, the sides slightly convex above then straight and tapering to a narrow base; body whorl with about 6-12 low spiral ridges above the base, otherwise smooth except for numerous fine spiral and axial threads that can be quite conspicuous; shoulder broad, angulate to carinate, slightly concave above; spire low to nearly flat, the sides deeply concave, sharply pointed; tops of whorls slightly concave, with about 4-5 indistinct low spiral ridges crossed by crowded prominent axial threads; early whorls not nodulose, the margins sometimes distinctly carinate. Body whorl usually some shade of bright orange-brown to reddish or occasionally yellow, often with several widely spaced spiral rows of small squarish brown dots, these seldom obvious in average specimens; often with wide spiral bands of somewhat darker or lighter shade than background color, the midbody area often pale; occasionally with indistinct whitish blotches at shoulder or midbody; base paler, sometimes pinkish; spire and shoulder about same shade as body whorl, with a few scattered reddish brown squarish spots; apex pinkish. Aperture narrow, uniform in width; outer lip thin, straight; mouth violet to pinkish, sometimes orange, the outer margin narrowly reddish brown in some specimens. Columella short, narrow. Length 25-56mm.

Comparison.—*C. daucus* is very close to *C. attenuatus* and in some patterns they are difficult to distinguish. However, *C. attenuatus* is relatively narrower and more elongated, often has the spire whorls forming a distinct dome, and apparently always (?) lacks small dots in spiral rows. Many *C. attenuatus* have heavy axial patterns, usually absent in *C. daucus*. *C. magellanicus* is smaller, lacks distinct spiral ridges on the spire whorls, and often has heavy axial flammules.

C. daucus is most likely to be confused with some variants of *C. amphiurgus* and *C. beddomei*. Both these species may be very similar in general appearance to *C. daucus*, but as a general rule the present species can be distinguished by the lower, deeply concave spire, the carinate or strongly angled shoulder, and the "trimmer" appearance of the body whorl with straighter sides. In pattern and color there is some overlap, but few *C. daucus* have more than a few spiral rows of rather squarish spots, not the numerous rows found normally in *C. amphiurgus*.

Variation.—Although usually some shade of orange or reddish, this is a very variable species. Solid yellow individuals ('luteus' Krebs, 1864, a nomen nudum) are found uncommonly, and solid white shells are known.

KERMADECENSIS.—Above: Left and Center: Kushi, Nada-cho, Wakayama Pref., Japan, 46.6 and 47.2mm; Right: Taiwan, 45.5mm (all GG). *Below:* Left and Center: SW Conducia Bay, Mozambique, 31.8 and 33.9mm (NM H157 & H158); Right: Jumpinpon, Queensland, Australia, 29.1mm (GG).

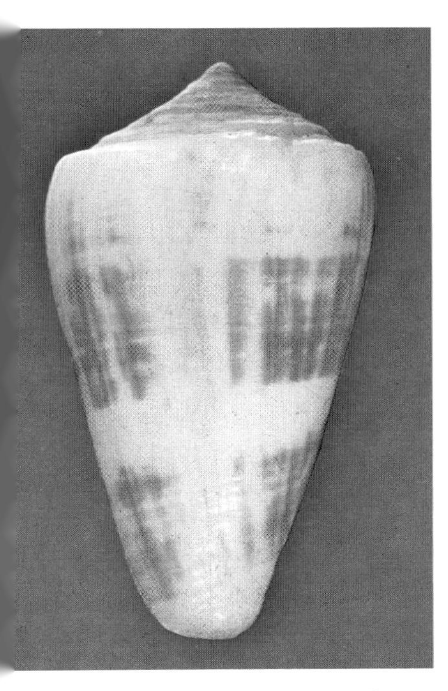
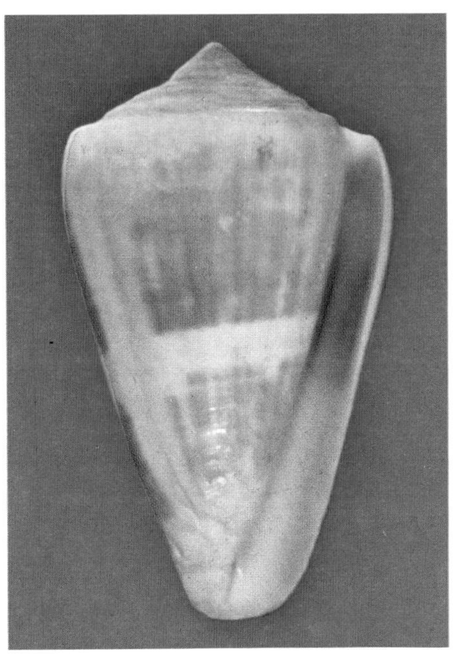

KERMADECENSIS.—Above: Point Mud, NW Western Australia, Australia, 37.9mm (GG). *Below:* NW Cape, Exmouth, Western Australia, Australia, 47.0 and 56.9mm (GG).

Generally the pattern is a simple one of several spiral bands of slightly different reddish tones on a reddish background, perhaps with 2-4 inconspicuous spiral rows of squarish brown spots hidden in the background and most conspicuous at midbody. The midbody area may be distinctly paler than the rest of the whorl, as may the shoulder area, both occasionally with indistinct white blotches. In some specimens the spiral rows of dots are heavier and more numerous, perhaps 8-12 over the body whorl. In beach specimens the reddish colors fade to brownish or yellowish and the spots usually hidden in the background become more obvious.

Distribution.—Widespread in the Caribbean from southern Florida and the Bahamas south through the Antilles and the Central American coast to the northeastern coast of Brazil. Seldom common.

Synonymy.—The variability in color, especially post-mortem changes, has led to many names founded on single specimens. There is little doubt that the listed names fall fairly easily within the synonymy of *C. daucus.* The relationship of this species and *C. amphiurgus* on the one hand and *C. attenuatus* on the other deserves careful investigation.

Notes.—Beware of yellow or whitish shells sold at high prices—many are just beach specimens that have faded. Spotted patterns are *not* rare, but are more conspicuous in beach specimens. The bright red or orange-brown patterns are most popular. The lip is thin and fragile, and there are often minor growth flaws.

DELESSERTII Recluz, 1843

1843. *Conus Delessertii* Recluz. *Rev. Mag. Zool.*, (*Soc. Cuvier*), 4: 2, pl. 42 (Red Sea, near Socotra).

1845. *Conus candidus* Kiener. *Species gen. et icon. des coqu. viv.*, 2 (*Conus*): Pl. 97, fig. 1. 1848, *Ibid.*, 2: 214 (Locality not known). Non *Conus candidus* Born, 1778.

1845. *Conus Largillierti* Kiener. *Ibid.*, 2(*Conus*): Pl. 98, fig. 3; pl. 101, fig. 1 (not fig. 1a). 1848, *Ibid.*, 2: 212 (Mexico coasts).

1939. *Conus sozoni* Bartsch. *Smithsonian Misc. Collection*, 98(1): 1, pl. 1, figs. 1-3 (Tarpon Springs, Florida).

Description.—Moderately heavy, with a high gloss; biconical, the sides straight or slightly convex posteriorly, tapering to a narrow base; body whorl with about 6-12 weak spiral ridges basally separated by very shallow grooves; numerous axial growth lines and threads over the body whorl, crossing inconspicuous spiral threads to produce a very faintly cancellate pattern; shoulder wide, carinate, weakly concave above; spire tall, sharply pointed, the sides slightly concave or almost straight; tops of whorls concave to almost flat, the margins carinate; first 3-5 whorls with small but distinct nodules; tops of whorls with numerous curved axial threads and 2-3 weak spiral grooves which are often not visible on later whorls. Body whorl white, with three salmon to sandy tan broad spiral bands, one below shoulder and the others above and below midbody; numerous spiral rows of large dark brown dashes over entire whorl, these often fused into short oblique axial flammules, especially below the shoulder within the salmon bands; base whitish to very pale tan; spire and shoulder white to pale salmon, usually with numerous orange-brown axial lines and some spots, the pattern less distinct on early whorls; nodules of early whorls separated by small brown spots; tip of spire often pale brown. Aperture rather narrow; outer lip thin, straight, sloping strongly below level of shoulder; mouth white. Columella internal. Length 40-100mm.

Comparison.—This beautiful cone is closely related, both structurally and geographically, to *C. centurio*. However, *C. centurio* is relatively wider, with a lower spire and only the first two or so whorls weakly nodulose; the pattern in *C. centurio* lacks the salmon bands of *C. delessertii* but instead has heavier and more distinct axial flammules of brown which are often partially fused into dark brown bands at the same position as the salmon bands of *C. delessertii*. In *C. centurio* the basal ridges are more distinct, rounded, and regularly spaced.

C. cingulatus has heavy spiral ridging over the whole body whorl in average specimens and lays brown clouds in axial fashion instead of spiral bands; the spire of *C. cingulatus* is largely dark brown with small white half-moons. *C. philippii* in its deep-water patterns often approaches *C. delessertii* closely; however, it is usually a darker chestnut or mahogany brown on yellowish, with the bands less distinct; the upper sides are often more convex and the spire lower; the sides of the spire are usually rather flat, the individual whorls flat above; finally, only the earliest 1-2 whorls are weakly nodulose, or none are.

Variation.—*C. delessertii* is rather constant in color, pattern, and shape, the major variation occurring in height and degree of concavity of the spire, the width and brightness of the salmon or tan bands, and the degree of fusion of the brownish dashes into axial flammules. Older specimens tend to have faded bands and develop some faint axial clouds of light brown or tan over the body whorl. In juveniles the shoulder is narrower, the salmon bands are indistinct or absent, the dashes are in definite spiral rows with little fusion, and there are weak axial blotches in the positions where the salmon bands will appear later. Juveniles have the spire whorls flat or only weakly concave above.

Distribution.—Uncommon to rare in moderately deep water from the offshore reefs of the Carolinas to southern Florida; also reported from off Texas and off the Yucatan coast, where it is apparently rare. Perhaps more widely distributed, but I know of no records from the Caribbean islands, only the continental mass.

Synonymy.—Although it has been repeatedly synonymized with *C. delessertii*, many collectors still prefer to use *C. sozoni* as the name of this species—certainly *sozoni* is easier to pronounce. However, the figure

KERSTITCHI.—Above: Isla Tres Marias, Nayarit, Mexico: Left (holotype): 31.8mm; Right: 38.8mm (photo by A. Kerstitch). *Below:* Cabo San Lucas, Baja California, Mexico; Left: 45.8mm; Right: 32.0mm (photos by A. Kerstitch).

***KIMIOI.**—Above:* Taiwan Straits, 22mm (DM) (photos by Jayavad). *Below:* Zenisu Bank, Izu-Shichito I., Japan, 19.9mm (photos by Dr. T. Okutani).

of Recluz and the good description certainly apply to the present species, and no species answering to this description has been found in the Indian Ocean—early locality data are best ignored in almost all cases as little can be trusted.

C. candidus appears to be based on a juvenile of this species. Although Clench applied the name *largillierti* to the species here called *C. cingulatus* (in his treatment *largillierti* is probably a composite of *C. cingulatus* and *C. lorenzianus* as treated here), Kiener's description and plate certainly seem to be a specimen of *C. delessertii*. Kiener distinctly states that sculpture is restricted to the base of the body whorl, which readily contrasts this species with *C. cingulatus;* in addition, his figures show a variegated species which closely matches the present one in pattern range.

Notes.—It is difficult to find a large, perfect specimen of this species as all average specimens have healed or fresh breaks on the body whorl or spire; the lip also chips very easily. Although decent specimens of rather large size are easily available, they will seldom grade better than fine or F+. Large true gems are rare and still expensive. Small juveniles look quite distinct from adults. The clearest patterns are usually found in specimens under 75mm in length.

The exact relationship of this species to *C. centurio* is uncertain. As a general rule, *C. delessertii* is a species of the North American continental shelf from the Carolinas to Yucatan, while *C. centurio* is found through the Antilles to the northern coast of South America. If this distribution is correct, they could easily be considered subspecies. However, I have seen a fairly typical specimen of *C. centurio* collected off southern Georgia by Mr. T. Yocius, well within the range of *C. delessertii*. Whether this is just a freaky *delessertii* or a true population of *C. centurio* is uncertain, but for the moment it is considered strong enough evidence to recognize the two as full species with overlapping (though rarely) ranges.

DIADEMA Sowerby i, in Sowerby ii, 1834

1834. *Conus diadema* Sowerby i, in Sowerby ii. *Conchological Illustrations:* Pt. 56/57, fig. 88 (Locality not stated).

1882. *Conus prytanis* Sowerby iii. *Proc. Zool. Soc.* (*London*), 1882: 117, pl. 5, fig. 1 (Galapagos Is.).

1910. *Conus brunneus* var. *pemphigus* Dall. *Proc. U.S. Natl. Mus.*, 38 (1741): 220 (Tres Marias Is.).

Description.—Moderately light in weight, with a good gloss; obconical to low conical, the upper sides convex and tapering to a relatively narrow base; body whorl with about 6-8 low, weak, widely spaced spiral ridges, these soon obsolete and interrupted; basal ridges sometimes granulose; rest of body whorl with fine axial growth lines and threads, rarely with spaced rows of large granules; shoulder angulate, broad, strongly coronate, with 2-3 low, irregular spiral ridges crossed by weak axial growth lines; spire low to moderately high, the whorls often stepped, the sides usually convex but sometimes concave, bluntly pointed; whorls strongly coronate, usually with 1-3 wide, weak spiral ridges and faint axial threads. Body whorl waxy brown, light or dark, usually with a narrow paler tan to dirty cream midbody band; base faintly stained violet; shoulder with or without an indistinct narrow pale band; spiral dashes absent on body whorl; spire cream to pale tan, the coronations often paler than rest and with dark brown line along the margin; tip of spire faintly pinkish. Aperture rather narrow, somewhat widened anteriorly; outer lip thin, convex; mouth violet throughout, usually with a darker violet band on posterior half. Columella long, very narrow and largely internal, set off posteriorly by a more-or-less distinct ridge. Length 25-53mm.

Comparison.—This species is very similar to the sympatric *C. brunneus*, but it is generally smaller, waxy in texture, lacks spiral rows of brown dashes or large white blotches on the body whorl, and has a violet aperture rather than a white mouth with a large yellow blotch. *C. balteatus* and *C. cernicus* can be similar in appearance, but they are covered with weak spiral ridges and are much rougher in feel.

C. diadema and *C. lividus* are extremely similar and there is little doubt that they are closely related. As in *C. abbas*-*C. dalli*, there is a Marquesan variant of *C. lividus* which closely approximates *C. diadema* in all features and is very difficult to distinguish from it. Typical *C. lividus* are usually coarse shells with the margin of the lip reddish brown and the interior pale violet with two wide dark violet bands; the spire is unmarked dirty white, the white extending well onto the body whorl below the shoulder; commonly widely spaced rows of granules extend over most of the body whorl. The Marquesan variant (sympatric with typical *C. lividus* and commonly sold under the name *C. prytanis*, here considered a synonym of *C. diadema*) is usually almost flat-spired and is more uniform in color and somewhat waxy, with very weak sculpture; the mouth is more subdued violet with a more distinct dark violet blotch posteriorly, as in *C. diadema*. Both these variants can be distinguished from *C. diadema* by the generally more waxy tones of *C. diadema*, not extending to reddish brown or bluish gray, the straighter sides in *C. lividus*, the absence of dark marks on the spire whorls (sometimes absent in *C. diadema*), the often flatter spire (highly variable), and perhaps a generally darker violet mouth (also very variable in *C. lividus*). Many *C. lividus* also have a strong tendency to develop darker axial flammules of body color along the numerous axial growth lines which are almost characteristic of that species (but distinct growth lines are not common in *C. diadema*).

Variation.—The body color varies from pale tan to deep brown but seldom has any reddish tones; the texture is distinctly smooth and waxy. The spire may be almost flat or strongly domed. Paler midbody and shoulder bands are often present, although the shoulder band may be

KINOSHITAI.—Above: Near Hong Kong, South China Sea, 56.9mm (GG). *Below:* Left: Taiwan Straits, 47.9mm; Right: Off southern Honshu, Japan, 44.6mm (MG).

KLEMAE.—*Above:* Yallingup, Western Australia, Australia, 28.3mm (MM). *Below:* South Australia, 27.5mm (GG).

very poorly defined and narrow. Spiral dashes are apparently constantly absent. The base may be brown or strongly toned with violet. Occasional specimens and populations are partially or wholly covered with large nodules on the basal ridges and even to the shoulder; this appears to be strictly an ecological character.

Distribution.—Moderately common in the eastern Pacific from the islands of the Gulf of California south to the Galapagos; apparently more common on offshore islands than on the mainland, but this may be partially because of confusion with *C. brunneus*. Shallow water.

Synonymy.—*C. prytanis* appears to be this species and not the variant of *C. lividus* mentioned earlier; the shape and pattern agree, as does the locality (if it can be trusted). *C. diadema* var. *pemphigus* is the name given to distinctly granulose individuals.

Notes.—Although not very subject to bad growth flaws, this species is very subject to erosion. Both the spire and the body whorl may be eroded enough to obscure the color. Mainland specimens are perhaps rougher than those from the islands. Very large specimens are not at all common.

DISTANS Hwass, in Bruguiere, 1792

1792. *Conus distans* Hwass, in Bruguiere. *Cone*, in *Ency. Method., Hist. Nat. des Vers*, 1: 634 (New Zealand). Holotype figured by Kohn, 1968.

1896. *Conus Waterhouseae* Brazier. *Proc. Linnaean Soc. New South Wales*, 10: 471 (Solomon Islands). Holotype (?) figured by Cotton, 1945, pl. 4, fig. 5.

1896. *Conus Kenyonae* Brazier. *Ibid.*, 21: 346 (Shark's Bay, West Australia). Holotype figured by Cotton, 1945, pl. 4, fig. 6.

1896. *Conus Kenyonae* var. *Arrowsmithensis* Brazier. *Ibid.*, 21: 346 (Arrowsmith Isl., Marshall Is.). Holotype figured by Cotton, 1945, pl. 4, fig. 4.

?1907. *Conus Bougei* Sowerby iii. *Proc. Malac. Soc. London*, 7: 299, pl. 25, figs. 1-2 (Monac Island, New Caledonia).

1970. *Conus* (*Rhizoconus*?) *chinoi* Shikama. *Venus*, 29(4): 115, figs. 1-4 (Ogokuda, Shionomisaki, Wakayama Pref., Japan).

Description.—Heavy, coarse, dull; low conical to obconical, the upper sides slightly convex then slightly concave; body whorl with several shallow grooves near the base then about 6 widely spaced grooves on the anterior fourth in adults (these same grooves extend almost to shoulder in small juveniles); grooving indistinct in adults; numerous coarse axial growth lines in adults and many healed breaks in typical specimens; shoulder broad, with very heavy rounded coronations; spire low, sometimes almost flat, the tip peculiarly flat, not at all pointed; spire whorls with traces of several indistinct spiral ridges, these obscured by the heavy coronations; coronations very large, closely spaced, rounded. Body whorl in subadults pale yellowish brown to dark brown, streaky because of axial growth marks, with three spiral bands of irregular flexuous axial white blotches, these most prominent at midbody, often indistinct at shoulder and base; base with purple spots and streaks; with growth the brown areas increase at the expense of the white zones, resulting in largely brown adults with indistinct but often wide white midbody band; base purple; with further growth, the brown begins to recede in many specimens, the areas below the shoulder turning dirty white and eventually joining the midbody band; spire whitish, with large brown spots between the coronations, the spots in regular radiating rows, the coronations conspicuously white. Aperture moderately narrow, slightly widened anteriorly; outer lip sharp, concave at the middle; mouth dark violet in small shells, whitening with growth until entirely pale bluish white; extreme anterior edge of lip dark violet or purple. Columella short, narrow, not very distinct. Length 50-123mm.

Comparison.—*Conus distans* is a species unto itself, being very distinct in pattern, the widely spaced spiral grooves, the great ontogenetic changes, the large and closely spaced coronations, and the curiously flattened spire tip. Only *C. moreleti* is similar, and it is more elongated, olive in color with very little white, much smaller in adult size, and has evenly spaced often granulose ridges over the anterior half of the body whorl; the spires of the two are very similar, however. *C. lividus* may be vaguely similar, but it has a solid white spire with more normal coronations, the apex is not flattened, and it commonly is bluish in tone or has heavy granulations.

Variation.—Because of the great changes in coloration with growth, it is possible to confuse the juveniles (under 40mm) with another and always seemingly 'undescribed' species. In juveniles the basal grooves are very distinct and extend almost to the shoulder; the color is a bright pinkish tan, while there are only a few curved white spots at midbody and above the base, as well as a few white spots on the violet base; the mouth is dark violet in two broad bands, Later changes are covered in the description. Curiously, the basal grooves vary little in actual position, remaining about the same distance apart, but they of course gradually come to occupy an increasingly more anterior position with growth and become less and less distinct. Almost all white individuals are usually old adults. The spire may be greatly stepped in old specimens and very much distorted.

Distribution.—A common species of shallow water across the entire Indo-Pacific from eastern Africa to Hawaii and French Polynesia.

Synonymy.—The Brazier names are based on worn individuals showing minor differences in sculpture and pattern because of slight size differences; none are based on adults. *C. chinoi* is similarly based on two very small specimens. *C. bougei* may not belong here, but it is somewhat

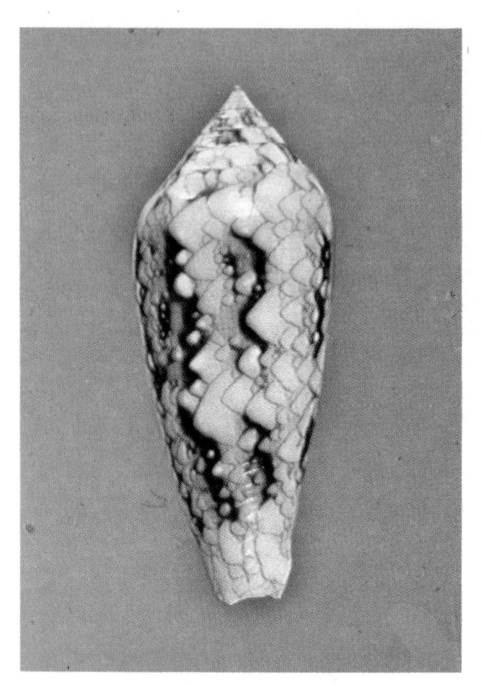
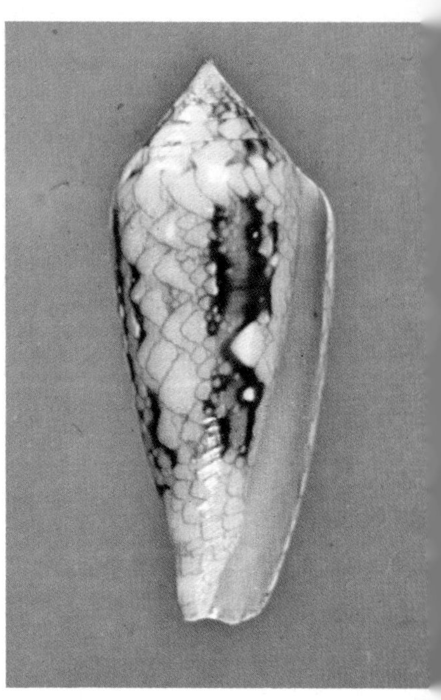

LEGATUS.—Kwajalein Atoll, Marshall Is. (GG). *Above:* 34.4mm. *Below:* 21.1-31.2mm.

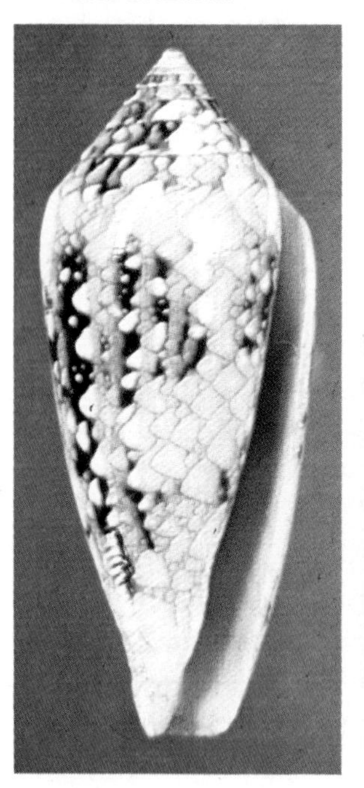
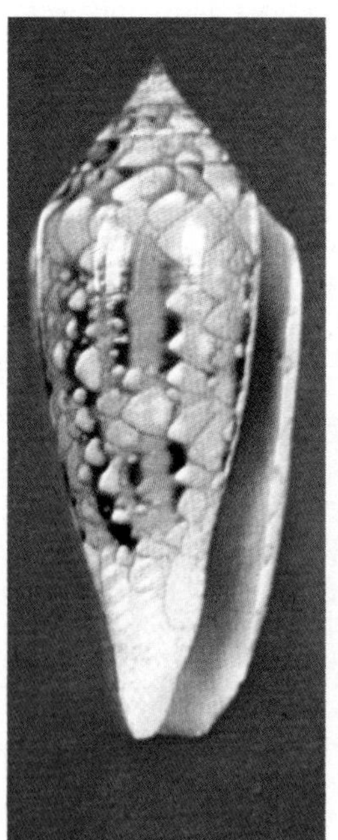
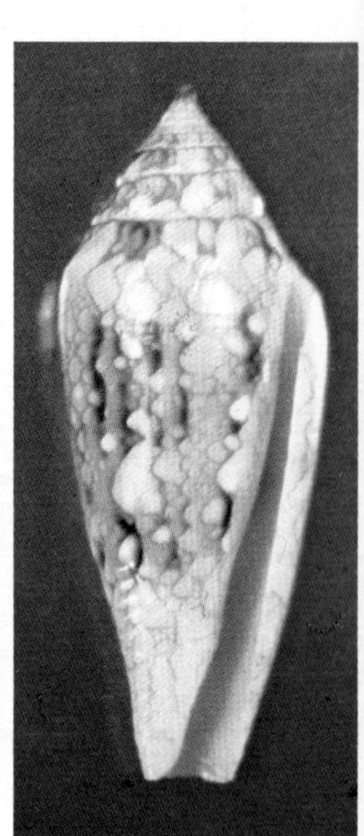

LEMNISCATUS.—Above: Left: Bombay, India, 29.5mm (GG); Right: Off Rameswaram, India, 32.8mm (K). *Below:* Left: Aden, 35mm (DM) (photo by Jayavad); Right: Gulf of Oman, 26.0mm.

similar in shape and sculpture to a juvenile of *C. distans*, although the color pattern is much more variegated white and brown than usual at a small size; specimens agreeing with Sowerby's description have apparently not been reported in the literature since.

The shell commonly sold as *C. kenyonae* from Western Australia is here considered to be *C. kermadecensis*.

Notes.—Small specimens are rare, apparently being found in deep water until they reach a length of 50mm or so; they would certainly be worth a premium if in good condition. Although truly gem adults are seldom seen, specimens in decent condition are not uncommon; the heavy flaws of most adults, however, make them unacceptable specimens.

DORREENSIS Peron, 1807

1807. *Conus Dorreensis* Peron. *Voy. Decouvertes Terres Australes*, 1: 120 (Terres d'Endracht). Note: I have not seen this reference; it is usually cited as Peron and Lesueur, but Tomlin cites only Peron. Tomlin also implies that this name may refer to *C. striatus*.

1810. *Conus pontificalis* Lamarck. *Ann. du Mus. Hist. Nat. (Paris)*, 15: 38 (Van Diemen Land).

1830. *Cucullus Doreyanus* Blainville. *Dict. Sci. Nat.*, 60: 119 (New Guinea). Tomlin considers this to be an emendation of *C. dorreensis* and possibly = *C. striatus*.

Description.—Light in weight but thick, with a low gloss; biconical, the sides convex; base usually with low and indistinct widely spaced spiral ridges; rest of whorl covered with closely spaced spiral rows of large deep punctations which form nearly perfect axial and spiral lines; shoulder rather narrow, with heavy, rounded coronations; spire moderately high, conical, the sides convex; whorls stepped and heavily coronated; tops of whorls covered by about 4 rows of deep punctations like those of body whorl; tip of spire sharply pointed but usually heavily eroded. Entire shell solid white, often with pinkish tints at base; periostracum very thin and adherent, tan with black spiral bands above base and on shoulder. Aperture narrow posteriorly, wide anteriorly; outer lip sharp, slightly convex; mouth white, sometimes with pinkish tinge. Columella internal. Length 20-40mm.

Comparison.—This is the only cone known to me which has a distinctly patterned periostracum, the black bands at the shoulder and base being very distinctive. With the periostracum removed, the shape of the spire combined with absence of pattern and the extremely regular and distinct punctations will separate it without problems. The relationships of this species are uncertain, but perhaps it is an extreme development of the *C. miliaris* line; certainly there is a similarity in general shape, and *C. miliaris* is covered with regular rows of small but distinct punctations.

Variation.—The spire is of variable height, and the whorls are variably stepped. Some specimens are distinctly stained or toned with pink or even violet, especially at the base. The periostracum is very adherent because of the many pits, and even after long soakings it may leave traces that look like an indistinct brownish spiral band near midbody and on the top of the spire whorls.

Distribution.—Confined to shallow water between Cape Leeuwin and Barrow Island, Western Australia, Australia. Common.

Synonymy.—This species was long known as *C. pontificalis*, but current practice is to call it *C. dorreensis*. I have not seen the original description of *C. dorreensis* and am not really sure whether the correct authorship is Peron or Peron and Lesueur. Tomlin seems to be uncertain if *C. doreyanus* is a new name or an emendation of *C. dorreensis;* he also seems to be of the opinion that *doreyanus* is a *C. striatus*, so the whole problem is badly muddled at the moment and deserves clarification by someone who can get the original references.

Notes.—This species is usually collected in pairs, one with and one without periostracum. The problem with this is that natural specimens tend to become covered with a greenish mold, and the periostracum is very difficult to remove without leaving traces. Many specimens have heavy axial growth lines and flaws.

DUPLICATUS Sowerby i, 1823

1823. *Conus duplicatus* Sowerby i. *Genera Recent and Fossil Shells*, 2(Pt. 16): Pl. 267, fig. 5 and caption (Locality not stated).
1965. *Asprella kuroharai* Habe. *Venus*, 24(1): 47, pl. 4, figs. 3-4 (Off Okinoshima, Kochi Pref., Shikoku, Japan).
1971. *Conus* (*Asprella*) *armadillo* Shikama. *Sci. Rpts. Yokohama Nat'l. Univ.*, Sec. II(18): 34, fig. 2 (Taiwan).

Description.—Moderately heavy, thick, with a low gloss; low conical, the upper sides distinctly convex then tapering to base; body whorl covered with narrow, widely spaced spiral grooves, leaving flat ribs between; whole shell covered with finer spiral and axial ridges and threads, giving a highly sculptured effect although normally the ribs are quite low; occasionally grooving is deeper and wider, leaving narrower ribs which are more widely spaced; basal ribs sometimes granulose on the margins; shoulder wide, roundly angled, often with a weakly defined carina; spire low to moderate, sharply pointed, the sides somewhat convex; first 2-5 whorls distinctly nodulose, rest with smooth and often carinate margins; tops of whorls nearly flat above, sometimes concave, with 1-3 distinct spiral ridges usually located on the anterior part of the whorl and often covered with curved axial threads. Body whorl whitish to straw, variably covered with numerous thin, flexuous, irregular axial

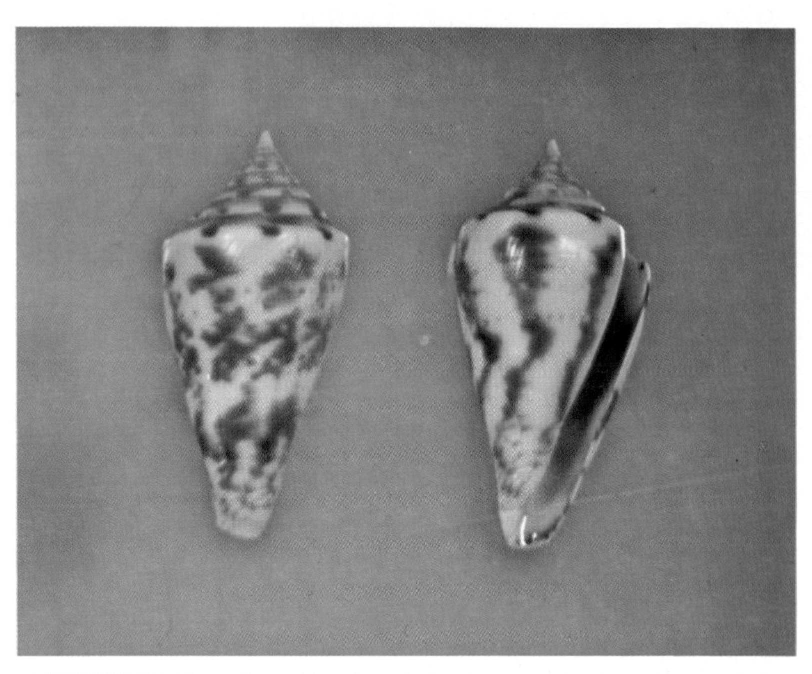

LENTIGINOSUS.—Above: Bombay, India, 31.4 and 31.7mm. ***Below:*** Left: Madras, India, 23.6mm; Right: Bombay, India, 24.5mm.

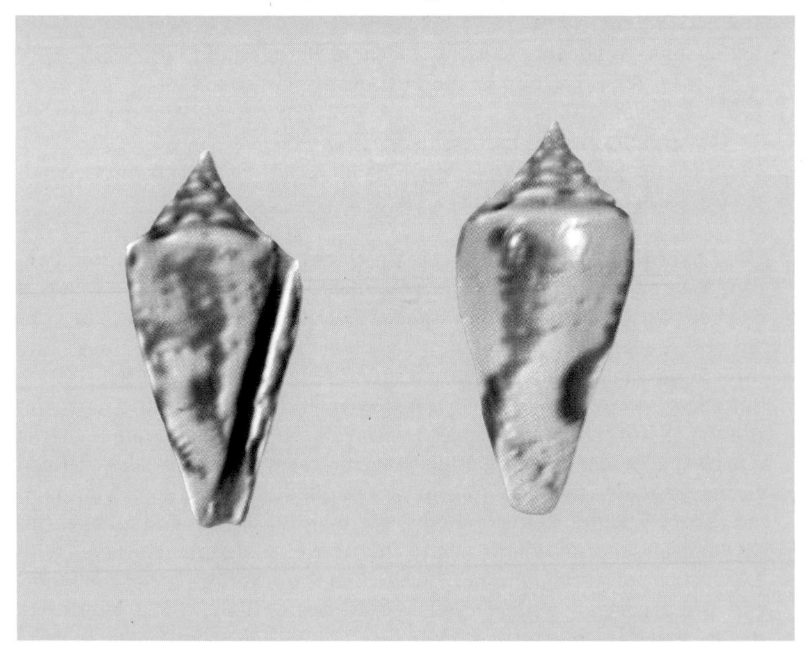

LEOPARDUS.—Above: Sulu, P.I., 77.7mm. *Below:* Kona, Hawaii, Hawaii: Left: 68.9mm; Right: 39.9mm.

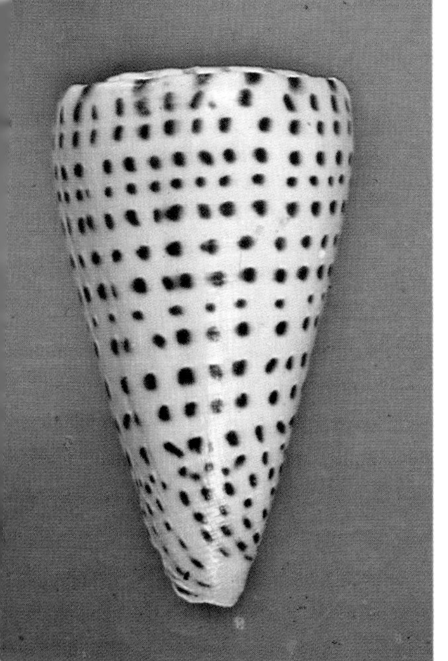

lines of reddish brown to yellowish brown, the lines long or short and heavily broken; tendency for lines to 'condense' (form closer bunches) into vague blotches or even spiral bands above and below midbody and above the base; in specimens with heavy sculpture the lines may be very short and of almost even length, following closely the width of the ribs; spire whitish and covered with long and short axial lines like those of body whorl; apex and base white. Aperture moderately wide, somewhat wider anteriorly; outer lip very thick in many specimens, weakly convex, sloping below level of shoulder; mouth whitish. Columella long and narrow, weakly set off posteriorly by a low ridge. Length 40-80mm.

Comparison.—This variable species has often been confused with *C. australis*, which is the only species it is likely to closely resemble. It is broader at the shoulder and distinctly more convex, with the spire lower and somewhat flatter sided. In typical specimens the spiral grooves on the body whorl are narrow, barely separating low but broad flat ribs that are weaker posteriorly; even in populations with heavy sculpturing the ribs are still flat, not the usual rounded ribs as in *C. australis;* most *C. duplicatus* are comparatively smooth and weakly sculptured compared to *C. australis.* In *C. duplicatus* the early whorls are distinctly nodulose, while in *C. australis* they are carinate. *C. duplicatus* has a much more open pattern of axial lines with little blotching present in typical specimens. The two species look quite different when several specimens of each are examined, and the difference in spire sculpture is the clencher.

Variation.—Two populations of this species have been examined, although at least one other one is known to exist. The Taiwanese (and presumably the Japanese) shells are somewhat differentiated from Solomons shells recently taken by Mr. Brian Bailey. The differences are summarized below.

Taiwan: Relatively broader; spire whorls flat above, the first 5 whorls nodulose; body sculpture of narrow grooves barely separating wide, flat, indistinct ribs; tendency for axial lines to cluster into indistinct blotches in spiral bands.

Solomons: Somewhat narrower; spire whorls concave above, the first 2-3 whorls nodulose; body sculpture often heavier, the grooves deeper

Heavily granulose *C. duplicatus* with distorted pattern, Solomon Is. (MG and BB).

and the ribs granulose on anterior and/or posterior margins; axial threads somewhat heavier on the body whorl; pattern open, the lines often very short and lost in the grooves so as to appear spotted.

Individual variation includes height of spire, width at shoulder, color of axial lines, density of spire pattern, and angulation of shoulder.

Distribution.—Presently this is a very rare species known from deep water off southern Japan, the Taiwan area, and the Solomon Islands. It is almost certainly more widely distributed but confused with *C. australis*.

Synonymy.—The figure and characters given for *C. duplicatus* by Sowerby are in direct comparison to *C. australis* and are in my opinion valid species characters. *C. kuroharai* is a rather typical pattern for this species and a close match with the Sowerby figure of *C. duplicatus*; in my opinion they are the same species. *C. armadillo* is an extreme variant, very large for the species (73mm), with the axial lines short, very regular, and condensed into two distinct spiral bands; a similar pattern has been seen in an extreme Solomons specimen and the beginning of a similar pattern is obvious in several Taiwanese specimens. I can only conclude that the holotype of *C. armadillo* is just an unusual *C. duplicatus*.

Notes.—This rare species is usually only seen as dead-dredged specimens from the northern China Sea; such specimens are often heavily eroded and covered with holes from burrowing sponges. Recently better specimens have been taken in the Solomons, although most of these have heavy growth flaws and broken lips. Perhaps in the near future a collectable population of this species will be found which will yield numbers of gem specimens.

Heavily sculptured specimens or those with very short and regular lines (almost spot-like) may appear very different from smoother specimens with long axial lines. The shape and spire sculpture will readily connect them with typical specimens, however.

DUSAVELI (H. Adams, 1872)

1872. *Leptoconus* (*Phasmoconus*) *du saveli* H. Adams. *Proc. Zool. Soc.* (*London*), 1872: 12, pl. 3, fig. 17 (Mauritius).

Description.—Light in weight but solid, with a high gloss; rather cylindrical, the upper sides convex below the shoulder; body whorl with several low spiral ridges above the base, these separated by narrow, shallowly punctate grooves; other spiral grooves sometimes extend over about half to two-thirds of body whorl, seldom distinct; numerous fine axial threads imparting a satiny sheen to shell; shoulder narrow, roundly angled, slightly concave above; spire moderate, sharply pointed, the sides straight or slightly convex; tops of whorls slightly concave, with about 3-4 weak spiral ridges; first 2-3 whorls weakly nodulose. Body whorl golden brown, the base light; covered with numerous spiral bands of narrow

LIENARDI.—Above: Mamie, New Caledonia, 36.8mm. ***Below:*** Left: New Caledonia, 36.0mm (GG); Isle des Pins, New Caledonia, 35.4mm (MM).

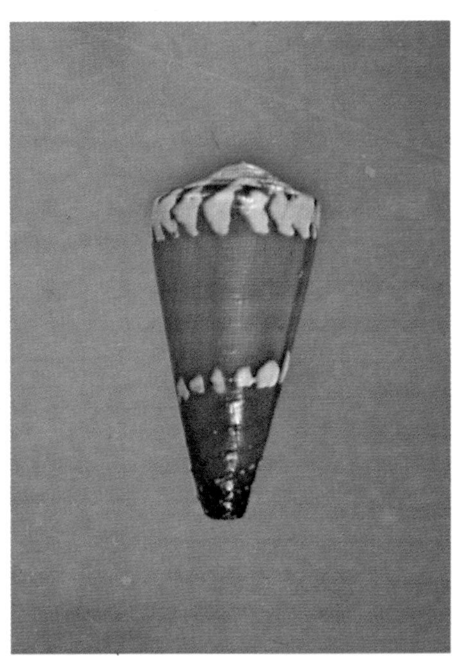

LITOGLYPHUS.—Above: Left: Okinawa, Japan, 39.0mm (TCG); Right: Tahiti, Society Is., 36.5mm. *Below:* Left: Dahlak Arch., Ethiopia, 43.4mm; Right: Cebu, P.I., 19.2 and 40.2mm.

bluish dashes alternating with white dashes and occasionally with dark brown dashes, most of each band white; three broad but variable bands of squarish chestnut blotches, one below shoulder, one above midbody, and one (narrowest) below midbody; the blotches making up these bands are somewhat axial, especially those below the shoulder, and may be divided by the dash bands which go through them; indistinct clouds of pale brown sometimes scattered over shell; shoulder and spire pale brown to white, covered with numerous fine axial lines of reddish brown; earliest whorls unmarked. Aperture wide anteriorly, narrower posteriorly; outer lip thin but sharp, convex, distinctly angled at posterior fourth; mouth whitish. Columella long, narrow. Length 40-60mm.

Comparison.—It is hard to determine the exact relationships of this still very rare and poorly known species. However, it is superficially similar to *C. cinereus* in general appearance and some aspects of pattern, although the shape of the mouth and the texture of the shell resemble *C. vicweei* and allies. From *C. cinereus* it is easily distinguished by the heavier spiral sculpture on the spiral whorls and the (weakly) nodulose early whorls, as well as the different coloration of the body whorl and mouth. From *C. vicweei* and other thin *bullatus*-like species (*C. cervus*, some *C. bullatus*) it differs greatly in color and pattern. From both it differs in the numerous fine axial reddish brown lines covering the spire and shoulder, a very distinctive color pattern that appears constant.

Perhaps the closest relative of *C. dusaveli* is *C. floccatus*, which is practically identical in shape of the body whorl and spire, has similar spire sculpture and columella, has the lip generally similar, and has the bright coloration and bluish dashes of *C. dusaveli*. In both species the texture is similar, with a strong satiny sheen over the body whorl. Of course there are obvious differences in pattern, especially the fine lines on the spire.

Also similar in some ways to *C. dusaveli* in pattern is an uncommon pattern of *C. monachus* in which the body whorl is encircled by heavy brown spiral lines which are interrupted by white dashes and blotches. Although this shell is broader at the shoulder and quite different in pattern, it should be mentioned to prevent possible misidentifications.

Variation.—I have seen too few specimens and photos to be sure just how variable this species is. The height of the spire varies from moderate to quite high, with the tops of the whorls almost flat to distinctly concave; the sculpture of the body whorl may be reduced to a few basal ridges or have weak ridging extending to above midbody. Of course, as would be expected, the pattern varies considerably in details, although blue-white-brown dash bands are fairly constant and the three brown bands are always represented, although their size varies. The background color may be very pale, almost yellowish, or quite golden tan. As far as I know, the narrow lines on the spire are constant and the easiest means of recognizing the species. An almost white shell probably belonging to this species has been examined courtesy of T.C. Good.

Left: Nearly colorless *C.* cf. *dusaveli,* Okinawa (photo TCG).

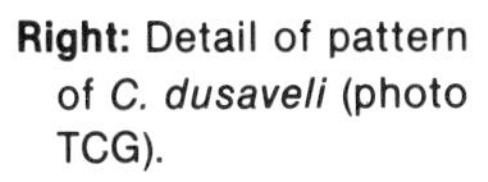

Right: Detail of pattern of *C. dusaveli* (photo TCG).

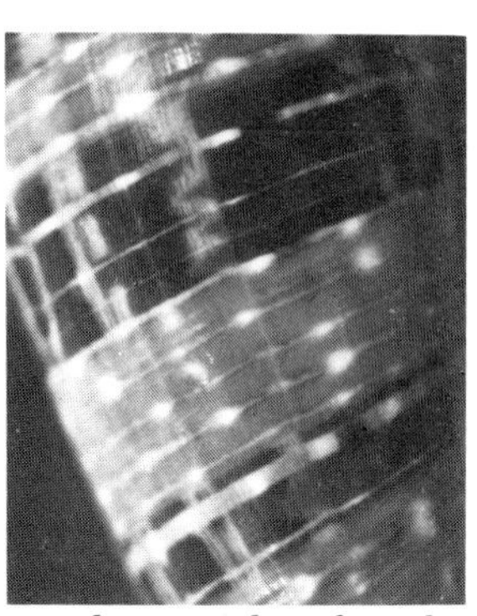

Distribution.—This species has only been rediscovered within the last few years, so it is very poorly known. The holotype was taken from the stomach of a fish taken off Mauritius, but perhaps this locality is incorrect. Most specimens taken lately have come from the Ryukyus, where it is apparently taken with some regularity by divers in deep water. Supposedly specimens have been taken in the Philippines. I have also identified, very tentatively, this species from a fragment dredged in deep water in the Solomons, but it is possible that the fragment represents an undescribed species. In all likelihood the distribution of this species in the western Pacific is wide and the species will prove to be merely uncommon once its ecology is better understood.

Notes.—This species currently has the most unstable market of any cone, not excluding *C. gloriamaris*. There is every indication that several dozen specimens are currently sitting in private collections around the world, yet every new specimen is said to be the 'second or third' specimen reported. Certainly *C. dusaveli* is not as rare as the extremely high prices (up to $5,000) which have been paid lately would indicate. Extreme caution is advised with this species unless you can afford to chance parting with several thousand dollars for a shell which in a couple of years might be worth only $500. To add to the uncertainty, there is a faint possibility that two species are presently confused under the name *C. dusaveli.*

EBRAEUS Linnaeus, 1758

1758. *Conus Ebraeus* Linnaeus. *Systema Naturae per Regna Tria Naturae*, ed. 10, 1: 715 (India). Lectotype selected and figured by Kohn, 1964. This name is commonly misspelled *C.* 'hebraeus'. and *C.* 'haebreus.'

1811. *Conus quadratus* Perry. *Conchology*: Pl. 24, fig. 5 (Africa).

Description.—Heavy, solid, with a low gloss; broadly conical to obconical, the sides convex; body whorl with low rounded spiral ridges basally, these often extending to at least midbody and sometimes bearing

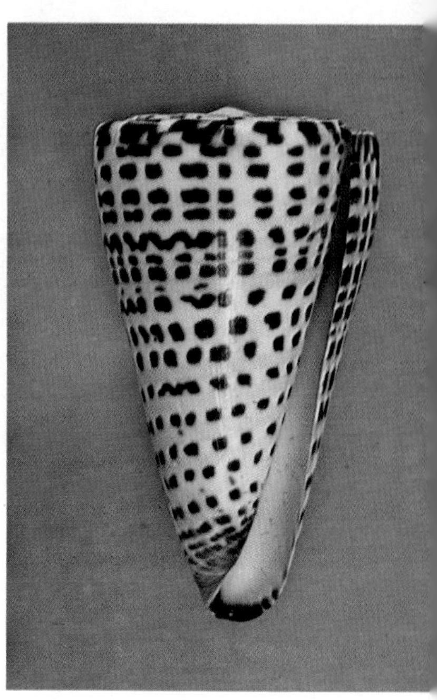

LITTERATUS.—Above: Sulu, P.I.: Left: 94.3mm; Right: 88.0mm. ***Below:*** Left: Honiara, Guadalcanal, Solomon Is., 47.8mm; Right: Zanzibar, 55.6mm.

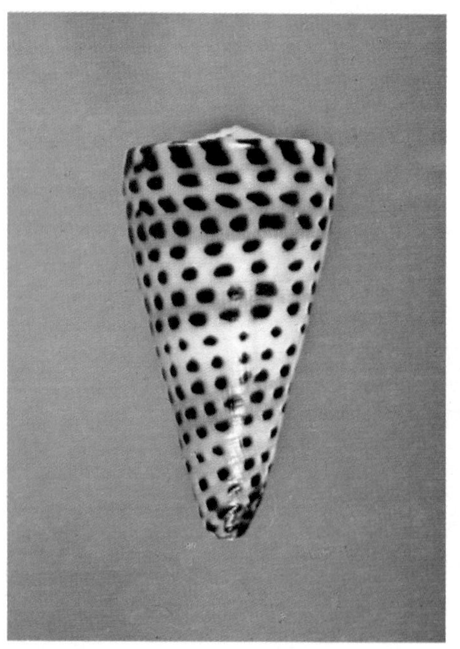

LIVIDUS.—Above: Boac, Marinduque, P.I., 31.6 and 39.6mm. ***Below:*** Tahiti, Society Is.: Left: 22.7 and 27.6mm; Right: 25.3 and 26.2mm.

large nodules; shoulder roundly angled, coronated, the coronations sometimes almost absent or very weak; spire low, convex, bluntly pointed; spire whorls weakly coronated and with 4-5 heavy spiral ridges on top. Body color white to pinkish, with three series of axially rectangular black spots; series below shoulder separated by a narrow and even band of background color from series above midbody; midbody band wide, clear; basal series of spots smaller, sometimes elongated and partially fused with the deep purple base; sometimes with secondary series of black dashes at midbody; shoulder with large squarish black spots alternating with light background; spire with irregular black blotches. Aperture narrow posteriorly, wide anteriorly; outer lip thick, sharp, convex; mouth pinkish white or white, the external pattern showing through as three broad blackish or brownish bars. Columella short, wide, bounded posteriorly by a heavy ridge. Length 20-62mm.

Comparison.—Although related to *C. sponsalis*, *C. ebraeus* is only likely to be confused with the very similar *C. chaldaeus*. From that species it differs in the series of black blotches being squarish or rectangular, sharply defined, and in distinct rows, not axially fused. The mouth in *C. ebraeus* is whitish with the external pattern showing through, while in *C. chaldaeus* the normal mouth pattern consists of only a large violet blotch on bluish white. Occasional specimens of *C. ebraeus* show partial fusion of the anterior two series of spots, or occasionally the series posterior to midbody and below the shoulder, but the resulting axial blotches are wider and more regular than in *C. chaldaeus*. The two species are not likely to be confused.

Variation.—The background color is sometimes distinctly pink and quite attractive; the shoulder and spire coronations may be quite heavy or almost obsolete. The tendency toward fusion of the spots into axial flammules has been mentioned. In some specimens the bands in the mouth are brownish and similar to those of *C. sponsalis*.

Distribution.—Widely distributed and abundant in shallow water throughout the Indo-Pacific, from the eastern African coast to Hawaii and French Polynesia. Also occurs along the Central American coast and the offshore island, but very uncommon there.

Synonymy.—The distinctive pattern is clearly shown in Perry's figure. The various alternate spellings are here considered errors, but they were probably actually emendations, addition of the beginning 'h' being considered better formation. The name should of course be spelled as Linnaeus spelled it.

Notes.—An abundant little cone which is still one of the cheapest species but quite attractive, especially those with bright pink backgrounds (apparently a strictly ecological feature) and sharply defined spots. The outer lip is commonly thickened at the middle like other small coronated cones. *C. ebraeus* often has heavy growth flaws, but the species is so common this seems to go unnoticed. Specimens over about 40mm are very uncommon.

EBURNEUS Hwass, in Bruguiere, 1792

1792. *Conus eburneus* Hwass, in Bruguiere. *Cone*, in *Ency. Method.*, *Hist. Nat. des Vers*, 1: 640 (East Indies). Lectotype figure (Kohn, 1968): *Tableau Ency. Method.*, Pl. 324, fig. 1.

1798. *Cucullus quadratulus* Roeding. *Museum Boltenianum:* 41 (Locality not stated). Lectotype figure (Kohn, 1975): Martini, Pl. 61, fig. 670.

1807. *Conus alternatus* Link. *Beschr. Nat.-Samml. Univ. Rostock*, 3: 101 (Locality not stated). Holotype figure: Martini, Pl. 56 (actually Pl. 61), fig. 670.

1857-1858. *Conus crassus* Sowerby ii. *Thesaurus Conchyliorum*, 3 (*Conus*): 25, pl. 12(198), figs. 254-255 (FeeJee Islands). Lectotype figure (here selected): Pl. 12, fig. 254, as fig. 255 is stated to be abnormal.

1874. *Conus eburneus* var. *polyglotta* Weinkauff. *JB. deutsch. Malak. Ges.*, 1: 244 (Pelew Is.). Based on *Thesaurus Conchyliorum*, P. 12 (198), fig. 248.

Description.—Moderately light in weight, with a high gloss; low conical to obconical, the upper sides distinctly convex to straight; body whorl with about 12 heavy, rounded spiral ridges on the anterior third, these separated by deep grooves which may be punctate; rest of body whorl smooth except for fine spiral and axial threads which are seldom conspicuous; shoulder weakly angled to rounded; spire low, sometimes almost flat, to moderate, sharply pointed, the sides straight to concave; spire whorls flat or slightly concave, but usually convex; tops of whorls with a heavy, rounded spiral ridge near the middle, flanked by two lower and sometimes inconspicuous spiral ridges; axial sculpture weak or absent. Shell glossy white with numerous spiral rows of black to reddish brown or orange rather large squarish to rectangular spots of various sizes; these are commonly in at least a dozen rows, but there may be several rows missing, especially in the midbody series, leaving white bands; usually at least three orange spiral bands developed in the background between rows of spots, one above midbody, one below midbody, and one in the basal area; the orange bands may be very weak, strong, or absent, and more or less than three in number; the base is white; series of spots show a strong tendency in some populations to fuse laterally into large blackish dashes; shoulder with regular black spots like those of the body, sometimes somewhat oblique and extending onto last spire whorl; other spire whorls with regularly revolving black to orange spots, these sometimes fused into axial lines; tip of spire white. Aperture fairly narrow, rather uniform in width and not greatly widened at the base; outer lip thin to thick, sharp; mouth shiny white. Columella short, broad, thick, bounded posteriorly by a heavy ridge. Length 30-65mm.

Comparison.—The contrasting blackish spots on a glossy white background, plus the usually present orange bands, allow this species to be confused with few others. *Conus leopardus* is much larger, coarser, with

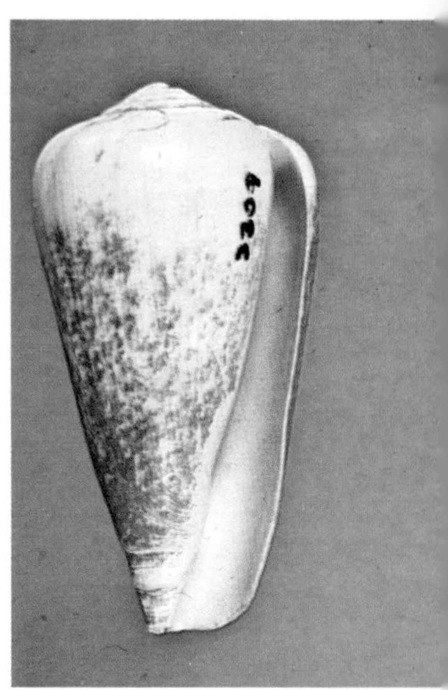

LOHRI.—Above and spire below (paratype): Scottburgh, Natal, South Africa, 56.7mm (NM 3309). *Below* left (holotype): Baia des Cocos, between Inhambane and Jangamo Beach, Mozambique, 40.2mm (NM F8701 (photo by R.N. Kilburn).

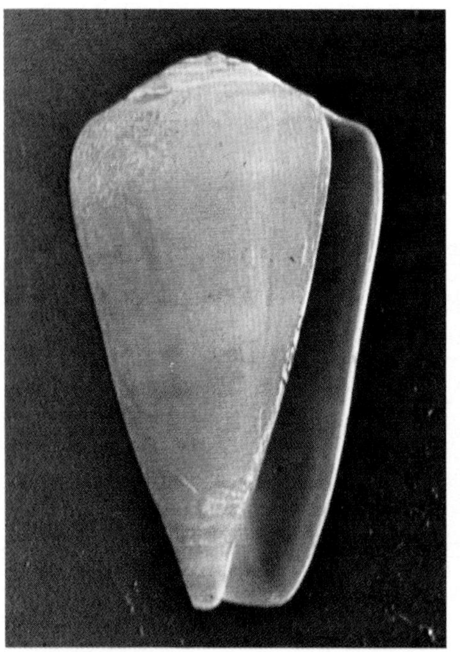

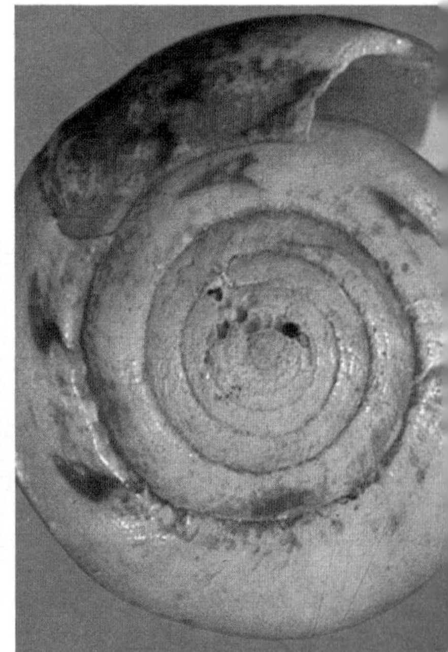

LORENZIANUS.*—Above:* Off Cartagena, Colombia, 39.0mm (EP). ***Below:*** Left: Off Cartagena, Colombia, 56.2mm (EP); Right: "West Indes", 24.6mm (PC).

the pattern consisting of rather small, often irregularly spaced spots and no orange bands; the spire whorls are deeply concave, as is the shoulder, and with weak spiral ridges; juvenile *C. leopardus* are distinctly narrower than *C. eburneus* of similar size and not as glossy. *C. litteratus* is more similar, looking like a gigantic *C. eburneus*, but the spotting is sometimes distinctly axially elongate, the shoulder is usually strongly concave above, the spiral ridges of the spire are weak and indistinct, the columella is curved so the outer lip and inner lip form a smooth and continuous curve, and the base has a large blackish or violet spot; these differences hold for juveniles as well. *C. tessulatus* is thicker and heavier than most *C. eburneus*, although this will not always be useful, with the base stained a deep pinkish violet that is almost always present; the spots are usually smaller, often pale orange or bright red but seldom reddish brown, and in less regular series. In *C. tessulatus* the spire whorls have a single very heavy spiral ridge at the middle, although sometimes a much weaker second ridge may be visible.

Variation.—Variation in *C. eburneus* involves both pattern and shape. The extremes of each type of variation have been given a distinct name, so it is possible to use these names as varieties, although the variants themselves are apparently just unusual individuals or small and scattered populations.

C. eburneus, typical: The most common pattern and shape, with the sides nearly straight or slightly convex posteriorly and the spire almost or completely flat except for the pointed cone of the early whorls. The rows of spots are fairly uniform and show little tendency to fuse, although they may occasionally be almost totally lacking in some specimens; the orange bands are narrow. Widely distributed throughout the Indo-Pacific except Hawaii.

C. eburneus var. *polyglotta:* Specimens of this variety are of the same shape as typical *eburneus*, but show a strong tendency to enlarge and fuse the black spots to produce long dashes or complete spiral bands; only 1-2 rows, especially those near the midbody area, may be fused, or all the rows may become band-like; the dark appearance of the shell is increased in some specimens by fusion of the orange bands into large orange areas of intensified color above and below the midbody area, leaving only the midbody area white. This very attractive and common variety seems to be best developed in the Philippines, although occasional specimens of the variety may show up almost anywhere.

C. eburneus var. *crassus:* Similar in color pattern to typical *C. eburneus*, but differs in shape and (sometimes) color. The sides of the body whorl are strongly convex below the shoulder then distinctly concave, producing a pyriform outline; the spire is high for the species and distinctly conical; the spiral ridges of the spire whorls are heavily developed, usually two but sometimes all three strong; the spots on the body whorl in the typical (most expensive) form of this variety are bright reddish brown or orange, but black spots also occur in association with the peculiar shape. Since the distinctive shape can be found with normally colored spots and since reddish spots can be found on shells of normal shape, I can see no way to retain *crassus* as a full species; it seems to be a rare or at least uncommon individual variant, although it may be a sizable part of some populations. Seemingly it is restricted to a belt in the southern Pacific from Fiji to New Britain.

Distribution.—Widely distributed and often abundant across the Indo-Pacific from East Africa to French Polynesia, but apparently not recorded from Hawaii. Shallow waters.

Synonymy.—Since the names *polyglotta* and *crassus* seem to refer to individual variants or at most small and scattered populations, they are discussed under *Variation. C. quadratulus* and *C. alternatus* seem to be based on the same figure, which is in all likelihood *C. eburneus;* the Link name was referred to *C. genuanus* by Tomlin, but this does not agree with the figure reference.

Notes.—A common and very attractive species. The *polyglotta* variety draws a small premium, although it is an extremely attractive shell when fully developed with all the spots fused and the orange bands greatly enlarged and widened. Of the typical variety, only the partially albinistic (actually, just reduced pattern) specimens draw a premium, although the price differential is often large; specimens without spots (or with only 3-4 or so) but with well developed orange bands are considered especially interesting. The *crassus* variety is least common, although specimens with the *crassus* shape and normal black spotting are found with normal *eburneus* at New Britain, where there is a great deal of commercial collecting; to a dealer or collector, however, *crassus* implies reddish spots. The combination of reddish spots and pyriform shape is uncommon to say the least, and large specimens in gem condition with this combination are distinctly rare.

C. eburneus is a glossy cone but it is subject to large breaks near fresh growth. Some specimens have obvious growth marks, but these are not really common. Since the typical variety is so common, only gem specimens are ever offered for sale.

ELEGANS Sowerby iii, 1895

1895. *Conus elegans* Sowerby iii. *Proc. Malac. Soc. London*, 1: 215, pl. 13, fig. 8 (Persian Gulf).

1904. *Conus aculeiformis* f. *torensis* Sturany. *Denkschr. K. Akad. Wiss. Wien, 74, Exped. "Pola" Rothe Meer:* 227, pl. 4, figs. 8a-8b (Station 58, Red Sea).

Description.—Very thin, light, and fragile, with a low gloss; elongate biconical; body whorl with anterior half to two-thirds covered with narrow spiral punctate grooves separating broad spiral flat ribs;

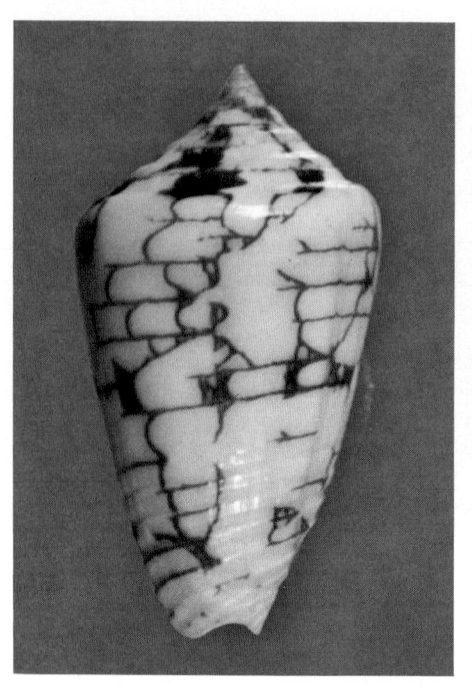

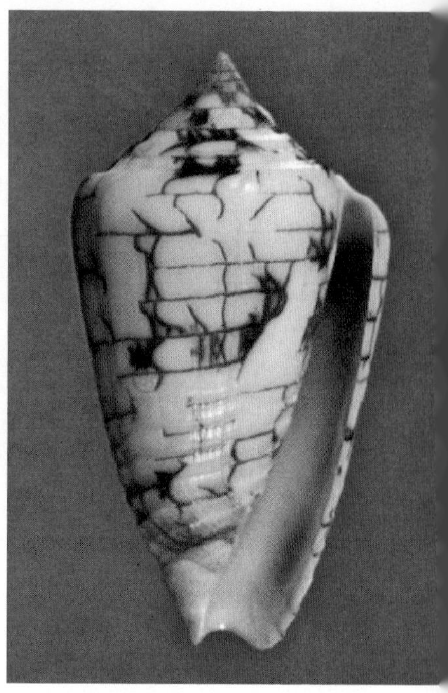

LUCIDUS.—Above: Santa Cruz, Galapagos Is., 29.1mm (GG). ***Below:*** Cabo San Lucas, Baja California, Mexico: 58.0 and 58.2mm (photo by A. Kerstitch).

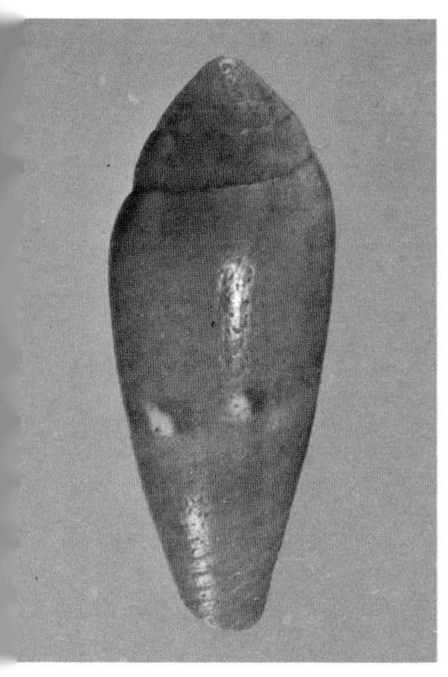

LUTEUS.—Above (*nucleus*): Okinawa, Japan, 31mm (DM) (photos by Jayavad). *Below:* Left: Koniya, Amami-Oshima, Ryukyus, Japan, 31.8mm (EP); Right: Tolwat Beach, E New Britain, Papua New Guinea, 18.4mm (B. Parkinson).

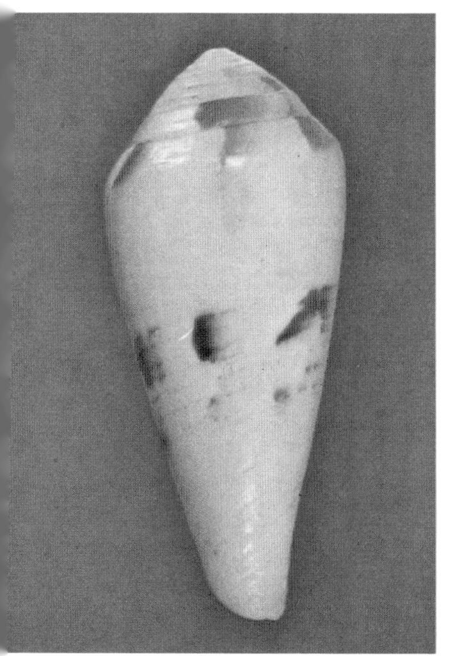

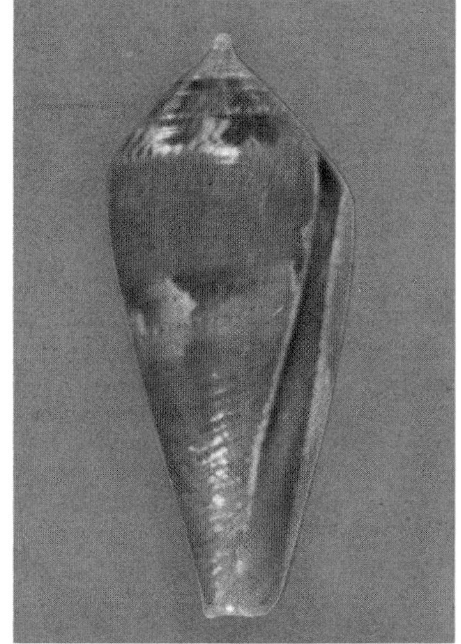

occasionally (small specimens) grooving may extend to shoulder; whole shell covered with very fine axial threads; shoulder rounded, narrow, merging into spire, weakly carinate; spire high, sharply pointed, the sides straight or slightly concave, the whorls stepped; first 0-5 whorls bearing strong rounded nodules, these usually extending to last whorl although commonly absent from shoulder; tops of whorls concave, with 1-2 narrow spiral ridges at the middle and broad flat ribs anteriorly and posteriorly; numerous fine curved axial threads heaviest in center of whorl. Body whorl white to cream, covered with widely spaced spiral rows of small squarish reddish brown to yellowish brown spots; occasional traces of very indistinct larger blotches; spire like body whorl, with small brown spots between nodules and scattered small blotches of yellowish brown. Aperture narrow, uniform; outer lip very thin and fragile; mouth dirty white. Columella not visible externally. Length 20-30mm.

Comparison.—This species, if it is a full species, is extremely close to *C. aculeiformis* and needs to be compared mostly with that species. The spiral grooving in *C. elegans* normally fails to extend to the shoulder, leaving the last third to half of the body whorl smooth; juvenile shells, however, have the grooving extending almost to the shoulder, as in *C. aculeiformis*. The pattern is fairly distinctive, the spots widely spaced and distinctly squarish, with only vague traces of other dark patterns; such a pattern is seldom found in *C. aculeiformis*, which commonly has the spots more dash-like, closer, and sometimes covered with heavier clouds and blotches. In most *C. elegans* the spire whorls are strongly nodulose to the last whorl, thus heavier than in *C. aculeiformis*.

The narrow variant of *C. acutangulus* found in the eastern Indian Ocean is very similar superficially, but it is wider at the shoulder, has the ribbing of the body whorl narrower and heavier, has stronger axial ridging on the spire whorls, and has the pattern largely axial clouds or heavy spiral bands of large spots and blotches.

Variation.—In fresh specimens the spotting is distinctly reddish brown, but this fades to yellowish brown. Grooving varies from over the anterior half of the body whorl to at least two-thirds, occasionally reaching the shoulder in small specimens. The nodules of the spire whorls are fairly heavy in the specimens examined, but in Sowerby's type (larger than any specimen I have seen) the nodules were described as minute and restricted to the first 4-5 whorls; the type of *C. torensis* apparently lacks nodules. Perhaps this is due to size or simply individual variation, as in *C. aculeiformis* and *C. insculptus*, which both vary in number of nodulose spire whorls. I am not at all sure that specimens could always be separated from some variants of *C. aculeiformis*.

Distribution.—Found in deep water from the Persian Gulf south to off Mozambique. Uncommon to rare, but this area is not heavily collected.

Synonymy.—*C. torensis* was well described and figured and certainly belongs here. The sculpture and pattern are typical, but the spire whorls all appear to be smooth. This variation in spire nodulosity from smooth to half nodulose to completely nodulose parallels the variation in *C. aculeiformis*.

Notes.—Other than the number of nodulose whorls, the present specimens agree with Sowerby's description and figure quite well. This is a very fragile species.

EMACIATUS Reeve, 1849

1849. *Conus emaciatus* Reeve. *Conchologia Iconica*, 1(*Conus* Suppl.): Pl. 5, sp. 248 (Philippine Islands).

Description.—Heavy, with a low gloss; elongate obconical, the body whorl tapered, the middle sides slightly to distinctly concave; body whorl with heavy, widely spaced, often granulose spiral ridges from base to posterior fourth of the shell, where they become obsolete; numerous fine axial threads over whorl; shoulder wide, weakly carinate, almost flat above; spire low, almost flat, the earlier whorls forming a small raised cone in the middle, blunt; spire almost always eroded, even in small specimens; tops of whorls with many very weak spiral and axial threads, often appearing smooth to the eye. Body whorl bright yellow to orange, usually with poorly defined lighter yellow to yellowish cream bands at the shoulder and midbody; base brownish purple to deep violet, sometimes very weakly marked; shoulder and spire yellowish cream, the tip of the spire, when visible, violet. Aperture narrow posteriorly, suddenly widened anteriorly; outer lip strongly to weakly concave, thin; mouth bluish white in adults, with a large to small violet blotch at the anterior margin, more strongly and widely toned with violet in juveniles. Columella very narrow. Length 30-56mm.

Comparison.—*C. emaciatus* is often considered to be a juvenile or dwarfed *C. virgo*, but they appear quite distinct. Few juvenile *C. virgo* are known, but they are said to be waxy orange with the spiral ridges crowded anteriorly on the body whorl (Cernohorsky, 1964). Small *C. emaciatus* tend to be yellowish, somewhat lighter than adults, with two even paler bands. Whether these differences mean anything or not is uncertain. A simple difference which is probably ecological in origin is that specimens of *C. emaciatus* even 35-40mm long are almost always heavily eroded on the spire, even more so than typical *C. virgo* 70-80mm in length. In *C. emaciatus* the sides of the body whorl are concave, but this is often difficult to see.

Also similar to *C. emaciatus* are some variants of *C. flavidus*, but *C. emaciatus* has the spiral ridges usually extending higher on the body whorl, is more slender with concave sides, and usually lacks the dark violet aperture typical of *C. flavidus*. *C. frigidus* is somewhat more similar in body sculpture, but that species has distinctly striate spire

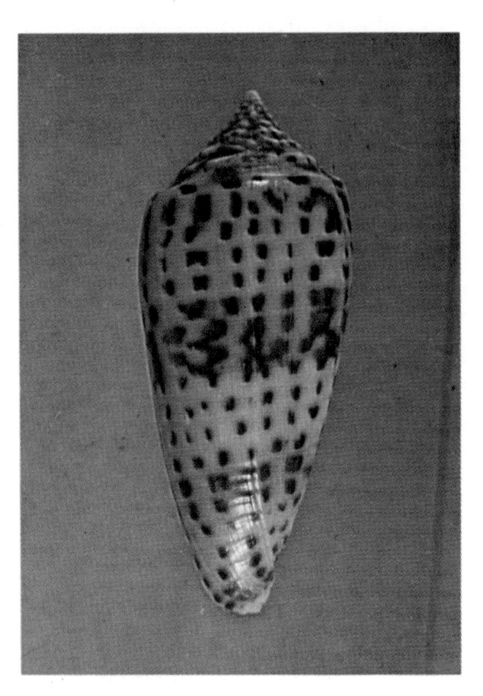

LYNCEUS.—Above: Rangoon, Thailand, 68.7mm (K). *Below:* Left: Off Kov Keow I., Andaman Sea, 83.6mm (MM); Right: Andaman Sea, 52.0mm.

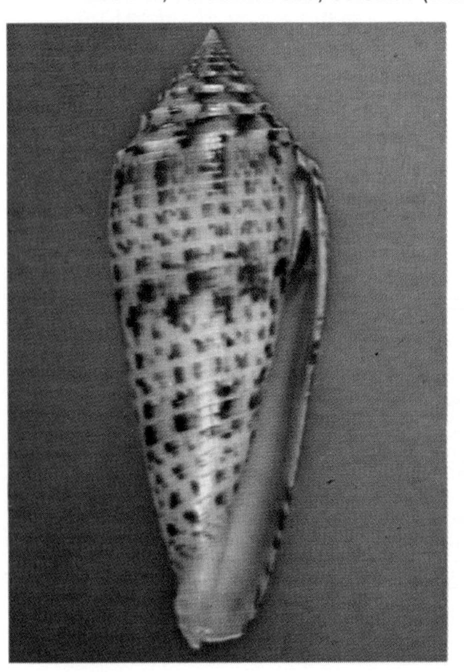

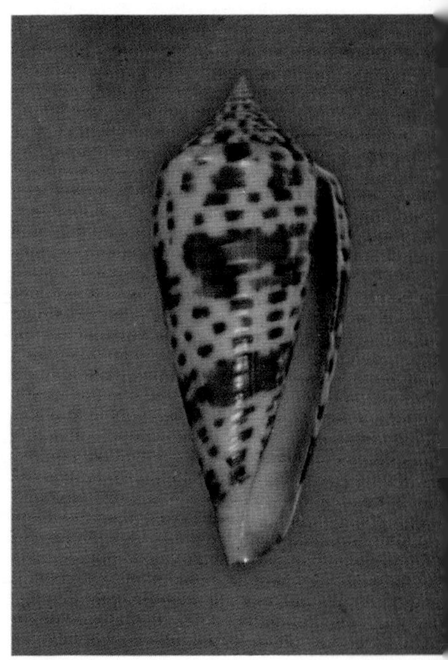

MAGELLANICUS.—Off Hollywood, Florida. *Above:* 23.4-26.3mm. *Below:* Left: 21.9 and 24.2mm; Right: 22.2 and 22.6mm.

whorls and a deep violet aperture. *C. terebra* is similar in general sculpture and size, but it has a higher spire, heavy striations to the shoulder, and is usually whitish or olive.

Variation.—There is some variation in intensity of the yellow color, the distinctness of the paler bands, the concavity of the sides, the height of the spire, extent of the spiral ridges, and amount of purple on the base, but these appear to be mostly individual and of little significance. The juveniles are somewhat paler than adults, not orange but yellow, have the same bands, and have the mouth more violet.

Distribution.—Widely spread over the Indo-Pacific from eastern Africa to French Polynesia, but not yet reported from Hawaii. Common in shallow water.

Notes.—Erosion of the spire seems typical of this species and should be expected, although it is not always very bad. The lip is fragile and easily broken. Some specimens become bright orange rather than yellow and are more attractive. Juveniles of this species are not uncommon, but I am not really sure they could be separated from those of *C. virgo*. Like *C. flavidus*, there are often distinct crowded axial growth marks over the body whorl.

ENCAUSTUS Kiener, 1845

1845. *Conus encaustus* Kiener. *Species gen. et icon. des coqu. viv.*, 2 (*Conus*): Pl. 14, fig. 2. 1846, *Ibid.*, 2: 54 (Locality unknown).
1848. *Conus praetextus* Reeve. *Conchologia Iconica*, 1(*Conus* Suppl.): Pl. 2, sp. 277 (Marquesas Is.).

Description.—Heavy, thick, with a high gloss; low conical or obconical, the sides distinctly convex to almost straight; body whorl with a few low spiral ridges, seldom granulose, above the base, these obsolete by midbody; rest of body whorl with widely spaced spiral rows of deep punctations, these continuing to shoulder; usually with several heavy axial growth marks and threads; shoulder broad, angulate, usually with heavy coronations; spire low or flat, sometimes weakly domed, the whorls weakly (early) to strongly (later) coronate; tops of whorls with 2-5 low, weak spiral ridges; spire bluntly pointed. Body whorl creamy gray with strong bluish tones, covered with numerous closely spaced spiral lines of long brown dashes (or small brown spots) alternating with long opaque white dashes; usually the opaque white dashes are more prominent than the brown ones, especially at midbody and below the shoulder, where whitish bands of variable width and development result; usually with several large or small irregular opaque whitish axial blotches or flammules, these seldom with distinct borders; a wide pale band below shoulder, only weakly marked with brown dashes; shoulder whitish, usually with a narrow dark brown line between the coronations and marking the margin of the whorl; spire whitish, similarly marked with brown at the margins; base of shell cream. Aperture narrow, uniform in width; outer lip fairly thick, often with a thickened ridge at middle; mouth mostly violet-brown, with paler bands at middle and below shoulder. Columella short, thick, bounded above by a heavy ridge. Length 20-40mm.

Comparison.—*C. encaustus* is closely related to *C. miliaris*, *C. fulgetrum*, *C. tiaratus*, and *C. abbreviatus*. From all of these it differs in the very fine but distinct spiral rows of brown and opaque white dashes which tend to form indistinct opaque white bands at midbody and below the shoulder. Such opaque white dashes are poorly developed in *C. miliaris* and widely spaced when present; the dashes in *C. miliaris* are usually brownish to bluish and wider. *C. fulgetrum* is usually reddish brown and has the opaque white arranged in constant axial flammules and cross-hatches over the body whorl; the brown spots between the coronations are usually reddish brown and formed of closely packed axial lines. *C. tiaratus* seems to lack the rows of punctations of the other species or they are indistinct; the spiral rows of dashes are larger and more widely spaced, the shell having little opaque white but tending toward blue to brown in general color. *C. abbreviatus* is very similar to *C. encaustus* in shape and spire, but the body whorl is covered with very constant squarish spots and no opaque white in normal specimens. Compared to all these species, *C. encaustus* is rather large, is very glossy and feels very nacreous to the touch, and has the whole pattern very subdued because of the predominance of opaque white over the whole shell.

Variation.—As might be expected, there is considerable variation in details of pattern depending on how much opaque white is present. Usually the whitish midbody and shoulder bands are distinct, often with a second and very broken band of white blotches below the constant shoulder band. The rest of the body whorl may appear brown, very finely specked with opaque white, or may appear mostly whitish with a number of nebulous brown-specked clouds. Axial blotches of opaque white may be present, but they are not constant in form, usually being just irregular blotches. The spire whorls may be heavily or weakly coronated, with the dark brown as a solid line or just spots between the coronations. Small specimens seem to have more brown dashes in the pattern relative to the opaque white than in larger shells.

Distribution.—Seemingly restricted to the Marquesas Islands, French Polynesia, where common in shallow water. *C. miliaris* apparently does not occur sympatrically.

Synonymy.—*C. praetextus* is a fairly normal-appearing *C. encaustus*, perhaps with the brown dashes a bit more obvious than normal. I certainly can see no basis for recognizing two endemic Marquesas species of this group.

Notes.—Large specimens tend to have the whorls of the spire eroded and the shoulder coronations low; there may be several heavy growth flaws over the body whorl, and the pattern may be very diffuse. Smaller

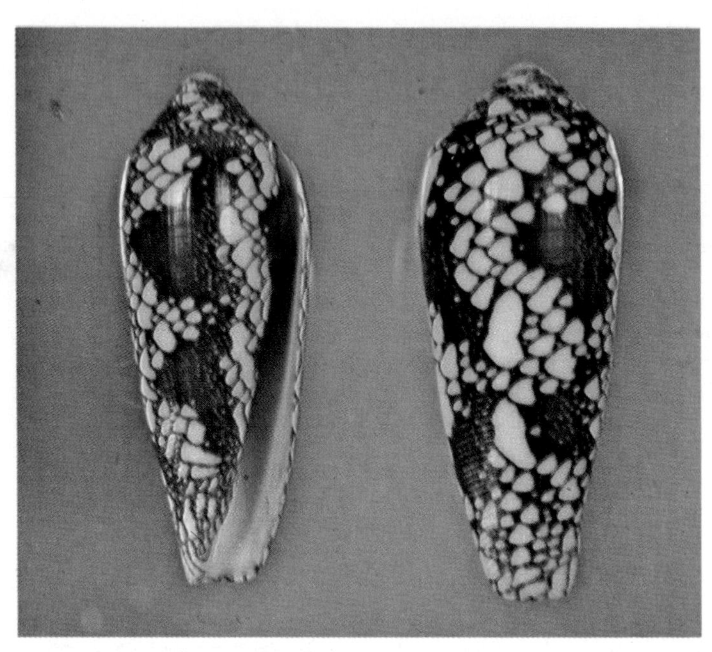

MAGNIFICUS.—*Above:* Boac, Marinduque, P.I., 60.0 and 63.0mm. *Below:* Left: Marau Sound, Guadalcanal, Solomon Is., 69.4mm; Right: Zanzibar, 91.8mm (MF).

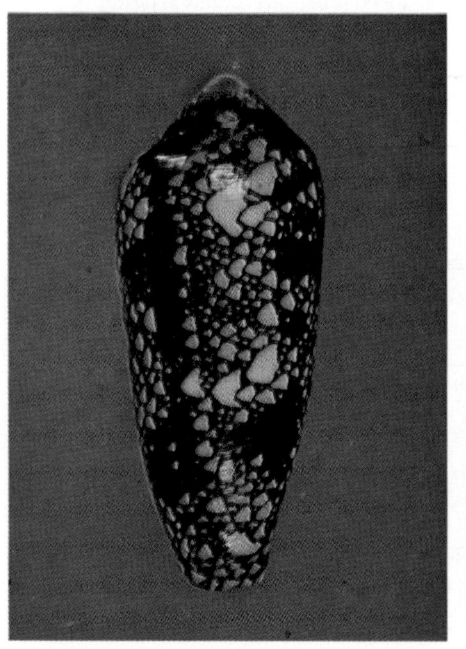

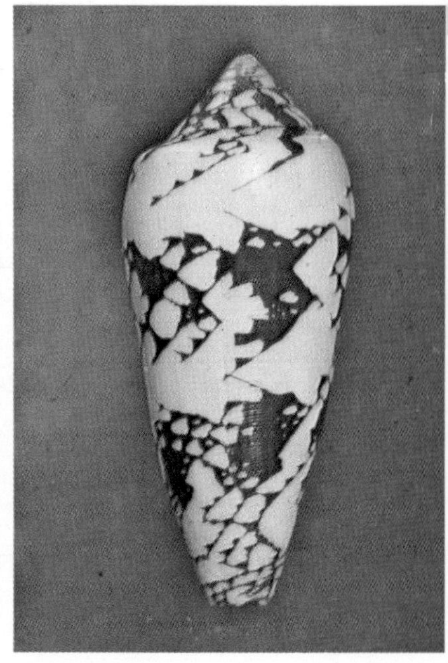

MAGUS.*—Above:* Left: Langa Langa, Guadalcanal, Solomon Is., 49.3mm; Right: Sulu, P.I., 63.9mm. *Below:* Boac, Marinduque, P.I., 42.1-50.3mm.

shells are more attractive in pattern and have better spires.

I have seen no tendency toward *C. miliaris* in specimens examined or mentioned in the literature, so there is no reason to think this is a variant or subspecies of *C. miliaris*. The closest species would seem to be the Hawaiian *C. abbreviatus*, which is very close in spire and pattern although lacking the opaque white. *C. miliaris* from Tahiti and the Tuamotus as well as Easter Island are extremely distinct from *C. encaustus* in pattern and texture; the closest approach to *C. encaustus* by *C. miliaris* seen by me was in specimens from India, and even they were very distinct.

ERMINEUS Born, 1778

1778. *Conus ermineus* Born. *Index Rerum Naturalium Musei Caesari Vindobonensis,* 1(*Testacea*): 141 (Indies). Lectotype selected and figured by Kohn, 1964.

1791. *Conus betulinus medusa* Gmelin. *Systema Naturae per Regna Tria Naturae,* ed. 13, 1: 3383 (Locality not stated). Based on Martini, Pl. 61, fig. 675.

1791. *Conus oculatus* Gmelin. *Ibid.*, 1: 3387 (Locality not stated). Based on Martini, Pl. 56, fig. 616.

1792. *Conus ranunculus* Hwass, in Bruguiere. *Cone,* in *Ency. Method., Hist. Nat. des Vers,* 1: 671 (Dutch Guiana). Lectotype figure (Clench, 1942): Seba, Pl. 43, fig. 36 (holotype designation of Kohn, 1968, is in error).

1792. *Conus testudinarius* Hwass, in Bruguiere. *Ibid.*, 1: 694 (Surinam). Lectotype selected and figured by Kohn, 1968.

1792. *Conus Guinaicus* Hwass, in Bruguiere. *Ibid.*, 1: 697 (Guinea). Lectotype selected and figured by Kohn, 1968.

1792. *Conus eques* Hwass, in Bruguiere. *Ibid.*, 1: 705 (America). Lectotype figure (Kohn, 1968): *Tableau Ency. Method.*, Pl. 337, fig. 3.

1792. *Conus Luzonicus* Hwass, in Bruguiere. *Ibid.*, 1: 706 (Philippines). Lectotype figure (here selected): *Tableau Ency. Method.*, Pl. 338, fig. 6 (incorrectly considered the holotype by Kohn, 1968).

1792. *Conus Portoricanus* Hwass, in Bruguiere. *Ibid.*, 1: 714 (Porto Rico). Holotype: *Tableau Ency. Method.*, Pl. 338, fig. 4.

1798. *Cucullus tigris* Roeding. *Museum Boltenianum:* 40 (Locality not stated). Same as *C. betulinus medusa* Gmelin, 1791.

1798. *Cucullus barathrum* Roeding. *Ibid.:* 43 (Locality not stated). Lectotype figure (Kohn, 1975): Martini, Pl. 55, fig. 607.

1798. *Cucullus crucifer* Roeding. *Ibid.:* 48 (Locality not stated). Based on Martini, Pl. 55, fig. 605.

1798. *Cucullus cutis anguina* Roeding. *Ibid.:* 48 (Locality not stated). Based on Martini, Pl. 55, fig. 605 (as above).

1807. *Conus leaeneus* Link. *Beschr. Nat.-Samml. Univ. Rostock,* 3: 103

(Locality not stated). Perhaps unrecognizable.

1810. *Conus narcissus* Lamarck. *Ann. du Mus. Hist. Nat.* (*Paris*), 15: 281 (American Ocean). Holotype figured by Kiener, Pl. 52, fig. 4.

1810. *Conus crocatus* Lamarck. *Ibid.*, 15: 424 (Mers des Grandes Indes). Probable syntype figured by Kiener, Pl. 52, fig. 3.

1817. *Conus coerulescens* Dillwyn. *Descr. Catalogue Recent Shells*, 1: 368 (St. Thomas, West Indies). Lectotype figure (here selected): Chemnitz, Pl. 182, figs. 1762-1763. Also spelled 'caerulescens.'

1833. *Conus aspersus* Sowerby i, in Sowerby ii. *Conchological Illustrations:* Pl. 28, fig. 16 (Locality not stated). Fide Reeve and Sowerby ii.

1838. *Conus caerulans* Kuester. *Syst. Conch. Cab.*, ed. 2, 6: 85, pl. 14, figs. 3-4 (St. Thomas, West Indies).

1844. *Conus Grayi* Reeve. *Conchologia Iconica*, 1(*Conus*): Pl. 46, sp. 258 (Locality unknown).

1849. *Conus inquinatus* Reeve. *Ibid.*, 1(*Conus* Suppl.): Pl. 5, sp. 251 (West of Africa).

1873. *Conus rudis* Weinkauff. *Syst. Conch. Cab.*, ed. 2, 216(*Conus*): 158, pl. 10, figs. 1-2 (Locality not stated). Based on Chemnitz, Pl. 144A, figs. e-f.

1879. *Conus carnalis* Sowerby iii. *Proc. Zool. Soc.* (*London*), 1878: 796, pl. 48, fig. 2 (Locality unknown). Almost colorless.

1942. *Conus verrucosus piraticus* Clench. *Johnsonia*, 1(6): 14, pl. 11, fig. 1 (Off Singer's Hotel, Palm Beach, Florida). Juvenile.

Description.—Moderately heavy, with a low gloss; low conical, the sides convex posteriorly; body whorl with about 6-10 low spiral ridges above the base, these obsolete before midbody and very rarely granulose; shoulder roundly angled, moderately broad to rather narrow, usually distinct from spire; spire low to moderate, sometimes almost flat or distinctly domed, sharply pointed if not eroded; spire whorls usually flat or slightly convex above, with about 2-3 low but distinct spiral ridges and often 2-3 weaker spirals also visible; axial threads weak and few; early whorls carinate, apparently not nodulose, although usually badly eroded. Body whorl pale bluish white to distinctly light blue; dark brown dashes and short lines in spiral rows are usually visible in the pale portion of the pattern, although the lines are usually inconspicuous and seldom a major portion of the pattern; usually with three broad bands of chocolate blotches, at shoulder and above and below midbody, these blotches variously fused spirally and axially but usually leaving the midbody area pale and sparsely marked; blotches often heavily infiltrated with large and small whitish blotches, spots, and triangles, these sometimes fusing to produce large whitish axial blotches above midbody and below the shoulder; sometimes blotching reduced, absent, or greatly irregular; in very pale shells the whole body whorl may appear pinkish or very pale violet; base usually white; spire mostly white with small to moderate brownish spots, the pattern reduced in shells with light body whorls; apex

MAGUS.—Above: Palawan, P.I., 60.6 and 64.4mm. *Below:* Left: Turtle Is., P.I., 49.0mm (Dayrit); Right: Milne Bay, Samarai I., Papua New Guinea, 42.1mm.

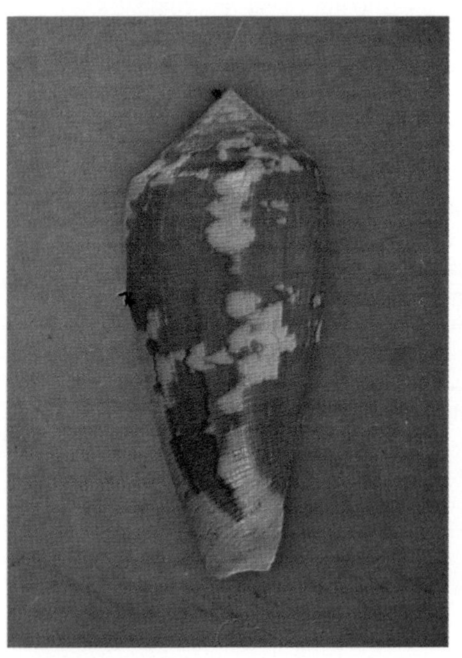

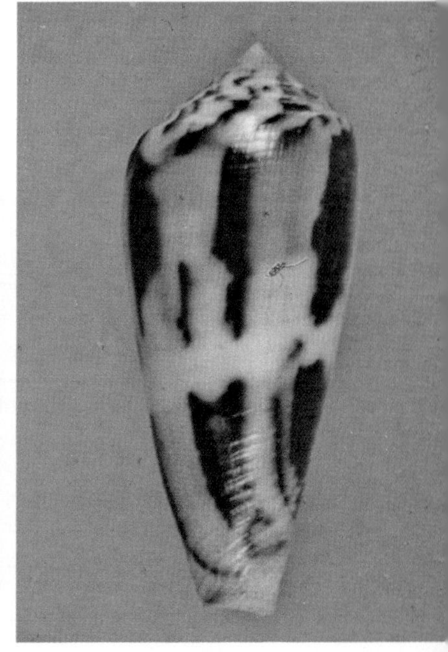

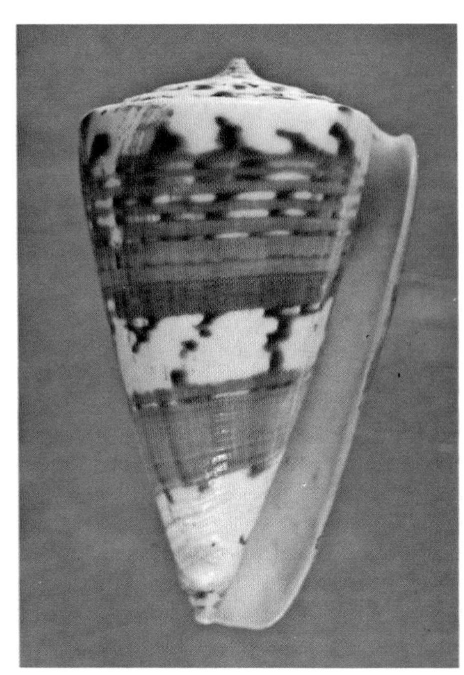

MALACANUS.*—Above:* Sri Lanka, 55.0mm (MG). *Below:* Off Madras, India: Left: 67.0mm; Right: 66.8mm (K).

often pink. Aperture moderately wide to distinctly wide, uniform in width; outer lip sharp, sometimes rather thick, slightly convex; mouth bluish white to very pale violet, occasionally pinkish in very light shells. Columella long, narrow, with a slight ridge posteriorly. Length 30-70mm.

Comparison.—The shape and pattern of this species, although highly variable, are fairly distinctive. It must be carefully compared with the African *C. taslei* and *venulatus* and the Pacific *C. purpurascens* and *C. monachus.*

African species: There is a general similarity to *C. aemulus* and *C. mercator*, but both these species have distinctive patterns or groups of patterns and have somewhat higher spires. *C. aemulus* in addition has broad violet-brown bands in the mouth. *C. venulatus* (and to a lesser extent *C. balteus*) is very similar in some patterns but is usually more strongly shouldered, has the spire sculpture very weak or absent, often has distinctly stepped whorls, has the dash pattern of a somewhat different type, and usually has a different range of patterns than in *C. ermineus. C. ermineus* populations sympatric with *C. venulatus* are round-shouldered with the dark brown blotches usually well developed and infiltrated with many white triangles. *C. taslei* as a rule has the shoulder almost bulbous and the spire flat laterally with a sharply pointed central cone; the pattern consists of very fine axial lines of brown to olive with a few scattered white blotches or tents. The very similar *C. cloveri* is like *C. taslei* in shape but has the axial lines heavy and often absent, leaving a mostly white shell.

Pacific species: The major problem in this area is *C. purpurascens*, which could easily be considered an eastern Pacific subspecies of *C. ermineus.* That species is more distinctly violet in most patterns, even the base being strongly tinted; the shoulder is usually broader than in *C. ermineus* and more strongly angled, with the spire more conical and the whorls often weakly stepped; there are usually long brownish spiral lines conspicuous in the pattern, and the blotches often have distinct 'shadows' of paler blue or whitish. Certainly the two are very close, and isolated specimens of *C. purpurascens* often would easily fall within the range of variation of Caribbean *C. ermineus.* The Indo-Pacific *C. monachus*, especially the *achatinus* variant with distinct spiral lines and dashes, is sometimes very similar also, although most specimens would present no problems in identification. *C. monachus* is more cylindrical than *C. ermineus* and has heavier basal sculpture that is often heavily granulose; the early 2-3 whorls of the spire are finely but distinctly nodulose; and the spiral grooves of the spire whorls appear punctate.

Variation.—That this is a highly variable shell is obvious from the long synonymy, although most of the names here are based on such minor differences that even today's collectors would ignore them. The basic shell is a bluish white with three series of brown blotches and a whitish base; this is overlaid by usually inconspicuous rows of spiral dashes or short lines; the midbody area is clear or weakly marked. Probably five basic patterns would cover most of the obvious variation likely to be found except for isolated individuals.

One: The common pattern is typical of most African populations and some American ones. In this pattern the brown blotches are very heavy, the ones below the shoulder and above midbody largely fused and these in turn fused spirally to produce a largely brown shell above midbody; the band below midbody is also spirally fused, and the midbody band is heavily infiltrated from above and below by axial flammules; the brown areas are variably infiltrated with whitish tents and blotches which are randomly arranged and often connect axially to produce larger blotches; over all this can be seen scattered short spiral lines and dashes, these seldom conspicuous except in the midbody area; the spire is heavily and regularly marked with axial brown blotches.

Two: Similar to One, but the pattern clearer, the white blotches so heavily developed that the brown areas are reduced to rather regular spiral brown bands, a narrow one below the shoulder, a wider but rather scalloped band above midbody, and a wide but heavily broken band below midbody. The mouth may have broad violet bands like *C. aemulus.* This is apparently a northern African form, common off Spanish Sahara but also found in various intergrading forms further south with pattern One.

Three: A rather plain pattern, the background pale bluish white with faint tan tones, lightly and irregularly covered with broad, wavy axial brownish bands, the bands thickened in the same positions as the usually three spiral bands; often the midbody band is crossed by thin sections of the blotches; dash bands are few and not conspicuous; the spire is lightly patterned. This pattern, often called *C. grayi*, is more characteristic of southwestern Africa, but similar patterns and intergrades can be found in large series that are mostly type One.

Four: Distinctly bluish shells with rather wide shoulders; the blotches are very small and often only distinct above midbody, the other two bands being just large jagged brown spots; usually the spiral rows of brown dashes and lines are fairly well developed and conspicuous over the body whorl, the basal ridges being especially heavily dashed; the spire in average specimens is lightly marked. This pattern is a rather extreme American type, but a similar pattern (seldom bluish, however) can be found in Africa.

Five: Rather round-shouldered, the spire moderately high; whole shell pale bluish white with pale pinkish tones, the blotches barely visible as tan areas above midbody and sometimes below the shoulder, the basal area often unmarked; dashes absent or very poorly represented. I have seen this pattern mostly from the southern Antilles.

Other patterns, more typical of individuals rather than populations, include solid whitish shell with no obvious pattern, almost totally dark shells with very little white, and bluish shells heavily covered with long

MARIELAE.—(Holotype, dorsal, and paratype, ventral) Off Baie Moto-Hee, Nuku Hiva, Marquesas Islands, 40.3 and 39.3mm (photos by USNM through Dr. H. Rehder).

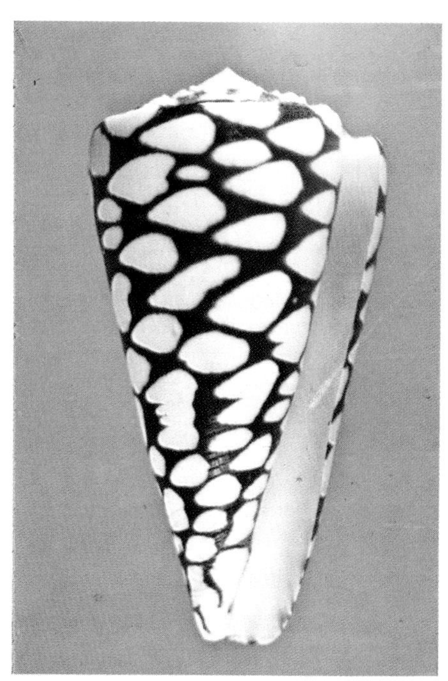

MARMOREUS.—*Above:* P.I., 91mm. *Below:* Left (*nocturnus*): Boop I., Moluccas Is., Indonesia, 52.1mm (EP); Right (*vidua*): Coron I., P.I., 38.0mm.

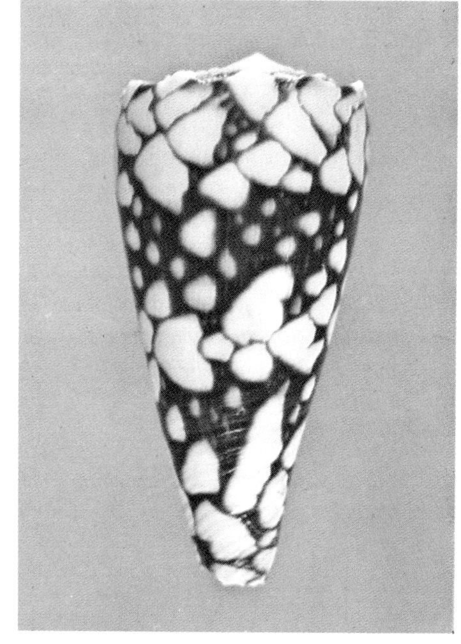

spiral lines and almost no blotching. These are not common.

Distribution.—Widespread and common on the western African coast from the Sahara south to Angola, including the offshore islands and the Atlantic islands; reappears on the Brazilian coast and is found through the Antilles and Central American Coast to the Gulf of Mexico off Texas and in Florida and the Bahamas; also found off South Carolina. Usually moderately common in shallow to moderately deep water.

Conus ermineus variants. Left to right: Ghana, 52mm; Off Cameroons, 42mm (EP); Bonaire, 15mm juvenile (R. Guest).

Synonymy.—As many as five distinct species are sometimes recognized within this species, but the patterns and shapes vary too much and have such total intergradation on both continents that I cannot see recognizing more than one variable species. The earliest name for the species is *C. ermineus*, here used. There is just not enough room to go through each synonymy individually, but most have been discussed in the literature and assigned to such taxa as *C. ranunculus*, *C. testudineus*, and *C. ermineus* by various authors. Most synonyms fall within pattern type One (common on both continents) and Four (especially common in America).

Two names deserve special mention, however. *C. grayi* is a rather normal *C. ermineus* with the pattern distinctly axial and thickened at the position of the normal three spiral bands. Full intergradation with pattern One is not uncommon, even within such limited populations as those of the Cape Verde Islands. *C. verrucosus piraticus* is of all things a juvenile *C. ermineus*; this is obvious from the bluish coloration, well developed spiral ridging on the spire, the bluish mouth, and the presence of small blotches in the proper positions for *C. ermineus.* Juveniles are apparently uncommon or rare in the Antilles, which may have led to this confused situation; I have examined juveniles and can see no real differences from Clench's description of *piraticus.* It might be mentioned that some juveniles of *C. ermineus* have faintly nodulose spire whorls, but this is apparently reduced and lost with growth.

Notes.—Certainly several of the synonyms will continue to appear on dealer's lists for years, although I can never find the basis by which dealers and many collectors assign these names to specimens—other than random guesses. None of these variants is really uncommon except for the individual patterns mentioned after pattern Five and perhaps small juveniles.

American populations are rather clean shells with only small flaws and little erosion of the spire whorls. African populations often are very rough, with heavy breaks and heavy spire erosion, as well as extremely irregular patterns. It is not uncommon in African shells to find pattern One on the dorsal side and pattern Two or Three on the ventral, which is pretty good evidence that the patterns are just minor ecological changes.

ERYTHRAEENSIS Reeve, 1843

1843. *Conus erythraeensis* Reeve. *Conchologia Iconica*, 1(*Conus*): Pl. 24, sp. 137 (Locality unknown).
1844. *Conus piperatus* Reeve. *Ibid.*, 1(*Conus*): Pl. 43, sp. 230 (Locality unknown). Non *Conus piperatus* Dillwyn, 1817.
1849. *Conus Dillwynii* Reeve. *Ibid.*, 1(*Conus* Suppl.): 2. Nomen novum for *Conus piperatus* Reeve, 1844.
1849. *Conus induratus* Reeve. *Ibid.*, 1(*Conus* Suppl.): Pl. 7, sp. 268 (Red Sea).
1858. *Conus Hamilli* Crosse. *Rev. Mag. Zool.*, (Ser. 2), 10: 122. Nomen novum for *Conus piperatus* Reeve, 1844.
1866. *Conus quadratomaculatus* Sowerby ii. *Thesaurus Conchyliorum*, 3(*Conus* Suppl.): 328, pl. 27(288), figs. 637-638 (Locality unknown).

Description.—Moderately heavy and solid for its size, with a good gloss; low conical to obconical, the sides nearly straight or slightly convex posteriorly; body whorl with a few broad and low spiral ridges near the base, these separated by shallow, weakly punctate grooves; occasionally grooving extends to above midbody but this is uncommon; rest of body whorl smooth except for fine axial threads; shoulder broad, sharply angled, weakly concave or flat above; spire low, sometimes nearly flat, the sides concave, sharply pointed; whorls often distinctly but weakly stepped, concave above, with two or three faint spiral ridges and faint axial threads; tops of the whorls occasionally appear smooth and are often eroded. Body whorl bluish white or white with pinkish tinge, covered with many rows (often 15-20) of squarish reddish brown spots, these usually very regularly arranged; spots have tendency to fuse into short spiral dashes near the base and into short narrow axial lines at the shoulder, as well as into larger axial blotches above and below the midbody area; spots occasionally absent over part of shell or obscured by small brown clouds; spire bluish white with regularly spaced large brownish squares especially prominent on shoulder and becoming smaller

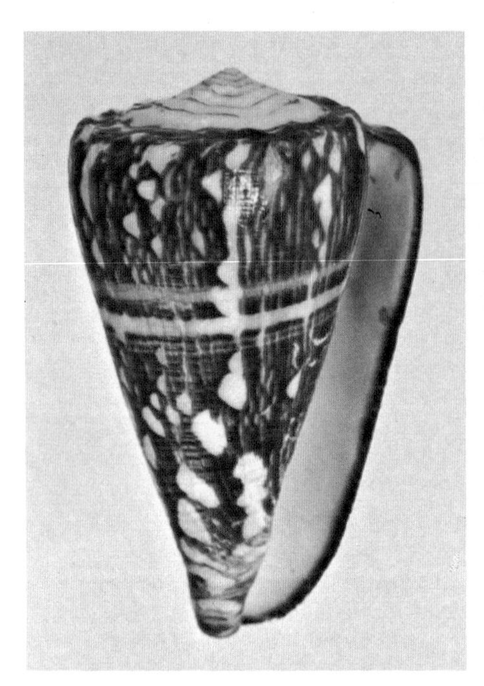

MARMOREUS.—*Above:* Left (*crosseanus*): New Caledonia, 55mm (DM) (photo by Jayavad); Right (*nigrescens*): Apia, Western Samoa, 32.9mm (MG). *Below* (*suffusus* and intermediate): New Caledonia: Left: 44.0mm (MM); Right: 46mm (DM) (photo by Jayavad).

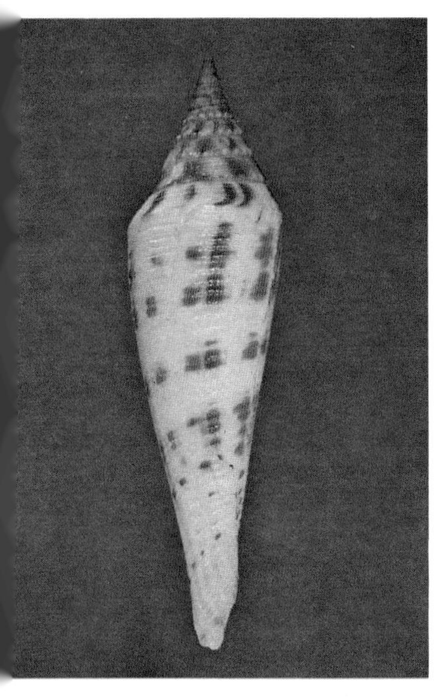

MAZEI.*—*Above: S of Pensacola, Florida, 52.0mm (GG). ***Below:*** 110 mi. W Egmont Key, Florida, 25.2mm (GG).

and more line-like on earlier whorls; tip of spire brownish. Aperture moderately wide, nearly uniform in width; outer lip sharp but solid, straight or weakly convex; mouth bluish white near lip but becoming dark violet deeper within in most specimens. Columella short, narrow, slightly indented from inner lip. Length 15-35mm.

Comparison.—The thickness of typical shells, even of small specimens, should prevent their being confused with juveniles of such things as *C. inscriptus* (which is heavily grooved both on the body whorl and the spire), *C. tegulatus* (like *C. inscriptus*), or *C. iodostoma* (shoulder rounded, the spots very small, body whorl more slender in juveniles). *C. musicus* is similar in some patterns but coronated and with a violet base (white in *C. erythraeensis*). *C. eburneus* has a white mouth and is wider at the shoulder as well as having orange spiral bands.

There is a strong superficial resemblance between *C. erythraeensis* and some patterns of *C. puncticulatus*, even down to the bluish gray or bluish white background colors and the regular series of spots. However, the spots in *C. puncticulatus* are usually smaller and less regular; *C. puncticulatus perplexus* is a closer approach to the pattern, but it lacks spiral sculpture on the spire whorls and is relatively larger and broader.

Variation.—The spire is usually low with weakly stepped whorls, but commonly it is almost flat or higher and more strongly stepped. The spiral grooving of the body whorl may be only on the base or may rarely extend almost to the shoulder. The shape is quite constant, except small specimens are a little convex posteriorly. The spots are usually very regular, but with various types of fusion and a few brown clouds rather different patterns can result. Commonly there are some small irregular blotches at random over the body whorl, and commonly several blotches may form above and below the midbody area. It is not uncommon for the blotches to become fairly large and distinctly axial, extending from the shoulder to about midbody and from about midbody to the anterior tip. Commonly the spots are fused into short spiral dashes basally. Rarely the pattern is reduced and spotting may be absent from sizable sections of the shell. The spire pattern is usually well developed, especially on the shoulder. The mouth must be subject to some fading, as the extent of the violet blotch deep within is very variable. Some shells appear distinctly pinkish with more reddish spots rather than bluish or blue-gray with dark brown spotting.

Distribution.—Apparently restricted to the Red Sea and adjacent areas. Common in shallow water.

Synonymy.—The type figure of *C. erythraeensis* is a very typical specimen with regular spotting, a very few small blotches, and short axial lines at the shoulder. *C. dillwyni* is distinctly red-spotted with a few irregular blotches. *C. induratus* has shallow grooving almost to the shoulder. *C. quadratomaculatus* has heavier blotching above and in distinct spiral bands above and below midbody and at the base. Obviously these all fall easily within the range of variation of a single species; the variants are easy to find in even small series of *C. erythraeensis*.

Notes.—This is a trim and distinct little species which is seldom injured except for minor erosion of the spire whorls. Large specimens (over 25mm) with heavy axial blotching look a bit different and are not as common as the smaller, regularly spotted specimens.

EUCORONATUS Sowerby iii, 1903

1903. *Conus eucoronatus* Sowerby iii. *Marine Invest. South Africa*, 2(3): 217, pl. 3, fig. 9 (Cape St. Blaize, Natal, South Africa).

Description.—Moderately heavy, with a low gloss; biconical, the upper sides distinctly convex then concave and tapering to a narrow base; body whorl covered with about 15 wide, flat spiral ridges, those above the basal 6 or so alternating with narrower and rounded ridges (about 10); major ribs covered with heavy, spaced axial grooves, these not present on minor ridges, so ribs look as though heavily beaded; sculpture weaker and more crowded below shoulder; shoulder broad, angled, with heavy, closely spaced rounded coronations, very numerous; top of shoulder with widely spaced low curved axial ridges crossing two low spirals, plus many very fine axial and spiral threads; spire moderate, sharply pointed, the sides straight or convex, the whorls weakly stepped; tops of whorls slightly concave, the margins heavily coronated with closely spaced rounded coronations; spire whorls sculptured with very fine axial and spiral threads and with widely spaced curved axial ridges crossing two wide spiral ridges. Body whorl white with pale violet tones, with three spiral rows (below shoulder, above and below midbody) of closely set, slightly axial, rectangular or squarish deep violet-brown blotches; blotches broken by white beads of sculpture, but may be almost continuous above midbody; sometimes with smaller spots between major bands; base white; spire white, the shoulder and later whorls with large brownish spots like the blotches of body whorl plus small spots between coronations; early whorls tan. Aperture moderately wide, uniform in width; outer lip sharp, thin, concave at the middle; mouth white with heavy violet tones under exterior pattern. Columella long but very narrow. Length 35-50mm.

Comparison.—This species is very distinctive in sculpture and pattern. It is somewhat similar to *C. orbignyi* or *C. acutangulus* at first glance, but it is much wider than the former and has much more pinched-in sides than the latter. From both it also differs in the heavily beaded body sculpture, the three regular bands of a peculiarly violet-brown tone on stark white, and the very broad spire.

Variation.—The sculpture varies a bit as might be expected, the two types of ribs and ridges not always completely distinct and the beading

MEMIAE.*—Above* (Lectotype of *C. adonis* Shikama, here selected): Kashiwajima I., Kochi Pref., Japan, 25.8mm (photos by Dr. T. Shikama). *Below:* Left (paratype of *C. adonis*): Kashiwajima I., Kochi Pref., Japan, 34.0mm (photo by Dr. T. Shikama): Right: Panglao, Bohol, P.I., 22.9mm (Dayrit).

MERCATOR.—Off N'Gor, Senegal. *Above:* 32.0mm (K). *Below:* 34.0 and 34.2mm (photo by A. Kerstitch).

sometimes weak, especially anteriorly and posteriorly. The spire may be quite tall or very low for this type of shell. The bands on the body whorl may be almost continuous or distinctly broken into large blotches; the broken bands may be a subadult character. The background color of the body whorl may be stark white, bluish white, or pale pinkish in tone.

Distribution.—Known only from deep water off northeastern South Africa and southern Mozambique. Uncommon to rare.

Notes.—Like many other heavily sculptured deep-water shells, this one is subject to bad breaks over the body whorl which heal poorly; the old varices may be especially badly broken and grown over. The lip is fragile and breaks easily. This is a very interesting and desirable species, although it is seldom colorful, only some specimens having attractive violet or pinkish tones.

EUGRAMMATUS Bartsch and Rehder, 1943

1943. *Conus eugrammatus* Bartsch and Rehder. *Proc. Biol. Soc. Washington*, 56: 85 (Off north coast of Molokai Island near Mokapu Islet, Hawaii).

1956. *Asprella (Conasprella) wakayamaensis* Kuroda. *Venus*, 19(1): 9, pl. 1, fig. 2 (Sea off Kii Peninsula, Japan).

Description.—Moderately light in weight, with a low gloss; biconical, the upper sides straight or nearly so, the base very narrow; body whorl covered with deep, axially lineate grooves from base to shoulder, separated by narrow to broad flat or weakly rounded spiral ribs; entire body whorl with fine axial threads; ribs immediately anterior to shoulder narrower; shoulder broad, carinate, concave to almost flat above; spire rather tall, sharply pointed, the sides concave, the whorls usually distinctly stepped; tops of whorls shallowly concave, covered with numerous strong curved axial ridges and traces of 1-3 very weak spirals; first 3-5 whorls weakly and sometimes indistinctly nodulose, the others smoothly carinate. Body whorl white to pale buffy tan, covered with numerous rows of large reddish brown to dark brown dashes in spiral lines; these may be concentrated into blotches in narrow spiral bands below the shoulder (least distinct) and above and below midbody, occasionally above the base, these spiral bands seldom complete; shoulder carina with squarish brown spots, these continuing onto spire; spire whitish, with radiating rows of brownish spots and blotches; early whorls white. Aperture rather narrow, uniform in width; outer lip thin, sharp, straight or slightly concave; mouth white, sometimes with faint pink or bluish tones. Columella internal. Length 20-40mm.

Comparison.—*C. eugrammatus* is extremely close to *C. acutangulus* but distinctive enough to be considered a full species. In shape it is almost identical, although the spire averages a bit lower than most *C. acutan-*

gulus populations; this is a variable character. The ribs on the body whorl of *C. acutangulus* are perhaps a bit broader and flatter than in *C. eugrammatus*, but this is also variable; in the same way, *C. acutangulus* tends to be covered with axial blotches and flammules instead of numerous dashes, but there are exceptions. The only really constant difference between the two species is spire sculpture: in *C. acutangulus* the spire whorls are all strongly nodulose, as is the shoulder in many populations (carinate shoulders are not uncommon); the tops of the whorls have heavy spiral ridges which distinctly cut through the axials, rather than just distinct axial over very weak spirals as in *C. eugrammatus*. Although the first 3-5 whorls are commonly the only ones nodulose in *C. eugrammatus*, some specimens seem to lack distinct nodules entirely.

C. comatosa is similar in general pattern, but it is much more elongate than *C. eugrammatus*. *C. otohimeae* has only weak grooving on the body whorl, has a different spire shape with only weak nodules and axial ridges, and has numerous spiral rows of small orange spots as well as one or two broad bands of orange blotches. *C. memiae* is like a slender, elongate, smoothish *C. eugrammatus* with rather similar spire sculupture; however, it has a distinct band of orange blotches at midbody and many large dashes of opaque white in the same area.

Variation.—For some reason Japanese specimens of this species have not normally been associated with Hawaiian specimens, although they are obviously within narrow limits of variation of a single species. The height of the spire varies a bit, as does the degree of concavity of the spire and the spire whorls. The nodules of the early whorls may be distinct, very weak, or apparently occasionally absent. The ribs on the body whorl may be narrow or relatively broader, rounded or flat. The pattern varies a bit, but usually at least indistinct blotches are formed in spiral bands above and below midbody, the bands at the base and shoulder much weaker and sometimes absent; scattered dashes between the bands are almost always present.

Occasional Japanese shells appear to be senile adults or perhaps a distinctive variety. These have the ribs of the body whorl flat and not very distinct, with the bands above and below midbody and below the shoulder very well defined; the spire whorls are almost flat above and only the first few whorls are weakly nodulose, with occasionally no nodules present. This is apparently the type of shell described as *C. wakayamaensis*; intergrades to the typical form are not uncommon.

Distribution.—Uncommon to rare in the China Sea from southern Japan (locally common) to Taiwan (uncommon) and in the Philippines (rare); also in deep water in the Hawaiian Islands (rare).

Synonymy.—Japanese shells are called various names, one of the most common being *C. japonicus*, here considered unrecognizable. *C. wakayamaensis* is also applied at random to Japanese specimens. There is no doubt that typical Japanese adults are virtually identical to adults from Hawaii (normally only juveniles are taken in Hawaii), and

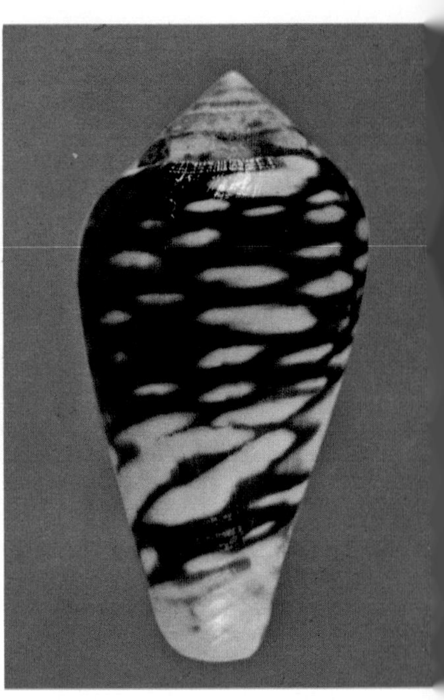

MERCATOR.—Off N'Gor, Senegal (K). *Above:* Left: 24.2mm; Right: 29.6mm. *Below:* Left:30.0mm; Right: 34.5mm.

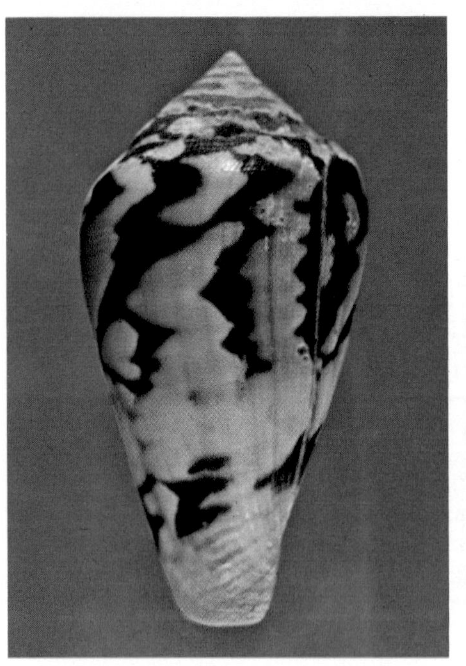

MERCATOR.--*Above* (*miser*): N'Gor, Cape Verde, Senegal, 39.3mm (EP). *Below:* Left (*miser*): N'Gor, Cape Verde, Senegal, 33.6mm (EP); Right (*lamarckii*): Belair, Dakar, Senegal, 17.3mm (PC).

C. wakayamaensis can be recognized only on a minor variety for smoother shells with weak sculpture. Hawaiian specimens over 25mm are perhaps a little more reddish than typically brownish Japanese specimens, but there are no obvious differences in pattern or sculpture not found in series of Japanese specimens. Only one species is involved.

Notes.—There is a gigantic price differential at the moment between Japanese adults and Hawaiian juveniles of this species. Specimens over 25mm in length have been taken both places, so do not believe anyone who says that *C. eugrammatus* from Hawaii is a smaller shell than Japanese specimens—it is just that mostly juveniles are taken in Hawaii, mostly adults in Japan. I have seen juveniles from off Taiwan which are like Hawaiian juveniles.

Heavy breaks and growth flaws are not uncommon in this species, and there are commonly breaks and bad erosion on the spire whorls. Occasionally there are burrowing sponge holes as well in Japanese material. The species is collected alive, so bad erosion should not be acceptable in specimen shells. The lip is fragile. Dead shells fade rapidly. The smoother variant is not as attractive of the typical form and is apparently not any less common in Japan.

EXCAVATUS Sowerby ii, 1866

1866. *Conus excavatus* Sowerby ii. *Thesaurus Conchyliorum*, 3(*Conus* Suppl.): 326, pl. 25(286), fig. 616 (Locality unknown).

1942. *Conus richardi* Fenaux. *Bull. l'Inst. Oceanogr.* (*Monaco*), No. 814: 4, fig. 11 (Madagascar).

Description.—Light in weight, with a low gloss; obconical or low conical, the sides nearly straight; body whorl with about a dozen low ridges at the base, these soon obsolete; rest of body whorl smooth except for numerous fine axial threads and a few scattered spiral ones, plus growth marks; shoulder broad, sharply angled, concave above and with only traces of axial and spiral threads; spire low, sometimes almost flat, the early whorls forming a sharply pointed cone; whorls slightly stepped, concave above, the margins thickened and raised, somewhat rude; first 3-4 whorls weakly undulate or vaguely nodulose; other whorls with straight margins; tops of whorls with about 2-3 narrow spiral ridges crossed by scattered and indistinct axial threads, sculpture all indistinct, especially on later whorls. Body whorl dark brown to yellowish, usually with a broad pure white midbody band and occasionally a second white band more posteriorly; midbody band often with small brownish spots at its margins; growth lines often with faint dashes of dark brown in axial rows, these sometimes going through midbody band; base not much darker than rest of body whorl; faint traces of brown dashes in spiral rows hidden in background; area immediately below shoulder white, with axial dark brown lines continued onto shoulder; shoulder and spire white, with heavy spots of dark brown in revolving pattern; early whorls faintly pinkish. Aperture moderately wide, slightly widened anteriorly; outer lip straight or slightly convex, sharp; mouth whitish with dirty violet tinge. Columella very narrow, largely internal. Length 35-55mm.

Comparison.—This seems to be a distinct species although it has been widely confused with *C. fumigatus* and perhaps even *C. capitaneus*. Specimens usually go under the name *C. coffeae*, a name here considered unrecognizable. From *C. fumigatus* it differs readily in the longer body whorl, larger size, thinner texture, and plainer pattern with few spiral lines visible; the white band at the shoulder is similar in both species; in *C. fumigatus* the shell is much thicker even in juveniles and the axial ridges on the spire whorls are heavy, even on the shoulder.

There is some resemblance to *C. argillaceus*, though that species is also relatively thicker, has a dark base, and has a much more complicated pattern of white blotches and spiral dark lines. *C. malacanus* is also a thick shell that is much wider across the shoulder, has somewhat different spire sculpture, a white base, and quite different pattern. *C. kermadecensis* has a more rounded shoulder, is coarser and heavier, and lacks the white band just below the shoulder. *C. xanthicus* has the early whorls strongly nodulose and lacks darker dashes on the body whorl.

Finally, there is a superficial resemblance to *C. capitaneus*, which may be similar in color and pattern but has a more rounded shape, punctate grooves at the base and on the spire, and numerous fine brown flecks basally; the same basic differences would also apply to *C. mustelinus*. *C. eximius* may be similar in shape and even texture, but it has heavier spire sculpture and basal sculpture as well as a spotted or blotched pattern on white. See also *Conus xanthicus*.

Variation.—The spire may be nearly flat or just low, with the whorls weakly stepped or strongly so. The midbody band is apparently usually distinct and clear except for spots on its margin and axially along growth lines when they are developed; although spiral dashes can be seen in some specimens through the brownish or yellowish background, especially near the midbody band and below the shoulder band, they are seldom conspicuous except in juveniles, where a few short blackish lines may be visible. The base may be paler than the body color, but not darker. Background color may be bright yellow to reddish brown, the color often appearing as though it has been applied in broad spiral bands of very slightly different shades. The spire and shoulder are heavily patterned brown and white, sometimes with more dark brown than white, or vice versa. Undulate early whorls appear the rule, occasionally with the other whorls irregular but not distinctly undulate. Secondary whitish spiral banding may be variously developed on the body whorl, as may occasional axial white blotches.

Distribution.—I have seen specimens only from the western Indian Ocean, particularly the Red Sea and vicinity; probably more widely dis-

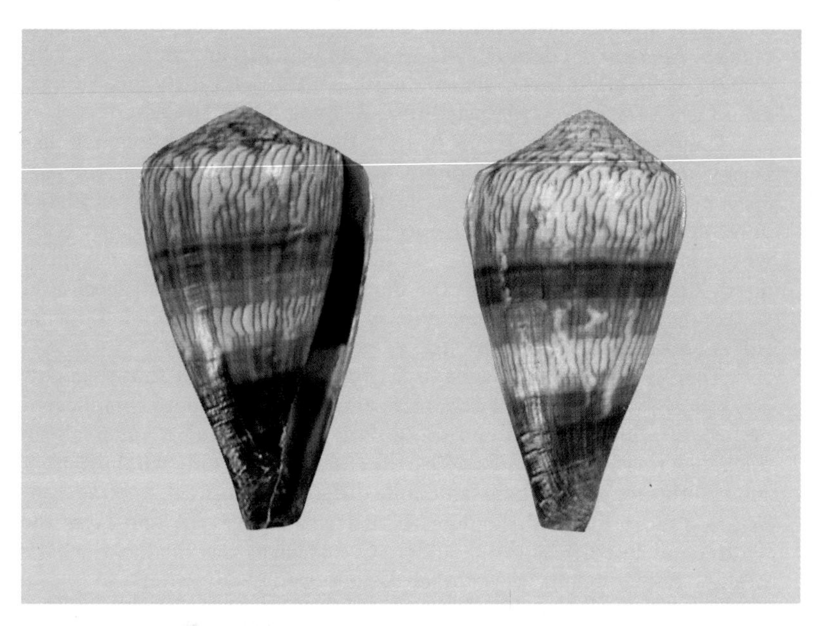

MILES.—Above: Boac, Marinduque, P.I., 55.5 and 56.1mm. ***Below:*** Tahiti, Society Is., 50.1 and 52.3mm.

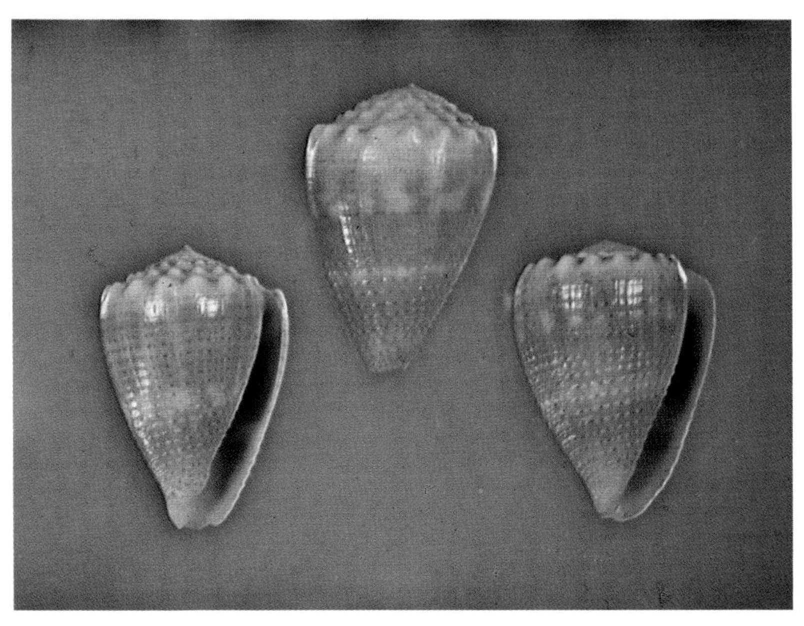

MILIARIS.—Above: Tahiti, Society Is., 24.2-25.0mm. *Below:* Left: Bombay, India, 22.0mm (GG); Right: Easter I., 28.0mm.

tributed but confused with other species.

Synonymy.—Although sometimes called *C. coffeae* or *C. fumigatus*, *C. excavatus* seems to be the earliest name for this species. *C. coffeae* is considered unrecognizable, while *C. fumigatus* is obviously distinct. The obscure *C. richardi* seems to be a pale specimen with much white. It is possible that an earlier name will turn up for this species.

Notes.—I have seen juveniles sold with *C. fumigatus*, although they are easily distinguished by pattern and thinness of texture. Apparently this is a rather rough cone which suffers small breaks commonly. The growth marks may be quite heavy, and the lip is commonly filed. This species is not rare, just poorly known and confused.

EXCELSUS Sowerby iii, 1908

1894. *Conus pulcherrimus* Brazier. *Proc. Linn. Soc. New South Wales*, (Ser. 2), 9(1): 187 (Tanna, New Hebrides, on beach after submarine volcanic eruption in 1878). Non *Conus pulcherrimus* Heilprin, 1879, a fossil. This name has been confused with *Kenyonia pulcherrima* Brazier; see *Conus chiangi*.

1908. *Conus excelsus* Sowerby iii. *Ann. Mag. Nat. Hist.*, (Ser. 8), 1: 465, fig. (New Caledonia ?).

1945. *Asprella tannaensis* Cotton. *Rec. South Australian Mus.*, 8(2): 270, pl. 4, fig. 3 (Tanna, New Hebrides). Based on same specimen as *C. pulcherrimus* Brazier; see Cernohorsky, 1974, *Rec. Auckland Inst. Mus.*, 11: 132-134, figs. 27-28.

1968. *Turriconus nakayasui* Shikama and Habe. *Venus*, 26(3/4): 57, pl. 6, figs. 1-4 (sea around Kashiwajima, a small island near Cape Ashizuri, Kochi Pref., Shikoku, Japan).

Description.—Heavy, with a good gloss when fresh; elongately biconical, the spire exceptionally high; body whorl slightly convex posteriorly and tapering to base; body whorl covered with narrow, deep spiral grooves separating broad flat ribs, the grooves sometimes with a weak spiral thread within; grooving tends to become weaker or obsolete below shoulder; number of grooves about 20 to 40, depending on method of counting; axial threads and wrinkles prominent, curved; shoulder weakly angled, rather broad, concave above; spire very high, pointed, the whorls strongly stepped; tops of spire whorls concave, with a single strong spiral ridge and sometimes traces of two or three weaker ones; margin of whorls angulate, sometimes with first eight or nine whorls bearing minute nodules. Shell creamy white, with two broad spiral bands of deep yellowish brown above and below midbody, these bands either nearly continuous or broken into isolated irregularly oval blotches; spiral lines of alternating brown and white spots visible within the brown spiral bands; bands with irregular extensions crossing midbody white band and more numerous finer lines extending to shoulder; base white with curved brown lines; shoulder white with numerous fine axial lines from brown bands; spire whorls with large to small brown areas on top and fine axial brown lines on sides; early whorls pale. Aperture wide, uniform in width; lip strongly convex on anterior margin, straight, and strongly sloping from shoulder posteriorly; mouth white, the margin with some brown marks from external pattern. Columella largely fused, indistinct. Length 60-102mm.

Comparison.—The only species which bears any resemblance to *C. excelsus* in size and shape is *C. milneedwardsi*. That species is of course immediately distinguished by the lack of deep spiral grooves over the body whorl (although a few shallow grooves may be present), the relatively shorter spire, and the color pattern of distinct triangular white spots on a reddish brown network. It is possible that the closest relative of *C. excelsus* is *Conus stupa*, which differs in shape and in lacking spiral grooving. There is also an unidentified small shell from the Japanese area which is similar in shape but with much narrower and more elevated spiral ribs on the body whorl and apparently lacking the color pattern except for faint spotting; the status of this species is deferred for the moment, although it seems to be a juvenile *excelsus*. Some deep-water *C. amadis* may be vaguely similar in shape and grooved, but it is a lighter and much lower spired species.

Left: Presumed juvenile of *C. excelsus*, Japan, 31mm (photo Okutani).

Right: Holotype of *C. excelsus*, 88mm (BM(NH)) (photo Abbott).

Variation.—Presuming that I am correct in placing *C. tannaensis* and *C. nakayasui* in this species, the shell is rather uniform in shape and color pattern. The shoulder may be relatively wide or narrower and the brown body pattern nearly continuous and very heavy or weak and blotchy; beach specimens are pale yellowish instead of deep yellowish brown. The number of spiral body grooves varies in the few recorded specimens, but this is often a hard character to count exactly because of secondary division. Both *C. tannaensis* and *C. nakayasui* seem to be lacking nodules on the spire whorls, but they could simply be overlooked or obsolete; they are strong on the type of *C. excelsus* but absent on a small Solomons specimen.

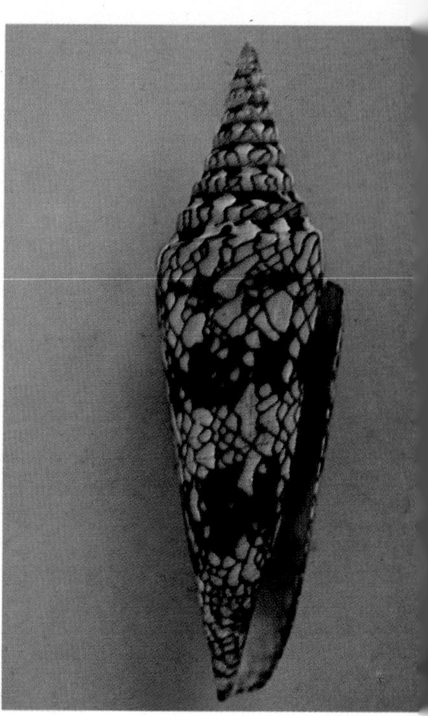

MILNEEDWARDSI.*—Above:* Mozambique, 118mm (photos by P. Clover). *Below:* Left: Mozambique, 96mm (MG); Center (*kawamurai*): Taiwan, 78mm (PC): Right (*kawamurai*): 53mm (photo by P. Clover).

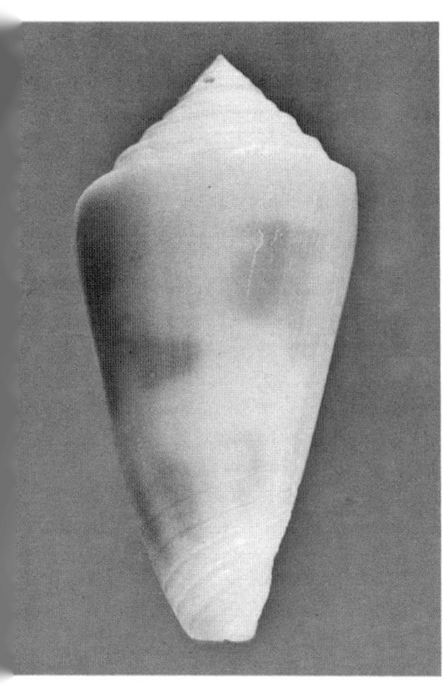

MINDANUS.—Herrington Sound, Bermuda (R. Jensen). *Above:* 45mm. *Below:* Left: 46.1mm; Right: 55.3 and 57mm.

Distribution.—A rare species in collections known with certainty from the New Hebrides to southern Japan; recently specimens said to come from the northeastern Indian Ocean have appeared. The species is probably very widely distributed in deeper water of the Indo-Pacific but seldom collected. I have just seen a specimen from the Solomons.

Synonymy.—The first description of this species was an untitled mention, in the *Proceedings of the Linnaean Society of N.S.W.*, of an exhibit by Brazier. The unique specimen was later described by Cotton, who was apparently unaware that it had been described earlier by Brazier. This specimen and the type of *C. excelsus* were illustrated by Cernohorsky and are virtually identical except that the type of *C. excelsus* is wider at the shoulder and has distinct but very small spire nodules apparently absent in the type of *C. pulcherrimus* (=*tannaensis*). The types of *C. nakayasui* are more weakly patterned and blotchy but otherwise have no significant differences from *C. pulcherrimus*(=*tannaensis*).

Notes.—This is an extremely rare shell in collections and commands a high price when offered. Rumor has it that the shell is not as rare as thought. At the moment even badly worn shells are considered collectable.

EXIMIUS Reeve, 1849

1849. *Conus eximius* Reeve. *Conchologia Iconica*, 1(*Conus* Suppl.): Pl. 6, sp. 256 (Moluccas).

Description.—Moderately heavy, with a good gloss; biconical, the sides nearly straight and tapering to a narrow base; body whorl with low, wide basal spiral ridges separated by shallow grooves; ridges usually obsolete before midbody, but sometimes the grooves are evident to the posterior third of the whorl; body whorl otherwise smooth except for axial threads and scattered spiral threads or a few weak ridges; shoulder sharply carinate, broad, flat above; spire moderately high, the sides concave, sharply pointed; early whorls not nodulose; whorls nearly flat above, covered with very fine axial curved threads and 2-3 narrow strong spiral ridges on posterior part of whorl, the anterior part without distinct spirals, whole top appearing finely cancellate; margins of whorls carinate. Body whorl glossy white, covered with two variable series of reddish brown to yellowish brown blotches, one above and one below midbody area; blotches usually fairly large but isolated, seldom reaching the shoulder or base, often axially rectangular; numerous fine dashed lined of brown and white visible in parts of pattern; often shell mostly white with just a few isolated brown blotches above and below clear midbody area plus a few small brown dashes in vaguely axial pattern visible on basal ridges and below shoulder; alternatively, the shell is almost all brown, except for the white midbody area which may be infiltrated by squarish brown blotches at its edges, the brown background with numerous white and darker brown dashes and occasional axial white areas; spire white, with numerous or few curved brownish axial lines; early whorls faintly pinkish. Aperture narrow posteriorly, widened anteriorly; outer lip thin, sharp, nearly straight; mouth white to pale bluish white or cream. Columella long and very narrow, sometimes weakly set off with a small ridge posteriorly. Length 30-55mm.

Comparison.—*C. eximius* is highly variable in pattern but has a distinct spire sculpture. Although it commonly resembles such species as *C. voluminalis*, *C. polygrammus*, or even *C. tribblei* in general shape and color pattern, all these species and their near relatives have distinctly nodulose early whorls. *C. monile* is much more elongate, lacks the faintly cancellate spire sculpture, and usually has a strong salmon tone. *C. excavatus* has a distinct white band below the shoulder and is more evenly colored, not blotched. Perhaps the closest relative of *C. eximius* is *C. malacanus*, which is somewhat similar in shape and patterns; however, *C. malacanus* has the spire whorls concave above, the early whorls faintly nodulose or undulate, and the spiral sculpture of the early whorls evenly distributed over the whorl. See also *C. lemniscatus*.

C. kermadecensis is quite similar to *C. eximius* in several ways, and some patterns of the two are not easy to distinguish. In *C. kermadecensis* the tops of the spire whorls are usually weakly sculptured or have the axial ridges strongly developed and the spirals almost absent; in some populations the whorls are distinctly concave and stepped. Almost all *C. kermadecensis* have numerous spiral rows of very small brown spots visible in the background, these not comparable to the dashes of *C. eximius*. *C. kermadecensis* is usually a coarser-looking shell, sometimes with a distinctly violet mouth. The two species are possibly sympatric in the western Indian Ocean and the China Sea.

Variation.—As indicated in the description, this shell may vary from mostly white with small brownish blotches to mostly brown with white dashes and blotches and a white midbody area. The three basic patterns are: 1) almost all white with a few faint yellowish brown spots above and below midbody; 2) white with two bands of isolated large brown axial blotches connected spirally by rows of brown spots and with a few faint dashes in the blotches; 3) mostly brown shells with axial flammules cutting across the midbody band and having many whitish dashes. The spire height varies a bit, as does the extent of the basal sculpture and the strength of the spire sculpture.

Distribution.—Commonly dredged in deep water of the China Sea (Taiwan), Philippines, Indonesia, Malaysia, and the Bay of Bengal. It probably occurs more widely in the Indian Ocean and in the western Pacific.

Notes.—Except for sometimes heavy axial growth marks, this is not a hard species to get in gem or near-gem condition. The lip chips easily and is often heavily filed, however. Specimens with heavy patterns seem to be

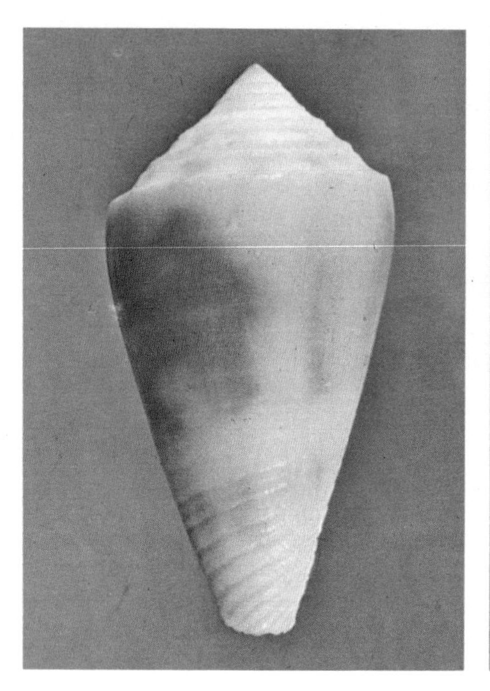

MINDANUS.—Above: Herrington Sound, Bermuda, 35mm (R. Jensen). *Below:* Left: Off Beaufort, North Carolina, 30.0mm (photo by A. Kerstitch); Right: Herrington Sound, Bermuda, 18.9mm (R. Jensen).

MINDANUS.—Above (*duvali*): Left: St. Vincent, 14.8mm (GG); Right: Guadaloupe, 18.3mm (TY). *Below:* Left: Bimini, Bahamas, 20.0mm (R. Guest); Right (*duvali*): Off Rio de Janeiro, Brazil, 15.8mm (photo by A. Kerstitch).

harder to find than those with type 1 or 2 patterns.

There is a chance that *C. eximius* of Reeve is not this species; the original figure is not very good and in some ways resembles the shell now called *C. kermadecensis*. However, the description correctly points out the general resemblance to *C. voluminalis*, the sharp shoulder, and the brown and white dashes, so it seems safe to continue using the name *C. eximius* in the sense used here.

FERGUSONI Sowerby iii, 1873

1873. *Conus fergusoni* Sowerby iii. *Proc. Zool. Soc. (London)*, 1873: 145, pl. 15, fig. 1 (Panama).

1873. *Conus fulvocinctus* Crosse. *J. Conchyl. (Paris)*, 20: 214 (West Africa). 1873, *Ibid.*, 21: 248, pl. 11, fig. 5. Crosse (21: 248) gives the date of the original Latin description as 1873, not 1872.

1880. *Conus consanguineus* E.A. Smith. *Proc. Zool. Soc. (London)*, 1880: 478, pl. 48, fig. 1 (Locality unknown).

Description.—Heavy, with a low gloss; low conical, the sides convex posteriorly, sometimes almost parallel for a short distance, then tapering to the base; body whorl with numerous low, crowded spiral ridges on the anterior third in adults, these rapidly becoming obsolete except for scattered short spiral ridges and many distinct axial threads and growth marks; juveniles with spiral sculpture extending above midbody; shoulder roundly angled in adults, but with a weak carina, convex above; spire low or moderate, in adults with straight or slightly convex sides and bluntly pointed; juveniles with somewhat relatively higher spire that is sharply pointed and has concave sides and weakly stepped whorls; early whorls distinctly nodulose, then becoming undulate, later whorls weakly carinate; in adults the nodulose early whorls seem to be eroded and appear only weakly undulate, then with rounded margins; tops of whorls with very faint traces of 3-5 spiral ridges crossed by numerous very weak axial threads, the sculpture all weak, even in juveniles, often not apparent in adults. Color of shell white in large adults; juveniles with two broad orange to orange-brown bands above and below midbody; juvenile pattern becomes less distinct with growth and usually by 75mm there is only a weak but broad orange band above midbody and traces of a narrower orange band below midbody; base white; juveniles and occasional small adults with scattered large rounded brown spots in vaguely axial series, the spots usually absent; spire white with orange tones, especially in juveniles, without distinct spots; early whorls faintly tan. Aperture wide, uniform; outer lip sharp, straight; mouth white. Columella long, narrow, mostly internal. Length 65-163mm.

Comparison.—Large adults that are all or mostly white could only be confused with *C. quercinus* or perhaps *C. patricius*. It is thinner than both these species and has a more convex-sided spire with indistinct sculpture; *C. quercinus* is yellowish with fine spiral brown lines in almost all specimens and has a nearly flat spire; *C. patricius* is strongly pyriform with deeply concave anterior sides, usually a sharply pointed but very low spire even in adults, and is distinctly nodulose over most early spire whorls even in adults.

C. xanthicus and *C. excavatus* are somewhat similar to the juveniles in shape and pattern and could be confused. *C. excavatus* is usually brownish or yellowish with a distinct white midbody band and indistinct brown dashes in the background; the early whorls are only undulate or nearly straight, and there is a heavy spire pattern which extends onto the body whorl just below the shoulder. In *C. xanthicus*, long considered a juvenile of *C. fergusoni*, the shell is thin, the outer lip sometimes raised above the level of the shoulder, and the early spire whorls strongly nodulose. The shoulder is sharply angled, the spire low, the whorls sometimes weakly concave above, and the spiral and axial sculpture heavier than in *C. fergusoni*. The spire is whitish and heavily marked with squarish brown spots, a feature not found in *C. fergusoni*; there are also several minor differences in color and pattern.

Variation.—There is considerable ontogenetic variation in shape and color. The juveniles are relatively slender, with a moderately tall and sharply pointed spire with concave sides and weakly stepped spire whorls; the nodulose early whorls are strong and distinct. The body whorl is white with two broad bands of dull orange or orange-tan above and below a whitish midbody band; there are sometimes rounded brown spots scattered over the whorl in vaguely axial or spiral series, but these are not constant. With growth the shell widens posteriorly, the pointed early whorls become eroded, and the spire assumes a rather broadly convex shape with indistinctly nodulose early whorls still visible. The pattern becomes weaker with growth, the anteriormost orange band disappearing first, while remnants of the posterior band may be seen in fairly large specimens. Full adults are solid white without pattern.

Distribution.—Common in shallow to moderately deep water from the Gulf of California south through the eastern Pacific to Ecuador

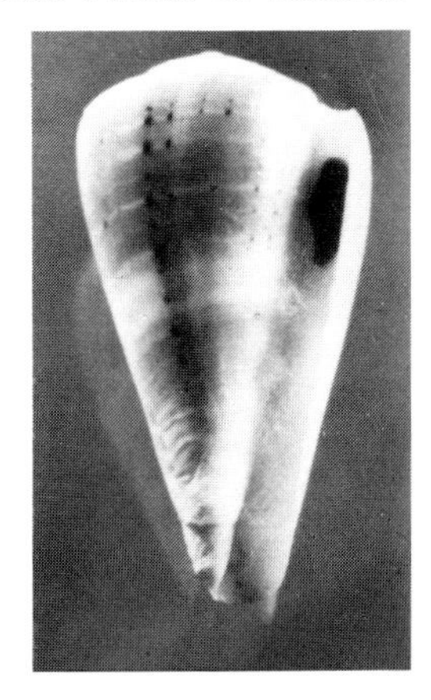

Spotted adult of *C. fergusoni*, Guaymas, Mexico, 88mm (photo Kerstitch).

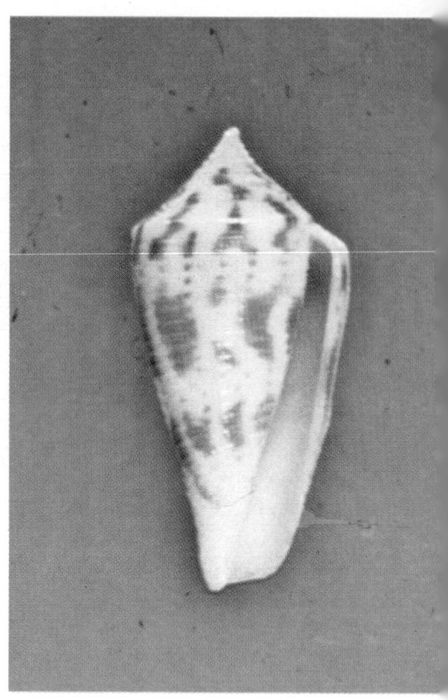

MINNAMURRA.—Above: Capricorn Channel, Queensland, Australia, 26.9mm (TY). *Below:* Left: Capricorn Channel, Queensland, Australia, 25.6mm (TY); Right: Off Terrigal, New South Wales, Australia, 18.1mm (EP).

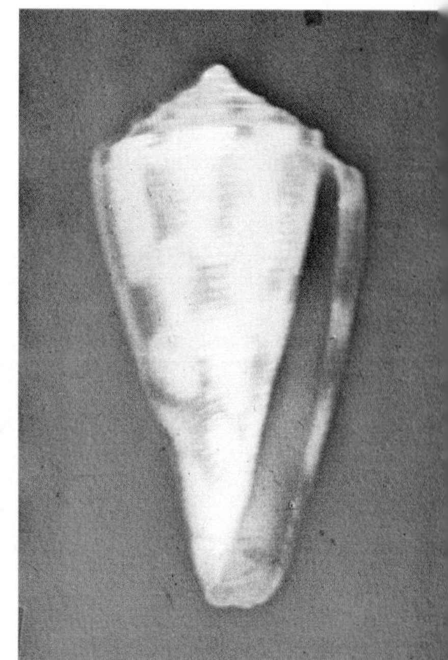

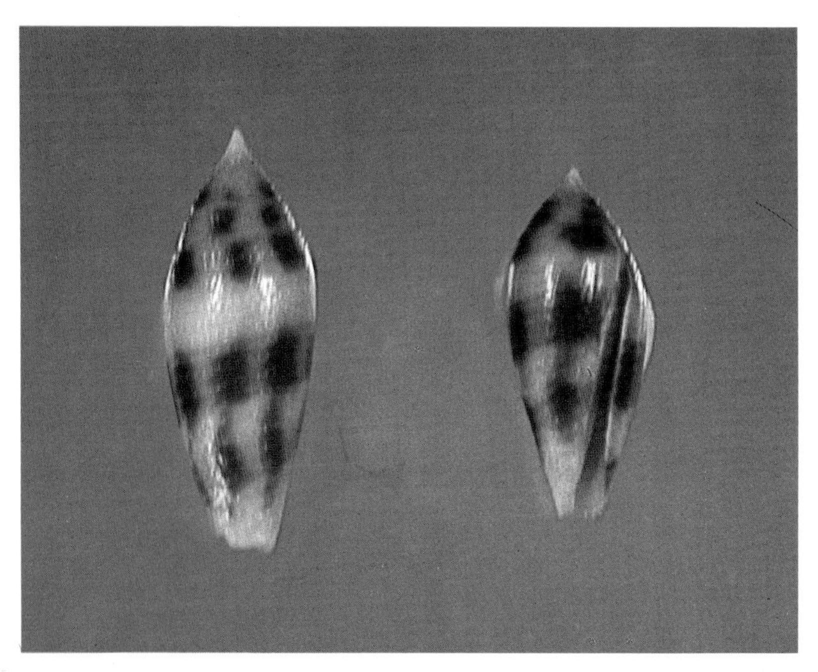

MITRATUS.—*Above:* Honiara, Guadalcanal, Solomon Is., 18.9 and 22.5mm. *Below:* Madagascar, 39mm (PC).

including the offshore islands.

Synonymy.—It has long been a debated wonder that this large and common shell was not described before 1873. However, between 1873 and 1880 it was described three times, each description being based on fairly large specimens. It is possible that *C. fulvocinctus* has a few month's priority over the familiar *C. fergusoni*, as the date is often given as 1872, while *C. fergusoni* was described in 1873. As with other French descriptions of this period, the shell was described as new in two papers, the first being a Latin diagnosis only, the second a fuller French description and plate. In the second paper, Crosse gives the date for his species as 1873, not 1872; I prefer to use 1873 as it at least gives a chance that *C. fergusoni*, the familiar name, is the correct one for matters of priority. Regardless, it would be senseless to change a familiar name at this date because of a few months. *C. consanguineus* is based on a badly distorted specimen.

Conus xanthicus has several times been cited as the juvenile of *C. fergusoni*, but there is no doubt that it is a distinct species. It has been suggested that the spotted specimens represent a distinct species, but Mr. Alex Kerstitch assures me that spotting appears to be just a juvenile character that is individually variable; spotted individuals are otherwise identical to non-spotted specimens. There is another *C. fergusoni*-like species being described from Central America, but the name is not available at the moment.

Notes.—Fully adult specimens are ugly white shells with little to recommend them. Subadults with good pattern traces are more presentable both color-wise and scar-wise. This species has heavy axial growth marks and is sometimes badly worn; the lip chips easily. Juveniles are not common, and spotted juveniles could be called rare.

FIGULINUS Linnaeus, 1758

1758. *Conus figulinus* Linnaeus. *Systema Naturae per Regna Tria Naturae*, ed. 10, 1: 713 (Locality not stated). Lectotype selected and figured by Kohn, 1963.

?1758. *Conus minimus* Linnaeus. *Ibid.*, 1: 714 (Locality not stated). Holotype figure: Argenville, Pl. 15, fig. A.

(1792. *Voluta fagina* Martyn. *Univ. Conch.*, 4: fig. 125 (Locality not stated). Non-binomial.)

1798. *Cucullus buxeus* Roeding. *Museum Boltenianum:* 42 (Locality not stated). Lectotype figure (Kohn, 1975): Martini, Pl. 59, fig. 656.

1845. *Conus Loroisii* Kiener. *Species gen. et icon. des coqu. viv.*, 2 (*Conus*): Pl. 65, fig. 1. 1847, *Ibid.*, 2: 91 (Indian Seas).

1933. *Conus figulinus violascens* Barros e Cunha. *Mem. Estud. Mus. Zool. Univ. Coimbra*, (Ser. 1), 71: 37 (Locality not stated).

1937. *Conus figulinus* var. *insignis* Dautzenberg. *Mem. Mus. Roy. d'Hist.*

Nat. Belg., 2(18): 108, pl. 1, fig. 6 (Amboine). Non *Conus insignis* Sowerby i, in Sowerby ii, 1833.

Description.—Very heavy, with a low gloss; pyriform, the upper sides very convex then concave and with a narrow base; anterior third to quarter of body whorl with variable spiral ridges, these sometimes heavy or very weak, the grooves between sometimes with axial threads within; rest of body whorl smooth except for numerous axial threads and growth marks, these sometimes conspicuous; shoulder rounded, flat or slightly convex above; spire very low or flat, the early whorls forming a small sharply pointed cone in the middle; tops of whorls with numerous curved, weak axial threads or wrinkles, sometimes with very faint indications of 2-3 spiral ridges, especially on the early whorls. Body whorl gray to dark tan, usually heavily covered with narrow darker spiral lines which are not interrupted except at major growth flaws; occasionally spiral lines reduced or absent, or broader and grouped into wider bands; often a clear midbody band, grayish, tan, or pinkish in color, and a narrower band below the shoulder of about the same color; top of shoulder and spire usually dark brown or grayish, contrasting with body whorl coloration in most specimens, sometimes almost black and with faint indications of lighter tan axial blotches; base of shell often dark brown. Aperture fairly wide, uniform or slightly widened anteriorly; outer lip thick, sharp, concave at middle; mouth bluish white to pale bluish, the outer edge often with a brown stripe. Columella long, narrow, set off posteriorly by heavy ridges in many populations. Length 40-97mm, probably larger.

Comparison.—This heavy, pyriform species closely resembles *C. betulinus* in shape and sometimes in color; rarely does *C. betulinus* have continuous spiral bands of brown and uniformly colored spire whorls, however, except for uncommon variants from deep water; the base in *C. figulinus* is usually narrower and longer than in similarly-sized *C. betulinus*. Actually, almost all *C. betulinus* are orange with spots or dashes on the body whorl and distinct dark spots or stripes on the spire; there should be no confusion. *C. suratensis* is narrower and more elongated, with the pattern of numerous small dashes on yellowish or orange. *C. glaucus* is smaller, actually thicker, bright bluish or bluish gray, including the mouth, and has a dashed pattern, the spire with heavy stripes. *C. patricius* lacks spiral bands or any other color pattern, has a nodulose spire, and is more elongated.

Variation.—There is little significant variation other than slight differences in tone of coloration, presence of clear bands, width of spiral lines, and convexity of spire. Juveniles are narrower and less pyriform than adults, with the spire more convex; they are sometimes pinkish.

One problematic taxon, *C. loroisii*, has often been considered a full species, but it is here treated as *C. figulinus* variety *loroisii*. First, the two shells are sympatric in the northern Indian Ocean and Philippines. Secondly, the major difference between the two forms is that *C. loroisii* is

MOLUCCENSIS.—Bolo Pt., Okinawa, Japan. *Above:* 44.0 and 47.0mm (TCG). *Below:* 45.5mm (K).

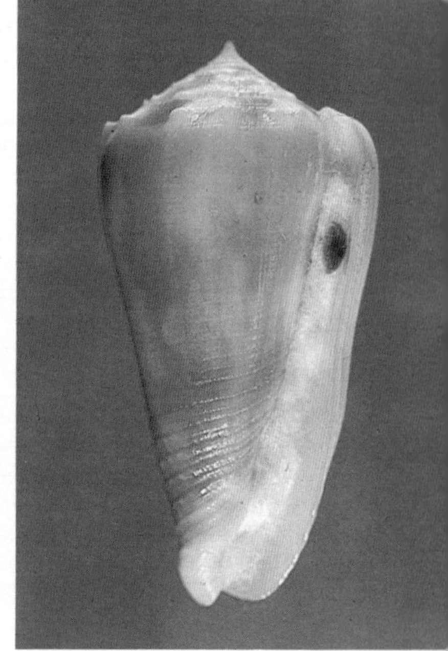

MONACHUS.—*Above:* Left: Andaman Is., 55.7mm; Right: Honiara, Guadalcanal, Solomon Is., 53.2mm. *Below:* Left: Matupit, E New Britain, Papua New Guinea, 42.4; Right: Broome, Western Australia, Australia, 39.1mm.

grayish, somewhat thicker than typical *C. figulinus*, and lacks the spiral brown lines. However, the absence of lines has long been known to be variable in shells called *C. loroisii*, and there have been several reported specimens with lines on one surface and none on the other; I have seen a Philippine specimen that was grayish like typical *loroisii* but had scattered indistinct pale brown lines anteriorly. It seems likely that *loroisii* is an ecological variant, often found in deeper water than typical *figulinus*. The variants can be distinguished as follows:

C. figulinus, typical: Usually tan or brownish with distinct spiral brown lines; base rather long and narrow; relatively lighter in weight; Indo-Pacific.

C. figulinus var. *loroisii:* Grayish, the spiral lines absent or very poorly developed; base shorter; relatively heavier and thicker; northern Indian Ocean (common) to the Philippines (rare) and Indonesia.

Distribution.—Widespread and common in the Indo-Pacific, from eastern Africa to French Polynesia, but not recorded from Hawaii. Shallow water.

Synonymy.—*C. minimus* is probably unrecognizable and based on a single bad figure; it has commonly been interpreted as *C. coronatus*, but it is best to treat it as a synonym of *C. figulinus* so there is no danger of the later *C. coronatus* being replaced. Some Japanese authors have used *C. minimus* as the correct name for *C. figulinus*, but there is nothing to be gained by this and a familiar and well-understood name would be lost.

The varietal names are meaningless, as they are based on individual variants which could appear in any population. Specimens with brightly marked central bands are uncommon.

Notes.—This common species varies considerably in pattern and size from area to area, although few of the variants are uncommon enough to be worth a premium. Those specimens brightly marked with rosy colors are uncommon and usually deserve a premium. The variety *loroisii* was once considered rare, but Indian Ocean specimens are now merely uncommon to locally common and only worth a small premium.

Note that *C. chytreus*, usually considered a dwarf variant of *C. figulinus*, is here treated as one of the many color patterns of *C. variegatus*, a West African species. This treatment seems obvious from the original description and later redescription, especially because of the small size and the erosion. Collectors would never mistake a specimen of *C. variegatus* for a *C. figulinus* as the textures and patterns are different in details.

FLAVIDUS Lamarck, 1810

1810. *Conus flavidus* Lamarck. *Ann. du Mus. Hist. Nat.* (*Paris*), 15: 265 (Locality unknown).

?1933. *Conus erythraeo-zonatus* Barros e Cunha. *Mem. Estud. Mus. Zool. Univ. Coimbra*, (Ser. 1), 71: 108 (Locality unknown).

Description.—Heavy, with a low gloss or dull surface; obconical, the sides generally straight or weakly concave at the middle; body whorl with narrow, widely spaced spiral ridges sometimes extending to above midbody, the ridges often heavily granulose; numerous prominent axial threads and growth lines; shoulder broad, roundly angled to carinate, not coronated; spire nearly flat or very low, bluntly pointed; spire whorls with 3-4 very low and weak spiral ridges, these sometimes obscured by axial growth lines. Body color usually orange or yellowish, sometimes tan or bluish gray; two narrow white or bluish white spiral bands present, one at midbody and one below shoulder; sometimes with a second narrow band above the midbody band; base broadly tipped with bright purple; shoulder and spire white or bluish white, unmarked. Aperture narrow, sometimes slightly widened anteriorly; outer lip thin, sometimes concave at the middle; mouth purple with pale bands at middle and below shoulder, occasionally paler anteriorly. Columella internal. Length 25-61mm.

Comparison.—*C. flavidus* is most often confused with *C. lividus*, which resembles it closely and is sympatric, but that species is heavily coronated on the spire whorls and shoulder and is commonly more brownish than orange or yellow; the two species are distinct although very close. *C. frigidus* is also very close to *C. flavidus*, but in that species the spire whorls are strongly striated with three spiral ridges, the spire is convex and the shoulder rounded, and the pale bands at midbody and shoulder are less distinct and generally yellowish or olive, not white. *C. emaciatus* commonly has the spiral ridges extending almost to the shoulder, is more elongated, lacks the sharply defined bands at midbody and shoulder but instead has pale yellow bands that are not sharply contrasting, and commonly has the mouth pale bluish in adults, not bright purple. *C. gilvus* differs in having the base brownish violet and not purple, has numerous fine brown dashes usually visible somewhere in the pattern, has a more distinctly sculptured spire, and has a pale violet mouth; in addition, most specimens of *C. gilvus* have small black spots scattered on the shoulder and on the margin of the spire whorls.

Variation.—The shoulder may be roundly angled or sharply carinate. Usually the sides of the body whorl are nearly straight, but in large specimens they are sometimes distinctly concave. The basal ridges usually extend to at least midbody but rapidly become obsolete. The ground color is usually some shade of orange or yellow, sometimes pale tan; some specimens are bright bluish gray; it is not uncommon for the colors to assume several different shades on the same specimen. The spire is always unmarked in specimens I have seen, although it may be toned bluish or tan. Some specimens are quite glossy and very clean.

Distribution.—Abundant in shallow water throughout the Indo-Pacific, from eastern Africa to Hawaii and French Polynesia.

Synonymy.—*C. erythraeozonatus* from the original description (it was not figured) sounds like a *C. flavidus* with bright colors and toned

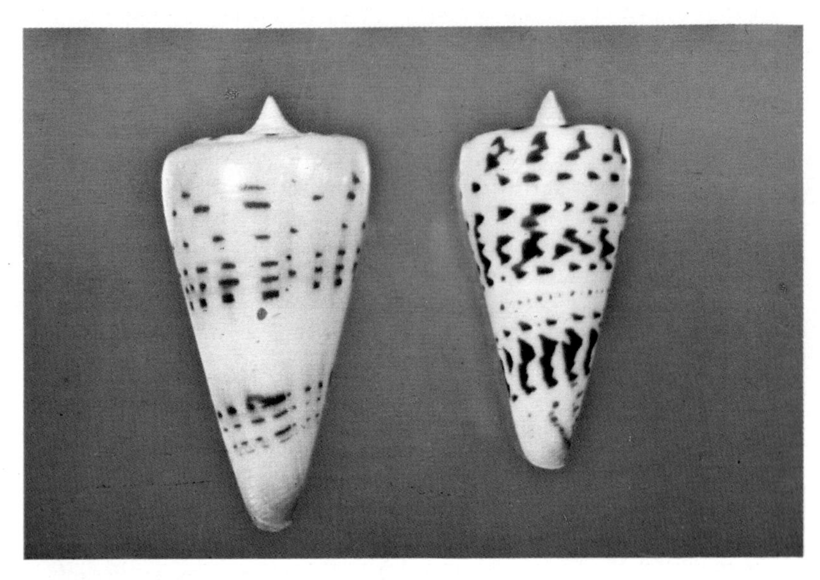

MONILE.—Above: Off W Thailand, 42.2 and 47.9mm (GG). *Below:* Left: Off Sri Lanka, 55.8mm (T.C. Lan); Right: Burma, 52.9mm.

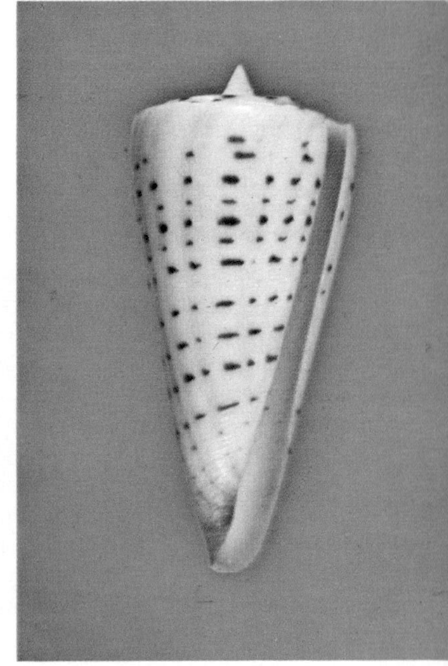

MORELETI.—Above: Honiara, Guadalcanal, Solomon Is., 35.3 and 37.9mm. *Below:* Left: Honiara, Guadalcanal, Solomon Is., 33.5mm; Right: E New Britain, Papua New Guinea, 24.6mm.

bands.

Notes.—Although often a very coarse and ugly cone, there are attractive specimens of this species. Usually juveniles or small adults are smoother and have some gloss, while the colors are often brighter. In some populations almost all individuals, regardless of size, have heavy erosion and persistent periostracum, making them very poor specimens. Specimens with bright bluish gray colors or actually reddish are uncommon compared to the usual orange or yellow specimens. Granulose specimens are not much less common than smooth specimens.

Although very like *C. lividus* in many ways, *C. flavidus* seems to be a full species and not just an ecological variant as has been suggested. The colors and patterns are different, and the two look quite different and have somewhat different textures.

Probably there are many *C. flavidus* in collections that are really *C. frigidus*—the two species seem to be readily confused.

FLOCCATUS Sowerby i, in Sowerby ii, 1839

1839. *Conus floccatus* Sowerby i, in Sowerby ii. *Conchological Illustrations:* Pt. 153/154, fig. 112 (Locality not stated).

1845. *Conus Magdalenae* Kiener. *Species gen. et icon. des coqu. viv.*, 2 (*Conus*): Pl. 69, fig. 4. 1849-1850, *Ibid.*, 2: 293 (Locality unknown).

1865. *Conus circumsignatus* Crosse. *J. Conchyl.* (*Paris*), 13: 311, pl. 10, fig. 4 (Locality unknown).

Description.—Moderately light but solid, with a high gloss; cylindrical, somewhat ovate, the sides slightly convex; body whorl covered from base to shoulder with fine, closely spaced spiral ridges, these heaviest at base and sometimes obsolete below shoulder; in many specimens the ridges are barely feelable and may disappear in beach specimens; shoulder roundly angled, narrow, slightly concave anterior to middle, with a weak groove at middle; spire low or moderate, bluntly pointed, the sides straight; spire whorls slightly depressed at the middle, sculpture consisting of a single median spiral groove and numerous fine axial threads. Body whorl pale pinkish white to deep tan, lavender, violet, or rosy; numerous spiral lines of brown and white dashes over body whorl, at least the brown dashes usually distinct and the white dashes often opaque white; commonly there are large to small reddish brown axial blotches above and below midbody and few to many narrow, wavy axial flammules of reddish brown; sometimes the flammules are in the form of very fine brown lines; lip often with reduced pattern, whitish; base pale, violet to white; shoulder and spire with large and small squarish blotches of pale brown and sometimes with small brown spots at the margins; early whorls pinkish. Aperture narrow above, widened anteriorly; outer lip slightly convex, sharp but thick in adults; mouth glossy white. Columella long, wide, thick, with heavy ridges posteriorly. Length 40-65mm.

Comparison.—At first glance this species reminds one of either *C. bullatus* and allies or *C. circumcisus* and allies. From both it is readily distinguished. The numerous very fine spiral ridges over the body whorl will separate it from *C. bullatus* and *C. julii*, which may have somewhat similar patterns. *C. circumcisus* in a smooth or nearly smooth phase is similarly sculptured, but it has very heavy blackish blotches on the shoulder and usually later spire whorls; it also has the banded pattern somewhat developed, although this is obscure or absent in *C. floccatus*. Perhaps the closest relative of *C. floccatus* is *C. dusaveli*, which is similar in general shape and texture but has the basal ridges wide and separated by distinct grooves; the spire pattern of *C. dusaveli* is composed of numerous fine axial brown lines absent in *C. floccatus;* there are of course many other differences in pattern.

Variation.—As indicated in the description, this is a highly variable species in color pattern. The shape and sculpture are very constant, however. The shell may appear pale pinkish white with just a few scattered squarish reddish brown spots visible and perhaps a few small blotches; commonly the shell is pale violet on pinkish with many spiral rows of brown and white dashes and some larger blotches, plus several axial flammules above and below the midbody area. Extreme specimens are very dark, sort of lavender-tan, have the spiral rows of dashes very heavy and conspicuous, and are covered with many blotches and axial flammules.

The unpigmented lip in adults is similar to the case in *C. bullatus* and allies and is a good argument for close relation with that group.

Distribution.—An uncommon to rare species apparently widely distributed in the western and southern Pacific from at least the Philippines to the Solomons and Fiji and north to the Marshall Islands. Apparently an occupant of offshore reefs in deep water, but commonly taken in fresh condition on beaches after heavy storms.

Synonymy.—*C. magdalenae* has long been recognized as being this species. *C. circumsignatus* appears to be a fairly typical but small specimen with a heavy spiral pattern and pale colors.

Notes.—This beautiful species is harder to get in really nice condition than many cones which sell for higher prices. Good specimens with bright patterns and heavy gloss are available at relatively low prices, but they usually have broken lips—these are typical of fresh-dead specimens taken after storms. Live-taken specimens are seldom offered, but I am not sure I could tell them from excellent fresh-dead, although there is a considerable difference in price. Specimens with dark patterns are more attractive than those with little flammulation; the most uncommon color appears to be yellow.

A thickened, partially unpigmented lip seems to be characteristic of adults of this species and should not be considered a flaw.

MUCRONATUS.*—*Above (typical): Tayabas Bay, Luzon, P.I., 37.1 and 39.8mm. *Below:* Taiwan: Left: 39.3mm; Right (*alabaster*): 43.8mm.

MURICULATUS.—Above: Left and Center (pustulose): Calapan, Mindoro, P.I., 20.3 and 28.8mm (GG); Right (pustulose): Off Russell I., Solomon Is., 34.1mm (EP). *Below:* Left and Center: Sulu, P.I., 26.2 and 27.6mm; Right: Tolwat Beach, E New Britain, Papua New Guinea, 22.1mm (B. Parkinson).

FRIGIDUS Reeve, 1848

1848. *Conus frigidus* Reeve. *Conchologia Iconica,* 1(*Conus* Suppl.): Pl. 3, sp. 284 (Locality unknown).
1873. *Conus Maltzanianus* Weinkauff. *Syst. Conch. Cab.*, ed. 2, 218 (*Conus*): 204, pl. 32, figs. 3-6 (Tahiti).

Description.—Moderately heavy, with a low gloss; low conical, the sides slightly convex posteriorly; body whorl covered on anterior half to three-quarters by wide spiral ridges, usually granulose, separated by shallow grooves with fine axial threads within; rest of body whorl covered with numerous axial and spiral threads; shoulder roundly angled, wide; spire low, bluntly pointed, the sides nearly straight; tops of whorls with 3-4 heavy spiral ridges separated by very finely axially threaded grooves; early whorls usually eroded. Body whorl olive tan to grayish tan, sometimes dirty cream, usually with indistinct yellowish tan bands at midbody and below shoulder; base heavily stained with purple; spire yellowish tan, without spotting. Aperture narrow posteriorly, slightly widened anteriorly; outer lip thin, sharp; mouth purple, with lighter bands at middle and under shoulder, sometimes entirely pale purple, outer edge reddish brown. Columella indistinct. Length 20-50mm.

Comparison.—*C. frigidus* belongs to a complex of usually small and dully colored cones including *C. flavidus, C. lividus, C. muriculatus, C. gilvus, C. emaciatus, C. sanguinolentus,* and perhaps other species. Of these *C. lividus, C. muriculatus,* and *C. sanguinolentus* are distinctly coronated although often patterned like *C. frigidus. C. emaciatus* is larger, more elongated, usually yellow or orange, and often has concave sides. *C. flavidus* is extremely similar, but it has a flatter spire with more sharply angled shoulder, weaker and indistinct spire sculpture, usually bright orange or yellowish colors with sharply defined bands of white at the midbody and below the shoulder, and the spire unrelieved whitish. In *C. gilvus* the shape is somewhat similar, as is the spire sculpture, but the base is more brown than purple, the spiral ridging on the body whorl is weak and sometimes indistinct, there are fine brown dashes over the body whorl usually visible, and the shoulder and spire margins usually have small scattered black spots.

Variation.—Other than color of the body whorl, this species is very constant—it almost has to be constant or it would be classified with another species! Body color is usually some shade of olive brown to grayish brown; juveniles and small adults of some populations are dirty white, but this is not true in all populations, as grayish juveniles have been seen from the Solomons. The spiral color bands are poorly defined and usually yellowish in tone. Apparently the spire is normally eroded.

Distribution.—Uncommon to locally common in the western and southern Pacific from the Philippines and Ryukyus south through the New Guinea-Solomons area to Fiji and French Polynesia. Usually in shallow water. Although usually stated to be Indo-Pacific, I have not seen Indian

Ocean specimens or records that definitely belong here rather than with *C. flavidus;* it probably does occur in the Indian Ocean, however.

Synonymy.—*C. maltzanianus* has long been placed with *C. frigidus*, of which it is certainly synonymous.

Notes.—Probably many *C. frigidus* have been sold as unusual *C. flavidus.* The two are certainly similar and I would not be surprised to find intermediates. The spire is usually eroded, the lip is fragile, and the body whorl often appears rough although not broken.

FULGETRUM Sowerby i, in Sowerby ii, 1834

1834. *Conus fulgetrum* Sowerby i, in Sowerby ii. *Conchological Illustrations:* Pt. 56/57, fig. 82 (Locality not stated).

1845. *Conus scaber* Kiener. *Species gen. et icon. des coqu. viv.*, 2(*Conus*): Pl. 100, fig. 1. 1849-1850, *Ibid.*, 2: 351 (Oceania). Non *Conus scaber* Link, 1807.

Description.—Heavy, thick, with a good gloss; low conical with the sides convex; body whorl with about a dozen low, widely spaced spiral ridges on the anterior half, these ridges usually heavily granulose; rest of body whorl usually smooth except for widely spaced spiral rows of small punctations and numerous fine axial threads; shoulder strongly angulate, heavily coronate; spire moderately high for this type of shell, pointed, with straight or convex sides; spire whorls strongly coronated and with about 2-4 heavy spiral ridges on top. Body whorl deep reddish brown to yellowish brown, usually without distinct darker spots; numerous spiral rows of small opaque white dots or dashes, these heavily fused to produce many small crosshatchings and axial zigzag flammules; opaque white markings somewhat concentrated into a vague band at midbody and another one below the shoulder; base white; spire and shoulder whitish, with large brown spots between coronations; spire spots with numerous fine axial lines within; coronations white. Aperture narrow posteriorly, widened anteriorly; outer lip convex, sharp but thick, often with a thickened band at the middle; mouth dark violet with paler bands at middle and under shoulder. Columella short, wide, thick, bounded posteriorly by heavy ridges. Length 15-35mm.

Comparison.—This species is obviously closely related to *C. miliaris* and *C. encaustus*, as well as closely to *C. tiaratus* and *C. abbreviatus.* Because of the large amount of opaque white in the pattern, *C. fulgetrum* is superficially closest to *C. encaustus*. However, it is readily distinguished by virtually lacking dark brown dots in spiral bands, having a higher and more sharply coronated spire, having the opaque white forming chevrons and x's as well as axial flammules, all these features usually absent in *C. encaustus* or at best poorly developed; in addition, *C. encaustus* generally lacks granulose spiral ridges at the base and does not approach the

MUS.—Above: Left: Riviera, Palm Beach Co., Florida, 26.4mm; Right: Aruba, Neth. Ant., 16.0 and 22.9mm (GG). *Below:* Left: Florida, 37.0mm; Right: Curacao, Neth. Ant., 20.0 and 20.5mm.

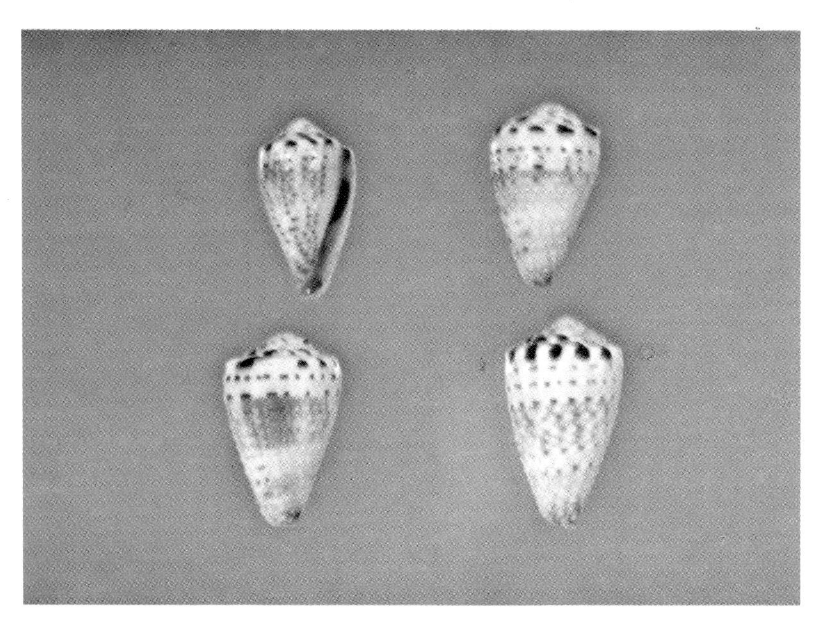

MUSICUS.—*Above* (Pacific form): Boac, Marinduque, P.I., 15.6-18.7mm. *Below:* Left (Indian form): Zanzibar, 17.4mm; Right (*mighelsi*): Eel Reefs, Cooktown, Queensland, Australia, 18.0mm (EP).

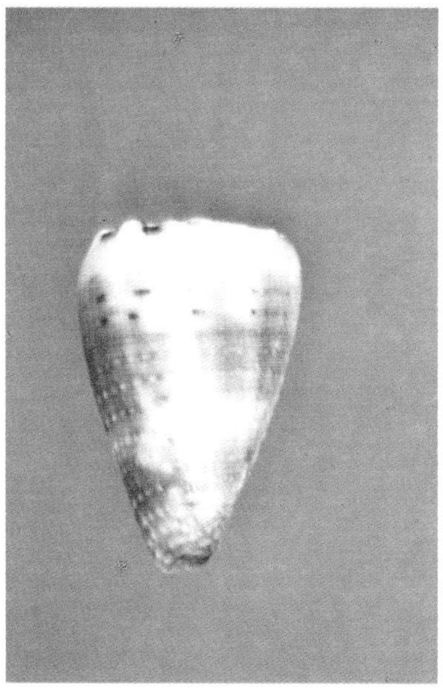

deep brown color of *C. fulgetrum;* the spots or stripes between the coronations are solid in *C. encaustus* rather than containing fine axial lines as in *C. fulgetrum.*

Certainly *C. fulgetrum* is actually closest to *C. miliaris,* and there is no doubt that some specimens are difficult to place. However, the two seem distinctive enough in pattern to be tentatively considered full species. Few *C. miliaris* have the large number of opaque white chevrons, although they may have small and irregular opaque white blotches; the opaque white pattern never approaches the complex patterns of *C. fulgetrum.* In addition, most *C. miliaris* have strong patterns of brownish dashes or lines in the background and are usually tan or creamy gray, sometimes pinkish, not deep brown like most *C. fulgetrum.* The combination of deep brown color, lack or reduction of brown spiral patterns, and heavy opaque white chevrons and flammules should distinguish *C. fulgetrum.*

Variation.—The shape and sculpture are rather constant, although sometimes the granulose ridges are heavily developed and extend to the shoulder. The color is also fairly constant, although there is of course a great deal of variation in the exact patterns of opaque white. The midbody band is usually visible, but the band below the shoulder may be so vague as to be absent to the eye. The shell may be almost covered with white chevrons, or there may be relatively few chevrons but several groups of axial flammules.

Distribution.—A moderately common shallow-water shell whose exact distribution is uncertain because of confusion with *C. miliaris.* It apparently ranges widely through the Indo-West Pacific, as specimens have been seen from localities from the Red Sea to the Philippines and Ryukyus; presumably it ranges into the New Guinea-Solomons area at least.

Synonymy.—*C. scaber* (preoccupied) is a specimen with extremely heavy granulations over most of the body whorl.

Notes.—I have little doubt that this is a distinct species although some *C. miliaris* do come close. It is not an uncommon shell, just poorly known. Most specimens come from the Philippines or Ryukyus. The spire is seldom badly eroded for a shell of this group, and there are seldom more than minor axial growth marks on the body whorl. The gloss is quite high, brighter than most *C. miliaris.*

FULMEN Reeve, 1843

1843. *Conus fulmen* Reeve. *Conchologia Iconica,* 1(*Conus*): Pl. 39, sp. 215 (Capul, Philippine Is.).

1956. *Conus fulmen kirai* Kuroda. *Venus,* 19(1): 7 (Tanoura, Kii, Japan).

?1970. *Conus* (*Chelyconus*) *wistaria* Shikama. *Sci. Rpts. Yokohama Nat'l. Univ.,* Sec. II(16): 24, text figs. 3-4 (Bungo Channel, Japan).

Description.—Fairly heavy, with a low gloss or dull; low conical, the sides convex; body whorl with a few strong spiral ridges above the base and numerous fine axial and spiral threads over the rest of the whorl; numerous growth lines usually present; shoulder rounded, moderately broad, slightly concave above; spire rather low, sometimes domed, the sides straight or convex, bluntly pointed; first 3-5 whorls weakly nodulose; early whorls with 2-3 weak spiral ridges and fine axial threads; later whorls with traces of spiral threads and numerous axial wrinkles, almost smooth. Body whorl pale violet or dull cream, the background color showing through only in a wide midbody band and at the shoulder; body whorl otherwise pale to dark brown, the intensity of color varying in broad spiral bands as follows: base whitish, then a pale brown band, a darker brown band, violet or cream midbody band, darker brown band, pale brown band, darker brown band, and pale violet or tan shoulder; this sequence is not always distinct; body whorl usually with large black irregular axial blotches, these continued onto shoulder; spire violet with large black blotches; early whorls tan and pink. Aperture wide above, much wider below; outer lip thin, sharp, evenly convex; mouth pale bluish white to dark violet. Columella long, narrow, weakly indented from inner lip. Length 30-70mm.

Comparison.—The shape and pattern on a violet background are fairly distinctive and not likely to be confused with any species other than *C. kinoshitai.* There is some similarity to various patterns of *C. monachus*, but that species is usually thicker, has strong spiral sculpture on the spire and body whorl, and has a different pattern. *C. kinoshitai* is usually obviously blotched in two or three spiral rows and the body whorl has much straighter sides; the spire is straighter-sided, shouldered more heavily, and the whorls have several distinct spiral ridges. The shoulder of *C. kinoshitai* usually has small black blotches instead of the large black blotches or no markings as in *C. fulmen.*

Variation.—The upper sides can be slightly convex to strongly convex, the sides of the spire nearly straight or distinctly domed. The body sculpture is usually very weak, but the axial threads are often conspicuous. Although the brown on violet or cream pattern is fairly constant, the large black blotches on the body whorl may be present or absent. Typical *fulmen* has the blotches large and deep black on a distinctly violet background; variety *kirai* (strictly ecological, especially depth) lacks the black blotches on the body whorls and sometimes the spire and often has a creamy background. Individuals with brown almost lacking and the violet showing through very strongly are not common. The mouth is usually dirty white with distinct violet tones, but it is sometimes deep violet.

Distribution.—Seemingly restricted to the northern China Sea from southern Japan to the Ryukyus; apparently not recorded south to Taiwan. Common locally in moderately deep water.

Synonymy.—*Conus fulmen kirai* is sympatric in slightly deeper

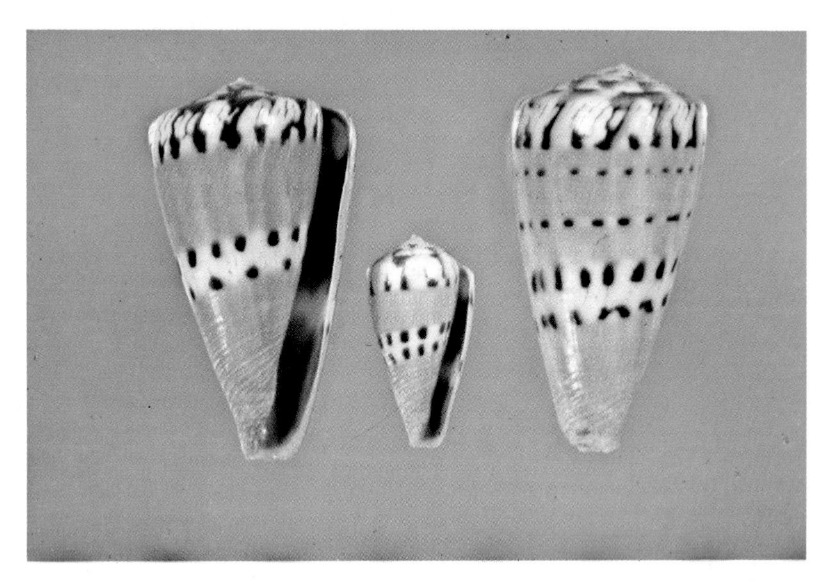

MUSTELINUS.—Above: Boac, Marinduque, P.I., 24.4-44.3mm. ***Below:*** Left: Southern Mindanao, P.I., 72.9mm (MG); Right (holotype of *C. melinus*): Arafura Sea, NW Australia, 71mm (photo by Dr. T. Shikama).

NAMOCANUS.*—*Above: Zanzibar, 47.0 and 49.7mm. ***Below:*** Left: Nossi Be, Madagascar, 42.9mm; Right: Zanzibar, 58.9mm (MF).

waters than typical *fulmen*, so the relationship is obviously varietal; intermediates are not uncommon. *C. wistaria* is a solid violet shell with only faint traces of spiral brown bands. Because the sides are convex and the shoulder is not especially well marked, it is assigned to *C. fulmen* as a freak individual, perhaps albinistic; there is a chance it might be a *C. kinoshitai*, but the shape is not quite right for that species.

Notes.—This apparently strictly Japonic shell is quite attractive in a subdued way. Individuals with the brown reduced and largely violet are quite pretty. The black blotches give the shell an especially distinctive appearance as they are usually pitch black. The growth marks are often conspicuous but seldom detract from the appearance of the shell.

FUMIGATUS Hwass, in Bruguiere, 1792

1792. *Conus fumigatus* Hwass, in Bruguiere. *Cone*, in *Ency. Method., Hist. Nat. des Vers*, 1: 704 (Mers d'Amerique). Holotype figured by Kohn, 1968.

1845. *Conus Blainvillei* Kiener. *Species gen. et icon. des coqu. viv.*, 2 (*Conus*): Pl. 111, fig. 1. 1849-1850, *Ibid.*, 2: 358 (Locality unknown). Non *Conus Blainvillei* Vignart, 1829.

1857. *Conus Pazii* Bernardi. *J. Conchyl.* (*Paris*), 6: 385, pl. 11, figs. 1-2 (Locality unknown).

1857-1858. *Conus adustus* Sowerby ii. *Thesaurus Conchyliorum*, 3 (*Conus*): 25, pl. 17(203), fig. 403 (Red Sea).

1857-1858. *Conus aureolus* Sowerby ii. *Ibid.*, 3(*Conus*): Pl. 17(203), fig. 395 (Locality unknown). Although cited in the index and plate caption as sp. 141 (*Conus coffea*), it is not discussed under the name *C. aureolus* in the text.

Description.—Heavy, thick, with a low gloss; broadly low conical, the sides nearly straight; body whorl with about 6-12 narrow, strong spiral ridges, these widely spaced and obsolete about midbody; rest of whorl with traces of fine spiral ridges and threads and numerous axial threads and growth marks; shoulder broad, sharply angled, concave above; spire low, the sides concave, bluntly pointed, the whorls concave above and weakly stepped with raised margins; tops of whorls with about 3-4 low but distinct spiral ridges crossed by numerous curved axial threads; early whorls perhaps weakly undulate or nodulose, but badly eroded in the specimens I have seen. Body whorl tan, dark brown, or grayish, covered with numerous heavy spiral rows of dark brown lines interrupted into long or short dashes or occasionally complete; when broken into dashes, they are often aligned axially as well as spirally; tendency for lines to be concentrated on either side of a clear white broad midbody band which is sometimes crossed by scattered axial rows of brown dashes; base about same color as body whorl; area just below shoulder with a narrow white band heavily marked with brown dashes, the pattern continued onto shoulder; spire and shoulder dirty white, heavily or sparsely marked with dark brown curved spots; early whorls tan, badly eroded. Aperture fairly narrow posteriorly, somewhat widened anteriorly; outer lip thick, sharp, straight or slightly convex; mouth brownish violet, with paler bands at middle and below shoulder. Columella long but very narrow, mostly internal. Length 15-35mm.

Comparison.—This species is like a small, thick *C. excavatus* with a short and wide body whorl and numerous spiral lines and dashes. Like *C. excavatus* it has a clear white midbody band and a band just below the shoulder, the spire with curved brown spots. Even juvenile *C. excavatus* are more elongate and somewhat thinner than juvenile *C. fumigatus* of the same length. From *C. miliaris*, which sometimes is very similar in size, shape, and pattern, it differs in lacking coronations and having the columella mostly internal, as well as lacking the rows of small punctations typical of *C. miliaris*. *C. erythraeensis* is a much smoother, narrower shell with the spotting distinct and squarish on a clear background.

Variation.—The color varies from dark brown to tan or even grayish, but the dashes and lines are dark brown; the midbody band is stark white and sharply contrasted even in grayish specimens. The brown lines may be long and almost continuous, especially in about 3-4 rows above and below midbody, or broken into short dashes in conspicuously axial rows. Usually the area above midbody is weakly marked compared to the anterior third. The band below the shoulder is narrow but usually sharply defined and with rather oblique brownish dashes extending onto the shoulder. The spire pattern is usually heavy, but sometimes it is almost absent above the shoulder and last 2-3 whorls.

Distribution.—Apparently restricted to the Red Sea and adjacent areas in shallow water, where not uncommon.

Synonymy.—The names listed have previously been referred to such species as *C. erythraeensis*, *C. boeticus*, or even *C. voluminalis*, but the shapes and patterns seem to all be in close agreement with the variation in *C. fumigatus*, so they are assigned here. *C. aureolus* is undescribed except for a single figure and the name, but it was apparently considered to be close to *C. coffeae*, a name which I consider unrecognizable but which has been used for *C. fumigatus* and *C. excavatus* at various times. *C. fumigatus* is more commonly known as *C. classiarius*, here considered a synonym of *C. capitaneus* but probably unrecognizable.

Notes.—Dealers at the moment fail to distinguish between *C. fumigatus* and juveniles of *C. excavatus*, at times selling them both as *C. coffeae* or *C. classiarius*. The two species seem quite distinct even as juveniles, however, and the first certainly recognizable names are those used here.

C. fumigatus is subject to large breaks on the body whorl and heavy erosion on the body whorl and spire, sometimes enough to obscure the pattern; the lip is fairly fragile. At best this is not an attractive cone.

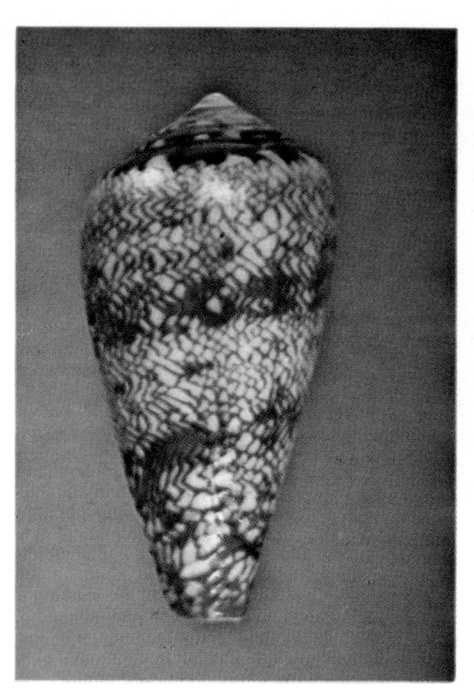

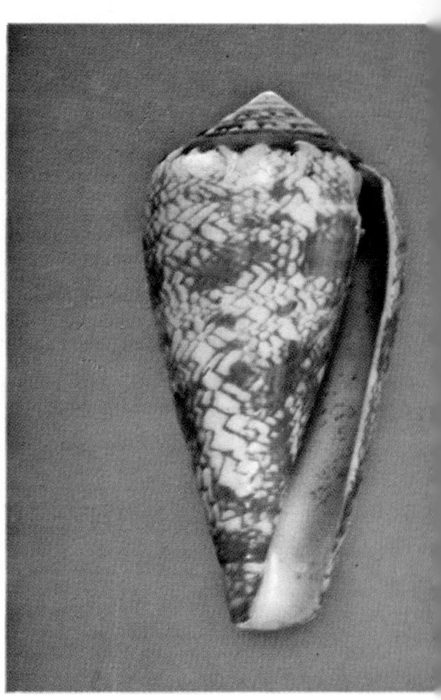

NATALIS.—Above: Jeffreys Bay, Cape, South Australia, 30.1mm (GG). *Below:* Left (*gilchristi*): Transkei Beach, South Africa, 32.9mm; Right: Gonubie, Cape, South Africa, 40.1mm.

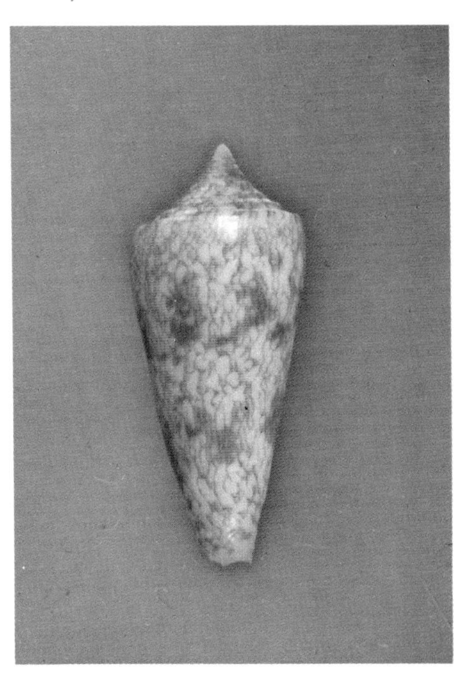

NEPTUNUS.—*Above:* Off Coron, Palawan, P.I., 42.2mm (Dayrit). *Below:* Russell I., Solomon Is., 38.4mm.

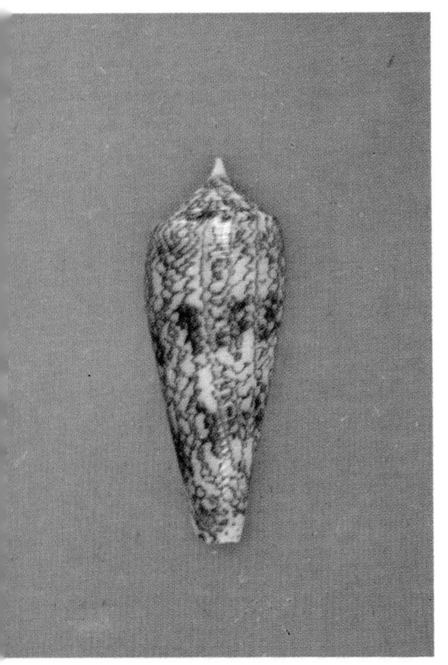

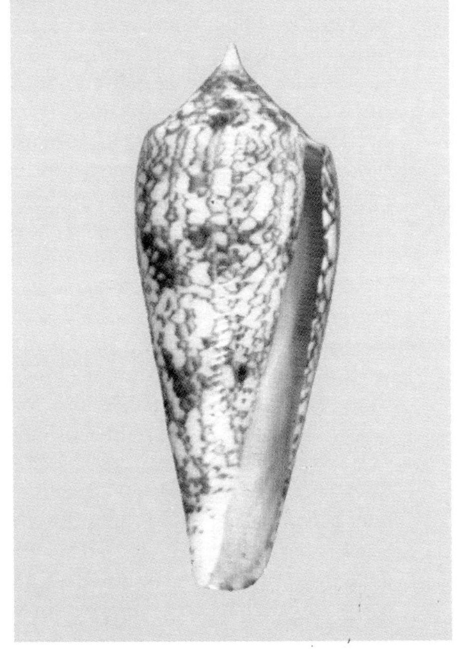

FURVUS Reeve, 1843

1843. *Conus furvus* Reeve. *Conchologia Iconica*, 1(*Conus*): Pl. 13, sp. 69 (Ticao and Masbate, Philippine Islands).
1843. *Conus lignarius* Reeve. *Ibid.*, 1(*Conus*): Pl. 24, sp. 136 (Port Sacloban, Leyte I., Philippines).
1844. *Conus crepusculum* Reeve. *Ibid.*, 1(*Conus*): Pl. 45, sp. 251 (Locality unknown).
1844. *Conus buxeus* Reeve. *Ibid.*, 1(*Conus*): Pl. 47, sp. 265 (Locality unknown). Non *Conus buxeus* (Roeding, 1798).
1845. *Conus Cecilei* Kiener. *Species gen. et icon. des coqu. viv.*, 2 (*Conus*): Pl. 98, fig. 4; Pl. 107, fig. 3 (var.). 1849-1850, *Ibid.*, 2: 286 (China Seas).
1845. *Conus striolatus* Kiener. *Ibid.*, 2(*Conus*): Pl. 105, fig. 1. 1849, *Ibid.*, 2: 266 (Locality unknown).
1849. *Conus aegrotus* Reeve. *Conchologia Iconica*, 1(*Conus* Suppl.): Pl. 5, sp. 250 (Locality unknown).
1849. *Conus granifer* Reeve. *Ibid.*, 1(*Conus* Suppl.): Pl. 7, sp. 272 (Philippine Is.).
1870. *Conus submarginatus* Sowerby iii. *Proc. Zool. Soc.* (*London*), 1870: 255, pl. 22, fig. 6 (Locality unknown).
1887. *Conus albus* Sowerby iii. *Thesaurus Conchyliorum*, 5(*Conus* Suppl.): 274, pl. 36(512*), fig. 761 (Locality unknown).

Description.—Moderately light in weight, with a low gloss or dull surface; rather elongately low conical, the sides straight or slightly convex posteriorly; body whorl with at least 6-10 narrow spiral ridges above the base, these usually continuing in weaker form to midbody and sometimes to shoulder, but just as commonly obsolete before midbody; occasionally some or all the ridges are finely granulose; shoulder sharply angled but not carinate, concave above; spire moderately high, bluntly pointed, the whorls often weakly stepped and concave above, the sides of the spire straight to slightly concave; tops of whorls and shoulder with two closely spaced narrow spiral grooves at the middle, dividing the whorls into two distinct parts and usually forming a shallow channel; early 3-6 whorls weakly nodulose in most specimens, the later whorls usually with simple margins; occasionally no spire whorls nodulose. Color and pattern variable; in the basic form the shell is creamy white or pale yellow, covered with numerous spiral rows of closely spaced small brown spots or dots, these tending to be thicker and forming two broad bands above and below midbody; the shoulder usually has short oblique brown lines and dots; from this basic pattern the shell can become suffused with brown thick enough to completely hide the dots except under a lens, or the dots may become paler and eventually disappear, leaving a white shell; typically the base is strongly stained brown or purplish brown, but white shells have the stain reduced or absent; spire tan to white, usually agreeing with body whorl color, the margins usually with small brown spots; tops of whorls variable, often tan, white with tan lines, or white; early whorls often eroded. Aperture moderately wide, slightly wider anteriorly; outer lip thin, sharp, straight or slightly concave at the middle; mouth white or bluish white (darkest in shells with dark body whorls). Columella long, very narrow, sometimes weakly indented from inner lip or mostly internal. Length 25-60mm.

Comparison.—This species is so variable in pattern, color, and shape that it must be examined in large series to show that extreme patterns intergrade. Typical patterns with brown base and numerous very fine brown dots in packed spiral rows are very distinctive and nothing else looks quite like them. In specimens which have lost most or all the pattern, the almost unsculptured spire which is concave at the middle of the whorls and the weak nodules on the early whorls are distinctive; usually even almost all-white shells show traces of the dots and brown base. Sometimes *C. ochroleucus* is sold as pale *C. furvus*, but that species is heavily grooved basally. *C. thalassiarchus* has fine axial lines in the pattern, plus many other differences.

Variation.—As indicated, this is an extremely variable and troubled species. The basic patterns can be briefly described as follows:

One: Dark brown, sometimes with fainter bands at midbody and shoulder, the dots almost completely hidden in the background; mouth dark, with a brown margin; spire tan, usually rather low and without nodules or very weak ones.

Two: Light brown, sometimes with fainter bands at midbody and below the shoulder, the base darker than rest of whorl; spots obvious, in distinct lines; mouth dark bluish white; spire tan, weakly nodulose.

Three: Yellowish, the base much darker and strongly contrasting; dots covering whole body whorl, but usually only conspicuous in two broad tan spiral bands above and below midbody and a narrower band below the shoulder; spire cream or yellowish with small brown dots and lines; spire whorls often nodulose and somewhat stepped.

Four: Shell dirty white with pale yellowish bands above and below midbody, these produced by faint dots; sometimes dots almost all absent; base pale violet; spire often taller and stepped, nodulose, without pattern.

Five: Solid white, without any pattern; base white; spire white, stepped, tall, strongly nodulose.

Six: Shell pale to dark tan, the dots conspicuous; spiral ridges granulose, especially on basal half but sometimes to shoulder; spire whorls coronated or not.

Additionally, any two or even three of the five color patterns can occur on a single shell, and the granulose pattern is not always restricted to brown patterns but may occur on white shells as well. Uncommonly the dots form heavy axial bands instead of or in addition to the spiral ones.

The body whorl varies from quite wide at the shoulder to narrow, the spire from low to high, the whorls flat-sided or strongly stepped, the

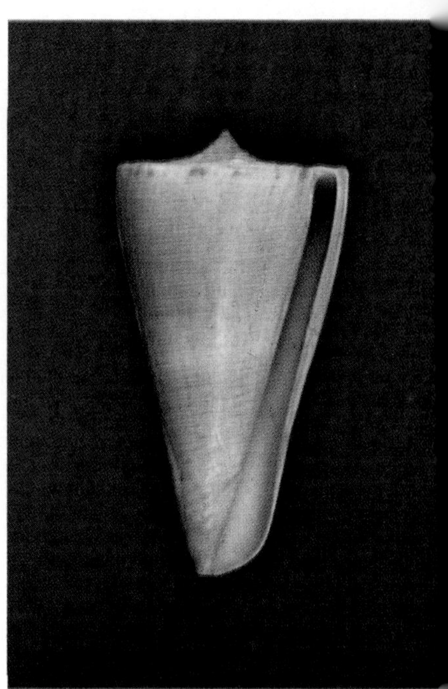

NIELSENAE.—Above: Moreton Bay, Queensland, Australia, 45.5mm (EP). *Below:* Left: Cape Bowling Green, Queensland, Australia, 32.1mm (MM); Right: Cape Keraunbren, Western Australia, Australia, 33.6mm (MG).

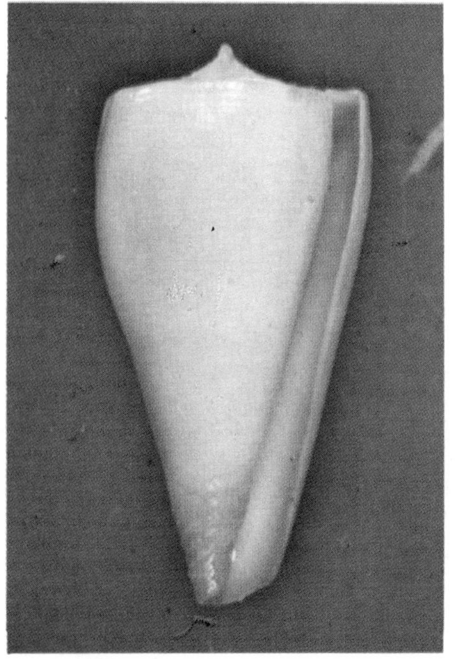

NIGROPUNCTATUS.—Sulu, P.I. *Above:* 27.2 and 31.3mm. *Below:* 28.2-30.9mm.

nodules absent or strong. As a general rule, with many exceptions, the paler shells have higher and more nodulose spires than the normal brown patterns. Deep-water variants may be very slender with strongly stepped whorls and granulose in addition.

Distribution.—Although once thought to be restricted to the Philippines, I have also seen specimens from the Ryukyus and from the northern Indian Ocean; unusually slender deep-water variants have been taken off Hawaii. In all likelihood *C. furvus* is widely distributed but very confused because of the complex variation and synonymy.

Synonymy.—Most of the names fall rather easily into one of the basic patterns. Since all these patterns intergrade and combinations may occur on single specimens, the names are obviously not worthy of recognition.

The names fall into the following patterns or close to them: One: *lignarius, buxeus;* two: *furvus, cecilei*; three: *crepusculum;* four: *striolatus, aegrotus;* five: *submarginatus, albus;* six: *granifer.*

Notes.—It may be hard to believe that the typical brownish *C. furvus* belongs to the same species as a solid white shell sold as *C. submarginatus*, but intergrades are very common and easy to obtain from any large unsorted shipment from the Philippines. Except for the granulose variant, which is not really rare but only uncommon, none of the variants are any less common in large series than the others. The most pleasing patterns are probably two-four, with the white shells often quite ugly. The spire is normally eroded or covered with calcium deposits. The lip chips easily in this species.

The slender deep-water variant from Hawaii probably belongs here; see Kohn, 1962, *Pacific Science*, 16(4): 356, as *C. granifer.*

GENERALIS Linnaeus, 1767

1767. *Conus generalis* Linnaeus. *Systema Naturae per Regna Tria Naturae*, ed. 12, 1(2): 1166 (East India). Holotype figured by Kohn, 1963.

1792. *Conus Maldivus* Hwass, in Bruguiere. *Cone*, in *Ency. Method., Hist. Nat. des Vers*, 1: 644 (Maldive Islands). Lectotype figured by Kohn, 1968.

1798. *Cucullus Dux* Roeding. *Museum Boltenianum:* 44 (Locality not stated). Lectotype figure (Kohn, 1975): Martini, Pl. 58, fig. 648. Non *Conus dux* Hwass, 1792.

1798. *Cucullus filosus* Roeding. *Ibid.:* 44 (Locality not stated). Lectotype figure (Kohn, 1975): Favanne & Favanne, Pl. 15, fig. C.

1798. *Cucullus locumtenens* Roeding. *Ibid.*: 45 (Locality not stated). Lectotype figure (Kohn, 1975): Martini, Pl. 58, fig. 645. Non *Conus locumtenens* Blumenbach, 1791.

1849. *Conus spiculum* Reeve. *Conchologia Iconica*, 1(*Conus* Suppl.): Pl. 7, sp. 266 (Cagayan, Mindanao, Philippines).

1863. *Conus spirogloxus* Deshayes. *Cat. Moll. de l'Ile de la Reunion:* 135, pl. 13, figs. 13-14 (Reunion).

1933. *Conus generalis Monteiroi* Barros e Cunha. *Mem. Estud. Mus. Zool. Univ. Coimbra*, (Ser. 1), 71: 79 (Locality unknown).

1937. *Conus generalis* var. *Regenfussi* Dautzenberg. *Mem. Mus. Roy. d'Hist. Nat. Belg.*, 2(18): 120, pl. 2, fig. 6 (Amboine).

1937. *Conus generalis* var. *subunicolor* Dautzenberg. *Ibid.*, 2(18): 120, pl. 2, fig. 7 (Amboine).

1937. *Conus generalis* var. *pallida* Dautzenberg. *Ibid.*, 2(18): 120, pl. 2, fig. 8 (Amboine).

Description.—Heavy, with a low to high gloss; elongate biconical, the sides straight and tapering to a narrow base; body whorl with a few low spiral ridges near the base, otherwise smooth except for sometimes conspicuous axial growth marks; shoulder broad, angulate to carinate, slightly concave above; spire tall if not eroded, deeply concave, turbinate; earliest 4-6 whorls with traces of small nodules; whorls with little or no sculpture except weak axial threads and traces of a single spiral ridge at the middle of the whorl; last 2-3 whorls concave above. Body whorl glossy white or chalky, usually with widely spaced spiral rows of small brown dashes or fine lines visible in some part of the pattern; two broad bands of orange, brown, or almost black covering most of body whorl except for clear bands at base, midbody and shoulder; brownish bands sometimes subdivided spirally by narrow white bands or axially by large and small white blotches and tents; heavy black axial flammules may be present in the white bands at shoulder, midbody, and base, but these seldom pass through the brown bands; base deep purplish brown, almost black; shoulder and spire white, with axial black flammules or spots. Aperture narrow, uniform in width; outer lip thin, straight; mouth white except where external pattern shows through, with a purplish blotch anteriorly. Columella long, narrow, dark like base. Length 40-90mm.

Comparison.—The elongate and strongly turbinate shape in combination with the blackish or violet base is very distinctive. Probably no one would ever have problems recognizing *C. generalis*. However, some specimens of *C. vitulinus* may have a body pattern similar to *C.g. generalis;* such shells have a low, non-concave spire and heavy black flammules covering most of the spire; the basal ridges are commonly granulose in *C. vitulinus*. *C. monile* is very similar in shape to *C. generalis* but differs in color; the base is usually salmon or pale violet, not almost black as in most *C. generalis;* there are large reddish or brownish dashes or lines over the body whorl, but these are conspicuous because there are no broad spiral bands of brown; the late whorls of the spire often appear convex.

Variation.—*C. generalis* is divisible into two distinct subspecies which must be discussed separately.

C. generalis generalis: Glossy, the spire tall in adults and usually not heavily eroded; body whorl covered with broad orange brown to almost black bands which are sometimes subdivided into narrower bands; heavy

NIMBOSUS.—*Above:* Left: Galle, Sri Lanka, 36.1mm; Right: Port Vila, New Hebrides, 39.2mm. *Below:* Left: Between Inhambane and Beira, Mozambique, 41.8mm (Meyer); Right: New Hebrides, 38.3mm (MM).

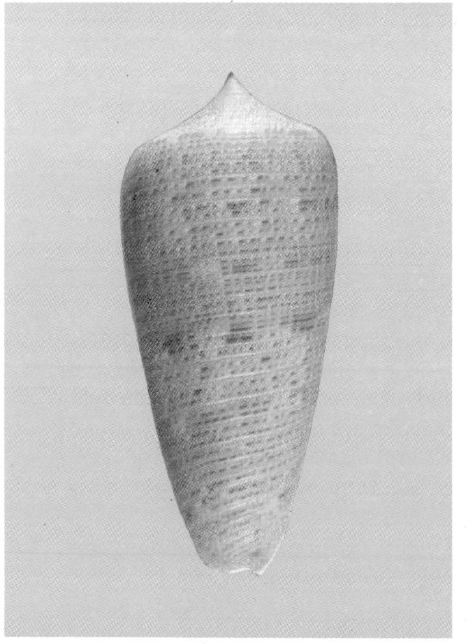

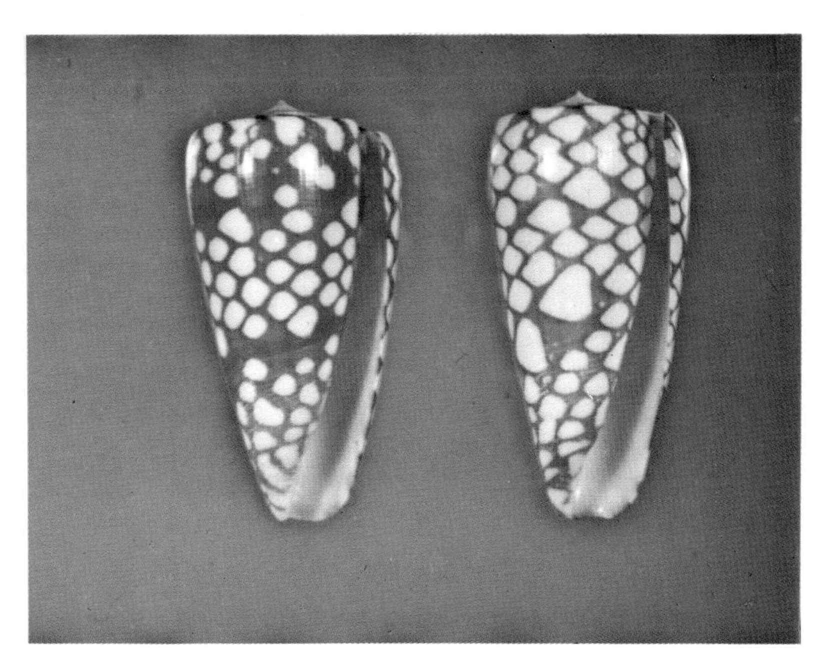

NOBILIS.—*Above* (*n. nobilis*): Balabac, Palawan, P.I., 43.1 and 43.9mm. *Below* (*n. marchionatus*): Marquesas, French Polynesia, 28.7 and 31.1mm.

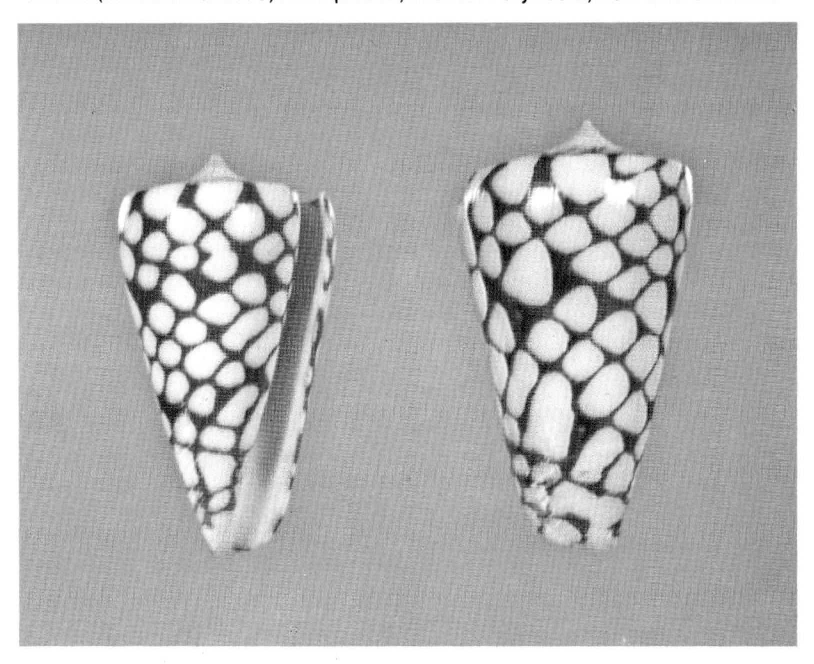

black axial flammules present, not passing through brown spiral bands so conspicuous only in white spiral bands at midbody, base, shoulder, and between subdivided brown bands; spiral rows of dashes usually very inconspicuous. Pacific Ocean.

C. generalis maldivus: Chalky, the spire lower and heavily eroded in even subadults; body whorl with conspicuous spiral lines and dashes partially covered with brown to orange-brown axial blotches which sometimes fuse into broad and almost complete spiral bands above and below midbody; the blotches usually have large and small white tents and blotches visible within them and do not hide the spiral rows of dashes; axial flammules are poorly developed or absent. Indian Ocean, intergrading with the nominate subspecies in the Philippines and perhaps Indonesia.

As a general rule, juveniles of both subspecies are more slender and have taller spires than adults. In *C. g. maldivus* the juveniles are glossy and have only a few small blotches on the body whorl other than the spiral lines of dashes.

C. g. generalis has a great assortment of patterns and tones of color; the bands are commonly subdivided or in some cases fused so the midbody band almost disappears. Many of these common variants have been named, but their recognition even at the varietal level seems unnecessary. In *C. g. maldivus* the brown bands may be almost continuous in adults or heavily broken with white, resulting in axial blotches with very irregular edges; even specimens with almost complete bands still have a few small white triangles visible and of course the spiral pattern of dashes can be seen.

Distribution.—Widely distributed and common through most of the Indo-Pacific in shallow water. *C. g. maldivus* ranges from the eastern African coast to the Bay of Bengal and often shows strong influence as far east as the Philippines and Indonesia. *C. g. generalis* occurs from the western Pacific east to at least Fiji, but seems to be absent from Hawaii and French Polynesia. *C. generalis* with very disrupted patterns occur in the Philippines and are interpreted as intergrades; these also have the black flammules reduced or absent.

Synonymy.—Almost all the synonyms are based on minor variations in *C. g. generalis*. Two names are based on juveniles: *C. spiculum* for *C. g. generalis* and *C. spirogloxus* for *C. g. maldivus*. Since the Dautzenberg names are still sometimes used, they are briefly defined: typical: heavy, long flammules, the bands moderately broad; *regenfussi:* two narrow bands, the midbody band very wide and with broken flammules; *pallida:* pale orange, otherwise about like typical except bands more subdivided; *subunicolor:* very dark, the bands broad and obscuring mid-body white band; these are all very minor variants, although some nearly black *subunicolor* are attractive shells.

Notes.—This familiar shell is usually perfect in the typical subspecies except for minor spire erosion in most populations; *maldivus* is a much rougher shell with a chalky finish, and heavy erosion of the spire is to be expected. Juveniles of either species are at least uncommon and not often seen. The lip is fragile in both species and easily broken. Minor axial growth flaws should be expected in most gem *C. g. generalis.* Of the several varieties, none really deserves a premium except perhaps nearly black *C. generalis generalis* var. *subunicolor.* Some very interesting Philippines specimens with bright orange bands on one side and nearly solid black on the other are occasionally available. Dwarf adults of the typical subspecies are found on certain New Guinea and Philippine islands.

GENUANUS Linnaeus, 1758

1758. *Conus genuanus* Linnaeus. *Systema Naturae per Regna Tria Naturae,* ed. 10, 1: 714 (Locality not stated). Lectotype figure (Kohn, 1963): Rumphius, Pl. 34, fig. G.
1798. *Cucullus papilio* Roeding. *Museum Boltenianum:* 42 (Locality not stated). Lectotype figure (Kohn, 1975): Martini, Pl. 56, fig. 623. Non *Conus genuanus papilio* Linnaeus, 1767.
1798. *Cucullus sphinx* Roeding. *Ibid.:* 42 (Locality not stated). Lectotype figures (Kohn, 1975): Martini, Pl. 56, figs. 624-625.
1811. *Conus fasciatus* Perry. *Conchology:* Pl. 24, fig. 3 (Locality not stated). Non *Conus fasciatus* Schroeter, 1803.

Description.—Moderately heavy, with a good gloss; low conical, the sides convex; body whorl smooth except for a few low spiral ridges at the base and sometimes heavy axial threads and growth marks; shoulder wide, round to roundly angled, flat or convex above, not very distinct from spire; spire low to moderate, sharply pointed, the sides straight to slightly concave; tops of whorls flat or slightly convex, covered with very fine spiral and axial threads, none especially distinct; early whorls eroded. Body whorl pinkish gray to bluish gray, usually with broad indistinct bands of olive above and below midbody; base whitish; whorl covered with about 10-20 widely spaced spiral rows of black dashes, bars, and dots, usually of at least two distinct widths, variously alternating; often the black bars alternate in a row with white bars containing small rounded black dots within; shoulder and spire whorls margined with row of alternating black and white dashes; spire like body whorl, with traces of indistinct axial olive blotches; early whorls eroded white. Aperture rather narrow above, wider basally; outer lip sharp, thin, evenly convex; mouth bluish gray to pinkish gray, usually pale; margin of lip often reddish brown. Columella rather long, narrow, slightly indented from inner lip. Length 35-71mm.

Comparison.—Few other cones have heavily dashed patterns over a rather pale body whorl. In *C. genuanus* the dashes are of particularly

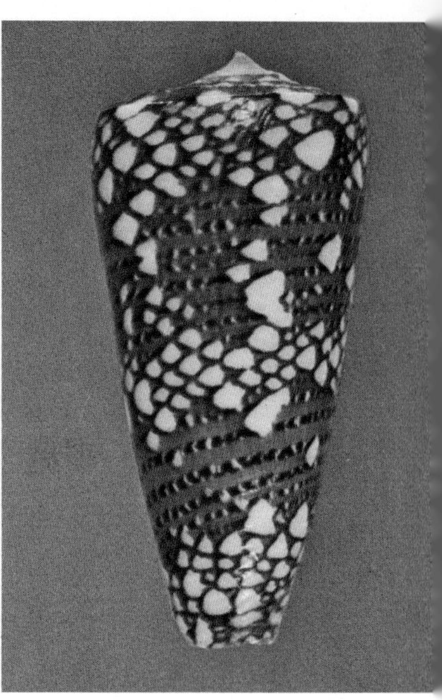

NOBILIS.—Above (*n. victor*): Left: Straits of Flores, Indonesia, 40mm (DM) (photo by Jayavad); Right: Nusa Tenggara Is., E of Bali, Indonesia, 35.6mm (EP). *Below:* Left (*n. victor*): Nusa Tenggara Is., E of Bali, Indonesia, 55.6mm (MM); Right (*n. lamberti*): Holotype from Souverbie, Pl. 2, fig. 7, Ovvea, 107mm.

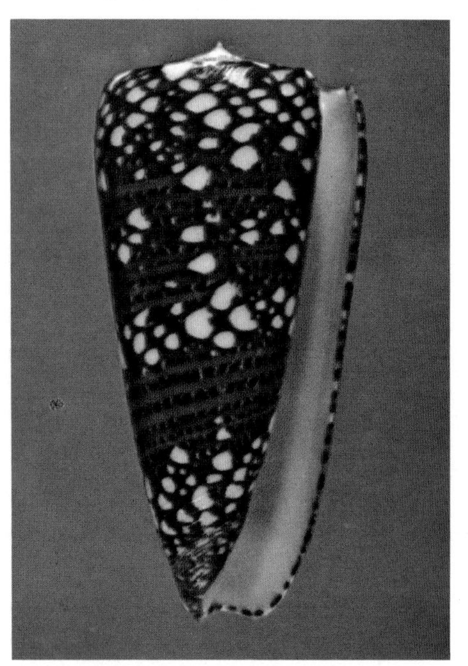

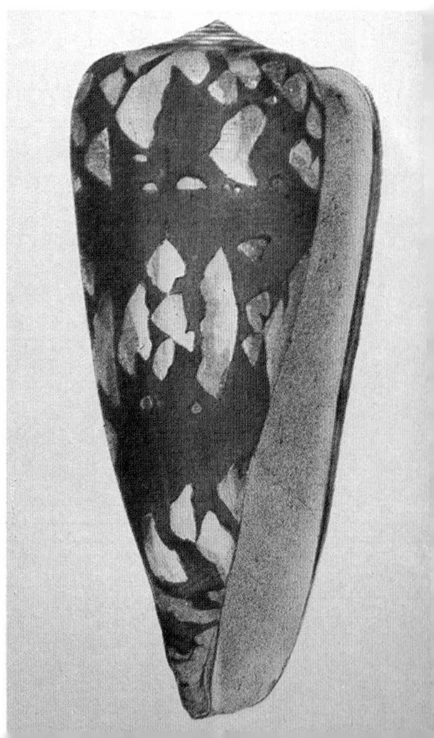

NUSSATELLA.—Above: Tahiti, Society Is., 38.3 and 40.8mm. ***Below:*** Left: Kona, Hawaii, Hawaii, 54.3mm; Right: Zanzibar, 56.2mm.

varied sizes, from small dots to large bars of black. *C. betulinus* may be somewhat similar in some patterns, but it is orange in color even in juveniles; patterns which have dashes instead of dots have the dashes of about the same size over the body whorl and have the spire marked with large black axial blotches. *C. taeniatus* is very similar at times but has a distinctly bluish tone, often sky blue; the spire whorls are weakly coronated or undulate, and the mouth is dark brownish purple in most specimens; adult *C. taeniatus* are usually smaller than subadult *C. genuanus*. *C. pulcher* and *C. byssinus* are white cones with the spiral banding much more varied and not uncommonly axially elongated; there are commonly orange bands in the background.

Variation.—There is some variation in the angulation of the shoulder and its width, with distinctly angulate shoulders less common than rounded shoulders. The intensity of background coloration varies considerably, from distinctly pinkish to distinctly bluish gray; usually the olive bands above and below midbody are very indistinct, as are the olive blotches on the spire. The pattern contains at least two sizes of dashes, some relatively small and squarish, others longer, wider, and bar-like; the basal ridges are usually covered with small dashes, and a few lines of small brown spots can usually be seen. The dashed lines around the spire whorls and shoulder are apparently constant and strongly developed. The mouth is darker in juveniles and adults with dark background colors.

Distribution.—Uncommon in shallow water along the western coast of Africa below the Bulge and on the offshore islands. Ranges south to Angola.

Synonymy.—The names are long associated with *C. genuanus*, although admittedly some of the old figures could just as easily be *C. betulinus* or even poor figures of *C. pulcher*.

Notes.—In fresh colors this is a very attractive cone, but it tends to fade somewhat. There are commonly small healed breaks over the body whorl, and some populations tend to be heavily eroded on the spire and posterior body whorl. Colorless axial growth marks sometimes detract from the appearance but are usually not conspicuous. The lip is fragile, especially in subadults.

GEOGRAPHUS Linnaeus, 1758

1758. *Conus geographus* Linnaeus. *Systema Naturae per Regna Tria Naturae*, ed. 10, 1: 718 (Indies). Holotype figured by Kohn, 1963.

1833. *Conus geographus* var. *rosea* Sowerby i, in Sowerby ii. *Conchological Illustrations:* Pt. 32, fig. 33 (Annaa I.). Non *Conus roseus* Fischer, 1807.

1843. *Conus intermedius* Reeve. *Conchologia Iconica*, 1(*Conus*): Pl. 23, sp. 129 (Annaa Island). Non *Conus intermedius* Lamarck, 1810, a fossil.

1858. *Conus mappa* Crosse. *Rev. Mag. Zool.*, (Ser. 2), 10: 200, 205. Nomen novum for *Conus intermedius* Reeve, 1843. Non *Conus mappa* Lightfoot, 1786.

1955. *Conus eldredi* Morrison. *J. Washington Acad. Sci.*, 45 (1): 32. Nomen novum for *Conus mappa* Crosse, 1858.

Description.—Very light in weight and thin in structure for its size, but solid and not fragile; gloss low or absent; rather ovately cylindrical, the sides nearly parallel and then suddenly constricted to a narrow, stem-like base; body whorl with a few very low and indistinct oblique ridges on the base, these seldom extending around the entire shell base, more like wrinkles than ridges; numerous axial threads over body whorl, sometimes conspicuous; shoulder wide, angled, with low and rounded coronations, concave above; spire low, the sides concave, the whorls stepped and with raised margins; spire whorls nodulose on the first 2-3, then becoming more distinctly undulate to coronate; tops of whorls concave, covered with about a dozen low and indistinct spiral ridges crossed by numerous fine axial threads. Body whorl creamy white, pinkish, or bluish white, heavily covered with a reticulated network of fine brownish lines from base to shoulder; reticulation may be distinct or very broken and appearing mottled; usually two spiral rows of large dark brown irregular blotches above and below midbody and smaller blotches below the shoulder and above the base; brown blotches sometimes with faint spiral rows of brown and white dashes visible within; inner lip usually without any pattern; base mottled; spire colored like body whorl, with large brownish blotches at the margins of the coronations and more-or-less axial; early whorls pink; coronations light, often with a small brown spot at their edge. Aperture very wide, at the base as wide as widest part of body whorl; outer lip thin, sharp, convex, not fragile; mouth bluish white to pale pinkish, the exterior pattern showing through. Columella long, narrow, strongly indented from inner lip. Length 70-153mm.

Comparison.—The very thin but solid structure and very wide aperture admit comparison only with the related *C. obscurus*, *C. tulipa*, and perhaps *C. cuvieri* and *C. vicweei*. The first two species are certainly very closely related, while the last two form a connecting link with the *bullatus* complex.

C. geographus is readily separated from *C. cuvieri* and *C. vicweei* by the distinctly coronate shoulder, reduction of spiral dashes in the pattern, and wider aperture basally. From *C. obscurus* it differs in having the late spire whorls distinctly undulate to coronate with the shoulder coronate (later whorls smooth to undulate in *C. obscurus*, the shoulder smooth), having few dashes in the pattern (dashes often heavy in *C. obscurus*), seldom being distinctly bluish, and being much larger. *C. tulipa* is a closer approach in size and in shoulder coronation, but it is more distinctly convex at the sides, has the shoulder only weakly coronate to undulate, has numerous large dashes in the pattern of most shells, and is usually distinctly bluish in tone; *C. tulipa* is somewhat thicker than *C. geographus*

OBSCURUS.—Above: Kona, Hawaii, Hawaii, 27.8 and 31.8mm. *Below:* Zanzibar, 38.3mm (MF).

OCHROLEUCUS.—Above: Boac, Marinduque, P.I., 55.6 and 61.6mm. *Below:* Batangas Bay, Luzon, P.I., 49.4-52.8mm (GG).

of similar size. Juveniles of *C. geographus* under 50mm in length have patterns like the adults and are similarly colored, with the shoulder distinctly coronate. Juveniles of *C. obscurus* are bright orange then change to bluish. Juveniles of *C. tulipa* are reddish violet with very heavy spiral rows of dashes and dots.

Variation.—The shape is constant, as is the sculpture. The pattern may be relatively open and regular or badly broken and heavily mottled and lineate; the blotches may be very irregular or fused into almost solid spiral bands. Dashes are seldom obvious, although they may be seen in the blotches.

The background color of the body whorl is usually creamy white, dirty whitish, bluish white, or even very pale tan. Occasionally, however, the shell assumes a distinctly pinkish or rosy tone, which appears to be an ecological variant and is not constant. There are no other differences between reddish specimens and normally colored ones, the usually stated differences (more slender, ridging at base more distinct) being normally variable in every population. I can see no reason to consider pinkish shells anything but individual variants.

Distribution.—Widespread in the Indo-Pacific and relatively common for a large carnivore. Not recorded from Hawaii. Shallow water.

Synonymy.—All the synonyms refer to the pinkish or reddish phase, which has had a complicated nomenclatural history. *C. eldredi* (the only valid name of the synonyms) is just an uncommon individual variant; I know of no evidence that it occurs in constant populations or ever occurs except in association with normal *C. geographus.*

Notes.—This large and familiar shell is extremely strong for its very thin structure. Along with *C. tulipa* and *C. obscurus* it has been placed in the genus *Gastridium*, which at first glance it seems to deserve. However, a study of the entire range of species of *Conus* reveals that this group of species is not as completely isolated as thought, being connected with the species clustered around *C. bullatus* through *C. cuvieri*, *C. vicweei*, and probably *C. cervus.* The relationship with *C. magus* and allies is obvious also.

It has been suggested on occasion that *C. obscurus* and *C. tulipa* are just variants or juveniles of *C. geographus*, but this is not possible as each species has distinctive juveniles of its own.

In *C. geographus* there are seldom major flaws, although occasional axial growth marks are found. The lip is sometimes chipped, but it is stronger than it looks and does not chip easily. Large specimens often have very mottled patterns and are not attractive. Juveniles and small adults are often more brightly colored and have sharply defined blotches. The spire is usually complete.

The pinkish variant is sometimes seen offered as *C. eldredi* at a large premium, although admittedly the only character by which it is identified is the rosy coloration.

GILVUS Reeve, 1849

1849. *Conus gilvus* Reeve. *Conchologia Iconica,* 1(*Conus* Suppl.): Pl. 6, sp. 255 (Saldanha Bay, South Africa).

Description.—Moderately light in weight, with a low gloss; low conical, the upper sides slightly convex to straight; body whorl with about 5-12 low, narrow spiral ridges at the base, these soon obsolete; grooves between ridges faintly axially striate; rest of body whorl with numerous often heavy axial threads and scattered spirals, plus often heavy growth marks; shoulder roundly angled, broad, not distinct from spire; spire low, the sides slightly convex, bluntly pointed; early whorls weakly undulate or rude, but with straight margins by the shoulder and last few whorls; tops of whorls flat or nearly so, covered with about 3-4 low, weak spiral ridges crowded onto the posterior half of the whorl, covered with numerous axial threads to produce a faintly cancellate effect; shoulder almost smooth. Body whorl light to dark olive brown with a faint bluish tone; bluish narrow spiral band at midbody and another one at the shoulder, these with indistinct margins; short brown lines and dashes usually present in background and often in midbody band, seldom conspicuous; base brownish, not much darker than rest of body whorl; spire light tan, paler than body whorl and with a faint bluish tinge; small reddish brown to blackish spots or oblique dashes usually widely scattered at shoulder and on spire margins, seldom conspicuous, usually only about 5-25 on most shells, sometimes absent but sometimes very regular. Aperture moderately wide, uniform in width; outer lip thin, fragile, straight or slightly convex; mouth pale bluish, with a narrow brown band at the margin. Columella very narrow, mostly internal. Length 25-40mm.

Comparison.—*C. gilvus* is very similar to both the dark phase of *C. concolor* and to *C. frigidus.* From *C. frigidus* it differs in having weaker spiral sculpture on the spire and weaker, non-granulose ridges on the body whorl; the band at midbody and shoulder is distinctly bluish, not yellowish as in *C. frigidus,* and *C. frigidus* lacks the small brown spots on the spire whorls and the dashes on the body whorl. *C. concolor* usually has straighter sides and lacks the distinct bluish bands at midbody and shoulder as well as the brownish spots on the spire whorls.

Variation.—The shape, color, and pattern seem to vary little in the specimens I have seen, although the spire may be almost flat or strongly convex. Spiral dashes seem to develop with growth as they are usually absent in the smallest specimens, but they may also be absent in large specimens. The dark spots on the spire may be very constant and equally spaced or may be represented by just one or two spots barely visible at the sutures or on the shoulder. The bluish bands are rather contrasty but have weakly defined margins. In some specimens the axial growth marks are darker than the rest of the body whorl, forming faint axial bars.

Distribution.—I have seen specimens only from New Guinea and the

ORBIGNYI.—*Above:* Off Tonkang, Taiwan, 54.1-59.6mm. *Below:* Left: Taiwan, 42.9mm; Right (*elokismenos*): N of Durban, South Africa, 56.0mm.

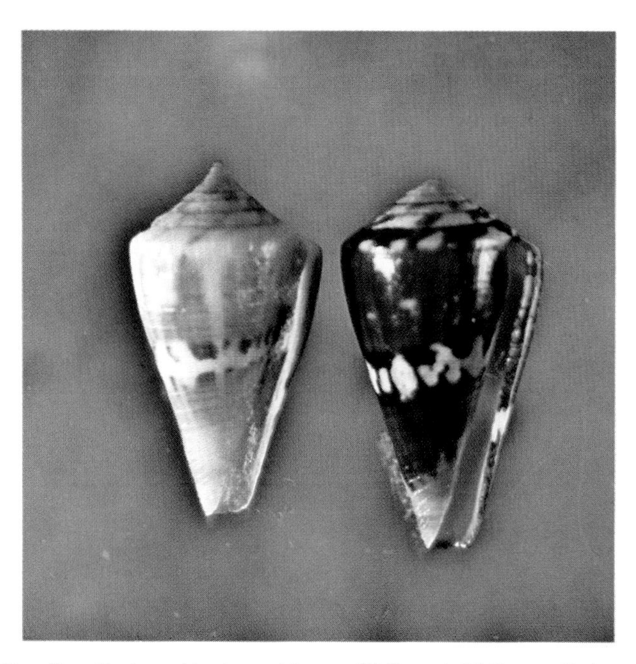

ORION.—San Carlos, Mexico. *Above:* 28.0 and 29.8mm. *Below:* 31.0-38.0mm (photos by A. Kerstitch).

Solomons, where it is rare or locally uncommon in moderately deep water. Probably more widely distributed in the western Pacific but confused with *C. frigidus* or even *C. flavidus*.

Synonymy.—*C. gilvus* has long been misapplied to an Australian cone related to *C. nielsenae* or considered a synonym of *C. tinianus* because of (incorrect) South African locality data given by Reeve. As treated here, the specimens I have seen agree very well with both the figure and description of Reeve in all respects, although only a few specimens have the spots regularly distributed on the shoulder as in Reeve's figure. *C. gilvus* as treated here is very close to *C. concolor* but the two seem distinct in pattern and some minor aspects of sculpture and shape; juveniles of the two species from the Solomons are quite different, the juveniles of *C. gilvus* being bright blue and bluish olive, with a good gloss, while those of *C. concolor* (from the same dredge sample) are more elongate, uniformly dark, and rather dull. I think they are distinct species.

Notes.—*C. gilvus* is a very rough species in the Solomons, at least, usually suffering from bad growth marks and unhealed breaks. The spire and shoulder are often eroded, and the lip is often chipped at the posterior margin. Large specimens have the color pattern very diffuse, although the bluish bands are still distinct.

This species has been sold under the name *C. pauperculus*, although it does not match the original figure of that name (here considered a doubtful synonym of *C. boeticus*).

GLADIATOR Broderip, 1833

1833. *Conus gladiator* Broderip. *Proc. Zool. Soc.* (*London*), 1833: 55 (Panama).

1865. *Conus tribunus* Crosse. *J. Conchyl.* (*Paris*), 13: 312, pl. 10, fig. 2 (California). Non *Conus tribunus* Gmelin, 1791.

?1882. *Conus evelynae* Sowerby iii. *Proc. Zool. Soc.* (*London*), 1882: 117, pl. 5, fig. 2 (Locality not stated).

Description.—Moderately light in weight, dull; low conical or obconical, the sides slightly convex; body whorl with several widely spaced spiral ridges above the base, usually almost smooth on posterior three-fourths of shell; numerous heavy spiral and axial threads on entire whorl, giving it a rough appearance; shoulder broad, angled, with low coronations; spire low to flat, bluntly pointed, the side concave; whorls slightly concave above, with 2-3 low and sometimes indistinct spiral ridges crossed by numerous curved axial threads. Body whorl whitish to pale yellowish tan, with weak to strong irregular axial brownish flammules above and below the usually clear midbody band; flammules may form broad irregular spiral bands or be almost absent; usually many short brownish dashes in the background, sometimes these conspicuous; base whitish; spire and shoulder tan to dark brown, the coronations whitish; tops of whorls may be uniform dark brown, light brown with dark brown spots between the coronations, or almost white with dark brown spots; early whorls pinkish. Aperture moderately wide, almost uniform or slightly wider anteriorly; outer lip sharp, slightly convex or straight; mouth white, sometimes with external pattern showing through as broad brownish bands. Columella indistinct. Length 20-37mm.

Comparison.—Although several small coronated Indo-Pacific cones resemble *C. gladiator*, these usually have violet or brownish bases or white shoulder bands. *C. gladiator* closely resembles *C. mus* of the western Atlantic, and the two may be distinct only at the subspecific level. As a general rule, *C. mus* has many spiral ridges over the entire body whorl, these usually absent in *C. gladiator*; in *C. mus* the spire has about 4 spiral ridges per whorl and they are usually more distinct than the 2-3 of *C. gladiator*; the spire of *C. mus* is usually taller and often has distinctly convex sides. Also, many *C. mus* have a bluish gray background color absent in most *C. gladiator*.

Variation.—The shape is rather constant, although some specimens are more slender than average; the spire is almost always low or even flat. The brown axial flammules may be very heavily developed or almost absent; occasionally the shell appears indistinctly mottled without any really distinct pattern. The spire may be very dark brown with strongly contrasting white coronations or mostly whitish with a few dark brown spots. Some specimens have the background distinctly bluish gray like typical *C. mus;* perhaps this type of specimen is the origin of reports of *C. mus* from the Pacific coast.

Distribution.—Common in shallow water from the Gulf of California south to Peru on the coast of western Central America.

Synonymy.—Although *C. tribunus* is certainly this species, it is quite possible that *C. evelynae* belongs elsewhere; it is more slender and has a higher spire than typical of *C. gladiator*, but it could be just an individual variant; perhaps it is a smooth *C. mus*.

Notes.—A common but very rough and unattractive little cone. Few specimens are anything to look at, the patterns being very irregular and often indistinct. Small growth flaws are common and to be expected.

GLANS Hwass, in Bruguiere, 1792

1792. *Conus glans* Hwass, in Bruguiere. *Cone*, in *Ency. Method., Hist. Nat. des Vers*, 1: 735 (Mauritius). Lectotype selected and figured by Kohn, 1968.

1807. *Conus violaceus* Link. *Beschr. Nat.-Samml. Univ. Rostock*, 3: 106 (Locality not stated). Based on Chemnitz, Pl. 143, fig. 1331(1). Non *Conus violaceus* Gmelin, 1791.

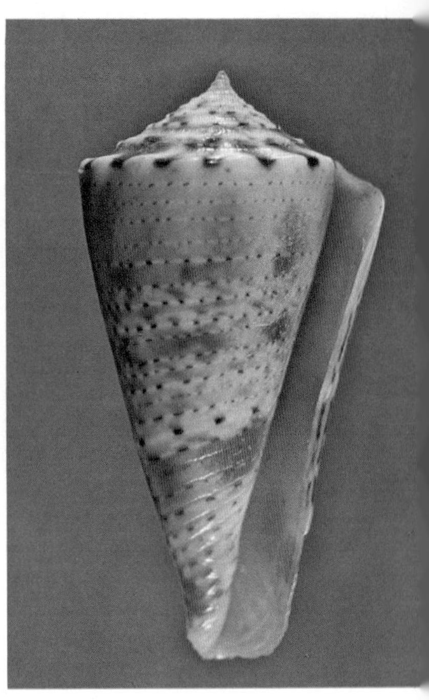

OTOHIMEAE.—Above: Off Kii Peninsula, Japan, 31.2mm (MM). ***Below:*** Left: Tiao-yu-tal I., Taiwan, 18.4mm (T.C. Lan); Right: Off Taiwan, 29.8mm (MM).

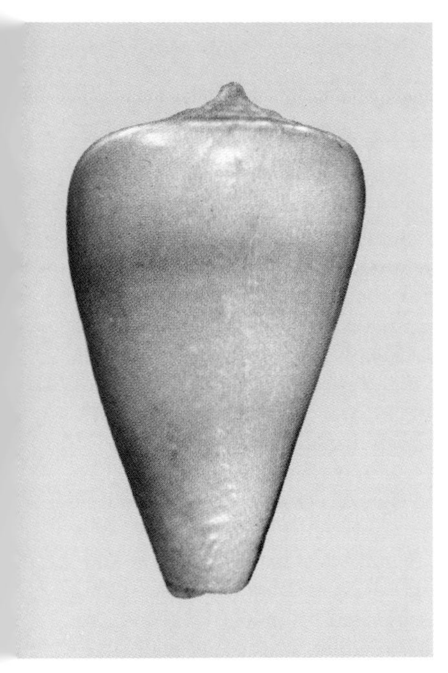

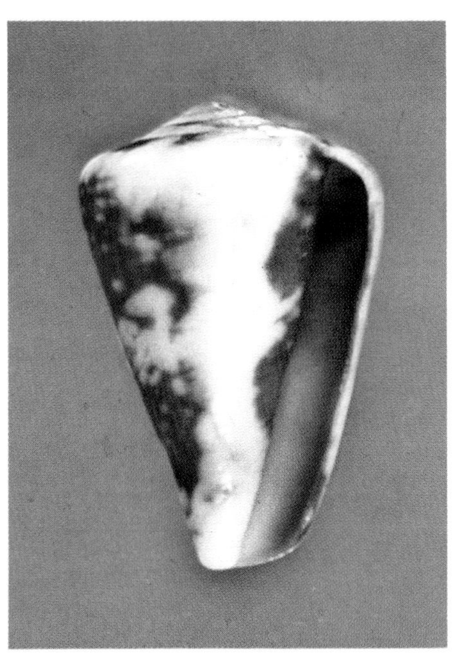

PAPILLIFERUS.—Above: Left: New South Wales, Australia, 21.2mm (GG); Right: Wooli, New South Wales, Australia, 21.9mm (GG). *Below:* Ilura, New South Wales, Australia, 31.3mm (JB).

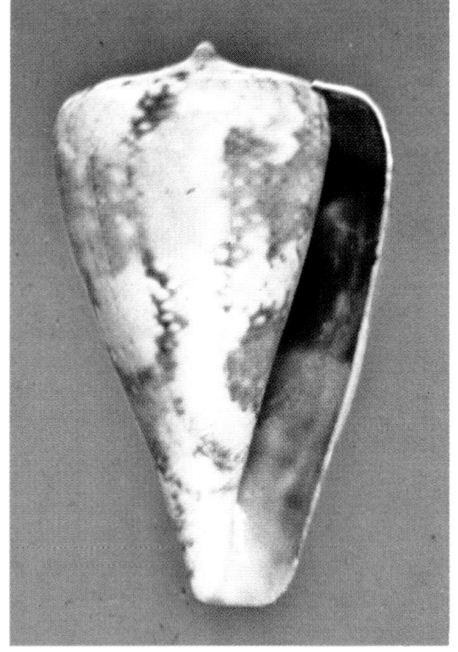

1937. *Conus glans* var. *tenuigranulata* Dautzenberg. *Mem. Mus. Roy. d'Hist. Nat. Belg.*, 2(18): 129, pl. 1, fig. 11 (Amboine).
1937. *Conus glans* var. *granulata* Dautzenberg. *Ibid.*, 2(18): 129. Based on *Tableau Ency. Method.*, Pl. 342, fig. 9. Non *Conus granulatus* Linnaeus.

Description.—Moderately heavy, thick, with a low gloss; cylindrical, the upper sides parallel, only the anterior third pinched in to the base; body whorl covered from base to shoulder with heavy, usually granulose, spiral ridges, these closely spaced and not noticeably weaker at midbody or shoulder; numerous axial threads; shoulder rounded, not distinct from spire; spire moderately high, sharply pointed, the sides convex; whorls with rather rude sutures, convex on top; first 4-5 whorls distinctly nodulose or undulate, later whorls undulate to smooth; tops of whorls with 4-5 heavy spiral ridges, these also on shoulder and much like those of body whorl; spiral ridges crossed by axial grooves to produce fine granules or a punctate appearance. Body whorl white to deep violet, with broad spiral bands of dark violet to dark brown above and below midbody; anterior band usually extends unbroken to base; posterior band often broken into two narrower bands separated by pale axial rectangles of body color; midbody band clear white to violet, crossed by wide axial bands of pale violet brown; base tinged violet; shoulder usually dark; spire violet to tan or sometimes whitish, with brown to violet squarish spots on the later whorls; early whorls whiter, earliest whorls pink. Aperture narrow posteriorly, wider anteriorly; outer lip sharp, thick, evenly convex; mouth uniformly dark violet. Columella narrow, long, mostly internal. Length 20-38mm.

Comparison.—*C. glans* is closely related to several very similar species which are sometimes hard to distinguish. It is often confused with *C. tenuistriatus*, but that species is more elongated, the sides not distinctly parallel; the body whorl appears a bit smoother, while the spire whorls are comparatively weakly striated and do not appear granulose; *C. tenuistriatus* is usually more uniformly tan or brown, not with the color in distinct spiral bands, while the aperture is paler violet. *C. scabriusculus* is basically a white shell with brown blotches tending to form spiral bands; the shape is distinctly bulbous below the shoulder, and the aperture has a broad white edge and often a white band at the middle; the entire shell is more weakly sculptured than that of *C. glans*. Less likely to be confused are *C. coccineus*, which has the shoulder distinctly undulate and a broad white midbody band as well as usually brighter colors, and *C. violaceus*, which is much more elongate and with a stronger axial and spiral pattern.

Variation.—When *C. tenuistriatus* is removed from *C. glans* as a full species, *C. glans* becomes a very sharply defined entity. As treated here, the major variation is in color, variable from brown to dark violet, the bands solid or broken by the more-or-less distinct axial bands above midbody. The entire basal third of the shell is dark. The heavy spiral sculpture of the body whorl extends almost full-strength over the shoulder and onto the later spire whorls. The sides are usually distinctly parallel, not convex.

Distribution.—Moderately common across the Indo-Pacific from eastern Africa to French Polynesia, but not recorded from Hawaii. Typically found in rather shallow water.

Synonymy.—*C. violaceus* Link could just as easily refer to *C. tenuistriatus*, although the violet color is usually better defined in *C. glans*. The two Dautzenberg varieties are very minor.

Notes.—This is a very solid shell that is seldom broken. The spire may be slightly eroded, and the posterior edge of the lip may be chipped on occasion. Calcium growths are not uncommon and are hard to remove because of the heavy sculpture. Brown and violet specimens are about equally common.

C. tenuistriatus has seldom been accurately separated from *C. glans* by dealers, so all collections should be checked.

GLAUCUS Linnaeus, 1758

1758. *Conus glaucus* Linnaeus. *Systema Naturae per Regna Tria Naturae*, ed. 10, 1: 714 (Asia). Based on Rumphius, Pl. 33, fig. GG.

1798. *Cucullus fraxinus* Roeding. *Museum Boltenianum:* 40 (Locality not stated). Lectotype figures (Kohn, 1975): Chemnitz, Pl. 138, figs. 1277-1278.

Description.—Heavy, thick, with a good gloss; pyriform, the sides very convex then slightly concave; body whorl with 10-15 narrow spiral ridges in the basal area, otherwise smooth except for numerous axial and spiral growth lines and threads, some conspicuous; shoulder rounded, not distinct from the spire; spire nearly flat or weakly convex, bluntly pointed; whorls smooth except for numerous crowded fine curved axial lines. Body whorl pale to dark bluish gray, the base and shoulder somewhat darker; covered with numerous short, closely spaced brown dashes alternating with inconspicuous white dashes; the dashes may form regular spirals or appear somewhat randomly spaced or even axial; the white dashes may be absent, but the brown dashes never form continuous spiral lines; a pale tan midbody band is often developed in larger specimens and may be conspicuous; shoulder and spire with broad black blotches alternating with lighter bluish gray areas, the blotches most heavily developed in larger specimens and sometimes extend onto body whorl as jagged axial flammules; base and apex sometimes brownish. Aperture moderately wide, slightly widened anteriorly; outer lip thick, often concave at middle and with a thickened ridge within; mouth bluish gray, darker in young specimens, the edge dark brown; sometimes mouth with indications of narrow light bands at middle and below shoulder. Columella long,

PARIUS.—Honiara, Guadalcanal, Solomon Is. *Above:* 29.9 and 31.4mm. *Below:* 30.9 and 31.8mm.

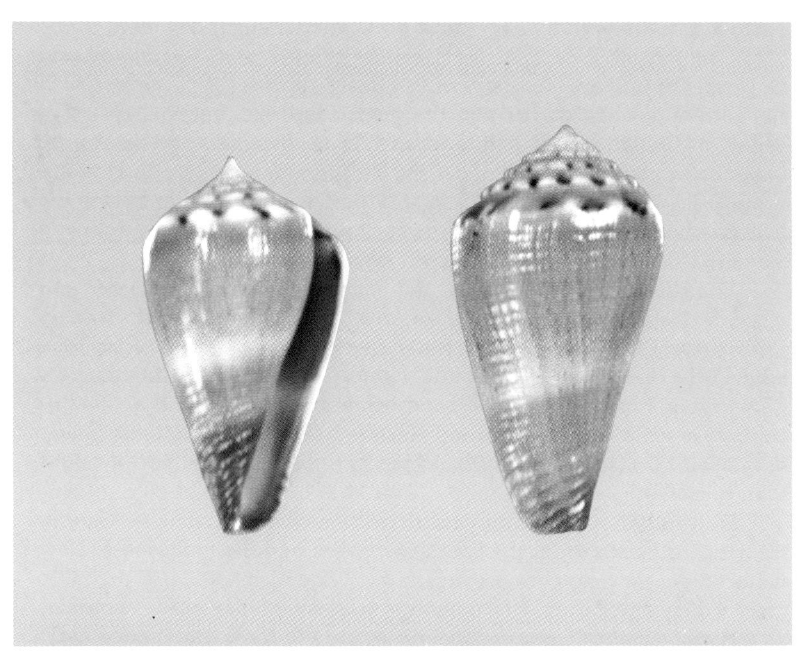

PARVULUS.—Above: Mackay, Queensland, Australia, 26.9 and 29.5mm. *Below:* Tayabas Bay, Luzon, P.I., 30.3 and 32.1mm.

narrow, usually with a heavy ridge posteriorly. Length 35-70mm.

Comparison.—*C. glaucus* is closely related to *C. betulinus* and *C. suratensis* in shape and pattern. It closely approaches *C. suratensis* in the pattern of small dashes and the small adult size, but that species is yellow or orange with a pale aperture; in *C. betulinus* the dashes are larger or reduced to squarish spots, the background is orange, and the size is much larger. *C. figulinus* is also very similar, especially in having the background color dark, but that species has the spiral lines complete and the spire almost unicolored, not heavily blotched.

Variation.—Specimens from the Solomons seem to be darker and more distinctly patterned than those from the Philippines, but this may just represent local populations. Small specimens tend to be darker than adults, the background becoming lighter with growth; intermediate growth stages have a brownish band below the shoulder and another at the base in some specimens. Juveniles have the black spire blotches poorly developed. The tan midbody band seems to be more distinct in adults than juveniles.

Distribution.—Apparently restricted to moderately shallow water in the western Pacific from the South China Sea and the Philippines, New Guinea, and the Solomons, as well as the Indonesian area; it is probably more widely distributed. Uncommon to rare, sometimes locally common.

Notes.—Except for occasional major growth flaws and eroded early whorls, *C. glaucus* is not a rough species. Fresh adults have a good gloss and bright colors. There are commonly axial colorless growth lines, but these do not detract from the appearance. The outer lip is sometimes thickened and distorted.

Although sometimes locally common in the Philippines and Solomons, this is not a common species over most of its range. It is sympatric with *C. suratensis*, and the two are not varieties of a single species.

GLORIAMARIS Chemnitz, 1777

1777. *Conus gloria maris* Chemnitz. *Besch. Berlin Ges. Naturf. Freunde*, 3: 321, pl. 8, fig. A (Locality not stated).

1801. *Conus gloria* Bosc. *Hist. Nat. Coqu.*, 5: 153 (Merdes Indes).

Description.—Heavy, with a good gloss; broadly cylindrical, the body whorl elongate, slightly convex posteriorly; body whorl smooth except for a few low spiral ridges at the base and spiral and axial threads and growth marks; shoulder moderately broad, rounded, not very distinct from spire; spire tall, broad basally, sharply pointed, the sides straight or nearly so; whorls slightly concave above, very weakly stepped; early whorls with fine nodules and 2-3 very weak spiral ridges; rest of whorls with faint traces of spiral ridges and weak axial threads. Body whorl white, bluish white, or pale golden tan, heavily covered from shoulder to base with fine axial zigzag lines of dull tan, golden brown, or reddish brown, these heavily overlapping to form many small tents of rather uniform size plus a few larger tents irregularly scattered over the body whorl; usually with three spiral rows of indistinct darker brown blotches with usually short and weak 'textile lines'; spire and shoulder heavily covered with fine tents and some dark brown axial lines much like body whorl, few large tents present; early whorls pale pink. Aperture rather wide, somewhat widened basally; outer lip sharp, rather thick, straight or slightly concave at middle; mouth white to pale bluish white. Columella long, narrow. Length 75-147mm.

Comparison.—Although this is a very distinct shell, it is often confused with such smaller species as *C. textile*, *C. telatus*, or even *C. neptunus*. From the first it is readily distinguished by the much more elongate shape and higher spire, the much heavier and finer tenting, and the nearly straight sides of the body whorl. *C. telatus* has the margins of the spire whorls raised and usually undulate, the body whorl usually distinctly spirally ridged, and the brown body whorl blotches larger and with longer 'textile lines.' *C. neptunus* and *C. lienardi* have stronger sculpture on the spire, more open patterns on the body whorl without distinct 'textile lines,' and usually obvious spiral grooves to about mid-body.

C. aureus, especially the form *paulucciae*, is sometimes more similar than the above species because the shoulder is wider and the spire taller, but that species is readily distinguished by the longer and heavier 'textile lines' which often extend over most of the length of the body whorl; the pattern is usually also much darker and more diffused; and the body whorl is covered with more-or-less distinct spiral ridges.

The closest relatives of *C. gloriamaris* are *C. bengalensis* and *C. milneedwardsi*. *C. bengalensis* has the spire somewhat higher, narrower, and a bit more stepped, with the body whorl more narrowly shouldered and straighter-sided. The tent pattern is more open and regular, with more scattered larger tents; the brown blotches are often reduced in size and axially elongate, the 'textile lines' often indistinct; on the spire the tents are large and white, separated by almost equal sized brown blotches. *C. milneedwardsi* is even taller spired and more strongly stepped, the sides of the body whorl often distinctly concave at the middle; the pattern of the body whorl is very open, the white tents often very large, and the brown blotches sometimes totally absent; the low-spired variant of *C. milneedwardsi* from the China Sea has the pattern like typical African specimens or even more open and has a similar narrow body shape.

Variation.—Considerable variation in background color and brightness of the pattern is obvious in different populations. The whole shell may appear very dull, as if covered with a faint golden glaze which diffuses the brown lines into a dull tan over a pale golden tan background; in more typical populations the background is white with well defined brown or pale reddish brown lines. In the most attractive specimens the

PATAE.—Above: Left (holotype): Off Pompano Beach, Florida, 24.3mm; Right (paratype): ½ mi. off Lauderdale-by-the-Sea, Florida, 24.5mm. *Below* (variant paratype): Ocho Rios, N coast of Jamaica, 20.7mm (all DMNH).

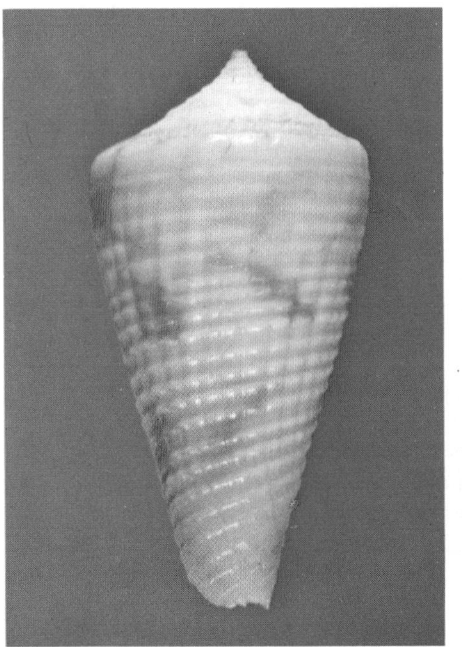

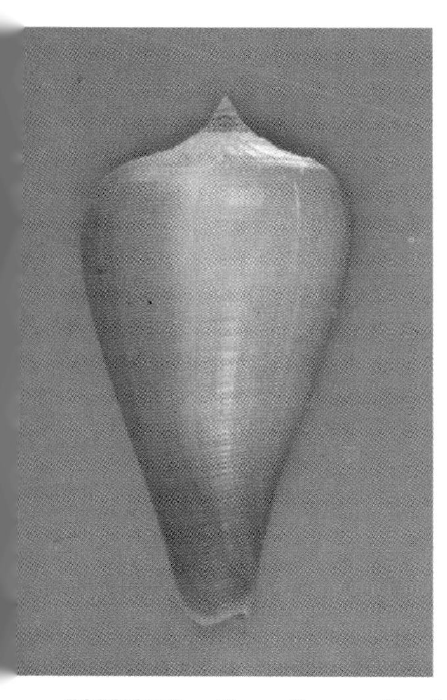

PATRICIUS.—Above: Panama Canal Zone, 44.6mm. *Below:* Left: Panama Canal Zone, 63.4mm; Right: Manzanillo, Colima, Mexico, 62.3mm (photo by A. Kerstitch).

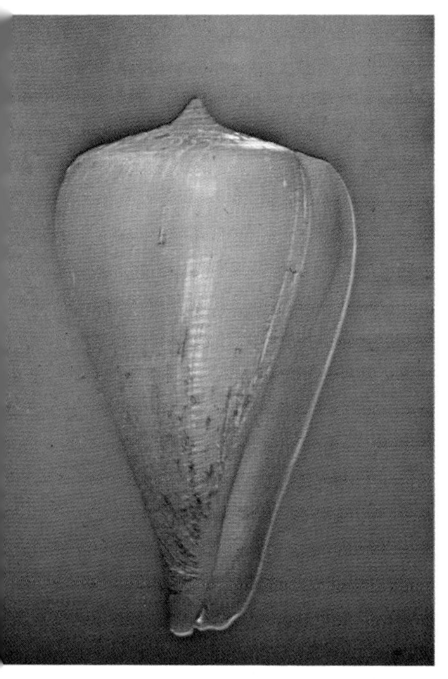

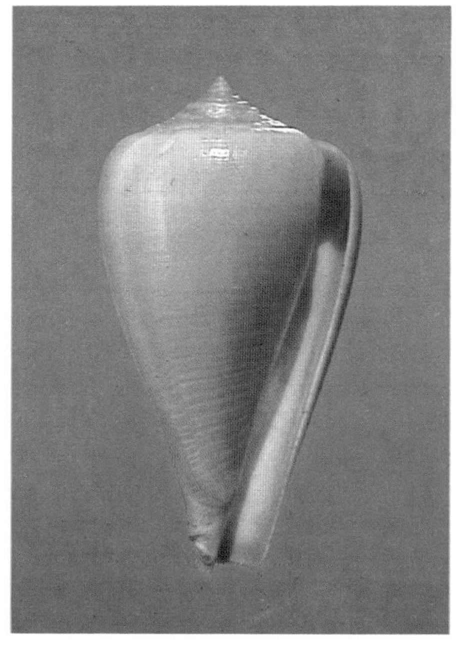

brown tenting is extremely sharply defined and bright chestnut, with a high contrast to white or pale bluish white tents. At the other extreme are specimens partially or entirely bluish in tone. The brown blotches and 'textile lines' may be sharply defined and squarish or occasionally almost fused spirally or they may be very indistinct and lost in the background; the blotches above and below midbody are usually present, but the band below the shoulder may be indistinct or absent. Width of the shoulder varies, as does its angulation, but this seems to be at the individual or local population level only.

Distribution.—Moderately rare but widely distributed through the western Pacific. Most specimens now come from the Philippines and Solomons, but sizable populations have also been found in New Guinea. Also recorded (sometimes doubtfully) from Flores Islands, Ceram, Fiji, and Yap Island (Western Carolines). Probably much more widely distributed, but populations appear to be local and of limited size in all cases. Found in deep water as the rule, but occasionally alive in shallow water after storms.

Synonymy.—This is the only Chemnitz name considered binomial, all the others proposed by him being invalid under modern rules. *C. gloria* is probably just an error in spelling. Note that *C. gloriamaris* to several early authors was just *C. textile*, not the real *C. gloriamaris* of Chemnitz.

Notes.—Specimens are often distinctly ugly for such a rare and famous shell; they have subdued patterns and may be covered with heavy patternless growth marks over the body whorl. Commonly the lip is chipped and heavily filed, and there are often bad and unhealed breaks on the body whorl and later whorls of the spire. Such specimens are, in my opinion, not suitable for collections and certainly not worth the inflated prices being asked. In the last few years *C. gloriamaris* has been collected regularly in the Philippines and often but irregularly in the Solomons. Hundreds of specimens are now on the market, many of which are true gems or at least very fine in condition with bright and distinct patterns. The price has dropped from about $1500 for a 100mm specimen in 1975 to under $800 at the time of writing, with smaller specimens considerably cheaper. Caution is suggested in purchasing this species until the market becomes more stable. This species would appear to be much more common than usually suggested, but very local in distribution in any area; once a colony is depleted, collectors must move to another colony, if one can be found.

The bite of this species is probably very dangerous because of the size of the radular teeth; caution is advised when dealing with living specimens.

The relationship of this species with *C. bengalensis* and *C. milneedwardsi* is obviously close. It and *C. bengalensis* are allopatric, while it now seems that *C. milneedwardsi* occurs in several populations in both the Indian and Pacific Oceans and might eventually be found sympatrically with both allied species. Some China Sea specimens of *C. milneed-*

wardsi have a spire no higher or more stepped than that of *C. gloriamaris.*

GRADATUS Wood, 1828

1828. *Conus gradatus* Wood. *Index Test., Suppl.:* 8, pl. 3, fig. 6 (California).

1832. *Conus scalaris* Valenciennes, in Humboldt and Bonpland. *Voy. Inter. Amerique—Recueil Observ. Zool. et Anat. Comp.,* 2: 338 (Acapulco).

1833. *Conus regularis* Sowerby i, in Sowerby ii. *Conchological Illustrations:* Pt. 26, fig. 29 (Locality not stated).

1833. *Conus Syriacus* Sowerby i, in Sowerby ii. *Ibid.:* Pt. 36, fig. 45 (Locality not stated). Non *Conus syriacus* (Roeding, 1798).

1833. *Conus dispar* Sowerby i, in Sowerby ii. *Ibid.:* Pt. 37, fig. 57 (Locality not stated).

1833. *Conus monilifer* Broderip. *Proc. Zool. Soc. (London),* 1833: 54 (Salango).

1834. *Conus ferrugatus* Sowerby i. *Ibid.,* 1834: 19 (Gulf of California and Guaymas Is.).

1843. *Conus gradatus* Reeve. *Conchologia Iconica,* 1(*Conus*): Pl. 25, sp. 140 (Salango). Non *Conus gradatus* Wood, 1828.

1854. *Conus angulatus* A. Adams. *Proc. Zool. Soc. (London),* 1853: 118 (Locality unknown).

1939. *Conus magdalensis* Bartsch and Rehder. *Smithsonian Misc. Coll.,* 98(10): 11, pl. 1, figs. 5-9 (Magdalena Bay, Baja California).

1955. *Conus gradatus thaanumi* Schwengel. *Nautilus,* 69(1): 15, pl. 2, figs. 12-13 (Bahia Salinas, Costa Rica).

1955. *Conus recurvus helenae* Schwengel. *Ibid.,* 69(1): 15, pl. 2, figs. 14-15 (Curu, Gulf of Nicoya, Costa Rica).

Description.—Moderately heavy, with a good gloss; low conical to strongly biconical, the sides usually straight or nearly so, sometimes weakly concave at the middle; body whorl with about 6-15 broad, low spiral ridges above the base, these seldom conspicuous or reaching mid-body; shoulder broad, angulate, sometimes carinate, seldom more than slightly concave above; spire low to very high, sharply pointed, the sides concave; first 2-3 whorls weakly nodulose; spire whorls flat to weakly concave above, often carinate, sometimes strongly stepped; tops of whorls covered with crowded axial curved threads and occasional traces of very fine spiral ridges (generally absent). Body whorl white to creamy or very pale tan, the usual pattern consisting of many spiral rows of squarish reddish brown to orange or yellowish spots; these are often covered by irregular axial flammules of reddish brown which may be poorly developed or very heavy and obscuring the spots; occasionally with broad spiral bands of yellowish tan between rows of spots and over them; base

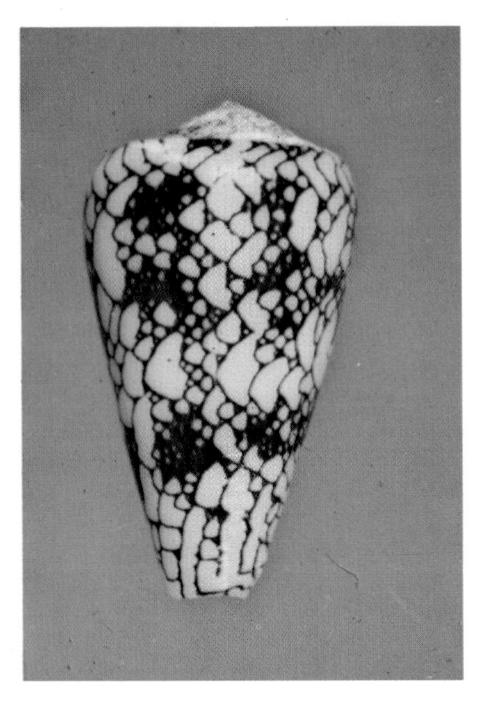

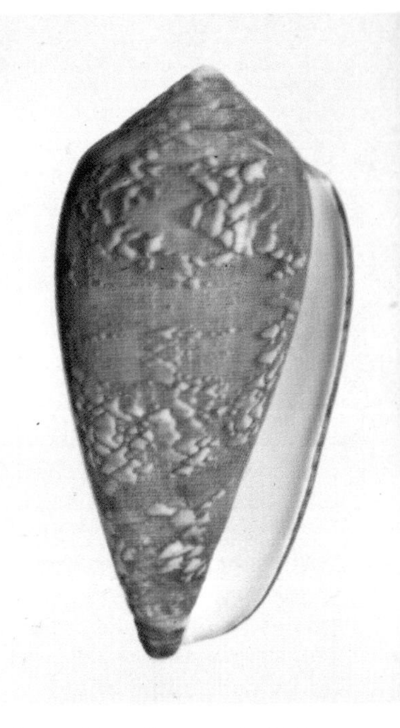

PENNACEUS.—Above: Left (*episcopus*): Kona, Hawaii, Hawaii, 26.7mm; Right (typical): Kenya, 53mm (DM) (photo by Jayavad). *Below* (*episcopus*): Left: Zanzibar, 58.9mm; Right: Maldive Is., 48.0mm.

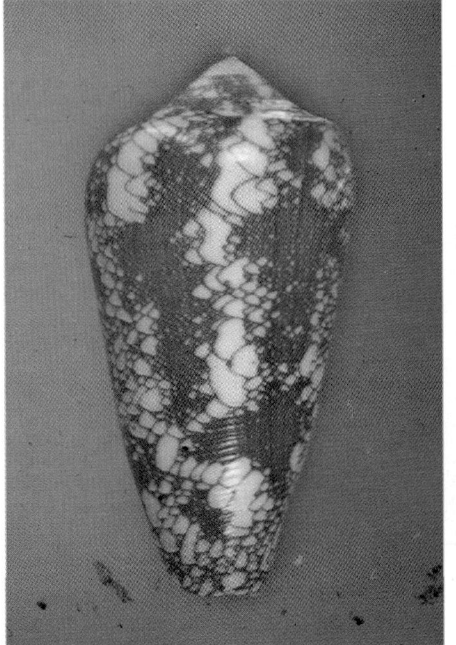

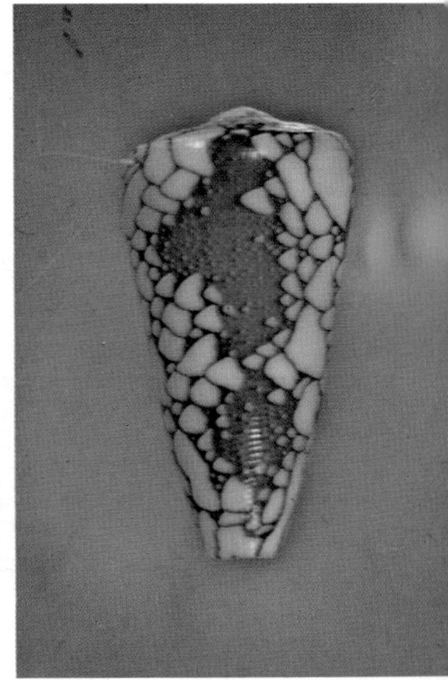

PENNACEUS.—Above: Left (*sindon*): Okinawa, Japan, 21.1mm (TCG); Right (golden *omaria*): Mahebourg, Mauritius, 26.8mm (EP). Below (*omaria*): Left: Ono, Okinawa, Japan, 42.8mm; Right: Kwajalein Atoll, Marshall Is., 51.6mm.

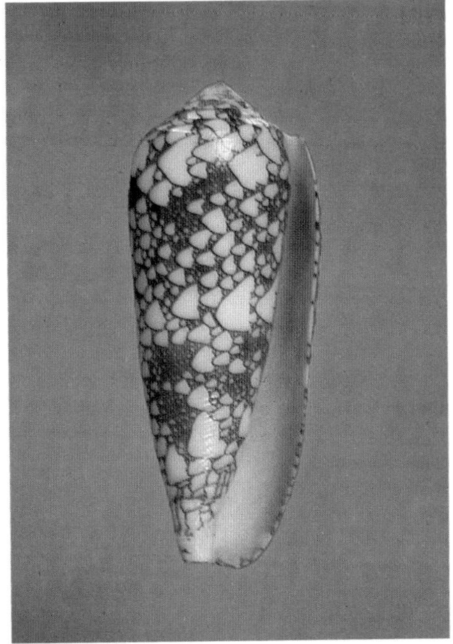

whitish; usually a pale midbody area that is weakly marked; spire with irregular reddish brown axial lines and blotches on white. Aperture moderately narrow, uniform in width; outer lip thin, straight or concave at the middle; mouth white to bluish white (paler and elongate shells) to heavily stained with brownish violet. Columella not visible externally. Length 30-84mm.

Comparison.—Although the great variability makes it difficult to compare this species with others, it is usually distinctive because the sculpture is very weak, the shape is elongate, and the pattern has the typical large squarish spots visible through the covering flammules and bands. The closest resemblance is to *Conus kerstitchi*, which is admittedly so close that it could be just an ecological variant. However, in *C. kerstitchi* the shape of the body whorl is quite distinctive, being wider at the shoulder and more sharply concave anteriorly than in any *C. gradatus* variant; the spire is also very concave and tall, a combination not usually seen in *C. gradatus*. Finally, the pattern of *C. kerstitchi* has the spots much smaller and has distinct (usually) zones of salmon at the shoulder and base (absent in all *C. gradatus* known to me). The variant of *C. gradatus* found with *C. kerstitchi* is a modified *scalaris*-type with broad pale yellowish tan bands.

Also very similar but fully distinct is *C. recurvus*, which is a much more fragile shell with deeply concave shoulder and spire whorls and the posterior edge of the lip sloping strongly below the level of the shoulder. The patterns are usually distinct, but some *scalaris*-types are covered with similar wavy axial stripes.

Such species as *C. poormani* and *C. virgatus* could be vaguely similar in shape and pattern to some variants of *C. gradatus*, although the similarity is never marked and should cause no confusion. I can think of no non-Panamic species which might be confused readily with *C. gradatus*.

Variation.—As indicated, *C. gradatus* is highly variable and has been split into many shape and pattern forms which are often considered full species (as they are sympatric, they cannot be subspecies). However, most of the common variation can be easily grouped into two extremes of shape and three basic patterns.

Shape: The spire is normally moderately high, concave, whorls weakly stepped or not stepped, and the shoulder is broad. The common variant type (form *scalaris*) has the spire whorls strongly stepped, sometimes slightly concave above, with the sides of the spire almost straight, the shoulder being relatively narrow.

Pattern: The typical pattern (1) (not common) is one in which the shell is covered with large pale reddish brown dash-like spots, often separated by rows of narrow dashes, the whole being overlaid with heavy orangish axial flammules which are rather regular and produce the appearance of large white spots where they fail to intersect the spiral rows of spots; this pattern is usually found in association with a quite low spire. The most common pattern (2) (found in both low-spired and *scalaris* shells) is of dark reddish brown spots variably covered with very irregular axial flammules of reddish brown to yellowish brown, these sometimes so short as to appear blotch-like or heavy enough to almost conceal the spots. The third pattern is usually associated with the *scalaris*-type of shape and consists of usually broad spiral broken bands of reddish brown to pale yellowish tan over rows of spots which are commonly narrower and more dash-like than typical; in this pattern the shell can be mostly yellowish tan with only a few spots showing through.

Occasional specimens are very waxy and have a high gloss. This situation is most likely to be found in the third pattern type and in small very elongated shells which are commonly associated with this species. The small shells are moderately high-spired, have an elongate body whorl, are very waxy, have no obvious spots showing, and are covered with a nearly solid pattern of broad pale brownish bands sometimes relieved by a paler midbody band; such shells closely resemble *C. anabathrum*, a doubtful name.

Distribution.—Common in shallow to rather deep water from the Gulf of California south to Peru; also found on the offshore Panamic islands.

Synonymy.—Many authors have suggested that *C. regularis*, *C. scalaris*, and *C. gradatus* might possibly be synonymous, and examination of an assortment of variants proves that all three forms intergrade broadly in shape, pattern, and coloration. Thus the high-spired type can occur in almost any pattern that the more typical lower-spired form does (*regularis*). The high-spired *scalaris*-type can also occur in pattern three, which is not usually found in the common *regularis*, but pattern three is uncommon at best. The typical *gradatus* is uncommon but not at all rare; the figure in Wood is small but easily recognizable as the first pattern.

Variant of *C. gradatus* often called 'dispar' or 'monilifer,' La Paz, Mexico, 31mm (EP).

Because the species is so variable, many will find it more convenient to use varietal names for the three major types: *gradatus* (low spire, pattern one), *regularis* (low to moderate spire, pattern two), and *scalaris* (high spire, pattern two or three). Of the synonyms, they fall most closely

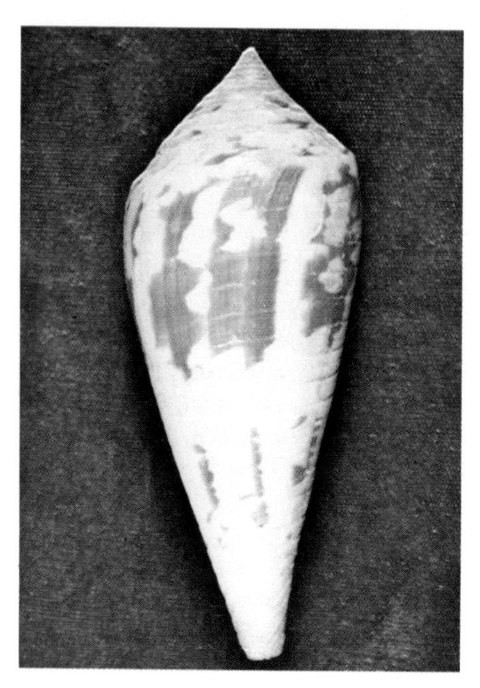

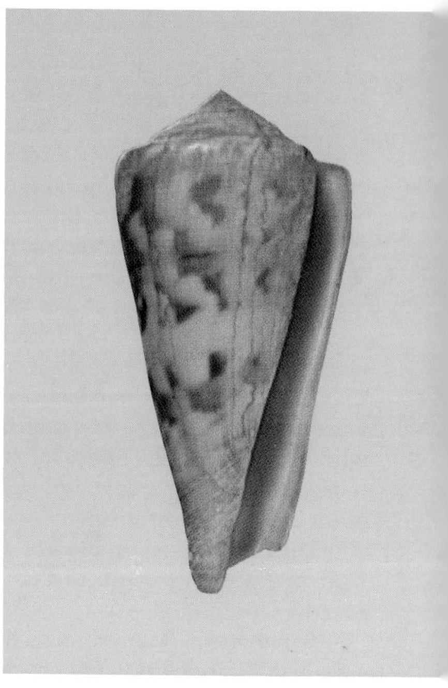

PERGRANDIS.—*Above:* Left (holotype of *C. fletcheri*): SE of Penghu Is., Taiwan, 105mm (photo by Ken Lucas at Calif. Acad. Sci.); Right (holotype of *C. tamikoae*): Eastern China Sea, 80mm (photo by Dr. T. Shikama). *Below:* S. Penghu Is., Taiwan, 133mm (photos by P. Clover).

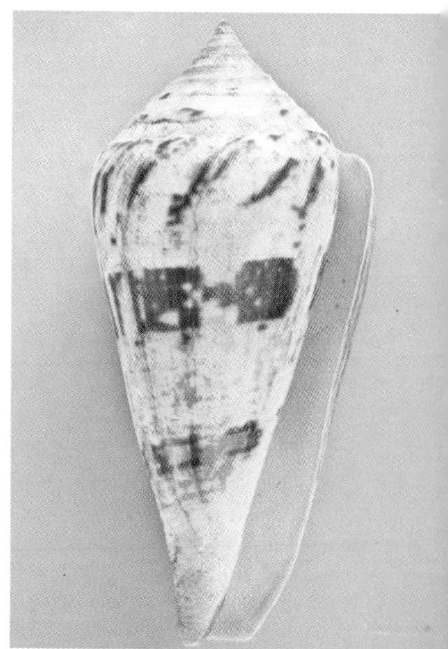

***PERONIANUS.*—**Thorny Pass, Port Lincoln, South Australia, Australia, 53.2 and 61.3mm.

into the following types: *syriacus—regularis; dispar—regularis; monilifer—regularis; ferrugatus—regularis; gradatus—scalaris; angulatus—regularis; magdalensis—regularis* (unusual); *thaanumi—regularis* (unusual); *helenae—regularis* (unusual). The last three names are based on rather blunt-spired shells with deeply concave spires, the body whorl being variously covered with an open reticulation of brown over the spots; *helenae* is strongly axially banded. *Dispar* is commonly used for the small shell with a waxy texture and reduced spotted pattern, probably incorrectly.

Notes.—I can see only one alternative to combining all the above names into a single species: the recognition of at least 10 poorly defined and broadly overlapping full species; certainly just three or four specific names cannot cover all the variation shown in the figures if one is a splitter.

The *scalaris* types and *gradatus* typical commonly sell for minor premiums. Pattern three is quite uncommon. Minor chips in the lip and eroded early whorls are to be expected, as are minor and major growth flaws.

GRANGERI Sowerby iii, 1900

1900. *Conus Grangeri* Sowerby iii. *Ann. Mag. Nat. Hist.*, (Ser. 7), 5: 441, pl. 11, fig. 5 (Locality unknown).

?1935. *Conus pseudosulcatus* Nomura. *Sci. Rpts. Tohoku Imp. Univ. Sendai*, 18(2): 108, pl. 7, fig. 3 (Taiwan). Fossil.

Description.—Heavy, with a low gloss; low conical, the upper sides distinctly convex and pinched-in to the base; body whorl with numerous broad, heavy, rounded or flat spiral ribs which are evenly spaced over the entire whorl, sometimes more crowded near the base; grooves between ribs shallow, usually smooth, with axial threads, or with weak, beaded ridges; major ribs usually smooth but occasionally granulose; shoulder roundly angled, weakly coronate or undulate; spire low to moderate, the sides slightly convex to straight or very slightly concave; sharply pointed; whorls concave above, with 4-6 distinct spiral ridges crossed by strong axial threads to produce a weakly cancellate appearance; spire whorls all weakly coronate to undulate. Body whorl and spire creamy white, sometimes very pale tan, with scattered reddish brown spots and interrupted rows of brownish dashes; usually the spots and dashes consolidate into two distinct brownish spiral bands, one above midbody and one immediately below the angle of the shoulder; occasionally there are weakly defined axial brown flammules between the two bands; base usually unmarked, while anterior half of body whorl is very weakly marked; spire with pale brown dots between coronations and sometimes faint brown lines on the earlier whorls. Aperture rather wide, uniform; outer lip thin, convex; mouth white. Columella long, narrow. Length 40-62mm.

Comparison.—The general appearance and pattern are very similar to some patterns of *C. mucronatus*, but in that species the spire whorls and shoulder are straight and not coronated. *C. sulcatus* is very similar also in shape and has similar sculpture in some variants; however, *C. sulcatus* usually has distinctly straighter and less convex sides, more concave spire, and lacks brown spots although it may be almost entirely brown. I have seen no intermediate of any type and think *C. grangeri* to be a quite distinct species.

Variation.—Although usually rather bulbous posteriorly, some specimens are just weakly convex and relatively slender. The spire may be distinctly convex or distinctly concave, with the height varying considerably. The ribs may be flat or rounded, smooth or lightly granulose, the grooves between with a finer ridge, axially threaded, or smooth, much the same variation seen in *C. sulcatus*. The pattern may be mostly spotted with faint traces of indistinct axial flammules posteriorly or with strong spiral bands; the pattern seems to be consistently weak anteriorly.

Distribution.—A poorly known species of deep water, but so far recorded definitely from the China Sea off Taiwan and from the Philippines.

Synonymy.—It seems as though *C. pseudosulcatus*, a Taiwanese fossil, is very similar to *C. grangeri* and probably the same species. The name is often applied to the Taiwanese variant of *C. mucronatus*, but it does not belong there.

Notes.—This uncommon species is apparently subject to damage frequently, as most specimens have large or small breaks on the body whorl or spire. Commonly the thin lip is chipped. The spire is often covered with asphalt. Specimens with two distinct spiral brown bands posteriorly are not uncommon and are nicer-looking than those with indistinct patterns.

The coronated spire in combination with the convex sides make this an easy species to recognize.

GRANULATUS Linnaeus, 1758

1758. *Conus granulatus* Linnaeus. *Systema Naturae per Regna Tria Naturae*, ed. 10, 1: 716 (Jamaica). Holotype figured by Kohn, 1963.

1791. *Conus laetus* Gmelin. *Ibid.*, ed. 13, 1: 3391 (Locality not stated). Lectotype figure (Kohn, 1966): Lister, Pl. 760, fig. 5.

1792. *Conus verulosus* Hwass, in Bruguiere. *Cone*, in *Ency. Method., Hist. Nat. des Vers*, 1: 719 (Locality not stated). Lectotype figure (Kohn, 1968): *Tableau Ency. Method.*, Pl. 341, fig. 7.

1798. *Conus Antillarum* Roeding. *Museum Boltenianum:* 47 (Locality not stated). Lectotype figure (Kohn, 1975): Knorr, Pl. 6, fig. 5.

1807. *Conus roseus* Fischer. *Museum Demidoff*, 3: 133 (Locality not stated).

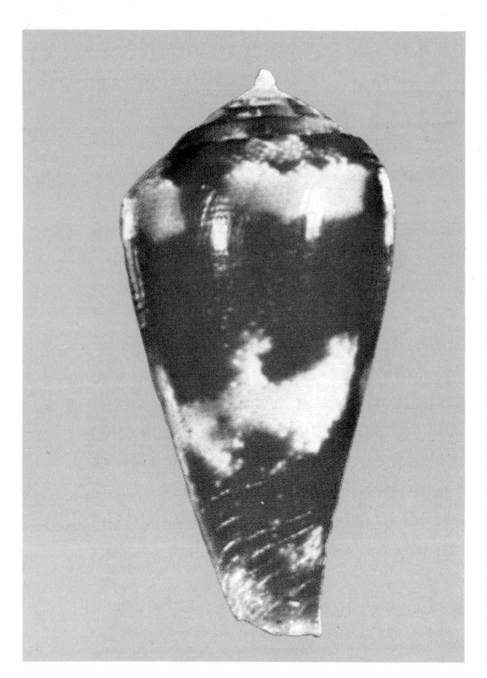
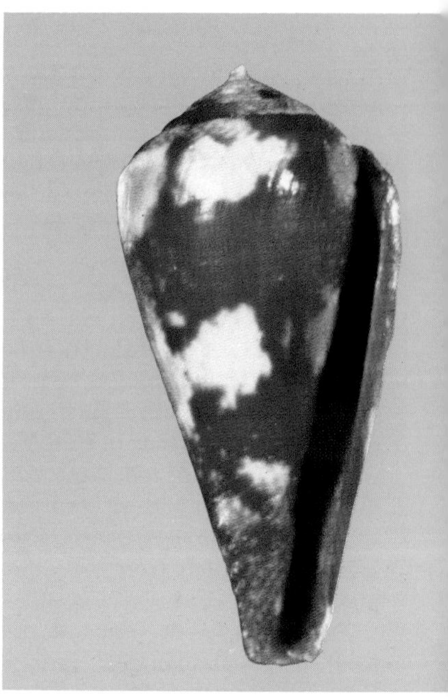

PERTUSUS.—Above: Samar, P.I., 25.5 and 25.6mm (GG). *Below:* Left: Kwajalein Atoll, Marshall Is., 33.2mm; Right (corded): Sulu, P.I., 28.6mm (GG).

PHILIPPII.—Above: Gullivan Bay, out of Woodland, Florida, 48.5-48.8mm (GG). *Below:* Off St. Augustine, Florida, 38.1-41.8mm (TY).

1877. *Conus inconstans* E.A. Smith. *Ann. Mag. Nat. Hist.*, (Ser. 4), 19: 224 (Locality unknown).

Description.—Moderately heavy, with a good gloss; rather cylindrical, the upper sides straight and nearly parallel or convex, anteriorly elongated; body whorl covered with broad spiral ribs separated by narrower grooves; the ribs usually seem to be made of two fused narrower ridges; numerous heavy axial threads over body whorl usually distinct; shoulder rounded, narrow, convex above, not distinct from spire; spire low to moderate, the sides straight to convex, pointed or eroded to blunt tip; early 4-6 whorls finely nodulose or undulate; spire whorls often distinctly but weakly stepped, convex above, with about 4-6 heavy spiral ridges and many heavy axial threads. Body whorl creamy pink to bright red, covered with many spiral rows of small brown dashes and often broad and heavy tan to dark brown axial bands and flammules; in fresh specimens the shell is brightest red with most distinct brown flammules and bands; usually with a well marked whitish or paler pinkish spiral band at midbody, this set off posteriorly by dark brown irregular squares; spire and shoulder pinkish to whitish, with dark brown to reddish spaced spots formed by the continuation of the brown flammules on the body whorl; early whorls pinkish. Aperture narrow posteriorly, wider anteriorly; outer lip thin, straight to slightly convex; mouth pink to white. Columella not visible externally. Length 35-61mm.

Comparison.—The shape and sculpture are signs that this distinctive species is allied to *C. artoptus* and *C. nussatella*. From both these it is readily distinguished by the broad ribs which seem to be formed by the fusion of two narrower ridges, the distinctly reddish color, the indistinctness of the spiral dashes in most specimens, and the rather distinct midbody band. From *C. vittatus*, which is similar in shape and often in the reddish color, it differs in the broad ribs (narrow ridges in *C. vittatus*), often distinctly stepped whorls (not stepped), and convex whorl tops (flat to concave); *C. vittatus* is also much more distinctly patterned and has the midbody band distinct and well defined.

Variation.—Fresh specimens are heavily flammulated and banded with brown over a bright reddish background, with the spire pattern and the brown above the midbody band distinct. With death the colors fade, yielding rosy to pinkish or even mostly yellowish shells with the spire pattern and midbody band becoming less defined. The convexity of the posterior part of the body whorl is variable, although it is usually rather parallel-sided; the shell of this species is usually much wider than equivalent specimens of *C. artoptus* or *C. nussatella*. The spire is low to moderate, usually bluntly pointed through minor erosion. Usually the spire is weakly stepped, but it may be almost flat-sided or distinctly turretted.

Distribution.—Usually rare to locally uncommon in deeper reefs of the Caribbean from southern Florida (rare) through the Antilles to northern South America; also the Central American coast. Seldom found in living condition, but sometimes locally not uncommon in the form of dead-dredged specimens with decent patterns.

Synonymy.—The early names are all assignable to this species with little doubt; *C. granulatus* has long been recognized as a rare and desirable Caribbean species. However, *C. inconstans* has for some reason often been considered to belong to such species as *C. miliaris* or *C. tiaratus*. The original description and the later figure by Sowerby in the *Thesaurus* leave little doubt that *C. inconstans* is closely related to *C. granulatus* in color and form and may very well be just a juvenile of that species; certainly it is not *C. tiaratus*.

Notes.—This is one of those species in which there is a tremendous price differential between live-taken specimens, which are currently very rare, and dead-dredged or beach specimens, which sometimes appear on the market. Dead specimens have the brown and red tones very faded and usually have bad lips which have been heavily filed. The spire pattern is seldom very strong in dead specimens, while the mouth often fades to white. It appears that *C. granulatus* is actually rare in live-taken condition and is seldom offered at any price. In fact, except for occasional small batches from the southern Antilles or bad beach specimens from Florida, the species is seldom available on the market with good patterns.

Like other cones of this general group, there is a tendency toward axial growth marks and spire blemishes. Apparently minor erosion of the spire is typical, with few specimens having perfect spires. The lip is quite fragile.

GUBERNATOR Hwass, 1792

1792. *Conus gubernator* Hwass, in Bruguiere. *Cone*, in *Ency. Method.*, *Hist. Nat. des Vers*, 1: 727 (Indian Ocean). Lectotype selected and figured by Kohn, 1968.

Description.—Heavy, with a high gloss; low conical, the upper sides broad at the shoulder then tapering to base and sometimes slightly concave at middle; body whorl with a few very low and weak spiral ridges at the base, sometimes with a more widely spaced narrow groove posteriorly, the rest of the whorl smooth except for faint spiral and axial threads which are not feelable with the fingertip; shoulder broad, roundly angled, concave above; spire low, bluntly pointed, the early whorls evenly rising from the later ones, not with a prominent mammillate protoconch; sides of spire concave, occasionally nearly flat; early 2-4 whorls very finely nodulose, sometimes inconspicuously so; early whorls flat above, with traces of 3-5 weak spiral ridges, later whorls slightly concave to deeply canaliculate, with the spiral sculpture indistinct and with weak axial threads. Body whorl white to pale pinkish white, usually covered with broad bands of pale rusty tan above and below midbody and often above the base and below the shoulder; this is usually (in fully patterned

PICTUS.—Jeffreys Bay, Cape, South Africa. *Above:* Left: 28.1mm; Right: 28.7mm (NM 5678). *Below:* 36.7mm (NM 5678).

***PILKEYI.**—Above:* Tolwat, E New Britain, Papua New Guinea, 47.7 and 49.4mm. *Below:* Left: Honiara, Guadalcanal, Solomon Is., 29.6mm; Right: Batangas Bay, Luzon, P.I., 46.4mm.

specimens) overlaid with weak, paler bands of the same color in an axial fashion, leaving large white rectangles and blotches; over this is often found a pair of broad spiral rows of large blackish curved and irregular blotches sometimes connected by weaker axial flammules; entire body pattern may be commonly absent; spire and shoulder with weak to heavy pattern of curved blackish flammules like the blotches of the body whorl on a white background, these often absent on the middle and early whorls; early whorls pinkish. Aperture wide, uniform in width or nearly so; outer lip thin, sharp, straight; mouth white, glossy. Columella very narrow, not very distinct, not indented from inner lip. Length 40-65mm.

Comparison.—This species has long been confused with *C. terminus* and has lately also been confounded with *C. barthelemyi.* From *C. barthelemyi* it is easily distinguished by the absence of a prominent mammillate protoconch, by a more irregular, blotch-like dark pattern on the body whorl instead of usually discrete spots as in *C. barthelemyi*, and by the more rusty coloration even in reddish specimens, not approaching very closely the colors of most *C. barthelemyi;* in addition, there are often very small brown dots in the background anteriorly in *C. gubernator*, these being absent in *C. barthelemyi;* the white dashes of *C. barthelemyi* are absent in *C. gubernator*. From *C. terminus* it is easily distinguished in most specimens by the broadly conical form, with the greatest width at the shoulder rather than anterior to it; the spire in *C. terminus* is usually higher and more deeply concave on top of the whorls, with the spire pattern usually heavy in *C. terminus* and often deep chestnut and contrasting with the body pattern. In *C. terminus* the body whorl is distinctly elongated, has the pattern very irregular and often strongly axial with deep bluish or violet tones over the whorl as shadows of the axial pattern; the columella is long and distinct, often set off posteriorly by a strong ridge absent in typical specimens of *C. gubernator;* and the entire body whorl of *C. terminus* is often covered with very fine brown dots in spiral rows, these often visible only under a lens but almost absent except anteriorly in *C. gubernator*.

C. striatus is very close to *C. gubernator* and *C. terminus* but usually can·be distinguished by the heavy spiral ridges over the body whorl and the much broader and sometimes heavier shell, as well as different colors and patterns on the whole. Some forms of *C. magus* from the Indian Ocean can approach *C. gubernator* very closely in form and pattern, but these have spire whorls which are not deeply concave above and have a more spotted spire pattern on the average.

Variation.—The shape is very constant, although the spire varies considerably in height, degree of concavity of the sides, and concavity of the whorls. In some specimens the shoulder and later spire whorls are almost flat above, although the distinctly raised margins are usually visible. The body pattern is extremely variable, going from solid white to almost completely rusty tan with heavy black blotches. Three patterns, plus intermediates, are common: solid white, white with two narrow rusty spiral bands above and below midbody and a few scattered half-moon black spots on these bands, and mostly rusty with heavy black crescents connected by axial flammules. The spire pattern is variable in the same way, from all white or with only a few blackish blotches in pale shells to heavy. Often the black blotches or half-moons on the body whorl are shadowed with white, but not with violet. A few brown dots in indistinct spiral rows are sometimes visible near the base, but they are never conspicuous and sometimes seem to be absent.

Intergradation of this species with *C. terminus* is not apparent, although admittedly some populations from the southern Indian Ocean are similar. These southern populations are for the moment referred to *C. terminus* because of the irregular pattern which is similar to that of *C. terminus* from the African coast. The shape is broader than African specimens, but the greatest width is below the shoulder, not at the shoulder as in *C. gubernator.*

Distribution.—As here restricted, *C. gubernator* is known only from the northern Indian Ocean, mainly the Maldives and possibly the adjacent Indian coast. Perhaps more widely distributed but confused with *C. terminus* and *C. barthelemyi.*

Synonymy.—Hwass' figure 4 on plate 340 of the *Tableau Ency. Method.* is certainly this species; figures 5 and 6 may be *C. gubernator* or possibly the broader southern Indian Ocean form of *C. terminus* because of the irregular patterns. Although Lamarck (1822, *Hist. Nat. des Anim. sans Vert.*, 7: 505) indicated figure 4 as typical of *gubernator* and considered the other two figures as variants, Kohn decided to select the specimen which was the basis for figure 6 as the lectotype. Unfortunately this specimen, as indicated above, is probably a variant of *C. terminus* and not *C. gubernator.* If the lectotype selection of Kohn is recognized, as apparently it must be, then *C. gubernator* and *C. terminus* would be synonymous. For the moment I cannot accept this and continue to recognize the two as distinct; perhaps it might be possible to accept Lamarck as the first restriction of the name *C. gubernator* to figure 4 of plate 340.

Notes.—Although *C. gubernator* has been badly confused with *C. terminus,* I feel they are distinct and constant enough to be considered full species. *C. gubernator* is uncommon compared to *C. terminus,* especially specimens with bright rusty patterns; specimens with reduced patterns are more common, and almost solid white specimens are not much less common. Except for a few minor axial growth marks and minor chipping of the delicate lip, this species is seldom badly flawed.

The market of the moment is highly unstable for this species, and caution is suggested before purchasing specimens at high prices. It has been widely called *C.* 'pramparti' in literature, a manuscript name that should not be used. Specimens of *C. terminus* from the southern Indian Ocean have been sold as this species, but they are easily distinguishable (usually) by the more irregular pattern and different shape.

PLANORBIS.—Above: Honiara, Guadalcanal, Solomon Is., 31.6-36.4mm. *Below:* Left (*chenui*): Thio, New Caledonia, 33.2mm (MM); Right: Nellie Bay, Queensland, Australia, 50.2mm.

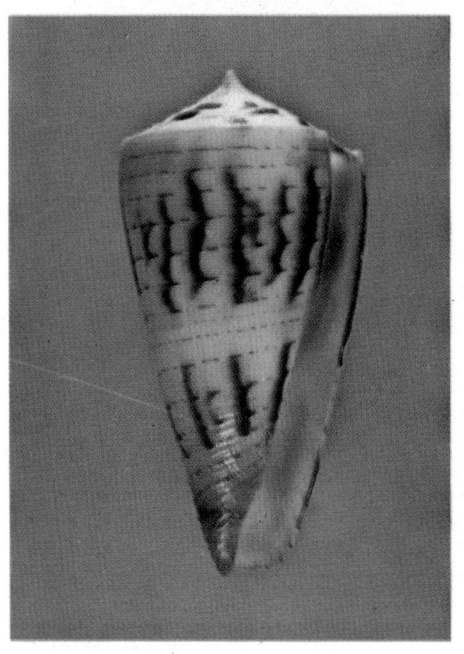

POHLIANUS.—*Above:* Tolwat, E New Britain, Papua New Guinea, 59.1 and 60.0mm. *Below:* Left (referred juvenile): Bai, E New Britain, Papua New Guinea, 30.6mm; Right: Matupit, E New Britain, Papua New Guinea, 46.4mm.

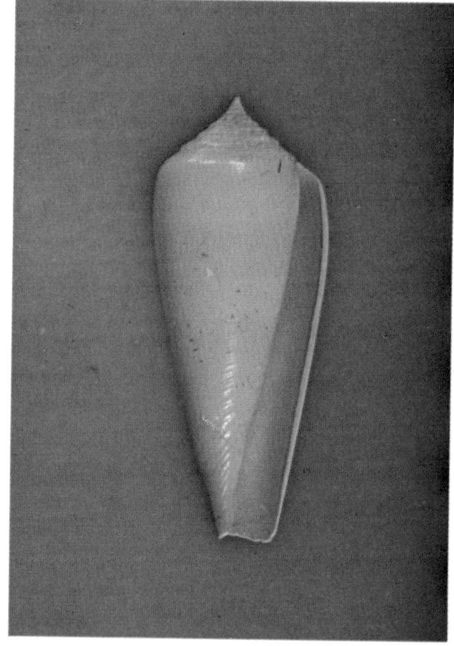

GUINEENSIS Gmelin, 1791

1791. *Conus Ammiralis Guineensis* Gmelin. *Systema Naturae per Regna Tria Naturae*, ed. 13, 1: 3380 (Locality not stated). Lectotype figures (here selected): Chemnitz, Pl. 144A, figs. i, k.
1792. *Conus Mozambicus* Hwass, in Bruguiere. *Cone*, in *Ency. Method., Hist. Nat. des Vers*, 1: 696 (Mozambique). Lectotype selected and figured by Kohn, 1968. (= g. *guineensis*)
1792. *Conus informis* Hwass, in Bruguiere. *Ibid.*, 1: 699 (America). Lectotype selected and figured by Kohn, 1968. (= g. *informis*)
(1802. *Conus elongatus* Holten. *Enumeratio Systematica Conchyl. beat. Chemnitzii:* 39 (Locality not stated). Based on Chemnitz, Pl. 144A, figs. i, k.)
1817. *Conus elongatus* Dillwyn. *Descr. Catalogue Recent Shells*, 1: 430 (coasts of Guinea). Lectotype figures (here selected): Chemnitz, Pl. 144A, figs. i, k. (= g. *guineensis*)
?1848. *Conus Caffer* Krauss. *Suedafr. Moll.:* 131, pl. 6, fig. 24 (Cape and Natal, South Africa). Non *Conus caffer* (Roeding, 1798). (= g. *informis*)

Description.—Light in weight but fairly solid, with a low gloss or dull finish; low conical, the upper sides convex; body whorl with a few weak spiral ridges at the base and numerous fine axial and spiral threads; shoulder rounded, slightly convex above, not very distinct from spire; spire low, blunt, the sides straight to convex, the whorls sometimes weakly stepped; tops of whorls convex and with 3-7 distinct spiral ridges crossed by weak axials. Body whorl bluish white to pinkish (redder when eroded), covered with heavy spiral and axial bands of dark brown, the brown often fading to orange with erosion; base and midbody area whitish; midbody band often with large dark brown spots posteriorly; usually many brown spiral dashes on basal third, these often continued over entire shell to shoulder; basal area sometimes with distinct axial brown streaks; shoulder whitish to pinkish, with scattered heavy or weak dark blotches; spire similar, often pinkish with little spotting. Aperture moderately wide, sloping below shoulder; outer lip thin, fragile, convex or nearly straight; mouth brownish to pale violet. Columella short, narrow. Length 30-70mm.

Comparison.—*C. guineensis* is very closely related to *C. tinianus*, *C. algoensis*, and *C. aemulus*. Like other South African cones, it is probably impossible to accurately identify many or even most of the beach specimens taken, yet live-taken shells are rare. Most *C. tinianus* have the columella free posteriorly and projecting into the aperture; they are also usually distinctly three-banded and seldom have many spiral brownish dashes. The shoulder ridges of *C. tinianus* are supposed to be finer and weaker than those of *C. guineensis*, but this character is very subject to erosion and individual variation. *C. algoensis* usually has a more distinctly straight-sided spire with regularly alternating dark and white blotches; the body pattern seldom has long brown dashes, although there may be short squarish spots; and the midbody band is seldom markedly distinct; the shoulder ridges are fine and indistinct, but again variable.

From *C. aemulus*, *C. guineensis* is most readily distinguished by the less regular pattern and paler background colors. The open three-banded appearance of *C. aemulus* is not found in *C. guineensis*, which is also typically with a more convex and bluntly pointed spire.

As with *C. algoensis*, the status of this species is doubtful, and it would not be implausible to consider *C. algoensis*, *C. guineensis*, and *C. tinianus* as a single variable species.

Variation.—There is considerable individual variation in height of spire, convexity of spire, and how distinctly stepped are the whorls. With erosion the browns change to orange, red, or yellow, as the background becomes more distinctly bluish. The spiral ridges on the spire whorls are apparently about 5-7 in fresh specimens, but these may appear as only 3 stronger ridges in beach specimens. Two variable and poorly distinguishable subspecies are recognized:

C. guineensis guineensis—Rather elongated, very thin and fragile, with a lower and more rounded spire; the body whorl is largely brownish with indistinct paler axial flammules; the body whorl usually has the spiral dashes poorly defined and restricted to the base; the midbody band is often indistinct. Southwestern South Africa and South-West Africa.

C. guineensis informis—Somewhat broader and shorter-bodied, a little thicker and less fragile; spire sometimes a bit higher; body whorl largely whitish with heavy and broken axial and spiral bands or flammules of brownish (eroded to orange or red); body whorl usually heavily covered with brown spiral dashes extending to shoulder; midbody band usually fairly distinct and with large brown blotches posteriorly. Southern and eastern South Africa.

These subspecies are quite possibly not distinct, the differences resulting from erosion often being greater than the supposed subspecific differences.

Synonymy.—The subspecies *C. guineensis guineensis* has generally been called *C. mozambicus*, which appears to be a synonym. *C. guineensis informis* has often been called *C. lautus*, although that name appears to be a synonym of *C. tinianus*, if recognizable at all. *C. informis* has often been considered a synonym of *C. ermineus* or *C. monachus*, but the lectotype and cited figures are of *C. guineensis*. *C. caffer* could be a yellowish specimen of the *informis* type with the pattern obscured or it could be a yellowish *C. tinianus*.

Notes.—This species is almost impossible to obtain in gem condition. *C. guineensis* is sometimes taken alive, but it is so fragile that there are always bad breaks over the body whorl. The nominate subspecies is taken alive most commonly but is a very ugly shell with distorted pattern and many axial breaks. *C. guineensis informis* is much rarer alive but no less ugly, although less subject to breaks. Beach specimens of both subspecies

POLYGRAMMUS.—Above: Left: Off Samarai I., Papua New Guinea, 36.4mm (EP). Right: Marau Sound, Guadalcanal, Solomon Is., 20.9mm (MG). *Below:* Passage between Basilaksi and Sidelia Is., Papua New Guinea, 24.6-27.2mm (B. Parkinson).

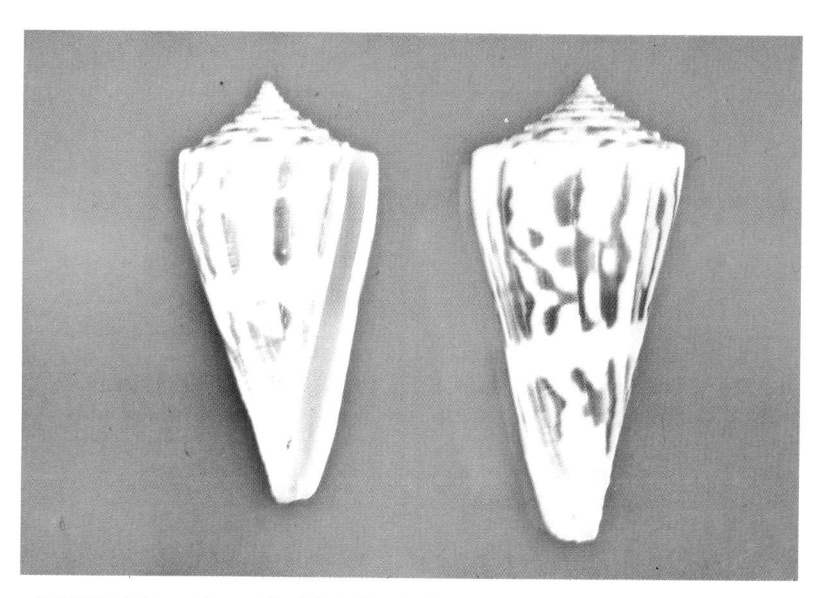

POORMANI.—Above: S of Del Rey I., Panama, 54.5 and 60.7mm. ***Below:*** Guaymas, Sonora, Mexico, 34.7-49.9mm (photo by A. Kerstitch).

become much brighter, especially *informis*. The coloration then is usually orangish or reddish, although the pattern is still distinct (perhaps even more distinct).

Beach specimens are common of both subspecies, although usually only the somewhat thicker *C. guineensis informis* survives in good enough condition to make a specimen.

HAVANENSIS Aguayo and Farfante, 1947

1947. *Conus havanensis* Aguayo and Farfante. *Rev. Soc. Malac. Carlos de la Torre*, 5: 11, figure (Arenas de la Chorrera, Habana, Cuba).

Description.—Heavy and thick for its small size, with a low gloss; low conical, the sides slightly convex; body whorl covered with narrow spiral ridges bearing strong pustules; the ridges are usually widely spaced but run from base to shoulder; shoulder angled, strongly nodulose or heavily undulate; spire low, bluntly pointed, the sides straight or slightly concave, sometimes very weakly stepped; tops of whorls with a few scattered axial threads; margins of whorls deeply undulate or nodulose. Whole shell pale yellow to white (dead), with about three spiral rows of yellowish brown blotches at midbody, below shoulder, and above base, the blotches seldom strong except at midbody; spire and shoulder with small brownish spots and streaks, these heavier toward tip, sometimes concentrated between the nodules. Aperture moderately narrow, uniform; outer lip thick, slightly convex; mouth pale violet to pinkish (faded). Columella very short, narrow. Length 14-21mm.

Comparison.—This species seems to be known only from a few dead specimens but seems fairly distinct although closely related to the *C. puncticulatus* assemblage. It is thicker and more strongly angled at the spire than most *C. puncticulatus*, but closer to *C. jaspideus* in shape and general appearance. From verrucose *C. jaspideus* it differs in the positioning of the granules on the ridges rather than between them, lack of deep spiral grooves at the base, the often heavier nodules on the spire whorls and shoulder, the concentration of the spire color on the early whorls, and the weak body pattern. The shell is very thick for such a small species. There is also a general resemblance to *C. cardinalis* in sculpture and thickness of the shell, but the reduced pattern of *C. havanensis* appears to be real; although most *C. cardinalis* have at least traces of spiral ridges on the spire whorls, some are nearly smooth like *C. havanensis*. Its actual relationships will have to wait for the examination of live-collected specimens.

Variation.—The few specimens known to me are quite similar in shape although the spire varies from very low to moderately high with slight stepping. The nodules of the spire whorls and shoulder are distinct, but those of the shoulder may be small or very large. The spiral ridges on the body whorl are always widely spaced and apparently normally extend from base to shoulder; the granulation may be fine or coarse. Background color varies from pale yellow with distinct irregular brown blotches in rows to solid white. Since living specimens are not known to me, it is possible that live-taken specimens would be much more heavily marked and of deeper colors. The spire spotting seems to concentrate posteriorly, with streaks and spots between the nodules; small specimens seem to have stronger spotting. Mouth color apparently fades from violet to pinkish.

Distribution.—Described from dead-dredged specimens from Cuba, this species is otherwise known to me only from beach specimens from the Bahamas. Probably much more widely distributed but not recognized.

Synonymy.—The Bahamas specimens have lower spires and heavier shoulder nodules than Cuban specimens, but they seem to be the same species.

Notes.—It is hard to place this species until live-taken specimens have been examined. It seems to be valid, but it is not impossible that it is a variant of either *C. jaspideus* or *C. cardinalis.* There is probably an older name for this taxon buried in the literature, but I have not been able to recognize it.

HIEROGLYPHUS Duclos, 1833

1833. *Conus hieroglyphus* Duclos. *Mag. Zool., Journal* 3: Pl. 23 (?California). Spelling often incorrectly emended to 'hieroglyphicus.'

1850. *Conus armillatus* C.B. Adams. *Contrib. Conch.*, 4: 59 (Jamaica). Lectotype selected and figured by Clench and Bullock, 1970, *Johnsonia*, 4(48): 374, pl. 176, fig. 5.

Description.—Moderately light, with a good gloss; low biconical, the upper sides strongly convex and almost bulbous; body whorl with crowded oblique ridges at the base, these changing to more normal narrow spiral ridges; ridges widely separated by flat grooves, usually about 15-20 in number, and extending from above base to shoulder; ridges bearing numerous spirally elongated beads; shoulder rounded, convex; spire moderately high to low, bluntly pointed, the tops of the whorls convex; spire whorls with about 3-4 narrow close set spiral ridges separated by deep grooves and covered with fine axial threads to produce a weakly cancellate appearance; whorls slightly stepped. Body whorl deep reddish brown to grayish black, with three spiral rows of large whitish blotches, at midbody, below shoulder, and above base; blotches rather irregularly rectangular, often fused spirally and sometimes connected axially as well; beads on spiral ridges white, and there may be fine white spiral dashes over the body whorl as well; base paler than rest of shell; shoulder and spire whitish to pale violet, with heavy radiating blackish or brownish lines, this pattern extending below the shoulder onto

PRAECELLENS.—Above: Tayabas Bay, Luzon, P.I., 43.4 and 47.0mm. *Below:* Left: Tayabas Bay, Luzon, P.I., 42.0mm; Right: S of Lubang, P.I., 30.9mm.

PRINCEPS.—Above: Mazatlan, Sinaloa, Mexico, 60.5mm (K). ***Below:*** Left (typical): Guaymas, Sonora, Mexico, 66.1mm; Center (*apogrammatus*): Pedro Gonzales, Panama, 54.0mm; Right (*lineolatus*): Pedro Gonzales, Panama, 44.9mm (photo by A. Kerstitch).

the body whorl. Aperture fairly wide, narrower posteriorly and widened basally; outer lip thin, convex; mouth pale violet with a faint paler band at middle. Columella short, very narrow. Length 13-22mm.

Comparison.—*C. hieroglyphus* is very close to *C. selenae* and I am not sure it would not be best to treat them as subspecies. However, *C. selenae* seems to be somewhat wider at the shoulder, is more heavily pustulose in most examples, often has the spiral ridges of the spire whorls strongly nodulose, and is a pale shell with small pale brown to lilac blotches and brown dots in typical patterns on a violet to pale pinkish background. The two seem to be separated geographically, but specimens answering very closely to *C. selenae* have been seen from southern Florida and the Central American coast.

There is a superficial similarity to *C. jaspideus*, but that species lacks spiral sculpture on the spire whorls, as does *C. puncticulatus*. Small *C. aurantius* and *C. regius* may be vaguely similar, but they have a more elongated body whorl or heavy spire nodules as well as different patterns. Some *C. sponsalis* may be similar but have the shoulder coronate and the base deep violet.

Variation.—The shape and sculpture of this species are very constant, but the pattern varies somewhat. Typically this is very dark reddish brown to nearly black shell with three rows of contrasting squarish white blotches. In some specimens the white blotches may be nearly absent except as whitish flecks or small spots; in others the blotches may be large and form broad bands and axial flammules as well. Sometimes the dark background bands may be secondarily divided by narrow brown and white dashed bands. Occasional specimens are pale brown with all pattern reduced, while rare specimens are mostly whitish with small brown blotches like those typical of *C. selenae.* Pale specimens may be distinctly pinkish, while dark specimens may be distinctly violet. The mouth color varies from deep violet to almost white, presumably from fading after death.

Distribution.—Rare to locally uncommon from Jamaica and the Virgin Islands south through the Antilles to Aruba and Bonaire. Found in shallow water and beach. Most specimens come from the Netherlands Antilles, where the species is locally uncommon.

Synonymy.—This species has long been poorly known and usually has been called *C. armillatus*. However, Clench and Bullock, 1970, have shown that *C. hieroglyphus* is the older name and also refers to this species.

Notes.—Typical Aruba specimens are the easiest to get, although the pale brown shells are also not uncommonly found in Aruba. Pale shells with reduced pattern are less common than dark shells, while those with the pattern almost totally absent are rare. Usually there is some erosion of the spire whorls and shoulder area, and commonly the lip is chipped. Large and small breaks in the body whorl are not uncommon. As mentioned in the comparison, specimens very much like *C. selenae* in color and pattern have been seen from southern Florida and the Central American coast. The status of these specimens is uncertain, but they could provide a convincing argument that *C. hieroglyphus* and *C. selenae* are conspecific.

HIRASEI (Kira, 1956)

1956. *Rhizoconus* (?) *hirasei* Kira. *Venus*, 19(1): 3, fig. 10 (Kashiwajima, Tosa, Japan).

Description.—Heavy, with a high gloss; low conical, the upper sides convex then tapering to a narrow base; body whorl with a few weak spiral ridges at the base, otherwise smooth except for scattered axial threads and scratches; shoulder roundly angled, broad, not very distinct from spire; spire low, sharply pointed, the sides straight or slightly convex; early 2-4 whorls weakly nodulose, next 2-3 weakly undulate; tops of whorls flat to slightly convex, with about six wide and distinct spiral ridges. Body whorl white to pale violet, covered with regular narrow spiral lines of deep reddish brown from base to shoulder, sometimes somewhat more widely spaced at midbody; base white; spire and shoulder whitish, regularly spaced small squarish dark brown spots at anterior margin of the whorls and on the shoulder. Aperture moderately wide, uniform; outer lip thin, sharp, straight or slightly concave at middle; mouth pale violet. Columella long, narrow. Length 40-60mm.

Comparison.—Although of simple sculpture, this is a very distinctive species because of the regular pattern on the body whorl and spire. *C. polygrammus* may be superficially similar, but that species has the later whorls nodulose or undulate and has the body pattern more irregular and dashlike in most shells; the spire pattern is much less regularly spotted and sometimes streaked. Some patterns of *C. typhon* and *C. caillaudi* are regularly striped over the body whorl, but these species have a more diffuse spire pattern without regular small brownish spots at the margins.

C. kimioi is smaller and thinner than *C. hirasei*, with a more angled shoulder. The body pattern has not only the spiral lines of *C. hirasei*, but indistinct brownish blotches between the stripes above and below midbody and below the shoulder; the spire pattern is similar to that of *C. hirasei* but somewhat more irregular. It seems quite possible that *C. kimioi* is a juvenile of *C. hirasei*, the brown blotching of the body whorl disappearing with growth and the shape changing slightly. Intermediate sized specimens have not yet been seen, however.

Variation.—Extremely constant in shape and pattern. The background color varies from white to distinctly pale violet. Occasional spiral lines are irregular and broken, but this seldom has any effect on the pat-

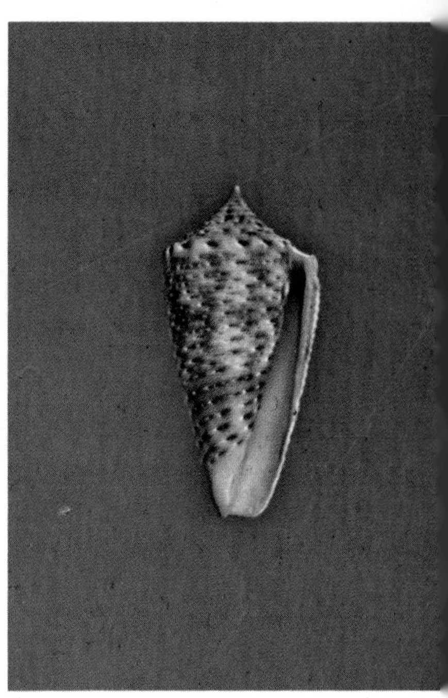

PROXIMUS.—Above: Ngo Bay, New Caledonia, 33.9mm (K). ***Below:*** Left: Siasi, P.I., 33.1mm (MM); Right: Russell I., Solomon Is., 30.3mm (EP).

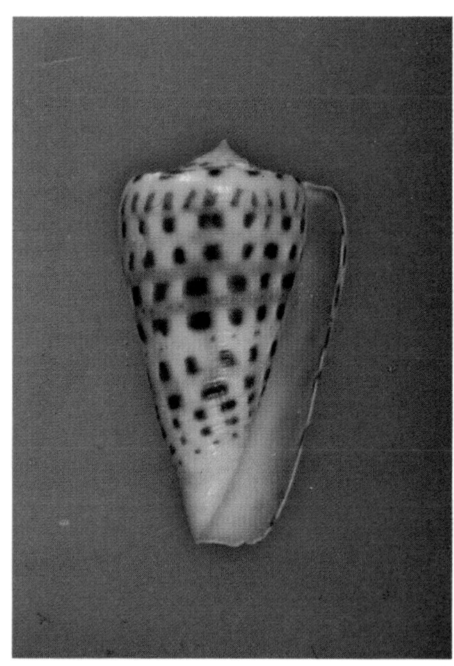

PULCHER.—Above: Left: N'Gor Bay, Cape Verde, Senegal, 58.8mm (EP); Right: Fernando Poo, 40.5mm. *Below:* Angola, 130mm.

tern. The spire pattern is constant, but the shoulder may be distinctly angled or only roundly angled.

Distribution.—Dredged in deep water from southern Japan to off Taiwan. Rare.

Notes.—This beautiful species is seldom badly scarred, although one or two minor growth marks are usually present on specimens. The lip is thin at the edge and chips readily. In some specimens there are poorly developed axial plicae visible. *C. hirasei* is uncommon or rare in any condition, and even faded dead-dredged specimens are still attractive.

HOWELLI Iredale, 1929

1929. *Conus howelli* Iredale. *Rec. Australian Mus.*, 17: 182, pl. 40, figs. 1, 8 (Off Montague Island, New South Wales, Australia).

Description.—Rather light, with a good gloss; biconical, the upper sides convex, nearly parallel, then concave to a narrow base; body whorl with weak spiral ridges over anterior quarter to half, sometimes with traces of obsolete ridges to the shoulder; rest of body whorl covered with sometimes heavy axial threads; shoulder strongly carinate, concave above; spire moderate to tall, bluntly pointed, the sides concave to nearly straight; tops of whorls concave, the margins with raised carinae; early whorls sometimes nodulose, the later whorls simple or undulate; tops of whorls with strong curved axial threads and sometimes with 2-3 fine spiral ridges. Body whorl opaque white, white, or pale yellow, usually covered with many axial wavy lines of reddish brown which are most distinct at midbody, above the base, and below the shoulder in broad spiral bands; usually with the rest of the body whorl pale yellowish brown to deep reddish brown, sometimes in broad spiral bands or as large squarish blotches; base pale; spire pale, whitish to yellowish tan, with large and small brownish blotches; carinae with rather regular alternating dark brown and white spots. Aperture wide, rather uniform in width; outer lip strongly sloping below level of shoulder, thin, evenly convex; mouth pink to white, the exterior pattern sometimes showing through. Columella not visible externally. Length 20-27mm.

Comparison.—Although this variable species is fairly distinct, at least some patterns are very similar to *C. teramachii* in shape, sculpture, and raised carinae, but *C. howelli* apparently always has a distinct color pattern involving wavy axial lines which are absent *C. teramachii*. The type specimen of *C. howelli* is virtually identical with *C. teramachii* except for the small size and presence of a pattern. Like *C. altispiratus*, the shape of the spire and pattern of the body whorl are variable. *C. howelli* with distinct spiral ridges on the spire whorls and nodulose carinae might be vaguely similar to *C. polygrammus* but would have more deeply concave tops of the whorls and a regular pattern of axial brownish lines.

Variation.—The type specimen of *C. howelli* has the early whorls weakly nodulose, the spiral sculpture absent on the spire and very weak on the anterior quarter of the body whorl, and the body whorl opaque white and covered with widely spaced axial lines and three spiral bands of pale brown squarish blotches mostly fused into regular bands. At the other extreme are mostly brown shells with high or low spire and white blotches restricted to the midbody area and above the base; such shells may have the spire whorls distinctly nodulose, weak spiral ridges may be present on the spire whorls, and the spiral ridges of the body whorl may extend (weakly) nearly to the shoulder and be faintly granulose anteriorly. Whether such extreme shells are really *C. howelli* or represent another species is uncertain because so few specimens are known; see Garrard, 1961, *J. Malac. Soc. Australia,* No. 5: 30, for a brief survey of variation. Perhaps the range of variation in this species is like that in *C. altispiratus,* which is closely related.

Distribution.—Seemingly restricted to deep water off the coasts of southern Queensland and northern New South Wales, Australia. Rare.

Synonymy.—The type specimen is so extremely similar to *C. teramachii* in shape and sculpture that later work may reveal that the two species are connected by intermediates in the western Pacific and are actually subspecies.

Notes.—This rare shell is poorly known and possibly two species are being confused here. Most specimens are quite dead and in bad condition. This is a small and fragile species with heavy growth marks even in good specimens. At the moment even very dead specimens are considered collectable, with few specimens found in private collections.

HYAENA Hwass, in Bruguiere, 1792

1792. *Conus hyaena* Hwass, in Bruguiere. *Cone,* in *Ency. Method., Hist. Nat. des Vers,* 1: 656 (West Africa). Lectotype selected and figured by Kohn, 1968.

1844. *Conus mutabilis* Reeve. *Conchologia Iconica,* 1(*Conus*): Pl. 45, sp. 249 (Locality unknown).

1845. *Conus Reevei* Kiener. *Species gen. et icon. des coqu. viv.,* 2(*Conus*): Pl. 44, fig. 2. 1847, *Ibid.,* 2: 115 (Locality unknown).

Description.—Moderately heavy, dull; low conical, the upper sides convex; body whorl with a few low spiral ridges above the base and scattered spiral threads; numerous heavy axial threads, growth marks, and breaks usually present; shoulder roundly angled, broad, sometimes with a weak carina, smooth or irregularly undulate; spire low, sharply pointed, mucronate, the sides straight to convex; early whorls weakly nodulose, later whorls smooth or irregular; whorls flat to slightly convex above, with about 5-6 low spiral ridges and weak axial threads, the ridges often

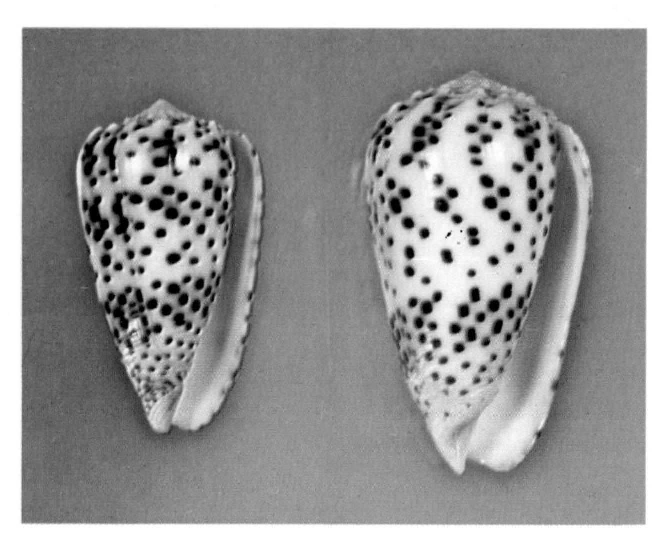

PULICARIUS.—Above *(p. pulicarius)*: Tahaa, Society Is., 40.0 and 48.4mm. *Below:* Left *(p. pulicarius)*: Tahaá, Society Is., 51.3mm; Right *(p. vautieri)*: Marquesas, French Polynesia, 26.5mm.

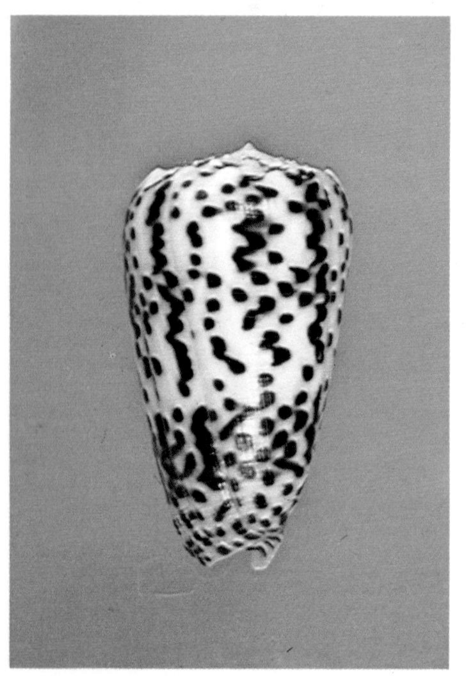

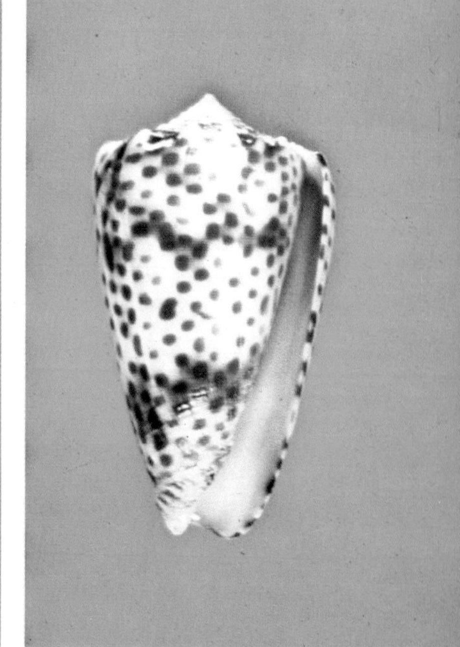

***PUNCTICULATUS.*—**(*p. puncticulatus*) *Above:* Amuay Bay, Venezuela, 21.9 and 22.7m (GG). *Below:* Left: Oranjestad Harbour, Aruba, Neth. Ant., 15.1mm (GG); Right (*columba*): Off Ft. de France, Martinique, 19.1mm (GG).

obsolete on late whorls and shoulder. Body whorl dirty white to yellow, heavily covered with broad reddish brown to dark brown flammules; usually a paler midbody area visible through the flammules; fine brown spiral lines or dashes visible, especially in and near the midbody area; base white to pale yellowish tan; spire and shoulder dirty white to pale tan, heavily covered with curved brown lines and blotches, about same color as body whorl; early whorls usually patternless. Aperture moderately wide, uniform in width; outer lip sharp, evenly convex; mouth clear bluish white or pale violet, sometimes with a darker blotch posteriorly deep within. Columella long, narrow, sometimes with a narrow ridge bounding it posteriorly. Length 40-65mm.

Comparison.—Superficially *C. hyaena* is closest to *C. concolor* in appearance, but it is distinguished from that species by the more irregular pattern which seems to be formed from broad axial flammules; the spire is more convex with a distinctly mucronate tip; the spiral lines of brown are usually more distinct in *C. hyaena* than in *C. concolor;* and the base is pale. *C. hyaena* sometimes has the early whorls distinctly nodulose, but the whorls do not appear finely cancellate as in *C. concolor.*

The closest relatives of *C. hyaena* appear to be *C. biliosus* and *C. lohri*. From *C. biliosus* it is readily distinguished by the broader form posteriorly, the more mucronate spire, lack of distinct coronations or undulations on the middle spire whorls, the reduced spiral pattern, the paler base, and, especially, the lack of distinct spiral ridges over the entire body whorl; there are also differences in mouth color, deep violet with a dark brownish violet stripe at the edge in *C. biliosus*, and in spire color, the blotches darker and more contrasting in *C. biliosus. C. lohri* looks much like an orange *C. hyaena*, but it lacks the distinct spiral sculpture of the spire, has almost no body whorl pattern visible, and has angular spots at the margins of the shoulder and later spire whorls.

Variation.—This cumbersome and often ugly species is not very variable in shape or sculpture, although some specimens (perhaps deep-water populations) are relatively more slender than those from tide pools. The early whorls may be distinctly nodulose or almost smooth. Body color usually is variably covered with broad brownish flammules or bands, sometimes resulting in a nearly all-brown shell; usually there is a distinctly streaky appearance in the pattern, however. The spiral lines or dashes may be very strong and visible over most of the whorl or greatly reduced and inconspicuous, even in the midbody area. The spire color usually matches the body color fairly closely. Spiral ridges are restricted to the base in all specimens seen. Occasional specimens are entirely pale orange in color, even the spire pattern being almost absent.

Distribution.—Found in shallow to moderately deep water around India and in the Bay of Bengal, perhaps more widely distributed; *C. lohri* from the Mozambique-Natal coast is certainly closely related. Common.

Synonymy.—*Conus mutabilis* of Reeve is the first acceptable use of a name from Chemnitz; since both Reeve and Hwass had the same shell in mind, the oldest available name is *C. hyaena* Hwass, as the *C.* 'mutabilis' of Chemnitz is not valid under present-day rules. *C. reevei* seems to be this species also.

Notes.—Until recently this was a rare species which was badly confused with *C. biliosus* and even *C. malacanus* (which is different in shape and pattern and has carinate spire whorls). However, the species is now commonly taken in tide pools in India and is available in at least two pattern types and two colors: a 'muddy' pattern with indistinct flammules and reduced spiral lines, a 'deep-water' pattern that is more slender with brighter colors and heavier spiral lines and dashes; the common color is some shade of brown or reddish brown, but orange specimens have lately been obtained and are not uncommon.

This is usually a very rough cone with heavy axial growth lines and flaws from the base to the spire; unhealed breaks are not uncommon, and the lip is delicate. The spire is commonly eroded and has small holes from boring sponges. Only a few specimens of this species could be considered attractive, and most are very ugly.

HYPOCHLORUS Tomlin, 1937

1877. *Conus croceus* E.A. Smith. *Ann. Mag. Nat. Hist.*, (Ser. 4), 19: 223 (Locality unknown). Non *Conus croceus* Sowerby i, in Sowerby ii, 1833.

1937. *Conus hypochlorus* Tomlin. *Proc. Malac. Soc. London*, 22(4): 206. Nomen novum for *Conus croceus* Smith, 1877.

Description.—Light in weight, with a high gloss and waxy appearance; elongate biconical, very slender, the sides almost straight and tapering to a very narrow base; body whorl covered with wide rounded spiral ribs separated by deep grooves; grooves with distinct axial threads so appear punctate; shoulder roundly angled, carinate; spire tall, sharply pointed, the sides nearly straight; tops of whorls nearly flat; early 2-4 whorls weakly nodulose, others carinate; tops of whorls with about 3-5 narrow but strong spiral ridges crossed by fine axial threads. Body whorl and spire totally deep yellow to deep orange-yellow, without traces of a pattern. Aperture narrow, uniform in width; outer lip thin, sharp; mouth pinkish tan. Columella not visible externally. Length 25-30mm.

Comparison.—This needle cone is closely related to *C. aculeiformis*, *C. insculptus*, *C. vimineus*, and the small form of *C. comatosa*. *C. aculeiformis* and *C. vimineus* are most similar superficially but have the spire sculpture of a modified type with the ridges restricted to a broad depression near the middle of the whorls; these species are seldom distinctly yellowish. *C. insculptus* in the common form has the posterior sides more convex, the whorls nodulose, small brown spots on the margins of the

PUNCTICULATUS.—(*p. perplexus*) *Above:* Left (granulose): No data, 18.5mm (GG); Right: Guayamas, Sonora, Mexico, 29.2mm (GG). *Below:* Bandaras Bay, Colima, Mexico; Left: 28.1mm; Right: 20.9mm (photos by A. Kerstitch).

PURPURASCENS.—Above: Pedro Gonzales, Perlas Is., Panama 49.3mm (GG). *Below:* Off Espinosa Pt., Isl. Fernandina, Galapagos Is., 31.0-40.3mm (GG).

whorls, and distinctly stepped whorls. *C. comatosa* is larger, has a distinct spotted pattern on the body whorl, and has strongly spotted carinae. From all these species *C. hypochlorus* is readily distinguished by its waxy appearance and uniformly deep yellowish to orangish color without spotting. Some *C. aculeiformis* may be mostly yellowish, but they have the modified spire sculpture and usually traces of spotting on the body whorl.

Variation.—This species seems to be constant in shape and sculpture as well as varying only in depth of color. The number of spiral ridges on the whorls varies from about 3 to about 5 or 6, but they are not placed in a depression and cover the entire top of the whorl.

Distribution.—Known to me only from a few specimens taken in deep water in the central Philippines, but probably more widely distributed.

Synonymy.—Since *C. croceus* was preoccupied, Tomlin renamed the species. It is still arguable whether it is distinct from *C. insculptus* and other needle cones, but this entire complex is very poorly known and species are hard to define.

Notes.—Although I have seen only a few specimens and know of no recent literature on the species, there is no guarantee that this is a rare or even uncommon species. Probably it is simply being confused with *C. insculptus* by collectors, although the waxy brightly colored shell is quite distinct at first glance. Like others of this complex, the lip is fragile.

ICHINOSEANA (Kuroda, 1956)

1956. *Asprella* (*Conasprella ?*) *ichinoseana* Kuroda. *Venus*, 19(1): 10, pl. 1, fig. 5 (Off Tosa, Japan).

1956. *Asprella* (*Conasprella ?*) *ichinoseana prioris* Kuroda. *Venus*, 19(1): 11, pl. 1, fig. 6 (Off Tosa, Japan).

Description.—Fairly heavy, with a good gloss; biconical, the sides slightly convex posteriorly, then tapered to a narrow base; body whorl with numerous wide spiral ribs on the anterior half or so, these separated by narrow, strongly punctate grooves; about midbody the grooves become irregularly punctate and sometimes obsolete; posterior half or third of whorl smooth; shoulder wide, sharply angulate, carinate; spire high, sharply pointed, the sides straight or nearly so, the whorls stepped; whorls concave above, with strong axial ridges and at most faint traces of spiral threads; early and middle whorls coronate, later whorls and shoulder simply carinate. Body and spire white, with reddish brown spots and dashes in about three spiral series, heaviest at midbody and below shoulder, weakest above base; these spots may be fused into large blotches and into axial flammules; spire and shoulder white, with brown spots at the anterior margins, widely separated by white areas. Aperture moderately narrow, uniform in width; outer lip strongly sloping below shoulder, thin, nearly straight; mouth white, sometimes with pink or violet tones. Columella internal. Length 50-69mm.

Comparison.—The slender and tall biconical shape immediately brings to mind *C. orbignyi*, which however is usually more convex posteriorly, has the ribs extending to the shoulder, and has strong spiral ridges on the spire whorls. In addition, *C. orbignyi* is seldom pure white, but is usually dirty cream or pale tan with the pattern more-or-less indistinct; typical *C. orbignyi* have all the spire whorls except possibly the last with strong rounded coronations. *C. comatosa* and *C. ione* have different shapes and have the carinae of the shoulder and spire whorls raised and rim-like. *C. ichinoseana* is not likely to be confused with any other species, although some Indian Ocean *C. acutangulus* may have a similar shape and pattern; such specimens would have the heavy spiral and axial sculpture of typical *C. acutangulus*, however.

Variation.—The grooving of the body whorl may extend over the anterior quarter, third, half, or three-quarters of the shell, seemingly as an individual variation. The shoulder area may be relatively slender or rather wide, although the sides of the body whorl are almost straight in all specimens. The spire may be relatively tall or moderate, but the whorls are always stepped and distinctly coronate on the early whorls at least. Although the axial sculpture of the spire is heavy, the spiral sculpture is commonly absent or extremely weak. The body pattern may be heavy blotches in spiral and axial bands or just separated dashes in indistinct spiral rows with a few pale blotches at midbody and below the shoulder. The spots on the spire may be strong and dark or light and indistinct.

Distribution.—Most common in deep water off southern Japan, but ranging south to Taiwan; also recently found in the Philippines. Uncommon.

Synonymy.—Since *C. i. prioris* was described from the same locality as the typical subspecies, it obviously cannot be a subspecies in the typical sense. Modern specimens with reduced sculpture are also found, so the subfossil subspecies is meaningless.

Notes.—Although apparently not rare, just uncommon, this species is seldom offered for sale at the moment. Except for minor erosion of the spire and lip chips, it is usually in good condition.

ILLAWARRA (Garrard, 1961)

1961. *Leptoconus illawarra* Garrard. *J. Malac. Soc. Australia*, No. 5: 31, fig. (East of Stanwell Park, New South Wales, Australia).

Description.—Very light and fragile, with a high gloss; low conical, the sides distinctly convex; body whorl with narrow oblique ridges at the base, these soon becoming rather broad spiral ridges that are flat on top; ridges separated by narrow, punctate grooves; spiral sculpture obsolete by

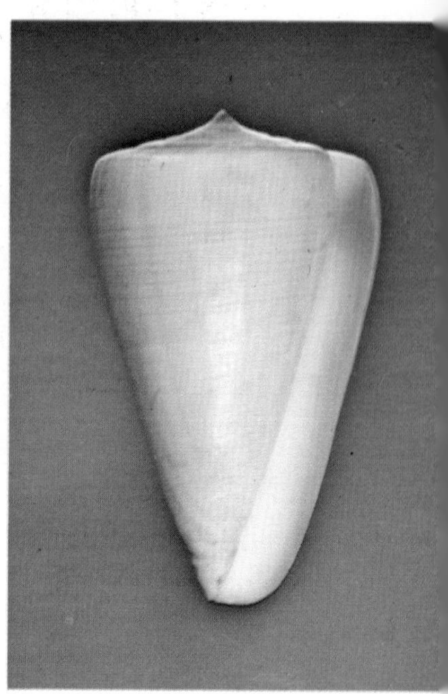

QUERÇINUS.—Above: Cebu, P.I.; Left: 72.2mm; Right: 67.5mm. ***Below:*** Left: Mombasa, Kenya, 47.5mm; Right: Keehi Lagoon, Oahu, Hawaii, 22.3 and 25.3mm.

RADIATUS.—Above: Manila Bay, Luzon, P.I., 61.3 and 66.4mm. *Below:* Mindanao, P.I., 47.8 and 51.5mm.

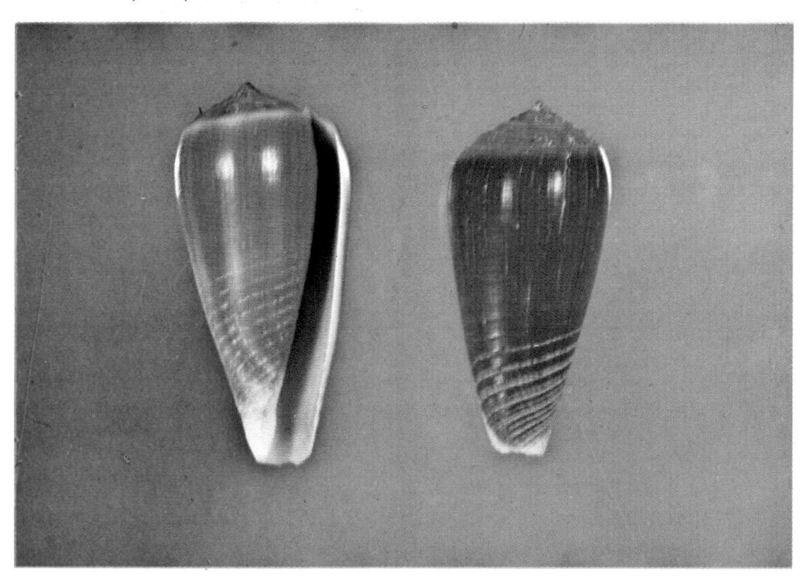

about midbody, with several rows of shallow punctations present just below the shoulder; shoulder roundly angled, sometimes slightly carinate; spire low or moderate, sharply pointed, the sides straight or weakly concave; spire whorls flat or slightly concave above, with about 2-3 narrow but strong spiral ridges crossed by numerous fine axial curved threads; no nodulose whorls. Body whorl white or cream, covered with irregular nebulous clouds and blotches of orange-tan; the blotches tend to form three broad but irregular spiral bands (below shoulder, at midbody, and above base), these sometimes connected by indistinct flammules; base white; spire and shoulder whitish with irregular orange-tan to yellowish tan axial blotches and streaks; early whorls whitish. Aperture wide, rather uniform in width; outer lip convex, very thin, sharp; mouth pinkish to pale dirty violet, sometimes with the external pattern showing through. Columella short and narrow. Length 25-30mm.

Comparison.—The pattern and general shape bear a close resemblance to *C. sibogae*, of which *C. illawarra* is perhaps just a southern terminal form. *C. sibogae* usually has the spiral ridges extending over the entire body whorl from base to shoulder, but sometimes the new growth of adult shells is almost smooth above the midbody area; in *C. sibogae* the shoulder is usually strongly carinate as opposed to rounded to very weakly carinate in *C. illawarra;* lastly, the spire whorls of *C. sibogae* have about 4-5 spiral ridges rather than the 2-3 of *C. illawarra*. The pattern of many *C. sibogae* has the orange-tan blotches mostly fused spirally into broad spiral bands absent in *C. illawarra*, but this is probably not constant. *C. wallangra* may be somewhat similar but is much smoother with a different shape and pattern and deep violet mouth in most specimens; it also has four or more distinct spiral ridges on the spire whorls. *C. minnamurra* has the early spire whorls nodulose, the later whorls undulate, and the body whorl covered with strong spiral ridges to the shoulder. *C. spectrum* has a much broader mouth, longer columella with often a strong indentation from the inner lip, and is usually covered with spiral rows of small brownish dots or dashes (sometimes visible only with magnification).

Variation.—This seems to be quite variable for a shell with a narrow range. The spire may be relatively low or moderate, with the sides distinctly concave to flat. The upper sides are distinctly convex, but the shoulder may be only rounded, weakly angled, or weakly carinate and concave above. The spiral ridges at the anterior part of the body whorl are usually wide and strong, extending to near midbody, but their width and extent vary considerably. The orange-tan blotches may be strong and dark with sharp borders or only nebulous clouds. Mouth color varies, but perhaps this is a factor of preservation; it is constantly pale, however.

Distribution.—Known only from deep waters off southern Queensland and northern New South Wales, Australia, where it is uncommon.

Notes.—This is not a very distinct species, and it would be easy to interpret some specimens as intergrades with *C. sibogae*, which occurs in adjacent areas of Queensland. Thus the variation of the shoulder from typically rounded or very weakly angled to weakly carinate shows a distinct trend toward the strongly carinate condition in *C. sibogae*. Probably study of large series of the two species from southern Queensland would reveal complete intergradation.

This is a very thin and fragile shell, with the lip especially subject to breaks and chips. Sometimes the pattern is very indistinct (as it sometimes also is in *C. sibogae*), making for an unattractive shell. Specimens with strong patterns of deep orange-brownish are quite pretty, however.

IMPERIALIS Linnaeus, 1758

1758. *Conus Imperialis* Linnaeus. *Systema Naturae per Regna Tria Naturae*, ed. 10, 1: 712 (Locality not stated). Lectotype (here selected): Linnaean Collection #291, incorrectly considered the holotype by Kohn, 1963, and figured by him.

1778. *Conus fuscatus* Born. *Index Rerum Naturalium Musei Caesari Vindobonensis*, 1(*Testacea*): 126 (Mauritius). Lectotype (here selected): N.M. Vienna #3929, incorrectly considered the holotype by Kohn and figured by him in 1964.

1798. *Cucullus regius* Roeding. *Museum Boltenianum:* 38 (Locality not stated). Lectotype figure (Kohn, 1975): Chemnitz, Pl. 139, fig. 1289. Non *Conus regius* Gmelin, 1791.

1798. *Cucullus corona ducalis* Roeding. *Ibid.:* 38 (Locality not stated). Lectotype figure (Kohn, 1975): Martini, Pl. 62, fig. 693.

1798. *Cucullus Imperialis* Roeding. *Ibid.:* 45 (Locality not stated). Lectotype figure (here selected): Martini, Pl. 62, fig. 691. This is probably not a new name.

1810. *Conus viridulus* Lamarck. *Ann. du Mus. Hist. Nat.* (*Paris*), 15: 31 (Southern Ocean). A type specimen was figured by Kiener, Pl. 7, fig. 1.

1906. *Conus queketti* Smith. *Ann. Natal Govt. Mus.*, 1(1): 22, pl. 7, fig. 1 (Isezela, Natal, South Africa).

1933. *Conus imperialis nigrescens* Barros e Cunha. *Mem. Estud. Mus. Zool. Univ. Coimbra*, (Ser. 1), 71: 17 (Locality not stated). Non *Conus nigrescens* Sowerby ii, 1859.

1933. *Conus imperialis flavescens* Barros e Cunha. *Ibid.*, 71: 18 (Locality unknown). Non *Conus flavescens* Sowerby i, in Sowerby ii, 1834.

1942. *Conus douvillei* Fenaux. *Bull. l'Inst. Oceanogr.* (*Monaco*), No. 814: 2, fig. 5 (Madagascar).

1942. *Conus dautzenbergi* Fenaux. *Ibid.*, No. 814: 2, fig. 2 (Madagascar).

Description.—Heavy, thick, with a low or moderate gloss, sometimes dull; elongate, obconical, the sides nearly straight and tapering to a narrow base; body whorl with several low, widely spaced spiral ridges on the anterior quarter, the ridges sometimes with low granules; in small

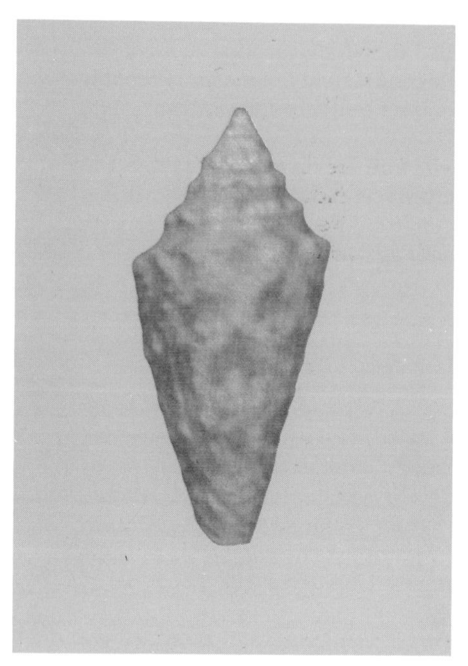

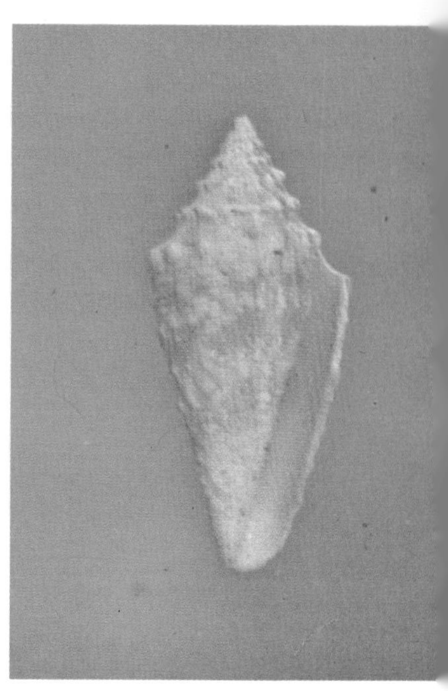

RAOULENSIS.—Sunday I., Kermadec Is., paratype, 19.6mm (Auckland Inst. and Dr. W.O. Cernohorsky).

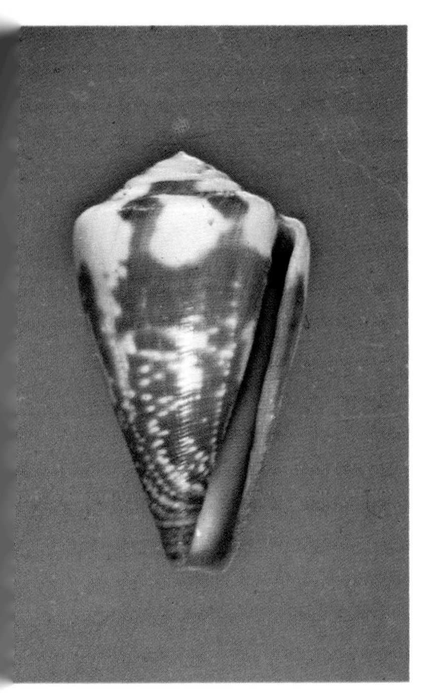

RATTUS.—Above *(r. rattus)*: Left: Zanzibar, 43.7mm (MF); Right (*r. taitensis*): Marquesas, French Polynesia, 38.9mm. *Below* (*r. rattus*): Tahiti, Society Is., 25.4-38.0mm.

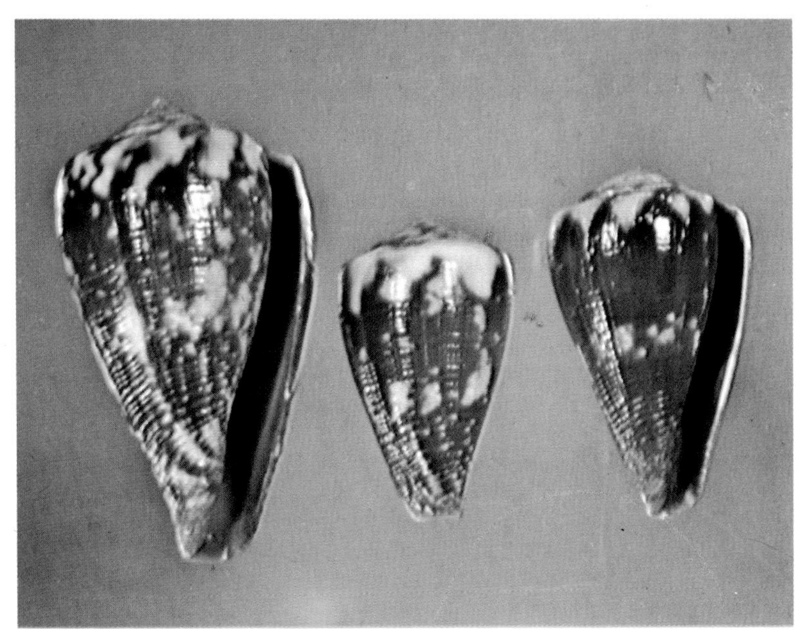

shells of some populations the body whorl may be covered with closely spaced spiral rows of large punctations, but these are usually obsolete in large specimens and in most populations; fine axial growth threads and marks usually present; shoulder wide, sharply angled, heavily coronated; spire low, sometimes flat, sometimes irregularly stepped, the tip rounded, blunt; spire whorls all heavily coronated and with indications of about 3-4 weak spiral ridges and irregular axial threads and wrinkles. Body whorl creamy white to grayish, covered with rather closely spaced spiral rows of brown and blackish dashes of variable length and small white spots; light brown irregular spiral bands often developed above and below midbody, sometimes blotchy or connected by irregular axial brownish flammules; pattern very variable; base of body whorl dark gray to violet; spire dull white with small brownish blotches and streaks; interstices of coronations often with darker and larger brown spots. Aperture narrow, widened slightly anteriorly; outer lip sharp, often concave at the middle, thick; mouth whitish to bluish white, the anterior edge with a dark purple stain. Columella long, narrow, often mostly internal and not visible in ventral view. Length 40-105mm.

Comparison.—This is one of the best known cones and the one least likely to be confused even by beginners. The pattern and heavy coronations bear some resemblance to *C. bartschi* and *C. regius*, but those species are much shorter bodied and wider at the shoulder as well as differing in many aspects of the pattern. The closest relative of *C. imperialis* is *C. zonatus*, with which it shares the elongate shape, heavy coronations, dark base, and narrow aperture. However, *C. zonatus* differs greatly in pattern, being heavily mottled and banded with bluish gray and having many fine but uniform reddish brown spiral lines over the body whorl; typical *C. zonatus* also have a higher surface gloss than *C. imperialis* (variable).

Variation.—*C. imperialis* is a highly variable shell which has been given a large number of names for the various patterns. Although some authors have given the better marked variants specific or subspecific status, I can see no reason for this as all possible intermediates are known between patterns. There is a tendency for darker specimens to come from the southern and central Indian Ocean, but typical *C. imperialis* is also found there, which would make subspecific status of dark shells very doubtful.

The basic patterns of *C. imperialis* may be summarized as follows (although many other patterns exist and are sometimes not uncommon). 1) Shell heavily covered with light brown blotches which have fused to produce very wide axial bands which are also partially fused spirally to leave only a few whitish blotches; 2) light brown color restricted to two rather wide spiral bands above and below midbody, with rather little other blotching (the typical and most common pattern); 3) like number 2, but with more axial flammules and tendency to break the bands into isolated blotches; 4) light brown spiral bands narrow, with straight edges, the anterior band wider than posterior one and often with several rows of dashes heavily developed and forming secondary bands that may be almost continuous spirally; 5) light brown blotches or bands absent or only vaguely indicated, the pattern consisting mostly of dark and light dashes; such shells may appear mostly yellowish. In addition, small shells often have more light brown blotches than larger shells. The background color is usually whitish, but not uncommonly it may be yellowish or pale bluish gray.

The shape also varies considerably, with the spire most commonly flat or depressed but also not uncommonly irregularly stepped and with straight sides. The coronations are almost always heavy, although those on the shoulder may be weak in occasional specimens. Some old specimens are distinctly inflated posteriorly and have very convex sides, while others have the sides distinctly concave near the middle. The base is usually dark brownish or dark purplish, but sometimes it is faded and not contrasting with the rest of the body color (uncommon condition).

All these variations seem to be individual or characteristic of small local populations and not worthy of recognition at any level.

Distribution.—Widespread and common in shallow waters of the Indo-Pacific from eastern Africa to Hawaii, including French Polynesia. Dark specimens and other less common variants usually come from the Indian Ocean islands.

Synonymy.—Typical *C. imperialis* has two broad brown spiral bands and a few brown blotches. In *C. fuscatus* the spiral bands are broader and may cover much of the body whorl; such a pattern is a minor development of the typical form and certainly not distinguishable. The names of Roeding are based on rather typical specimens, but *C. viridulus* of Lamarck is fairly distinctive, having many broad axial flammules as well as the usual spiral bands; such a pattern could easily develop into the largely dark shells found in the Indian Ocean and incorrectly called '*fuscatus*' by some dealers. Since the development of axial flammules and even of darkened background coloration is variable within many populations, even those of the Philippines, it is possible to find *viridulus*-like shells in many large series. Although often well marked, such dark shells are not worthy of recognition. *C. queketti* has long been considered a synonym of *C. imperialis*; the names of Barros e Cunha are unillustrated but there is nothing in the descriptions to discredit their placement in *C. imperialis*. *C. douvillei* seems to be a rather typical specimen with a pale base, a situation that is uncommon in *C. imperialis* but not unheard of. *C. dautzenbergi* seems to be a specimen of the number 1 pattern with very little light patterning. None of these names are recognizable at subspecific or specific level and all intergrade widely with typical specimens.

Notes.—*C. imperialis* is a common shell, but many specimens have heavy axial breaks and growth marks on the body whorl and badly eroded spires. True gems are not uncommon, however, although they sometimes have a very dull finish. The lip is rather fragile in some areas

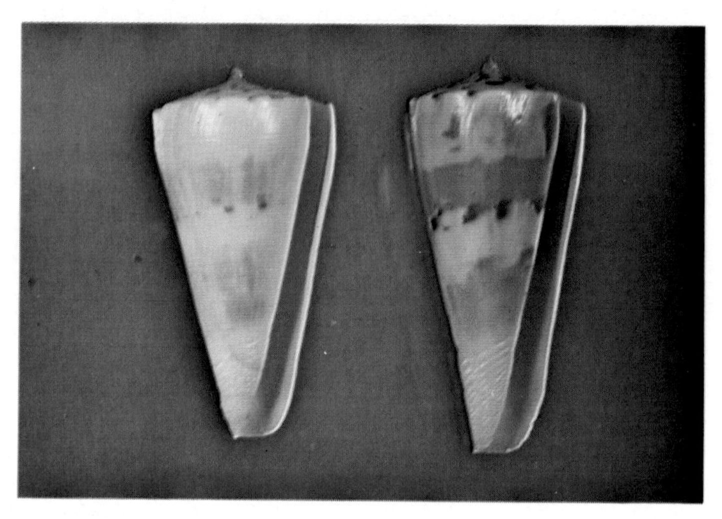

RECLUZIANUS.—Above: Left: Kusi, Nada-cho, Wakayama Pref., Japan, 45.7mm; Right: S of Straits of Malacca, Malaysia, 49.1mm (both GG). *Below:* Left: Off Tosa, Shikoku, Japan, 70.5mm (EP): Right: An Ping, Taiwan, 43.2mm.

RECURVUS.—*Above:* E of Isla Mina, Panama, 46.8 and 51.2mm. *Below:* Morro Colorado, Sonora, Mexico: Left (*scariphus*): 38.5mm; Right: 100.4mm (photos by A. Kerstitch).

and may be filed. Dark specimens of the number 1 pattern usually draw a heavy premium as they are not common in good condition; only pattern 2 is really common, with other patterns worthy of small to large premiums in good condition. Juveniles seem to be uncommonly collected.

INFRENATUS Reeve, 1848

1848. *Conus infrenatus* Reeve. *Conchologia Iconica,* 1(*Conus* Suppl.): Pl. 3, sp. 285 (Locality unknown).

?1974. *Conus* (*Rhizoconus*) *visagenus* Kilburn. *Durban Museum Novitates,* 10(6): 81, figs. I-III, text fig. I (East of Durban, Natal, South Africa).

Description.—Rather heavy, dull; low conical, the sides somewhat convex posteriorly; body whorl with a few weak spiral ridges above the base, otherwise smooth except for axial growth marks; shoulder roundly angulate, sometimes with a weak carina visible at the margin; spire low, the sides slightly convex, bluntly pointed; spire whorls slightly concave to slightly convex, usually too eroded to discern sculpture; fresher specimens show traces of at least three narrow spiral ridges crossed by wide and narrow axial threads; margins of whorls sometimes with very weak carinae visible, indistinct. Body whorl in fresher specimens creamy white, this rapidly eroding to pale violet or bright pinkish; about 6-20 spiral rows of large to small brownish dashes, these sometimes in pairs above and below midbody and containing clouds and blotches of brown; sometimes a secondary brown band of blotches at the shoulder; when spiral dashes are poorly developed, the shell appears to have three or two spiral bands of brown blotches only; base white; spire and shoulder whitish with curved and somewhat reticulated pale brown streaks and spots. Aperture rather wide, somewhat widened anteriorly; outer lip convex, thick; mouth whitish, violet, or pinkish depending on state of erosion. Columella long, narrow, oblique. Length 20-50mm.

Comparison.—If not too heavily eroded, *C. infrenatus* is likely to be confused only with *C. natalis.* In most *C. natalis* there are fine axial reddish brown wavy lines still visible and there are no spiral rows of dashes; even in heavily eroded *C. natalis* with the body pattern reduced to blotches, the tops of the spire whorls are largely dark brown, not just sparsely streaked and spotted as in *C. infrenatus. C. algoensis* and *C. guineensis* may be similar, but they are not so distinctly spirally dashed and do not have the brown blotching restricted mostly to three spiral bands. In *C. tinianus* the columella is distinctly cusped posteriorly and the pattern consists of large and irregular blotches in many specimens. *C. advertex,* which resembles the *visagenus* form in shape, has the columella with a heavy posterior cusp or talon and has the anterior quarter of the body whorl with heavy spiral dashes and lines of brown; there are fine punctate lines over at least the anterior half of the body whorl visible in many specimens. See also *C. bairstowi.*

Variation.—Fresh specimens are brownish with many spiral rows of brownish dashes from the base to the shoulder. With erosion the general color changes to pinkish or violet and often some of the lines of dashes may disappear. In extreme specimens the dashes are only visible above the base, near midbody, and below the shoulder in about 6-8 rows. Heavy blotching between pairs of dashed lines may be well developed or absent, perhaps in relation to erosion. Since I have seen no live-collected specimens, I do not know if the extent of spiral dashes and brown blotches is also variable in non-eroded shells, but probably it is.

Distribution.—Restricted to South Africa, but seldom found alive. Beach specimens are common on southern beaches, with living specimens recorded from Algoa Bay (very rare). Probably normally lives in rather deep water just offshore, as very fresh specimens are uncommon.

Synonymy.—*C. visagenus* is referred here with a query. The species was based on two specimens supposedly taken in 180 fathoms and was

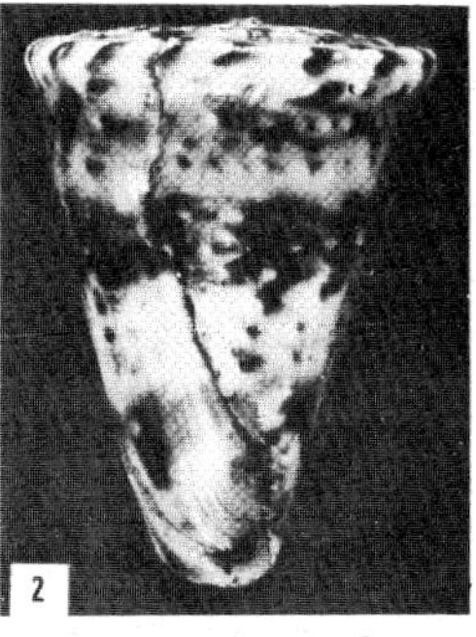

Left: Holotype of *C. visagenus*, 31.2mm (photo R.N. Kilburn).

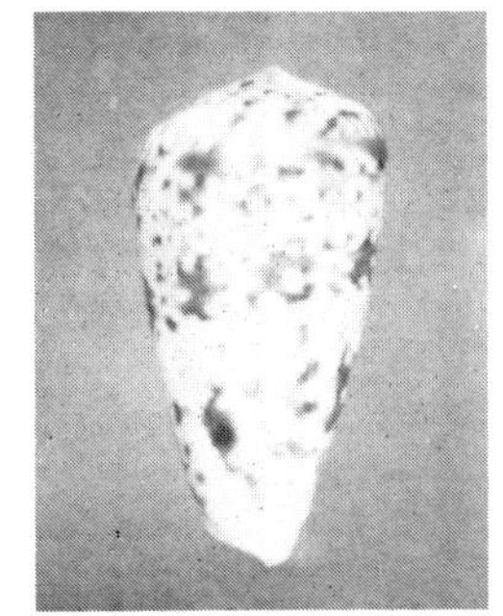

Right: *C. infrenatus*, eroded, 41mm.

compared mostly with *C. advertex*, which it admittedly resembles in shape. However, *C. advertex* differs in columella and color pattern, and I believe the two types of *C. visagenus* to be specimens of *C. infrenatus* with distorted shoulders. A similar flange-like distortion has been seen in some *C. altispiratus* and *C. natalis*, so perhaps there is some unknown ecological factor at work. The pattern of the types of *C. visagenus* can be matched quite easily in even small series of *C. infrenatus*, but I have not seen any *C. infrenatus* with as many spiral ridges on the spire whorls (8-10) as in *C. visagenus;* however, even my beach specimens sometimes show traces of at least three spiral ridges, so it is quite possible that the supposed differences in spire sculpture are insignificant. Certainly if the shoulder of *C. visagenus* is rounded off mentally, the specimens look like normal *C. infrenatus.* For the moment I think it best to consider *C. visagenus* a synonym of *C. infrenatus* until larger series of specimens agreeing with the types are available and can be compared with larger series of live-taken *C. infrenatus* (typical).

Notes.—*C. infrenatus* is a moderately common beach shell, subject to the usual erosion and bad lips of other South African beach cones.

REGIUS.—*Above* (*citrinus* and part *citrinus*): Left: Florida, 55.3mm (GG); Right: Long Reef, Florida, 40.4mm. *Below:* Bonaire, Neth. Ant., 29.8-45.5mm.

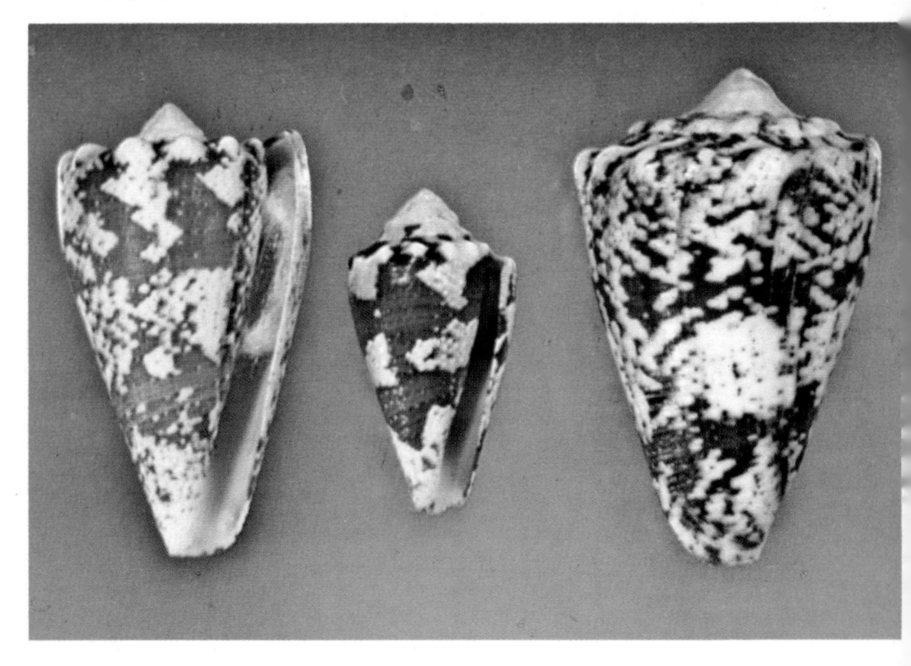

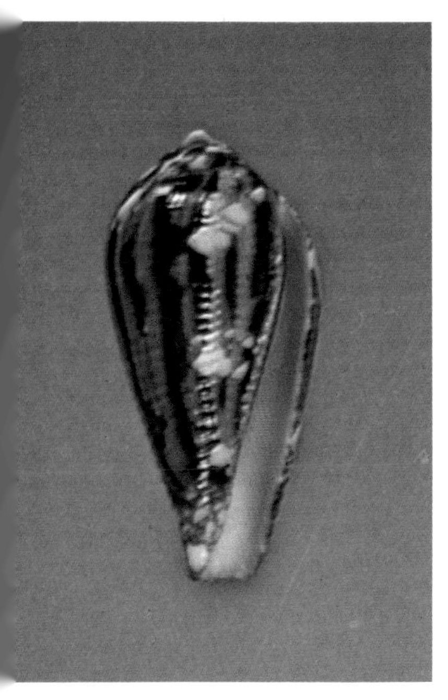

RETIFER.—Above: Left: Zanzibar, 35.3mm (MF); Right: Kona, Hawaii, Hawaii, 50.5mm. *Below:* Left: Quezon, P.I., 46.3mm (photo by A. Kerstitch); Right: Sorsogon, Palawan, P.I., 36.3mm (K).

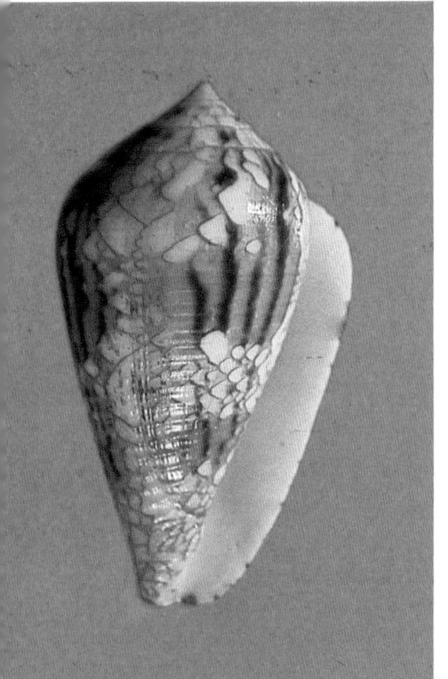

There are additionally often heavy axial breaks and growth marks visible. Bright pink or violet specimens are not uncommon, but neither are specimens still showing the more typical brown and white coloration. Specimens agreeing with *C. visagenus* are, as might be expected, very rare.

INSCRIPTUS Reeve, 1843

1843. *Conus inscriptus* Reeve. *Conchologia Iconica*, 1(*Conus*): Pl. 29, sp. 164 (Locality unknown).

1857-1858. *Conus Keati* Sowerby ii. *Thesaurus Conchyliorum*, 3(*Conus*): 34, pl. 20 (206), fig. 479 (Seychelles).

1870. *Conus planiliratus* Sowerby iii. *Proc. Zool. Soc. (London)*, 1870: 255, pl. 22, fig. 1 (Locality unknown). Non *Conus planiliratus* Sowerby ii, 1850, a fossil.

1891. *Conus adenensis* E.A. Smith. *Ibid.*, 1891: 401, pl. 33, fig. 1 (Aden).

1921. *Conus maculospira* Pilsbry and Johnson. *Proc. Acad. Nat. Sci. (Philad.)*, 73: 330. Nomen novum for *Conus planiliratus* Sowerby iii, 1870.

Description.—Moderately light in weight, glossy; low biconical, the sides distinctly convex to nearly straight posteriorly; body whorl with anterior half to two-thirds covered with broad, usually flattened spiral ribs separated by narrow grooves; rest of body whorl usually smooth except for fine axial threads; shoulder rounded to roundly angled, wide; spire low, sharply pointed, the sides concave to slightly convex; tops of whorls flat to slightly convex or concave, covered with about 6-10 narrow but sharp spiral ridges which appear to cover the entire whorl when viewed from posteriorly; axial threads very fine and not very distinct. Body whorl white, heavily covered with reddish brown to yellowish tan squarish spots in many spiral rows; the spots show a strong tendency to partially fuse into wide spiral bands above and below midbody, the blotches composing these bands with very irregular margins; below the shoulder the spots tend to fuse into axial flammules continued onto spire; base white; spire and shoulder whitish, covered with radiating brownish blotches continued from body whorl pattern; early whorls whitish. Aperture rather narrow, almost uniform in width; outer lip sharp, straight or slightly convex; mouth white to pale violet, uniform in tint. Columella narrow, long, sometimes set off from inner lip by a low ridge. Length 30-70mm.

Comparison.—*C. inscriptus* is closely related to the various species clustered around *C. stramineus*. *C. stramineus* has the mouth deep violet with distinct brown tones and should present no problems in identification. *C. subulatus* usually has the shoulder more strongly rounded, has the body whorl pattern less distinctly dashed with the spiral bands less distinct, has the spiral ridges of the spire whorls less conspicuous and perhaps fewer in number; the background color of the shell is often distinctly creamy white or bluish white instead of white. *C. janus* is more elongate and has the pattern tending to form irregular axial flammules; the mouth is often toned with yellow; otherwise *C. janus* is very like *C. subulatus. C. subulatus* and *C. janus* seldom have the spiral ribs as strong as in *C. inscriptus*, with the grooves not so distinctly unmarked white as in that species. *C. lynceus* is larger, more elongate, has a higher spire, and has a darker violet mouth than *C. inscriptus;* usually the pattern is much darker brown on off-white or orangish, the whole shell with reddish or violet tones. *C. tegulatus* is variable in sculpture but usually has the spiral ridges narrower and separated by wider grooves; commonly the ridging extends to the shoulder, although it may be very weak over the whole whorl; the shoulder is usually wider and more angulate, with the sides of the spire more distinctly concave than in *C. inscriptus.* In addition, the pattern of *C. tegulatus* is usually composed largely of broad wavy axial bands of brown to yellowish brown, the individual dashes comprising these flammules or bands being indistinct. *C. iodostoma* is smaller, has weaker sculpture, has a dirty violet mouth, and has the body whorl pattern very fine, either as dots or broken reticulations.

Variation.—This species seems to occur in two quite distinct forms which are partially separated geographically and perhaps ecologically. However, since the two types of shells both occur in the vicinity of Aden they cannot be subspecies and must be considered varieties, perhaps ecological in nature.

C. inscriptus typical: rather broad at the shoulder, with the spire often distinctly concave-sided and the whorls weakly stepped; the brown of the body whorl and spire is deep brown and very contrasting with the white of the body whorl background; the mouth is flushed with pale violet, at least deep within. Northeastern African coast and Red Sea to Aden and perhaps western India.

C. inscriptus variety *adenensis:* somewhat narrower and a bit more elongate, the shoulder roundly angled; spire with the sides more convex and not stepped; brown of body whorl usually pale tan to yellowish tan, often indistinct and leaving large white areas; spire bands more solidly fused in many specimens than in the typical variety; mouth glossy white, seldom with any violet visible or just very faint tones. Western Indonesian area and Bay of Bengal around India and Sri Lanka to Aden.

In addition to these differences, there are considerable variations in the extent of the ribbing on the body whorl, which may become obsolete below midbody or extend to shoulder, and in the convexity of the body whorl, which is usually almost straight but, especially in *adenensis*, may be distinctly convex. The grooves between ribs appear colorless, with this pattern often continued to shoulder even though grooves are absent.

Distribution.—Common in shallow to moderately deep water from the Red Sea and northeastern coast of Africa across the northern Indian Ocean to the Bay of Bengal and south along the western coast of Malaysia

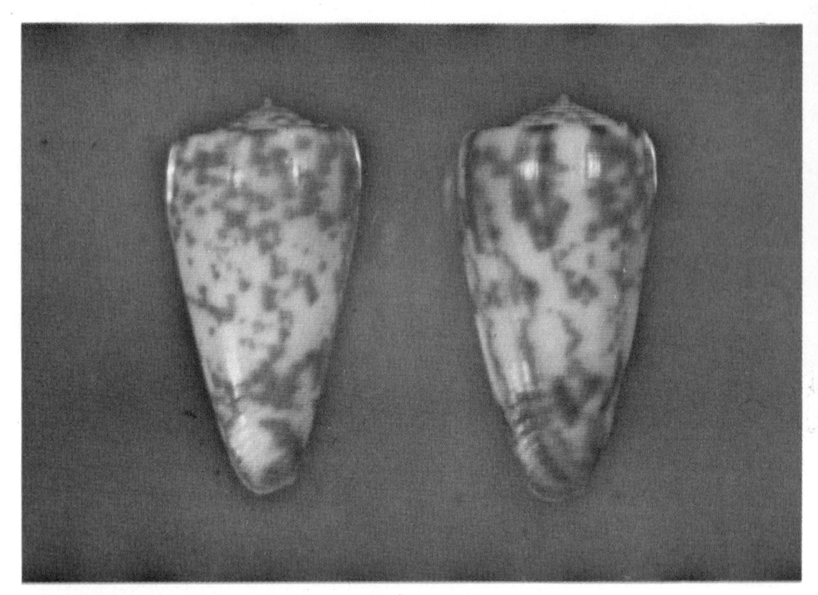

RUFIMACULOSUS.—Moreton Bay, Queensland, Australia, 40.5 and 41.4mm.

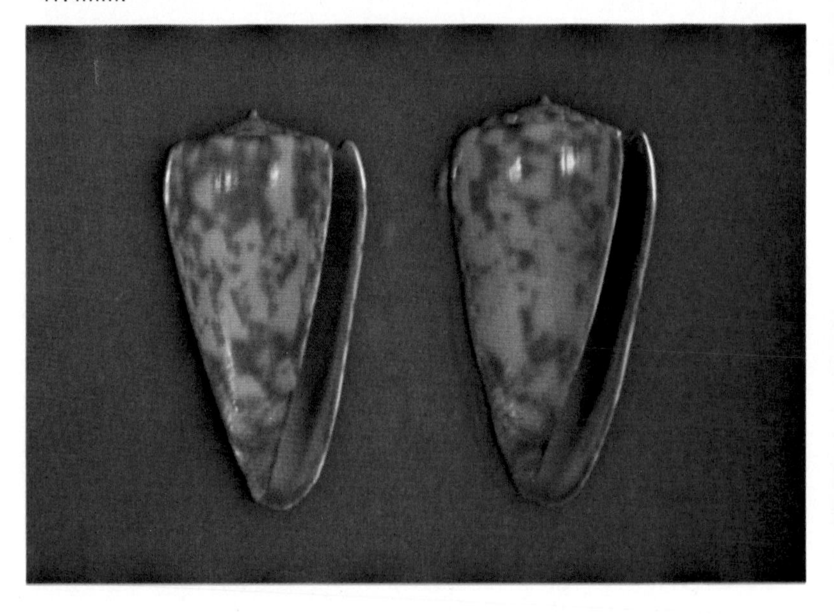

RUTILUS.*—Above:* Yallingup, Western Australia, Australia, 8.2 and 9.7mm (GG). *Below:* South Australia, Australia, 9.7 and 12.3mm (GG).

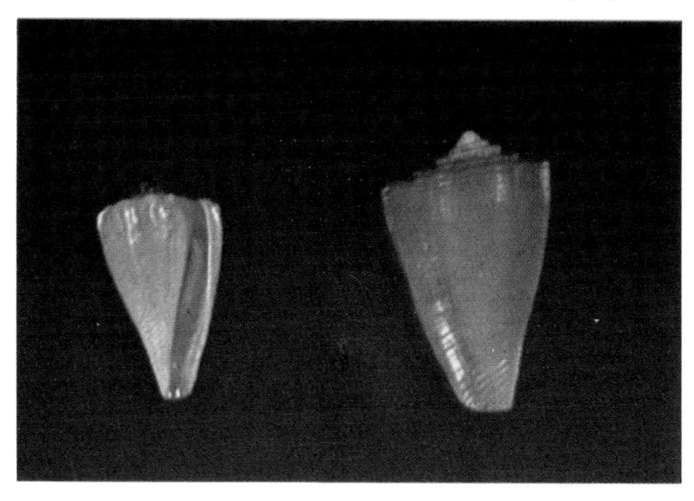

to Sumatra (Indian Ocean only, no valid Pacific records known).

Synonymy.—*C. inscriptus* of Reeve seems to be the common Red Sea -Arabian Gulf shell. *C. keati* is just a specimen of *C. inscriptus* with the spire whorls a bit more stepped than usual; it is not worthy of distinction. *C. planiliratus* is the first naming of the *adenensis* variety, but this preoccupied name was not replaced until 1921. *C. adenensis* looks very distinctive in typical form, with the brown distinctly yellowish and the mouth pure white; however, intermediate specimens with typical *inscriptus* are encountered not uncommonly; *C. adenensis* seems to be just an ecological variant of *C. inscriptus* although it is partially separated geographically.

Notes.—Good specimens of this species are now common, especially the *adenensis* form from southern India and Sri Lanka. Except for sometimes heavy axial growth lines which disrupt the pattern and minor chips on the lip, gems are not hard to find. Specimens from the Bay of Bengal are relatively large and sometimes tend to greatly reduce the pattern, eventually resulting in mostly white shells; such white shells are not rare, although they usually have some remnants of pattern.

Specimens of the typical form are sometimes sold as '*C. keati*' at large premiums—*beware*. There has been much confusion between this species and *C. stramineus* and *C. subulatus* in the past, although most specimens in collections are correctly identified. Lately *C. tegulatus* from the Bay of Bengal has entered the confusion, although misidentifications in this complex are more likely to occur between *C. subulatus* and *C. tegulatus* or *C. subulatus* and *C. stramineus* than between any of these and *C. inscriptus*.

INSCULPTUS Kiener, 1845

1845. *Conus insculptus* Kiener. *Species gen. et icon. des coqu. viv.*, 2 (*Conus*): Pl. 99, fig. 2. 1849-1850, *Ibid.*, 2: 309 (China Sea).

1854. *Conus sulciferus* A. Adams. *Proc. Zool. Soc.* (*London*), 1853: 116 (New Ireland). Non *Conus sulciferus* Deshayes, 1835, a fossil.

??1882. *Conus wilmeri* Sowerby iii. *Ibid.*, 1882: 118, pl. 5, fig. 5 (Port Blair, Andaman Islands).

1937. *Conus tmetus* Tomlin. *Proc. Malac. Soc. London*, 22(4): 206. Nomen novum for *Conus sulciferus* A. Adams, 1854.

Description.—Light in weight, with a low gloss or dull surface; elongate low biconical, the sides about straight posteriorly then tapering to a very narrow and long base, sometimes concave at middle; body whorl covered from base to shoulder with wide, flat or rounded spiral ridges separated by deeply punctate rather wide grooves; entire shell covered with very fine axial threads, these most distinct in grooves; shoulder wide, carinate, with or without nodules, concave above; spire tall, sharply pointed, the sides straight or concave; whorls distinctly stepped, concave above, the margins carinate and nodulose or not; tops of whorls with about 5-10 narrow spiral ridges crossed by heavy axial threads and distinctly cancellate. Body whorl pale brown to dark brown, sometimes rusty brown; occasionally with two spiral series of small brown blotches above and below midbody; spire paler brown to pale bluish tan, unmarked or with small brown blotches; carinae of nodulose specimens with small brown spots between nodules. Aperture narrow, uniform in width; outer lip very thin and fragile, concave at the middle; mouth whitish, pinkish, pale violet, or tan, probably due to shadow of external pattern showing through. Columella internal. Length 20-35mm.

Comparison.—This is another of the small needle cones which can be so difficult to identify. However, the presence of many distinct spiral ridges on the distinctly stepped spire whorls should distinguish this species from the sometimes similar *C. aculeiformis.* From *C. hypochlorus* it differs in being dull brownish, not waxy yellow, and in commonly having small brown spots on the carinae when nodules are present; *C. hypochlorus* is also narrower at the shoulder and less distinctly carinate, at least in some specimens. *C. comatosa* is larger, has the shoulder carina more ridge-like, and has some type of pattern over the entire body whorl, although it may be nebulous; the dark spots on the carinate spire margins are stronger and constant whether nodules are present or not. *C. schepmani* is very similar in shape to some *C. insculptus*, but it is distinctly patterned with opaque white and dark reddish brown spots near midbody.

Variation.—In color pattern and spire sculpture as well as general shape it would seem that this species breaks into two rather indistinct forms which further collecting might prove to be distinct species. Until then it is best to consider them just forms of a single species, although there is a valid name for each type.

C. insculptus typical: Rather wide at the shoulder and concave at midbody, the spire whorls without nodules; body whorl totally brown or with two spiral rows of darker brown blotches; spire plain or with brown blotches. Philippines to Fiji; rare.

C. insculptus variety *tmetus:* Narrower, the shoulder not as wide and the body whorl not as concave at midbody; spire whorls and usually shoulder with distinct nodules; body whorl all brown; spire without pattern except for small brown dots between nodules. Solomons and Philippines area west to Bay of Bengal; not uncommon.

Distribution.—Dredged in deep water from the Bay of Bengal through the China Sea and Philippines to the New Guinea-Solomons area; also Fiji. Moderately common to rare.

Synonymy.—I have no assurance that *C. wilmeri* belongs here; the figure is somewhat like *C. insculptus* in shape, although not as elongate; the spire is not nodulose, but Sowerby describes only 3 spiral ridges and apparently a spire sculpture not unlike that of *C. aculeiformis*. Probably *C. wilmeri* is just a poorly figured *C. insculptus*, but it could just as easily

SANGUINOLENTUS.—Above: Tahiti, Society Is., 37.2 and 37.5mm. *Below:* Queensland, Australia: Left: 41.2mm; Right: 24.8mm (R.W. Pitman).

SAZANKA.—*Above:* Waikiki, Hawaii, 23mm (DM) (photo by Jayavad). *Below:* Left (paratype of *C. kurzi*): Off Midway I., Hawaii, 26.3mm (EP); Right (holotype): SW Kochi Pref., Japan, 40.0mm (photo by Dr. T Shikama).

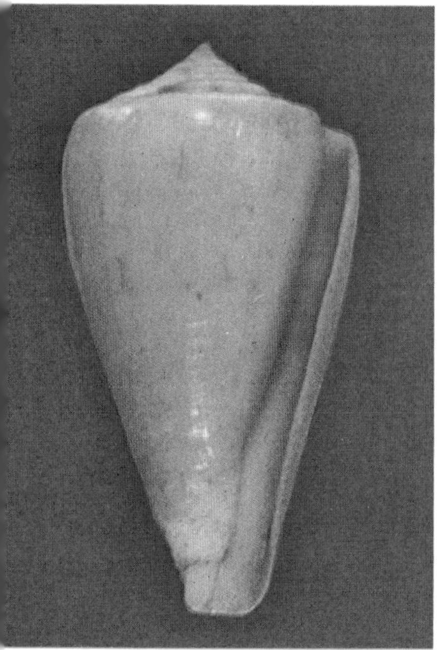

be a *C. aculeiformis* without pattern and with the anterior canal broken.

C. insculptus of Kiener is rather wide at the shoulder, uniformly brown, and has the spire whorls not nodulose. *C. sulciferus* (replaced by *C. tmetus* because of preoccupation) seems to represent with little doubt the form with distinctly nodulose spire whorls and somewhat narrower body whorl. The differences between the two types might very well be of specific value, but I have not seen enough specimens to be sure.

Notes.—The *tmetus* variety is not uncommon and is readily available, but the typical variety appears to be quite rare and seldom offered. Although a fragile shell, there is seldom any major damage except for a broken lip and occasionally heavy growth marks.

I am not sure of the relationship of this species to *C. hypochlorus*, the description and photos of which should be checked carefully.

IODOSTOMA Reeve, 1843

1843. *Conus iodostoma* Reeve. *Conchologia Iconica*, 1(*Conus*): Pl. 28, sp. 159 (Locality unknown).

Description.—Moderately heavy, with a good gloss; low biconical, the upper sides convex; body whorl with weak but wide spiral ridges on anterior fourth to half, these separated by shallow grooves; shoulder rounded to roundly angled, wide; spire moderately high, sharply pointed, the sides deeply to shallowly concave; spire whorls slightly concave, with about 3-5 wide spiral ridges crossed by fine axial threads; early whorls often heavily eroded. Body whorl creamy white with pale to distinctly violet tones, covered with numerous spiral rows of small orange-brown dots that are rather closely spaced; tendency to concentrate spots into two broad but indistinct bands above and below midbody area; often with small irregular orange-brown blotches also above and below midbody; base whitish; spire and shoulder whitish, with large to small brownish blotches and fine dots at the margins; early whorls white. Aperture moderately wide, a bit wider anteriorly than posteriorly; outer lip thin, sharp, straight or slightly convex; mouth dark violet deep within. Columella narrow, moderately long. Length 25-45mm.

Comparison.—The rather rounded shoulder, weak anterior sculpture, and dark violet mouth ally this species most closely to *C. subulatus* and less so to *C. stramineus*. *C. stramineus* is usually a larger and more cylindrical shell with the mouth even darker than in *C. iodostoma*; the spots on the body whorl are large and tend to form wide spiral bands in many specimens. *C. subulatus subulatus* is often similar in mouth color to *C. iodostoma* but has the spots distinctly larger and not dot-like; the spots of *C. subulatus* (both subspecies) tend to be axial or form wavy blotches. Actually, the fine spotting of *C. iodostoma* is very distinctive and identification of most specimens should cause no problems.

Variation.—There is the usual variation in height and concavity of the spire and to some extent the convexity of the posterior sides. The spire whorls have the ridges rather wide and usually about 3-4 in number, although some have as many as six distinct ridges. Generally the body whorl has a distinct violet tone, although this may be very pale. Intensity of spotting varies considerably; often the midbody area is almost devoid of spots.

A subadult specimen assigned to this species (from Zanzibar) differs somewhat from typical adults in being a bit more angled at the shoulder, having the spire comparatively heavily marked, and having the body whorl covered with remnants of a very open reticulated pattern and two series of brownish blotches (above and below midbody); new growth seems to be progressively less reticulated and more spotted, so the margin of the lip is only dotted. Certainly this specimen is *C. iodostoma*, but it also brings to mind the pattern of such species as *C. lienardi* and *C. wittigi*, although these of course differ in sculpture and other aspects of the pattern.

Distribution.—Apparently restricted to the western Indian Ocean from the African coast and Madagascar, probably including the smaller islands. Uncommon.

Notes.—Gem specimens of this species are not easy to obtain because there seems to be a tendency toward erosion of the spire and a few heavy axial growth marks. The lip chips easily. Most specimens come from Madagascar and Tanzania, so the distribution is poorly known. There exists the possibility that it is subspecifically related to *C. subulatus*, but it is sympatric with *C. janus*, a taxon even more closely related to *C. subulatus*.

IONE Fulton, 1938

1938. *Conus ione* Fulton. *Proc. Malac. Soc. London*, 23(1): 55, pl. 3, fig. 2 (Kii, Japan).

Description.—Light in weight, with a high gloss; low biconical, the upper sides almost straight then concave to a narrow base; body whorl with about a dozen low spiral ridges basally, these sometimes extending above midbody in a very weak way; grooves between ridges weakly punctate; axial growth line often heavy; shoulder wide, concave above, the margin strongly carinate and with a flat rim; spire moderately low to low, the sides concave to almost straight; first 2-3 whorls with very weak nodules, later whorls with carinate margins with flat rims; tops of whorls concave, with very heavy curved axial ridges and at best scattered traces of a single spiral thread; whorls distinctly stepped. Body whorl pale violet, covered with widely spaced spiral rows of small squarish reddish brown dots, these sometimes hard to see; usually two bands of large

SCABRIUSCULUS.—*Above:* Tahiti, Society Is., 33.4 and 35.1mm. *Below:* Marau, Guadalcanal, Solomon Is., 28.1 and 28.8mm.

SCALPTUS.—Above: Matupit, E New Britain, Papua New Guinea; Left: 32.9mm; Right: 37.8mm. *Below* (live-taken): Left: Off Deka-Deka I., China Straits, Papua New Guinea, 24.1mm (EP); Right: Tolwat, E New Britain, Papua New Guinea, 38.1mm.

irregularly squarish reddish brown blotches above and below midbody and occasional traces of smaller blotches above base; base whitish; spire and shoulder pale violet to whitish, with few to several radiating brownish lines. Aperture moderately wide, uniform; outer lip sloping a bit below level of shoulder, thin, concave at middle; mouth pale violet. Columella internal. Length 45-70mm.

Comparison.—*C. ione* is often confused with *C. sieboldii* when the reddish brown blotches are the only features noted. However, *C. ione* is broader in relation to width than *C. sieboldii*, with the body whorl less elongated and the spire lower. In *C. ione* the axial sculpture of the spire is heavier and the carinae are distinctly rim-like rather than with traces of obsolete nodules. The violet color of *C. ione* is constant though very pale and apparently not found in *C. sieboldii*, which also lacks the many rows of small dots. *C. otohimeae* is somewhat similar in general shape but is orange in tone and has the spire whorls flat or convex above and without rim-like carinae; although small spots are present on the body whorl, there is a wide band of large orange blotches below midbody that is quite unlike the reddish brown blotches of *C. ione*. *C. kermadecensis* may be similar in shape and pattern to some extent but is much thicker and heavier and not violet in tone as well as differing in details of pattern and sculpture.

Variation.—Because the shape and pattern are so simple there is little variation in this species. The spire may be low or moderate, the whorls weakly or strongly stepped and heavily or sparsely marked with brown. Commonly the shoulder carina has distinct larger spots contrasted with those of the body whorl. The larger blotches may be indistinct and pale or dark and large, occasionally partially fused spirally. The small dots and violet background are constant, although the dots may be absent in areas adjacent to breaks.

Distribution.—Uncommon in deep water off southern Japan; probably also found off Taiwan.

Notes.—This species is seldom offered today but is apparently not rare. It is easily mistaken for *C. sieboldii* on first glance, especially in specimens with the periostracum still present. The two are quite distinct in many features, however, and readily distinguished by careful observation. The tip of the spire may be broken or eroded in many specimens, and there are commonly bad growth marks on the body whorl and spire, as well as occasional unhealed breaks. The lip is very fragile and seldom complete.

JANUS Hwass, in Bruguiere, 1792

1792. *Conus Janus* Hwass, in Bruguiere. *Cone*, in *Ency. Method., Hist. Nat. des Vers*, 1: 690 (Indian Ocean). Lectotype selected and figured by Kohn, 1968.

Description.—Moderately heavy, with a good gloss; rather elongate biconical, the upper sides slightly convex; body whorl with about 4-12 spiral ridges at the base, these separated by shallow punctate grooves; rest of body whorl smooth except for fine spiral and axial threads; shoulder broad, roundly angled; spire moderately high, sharply pointed, the sides strongly concave; upper surface of spire whorls concave, with about 3-4 spiral ridges on the posterior part of the whorl, these crossed by a few weak axial threads. Body whorl glossy white with a variable pattern of yellowish brown to bluish brown wavy axial flammules which may be continuous from shoulder to base or be restricted to the vicinity of three bands of smudgy blotches; these bands are strongest at midbody and below the shoulder and also occur above the base; base often deep yellowish or yellowish brown; usually traces of spiral rows of brownish dashes visible somewhere in pattern; spire whitish with dark brown irregular radiating blotches and lines; tip brownish. Aperture moderately wide, especially anteriorly; outer lip thin, slightly convex or straight; mouth white, sometimes stained yellowish. Columella long, narrow, set off posteriorly by a low ridge. Length 50-76mm.

Comparison.—*Conus janus* is an old species which has seldom been correctly understood. It appears that it is very closely related to the shells here called *C. subulatus andamanensis*, which is quite possibly related subspecifically. *C. janus* appears to be distinctly more elongate than *C. s. andamanensis* of similar size, with a more tapered body whorl and relatively wider shoulder; the axial flammules are more commonly complete in *C. janus*, which has a sharper pattern that tends more toward pale colors than many *C. s. andamanensis*. Although both shells have a whitish aperture, that of *C. janus* is often distinctly yellowish, while that of some *C. s. andamanensis* is sometimes very pale violet. The two appear to be allopatric.

C. inscriptus has the mouth pale violet in some populations and has the pattern of brown in the form of strong dashes which tend to fuse into two broad spiral bands; the spire is usually lower than in *C. janus*. *C. jickelii* is a white shell with a very irregular pattern showing a weak tendency to produce larger blotches above and below midbody.

Variation.—This species is very constant in shape but variable in color and pattern. The brown on the body whorl ranges in tone from pale orangish brown to dark bluish brown, almost black. At the light extreme only three series of brownish blotches are present, these strongest below the shoulder and at midbody and connected by a few spots in axial series. At the other extreme the axial spots are fused into zigzag flammules and are continuous from shoulder to base, although commonly weak and broken at midbody. There are usually at least traces of spiral brown dash lines somewhere on the body whorl, usually near midbody, but they may be completely absent in occasional shells. Juveniles are relatively narrower and covered with widely spaced brownish dashes. In many adults there is a strong tendency toward the development of yellow stains at the

SCHEPMANI.—Above: Left: Russell I., Solomon Is., 23.9mm (MG and BB); Right: Okinawa, Japan, 24.8mm (TCG). *Below:* Left (holotype of *C. elegans*): Bougainville Straits, 21.0mm (Zool. Mus. Amsterdam) (photo by Dr. R.T. Abbott); Right: P.I., 31mm (Dayrit) (photo by W.E. Burgess).

SCOPULORUM.—Off Salinopolis, Brazil, 25.8mm (EP).

base and deep within the mouth.

Distribution.—Seemingly restricted to the islands of the southern and central Indian Ocean (Madagascar, Mauritius, Reunion, etc.) and the central African coast. Uncommon in shallow to moderately deep water.

Notes.—This species has recently entered the market in fairly large numbers and must be classed as currently uncommon instead of rare, as it was until recently. Gems are easy to find, as this is a clean species which seldom has more than very minor flaws in specimens that reach the market. Juveniles are uncommon. Of the various pattern types, probably the least common are those with very dark, almost black, coloration and those with complete axial flammules which extend unbroken from shoulder to base.

As mentioned in the comparison, *C. janus* is very close to *C. subulatus andamanensis* and differs only in minor tendencies of shape and pattern. Should it be determined that actual intergradation occurs, the complex would become *C. janus janus*, *C. j. subulatus*, and *C. j. andamanensis*.

JASPIDEUS Gmelin, 1791

1791. *Conus jaspideus* Gmelin. *Systema Naturae per Regna Tria Naturae*, ed. 13, 1: 3387 (Puerto Plata, Santo Domingo, designated by Clench, 1942). Lectotype figure (Clench, 1942): Martini, Pl. 55, fig. 612.

1792. *Conus verrucosus* Hwass, in Bruguiere. *Cone*, in *Ency. Method.*, *Hist. Nat. des Vers*, 1: 708 (Senegal). Lectotype figure (Kohn, 1968): *Tableau Ency. Method.*, Pl. 333, fig. 4.

1830. *Conus Pealii* Green. *Trans. Albany Inst.*, 1: 123, pl. 3, fig. 3 (Florida).

1845. *Conus nodiferus* Kiener. *Species gen. et icon. des coqu. viv.*, 2 (*Conus*): Pl. 100, fig. 4. 1849, *Ibid.*, 2: 228 (Mer des Indes).

1845. *Conus echinulatus* Kiener. *Ibid.*, 2(*Conus*): Pl. 105, fig. 2. 1849, *Ibid.*, 2: 270 (Locality unknown).

1854. *Conus stictus* A. Adams. *Proc. Zool. Soc. (London)*, 1853: 117 (Locality unknown).

1869. *Conus Stearnsii* Conrad. *American J. Conch.*, 5: 104, pl. 10, fig. 1 (Florida).

?1870. *Conus tenuisulcatus* Sowerby iii. *Ibid.*, 1870: 256, pl. 22, fig. 10 (Locality unknown).

1944. *Conus verrucosus vanhyningi* Rehder. *Nautilus*, 57(3): 105 (Off Pompano, Broward Co., Florida).

1953. *Conus jaspideus branhamae* Clench. *Johnsonia*, 2(32): 364, fig. 2 (Green Turtle Cay, Great Abaco, Bahama Islands).

1968. *Conus pseudo-jaspideus* Usticke. *Caribbean Cones from St. Croix and the Lesser Antilles:* 7, pl. 1, sp. 985 (St. Croix and Martinique).

Description.—Light in weight but solid, with a good gloss or dull;

biconical, the upper sides straight or gently convex; body whorl with broad, low, rounded spiral ridges over at least the anterior third of whorls, these sometimes continuing to shoulder; grooves between ridges finely axially threaded; ridges may have few to many low or heavy granules; many axial threads over whorl; shoulder broad, carinate, with indications of a flattened rim; spire moderate, sharply pointed, the sides straight or nearly so; spire whorls weakly stepped or not, the margins carinate and sometimes formed into an indistinct rim; tops of whorls concave or flat, bearing only fine curved axial threads and no spirals. Body whorl whitish to pale grayish brown, usually with indications of brownish to reddish brown clouds and blotches and spiral rows of brown and white dots and dashes; color and pattern highly variable; shoulder and spire about same color as body whorl, often with small spots of brown at the margins, sometimes with weak streaks and other blotches; tip pinkish or whitish. Aperture moderately wide, especially anteriorly; outer lip straight or slightly convex, thin; mouth white to pale brown. Columella very narrow, sometimes not visible externally. Length 15-31mm.

Comparison.—As restricted here, *C. jaspideus* is much easier to define and recognize than previously. It is very close to *C. puncticulatus* in size and sometimes pattern, but that species is recognized by the rounded shoulder, slender and deeply concave spire, and usually the violet mouth. *C. mindanus* (small tropical form) is very similar in having a dashed pattern and sometimes being pustulose, but it has the basal sculpture (grooves) heavier, the mouth wider, the spire more concave, and the shoulder more rounded; in addition, commonly the whorls of *C. mindanus* are rude and do not overlap evenly at the margins. *C. magellanicus* is sometimes similar but tends to be much waxier in texture, has only low spiral ridges at the base, is narrower, has different patterns as a rule, and has spire whorls and shoulder angulate but not with a carinate rim. Non-Caribbean species such as *C. clarus* and *C. cyanostoma* may be similar in shape and pattern but usually have distinct spiral ridges on the spire whorls.

Variation.—This highly variable species is here defined to exclude the round-shouldered shells of the Caribbean, but even so this is a confusing species. There is much localized variation resulting in shells which appear very distinct from small series but with intermediates found at other localities or in large series. The most confusing variant probably is the *verrucosus* form. This form, presumably of ecological origin although some populations appear genetically stable, is defined by having granules on some or all the spiral ridges and sometimes the margins of the whorls; the granules may be small and indistinct, restricted to the basal ridges, or very large and heavy and extending to the shoulder and spire. Since granules may occur in combination with any size, shape, and pattern of the species, their utility as a taxonomic character is nil.

Excluding granulation, variation in the species breaks into two very poorly defined type, which have been called subspecies. However, the

SCULLETTI.—Cape Moreton, Queensland, Australia. *Above:* 44.3mm. *Below:* Left: 42.8mm (K); Right: 42.2mm.

SELENAE.—Above: Brazil, 15mm (DM) (photo by Jayavad). *Below:* Left: Off Salinopolis, Brazil, 15.8mm (EP); Right: Off Fortalesa, Ceara, Brazil, 15.0mm (DMNH).

stearnsii type seems to be a local form which is barely distinguishable from typical shells and not restricted to a particular range although most common on the western coast of Florida. The two forms, here called varieties, can be distinguished as follows.

C. jaspideus typical: Shell relatively wide at the shoulder, not especially elongate; generally whitish with a variable pattern of reddish brown spots and blotches as well as dashes, sometimes the brown completely absent and shell entirely white; spiral ridges often extend to the shoulder or at least above midbody; *verrucosus* form common; ranging throughout the range of the species, although not common on the western coast of Florida.

C. jaspideus variety *stearnsii:* Shell relatively narrow and elongate; background color generally grayish or grayish brown with numerous rows of fine brownish or reddish dashes alternating with small white spots; axial flammules of reddish brown commonly present; spire dark like body whorl, often with white carinae bearing distinct spots of brown; spiral ridges seldom extending beyond midbody; *verrucosus* form rare; restricted in pure form to some populations on the western coast of Florida, but also found as part of more typical *jaspideus* populations in the northern Caribbean.

Distribution.—Widespread and common in shallow to moderately deep water from off North Carolina south to Florida and the Bahamas; throughout the Caribbean both in the Antilles and the Central American coast south to central Brazil; also in northern Mexico and off Texas in the Gulf of Mexico.

Synonymy.—Some of the synonyms listed here probably belong elsewhere, while probably there are several other names listed elsewhere that belong here. Certainly *C. verrucosus* and *C. stearnsii* belong here as explained above. *C. nodiferus* and *C. echinulatus* appear to be *verrucosus* forms, as does *C. vanhyningi* (a white *verrucosus* with a pink mouth). *Branhamae*, *pseudojaspideus*, and *pealii* also appear to fall easily within the normal range of variation of *C. jaspideus* and cannot be defined by any constant characters. *C. tenuisulcatus* and *C. stictus* are too poorly described to be certain of their identities, but they probably are *C. jaspideus.* Many of the above names are used interchangeably by some dealers for any exceptional forms of *C. jaspideus* or even of *C. mindanus.* Beware of any *jaspideus*-like shell offered at a high price.

Notes.—Except for occasional axial growth marks and minor spire erosion or growths, this is a clean species. Dealers sell the many common and uncommon variants under a variety of names without any real consistency of usage, so great caution must be exercised when buying variants sight unseen.

Several names commonly cited as synonyms of *C. jaspideus* have been removed to the synonymy of *C. puncticulatus* and *C. mindanus.* Doubtless the current treatment of all these shells as one variable species is incorrect, but I am not sure that treatment of them as three full species is the final answer. Certainly the small tropical *C. mindanus* represent a closer approach to *C. jaspideus* than to Bermudian forms of the species, while some forms of *C. puncticulatus* are also very close though distinguishable. Much more work and systematic collecting of all known populations in the Caribbean is necessary before a final solution will be possible.

It might also be pointed out that the name *C. jaspideus* for this species is probably incorrect. The lectotype figure is almost certainly a specimen of the very dark *C. ventricosus* known as *C. franciscanus*. Perhaps it would be best to call the species *C. verrucosus*, but *C. jaspideus* is now well established and it would serve little purpose to change it now on uncertain evidence of a very bad figure.

JICKELII Weinkauff, 1873

1873. *Conus Jickelii* Weinkauff. *Syst. Conch. Cab.*, ed. 2, 218(*Conus*): 206, pl. 32, figs. 11-12 (Massawa).

Description.—Light in weight, with a high gloss; low conical, the upper sides convex; body whorl with a few low flat spiral ridges on the anterior third to half, these separated by shallow grooves; fine axial and spiral threads over body, producing a rather silky appearance; shoulder roundly angled, wide; spire moderate, sharply pointed, the sides concave; tops of whorls slightly concave, with about three low spiral ridges clustered posteriorly on the whorls. Body whorl glossy white or cream, covered with many small squarish spots of dark reddish brown arranged in irregular spiral rows and vaguely axial rows; these usually heaviest anteriorly; usually with irregular larger blotches scattered in two indistinct spiral bands above and below midbody area; spire white, with scattered reddish brown spots and blotches. Aperture rather wide, somewhat wider anteriorly; outer lip thin, convex; mouth creamy white within. Columella long, narrow, indented from inner lip. Length 30-40mm.

Comparison.—*C. jickelii* is related to the group of species containing, among others, *C. janus* and *C. inscriptus*. Although it has few distinctive characters, it seems to fall outside the range of variation of these species and is probably distinct. The white mouth distinguishes it from violet-mouthed populations of *C. inscriptus* (as well as *C. stramineus* and *C. subulatus subulatus*, which have violet mouths). From the *adenensis* form of *C. inscriptus* it differs in having weaker sculpture on both body and spire whorls and in the much less distinct pattern of scattered spots. *C. janus* is more elongate and has the pattern heavily developed and often axially flammulate, with a heavy spire pattern and often yellowish staining at the base and in the mouth. *C. iodostoma* is more rounded at the shoulder, thicker, has the spots much smaller and more numerous, and has a violet mouth.

SENNOTTORUM.—*Above:* Campeche, Yucatan, Mexico, 31.9 and 34.2mm (DMNH). *Below:* Left: Campeche, Yucatan, Mexico, 28.4mm (PC); Right (aff. *sennottorum*): Off Arcas Cay, Campeche Banks, Yucatan, Mexico, 36.1mm (photo by A. Kerstitch).

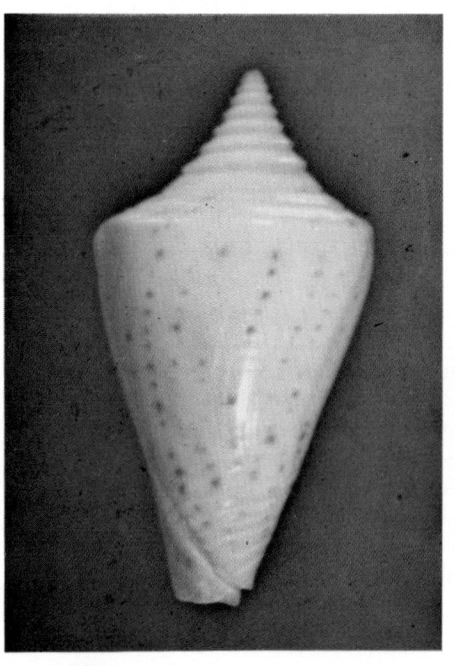

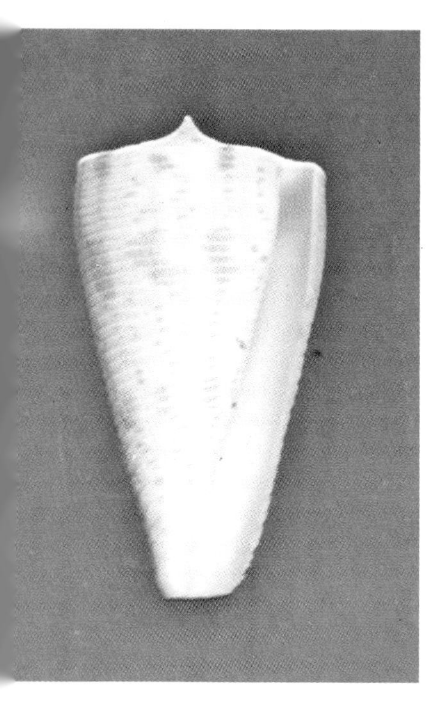

SIBOGAE.—*Above:* Left: Off Enu I., Aru Is., Indonesia, 39.2mm (EP); Right: Keppel Bay, Queensland, Australia, 38.7mm (GG). *Below:* Keppel Bay, Queensland, Australia, 29.5mm.

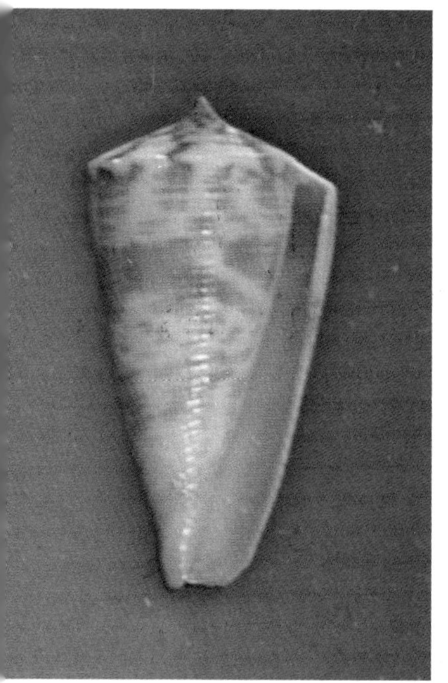

Variation.—The few specimens I have seen have been very constant in shape and pattern, the major differences being height of spire and number of spots on the body whorl. Although the spots are fairly large and cover the whole whorl, they tend to concentrate at the base and form spiral lines; in some shells they also are aligned in distinctly axial rows. The larger blotches are usually very irregular and sometimes absent. The spire pattern is usually weak.

Distribution.—Apparently restricted to the Indian Ocean, where it is rare or uncommon in deep water. There are apparently valid records for the Red Sea and Madagascar, but it is probably much more widely distributed.

Notes.—Although apparently related to *C. stramineus* and allies, there is also some resemblance to *C. spectrum* and allies. However, whichever group it really belongs to, it seems to be fairly distinct, although by rather negative characters. The specimens I have seen were all gems or near gems, the major flaws being filed lips.

JULII Lienard, 1870

1870. *Conus Julii* Lienard. *J. Conchyl.* (*Paris*), 18: 304 (Mauritius). 1871, *Ibid.*, 19: 17, pl. 1, fig. 6.

Description.—Solid, thick but light; glossy; oval, the greatest width well below shoulder, where a thickened cord may be developed; body whorl with a few widely spaced spiral grooves on the basal fourth, otherwise smooth or nearly so except for fine axial threads; shoulder rather wide, roundly angled; spire low, pointed, the sides straight or nearly so; tops of whorls flat or nearly so, with many fine axial threads and a few weak spiral ridges; a deep groove may divide the whorls at the anterior fourth; margins of early whorls sometimes undulate. Body whorl pinkish white or white, with many fine wavy axial brown lines tending to concentrate into blotches below the shoulder and above and below midbody; the blotches above and below midbody may be large and partially fused into spiral bands; lines and blotches often distinctly orange in tone; some opaque white spots and blotches hidden in background; spire whitish, with scattered orange-brown spots and blotches. Aperture wide, very wide anteriorly; outer lip oblique posteriorly then nearly straight; lip thick, rounded; mouth deep orange to rosy. Columella long, narrow. Length 35-65mm.

Comparison.—This poorly known species is quite distinct from most others in combining the shape of *C. bullatus* with the pattern (in part) of *C. floccatus*. *C. bullatus* is very similar to *C. julii* in shape but lacks the axial brownish lines. *C. floccatus* has similar lines but is usually a violet-toned shell with heavy axial flammules and large squarish brown spots in spiral rows as well; it also is finely spirally ridged.

Variation.—Although few specimens are known, this seems to be a uniform species. The spire varies from low to very low, while the density of the lines and size of the blotches of course vary a bit. Some shells are distinctly spirally banded, the blotches heavily fused. There is a considerable change from juvenile to adult, however. Large specimens are thicker than juveniles and have the widest portion of the body whorl prominent and sometimes corded. Juveniles have spiral grooves visible to the shoulder, although they are seldom conspicuous. In juveniles the pattern is poorly developed, with the blotches especially indistinct.

Distribution.—A rare shell known from moderately deep water on the coasts of Mauritius and Reunion. Probably more widely distributed but perhaps confused with *C. bullatus;* a photo of a specimen apparently of *C. julii* has been seen from Mozambique.

Notes.—Adults in good condition are beautiful shells toned with pink and orange. Like other species related to *C. bullatus*, there is a strong tendency to lose the pattern near the outer lip in adult shells; this should not be considered a flaw. At the moment *C. julii* is a desirable species in any condition.

KASHIWAJIMENSIS Shikama, 1971

1971. *Conus* (*s.s.*) *suturatus kashiwajimensis* Shikama. *Sci. Rpts. Yokohama Nat'l. Univ.*, Sec. II(18): 32, pl. 3, figs. 23-24 (Kashiwajima Island, S.W. Kochi Prefecture, Japan).

Description.—Heavy, with a good gloss; elongate obconical, the sides about straight, sometimes slightly convex posteriorly or weakly concave at the middle; body whorl with slightly enlarged spiral ridges at the base, these rather undulate; rest of whorl covered with sometimes strong spiral and axial threads, giving the whorl a satiny sheen; shoulder broad, roundly angled; spire low, almost flat, the sides straight or convex; tops of whorls slightly concave to flat, the early whorls with 1-2 low spiral ridges, these becoming narrower and less distinct on later whorls; last whorls and shoulder appear to be covered with fine spiral threads; all whorls with scattered axial threads; first 3-5 whorls weakly nodulose, later whorls with straight margins. Body whorl pale pink, covered with about four broad spiral bands of deeper pink and orange; midbody area usually very pale, surrounded by broad bands of deep orange or orange-pink; base and area below shoulder covered with broad bands of paler pinkish to pinkish violet or pale orange; base sometimes very pale; whole shell sometimes suffused with bright violet; spire pinkish orange to pinkish violet, with a few large, slightly darker orange blotches visible; tip pink. Aperture moderately narrow, slightly widened basally; outer lip straight or nearly so; mouth pale pink to pinkish violet; no dark staining anteriorly on shell. Columella long, very narrow. Length 35-90mm.

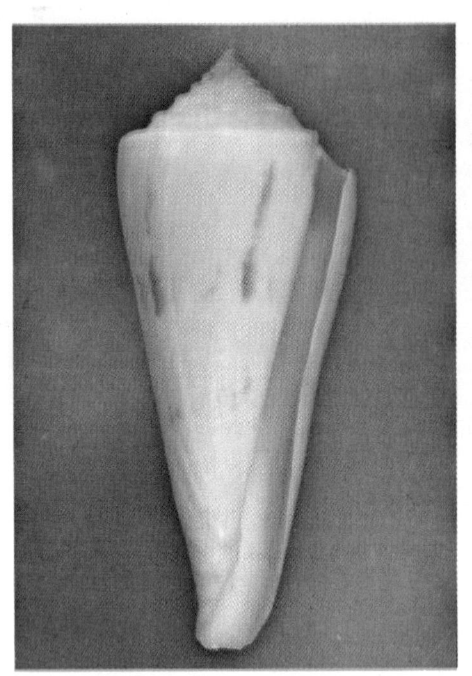

SIEBOLDII.—Above: Kii, Japan: Left: 92.1mm; Right: 71.5mm. *Below:* Mikawa, Aichi Pref., Japan: Left: 62.3mm; Right: 75.1mm.

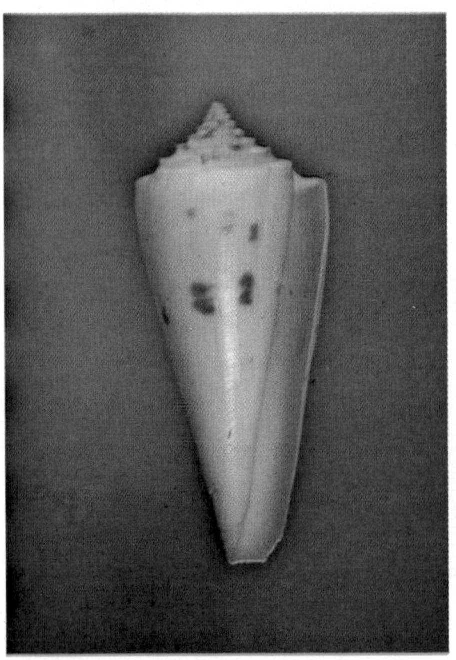

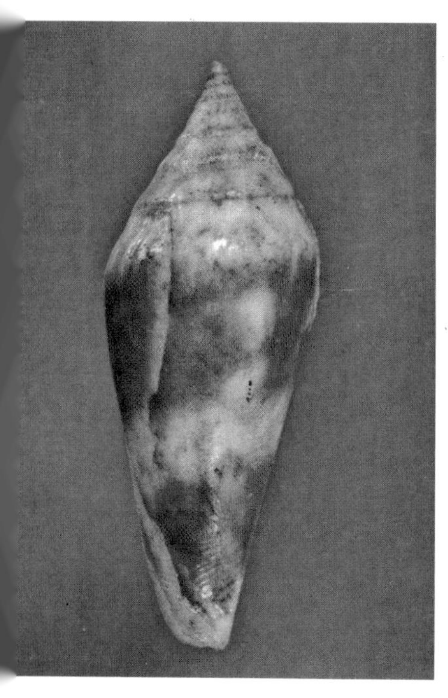

SMIRNA.—*Above* (*s. smirna*): SW of Midway I., Hawaii, 69.1mm (EP). *Below* (*s. profundorum*): Kii, Japan, 80mm (DM) (photos by Jayavad).

Comparison.—This species is very poorly known but appears quite distinct. The obvious comparison is with *C. suturatus*, at least the typical non-Hawaiian form, which has a very similar color pattern of broad pinkish or orangish bands. However, *C. suturatus* is much less elongated than *C. kashiwajimensis* and has the base of the shell stained with pale to deep violet. In addition, the spire sculpture of *C. suturatus* is constant and changes little with growth, featuring a broad median ridge sometimes with a very weak second ridge; this pattern is found in *C. kashiwajimensis* only on the early whorls (usually with the second ridge well developed), soon being replaced by more numerous but very narrow threads. Both species can have the early whorls nodulose, especially *C. suturatus sandwichensis* from Hawaii.

It is hard to compare this shell with any others because of the shape and bright banding. Perhaps there is a general resemblance to *C. coelinae berdulinus* in some patterns, but that shell is usually only a single color and not with variegated bands; in addition, the spire is very different.

Variation.—The shape of the sides seem to change with growth, being first rather convex in juveniles, straight in full adults, then slightly concave in larger specimens. The spire seems to also become lower with growth. In the 33mm type the midbody and basal bands are quite wide and very pale; in a very large specimen the whole shell is a very subdued orange-tan with even the midbody band partially interrupted. The spire pattern is always subdued and sometimes not developed, while the base is always pale, not stained with darker violet. Occasional shells are heavily suffused with pale violet, but in such cases the base does not contrast with the rest of the shell. In shells with heavy axial growth lines there are colorless areas next to the lines. Aperture color varies with the tone of the body whorl but is always pale.

Distribution.—Described from a beach juvenile from southern Japan, the species has been taken uncommonly in deep water off Taiwan as well. However, Taiwanese coral boats have taken this species in fair numbers from off Sri Lanka, presuming this locality data is correct.

Synonymy.—After checking over the various Reeve and Sowerby names applied to *suturatus*-like shells, none seems to fit this species. It would be very surprising if such a large and distinctive shell had escaped naming until 1971, but such seems to be the case. This is certainly not a subspecies of *C. suturatus*.

Notes.—Like other dredged shells, *C. kashiwajimensis* seems to be subject to lip damage, but it is otherwise a very clean species in the specimens I have seen. Axial growth lines and threads may be heavy and disrupt the pattern somewhat. There are commonly coralline encrustations on the spire, but nothing serious in better specimens. The color fades in dead-dredged specimens.

It is not yet certain whether this species is only moderately uncommon or moderately rare. At least a few specimens have reached the market recently.

KERMADECENSIS Iredale, 1913

1913. *Conus Kermadecensis* Iredale. *Proc. Malac. Soc. London,* 10: 227, pl. 9, figs. 15-16 (Sunday Island, Kermadec Group).
1961. *Conus* (*Rhizoconus*) *okamotoi* Kuroda and Ito. *Venus,* 21(3): 251, pl. 17, figs. 11-12 (Locality not stated). This may not be a valid description
?1973. *Rhizoconus yoshioi* Azuma. *Venus,* 32(1): 10, pl. 1, fig. 2, text figs. 9-9a (Off Kirime-zaki, Kii Peninsula, Japan).

Description.—Heavy, with a low gloss; low conical, the sides nearly straight or slightly convex, greatest width usually just below the shoulder; body whorl with about 4-10 low spiral ridges at the base, these rapidly replaced by weaker spiral threads that commonly undulate; numerous axial threads over entire whorl, sometimes conspicuous; shoulder roundly angled to (normally) weakly carinate, flat or concave above, broad; spire low, bluntly pointed, the sides straight or concave; whorls commonly stepped, at least the early ones; tops of whorls usually concave, the margins weakly carinate to roundly angled; tops of whorls with heavy curved axial threads or ridges and a variable number of spiral threads or low ridges, usually 1-3, sometimes none; early whorls may occasionally be weakly undulate. Body whorl white, covered with spiral rows of very small brown dots which are commonly visible only over new growth and may require a lens; numerous axial flammules of orange-brown to yellowish or reddish brown above and below a clear midbody area; commonly the flammules are partially or mostly fused into broken spiral bands; midbody band usually distinct, clear, sometimes with darker brown squares at the margins or with narrow continuations of the axial flammules; base pale; shoulder commonly with small dark brown spots or oblique blotches, usually in strong contrast to rest of whorl; spire whitish, heavily or sparsely covered with radiating reddish brown lines and streaks. Aperture moderately narrow, somewhat wided anteriorly; outer lip straight, usually thin; mouth white to bluish white, sometimes with a darker violet blotch within. Columella very narrow, mostly internal. Length 30-60mm.

Comparison.—As treated here, *C. kermadecensis* is a wide-ranging and somewhat variable species which is not always easy to distinguish. First, there is often a very close resemblance in shape and pattern to *C. eximius*, but that species usually has very distinctive spiral ridging on the spire whorls and deep grooves on the anterior part of the body whorl; *C. kermadecensis* is a coarse shell, while *C. eximius* seems more polished and attractive; although *C. eximius* may have small brown dashes in the pattern, these are not comparable with the very small brown dots of *C. kermadecensis*. Some *C. planorbis* and *C. vitulinus* may have similar patterns, but these species lack the strong axial spire sculpture of *C. kermadecensis*, have heavier basal sculpture that is often granulose, and usually have the base darkly stained. *C. capitanellus* may have a somewhat

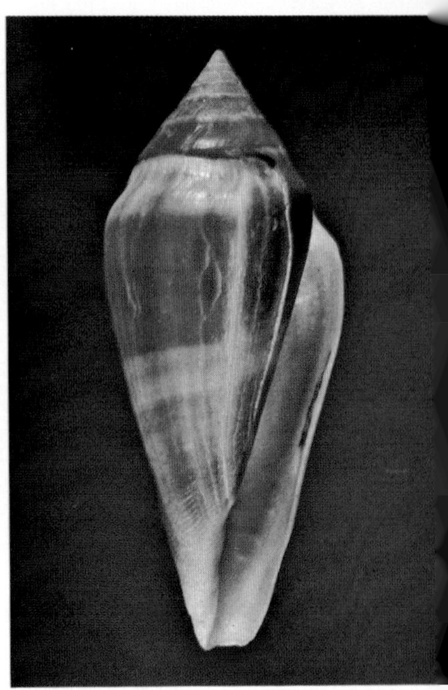

SMIRNA.—(*s. profundorum*) *Above:* Left: Off Tosa, Shikoku, Japan, 76.8mm (EP); Right (holotype of *C. soyomaruae*): 6 mi. off Hachijo I., Japan, 75.7mm (photo by Dr. T. Okutani). *Below* (holotype of *C. scopulicola*): Hyotanse Bank, Izu-Shichito I., Japan, 22.2mm (photos by Dr. T. Okutani).

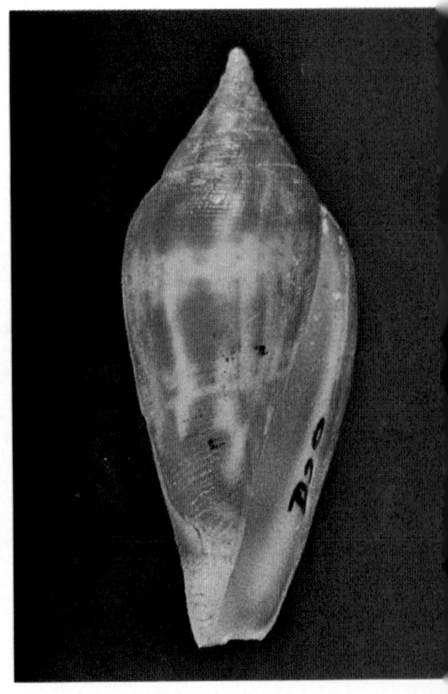

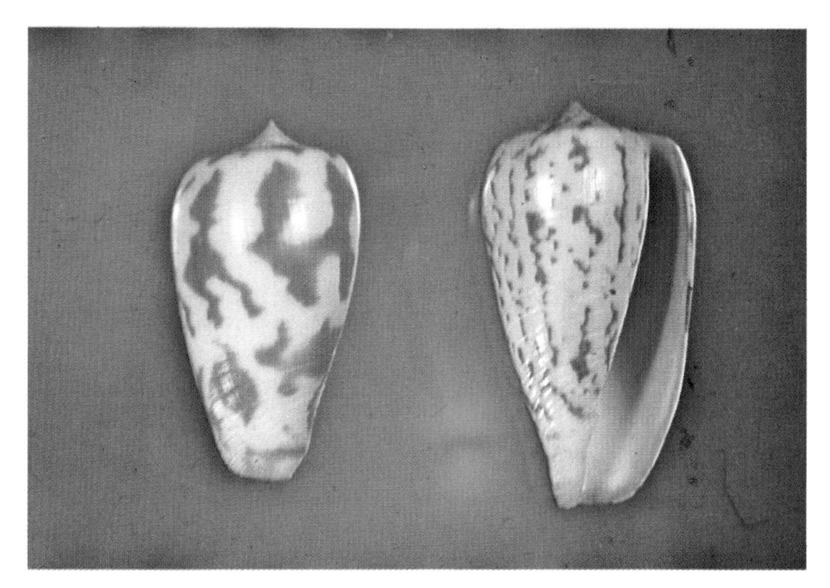

SPECTRUM.—Above: Subic Bay, Luzon, P.I., 37.1 and 41.4mm. ***Below:*** Left (*filamentosus*): Zamboanga, Mindanao, P.I., 37.6mm; Right (*pica*): Cebu, P.I., 34.3mm.

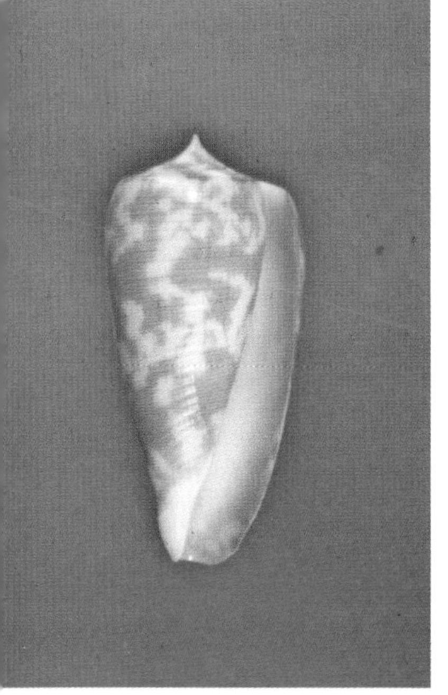

similar pattern to some *C. kermadecensis*, but it is lighter in weight, has nodulose early whorls, and has strong spiral sculpture on the spire whorls. *C. otohimeae* and *C. ione* are perhaps the closest relatives of *C. kermadecensis*, having very similar spire sculpture with heavy axials; both these species differ in color and texture, however, although they share the small spots in the body whorl pattern. *C. ione* is a very thin shell, violet in tone, with a very sparse pattern. *C. otohimeae* has a broad band of large orange blotches near midbody, not the irregular axial flammules of *C. kermadecensis*; the early whorls of *C. otohimeae* are strongly nodulose.

Variation.—There is considerable variation in spire shape and color pattern of the body whorl. The spire in China Sea specimens has distinctly stepped whorls which are concave above, while the specimens from Australia and South Africa have the whorls less concave and less distinctly stepped, the sides often distinctly concave. Smaller specimens of both types have the shoulder carinate, while in large specimens the shoulder may be just roundly angled. In specimens with sparse spire markings there are usually small brown dots demarking the whorl margins. Commonly the heavy axial threads are ridge-like and very conspicuous, while the spiral sculpture may be absent, represented by a single thread, or present as 2-3 weak ridges; not uncommonly all these situations may be found in a single specimen. The early whorls are usually eroded, so it is difficult to see the fine undulations if they are present.

Although the rows of small dots are present in every specimen examined, they are often very difficult to see. The shoulder spots are commonly strongly developed, while the midbody band is usually clear and distinct. The whole shell may be dark reddish brown or pale yellowish tan, or there may be only a few irregular axial flammules on a white ground. Occasional specimens are tinged with violet.

Four populations or population groups can be recognized, although they overlap strongly and individual specimens are seldom distinguishable with confidence. From the China Sea come specimens with distinctly stepped spire whorls that are concave above and have strongly carinate margins; the fine dots on the body whorl are usually numerous. Specimens from eastern Australia (and the Kermadecs) are lower-spired, the whorls only weakly stepped and often flat above; these specimens commonly have 1-2 weak spirals on the whorls, absent in most China Sea specimens; the pattern is often very weak and irregular. Specimens from Western Australia are very coarse in appearance, very wide at the rounded shoulder, have the spire poorly stepped, are dirty tan or pale reddish brown in color, and commonly have a large violet blotch in the aperture. The few South African and Mozambique specimens seen were very similar to those from eastern Australia, although with perhaps slightly higher spires and brighter patterns in small shells; small shells had a large violet blotch in the mouth.

Distribution.—Apparently antetropical, with populations in southern Japan and the China Sea near Taiwan; the Kermadec Islands;

Queensland and New South Wales, eastern Australia; Western Australia and perhaps the Northern Territory; and the Natal-Mozambique coast of Africa. It is seldom common at any locality and is usually dredged in moderately deep water, although the western Australia and South African specimens were collected in shallow water.

Synonymy.—*C. kermadecensis* is almost certainly not the earliest available name for this species, but early figures of non-descript shells are almost impossible to assign with certainty to this or any other species. See *C. fulmineus* and *C. lischkeanus* under dubious species. *C. okamotoi* is perhaps not validly described, but it is commonly used for Japanese specimens. *C. yoshioi* seems indistinguishable from *C. kermadecensis* by external feature, although the radular teeth are said to differ; however, I cannot be sure just what species were compared for this character because of confusion of nomenclature and the poor photos of specimens still covered with periostracum.

Notes.—Although some Japanese specimens are quite clean with distinct and bright color patterns, few specimens of this species are at all attractive. There is a strong tendency toward heavy axial flaws and breaks, with very irregular and muddled patterns. The spire is commonly eroded. The lip is thin in many specimens and chips badly. However, these faults are to be expected in such a species.

Japanese specimens are usually sold as either *C. okamotoi* or *C. fulmineus*, as are those from Taiwan. Specimens from eastern Australia have been sorted from specimens sold as *C. planorbis* and *C. vitulinus*, while those from western Australia are constantly offered as *C. kenyonae*—a synonym of the extremely different *C. distans!* Specimens from southern Africa are currently not available and, along with those from the inaccessible Kermadecs, would be the hardest populations to obtain.

KERSTITCHI Walls

Holotype: Delaware Mus. Nat. Hist.; 31.8mm; Mexico, Nayarit, off Isla Tres Marias (between Isla San Juanito and Isla Maria Madre).
Paratypes: Collections of Margaret Cunningham and Alex Kerstitch.
Diagnosis: A turnip-shaped cone related to *C. gradatus* but closely resembling *C. sennottorum;* creamy yellowish, suffused with salmon or salmon-tan at the base, below the shoulder, and often along axial growth lines; covered with about 10-15 regular spiral rows of small squarish deep brown spots that are smallest at base and shoulder. 32-46mm length.

Description.—Moderately light in weight, with a good gloss; broadly fusiform, the body whorl very wide anterior to shoulder and with nearly parallel margins, then strongly constricted to produce a turnip-like shape; body whorl with about 8-12 low, broad spiral ridges on the anterior third,

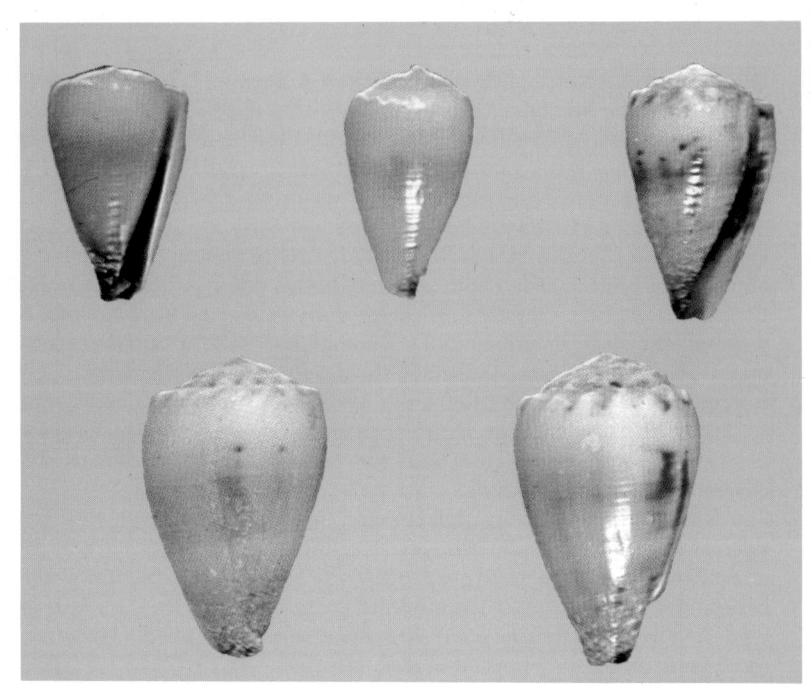

SPONSALIS.—Above: Tahiti, Society Is., 12.8-17.1mm. *Below:* Left: Tahiti, Society Is., 17.1-21.4mm; Right (*s. nux*): Guaymas, Sonora, Mexico, 16.2mm (GG).

SPURIUS.—Above: Off Marathon, Florida, 45.8 and 47.0mm. ***Below:*** Left
tan, Mexico (photo by A. Kerstitch);

the ridges separated by narrow shallow, slightly axially striated grooves; remainder of body whorl smooth except for numerous fine axial growth lines and some scattered nearly obsolete spiral threads; spire tall, sharply pointed, the sides deeply concave; first 6-7 whorls distinctly stepped, strongly carinate, the carinae close to mid-whorl; protoconch (possibly incomplete) glossy translucent white, of about 2½ unequal whorls; first 3-4 whorls weakly excavated to produce a weakly nodulose appearance; later spire whorls with smooth, carinate margins, their tops with a single visible spiral ridge near the carina and sometimes traces of 2-3 nearly obsolete other ridges; whorls concave both dorsal and ventral to carina; numerous fine curved axial threads and scratches over all whorls; late whorls not stepped, forming a rather flat plate for the raised portion of the spire; shoulder distinctly carinate, sometimes with a slight constriction anterior to carina. Body whorl glossy yellowish cream variously tinted with salmon or salmon-tan; usually the salmon is strongest at base and below the shoulder as well as along some heavier growth lines and the outer lip, but the entire shell may be suffused with color; about 10-15 widely spaced regular spiral rows of small squarish or slightly dash-like dark brown spots over body whorl from base to shoulder, those nearest midbody usually largest and most distinct, those below shoulder very small and sometimes obscured by salmon-tan axial streaks; spire and shoulder pale salmon or salmon-tan, the middle and late spire whorls covered with alternating salmon-tan and white rectangles of about equal size; shoulder with similar marks or not; early whorls faintly violet-brown, unmarked. Aperture narrow, of uniform width; outer lip thin, sharp, its margin following outline of body whorl, the posterior edge sloping below level of shoulder; anal notch moderately shallow, rather square; mouth pale pink to deep salmon, darkest deep within. Columella internal, not visible. Length 32-46mm.

Holotype of *C. kerstitchi*, Nayarit, Mexico.

Comparison.—There are obvious similarities between *Conus gradatus*, *Conus sennottorum*, and *Conus kerstitchi*. The similarity between the latter two is remarkable, both in shape and pattern, and one would be very tempted to consider them isolated subspecies of a single species. The only obvious differences between *C. sennottorum* and *C. kerstitchi* are the white color of *C. sennottorum*, the often paler orange-brown spots which are larger in size and more irregularly spaced, tending to be most obvious at base and shoulder, and the absence of salmon tinting at the base and shoulder. There are very minor differences in shape, with *C. sennottorum* being wider than *C. kerstitchi* at the shoulder and perhaps with a usually more concave-sided spire, but these differences are hard to see.

Although similar, *C. sennottorum* is related to *C. stimpsoni* and *C. lorenzianus* and I doubt that there is any close relationship between these species and *C. kerstitchi.*

In my opinion *C. kerstitchi* has evolved from *C. gradatus*, an extremely variable species that has a variety of very distinctive forms (or perhaps sibling species) evolved into several distinctive shape and color patterns. It has a similar reduced spire and body whorl sculpture to that found in *C. kerstitchi*, often has the same glossy texture, may have a similar spire pattern, and has a body whorl pattern of (usually) large spots. No variant of *C. gradatus* known to me has the distinctive shape of *C. kerstitchi* nor its relatively tall and concave-sided spire; the salmon tinting also appears to be absent in *C. gradatus*, and the spots are much smaller than in any *gradatus* form known to me.

The combination of differences in body whorl shape, spire form, pattern, and color tinting seems sufficient to distinguish *C. kerstitchi* from all variants of *C. gradatus*, although the possibility exists that it is just a very distinctive ecotype of *C. gradatus.*

Variation.—The few specimens available to me are very similar in shape and sculpture but differ in intensity of color and visibility of pattern. The holotype is a creamy yellowish shell covered with distinct spots and has the salmon restricted to the base and shoulder; the paratypes are more heavily suffused with salmon and salmon-tan, obscuring the spotted pattern.

Distribution.—As yet known only from a few specimens trawled in deep water or collected by SCUBA between Cabo San Lucas, Baja California del Sur, and off Isla Tres Marias, Nayarit, Mexico.

Notes.—One of the paratypes was found on a rock and sand substrate at 75 meters in a canyon rich in animal life. Collected with this paratype were *Conus bartschi*, *Conus ebraeus*, *Conus tessulatus*, *Cypraea rashleighana*, and more common Panamic cones and cowries. The other specimens were taken by shrimp trawlers.

This species is named in honor of Mr. Alex Kerstitch, University of Arizona, who made the types available to me and who has helped greatly in the preparation of this book.

KIMIOI (Habe, 1965)

1965. *Rhizoconus kimioi* Habe. *Venus*, 24(1): 47, pl. 4, figs. 1-2 (Off Okinoshima, Kochi Pref., Shikoku, Japan).

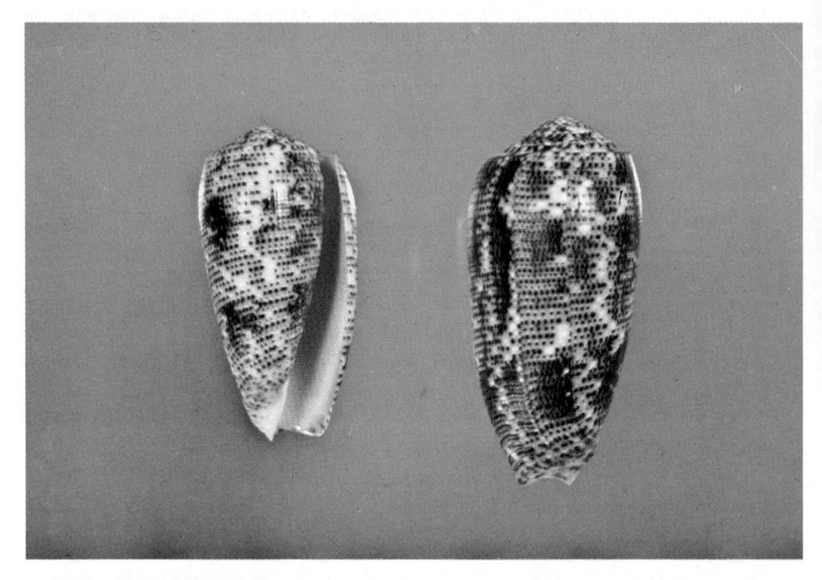

STERCUSMUSCARUM.*—Above:* Samar, P.I., 42.6 and 49.8mm. *Below:* Honiara, Guadalcanal, Solomon Is., 47.3 and 49.4mm.

STIMPSONI.—Above: Off St. Augustine, Florida, 30.1 and 31.4mm (TY). *Below:* 105 mi. WNW of Egmont Key, Florida, 13.1-30.2mm (GG).

Description.—Light in weight, fragile, with a high gloss; low biconical, the sides straight or slightly concave and tapering to a narrow base; body whorl with about 6-10 weak spiral ridges at the base and about 4 shallow spiral grooves just below the shoulder; fine axial threads over whorl, inconspicuous; shoulder sharply angled, slightly concave above; spire low to moderate, pointed, the sides concave; spire whorls slightly stepped, weakly concave above, with about 4-5 low spiral ridges and scattered axial threads. Body whorl light tan, covered with widely spaced spiral dark brown lines; a pair of lines below the shoulder and above and below midbody are slightly thickened and contain darker brown axial rectangles within, these separated by paler flammules or fine lines and producing a 'brick wall' pattern; rest of whorl with some nebulous axial tan clouds and irregular white blotches; shoulder with larger brown spots alternating with white ones, conspicuous; spire pale tan with squarish brown spots and blotches. Aperture narrow, uniform in width; outer lip thin, fragile, straight or slightly concave; mouth dirty white, the external pattern showing through. Columella mostly internal. Length 15-25mm.

Comparison.—In all likelihood this is the juvenile of another species, as the specimens certainly appear to be juvenile in shell characters. However, the pattern and shape are very distinctive and unlikely to be confused with any other species. *C. zonatus* also has a 'brick wall' pattern, but it is a large and thick cone with heavy coronations and an overall bluish gray coloration and a dark base. Possibly this is a juvenile of *C. hirasei*, although I have never seen a small specimen of that species. In his original description of *C. kimioi*, Habe indicated that young *C. hirasei* have only fine black lines on the body wall, presumably lacking the 'brick wall' effect. Large *C. hirasei* have the early whorls nodulose, while in *C. kimioi* they are apparently carinate; *C. hirasei* has the spire rounded, not stepped, but this could easily be explained by growth changes.

Variation.—This is a rare species with few good specimens known, but they appear very similar to each other in shape and pattern. The sides may be straight or weakly concave, while the spire may be low or moderate; stepping of the spire is more prominent in smaller specimens. There is some minor variation in number and spacing of brown spiral lines and the tone of the brick wall effect, but this appears negligible. The spots on the shoulder may be weak or strong.

Distribution.—A rare species dredged in deep water off southern Japan and Taiwan. Poorly known.

Notes.—This is a fragile deep-water species, so minor flaws should be expected. The lip is easily chipped, and dead specimens are covered with encrusting organisms. This is a rare species and acceptable in almost any condition that shows the pattern.

I find it very hard to believe that this is an adult shell. I am not aware of specimens over 25mm in length.

This species probably occurs in the Philippines, as I have seen a slide of a specimen supposedly collected there.

KINOSHITAI (Kuroda, 1956)

1956. *Chelyconus kinoshitai* Kuroda. *Venus*, 19(1): 6, fig. 7 (Kii, Japan).

Description.—Moderately light, with a good gloss; low biconical, rather cylindrical, the upper sides convex, widest just below the shoulder; body whorl with moderately heavy spiral ridges above the base, these soon becoming very fine spiral threads and widely spaced shallow grooves posteriorly; numerous axial threads, some heavy; shoulder angled, sometimes almost carinate, narrow; spire low, sharply pointed, the sides straight or weakly concave; tops of whorls slightly concave to flat, with about 4-6 spiral ridges; early 2-4 whorls weakly nodulose. Body whorl violet, usually with three bands of dark violet-brown irregular blotches below shoulder and above and below midbody; the blotches may be isolated and edged with axial lines and flammules over most of the whorl or mostly fused into nearly solid spiral bands; spiral brown dashes may be present in the blotches, especially basally; occasional shells are all pale violet with only small blotches above and below midbody area; midbody area usually paler violet than rest of whorl; shoulder with large blackish brown spots alternating with pale violet areas; spire violet with radiating brown blotches and wide lines; early whorls pale tan. Aperture moderately wide, slightly wider anteriorly than posteriorly; outer lip thin, sharp, straight or nearly so; mouth violet, fading to bluish white. Columella long and narrow. Length 40-90mm.

Comparison.—*C. kinoshitai* is very closely related to *C. fulmen* and *C. bruuni*. From *C. fulmen* it is distinguished by the more straight-sided shape and angled shoulder; the pattern of *C. fulmen* is usually in the form of large black axial blotches that are not in regular spiral bands. Furthermore, *C. fulmen* lacks nodulose early whorls and does not normally have the regularly alternating dark and light spotting along the margins of the shoulder and sometimes the spire whorls. *C. bruuni* has heavily granulose juveniles (presumably absent in *C. kinoshitai*) and usually retains a few granules on the basal ridges of even adult shells. In *C. bruuni* the shoulder is somewhat wider than typical of *C. kinoshitai* and apparently more concave above. *C. bruuni* is much more simple in pattern and lacks the numerous axial flammules usually present in *C. kinoshitai* as well as the spiral dashes normally found in that species.

Variation.—As usual, there is some variation in width and angulation of the shoulder, convexity of the sides, and height and shape of the spire, but they are minor. Color patterns in this species fall into two major categories. In the typical variety the blotches are isolated, usually with axial lines and extensions covering most of the shell; there are brownish dashes in vaguely spiral rows visible in the pale areas. In the other common pattern the blotches are fused into three fairly distinct and regular spiral bands. Occasionally the shell is almost without dark blotches except for small ones above and below the midbody area. Inten-

STRAMINEUS.—*Above:* Negros I., P.I., 39.4 and 40.1mm (Shell Cabinet). *Below:* Left: Natal, South Africa, 37.4mm; Right: Zaptos I., P.I., 43.9mm (Dayrit).

STRIATELLUS.—Above: Tanga, Tanzania, 52.2 and 59.4mm. *Below:* Left: Boac, Marinduque, P.I., 38.4mm; Right: Sulu, P.I., 45.0mm (GG).

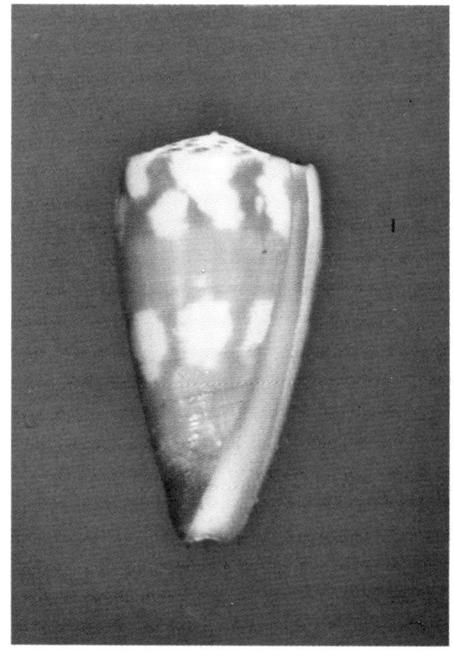

sity of the blotches and background colors and the presence or absence of light shadows on one edge of the dark blotches are all variable.

Although the basal ridges are often marked with spiral dashes of dark brown, these may be absent or extend much further up the shell and be easily visible in the dark blotches below the midbody area. The midbody area itself is usually a little paler than the rest of the shell but is commonly interrupted by many axial lines and spots. Some specimens have distinct spiral grooving, though shallow, over most of the body whorl, probably a remnant of juvenile sculpturing.

Distribution.—Normally taken by dredging in moderately deep water off southern Japan and Taiwan, and perhaps further south in the China Sea. Recently I have examined specimens dredged in the Solomons, and they differ little from typical Japanese specimens, so the range of the species is probably much wider in the depths of the western Pacific. It has recently been taken in the Philippines. Uncommon.

Notes.—Most specimens have traces of small breaks on the body whorl and spire, with occasional large and unhealed breaks visible. The lip is almost always broken and filed. Dead specimens fade to bluish white and lose most of the bright violet tones.

Live-taken specimens in good condition are very attractive, especially pale specimens with little blotching. The relationships of this species with both *C. fulmen* and *C. bruuni* are not satisfactory, although it is almost certainly a full species and distinct from both.

KLEMAE (Cotton, 1953)

1953. *Rhizoconus klemae* Cotton. *Trans. Roy. Soc. South Australia*, 76: 24, pl. 3, figs. 1-3 (South Australia: Corny Point, Eyre Is., Levens, Daly Head).

1970. *Rhizoconus coralinus* Habe and Kosuge. *Venus*, 29(3): 81, pl. 5, figs. 3-4 (Zamboanga, Mindanao, Philippines).

Description.—Moderately heavy, with a low gloss; biconical, the upper sides strongly convex then tapered to a narrow base; body whorl smooth except for a few low spiral ridges at the base and some axial and spiral threads; shoulder broad, angular, weakly concave above; spire moderate to high, bluntly pointed, the whorls distinctly but shallowly stepped; tops of whorls slightly concave, bearing 4-8 low spiral ridges which may become obsolete with growth; margins of whorls sometimes irregular or undulate, but usually too eroded to see clearly. Body whorl golden brown, dark brown, reddish, or even yellowish, the base white and strongly contrasting; above the base is a narrow band of dark brown spots; another white spiral band at midbody is usually heavily blotched with small brown triangles and axial lines; below the shoulder is a narrow white band containing axial brown lines continued onto spire; in some specimens the dark color of the body whorl may be interrupted by spiral rows of indistinct whitish spots and triangles sometimes separated by narrow brown dashes; axial flammules of dark brown may also be present; spire and shoulder whitish, covered with axial blackish or dark brown blotches and lines. Aperture fairly wide, uniform in width or widened anteriorly; outer lip strongly convex, thin; mouth pale reddish violet to dirty brown. Columella short, narrow, Length 20-75mm.

Comparison.—Although apparently derived from *C. anemone* and similar in spire shape to the *compressus* variety of that species, the smooth and very wide body whorl of *C. klemae* is very distinctive when combined with the normally three-banded appearance. *C. cardinalis* might possibly be confused with it, but that species is usually spirally ridged and granulose and does not have the strongly contrasting basal band.

Variation.—Juvenile specimens retain the fine spiral ridges over the body whorl that characterize *C. anemone*, but these become obsolete with growth. The height of the spire is very variable, as is the degree of convexity of the outer lip. Most variable, however, are the colors of the body whorl. Although commonly some shade of pale brown, specimens with bright red or yellow shells have also been found, as well as some with heavy axial flammules instead of the usual broad solid spiral bands. The amount of white triangles in the background is variable from specimen to specimen, but at least a few are usually present between midbody and the shoulder. The contrasting base is constant. Apparently living specimens have bright pink backgrounds which rapidly fade to white at death.

Distribution.—An uncommon species restricted to shallow to moderately deep water from South Australia west to Geraldton, Western Australia.

Synonymy.—*C. coralinus* does not differ in any way from reddish specimens of *C. klemae*, and I can only assume that the locality on the type specimen is in error. Although the early whorls are described as nodulose, these are normally eroded in *C. klemae*, but in some specimens they can be seen to be at least undulate.

Notes.—Only brown specimens are usually available, although most illustrations of this species show the rare bright colors. The spire is usually eroded, sometimes badly, while the body whorl is commonly very diffusely pigmented and dull. Axial growth marks and lines are not uncommon and may badly disrupt the pattern. The lip is fragile and chips easily as in most species related to *C. anemone*.

As mentioned, bright colors are uncommon to rare and certainly deserve a premium. It is always a disappointment to expect a bright red or yellow *C. klemae* and instead receive the usual dull brown one instead.

STRIATUS.—Above: Tahiti, Society Is., 65.7 and 66.2mm. *Below:* Zamboanga, Mindanao, P.I.: Left: 82.9mm; Right: 90.9mm.

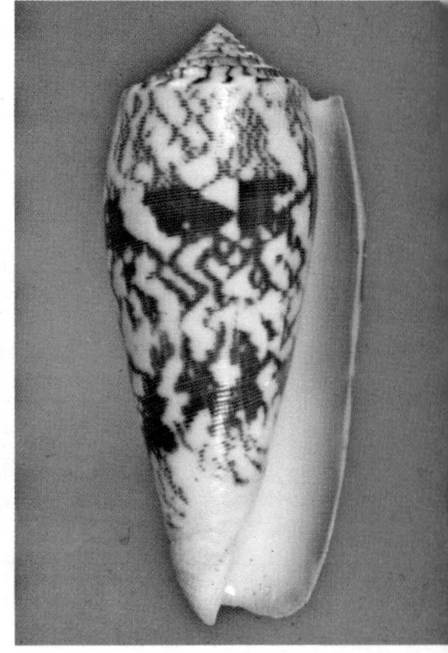

STUPA.—Above: Taiwan, 76.2mm (GG). *Below:* Left: Off Tosa, Shikoku, Japan, 93.5mm (EP); Right: NE of Taiwan, 91mm (DM) (photo by Jayavad).

LEGATUS Lamarck, 1810

1810. *Conus legatus* Lamarck. *Ann. du Mus. Hist. Nat.* (*Paris*), 15: 437 (Mers des Grandes Indes).

1833. *Conus musivum* Sowerby i, in Sowerby ii. *Conchological Illustrations:* Pt. 25, fig. 12 (Philippines).

Description.—Moderately heavy, thick, with a high gloss; cylindrical, biconical, the sides straight or slightly convex and suddenly constricted to the base; body whorl with low and obscure spiral ridges at the base, these sometimes visible to the middle of the shell but generally more restricted anteriorly; shoulder rounded, not distinct from spire; spire moderately tall, sharply pointed, the sides straight or nearly so; first 2-3 whorls with very weak nodules, easily eroded; tops of spire whorls smooth or with indications of a very weak spiral ridge at the middle. Body whorl deep pink or fleshy pink, covered with an open network of wavy axial brownish lines forming a few to many tents; 'textile lines' few, very contrasting blackish brown, usually fused into long axial stripes containing small to large starkly white tents; shoulder with a few dark blotches; spire mostly pink with scattered axial brown lines, a few dark blotches, and occasional white tents. Aperture narrow, uniform in width; outer lip thin, straight; mouth deep pink. Columella long and narrow, slightly indented from inner lip. Length 20-40mm.

Comparison.—*C. legatus* is certainly very close to *C. canonicus*, and occasional specimens are hard to place with certainty. However, the generally small size yet thick shells with the bright pink background are quite distinctive. In most shells the identification is further strengthened by the long blackish 'textile lines' containing contrasting white tents. Few *C. canonicus* are likely to have this combination of characters, but those that come close are usually larger and not as bright pink as *C. legatus.*

C. retifer is sometimes like *C. legatus* in general pattern and color, but it is much broader and more rounded in shape with strong spiral ridges over most or all the body whorl. *C. aureus* is sometimes misidentified as *C. legatus* in the Indian Ocean, but it is generally much larger, has a heavier tented pattern, seldom is bright pink, has strong spiral ridges over most of the body whorl, and often has a higher spire.

Variation.—Although constant in shape, there is considerable variation in pattern and some in sculpture. The body whorl may have only basal ridges or may be ridged to the shoulder, although weak ridging to the midbody area is most common. The black axial stripes formed from the 'textile lines' may be few and long or shorter and more numerous, forming an indistinct spiral band as in *C. retifer.* In some specimens the axial brown lines are almost absent so the pattern is virtually absent, while in the others the tents are heavier and more numerous. There are almost always contrasting white tents in the 'textile lines' that stand out in stark contrast to the bright pink background. This is always a relatively slender shell.

Distribution.—*C. legatus* is apparently restricted to the Pacific Ocean, where it is found from Okinawa and the Philippines south to Queensland and east to French Polynesia but not Hawaii. I am not aware of valid Indian Ocean records. Uncommon to rare in shallow water.

Notes.—This is one of the most attractive of the smaller cones, the bright pink contrasting greatly with the black axial stripes and the small white tents. After death the pinks fade somewhat and should not be exposed to light. The only major damage to most specimens is the presence of small chips in the fragile growing edge of the lip. Most gems sold are probably very fresh beach specimens, but it would be very difficult to tell these from live-taken material as long as the colors have not faded.

Specimens over about 30mm are especially uncommon.

LEMNISCATUS Reeve, 1849

1849. *Conus lemniscatus* Reeve. *Conchologia Iconica,* 1(*Conus* Suppl.): Pl. 5, sp. 246 (Locality unknown).

1865. *Conus sagittatus* Sowerby ii. *Proc. Zool. Soc.* (*London*), 1865: 518, pl. 32, figs. 8-9 (Locality unknown). Later illegally emended to 'sagittiferus' by Sowerby (*Thes. Conch.*, Suppl., Pl. 25 (286), fig. 608).

1875. *Conus Traversianus* E.A. Smith. *J. Conchol.* (*London*), 1: 107, text figure (Locality unknown).

Description.—Light in weight, with a good gloss; low biconical, the body whorl elongate and with straight or slightly concave sides tapering to a narrow base; body whorl with broad flat spiral ribs anteriorly separated by finely axially threaded deep grooves; above midbody the grooves become shallow and rapidly obsolete, although occasionally they reach to the shoulder; whole body whorl covered with fine but distinct axial and spiral threads, producing a finely cancellate appearance in fresh specimens (easily eroded); shoulder narrow, sharply carinate, concave above; spire moderate, sharply pointed, deeply concave; early whorls strongly stepped and with fine nodules; later whorls flatter and wider, with straight margins; tops of whorls with a strong median groove dividing the whorl into two parts, sometimes with 1-2 other fine grooves or spiral threads, and the whole covered with fine axial threads most prominent at the middle of the whorl. Body whorl pinkish to whitish, covered from base to shoulder with many spiral rows of fine orange to tan or yellow dashes alternating with white dots; over this is imposed darker clouds which form two broad spiral bands above and below midbody and several rather regular axial bands as well; when fully developed this pattern results in a bright orange shell with large axially rectangular pink areas and a pink midbody area, the whole covered with fine pale orange dashes; base pale tan; shoulder with larger dark brown to reddish squares at margin alternating with white areas; spire pinkish to tan, covered with

STUPELLA.—Above: South China Sea, 66.0mm (GG). *Below:* Off An Ping, Taiwan, 77.5mm (K).

SUBULATUS.—(*s. subulatus*) *Above:* Boac, Marinduque, P.I., 32.1 and 32.5mm. *Below:* Left: Marau Sound, Guadalcanal, Solomon Is., 29.7mm (GG); Right: Honiara, Guadalcanal, Solomon Is., 32.5mm.

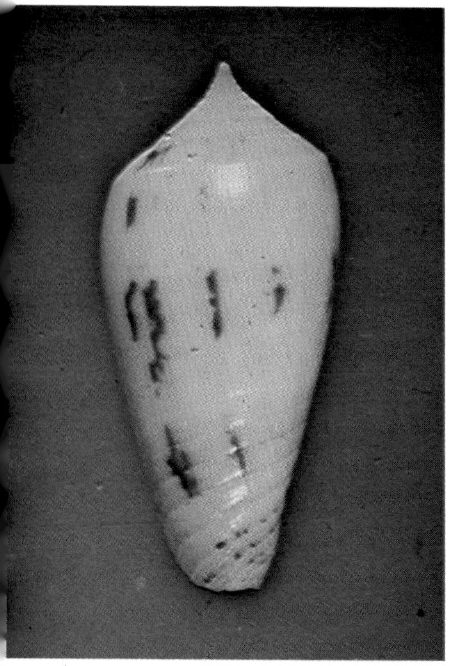

regularly spaced dark brownish spots; margins of spire whorls with fine brownish spots; early whorls tan. Aperture narrow, uniform; outer lip straight, fragile; mouth pink to bluish white. Columella internal. Length 20-50mm.

Comparison.—This distinctive little cone is very closely related to *C. lentiginosus* and shares with it most aspects of body and spire sculpture and some aspects of the pattern. However, *C. lemniscatus* is an extremely straight-sided cone with a very long body whorl and carinate or nearly carinate shoulder; the spire also appears much more concave-sided. In *C. lentiginosus* the posterior sides are convex and the shoulder is rounded, the spire considerably taller in proportion to the length of the body whorl than in *C. lemniscatus*. The colors of *C. lemniscatus* are pinkish and orangish, although sometimes bright tan might be a better description; *C. lentiginosus* is a bluish white or pale gray shell with dark brown axial and spiral flammules. The two are easy to contrast when examined together.

The distinctive spire sculpture of *C. lemniscatus* allies it to *C. aculeiformis* and *C. acutangulus* and allies, although it is obviously quite distinct from these species and not likely to be confused.

Variation.—The deep grooving of the anterior half of the body whorl is supposed to be continued to the shoulder according to literature, but in the specimens I have seen it stops at midbody and is virtually absent more posteriorly; this is probably individually variable or perhaps varies from area to area, but it seems of little importance since this whole complex of shells exhibits quite a bit of variation in grooving (see for instance *C. elegans*, which is probably not distinct from *C. aculeiformis*). The body whorl seems to become relatively more elongate with growth, or at least the contrast with the spire height becomes more obvious. The shoulder may be only sharply angled in large specimens. Color in life is apparently deep pinkish to pinkish white, the spiral and axial bands varying from bright yellow to orange or orange-red, sometimes to bright orange-tan. The spiral bands may be continuous or broken into blotches, while the axial bands may be incomplete or oblique, especially below the shoulder. The color fades rapidly.

Distribution.—Apparently endemic to the northern Indian Ocean from Rameswaram, India west to the Gulf of Oman and possibly to the African coast. Uncommon to rare in shallow to deep water.

Synonymy.—This seldom-seen cone is more commonly called *C. traversianus*. However, there is no doubt that *C. traversianus* is the same species as *C. sagittatus*, which was excellently figured by Sowerby ii. In addition, Sowerby later admitted that his species was the same as the type of *C. lemniscatus*, which was poorly figured and based on a poor specimen. Sowerby ii seldom synonymized his own species, so in this case he seems to be correct. Certainly the figure and description of *C. lemniscatus* mention the angled shoulder, long body whorl, numerous spiral rows of dashes, and the bright clouds. For this reason I believe the proper name for this species should be *C. lemniscatus* Reeve.

Notes.—Recently a few specimens dredged in deep water off India have come on the market, but previously the few specimens came from moderately shallow water in the Gulf of Oman and vicinity. This is still far from being a common species in collections. The tip of the spire is commonly broken off, perhaps because of the unusually stepped condition of the early whorls. Axial breaks and growth flaws are not uncommon, while erosion on the spire and body whorl are also found. The color fades rapidly, so dead-collected specimens have very subdued colors and an indistinct pattern.

LENTIGINOSUS Reeve, 1844

1844. *Conus lentiginosus* Reeve. *Conchologia Iconica*, 1(*Conus*): Pl. 44, sp. 245 (Locality unknown).
1854. *Conus optabilis* A. Adams. *Proc. Zool. Soc. (London)*, 1853: 116 (Locality unknown).
1855. *Conus selectus* A. Adams. *Ibid.*, 1855: 121 (Malacca).

Description.—Light in weight, with a good gloss; biconical, the upper sides convex then tapered to a narrow base; body whorl with about 6-12 wide spiral ribs anteriorly separated by narrow deep grooves which contain fine axial threads; rest of body whorl smooth except for fine spiral and axial threads; shoulder broad, rounded, concave above; spire moderately high, sharply pointed, the sides straight or slightly concave, the whorls weakly stepped; early whorls strongly nodulose, strongly stepped; later whorls with margins straight or weakly undulate, slightly concave on top; middle of whorls with a narrow groove dividing the whorl into two portions, and sometimes with second narrow groove or ridge; numerous fine axial threads over whorls, these most conspicuous near the middle on either side of the groove. Body whorl bluish white to pale bluish gray, covered with very irregular broad axial flammules of dark reddish brown, these usually few and formed by fusion of pre- and post-midbody series; the flammules usually are wavy and have heavily scalloped margins; a weak spiral band of brown blotches may form above midbody; there are also several spiral series of small brown dashes over the whorl, these seldom prominent and most common above midbody; spire bluish white to pale tan, with dark brown spots and radiating lines. Aperture moderately narrow, uniform in width; outer lip thin, convex, sloping below level of shoulder; mouth whitish, tinged with violet. Columella internal. Length 20-40mm.

Comparison.—The color pattern of broad axial flammules is found in several other species, such as *C. recurvus* and sometimes even *C. jaspideus*. However, regardless of pattern, the distinctive spire sculpture with a median groove will separate it from all similar but unrelated shells.

SUBULATUS.—(*s. andamanensis*) *Above:* Phuket, Thailand, 39.6 and 41.4mm. *Below:* Thailand, 44.7 and 45.4mm (Shell Cabinet).

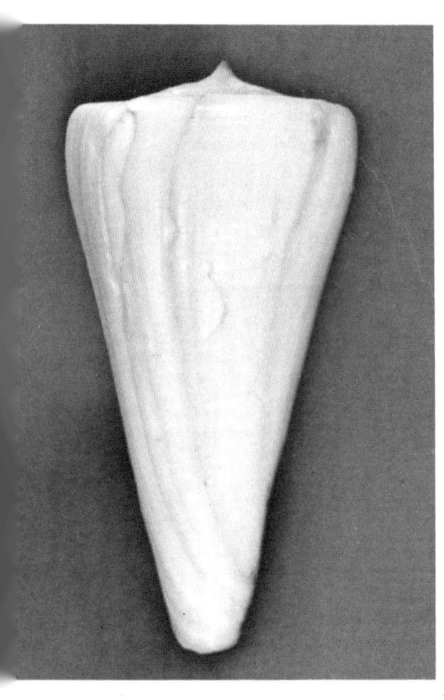

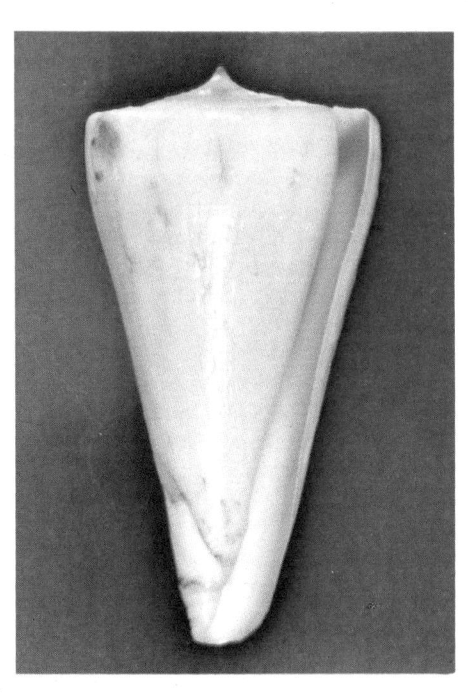

SUGIMOTONIS.—Above: Off Tosa, Shikoku, Japan, 79.4mm (EP). ***Below:*** Kii, Japan, 80mm (DM) (photos by Jayavad).

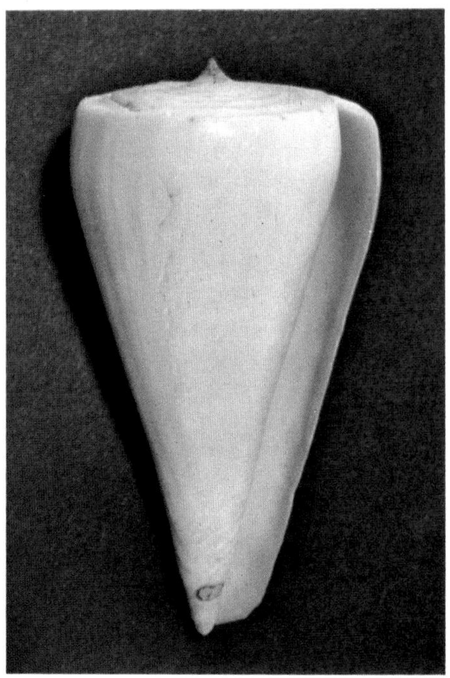

There is a close and natural resemblance to *C. lemniscatus*, which occurs in the same general area and is of about the same size. However, *C. lemniscatus* is a distinctly longer and more slender shell with straight sides and a much brighter pattern emphasizing two spiral bands which are seldom well developed in *C. lentiginosus. C. lentiginosus* has a relatively shorter body whorl with rounded sides and a relatively longer spire with straighter sides compared to *C. lemniscatus.*

Variation.—The shape of this species is rather constant, although the spire may be of variable height. The anterior sculpture is sometimes reduced to only 3-4 grooves, but at other times it extends to midbody. The axial flammules are always strong but seldom straight; often they seem to be formed by the fusion of not only a series above and below midbody, but also of a blotch at midbody, giving them a distinctly tripartite appearance. At other times the portions are all distinct, so there are flammules below the shoulder, a clear midbody area or one with a row of small blotches, and another set of flammules below midbody. The flammules sometimes partially fuse spirally at the base and below the shoulder. The spiral rows of dashes are often very inconspicuous but sometimes can be found over most of the body whorl, being especially prominent in the dark flammules where they alternate with small white dots.

Distribution.—Apparently restricted to the northern Indian Ocean, where it is not uncommon in southern Indian and Sri Lanka in shallow water. Its range outside the Indian area is uncertain.

Synonymy.—*C. optabilis* and *C. selectus* apparently belong here, but undoubtedly several other names also could be added here. The problem is that several poorly known names of this same general period are also based on cones of about the same shape with axial flammules. Chief of these is *C. emarginatus* Reeve, which looks very much like *C. lentiginosus* but is currently assigned to *C. recurvus.* Doubtless when the types of this and many other doubtful names have been examined the synonymy of this species will change.

Notes.—Recently large numbers of this species have entered the market from southern India. These are large and generally almost gem specimens having only very minor growth marks and minor chipping of the lip. This shell appears to be subject to bad breaks which do not heal well and to erosion of the spire and upper part of the body whorl. The pattern in most specimens is quite diffuse with little contrast.

LEOPARDUS (Roeding, 1798)

1798. *Cucullus leopardus* Roeding. *Museum Boltenianum:* 41 (Locality not stated). Lectotype figure (Kohn, 1975): Martini, Pl. 60, fig. 666.

1807. *Conus pardus* Link. *Beschr. Nat.-Samml. Univ. Rostock*, 3: 101 (Locality not stated). Non *Conus pardus* (Roeding, 1798).

1822. *Conus millepunctatus* Lamarck. *Hist. Nat. des Animaux sans Vert.*, 7: 461 (Asiatic Ocean). Lectotype figure (here selected): *Tableau Ency. Method.*, Pl. 323, fig. 5. Non *Conus millepunctatus* (Roeding, 1798).

1937. *Conus millepunctatus* var. *Aldrovandi* Dautzenberg. *Mem. Mus. Roy. d'Hist. Nat. Belg.*, 2(18): 171 (Locality not stated). Non *Conus aldrovandi* Risso, 1826, a fossil.

Description.—Very heavy and cumbersome, with a good gloss in smaller specimens; obconical, the sides straight or convex posteriorly, sometimes concave at the middle; body whorl smooth in adults except for a few low basal ridges, sometimes with shallowly punctate grooves between them; numerous minute axial and spiral threads sometimes visible, rather undulating; shoulder broad, sharply angled, concave above; spire very low or flat, the early whorls usually eroded and very blunt; whorls concave above and with about 2-4 low spiral ridges, these crossed by distinct spiral threads in juveniles, all sculpture sometimes obsolete in large adults. Body whorl white to cream, the base white; covered with many spiral rows of rounded or squarish blackish or deep reddish brown (juveniles) spots from base to shoulder, these sometimes in alternating large and small series; shoulder and spire with many narrow revolving black lines on white. Aperture relatively narrow, slightly wider anteriorly; outer lip thin, straight or slightly concave at the middle; mouth white. Columella long, relatively wide, weakly set off posteriorly by a low ridge. Length 75-222mm.

Comparison.—This, the heaviest and possibly the largest cone, is often confused with *C. litteratus* because both are large, spotted with black, and have names beginning with L (it is amazing how such a coincidence can confuse a situation). *C. litteratus*, however, is more brightly colored with one or two broad spiral bands of pale orange and elongate axial black stripes below the shoulder; it usually has a high gloss. More importantly, the base of *C. litteratus* is stained blackish or dark violet and the inner lip curved evenly into the outer lip instead of ending at the anterior edge of the columella; the spire whorls in *C. litteratus* have no distinct spiral sculpture. Very large *C. eburneus* are broader, lighter in weight, and have orange in the pattern. Juveniles of *C. leopardus* look much like the adults, just black or deep reddish brown spots on white or cream; juveniles of *C. litteratus* are bright orange and black on white with a high gloss. There should really be no confusion among these species.

Variation.—Although *C. leopardus* shows considerable local and individual variation in size and number of spots and their constancy of size on a single specimen, this variation is hard to categorize and apparently of no importance. Large specimens tend to reduce the number of spots and may be mostly or all white. The spire in large adults is commonly distorted and the sides concave. Juveniles have shallow grooving to

SULCATUS.—Above (*s. sulcatus*): Tayabas Bay, Luzon, P.I., 61.6 and 68.0mm. *Below:* Left (aff. *sulcatus*): Off Kavieng, New Ireland, Papua New Guinea, 78.4mm (EP); Right (*s. bocki*): Russell I., Solomon Is., 62.6mm (MG and BB).

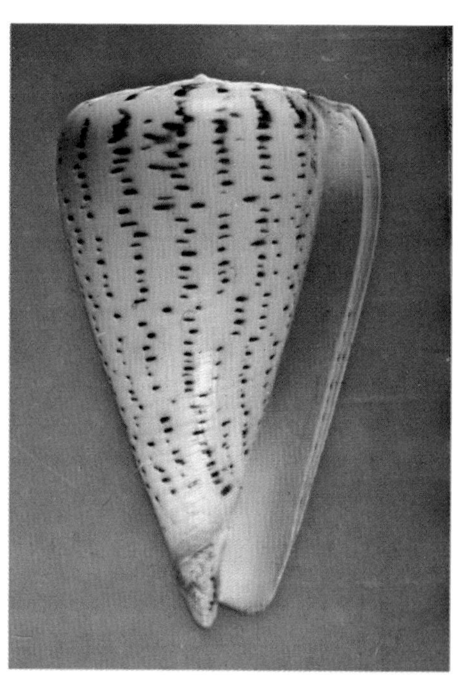

SURATENSIS.—Above: P.I., 89mm. *Below:* Left: Galle, Sri Lanka, 65.8mm; Right: Pangasinam, Luzon, P.I., 36.6mm.

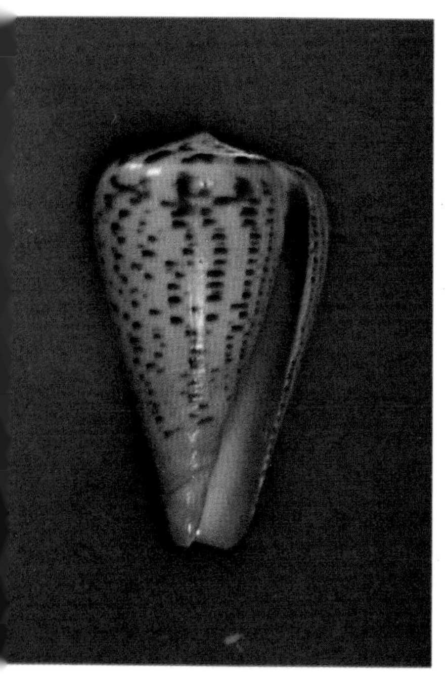
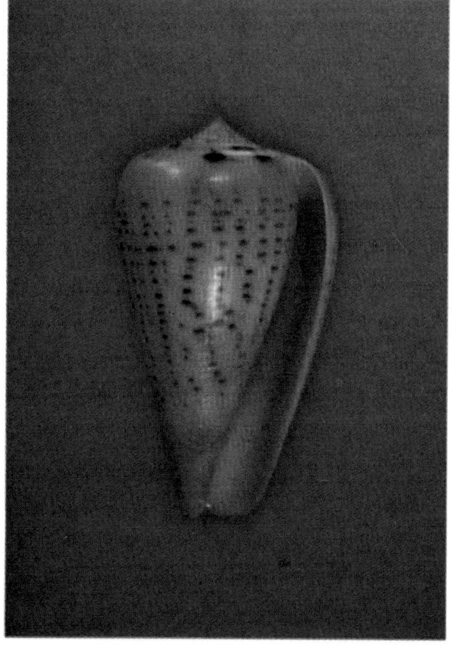

about midbody, a very blunt spire with deeply concave sides, dark oblique stripes on the base, and a tendency toward reddish brown instead of blackish spotting; the axial threads on the spire whorls are strong but widely spaced.

Distribution.—A common and widespread cone of the entire Indo-Pacific from the eastern African coast to Hawaii and French Polynesia. Shallow water.

Synonymy.—The first valid name for this species is apparently an invention of Roeding, but it has been in common use now for over a generation. For many years this species was called *C. millepunctàtus*, but that name is out of the running because of priority and because it is preoccupied. Dautzenberg's varietal name is meaningless.

Notes.—Except for erosion of the early spire whorls, which seems to be a constant character in this species even in juveniles, moderately large specimens are often perfect. Large shells (over 150mm long), however, tend to become distorted, covered with breaks, and have reduced spotting. The very largest shells look like rejects from a very heavy battle and are not suitable for a representative collection. But, as usual, those collectors who value their shells only for the number of millimeters in length will pay high premiums for exceptionally large specimens. Juveniles are very attractive and not common.

LIENARDI Bernardi and Crosse, 1861

1861. *Conus Lienardi* Bernardi and Crosse. *J. Conchyl.* (*Paris*), 9: 49, pl. 1, fig. 2 (New Caledonia).

Description.—Moderately light in weight, with a high gloss; biconical, the sides convex posteriorly and tapered to a narrow base; body whorl covered with crowded ridges near the base, these soon becoming wide flat spiral ribs separated by narrow punctate grooves; grooves usually extend above midbody and may reach the shoulder; very fine axial scratches over whorls; shoulder roundly angled, concave above; spire tall, sharply pointed, the sides concave; first 3-4 whorls with fine nodules, the later whorls with straight margins; whorls weakly stepped, concave above, with about 2-3 low and weak spiral ridges crossed by minute axial threads. Body whorl white to cream, covered with a variable pattern of wavy short axial lines of smudgy light brown spots arranged at random and sometimes overlapping to produce an indistinctly tented pattern; brown spots may be absent over most of shell or cover the entire whorl with loose tents; they may occasionally be produced into many irregular spiral rows of dashes that can further fuse to obscure most of the background except for large white triangles; in such very dark shells the color may be blackish instead of brown; spire whitish with a sparse or heavy covering of narrow brown lines and spots; early whorls pale tan.

Aperture moderately wide, slightly wider anteriorly; outer lip thin, convex; mouth white to pinkish, darkest in dark shells. Columella long, narrow, set off by a heavy raised oblique ridge. Length 25-45mm.

Comparison.—*C. lienardi* is closely related to *C. cinereus* on the one hand and especially to *C. neptunus* on the other. From *C. cinereus* as well as *C. spectrum* and allies it is readily distinguished by the presence of distinct nodules on the early spire whorls and the unusual pattern; the columella and general shape, however, are very close to *C. cinereus. C. wittigi* may have a similar pattern but lacks the nodules on the early whorls. From *C. neptunus* it differs in minor details such as the higher spire, more roundly angled shoulder, heavier ridge by the columella, and the usually more open or at least smudged pattern. Remember, however, that *C. neptunus* is poorly known and the extent of its variation is uncertain.

Variation.—Other than height of spire and extent of the spiral grooves on the body whorl, *C. lienardi* is fairly constant in shape. The pattern, however, varies considerably. At one extreme are dark brown, even blackish, shells with the white background showing through as large and small tents; the type specimen is of this form. The more commonly seen patterns, however, are loosely tented and reticulated with yellowish brown or dark brown, the lines formed by the fusion in a variable fashion of small triangular spots which are originally spiral in arrangement. Sometimes the spots concentrate into larger brown blotches above and below the midbody area, while at other times they are very reduced in number and the shell is almost white. The columellar area is often stained brownish like the early whorls.

Distribution.—At the moment *C. lienardi* is considered endemic to New Caledonia, where it is uncommon in shallow water. However, its relationship to *C. neptunus* is so close that it seems possible that it will be found in other areas as well, perhaps New Hebrides or Queensland.

Notes.—Dark specimens are not as common as reticulated ones and deserve a heavy premium at the moment. Except for chips in the delicate lip, specimens are usually in good condition. The common pattern is one in which most of the shell is covered with loose reticulations and some scattered lines and spots, as well as a few indications of larger blotches.

The relationship of this species to *C. neptunus* is very close, perhaps even at the subspecific level. It seems to differ enough in shape and sculpture to admit as a full species at the moment, but if further populations of *C. neptunus* are discovered it is quite possible that they might bridge the gap. At the moment the closest approach of *C. neptunus* to *C. lienardi* geographically is the Solomons, but there are old records for the *neptunoides* form from Australia (these are doubtful, however). Philippines *C. neptunus* show a closer morphological approach to *C. lienardi* than do those from the Solomons.

SUTURATUS.—*Above* (*s. suturatus*): Left: Pelican I., Queensland, Australia, 36.2mm; Right: Hope I., Queensland, Australia, 40.3mm. *Below:* Left (*s. suturatus*): Hope I., Queensland, Australia, 43.5mm; Right (*s. sandwichensis*) (holotype and paratypes): Pokai Bay, Oahu, Hawaii, 14.3-19.4.

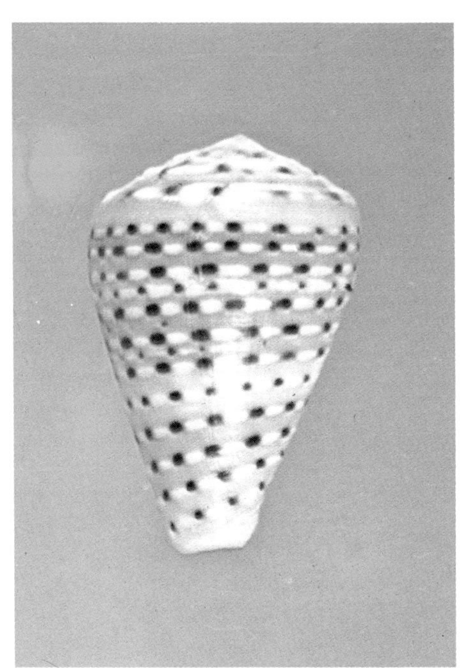

TAENIATUS.—Above: Left: Red Sea, 26.6mm; Right: Muscat, Oman, 26.7mm. *Below:* Red Sea, 34.3mm.

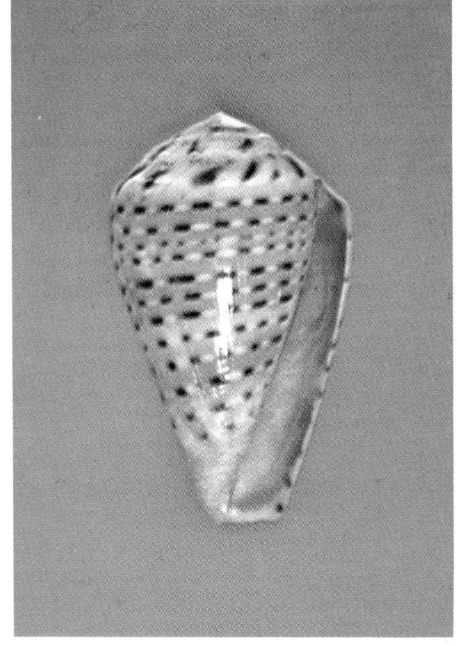

LITOGLYPHUS Hwass, in Bruguiere, 1792

1792. *Conus litoglyphus* Hwass, in Bruguiere. *Cone*, in *Ency. Method., Hist. Nat. des Vers*, 1: 692 (East Indies). Lectotype (here selected): specimen incorrectly cited and figured by Kohn (1968) as holotype.
1798. *Cucullus cinamomeus* Roeding. *Museum Boltenianum:* 43 (Locality not stated). Lectotype figure (Kohn, 1975): Martini, Pl. 57, fig. 631.
1798. *Cucullus Orleanus* Roeding. *Ibid.:* 44 (Locality not stated). Lectotype figure (Kohn, 1975): Chemnitz, Pl. 140, fig. 1298.
1807. *Conus subcapitaneus* Link. *Beschr. Nat.-Samml. Univ. Rostock*, 3: 103 (Locality not stated). Based on Martini, Pl. 57, figs. 630-631.
1833. *Conus bicolor* Sowerby i, in Sowerby ii. *Conchological Illustrations:* Pl. 24, fig. 2 (March 29) (Locality not stated).
1833. *Conus albomaculatus* Sowerby i, in Sowerby ii. *Ibid.:* Pl. 24, fig. 2. Nomen novum for *Conus bicolor* of March. Also spelled *Conus albimaculatus.*
1845. *Conus lacinulatus* Kiener. *Species gen. et icon. des coqu. viv.*, 2 (*Conus*): Pl. 108, fig. 2. 1849-1850, *Ibid.*, 2: 312 (Locality unknown).
1865. *Conus Carpenteri* Crosse. *J. Conchyl.* (*Paris*), 13: 302, pl. 9, fig. 1 (New Guinea).
?1874. *Conus* (*Rhizoconus*) *seychellensis* G. and H. Nevill. *J. Asiatic Soc. Bengal*, 43(2): 22 (Seychelles).

Description.—Heavy, with a good gloss; low conical, the sides nearly straight and tapering to a relatively narrow base; body whorl with about 4-10 low rounded spiral ridges basally, widely spaced; some of the basal ridges with heavy granules; rest of body whorl with sometimes heavy spiral and axial threads and a few scattered spiral ridges; shoulder broad, roundly angled; spire usually low, the sides straight or concave, sometimes high and with convex sides; whorls flat to slightly concave or convex above, with numerous fine curved axial threads and occasional traces of 1-2 indistinct spiral ridges; early whorls eroded. Body whorl dark reddish brown, yellowish brown, olive brown, or even black; base stained dark blackish violet; two spiral bands of large white spots, one at midbody and another below the shoulder; often these white spots are partially or entirely fused into solid white bands with scalloped edges; spire mostly brown like body, covered with large triangular white spots; early whorls whitish. Aperture narrow, of about uniform width; outer lip thin, straight; mouth white to pale brown, with a large brownish violet blotch anteriorly. Columella very narrow, sometimes internal. Length 35-70mm.

Comparison.—*C. litoglyphus* belongs to the group which includes *C. generalis*, *C. vitulinus*, *C. planorbis*, etc., but it is easily distinguished by the distinctive white spots at midbody and below the shoulder and the uniform dark body color without spiral dashes or lines. *C. vexillum* has a rounded shoulder and the white spots poorly defined. *C. pertusus* has a rounded shoulder, punctate grooves on the spire, and numerous spiral rows of small brown dots in the background.

Variation.—As indicated in the description, the dark color can vary considerably, although some shade of reddish brown is most common; occasional specimens have several regular spiral bands of different shades of the same color over the body whorl, with an especially conspicuous band below the shoulder. The white spots are often fused into narrow or broad spiral bands which are occasionally broken; very few specimens have the white spots reduced in size. Juveniles have the white spots much larger relative to body size. Although the spire is usually low, sometimes almost flat except for the median cone formed by the early whorl, some specimens have a quite high spire with convex sides. The spire may be mostly white or mostly brown.

Distribution.—Widespread and common in shallow water throughout the Indo-Pacific from the eastern African coast to Hawaii and French Polynesia.

Synonymy.—The names of Roeding and Link certainly apply here, but some of the later names are subject to doubt. *C. bicolor* appears to be based on a juvenile with very large spots. *C. lacinulatus* has the white bands very jagged, a condition sometimes found over part of some specimens but seldom seen over an entire shell's pattern. *C. carpenteri* appears to be a specimen with especially varied spiral banding of dark on the body whorl and reduced white. *C. seychellensis* has never been figured, but there is nothing in the original description to prevent tentative assignment with *C. litoglyphus.* In some older literature *C. litoglyphus* was called *C. ermineus*, a name now used for an Atlantic species.

Notes.—Brightly colored specimens of this species are very pretty and easy to find. Odd colors such as olive or black would be worth a small premium if they turned up more often. Juveniles are uncommon. *C. litoglyphus* is subject to bad reef breaks that sometimes heal poorly and distort the pattern and spire. The edge of the lip chips easily. The early whorls are almost always eroded.

LITTERATUS Linnaeus, 1758

1758. *Conus litteratus* Linnaeus. *Systema Naturae per Regna Tria Naturae*, ed. 10, 1: 712 (Asiatic Ocean). Lectotype selected and figured by Kohn, 1963.
1798. *Cucullus pardus* Roeding. *Museum Boltenianum:* 41 (Locality not stated). Lectotype figure (Kohn, 1975): Martini, Pl. 60, fig. 667.
1810. *Conus Arabicus* Lamarck. *Ann. du Mus. Hist. Nat.* (*Paris*), 15: 40 (Arabia, by implication).
1844. *Conus Gruneri* Reeve. *Conchologia Iconica*, 1(*Conus*): Pl. 43, sp. 231 (Java).

Description.—Very heavy, thick, with a good gloss in small and

TASLEI.*—Above:* Left (approaching *mercator*): M'Bur, Senegal, 42.4mm (GG); Right: Cape Verde Is., 19.5mm (MM). *Below:* Off Douala, Cameroons, 50.4mm (EP).

TEGULATUS.—*Above:* Trank, Thailand, 36.0 and 40.5mm (Shell Cabinet). *Below:* Left: Punta Engano, Cebu, P.I., 32.4mm (W. Lim); Right: Banda Sea off Timor, Indonesia, 27.6mm (GG).

medium specimens, duller with growth; obconical, the sides nearly straight or slightly concave at the middle in large adults; body whorl with a few low spiral ridges at the base, otherwise smooth except for numerous fine axial and spiral threads; shoulder broad, angulate, slightly concave above; spire low to flat, the early whorls eroded; spire whorls flat or slightly concave above, with curved axial threads only; juveniles may have traces of very weak spiral ridges as well. Body whorl white, with numerous spiral rows of large to small blackish spots and often three spiral orange or yellowish bands (below shoulder, at midbody, and above base); base blackish or dark violet; spots below shoulder axially elongated and extending onto spire; spire white with numerous revolving black lines. Aperture fairly narrow, wider basally; outer lip thin, sometimes concave at middle in adults; mouth white, the extreme anterior part with a blackish blotch. Columella short, wide, thick, with a heavy ridge above, shorter than inner lip and curved into the anterior profile of the outer lip. Length 75-150mm.

Comparison.—Adults of *C. litteratus* are often confused with *C. leopardus*, but are easily distinguished by the unique columella, the dark base, and the common presence of orangish bands on the body whorl. In addition, the shoulder blotches are generally elongated more than in *C. leopardus*, the surface gloss is much higher at equal sizes, and the spire whorls in even quite small specimens lack distinct spiral sculpture. Juveniles may be very like *C. eburneus* in color and pattern but are more elongated and have the base blackish.

Variation.—Although some changes occur with growth, there is never any doubt about the identification of this species. Juveniles are very brightly colored, with the black spots very dark and sharp and the orange bands wide and bright; the spots below the shoulder are very large and angular; and the spire has heavy axial threads crossing traces of fine spiral ridges. With growth the black spots become relatively smaller in many specimens and the orange bands fade, but the basic pattern remains; the spiral ridges on the spire whorls are soon lost. Large adults tend to reduce the number of spots, become quite concave at midbody, and have a much lower gloss than smaller adults. Occasional white specimens are found.

Distribution.—Widespread and common throughout the Indo-Pacific from the African coast to French Polynesia, but apparently absent from Hawaii. Shallow water.

Synonymy.—Although early records of this species are badly confused with *C. leopardus* and *C. eburneus*, the type specimen as restricted is the present species. *C. pardus* is still occasionally used for this species. *C. gruneri* is based on a small and very bright juvenile.

Notes.—Adults are common and easy to obtain in gem condition, although larger specimens, like those of *C. leopardus*, are often very ugly creatures and much battle-worn. Juveniles with very bright patterns are not commonly seen and are certainly worth a premium—they are much more attractive than even the most nicely patterned *C. eburneus*.

LIVIDUS Hwass, in Bruguiere, 1792

1792. *Conus lividus* Hwass, in Bruguiere. *Cone*, in *Ency. Method., Hist. Nat. des Vers*, 1: 630 (Antilles). Lectotype selected and figured by Kohn, 1968.

1798. *Cucullus monachus* Roeding. *Museum Boltenianum:* 39 (Locality not stated). Lectotype figure (Kohn, 1975): Martini, Pl. 63, fig. 694.

1807. *Conus plebejus* Link. *Beschr. Nat.-Samml. Univ. Rostock*, 3: 106 (Locality not stated). Based on Martini, Pl. 63, fig. 694.

1807. *Conus virgineus* Link. *Ibid.*, 3: 106 (Locality not stated). Based on Martini, Pl. 63, figs. 701-702.

?1849. *Conus primula* Reeve. *Conchologia Iconica*, 1(*Conus* Suppl.): Pl. 6, sp. 259 (Locality unknown).

Description.—Heavy, with a low gloss; low biconical, the upper sides somewhat convex, sometimes almost straight; body whorl with some crowded ridges at the base, these rapidly replaced by low, widely spaced spiral ridges that extend above midbody and are usually granulose; rest of whorl covered with often heavy spiral and axial threads and growth lines; shoulder angulate, with large rounded coronations; spire low, bluntly pointed, the sides usually concave; early whorls usually heavily eroded, later whorls with heavy rounded coronations; tops of whorls nearly flat, with about 2-4 low spiral ridges crossed by weak axial threads. Body whorl usually some shade of olive tan or bluish brown, only rarely yellowish; the base is heavily stained with dark purple that is usually visible only in ventral view; sharply defined whitish to yellowish spiral bands at midbody (sometimes double) and shoulder; spire white, sometimes with indistinct pale tan toning on the shoulder and last whorl; tip whitish to pale tan. Aperture moderately narrow, uniform in width; outer lip thin, with a marginal reddish brown band; mouth deep purple, with pale bands at middle and below shoulder; sometimes a thickened ridge at middle of mouth. Columella internal. Length 30-81mm.

Comparison.—*C. lividus* has sometimes been considered a coronated form of *C. flavidus* (or vice versa), but there are other differences besides coronation. In *C. flavidus* the usual body color is some shade of yellow or straw, the basal ridges often are not granulose, and the shoulder is more sharply angled. *Conus moreleti* is much more slender and has a more elongated body whorl at all sizes; the spire is roundly domed in many specimens, with the coronations very rounded and closely spaced, somewhat like those of *C. distans;* there are usually brown spots on the spire; and the basal ridges are seldom strongly granulose. *C. balteatus* and *C. cernicus* are covered with fine spiral ridges over the whole body whorl and have scattered white flecks. *C. muriculatus* has a distinctive pattern and often a paler violet mouth.

The only species likely to be confused with *C. lividus* is *C. sanguinolentus*. The two are very similar in shape, texture, and general pattern, and both have granulose basal ridges extending to midbody in most speci-

TELATUS.—Tayabas Bay, Luzon, P.I. *Above:* 51.1mm. *Below:* Left: 48.2mm; Right: 51.3mm (all K).

TENUISTRIATUS.—Above: Left: Kwajalein Atoll, Marshall Is., 16.0 and 20.5mm; Right: Boac, Marinduque, P.I., 36.9mm. *Below:* Mahe, Seychelle Is., 32.5mm.

mens. However, *C. sanguinolentus* is distinguishable by: the absence of a distinct pale band at midbody and a less sharply edged pale band at the shoulder; the purple staining of the base usually visible dorsally as well as ventrally; the spire strongly toned with pale tan, not solid white, especially on the later whorls; and the mouth without contrasting paler bands at middle and below shoulder. The two species are sympatric but their ecological relationship has never been clarified; perhaps they are a species pair like *C. ebraeus-C. chaldaeus*, with minor pattern differences and slightly different ecological preferences. Occasional specimens seem to be somewhat intermediate in one or more characters, but these cause little problem in identification. See also *C. diadema.*

Variation.—Other than differences in tone of background color and extent of the granulose spiral ridges, this species shows relatively little variation. The midbody band is sometimes double but there is normally a sharp edge to the band. Dark toning on the spire is usually very subdued or entirely absent. The mouth is only exceptionally all violet with the bands obscure.

It is possible that still another species is being confused with *C. lividus*, as occasional specimens from French Polynesia and other southern Pacific localities look much like smooth, small *C. lividus* but lack the midbody band or have it indistinct; these specimens usually do not have distinctly granulose basal ridges and may have a relatively high spire; the background color may be exceptionally blue. These are certainly not *C. sanguinolentus* or possible hybrids between that species and *C. lividus*, so they are for the moment considered variant *C. lividus.*

Distribution.—Widespread and common in shallow water throughout the Indo-Pacific, including Hawaii and French Polynesia.

Synonymy.—The names based on Martini figures almost certainly apply here, but the status of *C. primula* is uncertain. For one thing, the midbody band of the type is interrupted by large squares of dark color, a pattern not noticed in typical *C. lividus*. For another, the color is exceptionally reddish. Otherwise it fits *C. lividus* in shape and sculpture.

Notes.—This is not as rough a species as *C. flavidus*, but it is subject to minor and major axial growth flaws and to chips in the very fragile new growth of the lip. The early whorls of the spire are always eroded, sometimes quite badly, in adult shells. The smooth variant discussed earlier is not uncommon but is even more subject to axial flaws, although the spire is perhaps less subject to erosion. Yellowish specimens are not common in this species, while specimens with double midbody bands are moderately uncommon. Occasional specimens with reddish tones are found.

C. sanguinolentus seems to be specifically distinct from *C. lividus*, although the two are certainly close. For the moment, until good evidence of complete intergradation is forthcoming—if ever, they should be retained as species. There are probably many *C. sanguinolentus* in collections incorrectly identified as *C. lividus.*

LOHRI Kilburn, 1972

?1884. *Conus martensi* E.A. Smith. *Rept. Zool. Collections 'Alert' (1881-1882)*: 488, pl. 44, fig. A (Providence Reef, Mascarenes).

1972. *Conus lohri* Kilburn. *Ann. Natal Mus.*, 21(2): 428, figs. 8a, 14b-d (Baia dos Cocos, between Inhambane and Jangamo Beach, Mozambique).

Description.—Moderately heavy, with a low gloss; low conical, the sides convex; body whorl with a few low spiral ridges near the base, plus some scattered weak spiral and axial threads, otherwise smooth; shoulder rounded, broad, not very distinct from spire; spire low, rounded or bluntly pointed, the sides slightly convex; tops of whorls convex, the sutures rude and slightly overlapping; tops of whorls with faint traces of a single spiral ridge, almost smooth. Body whorl uniformly bright orange-tan or deep yellow, without obvious pattern or with two broad, slightly darker spiral bands above and below midbody; base same tone as rest of whorl; shoulder and spire same color as body whorl, the shoulder and last whorl with small deep reddish brown rather triangular spots that are widely spaced and not in uniform series; tip of spire eroded whitish. Aperture moderately narrow, nearly uniform in width; outer lip thick, convex; mouth faintly brownish violet. Columella long, narrow. Length 40-63mm.

Comparison.—This species is so non-descript that it is difficult to compare it with anything. However, the color of the body whorl is closely matched by some specimens of *Conus hyaena*, a species which sometimes shows similar brownish triangular spots on the shoulder as does *C. lohri*. However, *C. hyaena* is a much rougher cone with distinct (though irregular) body and spire sculpture, so the two are quite distinct. I can think of no other close relatives. *C. consors* is much more tapered in shape and has distinct nodules on the early whorls plus strong spiral sculpture on the other whorls.

Variation.—This species still seems to be known mostly by the type series, so little is known of the amount of variation to be expected. It would seem that larger specimens become duller in color with the already faint bands becoming even fainter. The sides of the body whorl may be more concave in larger specimens than in smaller ones.

Distribution.—So far known only from shallow reefs of the northern Natal coast, South Africa, and the southern Mozambique coast. Perhaps not uncommon but so plain that it is ignored by collectors.

Synonymy.—*C. martensi* was based on a single 24mm specimen from the central Indian Ocean. In shape and the plain color it is very close to *C. lohri;* however, the description mentions the spire as being three-grooved, which would probably correspond to having two spiral ridges. In the *C. lohri* so far known (none less than 40mm in length) the spiral sculpture is very weak or completely absent, although this could be a function of growth or erosion. Certainly it seems unusual for two species

TERAMACHII.—*Above:* China Sea, 95mm (DM) (photo by Jayavad). *Below:* Left: Mozambique Channel off Nossi Be, Madagascar, 53.8m (EP); Right: Off Durban, Natal, South Africa, 101.8mm (Meyer).

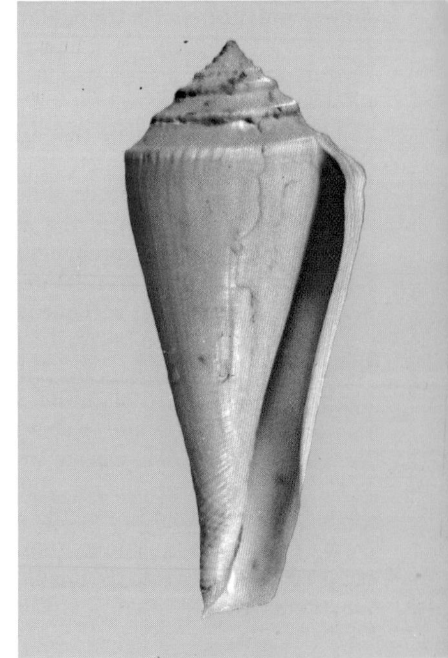

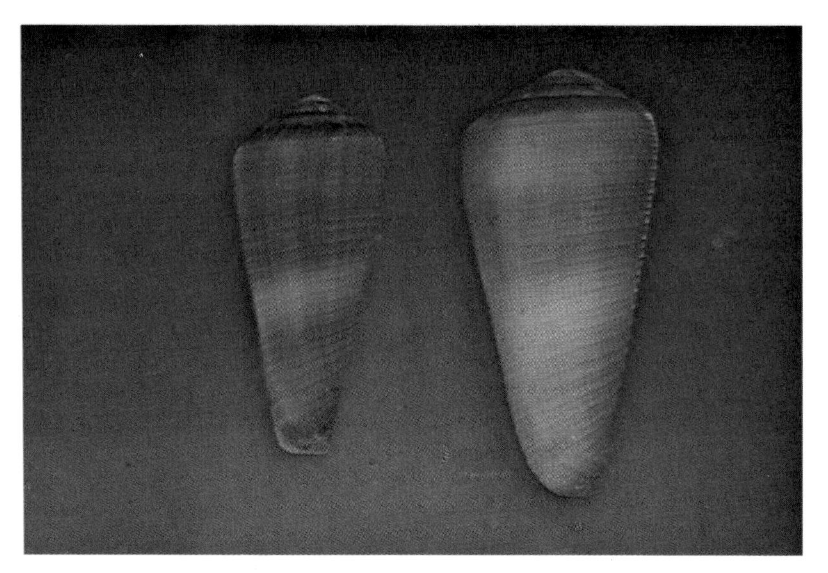

TEREBRA.—Above (*t. terebra*): Kakambaria, Guadalcanal, Solomon Is., 39.2 and 46.6mm. *Below:* Left (*t. thomasi*): Muscat, Oman, 49.3mm; Right (*t. terebra*): Tahiti, Society Is., 46.7mm.

of plain yellowish or orange cones with weak sculpture and similar shape to be described from the southern and central Indian Ocean. Until juveniles of *C. lohri* are known, nothing further can be said.

Notes.—The type series are not attractive cones, having heavy axial growth marks and eroded spires as well as damaged lips. Perhaps fresher specimens in good condition would be more attractive because of the bright orange or yellowish color, but the species has apparently not yet entered the market. It is, as mentioned earlier, possibly not rare, just ignored. I find it hard to believe the range is limited to the localities of the type series.

LORENZIANUS Dillwyn, 1817

1810. *Conus flammeus* Lamarck. *Ann. du Mus. Hist. Nat.* (*Paris*), 15: 279 (Mers d'Afrique). Holotype figure: *Tableau Ency. Method.*, Pl. 336, fig. 1. Non *Conus flammeus* (Roeding, 1798).
1817. *Conus Lorenzianus* Dillwyn. *Descr. Catalogue Recent Shells*, 1: 370 (East Indies). Based on Chemnitz, Pl. 181, figs. 1754-1755.
1937. *Conus phlogopus* Tomlin. *Proc. Malac. Soc. London*, 22(4): 206. Nomen novum for *Conus flammeus* Lamarck, 1810.

Description.—Moderately heavy, with a good gloss; low conical, the upper sides nearly parallel then tapering to the base, sometimes slightly convex posteriorly; body whorl with basal third to half covered with narrow raised spiral ridges, these usually close set and sometimes alternating narrow and wider; rest of whorl covered with quite distinct spiral threads; whole whorl crossed by sometimes heavy axial threads; shoulder broad, rounded to roundly angled, flat above in normal specimens but becoming deeply concave in old adults; spire moderate, sharply pointed, the sides weakly to deeply concave; tops of whorls flat to concave above, with numerous fine axial threads and sometimes traces of a single median spiral thread or low ridge; early whorls apparently not nodulose. Body whorl creamy white to pale grayish, covered with many spiral rows of small squarish reddish brown spots, these tending to be smallest at base and below the shoulder; spots become relatively smaller and more numerous with growth; narrow and scattered axial flammules, seldom long, may form by fusion of the spots, especially near midbody; numerous narrow colorless lines between series of spots, these often quite sharp and distinct, following path of spiral threads; angle of shoulder usually with many fine, short oblique reddish brown lines below, the angle itself with a distinct whitish band; spire whitish to grayish, covered with very fine short reddish brown lines, these appearing like spots at the angle of the whorls; early whorls usually eroded. Aperture moderately narrow, slightly widened anteriorly; outer lip sharp, thin, straight or slightly concave at the middle; mouth bluish white to pale grayish, sometimes with distinctly brownish tone. Columella long and very narrow, largely fused. Length 35-60mm.

Comparison.—Although *C. lorenzianus* has often been confused with *C. virgatus* in the literature, it is quite distinct from that species by the relatively wider body whorl and much narrower axial series of spots instead of strong flammules; the colors of the background are usually also different, often more salmon than the whitish or grayish of *C. lorenzianus*. The basal area with many small spots in *C. lorenzianus* is quite constant in that species and not found in *C. virgatus*.

C. lorenzianus is closely related to *C. spurius* and *C. sennottorum* as well as *C. cingulatus*. These species are all very close and occasional specimens are hard to place. However, as a rule the following characters are consistent enough to separate the species. In *C. spurius* the shell is thicker and heavier, somewhat more convex posteriorly; the body whorl is white with the spots large and well separated or fused into large blotches, not small and at best forming short axial flammules as in *C. lorenzianus*. *C. lorenzianus* is much more heavily sculptured on the basal third to half the body whorl than most *C. spurius* and has the spiral threads stronger over the remainder of the whorl, which is usually completely smooth in *C. spurius*.

In *C. sennottorum* the spots are sparse, the shoulder is carinate or very sharply angled, and the majority of the body whorl is smooth and almost sculptureless above the base. Juvenile *C. lorenzianus* are very similar at first glance to *C. sennottorum* of the same size, having the spire exserted and the spots relatively few on the body whorl, but the shoulder angle is rounded in these juveniles and the spots at the base are small, reddish brown, and squarish, much more distinct than in typical *C. sennottorum*. Typical *C. cingulatus* are almost covered with distinct, broad to narrow spiral ridges from the base to the shoulder; the shoulder is sharply angled, while the sides of the spire even in adults are almost flat or only slightly concave. The spots of *C. cingulatus* are large and dash-like in most specimens, tending to fuse into broad and heavy flammules. Commonly *C. cingulatus* has 2-3 weak but distinct spiral ridges on the spire whorls, while the spire itself is heavily marked with large rounded or crescent-shaped dark brown spots alternating with white. *C. cingulatus* is probably less closely related to *C. lorenzianus* than is *C. spurius*, but it is most commonly confused with *C. lorenzianus* in the literature.

Variation.—This species seems to change considerably with growth or perhaps as a response to different environmental condition. Small specimens have exserted spines with only slightly concave sides, rounded shoulders, and relatively large spots in fewer and more distinct rows. With growth the later spire whorls become broader, leaving the early whorls as a projecting cone, the shoulder becomes more distinctly angled, and the spots on the body whorl break into smaller and more numerous spots with a tendency to form short axial flammules. Old adults have the later spire whorls very broad and deeply concave, the shoulder angled

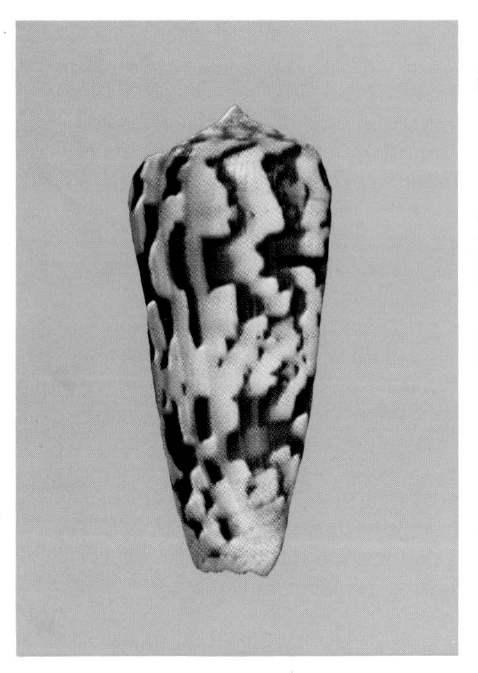

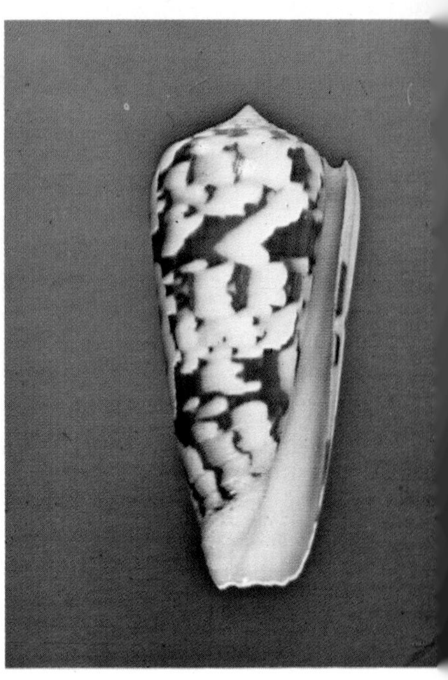

TERMINUS.—*Above:* Zanzibar, 68.5mm (MF). *Below:* Left and Center: Mozambique, 43.7 and 62.7mm (GG); Right: Trincomalee, Sri Lanka, 45.0mm.

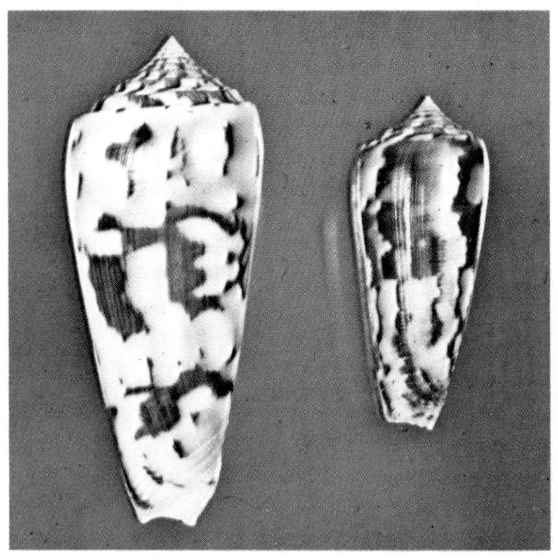

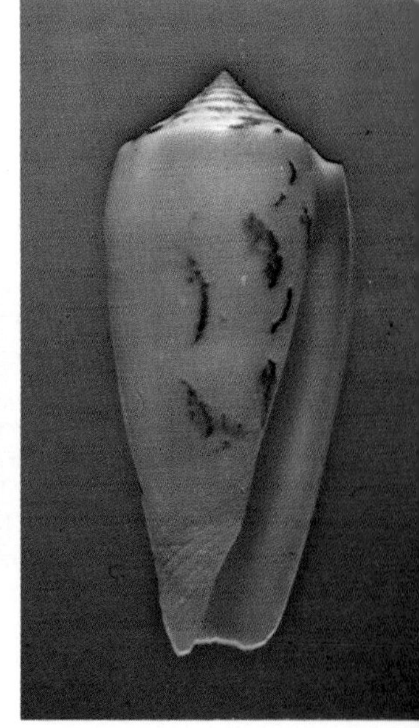

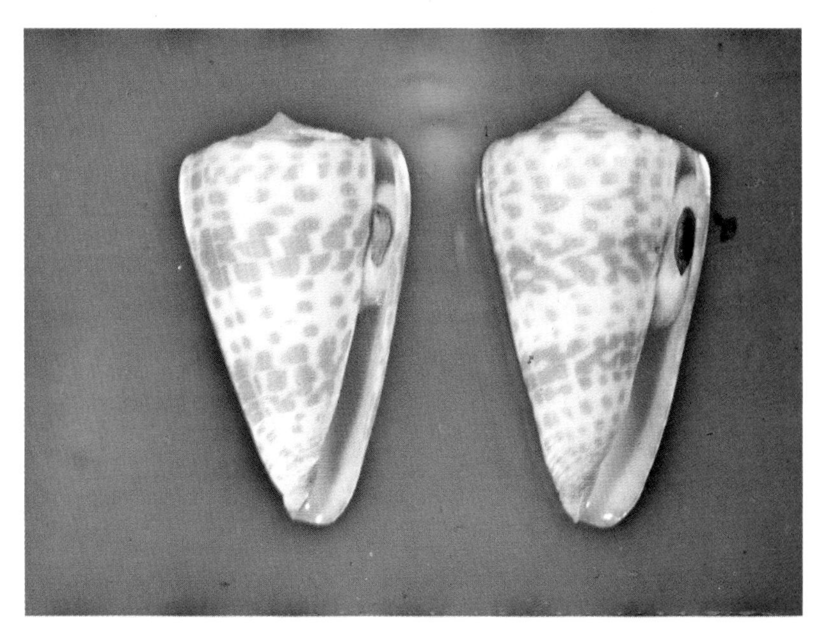

TESSULATUS.—Above: Mozambique, 55.9 and 59.2mm. *Below:* Left: Minabe, Wakayama Pref., Japan, 38.2mm; Right: Tahiti, Society Is., 26.2mm.

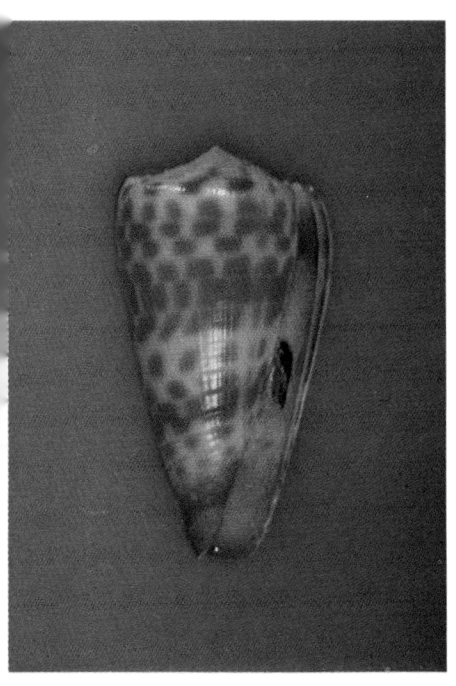

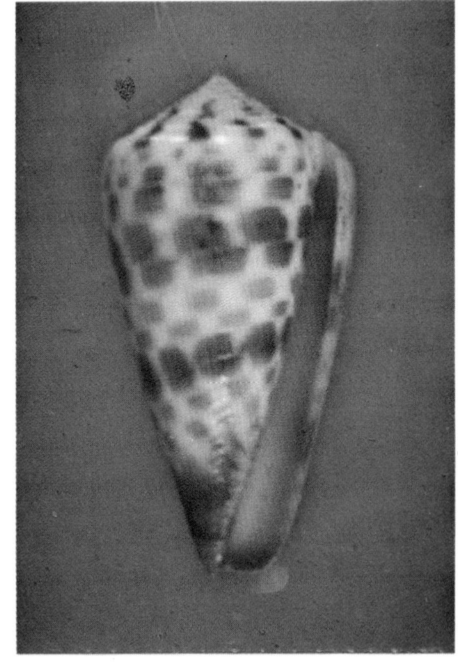

and concave, and the spots very numerous and small, forming axial rows and some flammules. The small spots at the base and the tendency to form oblique short stripes below the shoulder appear constant, as is the tendency to have the angle of the shoulder clear and unmarked. The spire whorls often appear spotted in larger shells because the margins have the same clear band as the shoulder that sets off the short brownish lines. Only occasionally are complete axial flammules formed over most of the body whorl, and even here there are commonly one or two fine clear spiral lines visible. The mouth may be pale violet in fresher specimens.

Distribution.—Seemingly restricted to moderately deep water off the northwestern coast of South America and adjacent Central America, from at least Panama and Colombia. Exact range unknown. Rare.

Synonymy.—When occasionally recognized as a full species in the literature, this species has been called *C. phlogopus*, a replacement for the preoccupied *C. flammeus* Lamarck. Comparison of the type figures of *C. flammeus* and *C. lorenzianus*, however, leaves little doubt in my mind that the two are the same species. Both figures show the small spots on the base, the spotting at the shoulder level, and the concave spire sides of this species.

Notes.—Adults of *C. lorenzianus* are rarely offered, although juveniles are sometimes sold as *C. sennottorum*. This species is usually found dead and heavily wormed and eroded, but it is still collectable in that condition. Large adults have a very unusual spire and very fine spots.

I have seen specimens from off Belize that could easily be considered intermediate between typical *C. spurius* and *C. lorenzianus* in both sculpture and pattern. Until more Central American material is available, these taxa should be treated as full species, but with the realization that they may actually be subspecies.

LUCIDUS Wood, 1828

1828. *Conus lucidus* Wood. *Index Test., Suppl.:* 8, pl. 3, fig. 4 (South Seas).

1834. *Conus reticulatus* Sowerby i, in Sowerby ii. *Conchological Illustrations:* Pt. 56/57, fig. 86 (Locality not stated). Non *Conus reticulatus* Born, 1778.

Description.—Fairly light in weight but thick, with a high gloss; low biconical, the sides convex; body whorl with anterior half or third with heavy narrow spiral ridges, widely separated, the grooves between minutely axially striated; rest of body whorl smooth except for axial growth marks in some shells; shoulder rounded or roundly angled, not very distinct from spire, sometimes slightly concave above; spire moderate, sharply pointed, the sides straight or concave; spire whorls flat or slightly concave above, without obvious sculpture or with fine curved axial threads. Body whorl white to pale cream, covered with numerous widely spaced narrow brown spiral lines, these often corresponding anteriorly with the spiral ridges; there are many irregular axial brown lines, long or short, straight or angular, connecting and breaking the spiral lines; commonly the lines define large white axial rectangles where the intermediate spiral lines are absent, but a vaguely tented pattern may also result; often there are large or small irregular dark blotches on the whorl, these most prominent near midbody and above the base; the blotches are composed of broad dark brown adjacent axial lines on a brown cloudy spot; shoulder and spire like body whorl, whitish with dark brown lines and small or large spots; early whorls whitish. Aperture moderately narrow posteriorly, widened anteriorly; outer lip thin to thick, convex or straight; mouth pale violet to whitish, but usually with a distinct violet blotch deep within. Columella short, narrow, projecting anteriorly, with low ridges posteriorly. Length 25-60mm.

Comparison.—The pattern of narrow brownish lines on white combined with the shape and generally small size are very distinctive. Only *C. boschi* is close in pattern, but it is a more obese shell with a dark violet aperture and more regular pattern; especially obvious is the more rounded spire with a bluntly mammillate protoconch—the spire of *C. lucidus* is sharply pointed. Some patterns of *C. puncticulatus*, which is probably the closest relative of *C. lucidus*, may have the spots almost fused into spiral lines, but of course such a pattern is much less distinct than that of *C. lucidus*; the background color of such shells is usually some tone of pale bluish gray instead of white. *C. scalptus* is more elongated with a narrower shoulder and commonly lacks the axial brown lines or has them very poorly developed; it has strong spiral sculpture on the spire whorls.

Variation.—As would be expected in a species with a lineate-blotched pattern, there is considerable variation in the intensity and concentration of the pattern. Some specimens are very densely marked, while others are mostly white with all the lines broken and irregular. There is also some variation in spire height and width of the shoulder. Mexican specimens are perhaps larger than those from the islands, with a denser pattern.

Distribution.—Uncommon to locally common in moderately deep water to shallow water from Baja California and the adjacent Mexican mainland south to Ecuador; most common on the offshore islands and the Galapagos.

Synonymy.—This species has long been called *C. lucidus*, although the 'description' consists of just the name and a very small figure.

Notes.—This is a very glossy shell, but there are commonly large axial growth marks over the body whorl; since these occur in almost every specimen and are straight and not unattractive, they do not affect the value of a specimen. Specimens from the islands are usually smaller than those from the mainland, but they are also more commonly collected and

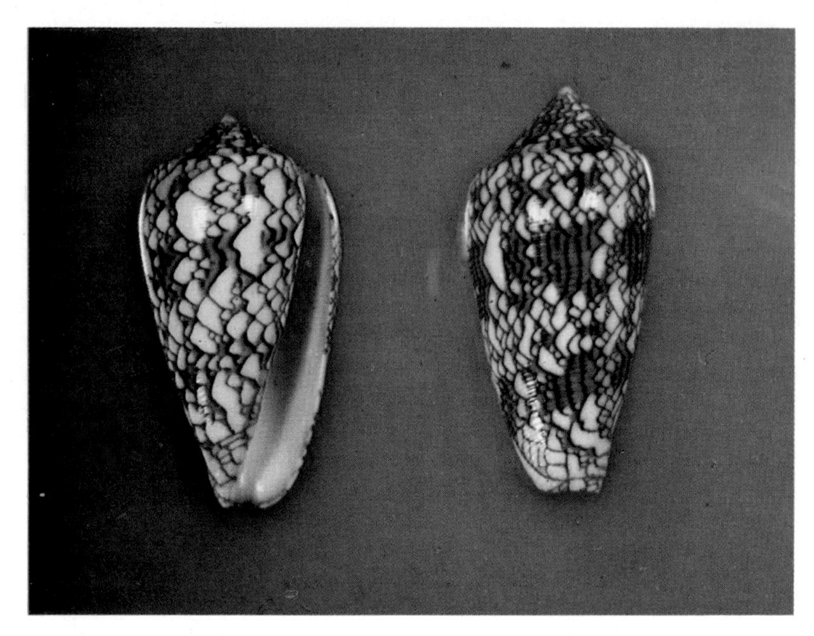

TEXTILE.—*Above:* Tahiti, Society Is., 50.4 and 52.5mm. *Below:* Left: Inhambane, Mozambique, 51.5mm (EP); Right: No data, 35mm (DM) (photo by Jayavad).

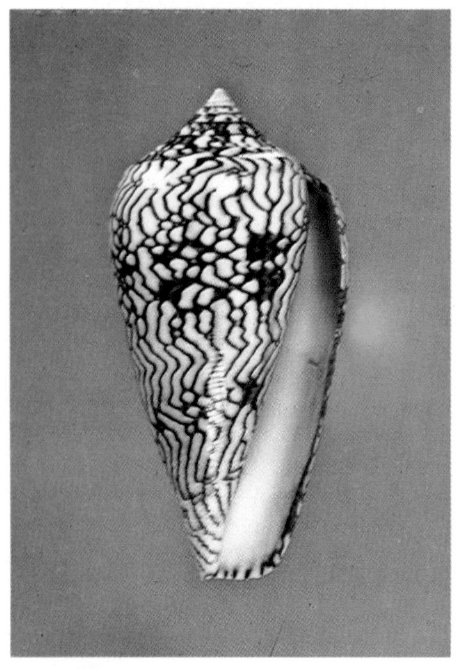

TEXTILE.—*Above* (*archiepiscopus*): Left: Umdloti, N of Durban, Natal, South Africa, 36.6mm (Meyer); Right: Zanzibar, 45.4mm (MF). *Below:* Left (*verriculum*): Zanzibar, 38.2mm (MF); Right (terminal *archiepiscopus*): Zanzibar, 58.0mm (MF).

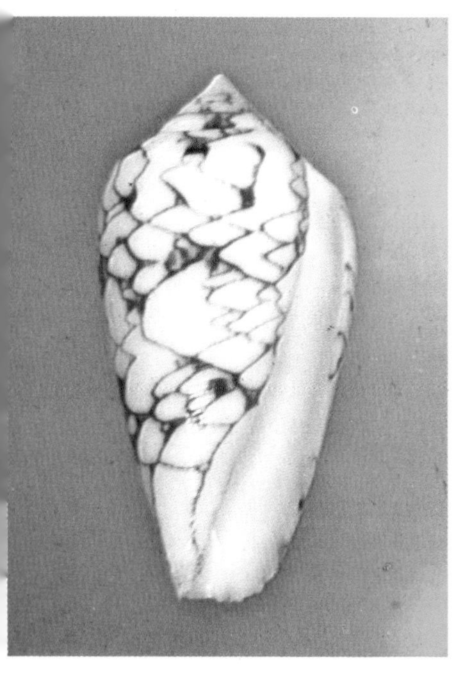

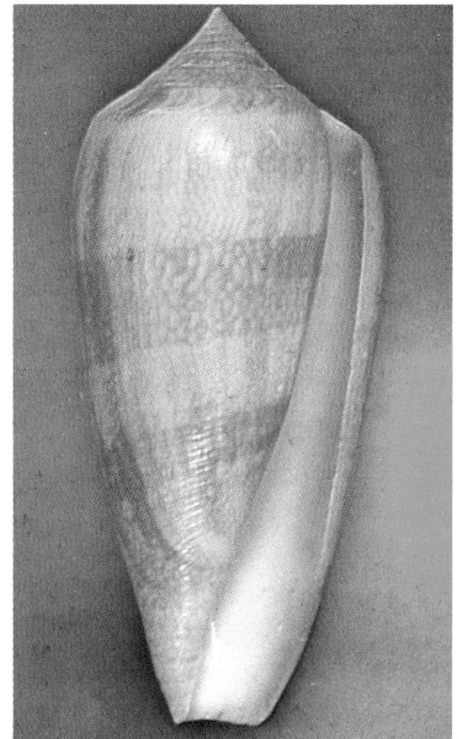

therefore more available.

Although the pattern suggests relationship with *C. textile* or some similar tented cone, the sculpture of the spire and the shape of the columella, as well as general shape, indicate a close relationship with *C. puncticulatus.* This is especially obvious when the Panamic *C. puncticulatus perplexus* is compared with *C. lucidus.*

LUTEUS Sowerby i, in Sowerby ii, 1833

1833. *Conus luteus* Sowerby i, in Sowerby ii. *Conchological Illustrations:* Pt. 25, figs. 8-8* (Locality unknown).
1845. *Conus corallinus* Kiener. *Species gen. et icon. des coqu. viv.*, 2 (*Conus*): Pl. 73, fig. 2. 1849, *Ibid.*, 2: 246 (Locality unknown).
1848. *Conus nucleus* Reeve. *Conchologia Iconica,* 1(*Conus* Suppl.): Pl. 3, sp. 280 (Matnog, Island of Luzon, Philippines).
1849. *Conus inaequalis* Reeve. *Ibid.*, 1(*Conus* Suppl.): Pl. 7, sp. 270 (Locality unknown).

Description.—Moderately light in weight, with a high gloss; low conical to almost cylindrical, the sides straight or slightly convex posteriorly, tapering to a narrow base; body whorl with about a dozen low spiral ridges anteriorly, rest of whorl with weak or distinct spiral and axial threads but appears glossy and smooth; grooves between spiral ridges very weakly axially threaded; shoulder narrow and rounded, indistinct from spire, to broader and roundly angled, fairly distinct from spire; spire low or moderate, bluntly pointed, the sides straight to convex, conical or domed; early whorls not nodulose; tops of spire whorls flat or convex, with about 3-6 fine but distinct spiral ridges crossed by very fine axial threads; sculpture sometimes almost obsolete on later whorls and shoulder. Body whorl reddish brown, pink, or pale violet, the base not contrasting with rest of whorl; usually several spiral rows of small brownish dashes around body whorl, these sometimes not very distinct, the rows usually widely spaced; midbody area paler to opaque white, blotchy, bounded posteriorly by large squarish blotches of dark brown; shoulder and spire same color as body whorl, with large to small curved blotches of dark reddish brown usually visible if not obscured; sometimes small brown dots along margins of whorls; early whorls whitish. Aperture narrow, about uniform in width; outer lip thin, fragile, straight or convex; mouth deep pink to pale violet or rosy. Columella internal or very narrow. Length 20-40mm.

Comparison.—This attractive little shell is highly variable in shape and to some extent pattern, which sometimes makes it hard to identify. However, the species most like it in shape (*C. mitratus* and *C. tenuistriatus*) have very different patterns of darker colors and are heavily ridged or granulose on the body whorl. *C. artoptus* sometimes is similar to the typical form of *C. luteus* in shape but is usually much darker in color, brownish instead of pinkish, has heavier basal ridges, and has strongly nodulose early whorls.

Conus pertusus is very similar in shape, pattern, and color to some *C. luteus* juveniles, but the two are probably not really closely related. In *C. pertusus* there are numerous fine brown dots over the base, while in *C. luteus* the brown dashes are larger, more widely spaced, and distinctly spiral even in the basal area. Even adult *C. pertusus* have traces of spiral rows of deep punctations over much of the body whorl and between the spiral ridges; such punctations are absent in *C. luteus* at any size.

Variation.—Although the sculpture of the body whorl and spire is fairly constant, the body shape and color pattern fall into two distinct groups which are possibly full species. For the moment, however, they are treated as variants of a single species. There seem to be intermediates known, but this species is simply too poorly known to make pronouncements on relationships.

C. luteus typical: Shoulder broader, roundly angled, sometimes a bit bulbous posteriorly; sides nearly straight, tapering strongly; color usually bright reddish brown or deep pink, sometimes bright yellow, with the spiral rows of brown dashes usually obvious at least near midbody; midbody area broad, pale, well demarked above by brown squares; spire with dark blotches usually fairly distinct. Uncommon to rare, from French Polynesia to the New Guinea-Solomons area, north to the Ryukyus.

C. luteus variety *nucleus:* Shoulder narrow, rounded, not distinct from spire; spire rather domed; sides of body whorl usually convex; color rather dull reddish brown, violet, or yellowish, with the spiral rows of dashes inconspicuous even under a lens; midbody area with large, widely spaced blotches of opaque white margined by similar blotches of dark brownish red; blotches on shoulder and spire usually indistinct. Uncommon, from Zanzibar east through the Philippines and Ryukyus to at least the New Guinea-Solomons area and the Marshall Islands.

Distribution.—Uncommon or rare over much of the Indo-Pacific, but apparently sometimes very local. Eastern African coast east to the Ryukyus, Philippines, New Guinea, Solomons, and Marshall Islands. French Polynesia population apparently isolated, as there are no records between the Solomons and Tahiti, but the species could have been overlooked. Dead shells recorded from off Hawaii. Usually dredged in moderately deep water or found beach.

Synonymy.—Except for *C. nucleus*, the other names all seem to apply to the typical variety of *C. luteus*. The more ovate shape and subdued pattern of *C. nucleus* appear very distinct at first, but there appear to be specimens of this shape with the pattern of typical *C. luteus* and vice versa, so it is retained as *C. luteus* variety *nucleus* for the moment.

Notes.—At best, this is an uncommon species. Most specimens in collections are probably of the *nucleus* variety, and many of these are slightly

THALASSIARCHUS.—Sulu, P.I. *Above:* 57.3 and 58.7mm. *Below:* 50.8-55.5mm.

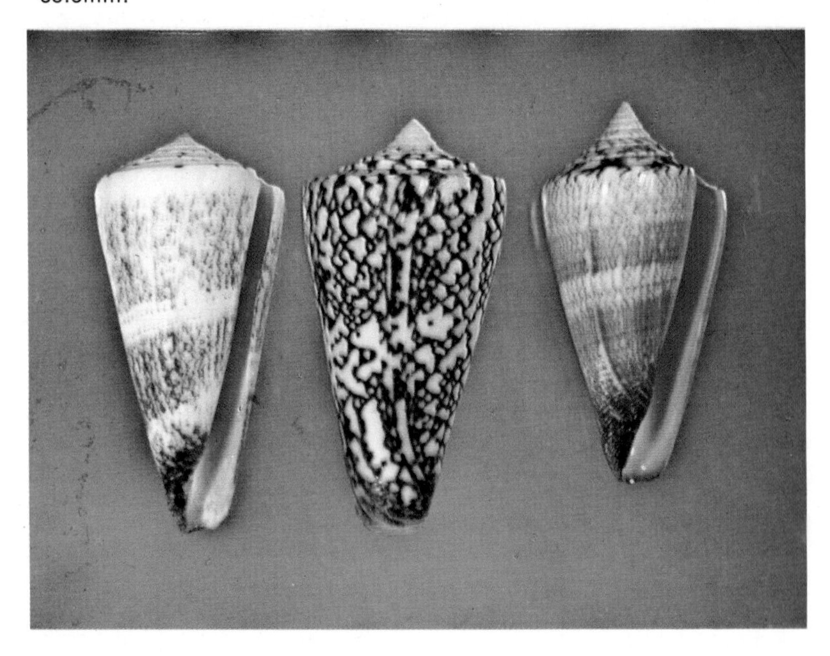

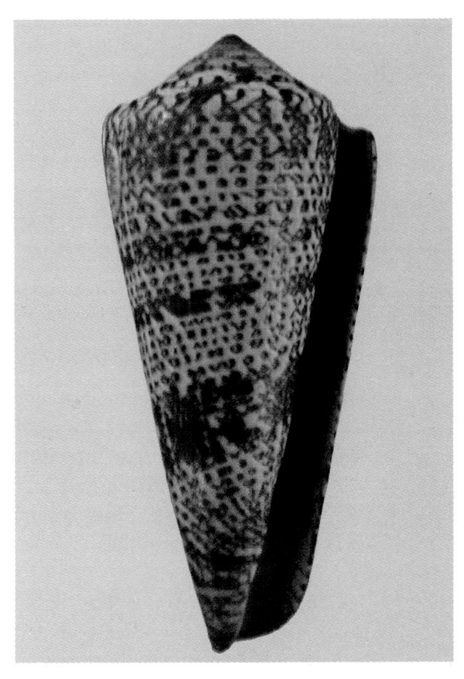

THOMAE.—Misool I., Indonesia. *Above:* 73.3mm. *Below:* 67mm (photos by P. Clover).

eroded beach shells. Although juveniles of the typical form are now being dredged in the New Guinea-Solomons area, they are still uncommon and adults are very uncommon. Large and brightly colored (pinkish or yellowish) specimens of the typical variety from either the Ryukyus or French Polynesia are sometimes offered but fetch high prices.

The lip of this species is apparently very fragile and easily chipped or completely broken; beware of heavy filing. Beach specimens appear weakly eroded under magnification and have subdued colors.

LYNCEUS Sowerby ii, 1857-1858

1857-1858. *Conus lynceus* Sowerby ii. *Thesaurus Conchyliorum*, 3 (*Conus*): 33, pl. 19(205), fig. 469 (Moluccas).

?1874. *Conus* (*Chelyconus*) *pretiosus* G. and H. Nevill. *J. Asiatic Soc. Bengal*, 43(2): 22 (Andaman Is.).

?1942. *Conus cavailloni* Fenaux. *Bull. l'Inst. Oceanogr.* (*Monaco*), No. 814: 4, fig. 1 (Bermuda).

Description.—Moderately light in weight, with a good gloss; elongate low biconical, the sides convex; body whorl elongate, covered from base to shoulder with shallow, widely spaced spiral grooves separated by broad flat ribs; grooves finely axially threaded or punctate; numerous fine axial threads over body whorl; shoulder narrow, roundly angled but often with a sharp edge, flat or slightly concave above; spire tall, sharply pointed, the sides concave to deeply concave, the whorls sometimes stepped; early whorls not nodulose; tops of whorls concave, covered with about 6-12 narrow but distinct spiral ridges crossed by distinct axial threads to produce a weakly cancellate effect, at least on the earlier whorls. Body whorl white to creamy white, covered with numerous large or small reddish brown squares, triangles, and dashes in roughly spiral and axial rows; these may be very regular over the whorl or poorly developed in some parts of the whorl and heavy on others; the spots show a strong tendency to fuse into large dark blotches above and below midbody, these blotches sometimes fusing into nearly complete spiral bands; base white; spire and shoulder whitish to cream, with scattered reddish brown blotches and spots; early whorls whitish. Aperture rather narrow, slightly widened anteriorly; outer lip convex, sharp; mouth pale to deep clear violet without brown tinges. Columella long, narrow, slightly oblique. Length 40-95mm.

Comparison.—This species looks very much like a gigantic *C. stramineus* but is relatively easy to distinguish from that species by the paler and clearer violet aperture and the more numerous and closely spaced spiral ridges on the spire whorls; it is also more elongate and has a higher spire. *C. inscriptus* is very similar in spire sculpture and sometimes aperture color, but it is marked with larger spots as a rule and is relatively broader at the shoulder; the early whorls of *C. lynceus* are more distinctly cancellate than those of *C. inscriptus*. *C. australis* is thicker, coarser, and has a different pattern of brown, not reddish brown, spots and streaks. *C. duplicatus* has the early whorls finely nodulose and is more convex-sided as well as being axially streaked or spotted with brown instead of reddish brown.

Variation.—There is considerable variation in color pattern in *C. lynceus*. Although the whole body whorl is usually covered with rather large and sharply defined spots of deep reddish brown and with two broad spiral belts of large blotches, occasional shells are very weakly patterned or may be almost white. Commonly larger shells have the spots reduced to smaller triangular spots and may show small spiral dashes in place of the larger spots. A row of small blotches may be developed below the shoulder or that area may be very weakly patterned. Occasionally the spotting is yellowish brown instead of reddish brown.

The spire whorls are not usually stepped, but in larger specimens the later whorls appear distinctly stepped and become relatively broader with sharper and almost carinate margins. The background color of the body whorl is usually white or cream, but it may on occasion be pale yellow or pinkish. The mouth is usually pale clear violet, rather uniform over the entire aperture, but may be deep violet; there is no brownish tinge in the mouth.

Distribution.—Uncommon to moderately common in the western Pacific from off Taiwan south through the Philippines to (rarely) Queensland, Australia. Also found in the eastern Indian Ocean off Malaysia north to off southern India. Possibly also further south in the central Indian Ocean.

Synonymy.—*C. pretiosus* is a very doubtful name which has never been illustrated. However, there is really nothing in the description to distinguish it from *C. lynceus* other than the larger than usual size; the proportions are the same, as is the general color and pattern. Very large *C. lynceus* from off Thailand appear to be this form and differ from typical, smaller (up to about 65mm) *C. lynceus* by the larger size (commonly over 80mm), the somewhat wider and more convex body whorl, slightly stepped later spire whorls, and reduced or very irregular pattern on a pinkish background. These would all appear to be characters depending on the large size and should fall within the normal range of variation of *C. lynceus*.

C. cavailloni possibly does not belong here, but it is also certainly not *C. gradatus*, to which synonymy it is commonly referred. It would seem to be a slightly distorted *C. lynceus* because of the general shape and pattern, but it could also belong to an allied species such as *C. inscriptus* or even *C. janus*. It is of course not from Bermuda.

Notes.—Most *C. lynceus* currently come from off Taiwan, the Philippines (both normal and very pale shells), and the Bay of Bengal. The large specimens over about 70mm are sold under various manuscript

TIARATUS.*—Above:* Left: Santa Cruz I., Galapagos Is., 25.6mm; Right: Cabo San Lucas, Baja California, Mexico, 38.1mm (photo by A. Kerstitch). *Below:* Pedro Gonzales, Perlas Is., Panama, 31.8mm (GG).

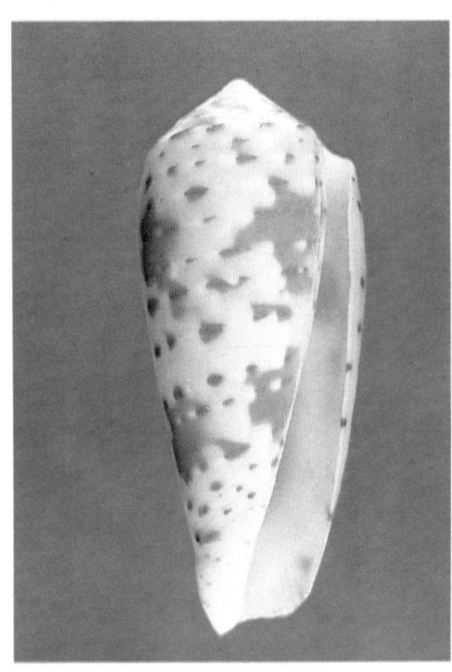

TIMORENSIS.—Above: Mauritius, 36.5mm (MM). *Below:* St. Brauson, Mauritius: Left: 36.9mm; Right: 31.8mm (both GG).

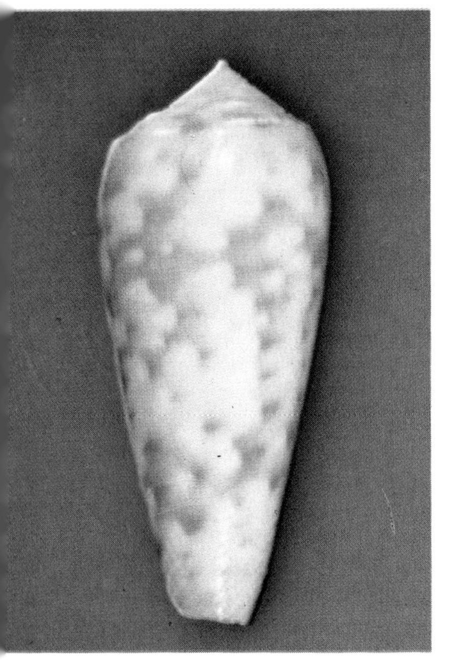

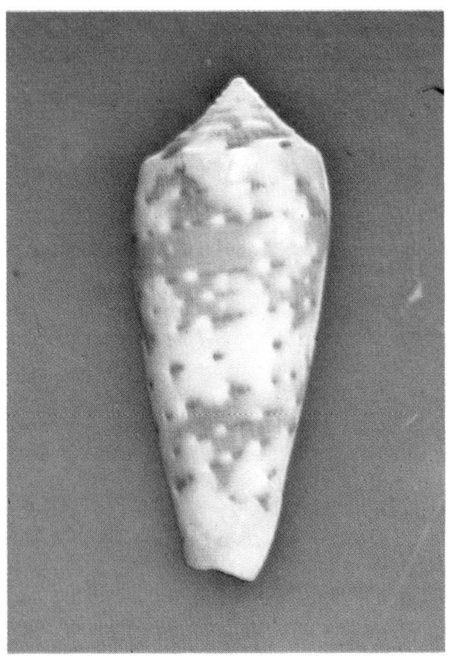

names at large premiums, but they are certainly not worth more than a moderate increase in price because of larger size.

White specimens assigned to this species possibly cannot be accurately identified unless taken with typical or intermediate specimens. The sculpture of the spire and body whorl is so simple that it is closely duplicated by several other species that sometimes differ radically in pattern. Probably *C. lynceus* is very widely distributed over the Indo-Pacific but difficult to identify if the pattern is absent or atypical, as may be the case in some populations.

MAGELLANICUS Hwass, in Bruguiere, 1792

1792. *Conus Magellanicus* Hwass, in Bruguiere. *Cone*, in *Ency. Method., Hist. Nat. des Vers*, 1: 633 (Strait of Magellan). Lectotype (here selected): the specimen figured by Kohn, 1968, as the holotype and by Kiener, Pl. 29, fig. 3.

1834. *Conus flavescens* Sowerby i, in Sowerby ii. *Conchological Illustrations:* Pt. 55, fig. 68 (Locality not stated).

1942. *Conus caribbaeus* Clench. *Johnsonia*, 1(6): 23, pl. 11, figs. 4-5 (Off Palm Beach Co., Florida).

(1968. *Conus caribbaeus* var. *circumpunctatus* Usticke. *Caribbean Cones from St. Croix and the Lesser Antilles:* 15, pl. 2, sp. 1002 (Anguilla). Infrasubspecific name that does not enter priority.)

Description.—Light in weight with a good gloss; low biconical, the sides straight or very slightly convex; body whorl elongate, with about a dozen low spiral ridges on the anterior third of whorl; rest of whorl covered with fine spiral and axial threads, often producing a satiny sheen; shoulder sharply to weakly angled, rather wide, distinct from spire; spire moderate, very sharply pointed with a persistent protoconch; sides straight or nearly so; last two or three whorls often distinctly stepped, concave above; all whorls with numerous curved axial threads, the spiral sculpture absent or sometimes with very weak median spiral threads; early whorls sometimes with rude sutures and may appear nodulose. Body whorl white to pale yellowish, covered with a very variable pattern of pale reddish brown or yellowish brown; the most familiar pattern consists of irregular axial flammules, few or numerous with a few or many spiral rows of brownish dashes; by further increases in the flammules and their fusion a nearly solid brownish shell with an interrupted midbody band may be produced; with reduction of the flammules a nearly all-white shell or one with faint traces of brown pattern above and below midbody may result; base pale; shoulder and spire whitish with radiating reddish brown to yellowish spots and streaks; white of shoulder sometimes extends onto body whorl as a jagged line. Aperture narrow, uniform in width;

outer lip straight, very thin; mouth pale pinkish. Columella very narrow or internal, slightly oblique and indented from inner lip. Length 15-35mm.

Comparison.—This little shell has been widely confused with *C. jaspideus*, but the two are quite distinct. In *C. jaspideus* the body whorl is covered with deep spiral grooves anteriorly, the shoulder is carinate with a weak but distinct rim usually present, and the spire is usually higher; there are usually differences in pattern, with the spiral rows of spots in *C. jaspideus* much smaller and more numerous than in *C. magellanicus*. In *C. mindanus* the shoulder is usually more rounded, the spire whorls fail to overlap slightly and have small spots at the margin, and the patterns are usually quite different; *C. mindanus* (large type) under 35mm would normally be covered with distinct spiral grooves which are commonly pustulose, while larger specimens have very strong grooving anteriorly. Small *C. gradatus* may have similar patterns but have the early whorls nodulose.

C. magellanicus is very closely related to *C. daucus* and *C. attenuatus* because of the shape of the body whorl and the spire, but it differs from both in a few characters. First, the lack of spiral sculpture on the spire whorls should separate it from typical examples of both species, although occasional *C. magellanicus* have a single spiral ridge and some *C. daucus* may lack obvious ridges. From *C. daucus* it differs in the smaller size, narrower and more elongate body whorl, and commonly in the pattern of heavy axial flammules. From *C. attenuatus* it differs mostly in the less bright colors, lacking the bright pinkish to salmon base and the bright aperture of that species; the spire of typical *C. attenuatus* is more domed than in *C. magellanicus*. Certainly the two species are closely related, however, and apparently derived from *C. daucus*.

Variation.—The spire may be rather low or rather high but usually has nearly straight sides and is always sharply pointed. In many specimens the last 2-3 whorls are distinctly but weakly stepped and concave above. Axial sculpture may be almost absent in some specimens, but rarely there is a single spiral ridge or thread present on part of the spire. The shoulder may be almost carinate to rather roundly angled. The greatest variation is in color pattern, which has been summarized above. Solid white shells are not uncommon, but neither are those that are all dark reddish brown with an interrupted white midbody band and pale base. The most common pattern consists of some variant of axial flammules and spiral dashes, although neither of these pattern elements is visible in many shells of the extremely light or dark patterns. Another common variation is the replacement of the reddish brown color with pale yellow.

Distribution.—Apparently this species is widespread in the Caribbean, being known from the Bahamas and southern Florida south to the Netherlands Antilles. It is seldom common, however, and seems to occur in small colonies. The distribution is incompletely known because of confusion with *C. jaspideus*.

TINIANUS.—Above: Jeffreys Bay, Cape, South Africa, 42.5 and 44.6mm. *Below:* Left: Jeffreys Bay, Cape, South Africa, 42.3mm; Right: Port Elizabeth, South Africa, 42.1mm (K).

TORNATUS.—Above: Perlas Is., Panama, 26.1 and 26.4mm. *Below:* Guaymas, Sonora, Mexico, 26.0-34.8mm (photo by A. Kerstitch).

Synonymy.—*C. magellanicus* was fairly well described and the type was well figured by Kiener. However, apparently the type is an unusual specimen in having the spire whorls very finely nodulose (actually the sutures are rude), a situation sometimes found in poorly developed form in modern specimens. *C. flavescens* is certainly this species from the figure and later descriptions, and the name has gained currency in Caribbean literature. *C. caribbaeus* is just a rather large and somewhat inflated specimen with a spiral ridge on the spire whorls, an uncommon but probably meaningless variation.

Notes.—Most of the patterns of this species can be found in any single large collection, so they are not worthy of any varietal names. Specimens from the Antilles might be somewhat larger than those from Florida and tend more to spotted patterns; they are slightly more inflated than more northern specimens, but this is probably a factor of size.

This is not a rare species in some areas, but it is apparently absent from several well-collected Caribbean islands (at least no specimens have been reported under any of the possible misidentifications). In southern Florida large series may be taken by night-diving.

This is a rather fragile species and gems are not common. Many specimens have broken lips and heavy growth marks, with occasional eroded spires or encrusting organisms. This is one species which must almost be collected in series of at least three patterns (very light, very dark, flammulated) before any concept of its nature can be formed.

MAGNIFICUS Reeve

1843. *Conus magnificus* Reeve. *Conchologia Iconica*, 1(*Conus*): Pl 6, sp. 32 (Matnog, Luzon, Philippines).

1937. *Conus episcopus* var. *elongata* Dautzenberg. *Mem. Mus. Roy. d'Hist. Nat. Belg.*, 2(18): 100, pl. 3, fig. 7 (Amboina). Non *Conus elongatus* Dillwyn, 1817. (This name is doubtfully established).

1942. *Conus episcopus* var. *oblongus* Fenaux. *Bull. l'Inst. Oceanogr. (Monaco)*, No. 814: 2. Nomen novum for *Conus episcopus* var. *elongata* Dautzenberg, 1937. Non *Conus oblonga* Bucquoy, Dautzenberg, and Dollfus, 1882.

Description.—Heavy, with a good gloss; elongate cylindrical, the sides slightly convex posteriorly then slowly tapering to the base, sides sometimes almost parallel; body whorl with numerous weak but distinct spiral ridges on the anterior third to half, these sometimes extending to the shoulder; ridges usually easily felt with the fingertip; shoulder rounded, narrow, not very distinct from spire; spire moderate, thick, bluntly pointed, the sides straight; whorls of spire flat to convex above, with traces of a few weak spiral and axial threads; early whorls not nodulose; tip of spire usually eroded, rounded, the protoconch small but sharp.

Body whorl white to pale pinkish, heavily covered with fine axial wavy or zigzag lines of deep reddish brown, golden tan, or blackish brown, these forming large to small tents; usually two spiral rows of large brown blotches above and below midbody, the blotches axially elongate and connected in small specimens; the blotches divide the tented areas into spiral bands at the base, at midbody, and below the shoulder as well as axial bands between the blotches; numerous spiral rows of small brown and white dots or dashes visible in the dark blotches, the white dots sometimes expanded into small triangles; shoulder and spire whitish or pinkish with large brownish spots and lines; early whorls white, protoconch pinkish. Aperture moderately narrow posteriorly, widened anteriorly; outer lip slightly convex to straight, often thick; mouth white. Columella long, narrow. Length 40-80mm.

Comparison.—The absence of nodules on the early spire whorls will distinguish *C. magnificus* from *C. textile* and allies. *C. magnificus* is very closely related to *C. pennaceus* (especially the *omaria* variety) and to *C. aulicus*, serving as something of a connecting link between the two. From *C. aulicus* it differs in having the spire blunt instead of sharply pointed, in having the spiral ridges of the body whorl usually weaker, and in being smaller at adult size; there are also differences in shape, but these are not constant. From *C. pennaceus* it differs in the relatively higher spire, the more rounded shoulder, and the stronger spiral ridges on the body whorl. Many forms of *C. pennaceus* are thicker and broader, with a low spire and distinctly angled shoulder; the variety *omaria* is more similar in body whorl shape to *C. magnificus*, but it has the low spire of *C. pennaceus*. In most patterns *C. magnificus* has larger and more distinct blotches and tents than many *C. pennaceus* patterns. Obviously all these species are very closely related, and occasional specimens are hard to place with certainty; however, series usually are fairly constant and present little identification difficulties.

Variation.—There is considerable variation in height of spire, convexity of body whorl, and strength of spiral ridges on the body whorl. The color also varies, the background from white to pink or even occasionally bluish, the pattern from pale golden tan through the usual reddish brown to almost black. These color variations seem to be dependent on local environmental conditions and have no taxonomic significance. The pattern itself varies in the size of both the blotches and the tents. In some specimens, including Reeve's type, the shell is largely white with two rows of brown blotches and remnants of the zigzag axial lines. Other specimens are covered with numerous small tents that invade the blotches.

Distribution.—Common and widely distributed over the Indo-Pacific in shallow water. Found from the eastern coast of Africa to French Polynesia, but apparently absent from Hawaii.

Synonymy.—This is the species long called *C. episcopus* in the literature. However, it seems certain that *C. episcopus* of Hwass is synonymous with *C. pennaceus* and cannot be used for this species. The next available

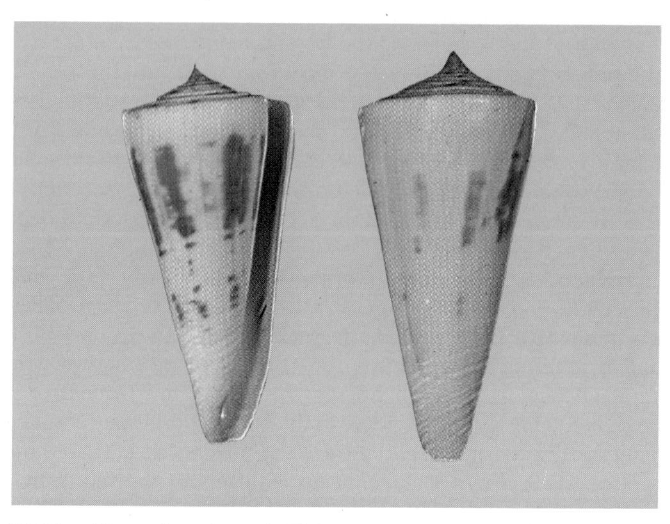

TRIBBLEI.—Above (paratypes): Taiwan, 71.1 and 76.4mm. *Below* (paratype): Marinduque, P.I., 48.4mm (AK).

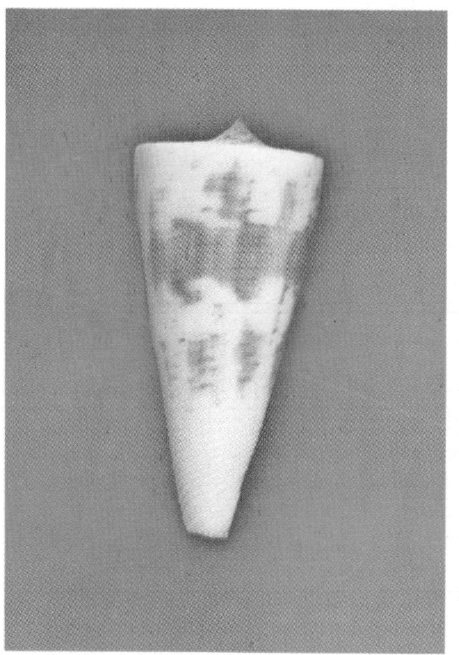

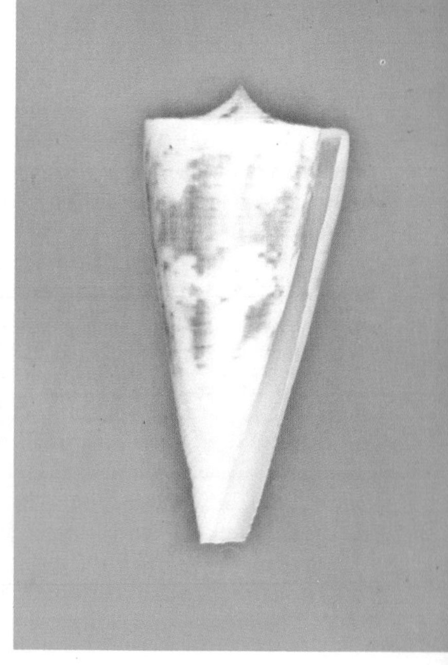

TRIGONUS.—Above: Sandy Reef, Dampier, Western Australia, Australia, 46.6mm (GG). *Below:* Left: "Cape Moreton, Queensland, Australia" (?), 64.8mm; Right: NW Cape, Western Australia, Australia, 39.6mm.

name seems to be *C. magnificus* of Reeve, which is based on an inflated specimen with reduced pattern; I have seen such specimens from Zanzibar and have no doubt of their identity with the narrower and more heavily patterned form better known to collectors. The varietal name of Dautzenberg is meaningless.

Notes.—This is a common species in the usual narrow, reddish brown form, but it comes in many less common patterns and colors. Often dealers have used the name *C. magnificus* to indicate any brightly colored specimens of *C. magnificus* (true), *C. pennaceus*, *C. aulicus*, or even *C. textile.* The golden tan color phase is sometimes sold as *C. auratus*, incorrectly; it also appears to be environmental in nature.

The tip of the spire is always rounded in this species, and there are commonly minor growth marks. Otherwise true gems are common.

MAGUS Linnaeus, 1758

1758. *Conus Magus* Linnaeus. *Systema Naturae per Regna Tria Naturae,* ed. 10, 1: 716 (Locality not stated). Neotype selected and figured by Kohn, 1963.

1792. *Conus raphanus* Hwass, in Bruguiere. *Cone*, in *Ency. Method., Hist. Nat. des Vers*, 1: 722 (Indian Ocean). Lectotype selected by Kohn, 1968.

?1798. *Cucullus cinereus* Roeding. *Museum Boltenianum:* 46 (Locality not stated). Based on Martini, Pl. 52, fig. 578. Non *Conus cinereus* Hwass, 1792.

1798. *Cucullus caesius* Roeding. *Ibid.:* 48 (Locality not stated). Lectotype figure (Kohn, 1975): Martini, Pl. 52, fig. 580.

1822. *Conus carinatus* Swainson. *Zool. Illus.*, 2: Pl. 112 (Locality not stated).

1844. *Conus epistomium* Reeve. *Conchologia Iconica*, 1(*Conus*): Pl. 42, sp. 227 (Mauritius).

1854. *Conus innexus* A. Adams. *Proc. Zool. Soc.* (*London*), 1853: 118 (Natal).

1857-1858. *Conus circae* Sowerby ii. *Thesaurus Conchyliorum*, 3(*Conus*): 39, pl. 21(207), figs. 513-514; pl. 22(208), fig. 525 (var.) (Philippines).

1864. *Conus consul* Boivin. *J. Conchyl.* (*Paris*), 12: 33, pl. 1, figs. 5-6 (Locality unknown).

1864. *Conus lictor* Boivin. *Ibid.*, 12: 36, pl. 1, figs. 1-2 (Locality unknown).

1865. *Conus Frauenfeldi* Crosse. *Ibid.*, 13: 307, pl. 10, figs. 1-1a (Madagascar). Lectotype selected by Tomlin, 1937, as specimen shown in fig. 1a.

1865. *Conus signifer* Crosse. *Ibid.*, 13: 308, pl. 10, fig. 6 (Locality unknown).

1865. *Conus Macei* Crosse. *Ibid.*, 13: 309, pl. 10, fig. 5 (Vizaga patam, Madras, British India).
1866. *Conus Tasmaniae* Sowerby ii. *Thesaurus Conchyliorum*, 3(*Conus* Suppl.): 328, pl. 27(288), fig. 636 (Tasmania).
1866. *Conus Borneensis* Sowerby ii. *Ibid.*, 3(*Conus* Suppl.): 329, pl. 28 (289), fig. 648 (Borneo).
1875. *Conus epistomioides* Weinkauff. *Syst. Conch. Cab.*, (Ser. 2), 233 (*Conus*): 315, pl. 57, figs. 5-6 (?East Africa).
1891. *Conus* (*Chelyconus*) *Worcesteri* Brazier. *Proc. Linnaean Soc. New South Wales*, 6: 276, pl. 19, fig. 4 (Mauritius).

Description.—Moderately heavy, with a low to moderate gloss; nearly obconical to low conical, the upper sides convex, the greatest width usually somewhat below the shoulder; sides variable, straight and little tapered to straight and tapered or even concave at the middle; basal third or so of body whorl with low spiral ridges which may be widely or narrowly spaced and even granulose, sometimes hardly visible; rest of whorl with numerous, often conspicuous, spiral and axial threads; shoulder broad, concave above, roundly angled to weakly carinate; spire low to moderate, the sides usually straight, sometimes concave, the point usually blunt; early whorls with small nodules; spire whorls flat or slightly concave above, with 3-6 distinct and close set spiral ridges crossed by fine axial threads to produce a minutely cancellate effect. Body whorl usually white, sometimes pinkish, pale tan, or even orange; pattern very variable, usually consisting of numerous spiral rows of small to large dots or dashes of brown crossed by narrow to wide axial flammules; the flammules vary from pale yellowish tan to black and may be fused into broad spiral bands above and below midbody or may be widely separated; midbody area usually pale; base usually pale; area below shoulder usually pale; spire and often shoulder with small dark brown to black squarish spots or blotches, these irregularly placed and often fused; sometimes the blotches are numerous, although they may be confined to the middle whorls; tip of spire usually pinkish to brownish. Aperture moderately narrow posteriorly, much wider anteriorly; outer lip usually thickish but may be sharp, rarely thin, straight to concave at middle; mouth usually white. Columella long, narrow. Length 40-75mm.

Comparison.—This is one of the most complex cone species and often one of the most difficult to identify. In almost all specimens there are at least traces of spiral rows of brown dots or dashes and a tendency to concentrate any remaining pigment into two blotchy or even-edged bands above and below midbody, with the background usually white or at least paler than the rest of the shell; the shoulder is also usually pale. It is useless to compare *C. magus* with unrelated species which it could resemble; from these species it can usually be distinguished by the striate spire whorls, nodulose early whorls, and weak body sculpture. Two species which are very close to *C. magus* are recognized here and are often diffi-

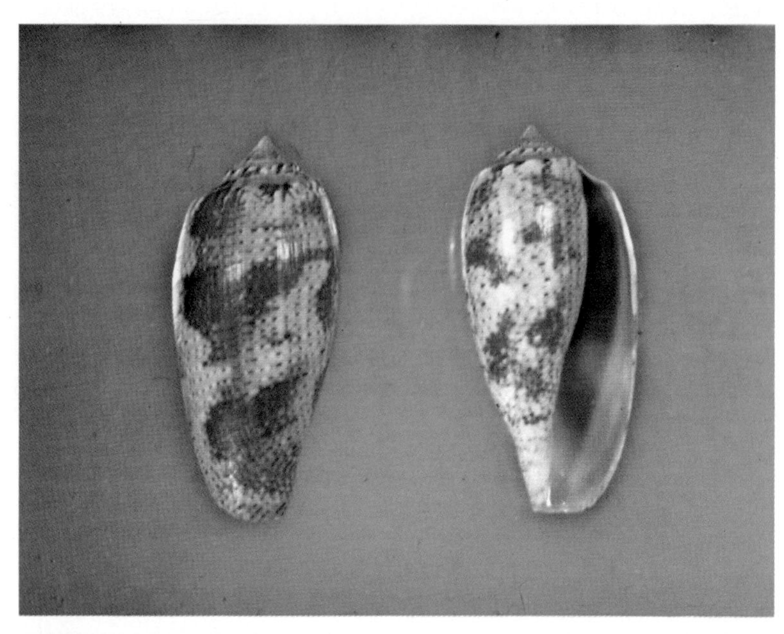

TULIPA.—*Above:* Zanzibar, 45.5 and 45.9mm. *Below:* Samar, P.I., 57.2 and 60.8mm.

TYPHON.—Above: Umdloti, N of Durban, Natal, South Africa, 32.9mm (Meyer). *Below:* Off Mozambique, 31.1mm (MM).

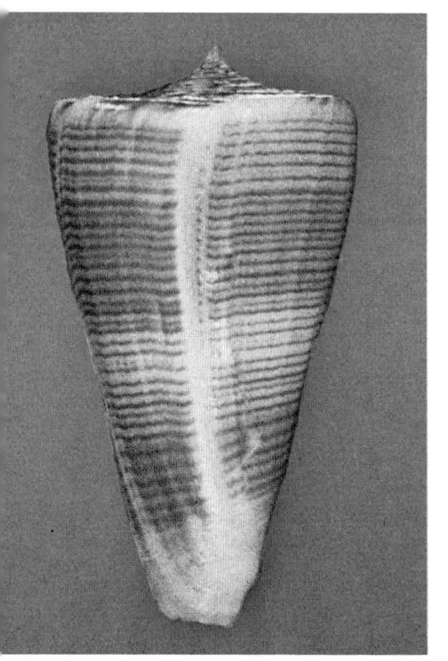

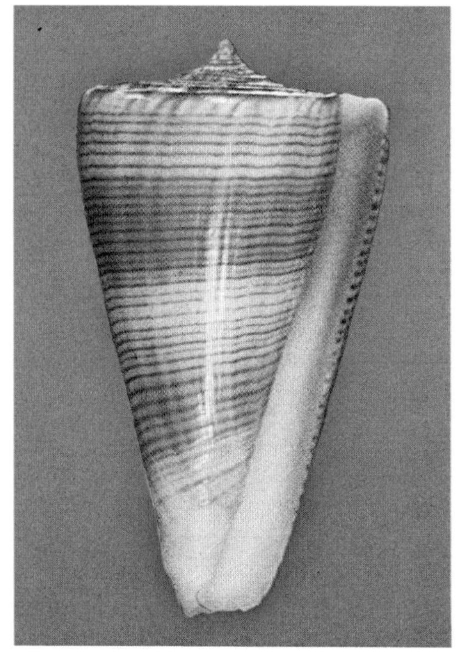

cult to distinguish with certainty from various patterns of *C. magus*; these are *C. consors* and *C. pohlianus*. *C. consors* usually lacks the spiral rows of distinct dots or dashes, has few or no dark spots on the spire, is often bright yellow with slightly darker spiral bands, lacks strong dark axial flammules, has more concave sides to the spire, and usually has distinctly convex sides of the body whorl below the shoulder. Of course all these characters can be found in individual *C. magus* from large series, but they are seldom associated. *C. pohlianus* is even more similar to *C. magus*, but it has the spire with convex sides, curved axial streaks on the spire and weak spotting, two brownish spiral bands above and below midbody, and a somewhat more elongate body whorl that is slender and convex posteriorly; juveniles may be violet in tone. Also related to *C. magus* is *C. nigropunctatus*, which is a smaller and thicker shell that is relatively wider; the shoulder is not concave above and never carinate; the basal ridges are more commonly granulose; and there are usually at least a few spots or blotches of opaque white on the body whorl.

Variation.—One of the most variable of cones and one of the most confused. For this reason I can see no reason at the moment to apply varietal names to any of the following variations, even though some are quite common and distinctive. All seem to intergrade broadly. The more common and distinctive variants are:

1) Entire shell heavily suffused with dark orange and overlaid by wide blackish brown axial bands; spire lines often solid, spire often nearly black; 2) shell heavily suffused with axial blotches and bands of dark blackish brown, the orange absent or restricted to two spiral bands above and below midbody; 3) as above, but axial bands orange, the spiral lines of dashes often reduced; 4) shell heavily suffused with orange-yellow, leaving only the shoulder, midbody, and basal areas white, without or with distinct spiral lines or dashes; 5) shell almost entirely orange from shoulder to base, with the spiral dashes nearly obsolete; a few white blotches may be present at midbody and shoulder; this shell is sometimes confused with *C. colubrinus;* 6) shell largely overlaid with broad spiral bands of orangish tan, some of the bands (usually total of four) darker than others; 7) as above, but the body whorl distinctly cylindrical, the sides nearly straight, the spire largely white with faint, curved spots, the shoulder marked with narrow white band; 8) like the last, but the spire more concave-sided and pointed; color dark tan to blackish, laid down in broad chevron-like axial bands above and below midbody, the white interstices often reduced or absent, as are the spiral brown dashes; 9) shell largely white, variably covered with smallish tan to dark brown or orange spots and blotches, the spiral rows of dashes usually distinct (common type). Other patterns occur but are seldom common. Also, there is much individual variation in height of spire, strength of sculpture, and convexity of the sides below the shoulder.

Distribution.—Very common and widely distributed in deep to shallow waters of the Indo-Pacific from eastern Africa to Fiji; apparently not recorded from Hawaii or French Polynesia.

Synonymy.—Trying to clarify the synonymy of this species is an almost impossible task as many of the type figures are not clear enough to be sure of the presence or absence of dots, the spire pattern, basal sculpture, etc. I believe that the names listed here all fall into *C. magus* as here defined, which is admittedly a broad definition. Many people recognize *C. epistomioides* as distinct (pattern 8), as well as *C. carinatus* (pattern 7), but I feel they are not distinctive enough for specific status and too many intermediates occur with other patterns of *C. magus.* Even *C. consors* and *C. pohlianus*, here recognized as valid, could be placed within the normal range of variation of *C. magus* with little trouble, but they are more constant with fewer intermediates.

Notes.—The only specimens of this species that truly deserve a premium are those with exceptionally bright colors and contrasting patterns. The other variants and patterns are all more-or-less about equally common in large series from heavily collected localities, and it is almost impossible to identify the variants in a price-list. Of course dealers will continue to ask for premiums for such shells as patterns 7 and 8.

There are often minor growth marks on the body whorl in this species, and the lip is commonly badly chipped even though it is often thick. The early spire whorls are generally eroded, and various types of encrusting organisms are common.

MALACANUS Hwass, in Bruguiere, 1792

1792. *Conus Malacanus* Hwass, in Bruguiere. *Cone*, in *Ency. Method., Hist. Nat. des Vers*, 1: 645 (Strait of Malacca). Holotype figure: *Tableau Ency. Method.*, Pl. 325, fig. 9.

1817. *Conus canaliculatus* Dillwyn. *Descr. Catalogue Recent Shells*, 1: 360 (Ceylon and the Nicobar Islands; Straights of Malacca). Lectotype figure (here designated): *Tableau Ency. Method.*, Pl. 325, fig. 9.

1865. *Conus subcarinatus* Sowerby ii. *Proc. Zool. Soc. (London)*, 1865: 518, pl. 32, figs. 12-13 (Nicobar Is.).

Description.—Heavy, with a low gloss; low conical to obconical, the sides straight; body whorl smooth except for a few low spiral ridges above the base, the grooves between weakly axially threaded or punctate; rest of whorl with occasional axial threads and growth marks; shoulder broad, carinate, concave to canaliculate above; spire low to almost flat, the sides convex above, early whorls forming a small projecting cone; first three or so whorls weakly nodulose and with 2-4 distinct spiral ridges; next 2-3 whorls faintly undulate, the spiral ridges less distinct and more numerous; later whorls with numerous fine spiral and axial threads; tops of most later spire whorls concave, the margins usually carinate. Body whorl white to pale violet, the base white; pattern very variable, but usually

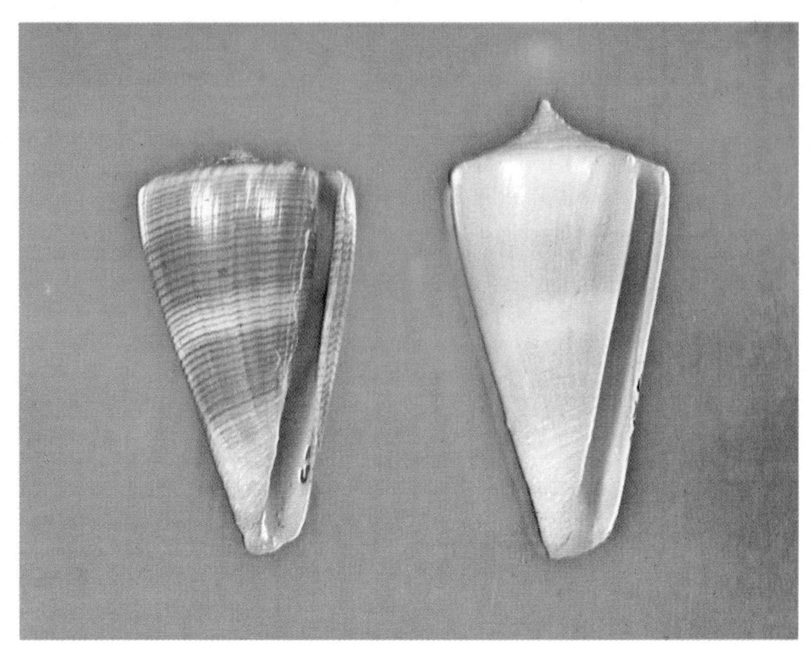

TYPHON.—*Above* and spire below (paratypes): N of Beira, Mozambique, 36.6 and 40.9mm (NM G2185). *Below* left: Between Inhambane and Beira, Mozambique, 46.4mm (Meyer).

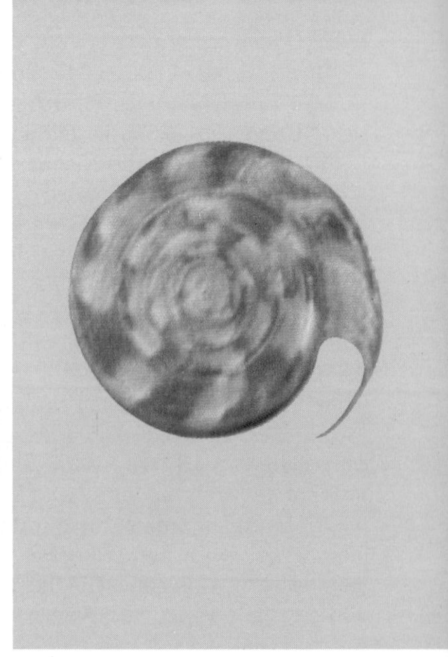

URASHIMANUS.—Nada-cho, Wakayama Pref., Japan, 49.6 and 63.8mm

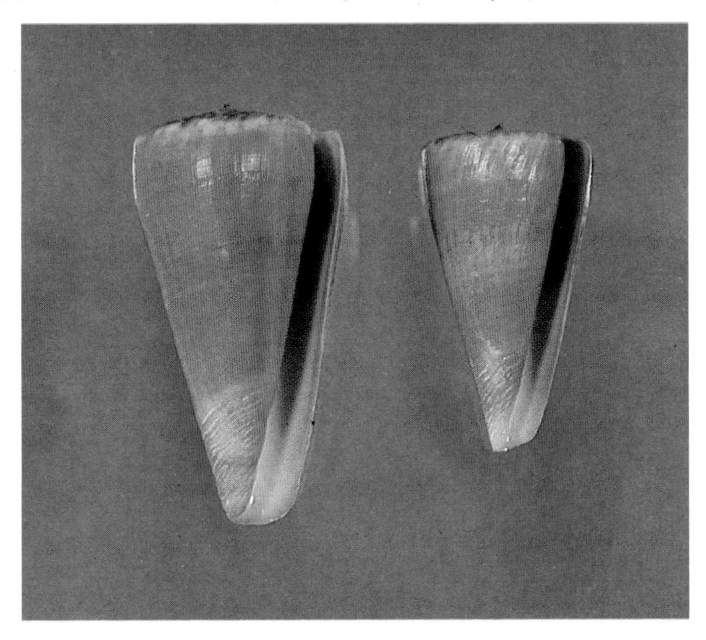

falling into two basic patterns and their intermediates: 1) several widely spaced narrow axial stripes which may be continuous from shoulder to base or interrupted at midbody, the stripes some shade of brown and often dotted with spiral rows of white dots and darker brown dashes or smudged spirally to produce narrow or broad spiral bands above and below midbody; 2) shell mostly pale brown, often darker above and below midbody and below the shoulder, with narrow dark brown spiral bands heavily interrupted with small white dashes, the midbody area usually white and with axial flammules crossing it; almost white to almost solid brown shells occur; spire and shoulder whitish, sparsely or heavily marked with brownish blotches and curved streaks; early whorls pale brown. Aperture moderately wide, uniform in width; outer lip thin, straight; mouth white, sometimes faintly bluish or creamy. Columella very narrow, indistinct, oblique anteriorly. Length 35-70mm.

Comparison.—*C. malacanus* is closely related to the broad form of *C. argillaceus* but is easily distinguished by color and shape. The body whorl is relatively wider, the shoulder more distinctly carinate and concave above; the shoulder is usually whitish and does not normally have the large black spots of *C. argillaceus*; the easiest character is the color of the base: violet-brown to blackish in *C. argillaceus*, white in *C. malacanus*. *C. fumigatus* and *C. excavatus* are sometimes somewhat similar, but the first species is much smaller, darker, has a rounded or at least not carinate shoulder, and is heavily covered with spiral lines and dashes; *C. excavatus* is more elongate, thinner in texture and less chalky, has a more nearly unicolored or at least weak pattern, and has the base pale but not white.

C. eximius has often been confused with *C. malacanus*, but the two are quite distinct. In *C. eximius* the early whorls are not nodulose, the

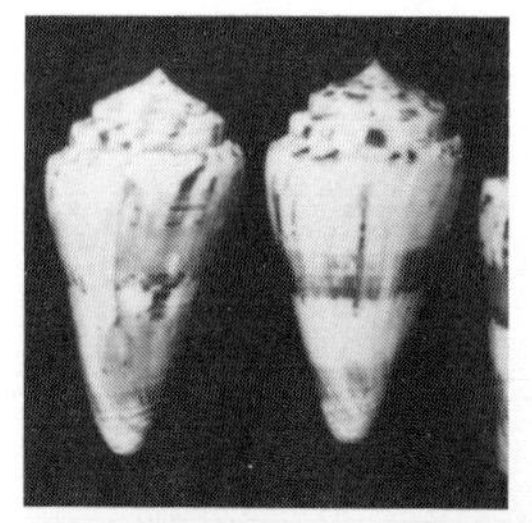

Variation in *C. malacanus* from Sri Lanka (photo Jonklass).

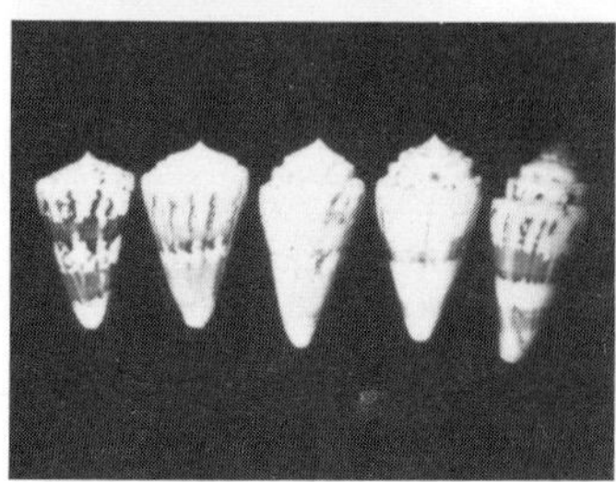

spiral ridges are more restricted to the posterior half of the whorls, the whorls are not deeply concave above, and the anterior half of the body whorl is covered with deep, widely spaced spiral grooves.

Variation.—The shape is quite constant, although this species, like *C. argillaceus*, tends to distort and irregularly step the last whorls of the spire. Generally the sides of the spire, excluding the early whorls, are slightly convex. The color pattern variation as described in the earlier paragraphs covers the extremes; although the pattern is variable, it falls within rather narrow limits: axially striped, spirally broadly striped, mostly dark, or mostly light. Generally there are darker brown squarish spots above midbody, but the entire midbody area may be covered with a broad brown belt.

Distribution.—This species has recently been found to be common in the northern Indian Ocean off southern India and Sri Lanka in moderately shallow water. It is also reported from off Burma and (a single specimen collected by Ed Petuch) from the New Guinea area. Probably widely distributed but only common in small colonies.

Synonymy.—The lectotype of *C. canaliculatus* is the same as the holotype of *C. malacanus*. The name *subcarinatus* has often been placed with *C. eximius*, but the description and figure are good; the name certainly applies here.

Notes.—This was until recently a rare and very poorly known species, but it is now easily available from the Indian area. The variants are about equally common and not worth premiums. This can be a very ugly shell when in a dark or light phase, only the broadly banded pattern being at all attractive. Heavy axial growth flaws are common and to be expected, as is erosion of the early whorls. This is often a very chalky shell without gloss.

MARIELAE Rehder and Wilson, 1975

1975. *Conus marielae* Rehder and Wilson. *Smithsonian Contr. Zool.*, 203: 14, frontis figs. 10-11, text fig. 10 (Off Baie Moto-Hee, Nuku Hiva, Marquesas Islands).

Description.—Moderately light in weight, with a good gloss; low conical, the sides nearly parallel or slightly convex posteriorly, then tapering to the base and sometimes slightly concave at the middle; body whorl covered with spiral ridges from base to shoulder, the ridges narrow and more widely separated anteriorly, wider, flatter, and less distinct posteriorly; fine axial threads and growth lines visible over shell and between ridges; shoulder sharply angled, coronate, the coronations erect and sharp; spire low to moderate, sharply pointed, the sides concave; whorls slightly stepped, concave above, the margins strongly coronate; tops of whorls with about three spiral ridges and numerous fine axial threads.

VARIEGATUS.—Above: Conchas Bay, Mocamedes, Angola, 26.2mm. *Below:* Left: Lucira Bay, Angola, 20.1mm; Right: Bala de Ste. Marta, Angola, 27.3mm (all GG).

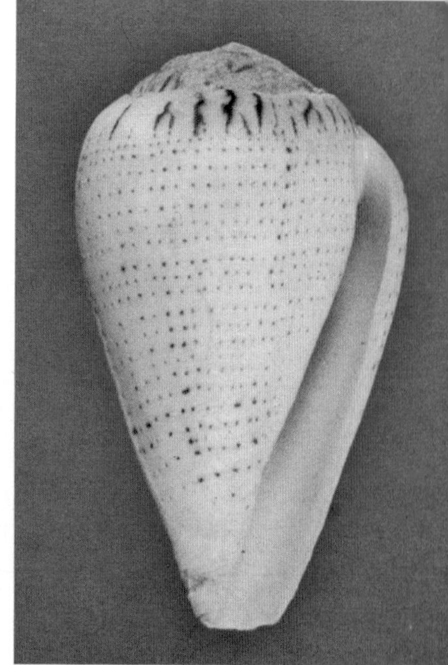

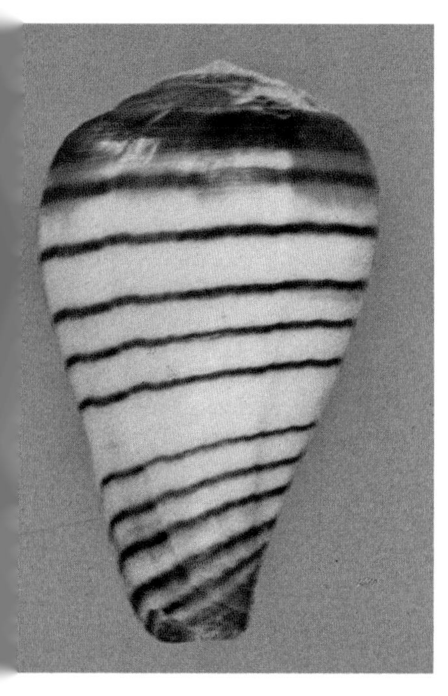

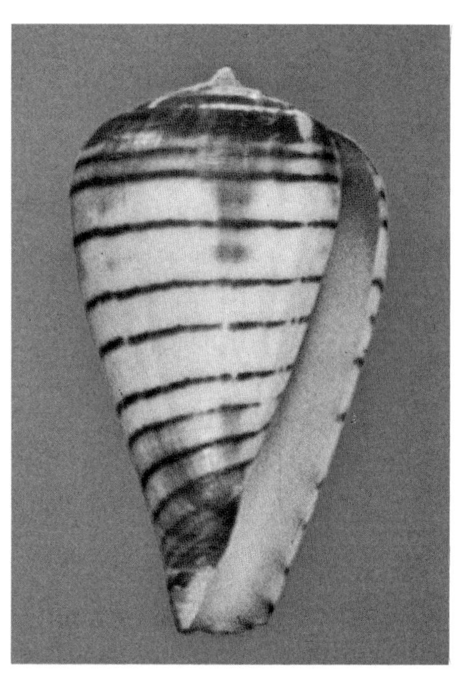

VARIEGATUS.—Above: Left: Baia de Elefante, Angola, 24.0mm (EP); Right: Lucira Bay, Angola, 23.2mm (GG). *Below:* Left: Mocamedes, Conchas Bay, Angola, 23.3mm (MM): Right: Baia du Chapen, Armado, Angola, 26.5mm (GG).

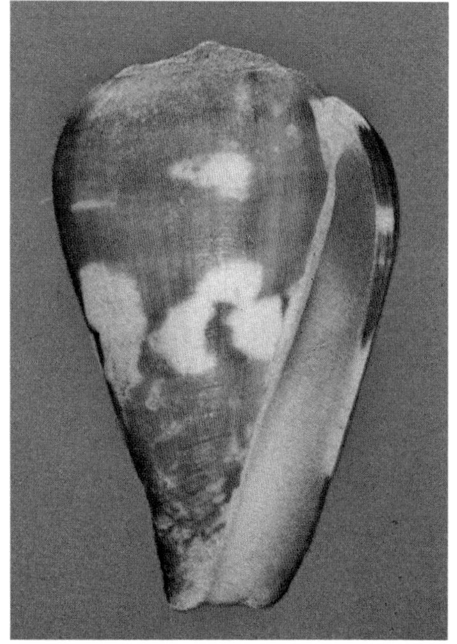

Body whorl white, covered with many small squarish or rectangular spots of reddish orange in spiral series over the ridges, these sometimes shaded with brown and with white dots within; spots may form larger blotches in two bands above and below the midbody area and a weak third band below the shoulder; base pale; spire whitish, with reddish orange spots between the coronations; early whorls white. Aperture moderately wide, slightly widened anteriorly; outer lip rather thin, straight; mouth white. Columella rather wide, long, strongly twisted. Length 34-44mm.

Comparison.—Frankly, I cannot separate this recently described taxon from Philippine *C. proximus*. Although *C. proximus* commonly has narrow, long brown dashes on the ridges, Philippine specimens are commonly covered with large squarish dashes of brown. In some specimens the brown becomes red. Sometimes brown and red, narrow and wide patterns are found in the same specimen. The spiral ridging is similarly variable, from narrow ridges to wide ribs on the same specimen. Other than the reddish color of *C. marielae* and its distribution, I can see no other character that would separate *C. marielae* from many *C. proximus*, and even the reddish color will not work. *C. proximus* also commonly lacks basal granules.

C. moluccensis differs in larger size, constricted outer lip, constriction at the shoulder, often heavy granulose basal ridges, and constriction at the shoulder level. The white mouth of *C. moluccensis* is found in both *C. marielae* and some *C. proximus*.

Variation.—Except for intensity of color, arrangement and size of spots and presence or absence of larger blotches, this shell seems rather constant in the few known specimens.

Distribution.—Apparently restricted to the Marquesas, but *C. proximus* occurs as far east as Fiji, so there may not be any geographic separation. Not uncommon in moderately deep water.

Notes.—Because I have not yet examined any Marquesan specimens, I retain this taxon as distinct. As noted, however, I really can see no way to distinguish it from some Philippine *C. proximus* of the same color, shape, and sculpture. It is probably not distinct even at the subspecific level.

MARMOREUS Linnaeus, 1758

1758. *Conus marmoreus* Linnaeus. *Systema Naturae per Regna Tria Naturae*, ed. 10, 1: 712 (Asia). Lectotype (here selected): specimen marked '290' in Linnaean collection, figured by Kohn, 1963, as the holotype.

1778. *Conus candidus* Born. *Index Rerum Naturalium Musei Caesari Vindobonensis*, 1(*Testacea*): 130 (Locality not stated). See Kohn, 1964. Perhaps unrecognizable.

1786. *Conus nocturnus* Lightfoot. *Cat. Portland Museum*: 156 (China).

Lectotype figure (Kohn, 1964): Martini, Pl. 62, fig. 687.

1792. *Conus Bandanus* Hwass, in Bruguiere. *Cone*, in *Ency. Method., Hist. Nat. des Vers*, 1: 611 (Banda Is., Moluccas). Lectotype (here selected): specimen MHN 1107/10, figured by Kohn, 1968, as the holotype.

1798. *Cucullus proarchithalassus* Roeding. *Museum Boltenianum:* 38 (Locality not stated). Lectotype figure (Kohn, 1975): Martini, Pl. 62, fig. 686.

1798. *Cucullus torquatus* Roeding. *Ibid.:* 38 (Locality not stated). Lectotype selected and figured by Kohn, 1975.

1798. *Cucullus equestris* Roeding. *Ibid.:* 38 (Locality not stated). Lectotype selected and figured by Kohn, 1975.

1811. *Conus maculatus* Perry. *Conchology:* Pl. 24, no. 4 (Locality not stated). Non *Conus maculatus* Bosc, 1801.

1839. *Conus marmoreus* var. *granulatus* Sowerby i, in Sowerby ii. *Conchological Illustrations:* Pt. 155/156, fig. 120* (Locality not stated). Non *Conus granulatus* Linnaeus, 1758.

1843. *Conus vidua* Reeve. *Conchologia Iconica*, 1(*Conus*): Pl. 8, sp. 45 (Capul, Philippines).

1857-1858. *Conus Deburghiae* Sowerby ii. *Thesaurus Conchyliorum*, 3 (*Conus*): 2, pl. 1(187), figs. 6-7 (Moluccas).

1859. *Conus nigrescens* Sowerby ii. *Proc. Zool. Soc.* (*London*), 1859: 429, pl. 49, fig. 2 (Locality unknown).

1861. *Conus Crosseanus* Bernardi. *J. Conchyl.* (*Paris*), 9: 168, pl. 6, figs. 3-4 (New Caledonia).

1870. *Conus suffusus* Sowerby iii. *Proc. Zool. Soc.* (*London*), 1870: 255, pl. 22, fig. 9 (New Caledonia).

1872. *Conus suffusus* var. *Noumeensis* Crosse. *J. Conchyl.* (*Paris*), 20: 155 (Noumea). *Ibid.*, 20: 350, pl. 16, fig. 2.

1875. *Conus pseudomarmoreus* Crosse. *Ibid.*, 23: 223, pl. 9, fig. 4 (Locality unknown).

1878. *Conus Crosseanus* var. *Lineata* Crosse. *Ibid.*, 26: 168, pl. 3, figs. 3-3a (New Caledonia). Non *Conus lineatus* Solander, 1766, a fossil. Probably meant as a descriptive term, not a new name.

Description.—Heavy, with a good gloss; low conical to obconical, the sides usually straight or only slightly convex posteriorly; body whorl generally with about a dozen or so weak and closely spaced spiral ridges above the base, occasionally these prominent over most of the shell and even granulose; strong axial growth lines and threads usually present; shoulder angulate, strongly coronate, slightly concave above; spire low to flat, the early whorls usually eroded and indistinct; whorls all strongly coronate and often concave above; spire whorls with 2-3 indistinct spiral ridges at the middle of the whorl, these often absent or not visible, and fine axial threads. Body whorl white or pale pinkish, the base usually heavily smudged with dark brown or blackish; whorl covered with a

VARIUS.—Above: Zamboanga, Mindanao, P.I., 34.7 and 34.8. *Below:* Left: Honiara, Guadalcanal, Solomon Is., 34.4mm; Right: Okinawa, Japan, 18.8mm (TCG).

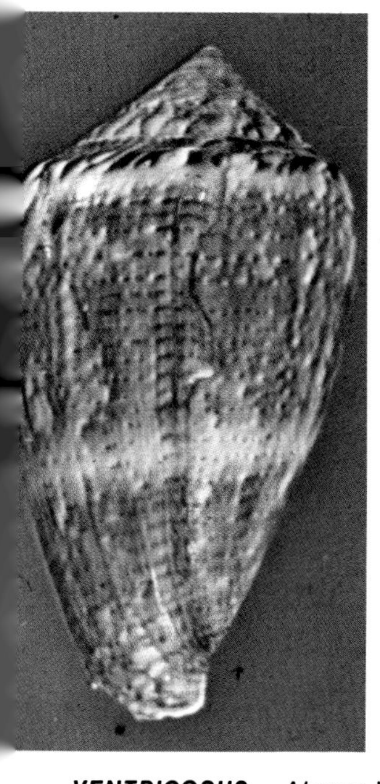
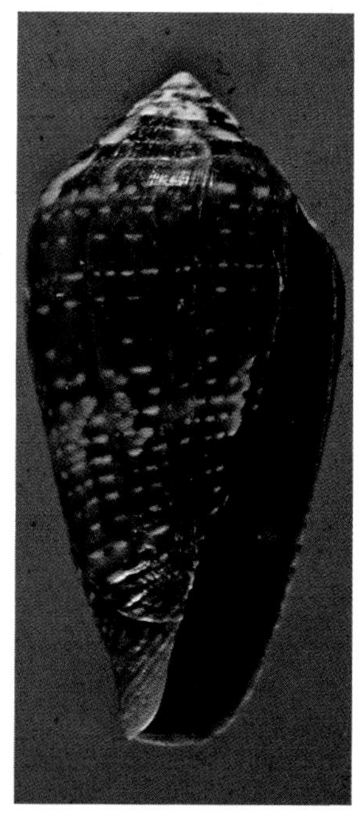

VENTRICOSUS.—Above: Left: Boulouris, France, 28.3mm (GG); Center: Dakar, Senegal, 25.1mm (MM); Right: N'Gor, Senegal, 26.9mm (EP). *Below:* Left and Center: Heraion Loutraki, Greece, 21.2 and 21.5mm (GG); Right: Boulouris, France, 22.7mm (GG).

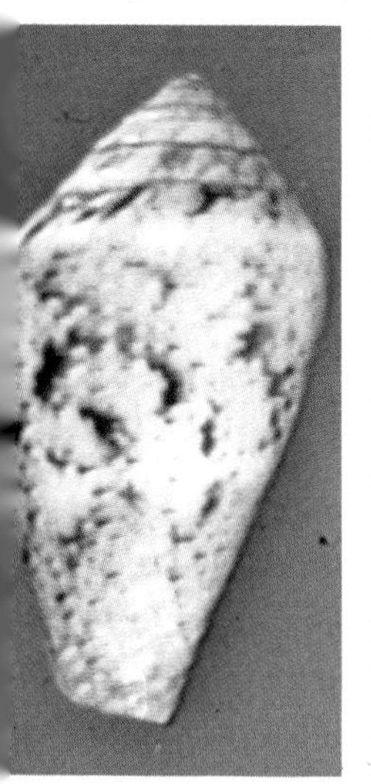
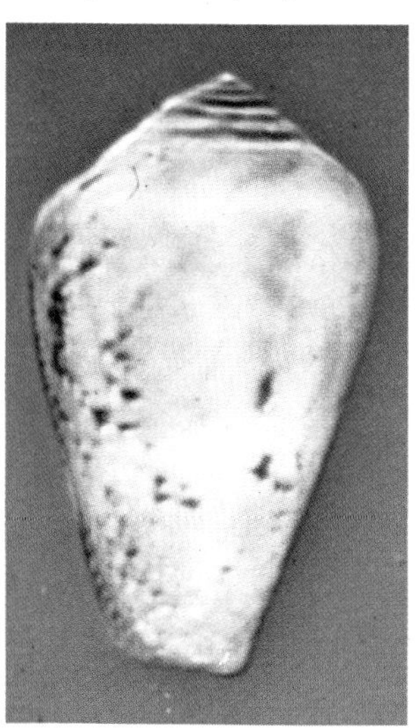
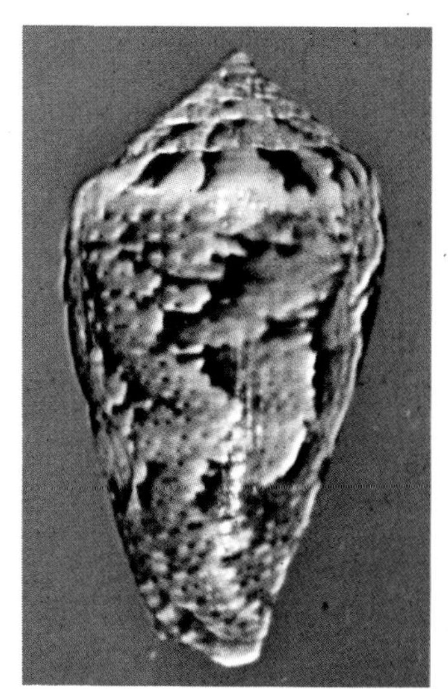

variable pattern of blackish brown axial zigzag lines which form some type of network of large and small tents; often two broad blackish spiral bands above and below midbody area marked with only a few white tents; shoulder and spire with very loose blackish brown reticulations on white; early whorls eroded white or pinkish. Aperture moderately wide, slightly wider anteriorly; outer lip moderately thick or thin, straight; mouth white, bluish white, or pale pink; extreme anterior edge of lip often blackish. Columella rather short and narrow. Length 50-150mm.

Comparison.—*Conus marmoreus*, although highly variable in pattern, is almost always easily recognized by shape. *C. araneosus* may be very similar in shape and pattern, but it usually has lower coronations, has a pale base, and has a large yellow blotch deep within the mouth. *C. imperialis* and *C. zonatus* are usually more narrow in shape, are very different in pattern, and, in *C. imperialis*, are often covered with coarse growth lines. *C. nobilis* in its various subspecies is orange or orange-tan in color, not black, and lacks the distinct shoulder coronations.

Variation.—The color patterns of *C. marmoreus* are legion but do tend to fall into well defined groups which can be dignified with varietal names as they are fairly constant and generally considered worth collecting. Individual variation usually includes height of spire, width of body whorl, occasional pinkish backgrounds, and height of the shoulder coronations. The following are the major varieties.

C. marmoreus typical: Often larger and with stronger tendency to conspicuous spiral ridging than the other varieties; pattern usually in the form of large black and white tents or rhomboids which show a little tendency to cluster into two broad bands above and below midbody; spire often heavily and evenly reticulated with black, the coronations white. Very widespread and common.

C. marmoreus variety *vidua:* Much like the typical variety, but often smaller and with a narrower body whorl; pattern clustered into two broad blackish bands above and below midbody, the white background showing through the bands as small white tents; dark lines reduced or absent at shoulder, midbody, and above base, producing nearly all-white bands with indistinct traces of fine brown lines. Most common in the western Pacific.

C. marmoreus variety *nigrescens:* Usually small (under 75mm) and relatively light in weight, often with conspicuous basal ridges; spire also commonly high and the coronations weak; tenting shows a strong tendency to fuse into axial dark lines at the varices and to smudge, sometimes resulting in nearly black shells with a few scattered white tents visible; when the smudging is weak and the white background is conspicuous, the variety *crosseanus* results; rare specimens have colorless spiral white lines near midbody, these being variety *lineata*. Sporadic, but usually New Caledonia to Samoa.

C. marmoreus variety *pseudomarmoreus:* A small and not very satisfactory variety with a tendency to a high spire and sometimes reduced coronations; often more convex-sided than normal; typical *marmoreus* pattern, but covered over at least half the shell with heavy spiral ridges that are commonly granulose. This is perhaps a remnant of the juvenile sculpture. Sporadic throughout the range.

C. marmoreus variety *suffusus:* Black tenting mostly or all absent, resulting in an all-white shell or one with only large or small triangular blackish spots (form 'batarde'). In all-white shells there are two broad and indistinct bands of pale pink to pale yellow above and below midbody. This is usually a small form with high coronations. Apparently restricted to New Caledonia, where not uncommon.

Holotype of *C. suffusus* Sowerby iii, 55.5mm (BM (NH)) (Abbott).

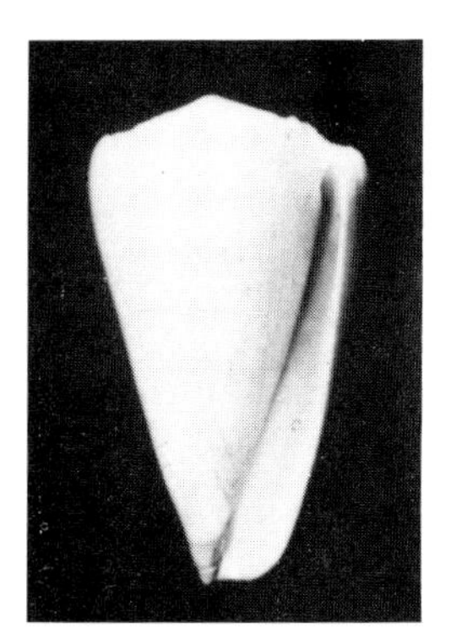

C. marmoreus variety *nocturnus:* The rarest and most attractive variety of *C. marmoreus;* the shell has two very broad spiral bands of black to deep reddish brown above and below midbody, these showing little or no white tents; sometimes the two bands are partially fused and cover the midbody area as well; the rest of the shell is white; the basal ridges may be granulose or not (form *deburghiae*). This shell is probably stark black and white when alive, but usually only old beach specimens are found, these with the black faded to reddish brown. Seemingly restricted to the Indonesian area, where rarely taken this century.

In addition, orange variants of the typical and *vidua* form are known, but the status of some of these specimens is doubtful as a very similar color can be created by heating normal shells in an oven. Such heated shells are not quite as glossy as those normally sold, but the difference is so small it is not worth the risk of paying the very high premium asked for orange shells.

Intermediates between *C. marmoreus* typical and *C. m.* var. *vidua* are common and are usually incorrectly sold as *C. m.* var. *nocturnus*; there is no comparison between these shells and true *nocturnus*.

Distribution.—Widespread and very common throughout the Indo-Pacific from eastern Africa to Hawaii and French Polynesia in shallow water. Not yet recorded from the Panamic area.

Synonymy.—Most of the synonyms are mentioned under *Variation* as the major names are still retained as varieties. Most of the old names

VENULATUS.—Above: Cape Verde Is., 27.5-33.7mm (PC). *Below:* Left: Sal I., Cape Verde Is., 42.2mm (EP); Right: Cape Verde Is., 33.1mm (PC).

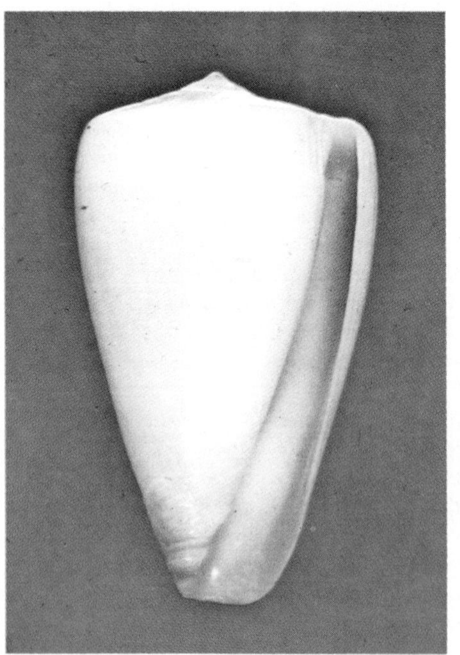

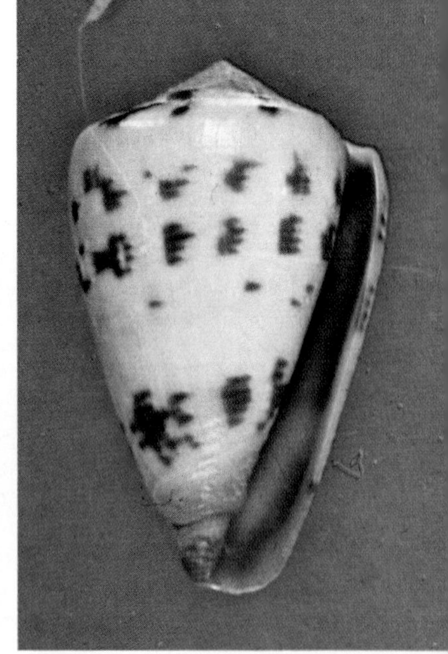

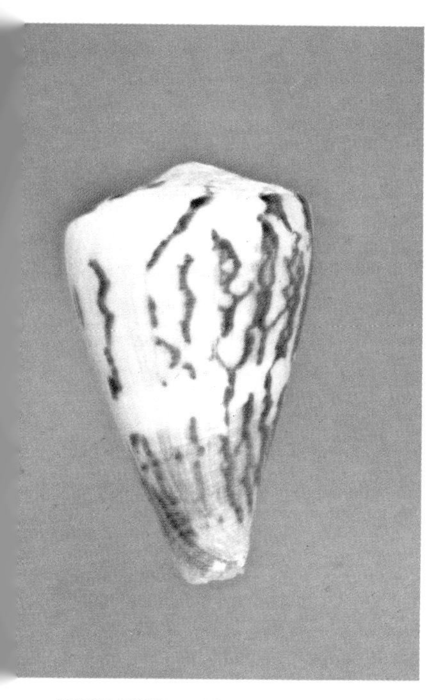

VEXILLUM.—*Above:* Left (*sumatrensis*): Dahlak Arch., Ethiopia, 60.4mm; Right (juveniles): Okinawa, Japan, 13.6 and 29.6mm (TCG). *Below:* P.I., 78mm.

provide no problems and can be applied to typical *marmoreus*. The name *bandanus*, however, has caused much problem; some authors have retained this as a full species or subspecies, although the type certainly looks much like a normal *C. marmoreus* and certainly falls easily within the range of variation of the species. Collectors and dealers often use the name *bandanus* for intermediates between typical *marmoreus* and *vidua* forms in which the banding is very poorly developed but the tenting is also not even over the whorl; this is based on the incorrect assumption that the name *bandanus* refers to a banded shell, while actually it refers to the Banda Islands.

Notes.—This common shell is easily available in gem condition, although commonly the coronations are eroded slightly and the early whorls may be badly eroded; such erosion is rather constant in the species and should not be considered a flaw. Heavy growth marks may occur on the body whorl, but they are seldom detracting. The lip chips easily in some populations.

Most of the varieties are worth a premium, small in the case of *vidua*, very large in *nocturnus*. Good *C. m.* var. *nocturnus* in fresh color is a major rarity. *C. m.* var. *nigrescens* and *suffusus* are often available and easy to obtain. *C. m.* var. *suffusus* is usually incorrectly sold as *C. caledonicus*, a very different although doubtful shell.

MAZEI Deshayes, 1874

1874. *Conus Mazei* Deshayes. *J. Conchyl.* (*Paris*), 22: 64, pl. 1, fig. 1 (Martinique).

1953. *Conus rainesae* McGinty. *Notulae Naturae, Acad. Nat. Sci. Phila.*, 249: 1, figs. (150 mi. NE of Progresso, Yucatan, Mexico).

1955. *Conus mcgintyi* Pilsbry. *Nautilus*, 69(2): 47, pl. 3, figs. 10-11 (Off Pensacola, Florida). Corrected to *macgintyi*.

Description.—Very thin and light in weight, fragile, with a good gloss; elongate biconical, the sides nearly straight or slightly convex posteriorly then strongly tapering to a very narrow base; body whorl with numerous flat spiral ribs extending from base to at least posterior third of whorl, often reaching the shoulder; ribs separated by narrower grooves containing fine axial threads or ridges; posterior third of whorl may be completely smooth or covered with heavy ribs; shoulder broad, carinate, smooth; spire tall, sharply pointed, the sides slightly concave; spire whorls slightly concave to flat above, weakly to strongly carinate, with traces of 3-6 fine spiral ridges and axial threads; early 3-5 whorls with fine nodules and heavy axial ridges posterior to the margin. Body whorl pale creamy white to pale straw or orange, covered with about 10-15 roughly spiral rows of squarish reddish brown spots, these about evenly spaced and showing a strong tendency for the spots to fuse into either axial flammules or broad bands below the shoulder and below midbody; base often stained pale brown; spire whitish, covered with large to small bright reddish brown spots in a regular arrangement and somewhat axially elongated; early whorls pale brown. Aperture very narrow, uniform in width; outer lip very thin and fragile, slightly convex; mouth white. Columella internal. Length 25-60mm.

Comparison.—*C. mazei* is closely related to *C. insculptus* and probably *C. aculeiformis* and *C. elegans*, being the only such shell yet known from the Atlantic. Like these species, it occurs in two sculpture and shape variants. From all other species of similar form it is distinguished by the distinct pattern on a whitish ground, the heavy spire pattern, the whorls with very fine spiral ridges, and only the first 3-5 whorls nodulose. *C. orbignyi* is larger, thicker, nodulose to at least the next to last whorl, and has heavier body sculpture. Identification should present no problems.

Variation.—As treated here, *C. mazei* occurs in two variants with somewhat different sculpture and pattern, these probably juvenile and adult. In addition, the type of *C. mazei* seems to represent a third very poorly understood variant.

C. mazei typical: Poorly known; large (over 50mm); the early whorls only weakly undulate, whorl margins not with projecting carinae and flat, not concave, above; pattern of about 9 rows of very distinct spots. Deep water off the Lesser Antilles.

C. mazei variety *rainesae:* Like the typical form in having the spiral ribs of the body whorl extending only to about midbody, rest of whorl smooth and polished; early whorls distinctly nodulose; whorls with projecting carinate margins; spotting distinct, in over 9 rows, showing some tendency to fuse or to be covered with larger clouds of pale reddish brown; size small, under 30mm; found in relatively shallow water (under 50 fathoms) from southern Florida to at least Yucatan.

C. mazei var. *rainesae*, Florida, 24-25mm (GG).

C. mazei variety *mcgintyi:* Larger and more elongate (to about 60mm) than *rainesae*, covered from base to shoulder with low but distinct spiral ribs; grooves between ribs with heavy axial ridges or threads; red-

VICTORIAE.—(*v. victoriae*) *Above:* Broome, Western Australia, Australia, 42.2 and 44.3mm. *Below:* Left: Exmouth Gulf, Western Australia, Australia, 36.4mm; Right: Broome, Western Australia, Australia, 43.1mm (K).

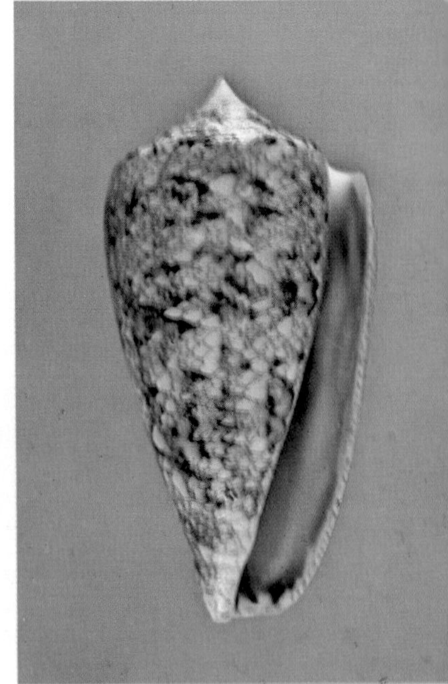

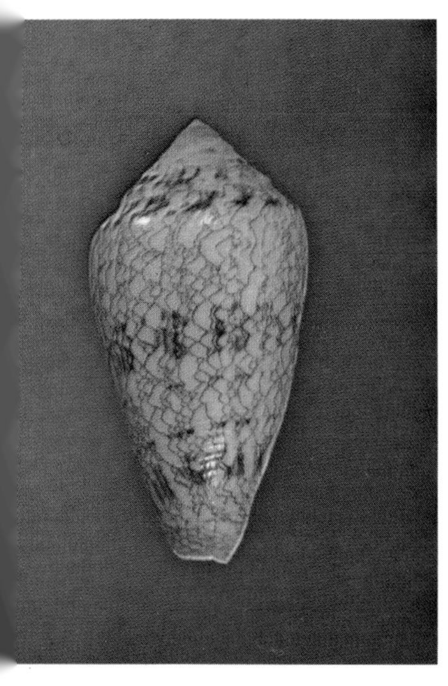

VICTORIAE.—*Above:* Left (*v. nodulosus*): Off NW Cape, Western Australia, Australia, 43.5mm (GG); Right (dark *v. victoriae*): Onslow, Western Australia, Australia, 53.4mm. *Below* (*v. nodulosus*): Geraldton, Western Australia, Australia, 30.2mm (MG).

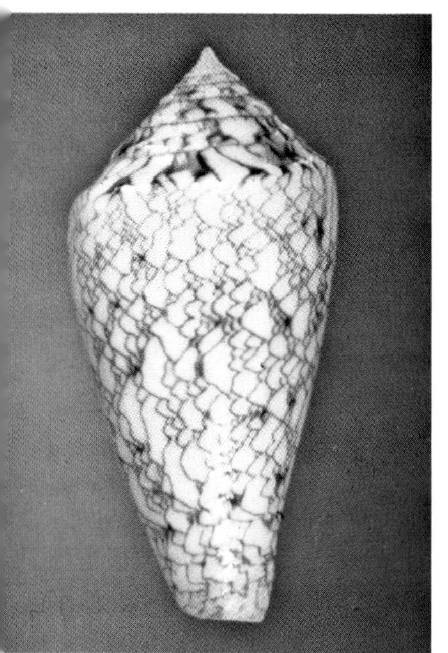

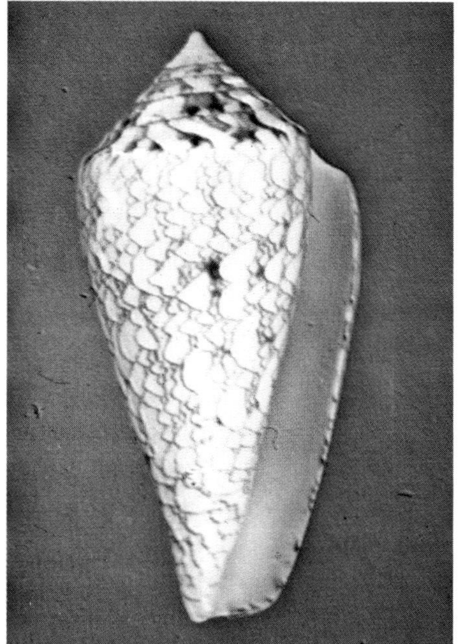

dish brown spotting tending to be connected into short axial flammules, especially below the shoulder and below midbody; deeper water (70-300 fathoms) from southern Florida to Brazil.

There is considerable individual variation in the last two varieties (the only ones known from good series) in strength of spotting and their tendency to fuse into spiral and axial bands or flammules. Additionally, each variety varies in strength of the sculpture on the body and spire.

Distribution.—Found in moderately to very deep water from off southern Florida, Cuba, the Virgin Islands, and the Lesser Antilles as well as off the Central American coast from Yucatan and Texas south to southern Brazil. Uncommon to rare.

Synonymy.—I can see no other recourse than considering *C. rainesae* the juvenile or shallow water ecotype of *C. mcgintyi*, as the two intergrade somewhat in sculpture with depth and no adults of the former or juveniles of the latter are known. The status of Martinique *C. mazei* is uncertain, although it certainly is almost identical with the *rainesae* form in general appearance, pattern, and reduced sculpture. However, it is much larger than any *rainesae* type known to me and apparently very rare. For the moment all are considered variants of a single species. Similar but less pronounced variation is known in related species.

Notes.—Of the three varieties, *rainesae* is moderately common off Florida, *mcgintyi* less so. Perfect specimens of either of these cones are hard to find, however, because they are very fragile; the lips break very easily, there are usually bad breaks on the body whorl of large specimens, and the tip of the spire may be broken. Commonly there is asphalt present on the spire. Of course such flaws should be expected in deep-water cones as delicate as these.

The status of typical *mazei* is very uncertain. For the moment, until this species is better known in the Antilles, it might be just as well to consider *C. mazei* a doubtful species restricted to specimens resembling the type, with *C. rainesae* used for the commonly seen forms.

MEMIAE (Habe and Kosuge, 1970)

1970. *Asprella memiae* Habe and Kosuge. *Venus*, 29(3): 82, pl. 5, figs. 1-2 (South China Sea).

1971. *Conus* (*Asprella*) *adonis* Shikama. *Sci. Rpts. Yokohama Nat'l. Univ.*, Sec. II(18): 33, pl. 3, figs. 25-28 (Japan, Kashiwajima I., SW Kochi Pref.).

Description.—Moderately heavy for its size, with a high gloss; biconical, the upper sides convex then concave and tapering to the base; body whorl covered with spiral ridges, these narrow and heavy anteriorly, wider and very low posteriorly; whole whorl covered with fine axial threads; shoulder broad, carinate; spire high, sharply pointed, the sides slightly to distinctly concave; spire whorls slightly stepped, flat or weakly concave above; first 3-5 whorls with fine nodules, later whorls carinate; tops of whorls with about 3-4 spiral ridges crossed by heavy curved axial threads. Body whorl pinkish white to bright pink or bright orange, covered with many spiral rows of small orange-brown dashes on the ridges; a row of dashes above and another below the midbody area enlarged and conspicuous; broad spiral band of darker orange blotches above the base usually conspicuous; base with many heavy spots; often with nebulous axial flammules obscuring or replacing part of pattern and giving the shell a duller tan look; spire pinkish to whitish, the margins of the whorls and shoulder with large dark brown spots alternating with white spots; sometimes a few axial blotches also present. Aperture narrow, uniform in width; outer lip thin, concave at middle; mouth pink to orange. Columella internal. Length 25-35mm.

Comparison.—*C. memiae* is related to *C. acutangulus*, from which it differs in lacking spire nodules on the later whorls and shoulder and in being pinkish with a different pattern. *C. eugrammatus* is more heavily sculptured, narrower at the shoulder, and lacks pink or orange in the pattern. Its closest relative, and the only species with which it is likely to really be confused, is *C. otohimeae*. It shares with that species the enlarged spiral dashes below the midbody and bordering a broad orange band, as well as the dark brown spots at the margins of the shoulder and spire whorls. However, *C. memiae* has a taller and more distinctly biconical shape, is covered with well defined spiral ridges lacking above the base in *C. otohimeae*, and has stronger spiral ridges on the spire whorls (weak or obsolete in *C. otohimeae*). The two are certainly close, however. *C. schepmani* is more elongated, wider at the shoulder, not pinkish, and has large opaque white spots at midbody and no orange bands.

Variation.—In addition to the usual variation in spire height and strength of spiral ridging, there are two basic patterns in this species. In the typical form the shell is pinkish with a broad orange blotch band above the base; the two enlarged rows of spiral dashes are not always very large. In the duller form, the color is more subdued, whitish or pale straw, with many axial flammules of pale pinkish or orangish straw, the orange band obscured; the dash rows may be greatly enlarged. Intermediates are known.

Distribution.—Uncommon to rare in deep water from southern Japan, the Philippines, and the Solomons. Probably widely distributed in the western Pacific but confused with *C. acutangulus*.

Synonymy.—*C. adonis* is obviously the same as *C. memiae*, but it was described almost one year later.

Notes.—Dead-dredged specimens have just become available of this recently described species. As might be expected, they are rather faded and may have bad lips and growth marks, but they are the best specimens currently available. Even live-taken specimens often have heavy growth

VICWEEI.—*Above:* Ranong, Thailand, 81mm (DM) (photos by Jayavad). *Below:* Left: Off Thailand, 75mm (La Cypraea); Right: Off Ranong, Thailand, 59.6mm (MM).

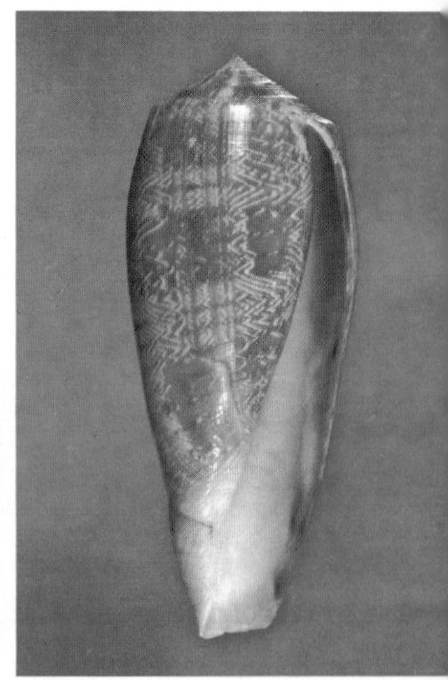

VILLEPINII.—Above: Left: Off Sand Key Light, Florida, 29.0mm (GG); Right: Off Hamilton, Bermuda, 50.7mm (photo by A. Kerstitch). *Below:* WNW Tampa, Florida, 59.1mm (MG).

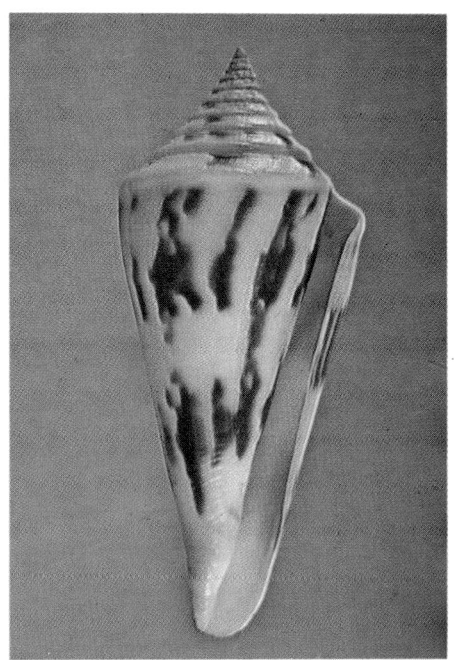

marks and scars. There has been some confusion between this species and a granulose variant of *C. otohimeae* from southern Japan, but the strongly biconical shape of *C. memiae* should be enough to identify any such specimens.

MERCATOR Linnaeus, 1758

1758. *Conus Mercator* Linnaeus. *Systema Naturae per Regna Tria Naturae*, ed. 10, 1: 715 (Locality not stated). Lectotype selected by Kohn, 1963.
1778. *Conus reticulatus* Born. *Index Rerum Naturalium Musei Caesari Vindobonensis*, 1(*Testacea*): 139 (Africa). Holotype figured by Kohn, 1964.
1798. *Cucullus aurelius* Roeding. *Museum Boltenianum:* 47 (Locality not stated). Lectotype figure (Kohn, 1975): Martini, Pl. 56, fig. 621.
1845. *Conus Lamarckii* Kiener. *Species gen. et icon. des coqu. viv.*, 2 (*Conus*): Pl. 83, fig. 4. 1849, *Ibid.*, 2: 240 (Locality unknown).
1845. *Conus unifasciatus* Kiener. *Ibid.*, 2(*Conus*): Pl. 110, fig. 4. 1849-1850, *Ibid.*, 2: 361 (Locality unknown).
?1864. *Conus miser* Boivin. *J. Conchyl.* (*Paris*), 12: 39, pl. 1, fig. 9 (Cape Verde).

Description.—Moderately heavy, with a low gloss; low conical, the posterior sides convex and then tapered to a relatively narrow base; body whorl with a few weak, crowded spiral ridges above the base; otherwise smooth except for sometimes conspicuous axial growth marks and threads; shoulder rounded, convex above, not very distinct from spire; spire low to moderate, bluntly pointed, the sides straight or concave; tops of whorls flat or slightly convex, with traces of very weak spiral and axial threads; erosion usually very bad. Body whorl in typical forms waxy yellowish, with two broad spiral bands of black reticulations leaving oval white spots, the widest band at and below the shoulder, the one at mid-body narrower; base paler; pattern very variable, sometimes of much finer reticulations, with larger reticulations that may cover most of the body whorl, or with broken reticulations to produce axial flammules; spire whitish or yellowish, later whorls covered with extension of body reticulation; early whorls usually eroded white or bluish. Aperture narrow posteriorly, considerably wider anteriorly; outer lip thin, slightly convex; mouth uniformly light bluish white, sometimes with external dark pattern showing through to produce pale brownish bands within. Columella narrow, mostly internal, somewhat indented from inner lip. Length 15-37mm.

Comparison.—In typical patterns with the heavy reticulations *C. mercator* is unmistakable, but some of the patterns are hard to place unless found in series with normal specimens. The entire series of species related to *C. mercator* is very confused and undoubtedly too many species are being recognized. However, of the various taxa, the closest is *C. taslei*. In *C. taslei* the body whorl is covered with very fine axial lines which commonly produce very small white tents in the midbody area and below the shoulder; broad spiral bands which are fairly solid may be present. In any event, the large and open reticulations of *C. mercator* are absent. Some *C. mercator*, however, show very fine lines in the midbody area in addition to a rather reduced reticulation. *C. cloveri* usually has the shoulder broader and more angled, the sides of the spire more concave; the pattern consists of long axial brownish lines which overlap only occasionally to produce indistinct tents. *C. balteus* is somewhat thicker and more angled at the shoulder, usually deeper brown or blackish in color, with at most heavy cross-hatching over the body whorl or between spiral dark bands; there are commonly large white blotches in the pattern.

Variation.—In addition to the usual variation in height and erosion of the spire as well as presence or absence of visible sculpture on the spire whorls, there is a fantastic array of pattern and color variants in this species. Fortunately most patterns are obviously individual variants, with no two exactly alike in large series. They seem to be mostly minor genetic variants perhaps due to local inbreeding or perhaps they are environmental in nature. Anyway, the patterns can be broken into about four types.

1) Typical, with distinct black and white reticulation on yellow or occasionally whitish ground color; the reticulation may be restricted to

Variation in *C. mercator* from a single Senegal collection (photo by Pete Carmichael through MG).

VIMINEUS.—*Above:* Tayabas Bay, Luzon, P.I., 32.3 and 36.8mm. *Below:* Boac, Marinduque, P.I., 41.4mm.

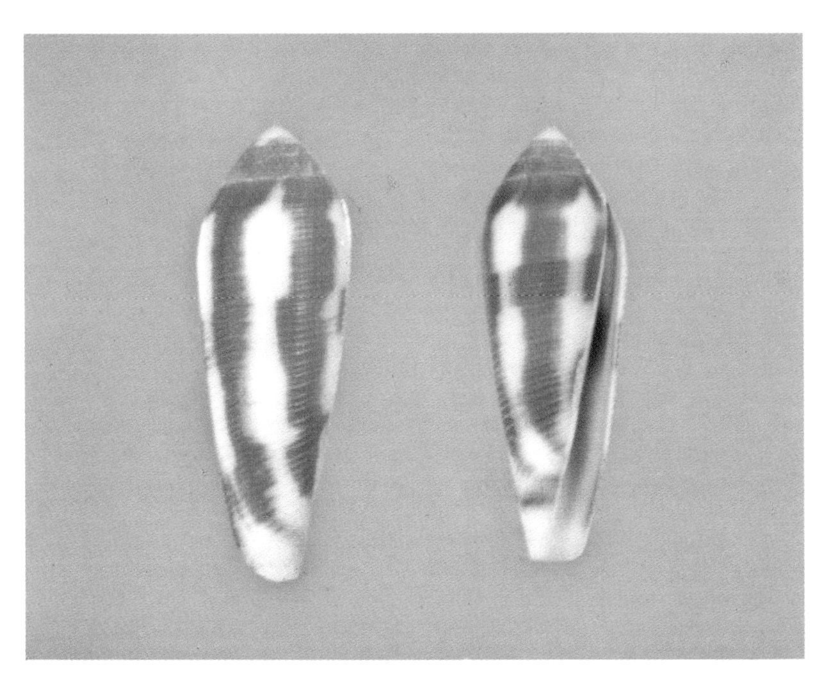

VIOLACEUS.*—*Above: Dar es Salaam, Tanzania, 42.8 and 45.1mm. ***Below:*** Zanzibar, 74.4mm.

two bands or extend over the entire whorl (uncommon); in some variants of this type the reticulation is filled in with blackish brown, eventually resulting in a solid brown shell or one with very large brown areas; 2) axially flammulate shells, usually on a white background on pale bluish; the axial flammules result from the irregular breaking of very heavy reticulations and loss of some of the spiral portions of the pattern; uncommon; 3) shell with pattern similar to pattern 1, but the reticulation very thin and more like the cross-hatching of *C. balteus*, sometimes with very fine and indistinct axial dark lines at midbody (rare); 4) very small, possibly dwarfed, shells with greatly reduced pattern consisting of faint and fine reticulations over much of shell and series of brown and white blotches near midbody (*lamarckii*) (uncommon). In addition to these basic types, solid brown shells (commonly called *miser*) exist and certainly belong to this species. All-white shells also are found.

Distribution.—Western African coast below the Bulge and in the Cape Verde Islands. Locally common, in very shallow to moderately deep water.

Synonymy.—Probably all West African cones that look anything at all like *C. mercator* could be safely placed in its synonymy, but that would be too drastic a step at the moment. *C. reticulatus* is a specimen with the reticulated network covering the entire shell, while *C. lamarckii* is a small shell with greatly reduced and somewhat indistinct pattern and white and brown blotches at midbody. *C. unifasciatus* is almost certainly the dark brown form from the Cape Verde Islands, but the name *miser* is commonly used for this form. Although included with *C. mercator* from tradition, it seems likely that the type specimen of *C. miser* is actually a specimen of *C. variegatus*, and as such should be included in the synonymy of that species.

Notes.—*C. mercator* is very common at several localities on the western African coast, and many specimens have reached the market in recent years. Included in large series are sure to be several freak variants which are sold at higher prices; these are not really uncommon, but it is probably true to say that each is unique. Specimens of *lamarckii* are very small and dull but sell at a premium. *C. 'miser'* specimens are very uncommon, regardless of their proper name.

C. mercator usually has the spire whorls eroded and encrusted and commonly has bad axial scars on the body whorl as well as the usual growth marks. The pattern of typical specimens is very different from that of most other cones.

As mentioned, it is not uncommon to find specimens of cones which cannot be satisfactorily placed in *C. mercator*, *C. taslei*, or allies.

MILES Linnaeus, 1758

1758. *Conus miles* Linnaeus. *Systema Naturae per Regna Tria Naturae*, ed. 10, 1: 713 (India). Lectotype selected and figured by Kohn, 1963.

Description.—Heavy, with a low gloss; low conical, the upper sides convex then tapered to a narrow base; body whorl smooth except for a few low and widely spaced spiral ridges above the base and often conspicuous axial growth lines; shoulder roundly to sharply angulate, broad, slightly concave above; spire low to moderate, bluntly pointed, the sides straight or slightly concave or convex, most convex in largest specimens; spire whorls with about two broad, low spiral ridges near the middle, these conspicuous only on early uneroded whorls, obsolete on later whorls and shoulder; occasionally the spire whorls and shoulder are weakly undulate. Body whorl white, densely covered with fine irregular axial lines of orange-brown that are more-or-less continuous from base to shoulder where not obscured by darker pattern; lines may occasionally be joined by small blotches of orange-brown; basal quarter to third of whorl deep solid brown, this followed by a narrower spiral band of lighter brown; a narrow to wide or occasionally double spiral band of dark brown near midbody, usually sharply defined; spire white, covered with fine orange-brown axial lines and regularly spaced small orange spots below the sutures; early whorls usually badly eroded white. Aperture moderately wide, uniform in width; outer lip thin, sharp, convex; mouth dark brown with wide white bands below the shoulder and at middle. Columella internal. Length 40-111mm.

Comparison.—There is almost never any doubt about the identity of *C. miles.* It is somewhat similar to *C. vexillum* in general appearance, but easily distinguished by the dark mouth, very broadly dark base, and very small spire spots.

Variation.—A very constant species. There is considerable variation in height and shape of the spire and amount of angulation of the shoulder, seemingly on an individual level. Juveniles may have large blotches of opaque white anteriorly. The major variation is in the development of the posterior brown band on the body whorl; this band may be wide or narrow, occasionally broken into blotches, more than one shade of brown, or even completely divided into two separate bands. Some specimens develop vague orange axial flammules above and below midbody.

Distribution.—Widely distributed and common throughout the Indo-Pacific from eastern Africa to Hawaii and French Polynesia. Found from shallow to moderately deep water.

Notes.—This very attractive and distinctive species is often marred by heavy reef breaks on the body whorl and conspicuous axial growth lines and flaws. The spire is sometimes very heavily eroded and encrusted. Specimens with double midbody bands might be worth a small premium.

VIRGATUS.—Guaymas, Sonora, Mexico. *Above:* 45.6 and 47.5mm. *Below:* 45.3 and 48.4mm.

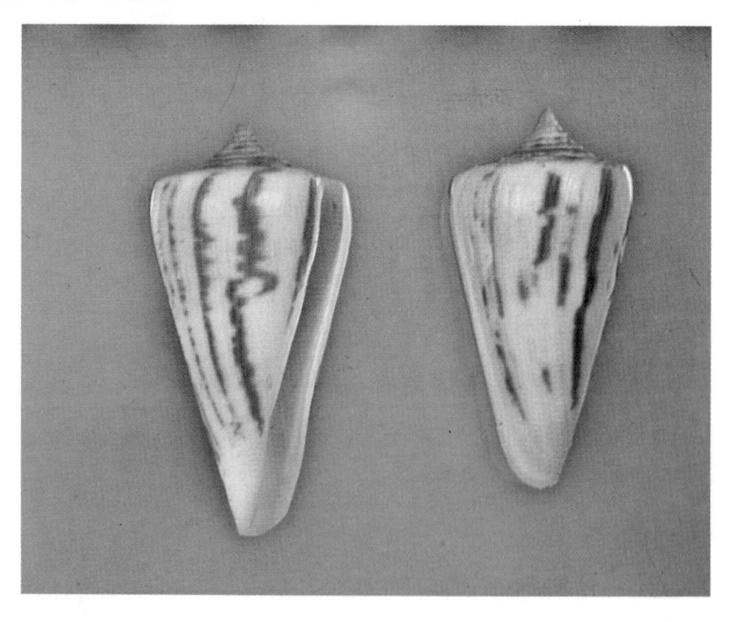

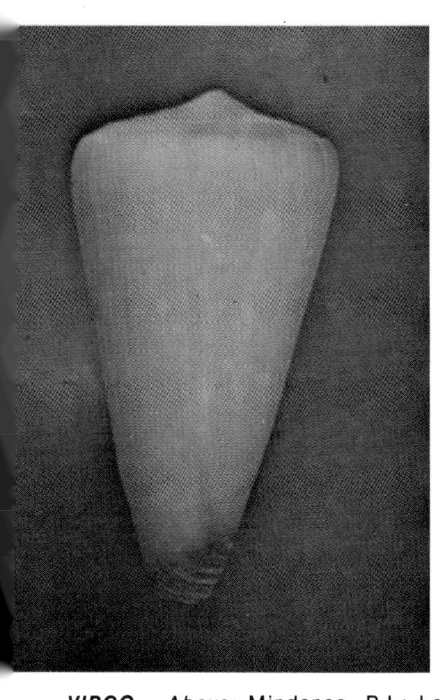

VIRGO.*—Above:* Mindanao, P.I.: Left: 78.0mm; Right: 81.8mm. *Below:* Marau, Guadalcanal, Solomon Is.: Left: 81.6mm; Right: 83.4mm.

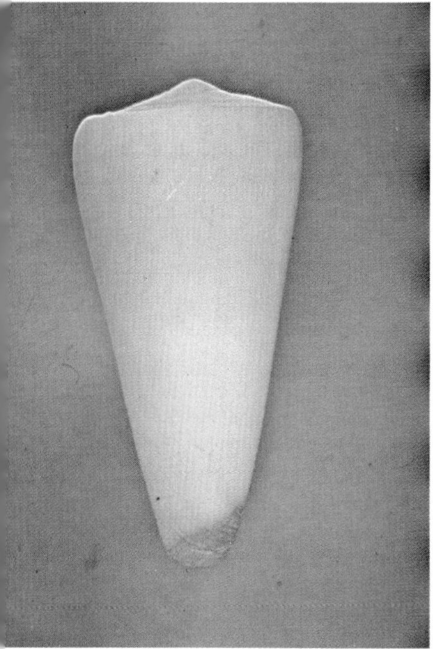
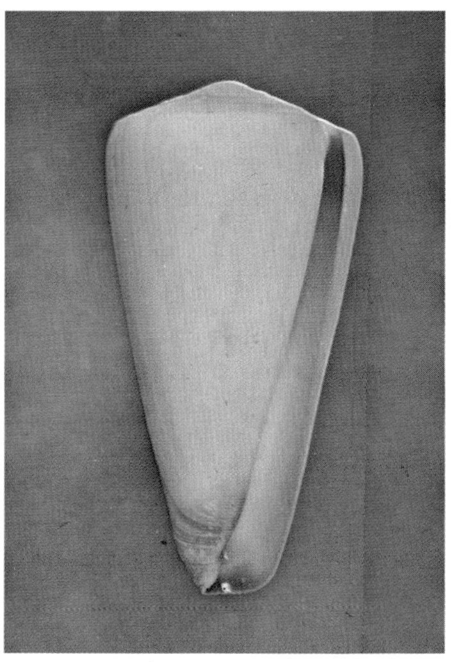

MILIARIS Hwass, in Bruguiere, 1792

1792. *Conus miliaris* Hwass, in Bruguiere. *Cone*, in *Ency. Method., Hist. Nat. des Vers*, 1: 629 (China). Lectotype selected and figured by Kohn, 1968.
1792. *Conus Barbadensis* Hwass, in Bruguiere. *Ibid.*, 1: 632 (Barbados). Lectotype selected and figured by Kohn, 1968.
1834. *Conus minimus* var. *granulatus* Sowerby i, in Sowerby ii. *Conchological Illustrations:* Pt. 56/57, fig. 81 (Locality not stated). Non *Conus granulatus* Linnaeus, 1758.

Description.—Heavy, thick, with a low gloss; obconical to low conical, the upper sides nearly straight and tapering to a narrow base; widest point about at shoulder; body whorl with anterior half or more covered with widely spaced spiral ridges, these usually granulose; grooves between ridges punctate, continued on rest of whorl as series of widely spaced rows of shallow punctations, these sometimes absent in large specimens; axial threads and growth lines often conspicuous; shoulder angulate, broad, heavily coronated; spire low, blunt, the sides nearly straight or concave; spire whorls strongly coronated, slightly concave above, with 2-4 low, wide spiral ridges. Body whorl whitish, usually heavily suffused and blotched with tan, apricot, orange, or bluish gray, the dark colors usually most prominent in two broad bands above and below midbody; opaque white lines and blotches, irregular in shape and size, may be present but are seldom prominent; few to many widely spaced spiral rows of dark brown dashes often present, these sometimes fused into nearly solid lines; base pale; shoulder and coronations whitish, the areas between coronations with small orange blotches formed of numerous very fine axial lines; spire usually whitish, the early whorls sometimes eroded to pinkish or violet. Aperture narrow, uniform in width; outer lip thickish, sometimes greatly thickened at the middle, almost straight to convex; mouth with two wide bands of brownish to pale reddish violet separated by whitish bands at middle and below shoulder. Columella short, wide, with heavy ridges posteriorly. Length 20-36mm.

Comparison.—*Conus miliaris* is the main species in a series containing *C. coronatus*, *C. abbreviatus*, *C. fulgetrum*, *C. encaustus*, and *C. tiaratus*. *C. coronatus* is chubbier, with the greatest width below the shoulder, the shoulder less angulate, the colors often bluish with darker axial flammules of brown, and the rows of punctations absent on the posterior part of the body whorl. Individual specimens of *C. miliaris* are sometimes difficult to separate from some *C. coronatus*, but they are quite distinct in series and after experience with both species.

C. abbreviatus is distinct in the usually bluish coloration and the prominent squarish reddish brown spots in many spiral rows; it is not sympatric with *C. miliaris*. *C. fulgetrum* is very close to *C. miliaris*, but it is usually bright brownish in color with numerous short opaque white oblique lines forming zigzag axial bands over much of the body whorl; the granulose ridges are also more heavily developed. *C. tiaratus* is very much like *C. miliaris* but has contrasting zones of brownish or bluish, has white midbody and shoulder bands in typical specimens, and has very distinct large brown dashes in spiral rows. *C. encaustus* is very finely flecked with brown and opaque white dashes and dots and heavily mottled with opaque white; it is usually smoother than typical *C. miliaris* and has a higher gloss. Some specimens of all these species can be very close to *C. miliaris*, and all are best separated in series if possible.

Variation.—This very variable species is easily recognizable once the collector becomes acquainted with its shape and color pattern. The variation is fairly well described earlier. This is usually a pale shell with a few dark aspects; bluish tones, when present, are subdued. The spiral rows of punctations are generally distinctive but appear to become obsolete with growth and perhaps do not appear in all populations; the punctations are often most visible just below the shoulder. Specimens from Easter Island have rather domed spires and continuous spiral brown lines, but they seem to intergrade with more typical *C. miliaris* of French Polynesia.

Distribution.—Very common throughout the Indo-Pacific in shallow water from eastern Africa to French Polynesia and Easter Island. Replaced in Hawaii by *C. abbreviatus* and in the Panamic region by *C. tiaratus*. Usually in shallow water.

Synonymy.—*C. miliaris* has been chosen over *C. barbadensis* as the proper name for this species; *C. barbadensis* should not be used.

Notes.—This little shell is sometimes very thick, with a thick ridge through the middle of the mouth. It is very close to both *C. abbreviatus* and *C. tiaratus*, which should perhaps best be considered subspecies; I find it especially hard to separate *C. tiaratus* from *C. miliaris* in some cases.

C. miliaris commonly has heavy growth marks and eroded spire whorls. The shell is so common, however, that such minor flaws usually go unnoticed. Lineate specimens from Easter Island are somewhat unusual and deserve a small premium, as do bright pinkish or rosy specimens from India.

MILNEEDWARDSI Jousseaume, 1894

1894. *Conus Milne-Edwardsi* Jousseaume. *Bull. Soc. Philom. Paris*, (Ser. 8), 6: 99 (Aden).

1899. *Conus (Cylinder) clytospira* Melvill and Standen. *Ann. Mag. Nat. Hist.*, (Ser. 7), 4: 461 (Arabian Sea, about 125 miles WSW of Bombay). Figured in *Proc. Zool. Soc. (London)*, 1901: Pl. 21, fig. 12.

(1905. *Conus aratispira* Pilsbry. *Proc. Acad. Nat. Sci., Phila.*, 57: 101, pl. 2, fig. 1 (Kikai, Osumi, Japan). Pliocene fossil.)

1962. *Leptoconus kawamurai* Habe. *Colored Illustrations of the Shells of*

VITTATUS.—Above: Pedro Gonzales I., Panama, 38.6mm. *Below:* Left: Bahia San Agustine, Sonora, Mexico, 47.9mm (photo by A. Kerstitch); Right: Nayarit, Mexico, 28.7 and 29.0mm (GG).

VITULINUS.—Above: Noumea, New Caledonia, 44.6 and 46.2mm. ***Below:*** Left: Tahiti, Society Is., 46.3mm; Right: Nellie Bay, Queensland, Australia, 48.4mm.

Japan, 2: 117, pl. 37, fig. 15 (Amami and Ryukyu Islands). 1964, *Shells of the Western Pacific in Color*, 2: 117, pl. 37, fig. 15 (English edition). See also *Venus*, 18(4): 301. I am not sure which of these, if any, is the original description.

Description.—Heavy, with a good gloss; elongate biconical, rather cylindrical, the sides straight or slightly concave at the middle; body whorl with a few weak spiral ridges at the base and often heavy axial growth marks and threads; occasional specimens with a few widely spaced shallow spiral grooves; shoulder narrow, weakly carinate, slightly concave above; spire very tall, about one-third or more of total length in average specimens, bluntly pointed, with straight or somewhat concave sides; protoconch large and rounded, persistent; early whorls with small marginal nodules, the later whorls smooth; spire whorls strongly stepped, concave above, usually with about 2-4 low but distinct spiral ridges and very weak axial threads. Body whorl white, covered with large and very open tents of dark orange brown to yellowish brown; three irregular and sometimes indistinct rows of large and small reddish brown blotches, the most conspicuous above midbody, a less obvious one below midbody, and a weak band below shoulder; tops of spire whorls with large reddish brown regular blotches, the sides of the whorls with axial orange-brown lines; tip of spire white if intact. Aperture narrow, slightly wider basally; the upper edge of outer lip strongly sloping below level of shoulder; outer lip rather thick, straight or concave at middle; mouth white to cream. Columella long, very narrow. Length 50-175mm.

Comparison.—*C. milneedwardsi* is related to *C. gloriamaris* and *C. bengalensis* in the tented pattern and elongate shape. However, it has a much longer and more stepped spire than either of these species, lacks 'textile lines' in the blotches, and has a blotched rather than distinctly tented spire pattern. Perhaps *C. milneedwardsi* is not actually very close to these two species but instead is related to *C. excelsus*. Both species certainly share the very elongated shape with strongly stepped spire whorls and distinct spiral ridges on the whorls; the pattern is similar in both species, just greatly reduced in *C. excelsus* so tenting is no longer apparent. *C. excelsus* has the body whorl covered with deep spiral grooves absent in typical *C. milneedwardsi*, but occasional *C. milneedwardsi* do have some spiral grooving on the body whorl.

Variation.—Specimens from the Indian Ocean are rather constant in shape and pattern, differing largely in details such as exact height of spire, how strongly tented is the body whorl, presence or absence of blotches below the shoulder, and tone of the brown coloring (from pale orange-tan to deep blackish brown). In the China Sea, however, are found specimens that vary greatly even among the relatively few specimens known. Although rather typical specimens have been collected near the Ryukyus, most specimens from this area have reduced patterns which may be almost absent except for axial lines or complete patterns that are very pale yellowish tan in color. In addition, specimens from this area often have shorter spires which are not strongly stepped and have projecting shoulders which give a concave appearance to the spire. Some specimens have very small tents, others very large. This is certainly an unusual situation and cannot be explained at the moment. Because Ryukyu specimens are so variable, they can be retained as a doubtful subspecies, *C. milneedwardsi kawamurai*, until the situation is better understood. Remember that only some specimens from this area are distinguishable from typical *C. milneedwardsi*, however.

Distribution.—Widely distributed throughout the Indian Ocean from the Mozambique coast to Aden, southern and western India, the southern Indian Ocean islands, and possibly the Bay of Bengal (the latter locality, where *C. bengalensis* is taken, is uncertain). There is then apparently a gap in the range until the species appears again in the China Sea off Taiwan and the Ryukyus. Although usually very rare, it may be locally uncommon; specimens have been taken by divers, so it is not a deep-sea cone, at least in some areas (Ryukyus, Mozambique).

Synonymy.—The status of the name *C. clytospira* has been discussed repeatedly, but it suffices to say here that it is a synonym of the poorly described (based on a single 46mm specimen) *C. milneedwardsi*. *C. kawamurai* is considered a subspecies, but admittedly its status is very uncertain; it could be a full species or a strict synonym—only more collections will tell. *C. aratispira* is a fossil name often allied with *C. milneedwardsi*, but since it has no color pattern it cannot be satisfactorily placed with any living species.

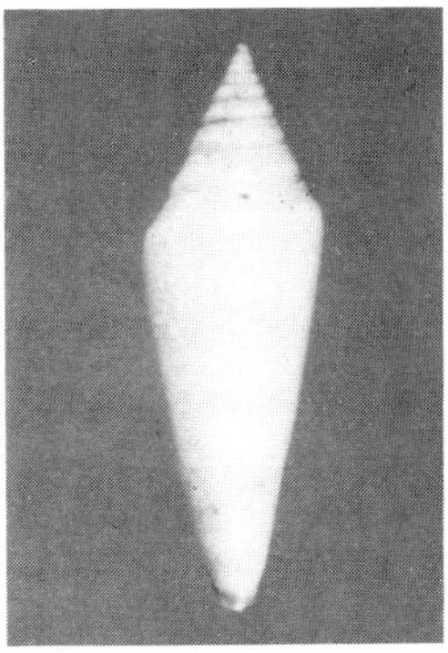

Possible juvenile of *C. milneedwardsi*, dredged dead, Ryukyus, 39mm (EP).

Notes.—This is of course a rare and desirable species, but it has recently come onto the market in some numbers. Unfortunately, the available specimens are usually either beach or juvenile (about 75mm) and probably not worth the prices placed on them. Supposedly good adults are still being taken off the Natal-Mozambique area, but unsettled political situations at the moment are preventing the specimens from being adequately distributed. Specimens are taken off India from time-to-time and off Mauritius-Reunion on occasion, but the source of supply in these areas is uncertain. Divers and dredgers take occasional specimens

VOLUMINALIS.—*Above:* Straits of Malacca, 41.5 and 42.1mm. *Below:* Andaman Sea off W Thailand, 41.7-46.0mm (GG).

WALLANGRA.*—Above:* Off Terrigal, New South Wales, Australia, 33.4mm (EP). *Below:* Queensland, Australia, 36mm (DM) (photos by Jayavad).

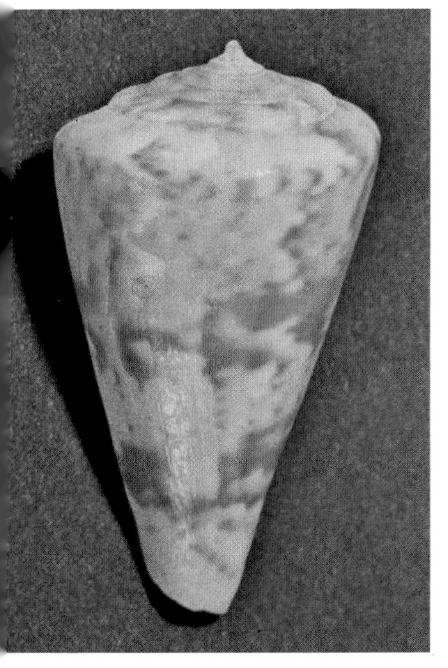
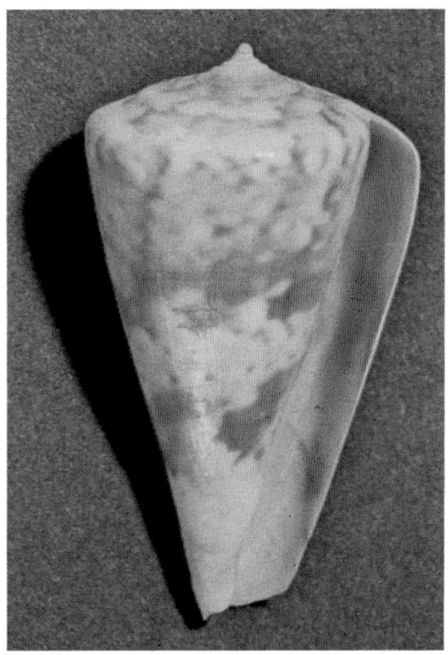

off Taiwan and the Ryukyus, but these are often either dead or badly scarred, or in some cases have very poorly developed patterns.

There seem to be two protoconch types in this species, one large and bluntly rounded, one narrow and rather pointed. I have seen too few specimens to understand what is happening here, but the presence of variant protoconchs in a single species would be very unusual.

The market for this species is unstable at the moment, and there are many rumors of large holdings of this species in dealers' basements. Certainly a large number of specimens have been taken in the last decade, but the price has not yet dropped to reflect this.

MINDANUS Hwass, in Bruguiere, 1792

1792. *Conus mindanus* Hwass, in Bruguiere. *Cone,* in *Ency. Method., Hist. Nat. des Vers,* 1: 711 (Philippines). Lectotype (here selected): specimen figured by Kohn, 1968, as holotype.

1833. *Conus elventinus* Duclos. *Rev. Mag. Zool.,* 3: Pl. 19 (Locality unknown).

1834. *Conus rosaceus* Sowerby i, in Sowerby ii. *Conchological Illustrations:* Pt. 54, fig. 55 (Locality not stated). Non *Conus rosaceus* Dillwyn, 1817.

1845. *Conus cretaceus* Kiener. *Species gen. et icon. des coqu. viv.,* 2 (*Conus*): Pl. 99, fig. 1. 1849, *Ibid.,* 2: 264 (Locality unknown).

1862. *Conus Duvali* Bernardi. *J. Conchyl.* (*Paris*), 10: 404, pl. 13, fig. 3 (Guadeloupe).

1865. *Conus anaglypticus* Crosse. *Ibid.,* 13: 314, pl. 11, figs. 8-8a (Antilles).

1866. *Conus acutimarginatus* Sowerby ii. *Thesaurus Conchyliorum,* 3 (*Conus* Suppl.): 328, pl. 27(288), figs. 640-641 (Locality unknown).

1870. *Conus corrugatus* Sowerby iii. *Proc. Zool. Soc.* (*London*), 1870: 257, pl. 22, fig. 7 (Locality unknown).

?1877. *Conus baccatus* Sowerby iii. *Ibid.,* 1876: 753, pl. 75, fig. 5 (Locality unknown).

1886. *Conus agassizii* Dall. *Bull. Mus. Comp. Zool., Harvard,* 12: Pl. 9, fig. 8. 1889, *Ibid.,* 18: 68 (Santa Cruz Island, Barbados, Bermuda).

1903. *Conus boubeeae* Sowerby iii. *J. Malac.,* 10: 76, pl. 5, fig. 5 (Locality unknown).

1942. *Conus bermudensis* Clench. *Johnsonia,* 1(6): 34, pl. 13, fig. 4 (Dyer Island, Bermuda).

1942. *Conus bermudensis lymani* Clench. *Ibid.,* 1(6): 35, pl. 13, fig. 3 (Off Nellies Point, South Lake Worth, Palm Beach Co., Florida).

1943. *Conus fulvus* Fenaux. *Bull. l'Inst. Oceanogr.* (*Monaco*), No. 834: 4, fig. 8 (New Guinea). Non *Conus fulvus* Schroeter, 1803.

Description.—Fairly heavy, with a good gloss; low conical to indistinctly biconical, the upper sides slightly convex to distinctly convex; body whorl with about 6-12 shallow, even, widely spaced spiral grooves anteriorly and numerous fine spiral and axial threads over the rest of the whorl; in some specimens, especially juveniles and those from the Caribbean, the grooves extend farther up the whorl and may reach the shoulder; the ridges are sometimes weakly granulose; shoulder roundly angled to sharp, usually concave above; spire low to high, usually pointed, the sides straight to concave; tops of spire whorls concave, with many weak axial threads and occasional traces of a spiral ridge; sutures often rude, may appear irregularly nodulose; sutures fail to meet evenly and leave a narrow ridge between the whorls. Body whorl pinkish to whitish or violet, sometimes pale golden tan; in complete patterns there are numerous spiral rows of small reddish brown and white dashes covering the body whorl; this is sometimes obscured by large or small clouds and blotches of reddish brown to deep pink; sometimes dots and blotches absent; shoulder with a series of larger dark brownish dots or dashes regularly alternating with white at the margins; spire whorls usually with similar alternating dots; there may be indistinct larger spots or blotches of brown on the spire. Aperture moderately wide, sometimes distinctly widened anteriorly; outer lip straight to slightly convex, thin and sharp; mouth pink to violet. Columella very narrow or internal. Length 15-60mm.

Comparison.—This species has long been very confused, although it appears to be quite distinct. The problem is that it occurs in two rather different sizes and has a variety of patterns. It is often considered a synonym of *C. jaspideus*, with which it shares the anterior grooving and (sometimes) the small dotting at the spire margins; however, the northern form is larger and broader than *C. jaspideus* while the southern form is more brightly colored and usually more round-shouldered. In both forms there is commonly a rudeness about the spire sutures which is not found in *C. jaspideus*, especially the strong tendency for the sutures not to overlap or even touch at the margins. *C. puncticulatus* differs in pattern in most cases, being duller in color than sympatric *C. mindanus*. *C. magellanicus* is much more slender with low ridges anteriorly instead of grooves.

Variation.—This is a very confusing species which may actually be a composite of two species, a large northern form and a small southern form. Both types occur off southern Florida and in the Bahamas, where they seem to intergrade. For the moment they are considered varieties, as I am uncertain of the distribution and do not feel that they are subspecies.

C. mindanus typical: Usually large, between 30 and 60mm in adults; broad, the shoulder sharply angled; usually pinkish with indistinct clouds and blotches of reddish brown, orange, or yellowish tan; in fully developed patterns there are many spiral rows of small orange or brownish and white dashes on the body whorl; the spots on the shoulder and spire are usually inconspicuous; juveniles (to about 25mm) may be spirally grooved to the shoulder and have small granules on the ridges; juveniles are commonly white. Bermuda and North Carolina south to

WITTIGI.—Lesser Sunda Is. N of Timor. *Above:* Left (paratype) 29.4mm; Right (holotype): 31.8mm. *Below* (paratypes): 24.9 and 27.0mm (all R. and E. Skinner).

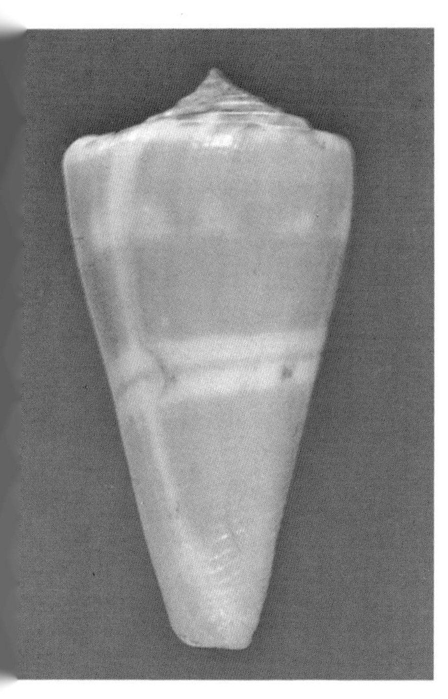

XANTHICUS.—*Above:* Off Los Frailes, Baja California del Sur, Mexico, 43.3mm (EP). *Below:* Left: Cabo San Lucas, Baja California, Mexico, 51.5mm; Right: Guaymas, Sonora, Mexico, 51.8mm (photo by A. Kerstitch).

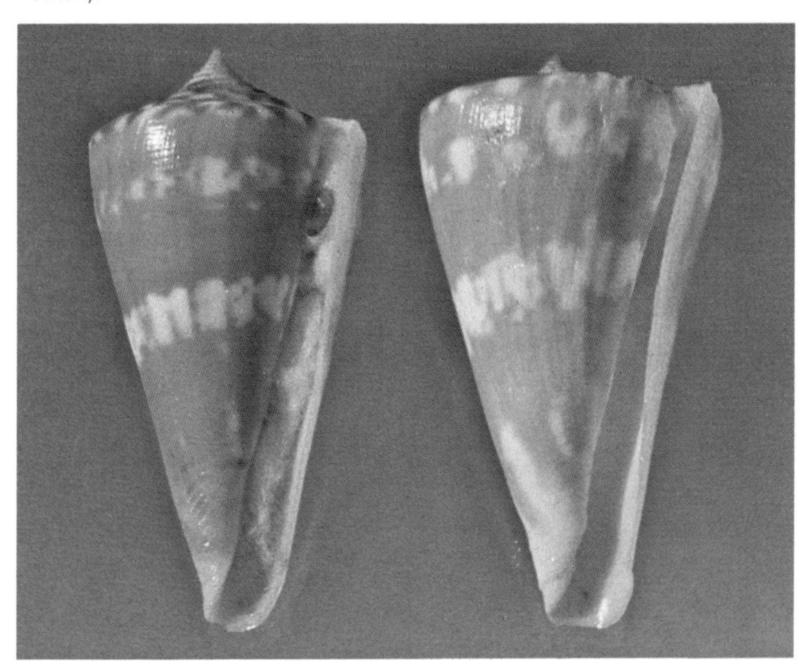

Florida and the Bahamas in moderately deep to shallow water.

C. mindanus variety *duvali:* Usually small, between 15-25mm in adults; shoulder usually rounded; usually pale tan or orange with very contrasting pattern of dark and light dashes in spiral rows over the body whorl; axial orange or whitish blotches may be present; spots on shoulder and spire margins usualiy very dark and distinct; sometimes with large black spire blotches; adults commonly grooved to shoulder, although the grooving may be absent above midbody; granules on basal ridges moderately common; juveniles like adults. Southern Florida and the Bahamas, Texas, Cuba, the Central American coast and Antilles, south to central Brazil in shallow to moderately deep water.

The *duvali* variety seems to have a wider mouth with brighter colors, but this is possibly a factor of size. In both forms the spire may be very tall and sharp or very low. Senile adults of the northern form are patternless and have heavy axial plicae on the body whorl.

Distribution.—Uncommon to locally common in shallow to deep water from Bermuda to central Brazil.

Synonymy.—It is often difficult to distinguish this species from *C. jaspideus* or *C. puncticulatus* in early descriptions, but I think all the names here listed apply to *C. mindanus*. Early authors seem to have had a fairly good concept of the large northern form, but repeatedly described the southern form. *C. duvali* appears to be the earliest name for the small form, but to be certain would require an examination of all the types. Of the various names listed, *C. anaglypticus* has recently been referred to the species here called *C. axelrodi*, but this seems incorrect on the basis of shape, pattern, and sculpture. *C. agassizii*, which is validly described and figured, seems to be a composite of both forms, including granulose juveniles. *C. boubeeae* is a bright and well marked specimen of the *duvali* form. For some unknown reason, Clench considered *C. agassizii* to be a synonym of the not at all similar *C. villepinii*; he described the northern form as a new species, *bermudensis*, with a subspecies *lymani* distinguished by trivial characters. Neither taxon can be recognized. *C. fulvus* was described from New Guinea but was compared with *C. magellanicus* and certainly appears to be the Atlantic species.

Notes.—It is currently easier to obtain specimens of the *duvali* variety than the large northern form. It appears that typical *mindanus* is quite rare or very uncommon over most of its range, while *duvali* may be locally common in shallow water. Neither form is cheap, however. This species, as might be expected, is sold under many different names by dealers.

The *duvali* variety is easy to obtain in gem, although sometimes the pattern is rather diffuse and not as distinct as one would wish. The northern form is often badly broken and scarred, with very indistinct patterns and very dull in appearance; bright patterns of this form are certainly worth a high premium.

MINNAMURRA (Garrard, 1961)

1961. *Mamiconus minnamurra* Garrard. *J. Malac. Soc. Australia*, 5: 32, pl. 1, figs. 4a-4b (E. of Botany Bay, New South Wales, Australia).

Description.—Light in weight, with a low gloss; low conical, the sides convex posteriorly then tapering to a narrow base; body whorl covered with narrow undulating spiral ridges from base to near shoulder, these crossed by light or heavy undulating axial threads and ridges; shoulder rather broad, roundly angled, sometimes undulate; spire low or moderate, bluntly pointed, the sides straight or slightly concave; early 3-5 whorls nodulose, rest with undulate margins; spire whorls slightly concave, the sutures overlapping, sometimes weakly carinate, with 2-4 low spiral ridges on top crossed by scattered axial threads. Body whorl pale straw or cream, with about three spiral rows of large and irregular darker brown axial blotches; blotches show a strong tendency to fuse into irregular axial flammules or into broken spiral bands; sometimes the blotches can be seen to be made of numerous brown spiral dashes on top of the ridges covered with brown clouds; shoulder and spire with broad brown bands and blotches, these continued from body pattern; often the spots are enlarged and darker at the angle of the shoulder; early whorls white. Aperture moderately wide, nearly uniform; outer lip thin, sharp, straight or slightly convex; mouth pale violet. Columella long and very narrow, mostly internal. Length 18-30mm.

Comparison.—This little cone is fairly distinct although apparently related to *C. sibogae*. From somewhat similarly patterned and sculptured species it differs in having the spire whorls weakly nodulose or undulate and by having the body sculpture very undulate and heavy for such a small shell. The dark spots at the shoulder, when developed, are quite distinctive.

Variation.—Height and degree of concavity of the spire vary considerably. In many specimens the spiral ridges of the body whorl appear to be in pairs separated by wider grooves; only the grooves may have strong axial sculpture, or the entire shell may be axially striate. The sculpture is usually distinctly undulate. Occasionally the spiral ridges are indistinct or obsolete just below the shoulder. Although the pattern is apparently made of fused dashes and brown clouds, the appearance is of three indistinct and very irregular bands of brown blotches that tend to fuse either axially or spirally. The pattern continues onto the spire but may be broken at the shoulder by larger and darker brown spots, sometimes with similar spots at the margins of the spire whorl.

Distribution.—Uncommon in deep water off southern Queensland and northern New South Wales, Australia.

Notes.—This species is usually sold under the name *C. lizardensis*, here considered a probable synonym of *C. arcuatus*. This is a fragile cone often showing distinct growth marks and scars adding to the irregularity

XIMENES.—*Above:* Guaymas, Sonora, Mexico, 36.8 and 42.5mm. *Below:* Cholla Bay, Sonora, Mexico, 24.1 and 24.8mm.

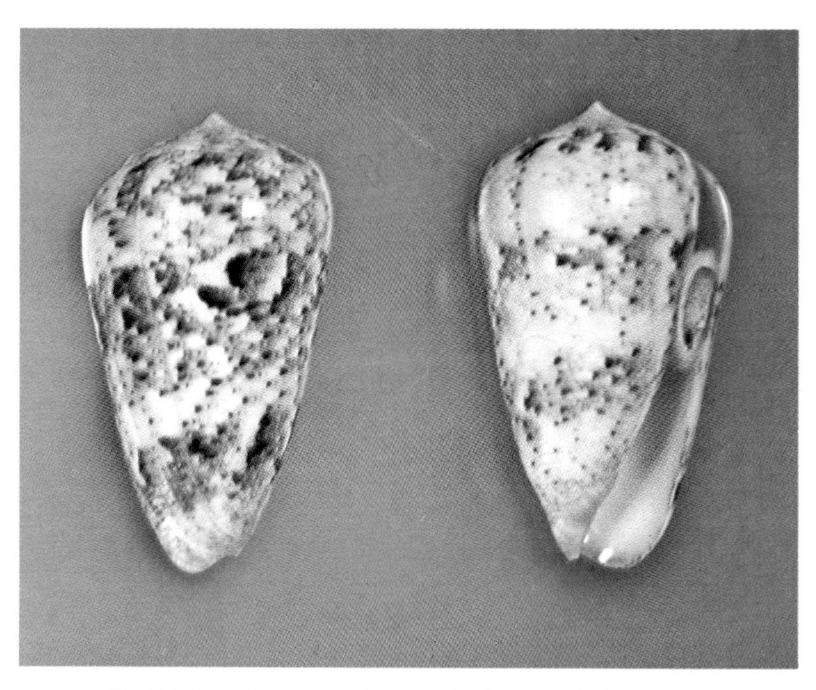

ZEYLANICUS.—Zanzibar. *Above:* both 46.0mm. *Below:* 49.1mm.

of the pattern. The lip is often broken. Specimens with heavy patterns over heavy sculpturing are quite attractive in a subdued way for a small cone.

MITRATUS Hwass, in Bruguiere, 1792

1792. *Conus mitratus* Hwass, in Bruguiere. *Cone,* in *Ency. Method., Hist. Nat. des Vers,* 1: 738 (Indian Ocean). Lectotype (here selected): specimen figured as holotype by Kohn, 1968.

1870. *Conus mitraeformis* var. *pupaeformis* Sowerby iii. *Proc. Zool. Soc. (London),* 1870: 256, pl. 22, fig. 2 (Mauritius). The term *'mitraeformis'* is certainly a lapsus.

Description.—Thick but rather light in weight, with a good gloss; elongate cylindrical, the sides convex; body whorl covered with widely spaced and weakly developed grooves, these separated by rows of low granules; granules may extend from base to shoulder or be restricted to the anterior half of the body whorl; fine axial threads usually present; shoulder indistinct, completely merging with spire; spire high, the sides convex to straight, sharply pointed or eroded blunt; earliest 2-3 whorls with fine nodules, the rest with straight and somewhat overlapping margins; tops of whorls slightly convex, with 2-4 broad and indistinct spiral ridges often bearing small granules. Body whorl yellowish tan to cream, with dark brown spots arranged in three broad bands on shoulder, at midbody, and above base, the spots often connected to form a solid set of broad spiral bands; lighter brown axial flammules may connect the large dark spots; base pale; spire and shoulder light yellowish with large squarish dark brown spots; tip white. Aperture narrow, especially posteriorly, somewhat widened anteriorly; outer lip thick, slightly convex; mouth white, the external pattern sometimes showing through at the margin. Columella internal. Length 15-35 (specimen with turreted spire 53mm).

Comparison.—The shape is very distinctive and shared with only a few species. Of these, only *C. cylindraceus* is at all close, but it differs in having closely spaced very weak ridges only basally, these seldom granulose, the rest of the body whorl smooth and with a high gloss; in *C. cylindraceus* the pattern is of broad brown axial stripes which fuse to leave only a little light background visible. *C. tenuistriatus, C. glans, C. luteus,* and even *C. artoptus* may somewhat resemble *C. mitratus* in shape, but all differ in sculpture (especially lacking the spaced granules on body whorl and spire) and pattern as well as color.

Variation.—This is a very constant species, the most obvious difference being the development of the color pattern, which may be of distinct rather oval dark brown spots or fused into solid spiral bands; the axial flammules may be light or dark. A few specimens seen have the granules restricted to the anterior half of the body whorl and barely apparent on the spire whorls. Some specimens have the spiral grooves on the body whorl deeper and more obvious than others.

Distribution.—Widely distributed over the Indo-Pacific from Africa to French Polynesia, but apparently not reaching Hawaii. Usually uncommon, but sometimes locally common when colonies are found. Shallow to moderately deep water.

Notes.—The specimens usually seen are small, under 25mm, and rather juvenile. The 53mm specimen has a badly distorted spire adding to the length, but the body whorl is 35mm in length (Zanzibar). A very interesting and unique species which is sometimes hard to find for sale at this time. Most specimens come from the Philippines or Solomons today.

MOLUCCENSIS Kuester, 1838

1838. *Conus Moluccensis* Kuester. *Syst. Conch. Cab.*, ed. 2, 8(*Conus*): 121, pl. 23, figs. 4-5 (Moluccas). This name was also used by Holten, 1802, and could be dated from then if desired.

1843. *Conus Stainforthii* Reeve. *Conchologia Iconica*, 1(*Conus*): Pl. 1, sp. 1 (Locality unknown).

Description.—Moderately heavy, thick, with a high gloss; rather cylindrical, the body whorl convex above then pinched-in at middle, also pinched-in at shoulder; body whorl with at least basal third covered by heavy spaced spiral ridges, these sometimes extending to shoulder; ridges anteriorly almost always with heavy granules; grooves between usually with heavy axial threads, appearing punctate; in smooth form the posterior half or more of body whorl is smooth except for axial and spiral threads and widely spaced rows of shallow punctations, these again deeper just below the shoulder; shoulder rounded, narrower than top of body whorl, heavily coronate with erect, sharp coronations; spire low to moderate, sharply pointed, the sides concave; whorls with distinct to heavy coronations at margins; tops of whorls concave, with 3-5 heavy spiral ridges crossed by curved axial threads. Body whorl golden tan, straw, cream, or pinkish, usually covered with very irregular axial flammules of deep brown to bright red, these sometimes partially fused into two blotchy spiral bands above and below midbody, with smaller blotches below shoulder; spiral dashes and spots of brown or red may be present on the spiral ridges in the strongly ridged form; columellar area paler, pinkish to violet; spire and shoulder whitish to straw, with large and small reddish or brownish spots and streaks, especially between coronations; tip white. Aperture moderately wide, slightly wider anteriorly; outer lip thick, straight but concave at middle; mouth white with pale violet tones anteriorly or deep within. Columella rather wide, long, twisted, with a heavy rounded ridge bordering it. Length 35-60mm.

Comparison.—This is a rare species which is still poorly known. It is

Continued to page 738

ZONATUS.—Above: Left: Hulule, Maldive Is., 50.3mm; Right: Phuket, Thailand, 56mm (DM) (photo by Jayavad). *Below:* Thailand: Left: 61.0mm; Right: 71.0mm (K).

About the Photos

Unless otherwise credited, all photos are by the author. They represent average specimens in most cases, with no attempt to illustrate exceptionally fine or large specimens. Although an attempt has been made to illustrate as many variations as possible, obviously most species present more color and pattern variants than shown; be sure to read the text carefully. Sizes of photographed specimens are indicated in the captions; no attempt has been made to keep the photos proportionate.

The many borrowed slides and specimens used have been credited individually in the captions by names or initials. Names or initials in parentheses after the data indicate the specimen illustrated is from the collection of the person or museum credited. Most specimens not otherwise credited are from my collection. The following abbreviations have been used for collectors and collections:

AK - Alex Kerstitch
BB - Brian Bailey
BMNH - British Museum (Natural History)
DM - A.J. da Motta
DMNH - Delaware Museum of Natural History
EP - Ed Petuch
GG - George Gundaker
Jayavad - Jayavad Mahathanaphanij
JB - Jim Bateman
K - Richard M. Kurz, Inc.
Meyer - Michael Meyer
MF - Misha Fainzilber
MG - Morrison Galleries, Inc.
MM - Mal de Mer Enterprises
NM - Natal Museum
PC - Phillip W. Clover
RJ - Rodney Jonklaas
TCG - T.C. Good
TY - Ted Yocius

probably most likely to be confused with *C. proximus*, although the two species are quite distinct. In *C. moluccensis* adults the shell is large with a thick lip and the body whorl is distinctly concave at the middle and pinched-in at the shoulder. In *C. proximus* the shell seldom exceeds 40mm length and is low conical, the upper sides nearly parallel and not pinched-in at the shoulder, then rather smoothly tapered to the base; the outer lip is straight or nearly so and thin. The columella in *C. moluccensis* is larger and bounded by a heavier ridge than in *C. proximus*. The supposed differences in sculpture between *C. proximus* and *C. moluccensis* do not hold up in large series, although as a rule the basal ridges of *C. moluccensis* are wider and more heavily granulose than in *C. proximus*. *C. proximus* is usually a whitish shell or pale straw, with a pattern of spiral dashes or spots of dark brown to red, the spots sometimes clustered into small blotches; there are seldom or never strong axial flammules as in most *C. moluccensis*.

Such species as *C. boeticus* or *C. muriculatus* could in theory be confused with *C. moluccensis*, but this is unlikely.

Variation.—This species seems to occur in two intergrading sculpture and pattern varieties, one variety or the other being dominant at a locality. In the typical variety the shell is covered from base to shoulder with heavy spiral ridges, those on the base heavily granulose, the granules sometimes occurring to the shoulder. In this form the pattern often is of dark brown or reddish brown irregular axial flammules and numerous long spiral dashes and spots. In the smooth form the spiral ridges and granules are restricted to the anterior third or less of the body whorl, the remainder being apparently smooth; however, under magnification can be seen numerous widely spaced spiral rows of punctations corresponding to the grooves between the obsolete ridges; the grooves become more distinct just below the shoulder. In this form the coloration is often much brighter, consisting of red or orange-brown axial flammules on golden tan or straw, the flammules tending to form indistinct blotches above and below midbody; the smaller spots and dashes may be totally absent.

Mouth color in this species varies more than stated in the literature.

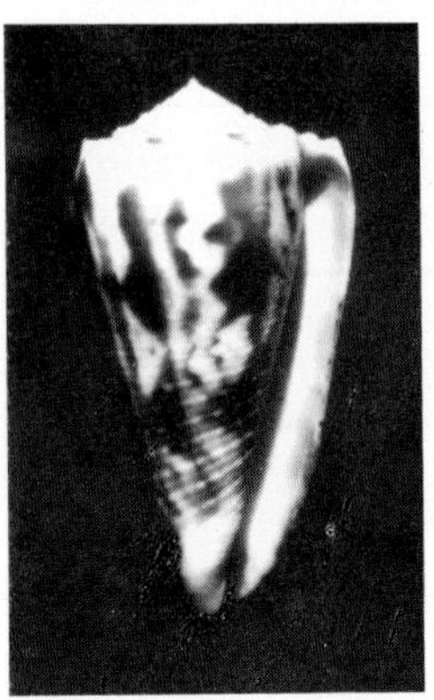

Axially flammulated *C. moluccensis*, Philippines, 36mm (Dayrit) (photo Burgess).

Although normally glossy white, it may also be distinctly pale violet or even pinkish, much the same variation found in *C. proximus*. The columella is often brightly colored.

The two varieties cannot be recognized as species or subspecies as they may both occur at the same locality (typical specimens have been taken with smooth specimens in Okinawa but do not reach the market), and often shells are distinctly intermediate between the two sculpture and color types. As a very general rule, the typical type is more likely to be found near the major island arc from the Philippines to the Solomons, while the smooth form occurs in outlying islands such as the Ryukyus.

Distribution.—Sporadically found along the western Pacific from the Ryukyus and perhaps southern Japan through the Philippines into the Indonesian area; also in the New Guinea-Solomons area, New Caledonia, and perhaps farther north. Rare or very uncommon, although locally large populations have been found and exploited.

Synonymy.—*C. stainforthii* is the typical, heavily ridged form. No formal name has been given to the smooth form, and in my opinion it does not need one.

Notes.—This is a rare and often very attractive species. It is subject to heavy growth marks and scars on the body whorl, which makes many offered specimens far from gems. The thick lip seems to shatter easily. The bright orange or red Okinawan shells are the ones currently most easy to obtain, but a few specimens from other localities are available from time to time. The typical form is not at all common and seldom offered.

Undoubtedly this species is very similar to *C. proximus*, but there should be little confusion identifying the two. Juveniles of *C. moluccensis* are slender and rather straight-sided with very large coronations for their small size; they are rare. The dealers' name *C.* 'merleti' (variously spelled) has been applied to New Caledonian specimens, but with no basis.

MONACHUS Linnaeus, 1758

1758. *Conus monachus* Linnaeus. *Systema Naturae per Regna Tria Naturae*, ed. 10, 1: 714 (Locality not stated). Lectotype figure (Kohn, 1963): Regenfuss, Pl. 12, fig. 68.

1791. *Conus achatinus* Gmelin. *Ibid.*, ed. 13, 1: 3386 (In Oceano Americano). Lectotype figure (Kohn, 1966): Chemnitz, Pl. 142, fig. 1317.

1791. *Conus nebulosus* Gmelin. *Ibid.*, ed. 13, 1: 3387 (Locality not stated). Lectotype figure (Kohn, 1966): Gualtieri, Pl. 25, fig. R.

(1792. *Voluta cosmographicus* Martyn. *Univ. Conch.:* iv, fig. 125 (Guinea coast). Non binomial.)

1798. *Cucullus cinerarius* Roeding. *Museum Boltenianum:* 46 (Locality not stated). Lectotype figure (Kohn, 1975): Knorr, Pl. 16, fig. 3.

1798. *Cucullus glaucus* Roeding. *Ibid.:* 48 (Locality not stated). Based on Martini, Pl. 55, fig. 613. Non *Conus glaucus* Linnaeus, 1758.

1798. *Cucullus guttatus* Roeding. *Ibid.:* 48 (Locality not stated). Based on Martini, Pl. 55, figs. 613-614.

1798. *Cucullus maculosus* Roeding. *Ibid.:* 48 (Locality not stated). Lectotype figure (Kohn, 1975): Knorr, Pl. 18, fig. 4.

1798. *Cucullus ventricosus* Roeding. *Ibid.:* 49 (Locality not stated). Lectotype figure (Kohn, 1975): Chemnitz, Pl. 142, fig. 1317. Non *Conus ventricosus* Gmelin, 1791.

?1807. *Conus pardalinus* Link. *Beschr. Nat.-Samml. Univ. Rostock*, 3: 104 (Locality not stated).

?1807. *Conus sardus* Link. *Ibid.*, 3: 105 (Locality not stated).

1854. *Conus assimilis* A. Adams. *Proc. Zool. Soc. (London)*, 1853: 118 (Australia).

1854. *Conus vinctus* A. Adams. *Ibid.*, 1853: 118 (Australia).

1898. *Conus Barbara* Brazier. *Proc. Linnaean Soc. New South Wales*, 22(1897): 781 (Solomon Islands). Holotype figured by Cotton, 1945, *Rec. S. Aust. Mus.*, 8(2): 242, pl. 4, fig. 2.

1937. *Conus achatinus* var. *infumata* Dautzenberg. *Mem. Mus. Roy. d'Hist. Nat. Belg.*, 2(18): 12 (Locality not stated). Lectotype figure (here selected): Chemnitz, Pl. 142, fig. 1320.

Description.—Moderately heavy, with a good gloss or dull; more-or-less cylindrical, the greatest width usually below shoulder, the sides convex; body whorl with usually heavy spiral ridges on basal third, these sometimes continuous to shoulder; ridges often granulose; grooves between ridges usually axially striate; upper half of body whorl sometimes very smooth except for weak spiral and axial threads; shoulder roundly angled, usually slightly concave above, not very distinct from spire; spire low or moderate, the sides usually convex, sometimes straight or weakly concave; early whorls with weak nodules or undulate; tops of whorls flat to concave with about 2-7 sometimes indistinct spiral ridges separated by distinctly punctate grooves and covered with axial threads. Body whorl with extremely variable colors and pattern; usually pale blue or even pale green, deep bluish gray, or gray, occasionally covered with darker olive brown to nearly black nebulous axial flammules and blotches which may fuse to leave lighter bands only at shoulder and midbody; in extreme examples the shell may be entirely ashy gray or blackish with a narrow midbody band of lighter bluish gray heavily reticulated with blackish; spiral lines of alternating brown and white dashes are usually present above the base and commonly extend to the shoulder; on occasion the dashes fuse into solid, rather broad spiral brown lines or may be entirely absent; spire white to bluish gray, heavily covered with revolving reddish to bluish brown lines and blotches; early whorls white to pale pink. Aperture wide, especially anteriorly; outer lip thin or thick, weakly to strongly convex; mouth white to deep blue, the outer edge often with the external pattern showing through. Columella long and narrow, with rather heavy ridges posteriorly. Length 30-65mm.

Comparison.—This is a very complex species which is usually treated as two species (*monachus* and *achatinus*) by collectors. Its great variability often makes it difficult to identify; *C. magus* is sometimes especially difficult, but it is usually less cylindrical in shape, has a wider shoulder that is more angulate, and tends to run toward yellows, oranges, and tans instead of blues and blue-grays in color. *C. nigropunctatus* is also very similar and possibly just a small variant of *C. monachus*; however, it is usually separable by its smaller size, thicker shell that is heavier and more solid than *C. monachus* of the same size, the somewhat less convex lip, and the usual presence of opaque white and orange in the pattern; the spire pattern is usually reduced to a few blotches. I am not at all certain that I can always distinguish *C. nigropunctatus* from juvenile *C. monachus*, even in series.

C. ermineus and *C. purpurascens* are often very similar in pattern and even color to some patterns of *C. monachus*, but they have a strong tendency to a wider shoulder and more conical shape, as well as usually differing in details of color and pattern; the grooves on the spire whorls are not punctate. *C. fulmen* and *C. kinoshitai* usually differ greatly in pattern and color.

Variation.—As indicated in the description, this is a very variable shell. The body whorl may vary from relatively slender with nearly straight sides to strongly convex and almost ovate. The sculpture may be weak and restricted to the extreme anterior part of the body whorl or may be very heavy, granulose, and extend to the shoulder. The number of spiral ridges on the spire whorl is unusually variable, even within small series, but generally the grooves between are distinctly punctate.

Collectors recognize two major groups of patterns as full species, here considered merely varieties because intermediates are not uncommon. *C. monachus* typical lacks spiral rows of brown and white dashes over the body whorl, except sometimes at the base; *C. m.* var. *achatinus* has such dashes.

Specimens of both varieties from the Indian Ocean often tend to have very light and clear background colors with contrasting narrow or wide axial clouds of brown or olive; specimens from Australia also commonly have contrasting patterns, but their colors are duller and more grayish. In the Philippines and New Guinea-Solomons area many populations are very dark, even almost entirely dark grayish brown with blue tinges; light specimens may have very blurred and indistinct patterns. The variation in color of a series from even one Solomons location may encompass the entire range of color variation typical of darker patterns of this species, with grayish dashless shells, shells encircled by dark lines, very dark blue and gray shells, and almost black shells found together. Certainly none of these are worthy of even varietal status.

One uncommon variant is sometimes very distinct. In this type the entire shell is pale or dark brown, encircled by broad spiral lines of darker brown or black; sometimes there are irregular white blotches at midbody

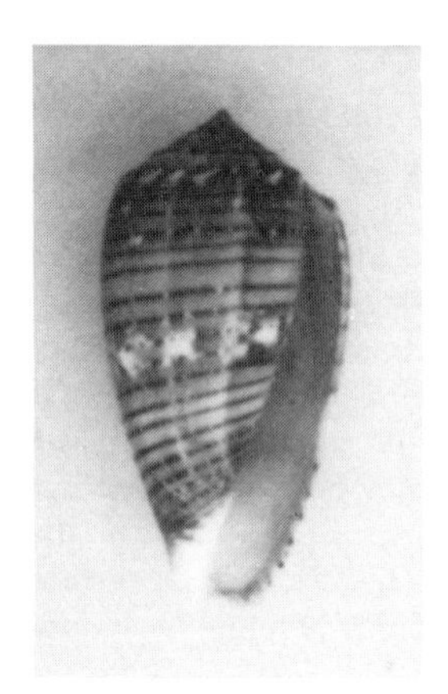

C. monachus var. *vinctus*, Indonesia, 46mm (photo Clover).

and shoulder. This type of shell can be called variety *vinctus* if desired, but it is connected to typical *C. monachus* by intermediates.

Distribution.—Widespread and sometimes common from the eastern African coast to the central Pacific. Apparently not recorded from Hawaii or French Polynesia, although it reaches to Fiji. Usually in shallow water.

Synonymy.—Some of the earlier names, especially those of Roeding and Link, are probably actually unidentifiable but fit here as well as anywhere else. *C. achatinus* intergrades commonly with typical *C. monachus*, with which it is often sympatric, so it cannot be retained; the *achatinus* pattern is usually more common than the reduced typical *monachus* pattern at most localities. *C. vinctus* is a highly developed *achatinus* pattern with fusion of the dashes. *C. barbara* applies to a fairly typical *C. monachus*, although the name is usually used by dealers (incorrectly) for very dark bluish gray or blackish *C. monachus* with otherwise reduced or obscure pattern; this dark pattern is not uncommon in some Solomons and New Guinea localities and appears to be environmentally induced.

Notes.—The finest specimens with brightest colors probably come from the Indian Ocean and Australia. Specimens from the Philippines to Solomons are often very dark with very indistinct patterns and low gloss. Heavily granulose specimens are not uncommon at some localities. Some patterns probably deserve small premiums, but except for the *vinctus* type all are relatively common locally.

C. monachus is subject to bad growth flaws and scars on the body whorl. It is often eroded on the spire and body whorl, with a very dull finish.

MONILE Hwass, in Bruguiere, 1792

1792. *Conus monile* Hwass, in Bruguiere. *Cone*, in *Ency. Method.*, *Hist. Nat. des Vers*, 1: 646 (Indian Ocean). Lectotype selected and figured by Kohn, 1968.

1798. *Cucullus ornatus* Roeding. *Museum Boltenianum:* 44 (Locality not stated). Lectotype figure (Kohn, 1975): Chemnitz, Pl. 140, fig. 1303.

1798. *Cucullus cereolus* Roeding. *Ibid.*: 44 (Locality not stated). Lectotype figure (Kohn, 1975): Chemnitz, Pl. 140, fig. 1301.

1811. *Conus marmoreus* Perry. *Conchology:* Pl. 25, no. 3 (Asiatic Seas). Non *Conus marmoreus* Linnaeus, 1758.

1845. *Conus fasciatus* Kiener. *Species gen. et icon. des coqu. viv.*, 2 (*Conus*): Pl. 109, fig. 2. 1849-1850, *Ibid.*, 2: 311 (Locality unknown). Non *Conus fasciatus* Schroeter, 1803.

1937. *Conus anadema* Tomlin. *Proc. Malac. Soc. London*, 22(4): 206. Nomen novum for *Conus fasciatus* Kiener, 1845.

Description.—Heavy, with a high gloss; turbinately obconical, the sides nearly straight and tapering to a narrow base; body whorl with a few weak spiral ridges above the base and sometimes axial threads and flaws that may be conspicuous; shoulder broad, carinate to angulate, concave above; spire with later whorls nearly flat or depressed, the early whorls forming a tall raised cone, the sides thus deeply concave; early whorls with fine nodules; tops of later whorls concave to flat, with only numerous curved axial threads, no distinct spirals. Body whorl glossy white, covered with a variable pattern of spiral rows of dark reddish brown squares and triangles, these usually heaviest above midbody; midbody area commonly unmarked; the banded areas may be suffused with pale to dark salmon, either in distinct narrow bands or broad clouds that may include the entire whorl; the dark reddish brown spots show a tendency to fuse both spirally and axially and to form irregular blotches; all-white shells probably occur, as the pattern is commonly very reduced; base of whorl usually tinged with very dark salmon; spire whitish, with reddish brown spots in radiating rows, sometimes tinged with salmon; early whorls usually white, chalky. Aperture narrow, uniform in width; outer lip thin, sharp; mouth white. Columella very narrow, mostly internal. Length 40-70mm.

Comparison.—*C. monile* is very closely related to *C. generalis* and *C. bayani* in shape, nodulose early whorls, and lack of spiral sculpture on the spire whorls. From all patterns of *C. generalis* it is easily distinguished by the lack of a dark violet to black base and the rows of reddish brown squares instead of broad spiral bands and axial flammules of color. *C. bayani* lacks the salmon tints of *C. monile* and is covered with a streaky brownish pattern in long axial lines and some weak dashes in spiral rows; it is a thinner shell than *C. monile* and seldom as glossy. There is also a vague resemblance between *C. monile* and *C. sculletti*, but the latter species is much lighter in weight, has a rather domed spire with heavy axial sculpture, and has a banded pattern. *C. rufimaculosus* has a lower spire with carinate margins, is not as elongated, and has a mottled pattern. *C. voluminalis* is much lighter in weight, is differently colored and patterned, and has spiral sculpture on the spire whorls.

Variation.—Although rather constant in shape, there is considerable variation in pattern. In some specimens the spire is rather convex on the later whorls and the sides of the body whorl may be weakly convex. As far as pattern goes, at one extreme the whole body whorl is covered with large squares or triangles of reddish brown, the whole shell also being suffused with salmon. At the other extreme are nearly white shells with only a few scattered reddish brown spots and a touch of salmon at the base. Commonly axial flammules connect some of the rows of squares on part of the pattern, but these are usually very short. In uncommon patterns the squares are replaced with irregular blotches in roughly the same position as the normal spots and there may be narrow spiral bands of salmon separating the rows of blotches. It is probably an extreme variant of this type which was named *C. fasciatus* by Kiener.

Distribution.—Uncommon to locally common in moderately deep water of the northern Indian Ocean to the Philippines. Most specimens now come from the Bay of Bengal and Sri Lanka.

Synonymy.—*Conus marmoreus* Perry is usually placed here, but if so it applies to the *anadema* pattern; possibly it is unrecognizable. *C. anadema* and *C. fasciatus* are the pattern with the spotting heavily fused into blotches in two broad spiral bands.

Notes.—Good specimens of this species are not hard to obtain, although the lip is rather fragile and there may be heavy axial growth marks. *Anadema* patterns are very uncommon and seldom seen; partial fusion of the spots axially, however, is not uncommon.

For some reason this species has been confused with *C. sculletti* and *C. bayani*, both very distinct species.

MORELETI Crosse, 1858

1843. *Conus elongatus* Reeve. *Conchologia Iconica*, 1(*Conus*): Pl. 27, sp. 157 (Locality unknown). Non *Conus elongatus* Dillwyn, 1817.

1849. *Conus oblitus* Reeve. *Ibid.*, 1(*Conus* Suppl.): 1. Nomen novum for *Conus elongatus* Reeve, 1843. Non *Conus oblitus* Michelotti, 1847, a fossil.

1858. *Conus Moreleti* Crosse. *Rev. Mag. Zool.*, (Ser. 2), 10: 122. Nomen novum for *Conus elongatus* Reeve, 1843.

Description.—Moderately heavy, with a low gloss; obconical, the body whorl elongated and with straight or nearly straight sides that taper to a narrow base; body whorl with about a dozen spiral ridges above the base, these most widely spaced posteriorly and often coarsely granulate; occasionally the ridges extend to above midbody; many spiral and axial threads and growth marks usually present; shoulder angulate, broad, with close set rounded coronations; spire low, blunt, with convex or straight sides; spire whorls heavily coronate, the coronations numerous and rounded, close set; traces of 2-3 low spiral ridges on top of the whorls.

Body whorl dark to rather pale olive brown, often with indistinct axial streaks of a darker or lighter shade; base violet-brown; a yellowish or whitish spiral band at midbody, often interrupted and indistinct; shoulder area whitish, as is spire; spire with brown streaks between coronations; tip eroded violet. Aperture narrow, uniform in width; outer lip thin, straight; mouth deep violet with lighter bands at middle and below shoulder. Columella long but very narrow. Length 30-50mm.

Comparison.—The brown coloration and whitish spire resemble *C. lividus* or *C. balteatus*, but *C. moreleti* has a more elongated body whorl and more rounded, crowded coronations than these species, as well as differing in details of color and sculpture. The nature of the coronations reminds one of *C. distans*, but that species is much larger, has widely spaced grooves anteriorly, and has the base and midbody mottled with white.

Variation.—Other than intensity of color and its exact shade, there is little variation in this species. Older specimens may be paler than small ones and have indistinct axial flammules. The midbody band may be quite distinct or apparently absent, as may the white area below the shoulder. Occasional specimens have granulose ridges extending above midbody, while others have only a few weak ridges at the base and no granules.

Distribution.—Widespread and locally common from the African coast to Hawaii and French Polynesia. Shallow water.

Synonymy.—The first two names by Reeve were unfortunately preoccupied.

Notes.—This is usually a very rough and ugly cone with bad axial growth marks, irregular patterns, and often heavy erosion on the spire. The lip is rather fragile. Good specimens are found, however, some even true gems. Although often locally common, *C. moreleti* seems to be rather sporadically distributed and not found uniformly over the Indo-Pacific like some of its relatives; perhaps it must compete with *C. lividus* or some other species.

MUCRONATUS Reeve, 1843

?1843. *Conus orbitatus* Reeve. *Conchologia Iconica,* 1(*Conus*): Pl. 27, sp. 156 (Locality unknown). October.

1843. *Conus mucronatus* Reeve. *Ibid.,* 1(*Conus*): Pl. 37, sp. 204 (Islands of Burias, Siquijor, Penay, etc., Philippines). December.

1849. *Conus alabaster* Reeve. *Ibid.,* 1(*Conus* Suppl.): Pl. 6, sp. 257 (China Sea).

Description.—Moderately light, with a low gloss; low biconical, the sides usually straight or nearly so; body whorl heavily sculptured with many rounded narrow spiral ridges from base to shoulder, the intervening grooves deep, wide, and with heavy axial ridges and sometimes secondary spiral ridges within; ridges tend to be widest anteriorly and more widely spaced posteriorly; ridges often appear to be in pairs and may be granulose; shoulder broad, carinate; spire moderate, sharply pointed, the sides concave; early whorls not nodulose; whorls slightly concave above, with about 4-6 strong spiral ridges crossed by weak axial threads; early whorls rather distinctly cancellate posteriorly. Body whorl dirty white, without pattern or with irregular light brown spots and streaks tending to form broad indistinct spiral bands below shoulder and above midbody, sometimes also above base; whole pattern when present nebulous and mottled; base pale; spire heavily or weakly reticulated with light brown spots and streaks, these sometimes absent; early whorls tan. Aperture moderately wide, rather uniform in width; outer lip thin, straight; mouth white, sometimes with very pale violet tone. Columella internal. Length 35-50mm.

Comparison.—Although vaguely similar to *C. sulcatus* and *C. grangeri* in sculpture and sometimes pattern, these species are larger, heavier, and have coronate or undulate spire whorls. In *Conus australis laterculatus* the ribs are flattened, closely spaced, and usually have large brown dashes on them; the body whorl is more elongated and the spire whorls usually have only 3-4 spiral ridges on top; the mouth is usually distinctly violet. *Conus tegulatus* variety *batheon* is very similar to *C. mucronatus* and sometimes difficult to distinguish if specimens intermediate to normal *C. tegulatus* are not present. However, in *C. tegulatus* var. *batheon* the grooves seem to lack axial threads and are almost smooth; the body whorl is more convex posteriorly and sometimes distinctly inflated; and the pattern, when present, is of dark brown streaks and spots much more distinct than in *C. mucronatus*; in addition, the spire of *C. tegulatus* is more slender and sharply pointed.

Variation.—There is considerable variation in the strength and relative sizes of the spiral ridges, but they are generally narrow and rounded, sometimes paired. Axial threads are usually conspicuous. The shoulder may be very broad and with a projecting carina or narrower and weakly carinate. In typical specimens from the Philippines the body whorl is rather heavily covered with pale brown blotches and lines which cover most of the whorl. In the China Sea the brown is commonly restricted to a broad band above midbody and below the shoulder. Solid white specimens are common in the China Sea but occasionally are found in the Philippines.

If desired, the typical Philippine shells which are more uniformly covered with brown and rather narrower and more elongate than China Sea specimens could be recognized as *C. mucronatus mucronatus*, with the white or reduced pattern shells from the China Sea called *C. mucronatus alabaster;* such shells are a bit wider and heavier than Philippine specimens. The distinction is hardly worth making, however.

Distribution.—As far as known, restricted to the China Sea near

Taiwan and probably southern Japan and to the Philippines. Usually dredged in moderately deep water. Common. Probably more widely distributed.

Synonymy.—*C. orbitatus* probably belongs here but is poorly figured and apparently based on a juvenile; it is older than *C. mucronatus* by two months. There seems little doubt that the brownish *mucronatus* intergrades fully with the white *alabaster* and the two are conspecific. As the patterns seem to be rather constant in the China Sea and Philippines and the shell shape is slightly different, they can be retained as subspecies.

For some reason the nomenclature of this species has been very confused. *C. mucronatus* is often applied to the shell here called *C. scalptus;* although belonging to the same group of shells, the two are very different in shape, sculpture, and color pattern. *Alabaster* was described as a solid white shell, yet the name has been applied to the mottled Philippine specimens; more typical *C. alabaster* from the China Sea have been called *C. pseudosulcatus*, a fossil name given to a shell which was probably very close to *C. grangeri.*

Notes.—This is a common shell and is usually in good condition although there may be minor scars on the body whorl and spire and the lip may be chipped; commonly there is asphalt on the spire. Solid white shells are not uncommon off Taiwan but are seldom sold for some reason. *C.* 'alabaster' from India is *C. tegulatus* variety *batheon.*

MURICULATUS Sowerby i, in Sowerby ii, 1833

1833. *Conus muriculatus* Sowerby i, in Sowerby ii. *Conchological Illustrations:* Pt. 24, figs. 1-1* (Masbate, Philippines).

1833. *Conus muriculatus* var. *laevigata* Sowerby i, in Sowerby ii. *Ibid.:* Pt. 24, fig. 1 (Masbate, Philippines). Non *C. laevigatus* Link, 1807.

1844. *Conus sugillatus* Reeve. *Conchologia Iconica*, 1(*Conus*): Pl. 45, sp. 247 (Locality unknown).

1848. *Conus floridulus* A. Adams and Reeve. *Zool. Voy. Samarang, Moll.*, 1: 18, pl. 5, figs. 9a-9b (Locality unknown).

1857-1858. *Conus tenuis* Sowerby ii. *Thesaurus Conchyliorum*, 3(*Conus*): 3, pl. 14(200), fig. 314 (Locality unknown). Non *Conus tenuis* Sowerby i, 1833.

Description.—Moderately light in weight with a good gloss; low conical to low biconical, the sides usually nearly straight or slightly convex posteriorly; body whorl often with just a few weak spiral ridges at the base, but sometimes covered to shoulder with low ridges; ridges often heavily granulose at base and sometimes to shoulder; granules widely spaced both spirally and axially; often fine spiral and axial threads over most of body whorl, these seldom conspicuous; shoulder wide, carinate or sharply angled, usually low coronate or undulate; spire low to moderate, sharply pointed, the sides slightly concave to straight; whorls with weakly coronate or undulate margins, sometimes almost smooth; tops of whorls slightly concave, with 2-4 weak spiral ridges crossed by heavy curved axial threads, the sculpture sometimes obsolete on later whorls. Body whorl whitish to cream or very pale straw, the base broadly stained dark violet to black both dorsally and ventrally; usually two broad spiral bands of olive brown to bluish olive, the one above midbody almost twice as wide as the one below; narrow clear bands usually present above basal staining, at midbody, and at the shoulder; brownish spiral bands often broken into axial blotches or clouded with lighter and darker brown flammules; most shells show at least traces of numerous spiral rows of fine dark brown dashes or interrupted lines, these sometimes very conspicuous or completely absent; shells often with strong bluish tones replacing the browns; spire and shoulder whitish, the shoulder and last whorl often heavily marked with large curved blotches of dark brown, but these may be almost absent; early whorls whitish. Aperture moderately narrow, widened anteriorly; outer lip thin, sharp, almost straight; mouth violet to brownish violet, usually with wide white bands at midbody and below shoulder, occasionally the violet reduced and the light bands joined to produce a white mouth with two violet spots deep within. Columella internal. Length 20-50mm.

Comparison.—If the pattern is well developed, this is a quite distinctive shell. It bears a resemblance to *C. flavidus* and *C. lividus*, but these species usually lack the carinate shoulder and are either smooth or heavily coronated, not weakly coronate to undulate as in *C. muriculatus*. *C. frigidus* has a more rounded spire which is not undulate; it also lacks the fine spiral lines. *Conus parvulus* has more convex sides, heavy spiral ridges on the body whorl, and large, rounded coronations; it is covered with fine brown flecks, not interrupted lines. *C. polygrammus* is lighter in weight, lacks the dark brownish violet base (although it may be pale violet), is more pastel in color, and has a very high gloss; the spire whorls are usually only weakly undulate.

Variation.—In *C. muriculatus* variation must be broken into variation of shape, sculpture, and pattern. The shape is usually rather broadly low conical with the shoulder carinate, but the spire varies from almost flat to quite high and stepped; specimens with very high spires give the impression of being distorted and may occur with otherwise normal specimens. Large specimens seem to have the body whorl wider relative to length than smaller shells, which are somewhat elongate. The sculpture is usually weak and restricted to basal ridges, but there may be granulose ridges at the base; the granules may extend to the shoulder but are not uncommonly present only ventrally or dorsally and seldom cover the entire shell evenly.

The banded pattern is usually fairly distinct, although the shade of brown varies from bright reddish brown to pale bluish olive; usually it is rather yellowish olive. The base is broadly stained with dark violet or brownish violet, but even this is sometimes greatly reduced and very pale. It is not uncommon for the broad spiral band above midbody to have axial brownish flammules, and the band may break into long wide flammules in some shells. Occasional shells are mostly bluish. The spiral lines of dashes are usually visible somewhere on the body whorl and are often conspicuous and almost complete lines; often, however, they are very weak or apparently absent.

Some people recognize three species or varieties based on some of the above variations. In *C. muriculatus* typical the shell has some granules on the body whorl; these are absent in the variety *sugillatus*. Shells with higher than normal spires and/or the posterior spiral band broken into axial blotches are referred to *C. floridulus*. In my opinion these variants are not worthy of recognition as they are either insignificant or inconstant. Thus many other shells occur in granulose and non-granulose variants, but in *C. muriculatus* the granules are often very poorly and sporadically developed, individually variable in shells from the same locality. The recognition of *floridulus* must depend on the correlation of a high spire with a blotchy posterior band; however, it is generally conceded that blotches are sometimes found in shells with low spires, with some high-spired shells having continuous and regular posterior bands. Admittedly, there are indeed two types of shells in *C. muriculatus:* one is a small, rather elongate shell generally with a low spire and heavily developed brownish bands; the other is larger, wider, sometimes with a moderately higher spire, the bands sometimes poorly developed. It seems that both these types can occur with or without granules.

In my opinion, if varieties must be recognized, they should be listed as *C. muriculatus*, shallow form, for the smaller elongate shells, and *C. muriculatus*, deep form, for the wider and larger specimens. Beyond this one is dealing with individual variants.

Distribution.—Moderately common in the western Pacific and apparently also found in the eastern Indian Ocean, where not common. Found in shallow to moderately deep water.

Synonymy.—See discussion of variation for the reasons for synonymizing *C. sugillatus* and *C. floridulus* with *C. muriculatus*.

Notes.—The shallow form is easily available from several localities, but good specimens of the deep form are rare. Occasionally specimens from shallow water in the Philippines that have very high and stepped spires are sold at a high premium. Many of the more expensive specimens of this species are really very ugly shells with poor patterns and distorted shapes. Heavy axial growth marks are common, as is erosion of the spire.

MUS Hwass, in Bruguiere, 1792

1792. *Conus mus* Hwass, in Bruguiere. *Cone*, in *Ency. Method., Hist.*

Nat. des Vers, 1: 630 (Guadeloupe). Lectotype (here selected): the specimen figured by Kohn, 1968, as the holotype.

Description.—Moderately heavy, dull; low conical, the upper sides convex, rather inflated, then pinched-in and rapidly tapering to the base; body whorl covered with closely spaced low spiral ridges, somewhat undulating, heaviest at base and sometimes obsolete below the shoulder; fine axial threads also present over the whorl; shoulder roundly angled, with low coronations, flat or concave above; spire low, sharply pointed, the sides concave to convex; spire whorls concave or flat above, the margins with rounded coronations; sculptured with about 4 strong spiral ridges crossed by numerous fine axial threads. Body whorl usually bluish gray or brownish gray, dirty white at the base, midbody, and shoulder; usually many brownish and white spiral dashes over whorl, these sometimes forming heavy axial flammules of brown and white, the flammules usually broken at midbody; shoulder whitish with heavy brown blotches on top and dark brown spots between coronations; spire dirty whitish with scattered brown lines and spots and larger spots between coronations; early whorls white to pinkish. Aperture moderately narrow, of about uniform width; outer lip thin, concave at middle; mouth white with two broad bands of dirty brownish violet, these sometimes reduced or absent. Columella not visible externally. Length 20-45mm.

Comparison.—Although an undistinguished shell, *C. mus* is usually readily recognizable. *C. gladiator* from the Panamic area is perhaps its closest relative, but that species differs in numerous details such as the weaker sculpture, broader and more angled shoulder, straighter sides, and general absence of bluish gray tones. *Conus parvulus* is similar in shape but has the base deep blackish violet and has numerous fine brown flecks over the body whorl. *C. biliosus* is vaguely similar but is more elongate, has a different spire outline, and has several different aspects of color and pattern.

The relationship of *C. patae* with *C. mus* is undetermined. The two are similar in general shape and at first glance the sculpture appears similar. However, in *C. patae* there are many heavy axial plicae over the body whorl and the shoulder and whorl margins are at most undulate, not coronate. The yellowish color of *C. patae* should be distinctive, but I have seen yellowish juveniles of *C. mus*.

Variation.—Although quite variable in color, the pattern of *C. mus* is usually of indistinct axial flammules and some dashes. Usually bluish gray or brownish gray in tone, the flammules are commonly darker brown or sometimes olive. Some specimens are uniformly gray with only a poorly defined midbody area of white, while others are heavily flammulate. I have seen yellow specimens of this species, but they are apparently not common. The dark spots between coronations are sometimes absent. The spiral ridges on the body whorl are usually distinct. Specimens from Bermuda may be very large and inflated.

Distribution.—Widespread and locally common in the Caribbean from southern Florida and the Bahamas south at least to Trinidad. Also in Bermuda. Shallow to moderately deep water.

Notes.—This is usually an ugly shell with eroded and chalky spire, bad axial breaks and growth marks, irregular and disrupted patterns, and chipped lips. Florida specimens seem to be smaller now than previously. *C. citrinus* was once used for this species but is now referred to *C. regius*, probably incorrectly.

MUSICUS Hwass, in Bruguiere, 1792

1792. *Conus musicus* Hwass, in Bruguiere. *Cone*, in *Ency. Method.*, *Hist. Nat. des Vers*, 1: 629 (China). Lectotype figure (Kohn, 1968): *Tableau Ency. Method.*, Pl. 322, fig. 4.
1792. *Conus Ceylanensis* Hwass, in Bruguiere. *Ibid.*, 1: 636 (Ceylon). Holotype figure: *Tableau Ency. Method.*, Pl. 322, fig. 10.
1845. *Conus Mighelsi* Kiener. *Species gen. et icon. des coqu. viv.*, 2 (*Conus*): Pl. 103, fig. 1. 1849, *Ibid.*, 2: 352 (Mer des Indes).
1857. *Conus acutus* Sowerby ii. *Thesaurus Conchyliorum*, 3(*Conus*): 16, pl. 6(192), fig. 142 (Ceylon).
?1887. *Conus exquisitus* Sowerby iii. *Ibid.*, 5(*Conus* Suppl.): 274, pl. 36 (512*), fig. 757 (California).

Description.—Heavy for its small size, with a low gloss; low conical, the sides nearly straight; body whorl with a few ridges above the base, these sometimes continued over body whorl to midbody or even shoulder as widely spaced spiral lines bearing heavy granules; shoulder broad, angulate, strongly to weakly coronate; spire low, blunt, the sides straight or convex; spire whorls coronate though often eroded, and with traces of two low spiral ridges. Body whorl white, the base purple; numerous spiral rows of large to moderate or small dark brown to blackish spots over rest of whorl, these variously developed and often partially absent; often a blotchy or complete pale reddish band at midbody and axial flammules of pale tan developed; shoulder with white coronations and large black blotches or smaller brown spots between; spire whorls white with similar spotting between coronations, sometimes eroded white. Aperture narrow, uniform in width or very slightly wider basally; outer lip thin, straight; mouth white toward margin, purple inside with two broad purple bands running toward edge above and below middle. Columella not visible. Length 10-29mm.

Comparison.—This attractive little shell has often been considered a subspecies or variety of *C. sponsalis*, but the two are broadly sympatric and quite distinct. *C. sponsalis* is a larger, more convex-sided shell that lacks the strong spotting of *C. musicus* and instead has (in fully developed specimens) broad wavy reddish axial flammules. I know of no intermediates. Also similar is *C. axelrodi*, which differs greatly in structure; in that species the body whorl is deeply grooved from base to shoulder, with the granules on the flat ridges between grooves; the spire whorls are weakly stepped and have about four distinct spiral ridges on top crossed by many fine axial threads; the pattern may be similar to some *C. musicus*, with scattered, rather axial rows of small brown spots, but the background color varies from straw to yellowish or pinkish, rarely white, and the base is not purple.

Variation.—Although quite variable, only two or three variants are at all commonly seen. In one variant, called *mighelsi*, the axial reddish flammules tend to be fused into a broad reddish band at midbody; intermediates with typical *C. musicus* are common in the Pacific.

It seems probable that *C. musicus* can be divided into an Indian Ocean subspecies and a Pacific subspecies. At the moment there seems to be no available name for the Indian Ocean subspecies, but it seems likely that one will eventually be discovered in the doubtful names. For this reason the subspecies are not formally named here.

C. musicus (Pacific): Large black spots between coronations on spire and shoulder; body spots in many spiral rows, rather small, regular; often a reddish band or row of flammules at midbody and another above the base, these sometimes connected to form indistinct axial bands; spire sides straight to slightly concave; often granulose to midbody or even shoulder. Pacific Ocean, excluding Hawaii and French Polynesia.

C. musicus (Indian): Small pale brown spots between coronations, these sometimes absent; body spots large, in relatively few rows and sometimes vaguely axial in alignment (spots smaller at base); reddish midbody band not developed; spire often with convex sides and flat on top; granules rare beyond the basal ridges. Indian Ocean from Natal coast to at least Sri Lanka.

Indian Ocean subspecies of *C. musicus,* Natal, South Africa, 15-18mm.

There is considerable individual variation in each subspecies, mostly in size of spots, their intensity, and markings on the spire. The Pacific form often has axial brownish flammules developed but the spots or dashes are still present. Some populations lack either reddish or brownish flammules and have only dashes for a pattern.

Distribution.—Found in shallow water from the Natal coast through the Indian Ocean and the western Pacific to Fiji; apparently absent or overlooked in Hawaii and French Polynesia. Sometimes locally common, but apparently rare or at least uncommon in many parts of its range.

Synonymy.—Although *C. ceylanensis* is sometimes used as the name of the Indian subspecies, the type figure shows a fairly normal Pacific *musicus* with wavy axial flammules, a not uncommon pattern. *C. pusillus* Lamarck is also applied to the Indian shell, but that name is possibly based on a juvenile of *C. catus* or a similar species and is not *C. musicus*.

C. mighelsi appears to be strictly an individual variant and occurs sporadically in the western Pacific. *C. acutus* is probably a Pacific *C. musicus* with an unusually high and concave-sided spire. *C. exquisitus* is covered with two series of angular flammules and could just as easily belong with *C. sponsalis* as with *C. musicus*. It is certainly not from California.

Notes.—Except for specimens from the Philippines, this is not an easy species to obtain. Even when it is common, few collectors waste their time with it because of its small size and supposed abundance throughout the Indo-Pacific. It appears to be rare in some areas and is certainly not common in many others. The Indian Ocean subspecies is especially difficult to obtain.

The lip is thin and breaks easily. In the Pacific form the shell is usually clean and the spire is not badly eroded. In the Indian form there are often major breaks on the body whorl and the spire is commonly heavily eroded. Specimens of the *mighelsi* variety with bright red markings commonly sell for a small premium. I have not seen any specimens over about 25mm and wonder if the 29mm shell might be *C. sponsalis*, which is usually larger than *C. musicus*.

MUSTELINUS Hwass, in Bruguiere, 1792

1792. *Conus mustelinus* Hwass, in Bruguiere. *Cone*, in *Ency. Method., Hist. Nat. des Vers*, 1: 654 (Indian Ocean). Lectotype figure (Kohn, 1968): *Tableau Ency. Method.*, Pl. 327, fig. 6.

1964. *Conus* (*Rhizoconus*) *melinus* Shikama. *Venus*, 23(1): 36, pl. 3, figs. 3-6 (Arafura Sea).

Description.—Moderately light in weight, with a good gloss; low conical, the upper sides nearly straight; base of body whorl with about a dozen low spiral ridges, these heaviest anteriorly, the grooves between crossed by axial threads to produce large punctations; more posteriorly the ridges become flat but the punctations between remain visible over the anterior fourth or so of the adult shell; remainder of body whorl smooth except for weak spiral and axial threads and occasional small punctations; shoulder wide, roundly angled, not distinct from spire; spire low to moderate, with straight or convex sides, bluntly pointed; tops of whorls flat or slightly convex, with about 4-6 spiral ridges separated by continuous or weakly punctate grooves. Body whorl pale yellowish tan to greenish olive, with a distinct white midbody band and a broad white band below the shoulder; the midbody band is usually marked with small, rather oval black spots which are evenly spaced and about equal in size and without connecting axial flammules; the shoulder band commonly has a series of larger black axial blotches which lead into the large reddish or blackish brown blotches on the shoulder; there are often narrow and incomplete oblique brown lines between the blotches below the shoulder; base about same color as remainder of shell, with only a few or no fine brown flecks; a row of large black dashes may be present below the midbody band, and sometimes there are 2-3 similar bands below the shoulder, but these are not connected by flammules or spirally fused; spire white, the tip yellowish, covered with large squarish reddish brown to blackish blotches equal to or larger than the white areas between them. Aperture moderately wide, uniform; outer lip thin, sharp, nearly straight; mouth purplish with two wide whitish bands below shoulder and at midbody, sometimes evenly suffused with violet or just pale bluish white. Columella long and very narrow, oblique. Length 40-75mm.

Comparison.—*C. mustelinus* could be confused with small *C. vexillum*, which, however, are largely lacking in punctate sculpture, have only a few dark blotches on the spire, and have a less regularly spotted midbody band. The real confusion occurs when trying to separate *C. mustelinus* from *C. capitaneus*. Undoubtedly the two are very similar and occasional specimens are hard to place, but there is no reason to consider the two taxa as conspecific. As a rule, *C. mustelinus* is narrower and more elongated with a distinctly narrower shoulder; the spire is commonly more convex than in *C. capitaneus*. Juveniles of *C. mustelinus* have the punctations poorly developed or absent above midbody and not reaching the shoulder as in small or subadult *C. capitaneus*. The base of *C. mustelinus* commonly lacks brown flecks or has them reduced in number compared to *C. capitaneus*.

Perhaps the easiest way of separating the two is by the clean appearance of *C. mustelinus* compared to the smudged and axially marked *C. capitaneus*. In *C. mustelinus* the spots in the midbody band are sharp and regular, without connecting lines or flammules, while the dashed bands above midbody are few in number, sharp, and not connected by axial lines as is the common situation in *C. capitaneus*. The background color is not a constant difference between the two, although *C. mustelinus* is usually a clear pale yellowish brown to distinctly greenish shell, seldom or never the darker tans common in *C. capitaneus*.

Variation.—Other than height and shape of spire, as well as shade of background color, this is a constant species. There is a good deal of variation in sculpture with age. Juveniles have the heavy basal ridges extending well up the shell and followed by rows of small punctations which do not reach the shoulder as distinct rows. With growth the ridges and punctations become relatively more restricted until they are found only on the basal area. Juveniles also seem to have a more distinctly purplish and white-banded mouth which in large adults may become a dirty bluish white. The yellow and green shades are apparently not related to age, and neither is the shape or height of the spire.

Distribution.—Widely distributed in the Indo-Pacific but seldom common. Apparently not reaching Hawaii and French Polynesia. Usually found in shallow water.

Synonymy.—*C. melinus* appears to be a rather normal *C. mustelinus*, with which it was not compared by Shikama.

Notes.—The lip is fragile and the spire commonly eroded or encrusted, but this is still a very attractive shell, especially when in a bright greenish phase. Juveniles are common. Most specimens are rather brownish mustard in tone rather than the more distinctive green.

NAMOCANUS Hwass, in Bruguiere, 1792

1792. *Conus Namocanus* Hwass, in Bruguiere. *Cone*, in *Ency. Method., Hist. Nat. des Vers*, 1: 712 (Namoca I., Pacific Ocean). Holotype figured by Kohn, 1968.

1810. *Conus nemocanus* Lamarck. *Ann. du Mus. Hist. Nat.* (*Paris*), 15: 422. Emendation or misspelling.

1846. *Conus badius* Kiener. *Species gen. et icon. des coqu. viv.*, 2 (*Conus*): Pl. 33, fig. 3. 1847, *Ibid.*, 2: 89 (Locality unknown).

1857-1858. *Conus laevigatus* Sowerby ii. *Thesaurus Conchyliorum*, 3 (*Conus*): 27, pl. 7(193), figs. 149-150; pl. 9(195), fig. 207 (Mauritius). Non *Conus laevigatus* Link, 1807.

Description.—Rather light in weight for its size, with a good gloss; low conical, the upper sides slightly convex or straight and tapering evenly to the base; body whorl with a few low and sometimes indistinct spiral ridges above the base and numerous fine spiral and axial threads and growth marks, some conspicuous; shoulder roundly angled, sometimes more sharply angled, flat or slightly concave or convex above; spire low, the sides convex, the whorls sometimes weakly stepped; bluntly pointed; tops of whorls slightly concave to slightly convex, earlier whorls with about 3-5 distinct spiral ridges crossed by numerous fine axial threads, sculpture usually indistinct by shoulder. Body whorl whitish to yellowish, heavily overlaid and suffused with olive or orange-tan, either as uniform broad spiral bands or as irregular axial flammules; usually a white or at least pale midbody band bordered above and below by dark squarish spots; base often clear pale orange-tan; axial flammules sometimes bright reddish brown; fine spiral rows of brown dots usually present at the base, these becoming fine undulating brown lines farther up on the whorl; lines often indistinct except on new growth, but usually present somewhere in the pattern; shoulder and spire white with large revolving blotches of reddish brown; early whorls pale yellow or white. Aperture moderately wide, uniform in width; outer lip straight, very thin and fragile; mouth pale to dark violet, usually with the white midbody band showing through and the new growth at the edge of the lip bright yellow. Columella long, very narrow. Length 40-100mm.

Comparison.—Although similar to *C. vexillum* in general shape and some aspects of the pattern, *C. namocanus* is usually easily distinguishable by the very thin and rather fragile shell for a large cone, the very blurred color pattern, and the normal presence on at least part of the shell of fine spiral brown lines; *C. vexillum* is also commonly dark at the base, *C. namocanus* light. *C. capitaneus* in some patterns might possibly be confused, but it has a broader shoulder in many populations and generally has remnants of the spiral rows of punctations on the body whorl, as well as many fine brown flecks basally.

Variation.—Although the body whorl pattern is very indistinct and variable, it usually falls into a few categories. In the best marked phase the body whorl is largely olive tan to orange brown or rather yellowish, sometimes reddish brown, with a narrow midbody white band and a pale base; there are numerous fine brown spiral lines over most of the whorl. More commonly, however, the lines are indistinct or absent except on new growth or at the base, and the straw to whitish background is heavily covered with wide axial flammules of olive tan, orange tan, yellow, or faded reddish brown; the midbody band is more-or-less scalloped and margined by rather irregular rectangular or squarish spots that are continuations of the flammules; there may be light shadows along axial growth lines all over the body whorl; the base is light. At the pale extreme are white or pale yellow shells with narrow orange or tan axial flammules which go through the midbody area; much white shows between the flammules, and the spiral lines are very indistinct or even absent. In all patterns the spire pattern consists of rather large reddish brown blotches.

Distribution.—Moderately common in shallow water throughout the Indian Ocean from the Bay of Bengal and Burma south to the Indian Ocean islands and over the African coast. There are repeated records for this species from the entire Pacific Ocean except Hawaii, but I have never seen specimens from any Pacific locality; sometimes small *C. vexillum* are called *C. namocanus.*

Synonymy.—The correct spelling of the name is 'namocanus', with an a, not an e. There is no problem putting *C. badius* and *C. laevigatus* within the normal range of variation of *C. namocanus.*

Notes.—Fully patterned specimens with distinct striping are not especially common. The spire is often badly eroded and encrusted, while in many populations the lip is extremely fragile and breaks into large slivers if touched. It is not an especially distinguished looking cone, although specimens with brighter orange-tan patterns are attractive.

NATALIS Sowerby ii, 1857-1858

?1854. *Conus succinctus* A. Adams. *Proc. Zool. Soc. (London)*, 1853: 118 (Natal).

1857-1858. *Conus Natalis* Sowerby ii. *Thesaurus Conchyliorum*, 3 (*Conus*): 31, pl. 13(199), figs. 292-293 (Cape Natal).

1892. *Conus natalensis* Sowerby iii. *Marine Shells South Africa:* 29. Emendation of *Conus natalis* Sowerby ii.

1903. *Conus Gilchristi* Sowerby iii. *Marine Invest. South Africa,* 2(3): 217, pl. 3, fig. 8 (Umhlangakulu River mouth, Natal, South Africa).

Description.—Moderately light in weight, solid, with a good gloss; low conical, the sides slightly convex; body whorl with a few (6-10) low spiral ridges at the base and axial threads; shoulder roundly to narrowly angulate, broad, not very distinct from spire; spire low to moderate, bluntly pointed, the sides straight or convex; whorls occasionally slightly stepped; spire whorls slightly concave or flat above, with about 2-3 narrow spiral ridges usually visible; whorls occasionally rude. Body whorl white to pinkish or straw, covered from shoulder to base with fine axial zigzag lines which overlap to produce many small tents; usually two broad spiral bands of pale brown blotches which are commonly connected to produce solid bands, one above midbody, one above base; other narrow and broken spiral brown bands usually present, usually 1-3 above the midbody area, at midbody, and (indistinct) at the base; base netted like rest of body whorl; strong tendency to reduce all pattern, resulting in pinkish or ivory shells with two spiral bands of oval reddish brown spots above and below midbody, the remnants of the spiral bands; in such reduced patterns there may be remnants of the axial lines or even large axial dark bands; shoulder and spire whorls white, heavily covered with large brown blotches so often only small white triangles are visible; brown of spire does not erode easily and remains visible after most of the body pattern is lost; early whorls white. Aperture moderately wide, slightly widened anteriorly; outer lip straight or slightly convex; mouth white to pale bluish white. Columella narrow, short. Length 25-52mm.

Comparison.—This distinctive species has sometimes been confused with the *textile* or *pennaceus* cones, but its lack of relationship is shown by the lack of nodules on the spire and the lack of a projecting protoconch. It is closely related to *C. infrenatus*, which usually appears to have a pattern of numerous spiral rows of dots and dashes of brown to violet or pink and of various sizes; the spire whorls are largely whitish, not brownish as in *C. natalis*. The brown spire whorls should distinguish even bad beach *C. natalis* from most other South African beach cones.

Variation.—There is considerable variation in width at the shoulder, from rather broad with convex sides to narrower with concave or straight sides. The spire is usually low but is occasionally tall and irregularly stepped. The tented pattern is very variable in density, as are the large and small spiral bands, which is to be expected with such a pattern. The major variation occurs in the reduction of the pattern from the heavily tented one to pale pinkish or ivory shells with two spiral rows of large oval brownish spots. Fortunately there are many intermediates between the tented and spotted patterns. However, there is certainly no harm in recognizing such reduced pattern shells as *C. natalis* variety *gilchristi; gilchristi* is often somewhat narrower and higher-spired than typical *C. natalis*, but this probably just represents the bias of the sample.

Distribution.—Restricted to South Africa from the Natal coast south to Port Alfred. Although probably a shallow-water species, it is seldom taken alive. Common, but less so than *C. tinianus*.

Synonymy.—*C. succinctus*, an older name than *C. natalis*, is quite possibly this species, but if so it represents an extreme shell of the *gilchristi* type, being pinkish with two spiral rows of small oval reddish brown spots. *C. gilchristi* almost certainly belongs here, but it must be pointed out that the type of *C. gilchristi* was a small shell dredged in deep water and still having periostracum; the pattern is heavier than the beach shells usually called *gilchristi* and tends to be rather axial; the shell is also rather wide for *C. natalis*, but probably falls within the normal range of variation of the species.

Notes.—Good beach specimens of both varieties are not hard to get, although the usual broken lips must be expected. Beach specimens commonly still have a good gloss and distinct spire sculpture. There may be axial growth marks which disrupt the pattern, but this is usually a very clean and attractive cone. The brown spire whorls are distinctive.

The *gilchristi* variety usually draws a small premium, and it does seem to be less common than the typical variety. Live-taken specimens seem to be rare, although the species must live in rather shallow water as the beach specimens are very well preserved.

NEPTUNUS Reeve, 1843

1843. *Conus Neptunus* Reeve. *Conchologia Iconica*, 1(*Conus*): Pl. 6, sp. 30 (Jacna, Island of Bohol, Philippines).

1880. *Conus neptunoides* E.A. Smith. *Proc. Zool. Soc.* (*London*), 1880: 479, pl. 48, fig. 2 (Australia).

Description.—Moderately light in weight, with a high gloss; low biconical, the sides nearly straight or slightly convex posteriorly, tapering; body whorl with deep spiral grooves over entire body whorl or at least to midbody area, the grooves with very fine axial threads; whole whorl with very fine axial scratches; shoulder moderately broad to quite broad, rather sharply angled, with a weak carina; spire moderate, sharply pointed, the sides shallowly concave; early four whorls with fine nodules, the later ones carinate; tops of whorls flat or slightly concave, with about four distinct spiral ridges crossed by fine curved axial threads. Body whorl creamy white with distinct yellowish to pinkish tinges; numerous wavy and zigzag fine reddish brown lines from shoulder to above base, these forming many small but very irregular and open tents; two spiral rows of brown blotches above and below midbody, these sometimes weakly developed or fused into nearly complete spiral bands; base stark white, contrasting; shoulder and spire colored like body whorl, covered with many fine reddish brown lines and rather small blotches; early whorls white. Aperture moderately narrow posteriorly, widened anteriorly; outer lip thin, fragile, slightly convex; mouth pale to deep pink. Columella long, narrow, straight. Length 30-50mm.

Comparison.—*C. neptunus* is a very attractive cone combining a weakly tented pattern with distinct spiral grooving on the body whorl. The presence of nodules on the spire whorls will readily distinguish it from *C. wittigi*, which has a very similar pattern. *C. amadis* is perhaps related, but that species is larger, wider, has the spiral sculpture of the spire whorls weaker, and has a wider shoulder, as well as having a less delicate pattern with larger blotches and spiral dashes usually present. *C. textile* and allies are thicker, heavier, usually of different pattern and shape, and are seldom distinctly grooved.

The closest relative of *C. neptunus* is certainly *C. lienardi*, a species which shares nodulose early whorls and very open and delicate pattern. However, in *C. lienardi* the shoulder and sides of the body whorl are more rounded, the spiral sculpture of the whorls is weaker, the columella is bounded by a heavy ridge and rather twisted, and the pattern is more likely to be partially absent or greatly darkened. *C. lienardi* gives the impression of a cone generally similar to *C. cinereus* in shape and texture.

Variation.—I have seen only a few specimens of this species, but it seems to be very constant in pattern. The axial lines are always well developed and very delicate, with the two rows of brown blotches (actually just darker brown clouds only partially obscuring the lines beneath) well developed on at least one surface; they may even be partially fused and form rather loose bands. The base is strongly contrasting and unpatterned. Juveniles have much the same pattern as adults. The tone of the shell varies from white with pale straw or yellowish tints to distinctly pinkish.

There is some difference in shape between apparent adults from the Philippines and those from the Solomons. Philippine juveniles are grooved to about the shoulder and have a narrow and rather weakly carinate shoulder. In adults the grooving is restricted to the anterior half of the body whorl and the last two whorls have become widened, giving a broader outline to the shell; the shoulder appears more roundly angled than in juveniles. Adults from the Solomons are much like Philippine juveniles in sculpture and shoulder shape, but the sides are somewhat more convex posteriorly.

It is possible that when further island populations of this species are discovered in the western Pacific (they are almost certain to exist), they will differ slightly from both the Philippine and Solomons specimens.

The maximum size of this species is unknown, but I have examined body whorl fragments from the Solomons with a length of 45mm, so a total length of at least 50mm must be attained in that population.

An extremely beautiful, elongate shell closely resembling *C. neptunus* in sculpture and pattern was also examined from the Solomons and is tentatively assigned to that species. It is very thin and fragile, over 77mm in length, and has the pattern more open than in typical *C. neptunus* but otherwise differs little from that species. The base is white and starkly contrasting as in typical *C. neptunus* from the Solomons.

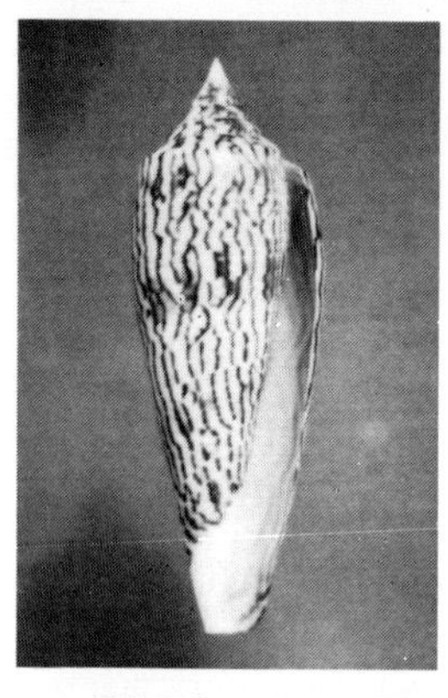

C. cf. *neptunus*, Solomon Is., 77mm (MG and BB).

Distribution.—Presently known to me only from deep water in the Philippines and Solomons; apparently quite rare. *C. neptunoides* was based on specimens from the Brazier collection; Brazier had much material from the Solomons, but the specimens are not the typical Solomons type; Brazier also had much material from the New Hebrides, and it is possible the species occurs there as well.

Synonymy.—*C. neptunoides* seems to belong with this species, but it does not match either population known to me. It does not appear to be *C. lienardi* in shape or pattern.

Notes.—This is a delicately beautiful shell which seems to be quite rare. Probably with further deep dredging more specimens and populations will be discovered in the western Pacific. At the moment specimens are worth whatever the market will bear. The lip is quite thin and chips easily.

NIELSENAE Marsh, 1962

1962. *Conus* (*Leptoconus*) *nielsenae* Marsh. *J. Malac. Soc. Australia*, No. 6: 40, pl. 4, figs. 1-2 (NE of Cape Bowling Green, near Townsville, Queensland, Australia).

Description.—Light in weight, with a high to moderate gloss; obconical or low conical, the sides slightly convex posteriorly but nearly straight to base; body whorl with about 4-10 low and indistinct spiral ridges at the base, otherwise smooth except for fine spiral and axial threads; shoulder broad, sharply angled, slightly concave above; spire nearly flat, the early whorls forming a slightly projecting cone in the middle of a nearly flat plate; tops of whorls slightly concave, with about 1-3 narrow spiral ridges crossed by many or few curved axial threads. Body whorl white, pale yellow, or tan, often covered with many closely to widely spaced spiral threads of brown, these quite regular; base occasionally uniformly pale tan, without lines, and lines may be absent from the entire shell; spire with few to many axial blotches of pale tan; early whorls white. Aperture moderately narrow, slightly widened anteriorly; outer lip straight or slightly concave, thin and sharp; mouth white to pale lavender. Columella long, very narrow, mostly internal. Length 30-61mm.

Comparison.—As treated here, *C. nielsenae* is a composite of three species of collectors: *C. nielsenae* from Queensland, *C. 'gilvus'* from southern Western Australia, and *C. 'clarus'* from northern Western Australia. The closest relative is apparently *C. typhon* from southern Africa. The two species differ in the more numerous spiral ridges on the spire whorls, the coarser axial threads of *C. typhon*, and the tendency of *C. typhon* to show two broad bands of color above and below midbody whether lines are present or not; the spire of *C. typhon* is higher and more prominent than in most *C. nielsenae* except those from Western Australia. In *C. nielsenae* the base is commonly pale tan if there is any pattern present, but in *C. typhon* the base is white or very pale tan regardless of whether other pattern is present or not; the spire markings in *C. typhon* are heavier, also.

Variation.—*C. nielsenae* is here considered to be a species with rather straight or sometimes concave sides to the body whorl, very low spire without strong sculpture, and a sharply angled shoulder. These characters hold constant regardless of pattern, although the three groups of populations differ somewhat in shape and size. They also differ in color, but for the moment they are considered to be conspecific. They are best defined separately.

Queensland (typical): Shell rather large, usually with at least some very indistinct to obvious spiral brownish threads over the body whorl; base commonly with a pale tan stain; sides often concave at middle; moderately deep water off Queensland, northern New South Wales, and the Kermadec Islands; to at least 61mm, commonly 45-55mm.

Northern Western Australia: Shell moderately large, without pattern; body whorl uniformly white or pale yellow; sides somewhat convex, the spire very low and even depressed; shallow water from the northern Kimberly region to Northwest Cape; usually 30-45mm, to 55mm.

Southern Western Australia: Shell moderately large, somewhat

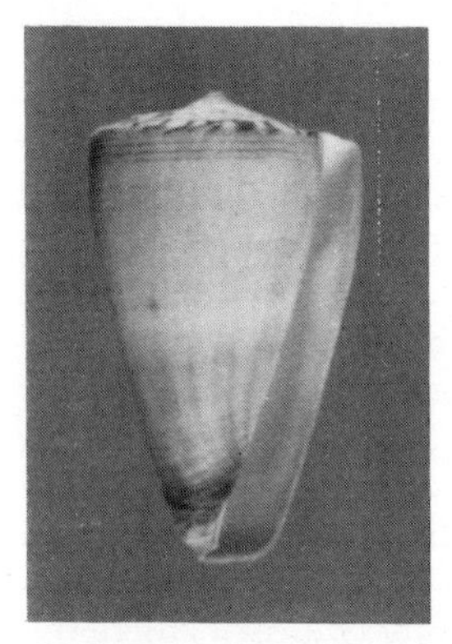

Southern Western Australia variant of *C. nielsenae,* lacking spiral lines, Broome, Western Australia, 37mm (K).

thicker than in the other forms; usually yellowish, covered with fine brown spiral lines over most of whorl or sometimes only posteriorly; basal half of whorl often uniformly pale tan, sometimes shell whitish rather than yellow; spire heavily marked, not depressed; shallow water from Albany to at least Geraldton; usually 25-40mm.

I believe these populations to represent a chain of slightly separated forms much like the variation shown by *C. kermadecensis.* Certainly there are no valid morphological characters to separate them, as they are commonly as variable within a population as between populations in number and strength of basal sculpture and spiral ridges. Probably all three forms are deserving of subspecific recognition at a later date when the distribution is better known and larger series are available.

Distribution.—Moderately uncommon along the coast of Western Australia, Queensland, and northern New South Wales, Australia; also in the Kermadec Islands. Shallow to moderately deep water.

Synonymy.—The use of the names *gilvus* and *clarus* for forms of this species seems incorrect; *C. gilvus* of Reeve appears to be a small shell with black spots at the shoulder and closely related to *C. frigidus; C. clarus* appears to be the proper name for the species usually called *C. segravei.*

Notes.—The typical form is very thin and fragile, commonly distorted and with broken lips and eroded spire whorls; the pattern is seldom well developed. The northern WAust. form is cleaner but also has a very thin lip and may have some axial growth marks. The southern WAust. form often has heavy axial marks and scars and may be badly eroded on the spire and body whorl; it is commonly found with three different patterns on various surfaces: solid pale yellow, yellow with spiral lines, and half lineate yellow with a tan base.

Of the various forms, only the typical form is hard to obtain; it is more expensive than the other forms.

NIGROPUNCTATUS Sowerby ii, 1857-1858

?1843. *Conus Metcalfii* Reeve. *Conchologia Iconica*, 1(*Conus*): Pl. 36, sp. 192 (Locality unknown).

1857-1858. *Conus Adansoni* Lamarck, Sowerby ii. *Thesaurus Conchyliorum*, 3(*Conus*): 38, pl. 13(199), figs. 286-289 (Red Sea). Misidentification of Lamarck's species.

1857-1858. *Conus nigro-punctatus* Sowerby ii. *Ibid.*, 3(*Conus*): 38, pl. 15 (201), fig. 342 (Locality unknown).

1860. *Conus Rollandi* Bernardi. *J. Conchyl.* (*Paris*), 8: 322, pl. 12, fig. 4 (Locality unknown).

?1898. *Conus Frostiana* Brazier. *Proc. Linnaean Soc. New South Wales*, 22: 781 (Solomon Islands).

1910. *Conus magus* var. *decurtata* Dautzenberg. *J. Conchyl.* (*Paris*), 58: 26 (Rua-Sura, Solomon Is.). Possibly intended as a replacement name for *C. adansoni* of Sowerby ii, 1857-1858.

Description.—Moderately light in weight but thick, with a good gloss; low conical, the upper sides convex; body whorl with heavy spiral ridges above the base, these usually obsolete before midbody but sometimes continued to near shoulder; basal ridges sometimes granulose; numerous spiral and axial threads over whorl; shoulder rather narrow, roundly angled, slightly concave above; spire low, sharply pointed, the sides slightly convex; tops of whorls slightly concave, with about 3-5 broad spiral ridges separated by punctate grooves and crossed by scattered axial threads. Body whorl opaque white to bluish white, covered with numerous spiral rows of fine dark brown dots and dashes; often with two broad yellow spiral bands of blotches above and below midbody area or with broad jagged-edged dark brown blotches above and below midbody area; brown or orange axial flammules may connect the two rows of blotches; base of shell whitish; spire mostly white with large or small dark brown to orange spots and blotches, sometimes the spots small and regularly spaced at the margins of the whorls; tip pink. Aperture moderately wide, especially anteriorly; outer lip thick but sharp, convex; mouth white or bluish white. Columella short and narrow. Length 20-41mm.

Comparison.—This little species is very close to *C. monachus* var. *achatinus* and I am not sure I can always separate the two, especially when compared with juvenile or small *C. monachus*. As a general rule, it seems to differ in its smaller adult size (usually about 30mm long) with thick and adult feel, the frequent presence of yellow or orange blotches in the color pattern or margining the dark blotches, the somewhat lower and more convex spire, and perhaps a narrower body whorl. As might be imagined, these differences are subject to interpretation and require series for accurate comparison; many shells sold as *C. nigropunctatus* are apparently juveniles of *C. monachus*.

This species is also very close to some variants of *C. magus*, with which it would share the orange blotching. Small *C. magus* are commonly sold as brightly colored *C. nigropunctatus*, but they can usually be distinguished by the broader form, much smaller dotting, and lack of punctate grooves between the spire ridges.

Variation.—This species is quite variable in color and pattern although fairly constant in shape and size. Very light shells are distinctly pearly white with the dashes rather pale; blotches are usually orange or bright yellow. Darker shells are bluish white with dark brown blotches that often are edged with a narrow orange or yellowish line. The spire blotches may be very irregular and large, or small and regularly spaced at the margin of the whorls; commonly they are either rusty brown or dark brown with rusty edges. Some specimens are completely covered with weakly granulose spiral ridges, but usually the ridges are not granulose and are restricted to the anterior end of the shell. Some specimens may be heavily flammulated with brown or orange obscuring the normal pattern, but all specimens seen have at least some spiral dashes.

Distribution.—Widely distributed and often common in shallow waters of the Indo-West Pacific. Exact eastern limits unknown, but probably extends to at least Fiji.

Synonymy.—If *C. metcalfii* is this species, it represents a specimen with granulose ridges to the shoulder, an uncommon occurrence in *C. nigropunctatus. C. decurtata* seems to be based on the concept of *C. adansoni* of Sowerby, and probably applies to this species; the name would fall as a synonym of *C. nigropunctatus* because it is based on Solomons specimens. Red Sea specimens figured by Sowerby possibly represent a distinct species or subspecies because the spire markings are small and regular, but no name appears to be available for such a shell. *C. rollandi* and *C. frostiana* are probably this species, but they could just as easily be small *C. monachus* or even *C. magus.*

Notes.—Probably most of the shells sold under this name are misidentified unless they have distinctly yellowish or orange markings. This is a very clean shell in most areas and easy to find in gem. Be sure to double-check all specimens purchased for juveniles of *C. magus* and *C. monachus.* Red Sea specimens are worth a premium.

NIMBOSUS Hwass, in Bruguiere, 1792

1792. *Conus nimbosus* Hwass, in Bruguiere. *Cone,* in *Ency. Method., Hist. Nat. des Vers,* 1: 732 (East Indies). Lectotype selected and figured by Kohn, 1968.

(1802. *Conus tenellus* Holten. *Enumeratio Systematica Conchyl. beat Chemnitzii:* 39 (Locality not stated). Holotype figured by Cernohorsky, 1974, *Rec. Auckland Inst. Mus.,* 11: 186-188, fig. 65, and the status of the name discussed.)

1817. *Conus tenellus* Dillwyn. *Descr. Catalogue Recent Shells,* 1: 417 (Moluccas). Based on Chemnitz, 11: Pl. 183, figs. 1782-1783.

Description.—Moderately light in weight, with a good gloss; low conical, the upper sides slightly to distinctly convex; body whorl covered from base to shoulder with numerous flat spiral ribs, these closely spaced and most distinct basally, weaker below the shoulder; grooves narrow, deep, with very fine axial threads within; sometimes ribs appear doubled; shoulder sharply angled, sometimes weakly undulate; spire low, sharply pointed, the early whorls forming a projecting cone on a flat plate formed by the last whorls; tops of whorls flat or slightly concave, with about 4-6 strong spiral ridges crossed by very fine axial threads. Body whorl white to cream or very pale straw, usually with large pinkish and tan clouds; clouds thickened into two broad bands of pinkish to tan blotches above and below midbody, these sometimes very dark and forming solid bands; numerous long and short dark brown dashes on the ribs, plus many small spots; some dashes darkened and thickened, forming indistinct axial flammules; shoulder with large dark brown, almost black, dashes at angle; spire creamy white or pinkish with regularly spaced large rectangular pale brown squares; many small spots and dashes of brown on the ridges; early whorls white. Aperture relatively narrow posteriorly, widened anteriorly; outer lip sharp, nearly straight or slightly convex; mouth white with a strong pink tone. Columella long, narrow, twisted, set off by a ridge posteriorly. Length 30-65mm.

Comparison.—*C. nimbosus* is very distinctive in shape, sculpture, and pattern. Perhaps it could be confused with *C. australis laterculatus*, but that species has a higher and more conical spire, lacks pink clouds, and is different in texture. There should be no problem identifying this species.

Variation.—The cloudy pattern defies description but is rather constant. The brown dashes are usually darker and more regular above and below midbody and very dark at the edge of the shoulder. The tan clouds may be very dark above and below midbody as well, or they may be absent or replaced by bright pink. Occasional Pacific shells are very dark, mostly dark brown, with only a few white dashes showing through. The convexity of the sides varies somewhat, as does the convexity of the spire whorls.

Distribution.—*C. nimbosus* was once thought to be restricted to the Indian Ocean (Natal to Sri Lanka and Burma), but it has now been found to be widely distributed in the western Pacific from the Philippines to the New Guinea-Solomons area and the New Hebrides and New Caledonia. It is uncommon in shallow to moderately deep water.

Synonymy.—*C. tenellus*, long used for the species here called *C. artoptus*, has been shown to be a synonym of *C. nimbosus*.

Notes.—Pacific specimens are available on the market about as commonly as Indian Ocean specimens and are not worth a premium. Very dark brown specimens from the New Hebrides and New Caledonia sell for about 2-3 times the price of normal specimens, but they are very unattractive specimens in my opinion.

There is commonly minor erosion on the spire in this species, and the whole shell may appear very dull. The pattern fades quickly from pinkish to pale tan. Pacific specimens are perhaps darker than those from the Indian Ocean, but the differences are minor.

I am far from certain about the relationships of *C. nimbosus*, although it might be related to *C. australis* and *C. mucronatus* because of the sculpture; the heavy spire sculpture and general shape are very distinct from these species, however, as are the pattern and coloration.

NOBILIS Linnaeus, 1758

1758. *Conus nobilis* Linnaeus. *Systema Naturae per Regna Tria Naturae*, ed. 10, 1: 714 (Locality not stated). Lectotype (here selected): specimen figured as holotype by Kohn, 1963.

1842. *Conus Victor* Broderip. *Proc. Zool. Soc. (London)*, 1842: 54 (Locality unknown).

1843. *Conus marchionatus* Hinds. *Ann. Mag. Nat. Hist.*, 11: 256 (Port Anna Maria, Nuhuhiva, Marquesas).

1866. *Conus cordigera* Sowerby ii. *Thesaurus Conchyliorum*, 3(*Conus* Suppl.): 329. Based on *Thesaurus Conchyliorum*, 3: Pl. 21(207), fig. 498.

?1877. *Conus Lamberti* Souverbie. *J. Conchyl. (Paris)*, 25: 71, pl. 1, fig. 1; pl. 2, fig. 7 (Ovvea (Loyalty), archip. Caledon.).

1883. *Conus marchionatus* var. *eudoxus* Melvill, in Tryon. *Manual of Conchology*, 6(1): 10, pl. 27, fig. 3 (Marquesas Islands).

Description.—Moderately heavy, with a high gloss; obconical to low conical, the sides straight or gently convex, the body whorl rather elongate; body whorl with a few low spiral ridges on the anterior third or so, rest of whorl smooth except for fine spiral and axial scratches and threads; shoulder moderately wide, sharply angled, concave on top; spire very low, the sides straight or concave, with projecting protoconch; early whorls finely nodulose to undulate, usually eroded, later whorls with straight margins; tops of whorls concave, with about 3-5 very fine and sometimes indistinct spiral ridges crossed by very fine axial threads. Body whorl dark brown, golden brown, or dark reddish, with many large to small white triangles, rhomboids, ovals, and hearts scattered evenly over whorl or clustered into spiral bands at midbody, below shoulder, and above base; white areas sometimes outlined with dark brown; sometimes with few to many spiral rows of brown to black dots or larger squares over whorl or clustered above and below midbody area; shoulder and spire with regular pattern of large oval white spots separated by narrower lines of body color; early whorls white. Aperture moderately narrow, wider anteriorly; outer lip thin, sharp, straight or slightly convex; mouth white.

Columella long and narrow, somewhat indented from inner lip. Length 30-62mm.

Comparison.—The pattern immediately brings to mind typical specimens of *C. marmoreus*, which may even occasionally have similar reddish brown colors. However, *C. marmoreus* has the spire whorls distinctly coronate, a situation which does not occur in *C. nobilis*, which is at most nodulose or undulate on the early whorls. Some large *C. pennaceus* or *C. colubrinus* may be vaguely similar, but these species lack the spiral ridges on the spire whorls and are not undulate on the early whorls; those species are usually also finely spirally striate over much of the body whorl and have very small brown dashes over the entire whorl, these not the same as the occasional dashes of *C. nobilis*. *C. ammiralis* is almost certainly the closest relative of *C. nobilis* and shares with it some aspects of color pattern, especially the regular pattern of the spire; however, *C. ammiralis* is much broader, has a high spire, and usually has many spiral brown lines through the pattern.

Variation.—Van Benthem Jutting and Van Regteren Altena (1965, *Basteria*, 29(5): 81-88) have discussed in some detail the variation in *C. n. nobilis* and *C. n. victor*; their paper should be seen for a better idea of the variability of this species.

The species can be divided into three subspecies which differ in color and pattern but are closely related; an additional subspecies is tentatively placed here.

C. nobilis nobilis: Ground color usually golden yellow to pale tan, seldom dark brown; sides straight or slightly convex; dark ground color may be uniformly netted or intensified into two spiral bands above and below midbody; spiral dashes, when present, small and light brown, not conspicuous; white markings relatively large. Nicobar and Andaman Islands, Sulu Sea, northern Borneo, and perhaps Indian Ocean shore of Sumatra.

C. nobilis victor: Similar in general shape to typical *nobilis*, but sometimes with a slightly higher spire and weakly concave at midbody; white spotting smaller and more clustered at midbody and shoulder on the average, but very variable; background color very dark brown, sometimes paler; spiral rows of dashes very dark, almost black, in many rows, very conspicuous, especially in shells with tan backgrounds. Restricted to islands of the Bali-Flores Strait area of Indonesia.

C. nobilis marchionatus: Usually with straighter sides and a wider shoulder than the other subspecies, more tapered; background color usually distinctly reddish, sometimes yellow, often with very large white spots and seldom with broad dark bands above and below midbody area; spiral dashes apparently absent; spire perhaps more concave than normal, and a bit more heavily nodulose; seldom exceeding 40mm in length. Restricted to the Marquesas Islands.

In addition, *C. lamberti* is tentatively placed here although most people have used the name for a shell like *C. colubrinus*. The type was described as being concave on the shoulder and having nodulose early whorls. It is very large (107mm), deep orange brown, with three spiral bands of large jagged white tents and scattered smaller white tents; the base is relatively light. The type was supposed to have come from the Loyalty Islands and is possibly still unique. Certainly the concave whorls and shoulder, the nodulose early whorls, the color, and the placement of the white spots resemble very closely *C. nobilis* and should separate this from *C. colubrinus*. Until it is better known there is no harm in considering it subspecifically related to *C. nobilis*.

Distribution.—Sporadically distributed from the eastern Indian Ocean (Andamans and Nicobars, perhaps coast of Sumatra), the Sulu Sea of the Philippines and northern Borneo, the Flores area of Indonesia, the Marquesas, and perhaps the Loyalty Islands. Obviously this was once a widespread species which has become extinct at many former localities or has yet to be discovered from many islands between the known populations. Rare to uncommon, shallow to moderately deep water.

Synonymy.—The differences between subspecies do not appear to reach the specific level, although *C. marchionatus* is found so far from known populations of the other subspecies that there would be no harm in calling it a full species; other than color, however, it shows very few and minor differences. *C. lamberti* is so poorly known that it cannot be placed anywhere with certainty. *C. cordigera* is a fairly typical *nobilis nobilis* with minor differences that occur at random in the species, while *C. eudoxus* is just a very large *marchionatus.*

Notes.—This is one of the most attractive cones, especially good specimens of *C. n. victor*. Of the various subspecies, the most common and easiest to obtain is *marchionatus*. Next is *C. nobilis nobilis*, which has apparently become scarce because of over-fishing; it has seldom been offered recently. *C. n. victor* has recently reached the market in fair numbers because of the collecting efforts of Renate Wittig Skinner, but it is still very uncommon; light tan specimens with very large black dashes are especially attractive but rare. *C. lamberti* is only known for certain from the holotype, as most or all the shells recently collected in New Caledonia and assigned to this name appear to be large *C. colubrinus*; its range and status are uncertain.

This is usually a clean and bright cone, although small chips in the lip are to be expected and the early spire whorls are usually weakly eroded. Minor axial growth marks are not uncommon but seldom conspicuous.

NUSSATELLA Linnaeus, 1758

1758. *Conus nussatella* Linnaeus. *Systema Naturae per Regna Tria Naturae*, ed. 10, 1: 716 (Nussatello Island, Asia). Lectotype (here selected): specimen figured as holotype by Kohn, 1963.

1834. *Conus nussatella* var. *tenuis* Sowerby i, in Sowerby ii. *Conchological Illustrations:* Pt. 54, fig. 62 (Annaa). Non *Conus tenuis* Sowerby i, in Sowerby ii, 1833.

1970. *Hermes kawanishii* Shikama. *Sci. Rpts. Yokohama Nat'l. Univ.*, Sec. II(16): 26, pl. 1, figs. 28-29 (Locality unknown).

Description.—Moderately heavy, with a good gloss; elongate, narrowly cylindrical, the sides straight; body whorl covered with narrow spiral ridges separated by about their own width and continuous from base to shoulder; these ridges may sometimes be weakly granulose basally; fine axial threads may be conspicuous or virtually absent over the whorl; shoulder rounded, not distinct from spire, with about 3-5 narrow spiral ridges like those of body; spire low to moderate, sharply pointed, the sides convex to straight; first 5-6 whorls with strong nodules, the later whorls plain; tops of whorls about flat, with 3-5 weak or well defined spiral ridges and very fine axial threads. Body whorl white to pale straw or yellow, covered with many small light and dark brown dots and dashes on the ridges, these usually widely spaced and alternating in size with every other ridge; spots on margin of shoulder larger and darker than those on body whorl; usually there are irregular axial clouds of yellowish to dark reddish brown over part of the whorl, especially above and below the midbody area; in very dark shells the dots on some ridges may be enlarged and very dark as well as closely spaced, presenting a very contrasting appearance; spire and shoulder whitish, usually with large and small blotches of pale orange or reddish brown, plus small dots and dashes of brown on the spiral ridges; margins of whorls dotted with brown. Aperture narrow, slightly wider anteriorly; outer lip straight, rather sharp and thick; mouth white. Columella very narrow. Length 30-83mm.

Comparison.—The narrowly cylindrical shape is found only in *C. artoptus* and *C. auricomus* of similar species. From both these *C. nussatella* differs in pattern and the strong, narrow spiral ridges. *C. auricomus* is a tented shell, presenting a pattern not found or approached by *C. nussatella*, and also has weaker ridging. *C. artoptus* may have a vaguely similar pattern containing both clouds and brown dashes, but it also tends to have large dark spots above the midbody area, may be violet in tone, and lacks the numerous dots and dashes on the ridges of the spire whorls and shoulder characteristic of *C. nussatella.*

Variation.—*C. nussatella* is very constant in shape and pattern. Occasional specimens have only very regular dots over the body whorl, without clouds, while occasional specimens may be largely dark reddish brown or yellowish brown with very contrasting dark dots and dashes. The more typical specimens, however, have at least a few clouds above and below midbody and only a few rows of dots enlarged. The dots may be widely spaced and roughly alternating large and small from ridge to ridge, or closely spaced spirally and all about the same size. Very contrasting patterns are not common. The spire whorls are usually spotted, at least on the margins, and may be lightly blotched or dark.

Distribution.—Widespread and moderately common throughout the Indo-Pacific from the eastern African coast to Hawaii and French Polynesia. Usually in shallow water.

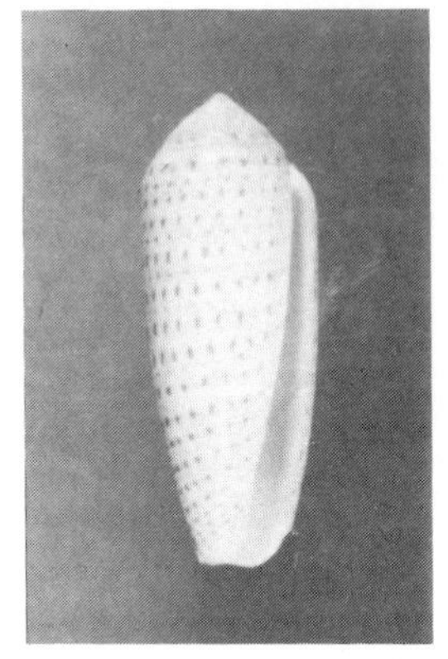

Holotype of *C. kawanishii* Shikama = *C. nussatella* (photo Shikama).

Synonymy.—*C. kawanishii* seems to be a single unusual individual in which clouding is absent and the dots are unusually constant in size over the whole body whorl. Both characters occur in occasional specimens of *C. nussatella*, but I have never seen them in combination. It seems probable that the type of *C. kawanishii* is just an unusual individual that happens to combine the two characters.

Notes.—This is a clean species which seldom has more than minor erosion of the spire whorls and minor growth flaws. Usualiy specimens are rather pale and do not have contrasting patterns, but bright reddish brown specimens are not uncommon. Very large specimens (over 60mm) are uncommon.

OBSCURUS Sowerby i, in Sowerby ii, 1833

1833. *Conus obscurus* Sowerby i, in Sowerby ii. *Conchological Illustrations:* Pt. 29, fig. 26 (Arabia).

1943. *Conus halitropus* Bartsch and Rehder. *Proc. Biol. Soc. Washington*, 56: 88 (Honolulu, Hawaiian Islands).

Description.—Extremely thin, light in weight and fragile, with a good gloss; elongate oval, the sides evenly but slightly convex; body whorl smooth except for weak and indistinct spiral ridges at the base and many axial and spiral threads that are seldom conspicuous; shoulder roundly angled, nearly flat above, rarely with undulate margins; spire low to moderate, blunt, the sides slightly convex to slightly concave; first 2-3 whorls with fine nodules, the later whorls straight or weakly undulate; tops of whorls nearly flat, with about six spiral grooves and numerous axial threads that produce a cancellate effect; protoconch usually persistent. Body whorl usually light or dark violet, sometimes very dingy, usually heavily clouded with large axial brownish flammules that may extend from base to shoulder or be more-or-less restricted to bands below the shoulder and above and below midbody; flammules or blotches usually with very irregular edges; many spiral rows of very fine dark brown dashes and light spots usually present, but sometimes visible only in dark blotches or on lip; spire mostly dark brown on later whorls, becoming bluish white on early whorls; early whorls with small brown dots at margins; protoconch bright pink. Aperture narrow posteriorly, becoming very wide anteriorly, usually over half greatest width of body whorl; outer lip very thin, slightly convex; mouth violet, external pattern showing through. Columella very long, narrow, deeply indented from inner lip. Length 20-41mm.

Comparison.—The great width of the aperture and the thin shell allow comparison only with *C. geographus* and *C. tulipa*. *C. geographus* is easily separated even in juveniles by the presence of coronations or nodules on all whorls, not just the early ones, and the usual absence of spiral lines or dashes over the body whorl, having instead a vaguely tented pattern; the shoulder and spire whorls are usually concave above. *C. tulipa* is closer to *C. obscurus* but still quite distinct. Adult *C. tulipa* have the sides more convex and are less clouded with brown on paler violet with the spiral rows of dashes very strong; the shoulder and late whorls are weakly but distinctly coronate or undulate; the shoulder and spire whorls are usually lightly spotted and concave above. Juveniles of these two species are very distinct. In *C. obscurus* the juveniles are bright orange with many rows of fine brown dashes and white blotches; the fourth and later spire whorls are dark on top. Juveniles of *C. tulipa* of equal size have higher spires that are pale and have the body whorl bright red or violet (sometimes with pale bands) and covered with a few rows of very dark large brown dashes; the spire whorls are heavily spotted on the margins; the basal sculpture is quite heavy.

Variation.—As with most shells with nebulous markings there is considerable variation in the tone and size of the brownish blotches and flammules. The intensity of the background color also varies from very dark dirty violet to almost bluish white. Occasional specimens are very weakly patterned, while others are almost solid pale brown with the violet restricted to a scalloped band at midbody and violet at the base. The shoulder is sometimes weakly undulate.

Distribution.—Moderately common throughout the Indo-Pacific, including Hawaii and French Polynesia. Specimens from the Indian Ocean appear to be darker brown and perhaps average larger at adult size. Shallow to moderate water.

Synonymy.—Although Hawaiian specimens of *C. obscurus* are sometimes more highly contrasted brown and pale violet than specimens from other localities, they are easily matched by series from many other localities; there is no basis for recognition of *C. halitropus* at the subspecific or even varietal level.

Notes.—This is a fragile species that often has the lip broken and may show heavy scars and pattern disruption on the body whorl. The spire is commonly encrusted and weakly eroded. The whorls are often irregularly stepped, perhaps from damage and repair early in life.

Juveniles are not commonly seen and at first glance do not resemble the adults because of the bright orange coloration. However, the change to adult colors and pattern apparently occurs abruptly, as the ventral surface of a juvenile is bright orange but the dorsal surface is typical of the adult. The orange fades after death.

OCHROLEUCUS Gmelin, 1791

1791. *Conus ochroleucus* Gmelin. *Systema Naturae per Regna Tria Naturae*, ed. 13, 1: 3391 (Locality not stated). Lectotype figure (Kohn, 1966): Martini, Pl. 52, fig. 573.

1792. *Conus praefectus* Hwass, in Bruguiere. *Cone*, in *Ency. Method., Hist. Nat. des Vers*, 1: 734 (American Seas). Lectotype figure (here selected): *Tableau Ency. Method.*, Pl. 343, fig. 6 (called the holotype by Kohn, 1968).

1798. *Cucullus eburneus* Roeding. *Museum Boltenianum:* 45 (Locality not stated). Lectotype figure (Kohn, 1975): Martini, Pl. 52, fig. 573.

1857-1858. *Conus fasciatus* Sowerby ii. *Thesaurus Conchyliorum*, 3 (*Conus*): 35, pl. 20(209), figs. 487-489 (Locality not stated). Non *Conus fasciatus* Schroeter, 1803.

Description.—Moderately heavy, with a low gloss; low biconical, the upper sides slightly convex then tapering to the base; body whorl with broad flat spiral ribs on the lower half, the ribs separated by deep grooves containing weak spiral and axial threads; upper half of whorl smooth or nearly so, but with many fine spiral and axial scratches; shoulder sharply angled, weakly carinate, concave above; spire moderate or tall, sharply pointed, the sides concave; spire whorls weakly stepped, with carinate margins, slightly concave to flat above; sculptured with about five spiral ridges crossed by numerous very fine axial threads. Body whorl mostly pale yellowish to orange tan, usually darker on posterior half than anteriorly and with two narrow pale bands above and below an indistinct darker tan midbody band; axial growth marks sometimes paler; margin of shoulder pale; spire mostly dark to pale tan, usually darker than body whorl and not distinctly flammulated with dark axial lines; early whorls darker than later ones; margins of later whorls inconspicuously dotted dark and light. Aperture moderately narrow, widest basally; outer lip sloping below level of shoulder, angled posteriorly then nearly straight, sometimes weakly concave at middle, thin; mouth creamy white with pale yellow tones, darker deep within. Columella long, narrow, bounded with a very heavy ridge posteriorly and laterally. Length 40-65mm.

Comparison.—*Conus radiatus* and *C. pilkeyi* are both similar to *C. ochroleucus*, especially the latter species. *C. radiatus* may be similar in basic pattern but has a lower spire with often depressed whorls, a more rounded shoulder, has a bluish white to grayish aperture, and is usually dark brown or gray; the grooving on the anterior part of the body whorl is very deep. *C. pilkeyi* is very similar, especially when the non-Philippine form is compared with *C. ochroleucus*. It is usually dark brown, almost black, or dark orange, commonly with indistinct darker spiral bands over the whorl; the mouth is deep orange, sometimes pale rusty red; and the spire is heavily marked with dark brown axial flammules that are well contrasted. The Philippine form of *C. pilkeyi* is usually easier to distinguish, as it has a lower spire than sympatric *C. ochroleucus*, has the shoulder more rounded, is less elongated, may have weak indications of spiral banding on the body whorl, and commonly has a much more orange mouth.

I am not sure I can distinguish the juveniles of *C. pilkeyi* and *C. ochroleucus*, but it seems that juveniles of *C. pilkeyi* have strongly contrasted spire markings and usually very dark spots or flammules on the body whorl compared to *C. ochroleucus;* juveniles of *C. pilkeyi* are apparently smoother than those of *C. ochroleucus* of the same size.

Variation.—Adults show considerable variation in tone and contrast of the body pattern, but this is very minor and should present no problems. Usually the early whorls are distinctly dark, the later spire whorls mottled tan and white, the edges of the whorls more-or-less spotted with dark and light. The aperture may be deep yellow, orange, or almost white. Grooving may be restricted to the anterior third of the body whorl or extend well above midbody.

There is considerable variation with growth, but I am not sure that I have correctly separated juveniles of this species from those of *C. pilkeyi*. However, the entire body whorl is rather deeply grooved, the ribs narrow and rounded; the shoulder is angled; and the body whorl is narrower than in adults. The color pattern consists of many large reddish brown or tan dashes forming heavy axial blotches or flammules on a pale background; the spire whorls are mostly dark brown with some white mottling. The change to adult coloration and sculpture is gradual, as specimens 25mm long are still mottled.

Distribution.—*C. ochroleucus* seems to be restricted to the Philippines and New Guinea area, where it is common in moderately deep water. Records from the Solomons to New Hebrides and east appear to be based on *C. pilkeyi*.

Synonymy.—*C. praefectus* is certainly this species. *C. fasciatus* of Sowerby is just the first validation of a non-binomial name by Martini.

Notes.—This is an attractive species although very subdued in color. The spire may have some scars and there may be heavy axial growth marks on the body whorl which disrupt the pattern with lines of lighter color. The lip is commonly filed.

ORBIGNYI Audouin, 1831

1831. *Conus Orbignyi* Audouin. *Mag. Zool.*, 1: Pl. 20 (?China).
1833. *Conus planicostatus* Sowerby i, in Sowerby ii. *Conchological Illustrations:* Pt. 28, fig. 15 (Locality not stated).
1874. *Conus d'Orbignyi* Weinkauff, *Syst. Conch. Cab.*, (Ser. 2), 227: 258. Emendation.
(?1963. *Conus emersoni* Hanna (in part). *Occas. Papers California Acad. Sci.*, No. 35: 25, pl. 1, fig. 2 (Off Los Frailes, Cape San Lucas, Baja California).)
1973. *Conus orbignyi aratus* Kilburn. *Ann. Natal Mus.*, 21(3): 575 (Off Tongaat, Natal, South Africa). Non *Conus aratus* Gabb, 1873, a fossil.
1975. *Conus orbignyi elokismenos* Kilburn. *Nautilus*, 89(2): 50. Nomen novum for *Conus orbignyi aratus* Kilburn, 1973.

Description.—Moderately light in weight, with a low gloss; elongate biconical, the upper sides slightly convex then strongly concave and tapering to a long base; body whorl covered with broad flat spiral ridges separated by narrow to wide spiral grooves containing axial riblets; entire shell covered with fine axial threads; ridges may be weak below shoulder; shoulder carinate, usually with large rounded nodules, narrow, very oblique; spire tall, sharply pointed, the sides straight or slightly concave; spire whorls stepped, with large rounded nodules on all whorls, although sometimes obsolete on last whorl and shoulder; top of whorls flat but steep, with about 3-6 narrow spiral ridges crossed by heavy axial threads so distinctly cancellate; sculpture sometimes weak or obsolete on late whorls and shoulder. Body whorl creamy white to pale straw, covered with many dark brown large dashes and spots which tend to form axial flammules and three spiral bands of darker brown blotches below shoulder and above and below midbody; pattern may be very heavily developed or almost absent; spire straw to whitish, with few to many axial spots and lines of same color as on body whorl; small spots between nodules; early whorls brown to white. Aperture narrow, uniform in width; outer lip strongly sloping below shoulder, thin and fragile, following outline of body whorl; mouth white, sometimes faintly salmon. Columella internal. Length 45-70mm.

Comparison.—*C. orbignyi* is one of the most distinctive cones in combining the elongate shape and heavy sculpture with relatively large size. *C. ichinoseana* is somewhat similar but has straighter sides, weaker body sculpture, very little spiral sculpture on the spire, and reduced coronations. *C. mazei* var. *mcgintyi* is very similar in all aspects but is much more fragile, has the later spire whorls and shoulder carinate and not nodulose, has a sparser pattern, and is relatively more slender. *C. comatosa* has straighter sides and only the first few whorls nodulose, the later ones with a rim-like carina. Smaller species such as *C. aculeiformis* and *C. insculptus* or *C. schepmani* are more narrow, have different patterns, or have different spire sculpture.

Variation.—The spire may be very high or quite low and squat. Although the nodules are strong on the spire whorls, they are occasionally obsolete on the last whorl and shoulder. The ridges of the body whorl may be rather narrow, with the grooves wider than the ridges, or wide and flat, the grooves very narrow; occasionally the ridges become very indistinct below the shoulder. Such differences in sculpture as exist between South African specimens and those from Taiwan can all be matched in series from Taiwan, although there definitely is a tendency for South African specimens to have wider and lower ridges that are commonly obsolete below the shoulder. The pattern varies from very dark to almost absent, with such differences common within even small series.

One interesting variant is a shell from deep water in the Philippines which certainly is *C. orbignyi* but differs in being somewhat thicker, smaller, having more strongly angled shoulder and spire margins that are not as oblique as in typical *C. orbignyi*, and commonly having the nodules reduced and low on the later whorls and spire. In such specimens the pattern tends to be clustered into two broad spiral brown bands with few other brown dashes present. It occurs with typical *C. orbignyi*, and there are intermediates.

The smallest *C. orbignyi* I have seen was about 30mm in length; it was identical with adults except that dark brown axial spots could be seen in three spiral bands under the overlying more typical brown pattern.

Distribution.—Very widespread over the Indo-Pacific and usually common when discovered. This is a deep-water cone, so its range remains largely unexplored except where there is commercial fishing. Known from off the Natal-Mozambique coast of southern Africa, the deep channel between Africa and Madagascar, deep water off Aden, common in the northern China Sea off Japan and Taiwan, found in the Philippines, and reported from off Queensland, Australia. Mr. Alex Kerstitch has taken quite typical specimens in deep water off Baja California, where the type of *C. emersoni*, which is possibly this species, was also taken. Obviously much further work is needed to fill the gaps in the range of *C. orbignyi.*

Synonymy.—*C. planicostatus* seems to be this species. Although South African specimens are somewhat different from the average Taiwan specimen, I can see no reason to recognize them at the subspecific level until specimens from the intervening areas have been examined; certainly the African populations are not isolated. The status of *C. emersoni* is very doubtful, and it is not certain if the holotype is a Recent or fossil specimen; the paratype is a fossil and probably belongs to another species.

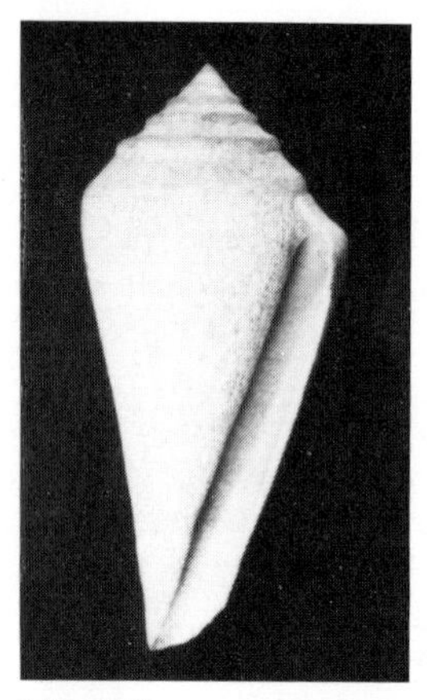

Paratype of *C. emersoni* (fossil) (photo Ken Lucas at Calif. Acad. Sci.).

The holotype of *C. emersoni* looks like a fairly normal *C. orbignyi* with a squat spire; such specimens are occasionally seen in Taiwanese material and are certainly not specifically distinct; add to this the fact that typical *C. orbignyi* have been recently taken (alive) in the Baja California area, and I can see no reason to recognize the name as valid until further specimens are available.

Notes.—This is a fragile deep-water cone, but it seldom has more damage than minor growth flaws and chipped lips. The spire is usually intact and uneroded. Specimens with heavy patterns are most attractive, but they are not as common as those with reduced or very irregular patterns. South African and Philippine material is less common than specimens from Taiwan and southern Japan; American material is very rare at the moment.

ORION Broderip, 1833

1833. *Conus Orion* Broderip. *Proc. Zool. Soc.* (*London*), 1833: 55 (Real Llejos, Central America).

1955. *Conus drangai* Schwengel. *Nautilus*, 69(1): 13, pl. 2, figs. 1-7 (Bahia Salinas, Costa Rica).

Description.—Moderately heavy, with a high gloss; low conical, the sides nearly straight or slightly convex; body whorl with low spiral ridges at the base, these almost obsolete below midbody, but sometimes with weak and widely spaced ridges scattered almost to shoulder; shoulder wide, sharply angled, flat or convex above; spire moderately low, sharply pointed (blunt in large specimens), the sides straight to slightly concave; spire whorls flat or slightly convex above, with traces of very weak spiral threads most distinct on the earlier whorls, about 3-5, crossed by fine axial threads, sculpture almost obsolete on later whorls. Body whorl glossy opaque and normal white, heavily covered with reddish brown to blackish brown in two heavily scalloped spiral bands above and below an almost unmarked or flammulated midbody band; from both dark bands extend heavy axial flammules; occasionally the dark bands are broken into irregular blotches; area below shoulder narrowly white, checkered with dark; often the shell gives the appearance of a dark brown shell with three spiral bands of irregularly axial rectangles of white above the base, at midbody, and below the shoulder; a few spiral rows of small squarish brown spots may be visible in the white areas; base pale to bright salmon, sometimes white; spire white, with large revolving brown blotches giving a checkered appearance; early whorls whitish. Aperture moderately narrow, uniform in width; outer lip thin, sharp, straight; mouth pale bluish white. Columella internal. Length 20-41mm.

Comparison.—The very glossy white background with opaque white blotches and a salmon base is quite distinctive when combined with the irregular pattern. *C. vittatus* is sometimes confused with *C. orion* because it also has a salmon base, but that species is more cylindrical, covered with distinct spiral ridges, has distinct though fine spire sculpture, and has distinct small spots on the ridges. *Conus beddomei* has a similar pattern in some forms and is similar in shape, but it has 3-4 distinct spiral ridges on the spire whorls, seldom has a salmon base, and has a deep pink to violet mouth. *C. purpurascens* is larger, often has a bluish background, and has distinct spiral rows of brown dashes.

Variation.—Large specimens have slightly convex sides and a blunt spire with a more rounded shoulder than smaller specimens. The pattern is fairly constant, but varies widely within the range described earlier; the major problem is that axial growth marks cause pattern disruptions and large colorless areas so the pattern may be very irregular. The spiral rows of dots are seldom conspicuous. The opaque white may be extensive over the shell or restricted to large blotches at midbody. Very light and very dark shells are known. The salmon base is normally present and is a good identification mark, but it is occasionally absent or reduced.

Distribution.—An uncommon to rare species in moderately shallow water of the eastern Pacific from central Mexico to Colombia.

Synonymy.—*Conus drangai* is very well described and certainly this species. Unfortunately the same thing cannot be said about Broderip's original description. There is no mention of the salmon base, and very little else of use in the description, but Reeve later figured a specimen collected by Cuming (as was the type) that seems to be this species. Unfortunately we are stuck with *Conus orion* for this shell.

Notes.—The specimens usually available are rather small but usually with good markings. The lip is fragile. Many freak specimens with odd color patterns have been collected, these patterns usually the result of disruptions by growth marks. The species is far from common at any locality, although not rare.

OTOHIMEAE Kuroda and Ito, 1961

1961. *Conus* (*Endemoconus*) *otohimeae* Kuroda and Ito. *Venus*, 21(3): 250, pl. 17, fig. 8 (Nado, Gobo, Kii Peninsula, Japan).

Description.—Moderately light in weight, with a good gloss; low conical, the upper sides slightly convex; body whorl with about 6-12 low spiral ridges at the base, these separated by weakly punctate grooves; rest of whorl smooth except for sometimes obvious axial threads; shoulder sharply angled, slightly convex or flat above; spire moderate or low, the sides straight or slightly convex; first 2-4 whorls very weakly nodulose; whorls flat or slightly convex above, covered with heavy curved axial threads and no obvious spiral sculpture. White to pale straw, covered with many spiral rows of small orange to brown dots, these most conspicuous on the basal ridges and as somewhat enlarged dots in single bands above and, especially, below midbody; a broad band of bright orange axial blotches above the base, these often partially fused into a nearly continuous band; bands of smaller orange-brown blotches may be present above midbody and below the shoulder but are not as well developed as the basal blotches; sometimes with indistinct tan axial clouds weakly developed; shoulder with large blackish brown spots alternating with white; spire checkered white and brown. Aperture moderately wide, uniform in width; outer lip straight or slightly concave at middle, thin; mouth pinkish with the orange basal band showing through. Columella very narrow, sometimes greatly indented. Length 25-32mm.

Comparison.—*C. otohimeae* is very close to *C. memiae* in pattern and general appearance but differs consistently in sculpture. In *C. memiae* the body whorl is covered with rather deep spiral grooves separating flat spiral ridges; the shoulder is carinate; and the spire is higher, more strongly nodulose on the early whorls, the slightly concave whorls with both heavy axials and 2-3 weak but distinct spirals. In addition, *C. memiae* is more elongated relative to width and has smaller but darker spots on the margins of the spire whorls and shoulder than in most *C. otohimeae*.

Conus ione juveniles are vaguely similar in having rows of small spots and some blotches, but that is a violet shell with strongly carinate spire whorls and of a different texture and overall appearance.

The relationship of this species to the dubious *Conus henoquei* is uncertain. The description of *C. henoquei* and of its probable synonym *C. nadaensis* both stress the presence of a single spiral ridge at the middle of the spire whorls; the shells are reddish and lack the distinct orange basal band. See *C. henoquei* in the *Doubtful Species* section.

Variation.—There is some variation in the height of the spire and the straightness of the sides. The body whorl varies from white to almost a pale grayish tone, sometimes heavily tinted with pink. The pattern is fairly constant in the presence of the basal orange band and at least one row of smaller spots near midbody, but the development of the smaller spots is quite variable. Juveniles are more like *C. memiae* in shape but still smooth; the pattern is subdued. Specimens covered with rows of large granules have been reported.

Granulose variant of *C. otohimeae,* Japan, 19mm (photo Okutani).

Distribution.—As yet known only from moderately deep water off southern Japan and Taiwan. Uncommon to rare.

Notes.—*C. otohimeae* is very similar to *C. memiae* but the two appear to be distinct. I have seen no intermediates, but admittedly both species are rare in collections as yet. The two appear very different in texture and size.

Conus henoquei and *C. nadaensis* are a thorn in my side; possibly they belong here (*henoquei* is by far the oldest name), but they seem distinct in shape (slight), sculpture (slight), and color (major). They do not seem to fall within the range of variation of *C. otohimeae*, but it would not be hard to bridge the gap between the type of *C. henoquei* and *C. nadaensis*, then between *C. nadaensis* and *C. otohimeae*.

PAPILLIFERUS Sowerby i, in Sowerby ii, 1834

1834. *Conus papilliferus* Sowerby i, in Sowerby ii. *Conchological Illustrations:* Pt. 56/57, fig. 79 (Locality not stated).
1848. *Conus Jukesii* Reeve. *Conchologia Iconica,* 1(*Conus* Suppl.): Pl. 2, sp. 278 (North Australia).
1860. *Conus Couderti* Bernardi. *J. Conchyl.* (*Paris*), 8: 212, pl. 4, fig. 3-4 (Locality unknown). Misspelled *C. conderti* by Sowerby ii, 1866.

Description.—Light in weight, with a low gloss; low conical or obconical, the sides nearly straight or distinctly convex; body whorl with about 6-12 low and sometimes indistinct spiral ridges at the base, rest of whorl with fine widely spaced remnants of spiral ridges and many axial threads; shoulder roundly angled, very broad, flat or slightly concave above; spire flat to low, bluntly pointed with persistent protoconch, the sides straight or slightly convex; tops of whorls flat or slightly concave, with about 6-12 fine spiral ridges crossed by fine curved axial threads, faintly cancellate; sutures sometimes rude. Body whorl bluish gray to cream, covered with often indistinct small brownish dots in spiral rows; usually with two spiral series of large brownish blotches above and below midbody, these blotches either isolated or extended into broad axial flammules that might cross the midbody area, sometimes weakly fused into broken spiral bands; numerous other spiral and axial dashes and lines may be present, and the pattern may even assume a finely crosshatched appearance near large axial flammules; color of flammules variable from pale brown to bright reddish brown or almost black; base commonly paler, white or tan; shoulder and spire whitish with many to few dark brown radiating blotches and lines; early whorls whitish. Aperture broad, uniform in width; outer lip straight, thin; mouth deep dark violet or brownish violet, often with two paler bands at middle and under shoulder. Columella long, narrow, partially free posteriorly and projecting into the mouth. Length 20-45mm.

Comparison.—For some reason *C. papilliferus* has been badly confused with *C. anemone* and *C. aplustre* in the literature. It appears to be sympatric with both these species and is quite distinct. From *C. anemone* it differs in the straighter sides and greater relative width at the shoulder as well as the absence of spiral ridges above the base. *C. peronianus* lacks such ridges also but is more inflated and different in pattern and color than *C. papilliferus*, as well as looking much like a smooth *C. anemone* in shape. *C. aplustre* is a small but thicker cone with a different shape, heavily covered with small brown dots, and often banded with pinkish on blue.

The real problem is separating *C. papilliferus* from the species here tentatively recognized as *Conus wallangra*. I strongly suspect that the two are conspecific, *C. wallangra* being just a deeper water ecotype of the tidal *C. papilliferus*, so the following differences are probably purely artificial and may not hold. In *C. wallangra* the spire is higher and more pointed, often with the whorls very slightly stepped; the pattern tends to form two broad belts above and below midbody, with the rest of the shell heavily flecked and dotted; the mouth is often bright pink but may also be brownish violet; and the spire sculpture is heavier. Certainly the two are very, very close.

Variation.—As indicated in the description, the pattern and coloration are variable and seldom very distinguished. This is a mottled and blotched cone with scattered small spots and dashes. Sometimes it is covered with very broad chocolate brown axial bands edged with fine crosshatched lines, other times it is pale bluish with only small spots above and below midbody. The columella seems to be constantly free posteriorly, but not as strongly as in *C. angasi* or *C. trigonus*.

Distribution.—Common in shallow water of New South Wales, Australia, but probably also extends into southern Queensland and Victoria.

Synonymy.—*Conus jukesii* and *C. couderti* apparently belong here because of the general appearance, but either could belong to various other undistinguished species as well.

Notes.—This is a very common species but not a popular one because of its generally ugly appearance. The spire is commonly eroded, and there are usually heavy axial growth marks and scars on the body whorl. Some specimens from slightly deeper water may have quite attractive patterns and be in better condition, but these are not normally sold.

The columella indicates a closer relationship with *C. angasi* and allies than with *C. anemone*, which would not be expected from the mottled pattern. Some specimens have patterns that closely match patterns of *C. tinianus*, another species with a partially free columella.

PARIUS Reeve, 1844

1798. *Cucullus lineatus* Roeding. *Museum Boltenianum:* 47 (Locality not stated). Lectotype figure (here selected): Chemnitz, Pl. 140, fig. 1304. Non *Conus lineatus* Solander, 1766, a fossil.

1810. *Conus lacteus* Lamarck. *Ann. du Mus. Hist. Nat.* (*Paris*), 15: 274 (Indian Ocean). Holotype figured by Kiener, Pl. 60, fig. 4. Non *Conus lacteus* (Roeding, 1798).

1844. *Conus parius* Reeve. *Conchologia Iconica*, 1(*Conus*): Pl. 43, sp. 235 (Locality unknown).

?1848. *Conus contusus* Reeve. *Ibid.*, 1(*Conus* Suppl.): Pl. 2, sp. 276 (Moluccas).

Description.—Moderately light in weight, with a good gloss; low conical, the upper sides convex; body whorl with a few spiral ridges at the base, these rapidly followed by broad flat ribs separated by deep grooves that have fine axial threads; grooves usually extend to about midbody or slightly higher; rest of whorl smooth except for fine spiral and axial scratches; shoulder rounded, narrow, convex above, not very distinct from spire; spire low to moderate, sharply pointed, the sides concave; tops of whorls convex, with fine spiral threads and curved axial threads, no sculpture very distinct; early whorls may be carinate. Shell variable in color from pale brown to waxy yellow or pale grayish blue, often two or more shades on the same shell; base generally pale; spire same shade as body whorl or different, often bluish gray, the early whorls dark brown when not eroded. Aperture moderately wide, somewhat widened anteriorly; outer lip sharp, evenly convex; mouth bluish white to violet, often with the violet as a large blotch at the inner posterior half. Columella long and narrow, indented from inner lip and sometimes bordered with a low ridge posteriorly. Length 25-35mm.

Comparison.—This species is very similar to *C. cinereus* and *C. radiatus* in shape, differing mostly in color being paler, lack of any distinct pattern, and having the sharply pointed early whorls often dark brown. In adition, *C. radiatus*, of which *C. parius* has sometimes been considered a dwarf pale variety, has several strong spiral ridges on the spire whorls. In sculpture it is closest to *C. cinereus*, but it lacks all pattern, while even very pale *C. cinereus* have traces of dots on the body whorl.

Variation.—Although constant in shape (some specimens have a more sharply angled shoulder than normal), sculpture, and texture (waxy), there is considerable variation in color from population to population. Generally specimens are pale bluish gray over at least the shoulder and lower spire area, often over all the shell. Other specimens are bright waxy yellow or pale yellow. Some specimens are brown or tan on the anterior half of the body whorl, grayish posteriorly. New growth is usually bright yellow. The brown early whorls are commonly obscured by erosion. The aperture may be entirely violet or largely bluish white with the violet restricted to a large blotch posteriorly deep within.

Distribution.—*Conus parius* seems to be restricted to the western Pacific from the Philippines through the New Guinea-Solomons area to the New Hebrides.

Notes.—*C. parius* seems to be subject to bad breaks and growth flaws on the body whorl; the lip is quite fragile. Most specimens I have seen were yellowish with grayish spires, and I have not seen any that were all brown. This species has been badly confused with *C. radiatus* without any real reason; although the patterns and sculpture of the body whorl may be similar, the spire sculpture is very different. The species has also been confused nomenclaturally with the white form of *C. puncticulatus* usually called *columba*, but restriction of the type of *C. columba* to a figure resembling the Caribbean shell leaves no doubt that *columba* must be applied to the Caribbean species, leaving *parius* for the Pacific species.

PARVULUS Link, 1807

1807. *Conus parvulus* Link. *Beschr. Nat.-Samml. Univ. Rostock*, 3: 106 (Locality not stated). Based on Martini, Pl. 63, fig. 707.

1810. *Conus roseus* Lamarck. *Ann. du Mus. Hist. Nat.* (*Paris*), 15: 37 (Antilles). Martini, Pl. 63, fig. 707 mentioned. Non *Conus roseus* Fischer, 1807.

1956. *Virroconus imperator* Woolacott. *Proc. Roy. Zool. Soc. New South Wales*, 1954-55: 72, fig. 3 (Trinity Bay, Queensland, Australia).

Description.—Moderately heavy, with a good gloss; low conical, the upper sides convex then pinched-in and tapered to the base; body whorl covered with narrow spiral ridges from base to below shoulder, weaker or obsolete on posterior fourth of whorl; fine axial threads usually present, and there may be axial plicae as well; shoulder angulate, heavily coronate with rounded coronations, flat above; spire low to moderate, blunt or sharp, sides nearly straight or slightly concave; spire whorls heavily coronate, flat or slightly concave above, with 2-4 narrow spiral ridges and many curved axial threads. Body whorl pale straw to deep bluish brown or reddish brown, usually with two broad darker bands above and below a paler midbody band; base deep blackish violet dorsally and ventrally; area below shoulder commonly paler than rest of whorl; numerous small brown flecks in vaguely axial rows over body whorl, these sometimes inconspicuous; spire and shoulder whitish to bluish or tan, the coronations white, usually with brown to black spots or short lines between the coronations; early whorls pinkish. Aperture moderately wide, wider anteriorly; outer lip straight or concave at the middle; mouth pale violet, often with white bands at middle and below shoulder, the growing edge yellowish. Columella rather short, narrow. Length 20-43mm.

Comparison.—*Conus parvulus* has been confused with *C. biliosus*, a species to which it is related. *C. biliosus*, however, is a larger and more elongated shell with nearly straight sides; the coronations are lower, especially on the shoulder, and the shoulder is narrower. In addition, *C. biliosus* has the base the same color or paler than the rest of the body whorl, is usually flammulated with dark brown, and has the spire heavily or sparsely streaked and blotched with dark brown. *Conus mus* may be very similar to *C. parvulus* in general shape and sculpture as well as color, but it lacks the blackish base and brown flecks and is somewhat lighter in weight. *Conus frigidus* and *C. gilvus* may be generally similar but lack coronations and are much more weakly sculptured.

Variation.—There is considerable variation between northern and southern populations of this species, but it is possibly clinal in nature rather than at the subspecific level. Specimens from the Queensland to New Guinea area are often brightly colored reddish brown or even bluish brown, the base very contrasted blackish and the spire generally dark; shells from this area have a higher gloss and commonly a higher spire as well. Specimens from the Philippines are paler, usually straw or pale tan, sometimes with pinkish tones, the base less contrasted and the spire whitish; the spire is lower, the sculpture somewhat heavier, and the gloss low. It is possible to find intermediate specimens in both areas, however.

The width across the shoulders varies a bit, as does the weight of the shell and the strength of the coronations. Commonly the sculpture is very weak or obsolete below the shoulder, and it may be weak over the entire body whorl. The brown flecks are sometimes very heavy and conspicuous, but usually one must look for them because they are small. Juveniles seem to be brighter in color, have higher spires, and may be more weakly sculptured.

Distribution.—Widely distributed and moderately common in shallow waters of the western Pacific from the Ryukyus and Philippines to the Solomons and Queensland, Australia.

Synonymy.—If desired, *C. imperator* could be retained as a subspecies for the darker, glossier specimens from New Guinea to Queensland, but the differences are probably not really at the subspecific level.

Notes.—The *imperator* form is much more attractive than more typical specimens from the Philippines, but both are about equally common. The spire is sometimes rather limy and there may be small scars on the body whorl, but this is usually a rather clean species. Philippines specimens often appear faded, but possibly these are beach.

PATAE Abbott, 1971

1971. *Conus patae* Abbott. *Nautilus*, 85(2): 49, figs. 1-6 (Off Pompano Beach, Broward Co., Florida).
1971. *Conus rudiae* Magnotte. *Busycon Bugle*, 6(10): 11, figs. 1-2 (13 fms. off Pompano Beach, Florida).

Description.—Moderately heavy, with a low gloss; low conical, the upper sides slightly convex; body whorl covered with narrow undulating spiral ridges and weak axial threads; conspicuous axial plicae abundant, giving whole shell a cancellate appearance; shoulder narrowly angled, sometimes undulate, broad; spire low or moderate, sharply pointed, the sides straight or slightly concave; tops of spire whorls slightly concave, with about six spiral ridges crossed by fine axial threads. Body whorl ivory white, with three indistinct bands of yellow, one above base, one at midbody, and an irregular band below shoulder; in life these colors might be pinkish or pale violet; occasionally with scattered small darker blotches above and below midbody; spire white with yellowish blotches. Aperture fairly narrow, uniform; outer lip straight or slightly convex; mouth pinkish, fading to white. Columella internal. Length 18-32mm.

Comparison.—This little shell immediately reminds one of *Conus mus*, and it seems possible that it is closely allied. However, it differs in having the shoulder and spire whorls straight or only undulating, not distinctly coronate, and in having heavy and numerous axial plicae. The yellowish color possibly does not mean much, as yellow *C. mus* occur. The axial plicae and to some extent the shape and pattern are also shared with *C. stimpsoni*, but that species has a more sharply angled shoulder and very weak body sculpture other than the plicae.

Variation.—The body whorl always appears cancellate because of the spiral ridges and plicae, but the shoulder may be straight or weakly undulate. All the specimens I have seen (the types) were collected dead and are yellow on white, except a paratype from Jamaica which is distinctly pinkish with darker pinkish brown blotches. The base and shoulder of yellow specimens may be tinted with rose or pale violet, and it is possible that living specimens are at least sometimes toned with pink.

Distribution.—Uncommonly dredged dead off southern Florida and Jamaica, certainly more widely distributed. Rare in live condition.

Synonymy.—*C. patae* is a few months older than *C. rudiae*, but the latter description fails to include a comparison and is thus technically invalid.

Notes.—This shell certainly looks like a yellowish *C. mus* at first glance, and I cannot help but think they are closely related. Living specimens are seldom offered at the moment, and even dead-dredged specimens are very uncommon. Probably this shell will prove to be widely distributed in the Caribbean and possibly even locally common. There is quite possibly an older name for this species buried somewhere in the literature.

PATRICIUS Hinds, 1843

1843. *Conus patricius* Hinds. *Ann. Mag. Nat. Hist.*, 11: 256 (Gulf of Nicoya).

1843. *Conus pyriformis* Reeve. *Conchologia Iconica*, 1(*Conus*): Pl. 13, sp. 70 (Bays of Caraccas and Montija, West Colombia).

Description.—Very heavy, with a low gloss or dull; obconical, the upper sides very convex in adults then tapering to a narrow base; body whorl with low rounded ridges above the base, the juveniles with ridges extending to midbody or beyond; numerous fine spiral and axial threads and scratches over rest of whorl; shoulder broad, smooth and rounded in adults, weakly angulate and with low coronations in juveniles; spire low, deeply concave, the early whorls forming a small pointed cone, otherwise nearly flat; spire tip usually eroded in adults; spire whorls slightly concave above, their margins deeply cut to produce a cogwheel-like pattern of nodules on the early and middle whorls; whorls covered with fine spiral and axial threads, none conspicuous, especially in adults. Body whorl variable in color with growth, waxy tan in juveniles, becoming duller and paler with growth, eventually whitish or creamy; spire like body whorl, the whorls commonly eroded whitish. Aperture moderately wide, nearly uniform in width; outer lip slightly concave at middle but not as much as inner lip, sharp and fragile; mouth marginally tan or creamy, interior bluish white to white in adults. Columella very narrow, long, bounded posteriorly by a heavy, twisted ridge. Length 70-140mm.

Comparison.—Although there are several species of large whitish or yellowish cones, none approach *C. patricius* very closely. The closest approach is perhaps *C. fergusoni*, which has nodulose early whorls, but it differs in having angulate shoulders and straighter sides as well as being banded with orange-tan or yellow on white. *C. figulinus* var. *loroisii* may be similar in color but is less elongate and lacks coronations on the early whorls. *C. quercinus* and *C. coelinae* have nearly straight sides, angled shoulders, and lack coronations. The coronate juveniles may be vaguely similar to *C. armiger* or *C. atractus*, but these species are much more heavily spirally ridged on both the body whorl and spire and are not waxy tan.

Variation.—The change in shape and color with growth is unusual, and the juveniles (*patricius*) were once considered a distinct species from the adults (*pyriformis*). Juveniles are waxy tan, spirally ridged to about midbody, and have somewhat angulate, coronate shoulders; the spire is rather high, the whorls stepped and concave above, with the margins strongly cogwheel-coronated. With growth the brown fades, the coronated whorls become relatively inconspicuous and eroded, and the ridging is restricted to the base of the body whorl. Juveniles may be distinctly dark brown at the base and have indistinct darker and lighter tan bands. Small adults are still a bit waxy tan, with a bluish tone, but grow up into dirty white rugged shells. The mouth tends to be bluish in small shells, white in adults.

The coronations are not typical of other cones, and Reeve called them plaits, which is perhaps a good description. In adults the early whorls look distinctly like cogwheels with heavy teeth.

Distribution.—A Panamic species ranging in shallow to moderately deep water from the Gulf of California and the offshore islands (rare or uncommon in both areas) through Central America to Ecuador (common south of Mexico).

Synonymy.—The juvenile was described a few months before the adults, so the name *patricius* takes priority over the once more familiar name *C. pyriformis*.

Notes.—Juveniles are seldom collected or offered and should be worth a premium over adults. The most presentable specimens are young adults under about 100mm in length, as older shells are very chalky, dull, and scarred. Shells at least as small as 50mm are adult in shape and sculpture but retain some of the waxy tan coloration of juveniles.

PENNACEUS Born, 1778

1778. *Conus pennaceus* Born. *Index Rerum Naturalium Musei Caesari Vindobonensis*, 1(*Testacea*): 152 (China). Lectotype (here selected): specimen figured by Kohn, 1964, as holotype.

1792. *Conus omaria* Hwass, in Bruguiere. *Cone*, in *Ency. Method., Hist. Nat. des Vers*, 1: 743 (Indian Ocean). Lectotype figure (Kohn, 1968): *Tableau Ency. Method.*, Pl. 344, fig. 3.

1792. *Conus rubiginosus* Hwass, in Bruguiere. *Ibid.*, 1: 744 (Mindanao, Philippines). Lectotype selected and figured by Kohn, 1968.

1792. *Conus praelatus* Hwass, in Bruguiere. *Ibid.*, 1: 746 (East Indies). Lectotype selected and figured by Kohn, 1968.

1792. *Conus episcopus* Hwass, in Bruguiere. *Ibid.*, 1: 748 (East Indies). Lectotype selected and figured by Kohn, 1968.

1798. *Cucullus aureus* Roeding. *Museum Boltenianum:* 49 (Locality not stated). Lectotype figure (here selected): Martini, Pl. 54, fig. 601. Non *Conus aureus* Hwass, in Bruguiere, 1792.

1798. *Conus gentilis* Roeding. *Ibid.:* 50 (Locality not stated). Lectotype figure (here selected): Martini, Pl. 53, fig. 591.

1844. *Conus sindon* Reeve. *Conchologia Iconica*, 1(*Conus*): Pl. 43, sp. 233 (Locality unknown).

1845. *Conus Elisae* Kiener. *Species gen. et icon. des coqu. viv.*, 2(*Conus*): Pl. 64, figs. 1-1a. 1849-1850, *Ibid.*, 2: 341 (Locality unknown).

1845. *Conus stellatus* Kiener. *Ibid.*, 2(*Conus*): Pl. 99, fig. 3. 1849, *Ibid.*, 2: 225 (Locality unknown).

1857-1858. *Conus Madagascariensis* Sowerby ii. *Thesaurus Conchyliorum*, 3(*Conus*): 43, pl. 24(210), fig. 582 (Madagascar).

1857-1858. *Conus convolutus* Sowerby ii. *Ibid.*, 3(*Conus*): 44, pl. 23 (209), fig. 564 (Locality not stated).

1874. *Conus racemosus* Sowerby iii. *Proc. Zool. Soc.* (*London*), 1873: 721, pl. 59, fig. 11 (Sandwich Islands).

1900. *Conus omaria* var. *magoides* Melvill. *J. Conchyl.* (*London*), 9: 310 (Locality not stated).

1900. *Conus omaria* var. *marmoricolor* Melvill. *Ibid.*, 9: 310 (Locality not stated).

Description.—Moderately heavy to very heavy, with a good gloss; low conical, the sides slightly convex to straight; body whorl with a few spiral ridges at the base, these obsolete or nearly so on posterior two-thirds of whorl; shoulder narrow to broad, rounded to narrowly angulate, flat or slightly concave above; spire low to moderate, the sides straight to slightly concave, with a sharp and persistent protoconch on a bluntly rounded tip; spire whorls slightly concave above or flat, with faint traces of 3-4 spiral threads and scattered axial threads; no nodulose whorls. Body whorl white to pinkish or occasionally pale bluish gray, covered with narrow golden tan to blackish axial wavy or zigzag lines that commonly overlap to produce large and small tents; usually two spiral rows of large to small dark brown rather axial blotches above and below midbody, these variously developed from very large and covering much of the shell to almost absent; whole shell covered with spiral rows of very small and usually inconspicuous brown dashes and white spots, these best seen in the blotches; occasionally entire shell suffused with grayish or bluish or pattern virtually absent, sometimes entire shell covered by very fine, nearly parallel lines which do not produce tents and form two slightly thickened bands above and below midbody; spire and shoulder whitish, lightly to heavily covered with white tents and brownish blotches; early whorls eroded white; protoconch pinkish. Aperture moderately wide, uniform in width or wider basally; outer lip sharp but thick, straight or slightly convex; mouth white, sometimes with pale pink or bluish white tones. Columella long and narrow, sometimes bounded by a ridge posteriorly. Length 35-85mm.

Comparison.—From *Conus textile* and allies this species is readily distinguished by the absence of small nodules on the early whorls, by the persistent and sharp protoconch, and by the numerous spiral rows of small brown and white dots and dashes. From various other tented cones it can usually be distinguished by at least the combination of the small dashes and the protoconch (such species include *C. neptunus*, *C. lienardi*, *C. wittigi*, etc.). From *Conus aulicus* and *C. auratus* it is distinguished by the lower spire, lack of distinct spiral ridges over the body whorl, and the bluntly rounded spire, as well as the smaller size and usually less cylindrical shape.

Conus magnificus is very similar to *C. pennaceus* in pattern, protoconch, blunt spire, and general appearance, and certainly there are specimens that cannot be easily placed in either species with certainty. However, as a general rule *C. magnificus* is rather cylindrical with a distinctly higher spire than any common variant of *C. pennaceus;* it is narrower at the shoulder than most *C. pennaceus*, but this overlaps broadly. As a general rule, the spiral sculpture of the body whorl is stronger and extends farther up the body whorl than in most *C. pennaceus*, which are nearly smooth above the base. *C. magnificus* juveniles tend to have the large axial blotches distinctly elongate and often extending through the midbody area, while they are smaller and more rounded in many *C. pennaceus* juveniles.

Variation.—Like other tented cones, *C. pennaceus* is very variable in pattern and color, but it is also very variable in shape as well. This has led to a very complicated synonymy and much controversy over characters and range of the species. It is probably best to first isolate three distinctive varieties which are at least partially sympatric with typical *C. pennaceus* (which is itself an uncommon shell) and recognized as full species by several authorities. In my opinion these cannot be consistently separated, and many intermediates exist.

C. pennaceus typical: A rather convex-sided shell with low spire and rounded shoulders; the whole shell is suffused with dark bluish gray and the tents are more-or-less restricted to narrow axial and spiral bands between very large and irregular dark blotches showing distinct spiral dots and dashes. This is the shell commonly called *C. praelatus*, but actually the type of *C. praelatus* agrees very closely with the type of *C. pennaceus*.

C. pennaceus variety *omaria:* Rather cylindrical, the shoulder rounded and the spire usually rather low and very blunt; pattern usually very open with small blotches, but variable. This is commonly considered a full species distinguished by a somewhat narrower body whorl, rounded shoulder, and low spire, but too many specimens and series are intermediate.

C. pennaceus variety *episcopus:* The most common concept of *C. pennaceus*, with the body whorl broad and the shoulder rather narrowly angled; the spire is moderately high to flat, and the pattern is quite variable but usually has distinct large blotches. Unfortunately this name has been badly confused with *C. magnificus.*

C. pennaceus variety *sindon:* A rather small and cylindrical *pennaceus* covered with nearly parallel dark lines which seldom form distinct tents but are closely spaced and thickened into two broad spiral bands above and below midbody. This uncommon variety is apparently a marginal form that may result from harsh environmental conditions; it has been recorded from the eastern African coast and Madagascar, the Ryukyus, and Hawaii. Collectors usually call this form *C. elisae.*

Individual and local populational variants are abundant and just cannot be covered in a limited space. The height of the spire is very variable, from flat to quite tall, sometimes deeply concave or convex; although usually white, pinkish tints are common, while bluish or grayish tints are less often found. The blotches may be absent, small, distinct, very ragged, spirally fused, axially fused, or cover most of the body whorl; occasional specimens are blackish lined and look like *C. marmoreus.*

Distribution.—Widely distributed and usually common throughout the Indo-Pacific from the African coast to Hawaii and French Polynesia. Usually found in shallow water, but sometimes dredged in moderately great depths. Some of the variants are uncommon.

Synonymy.—The major synonyms are covered with variants. I cannot see any reason to recognize the variants as full species or subspecies as they intergrade fully and are usually sympatric but sometimes ecologically separated from *C. pennaceus* var. *episcopus.* It is unfortunate that the varietal names must be changed from what is familiar to collectors, but I see no other choice. The following summary should help with the changes:

C. praelatus of collectors = *C. pennaceus* typical
C. episcopus of collectors = *C. magnificus*
C. pennaceus of collectors = *C. pennaceus* var. *episcopus*
C. omaria of collectors = *C. pennaceus* var. *omaria*
C. elisae of collectors = *C. pennaceus* var. *sindon*

Of the other names, most apply without doubt here and fall into the three major variants and occasionally into *sindon. C. stellatus* seems to be a *sindon*-like shell with greatly reduced pattern. *C. convolutus* is a pinkish *omaria*-type with a tall and concave-sided spire; similar shells have been taken in Zanzibar and Madagascar. *C. marmoricolor* is described as being netted with black and having the white areas placed as in *C. marmoreus,* but not coronated; I have never seen a specimen or figure of this form; *C. magoides* is similarly described as looking like *C. magus*; the status of both these names is doubtful but they presumably apply here.

Notes.—This is often a very common shell. Hawaiian specimens have sharply angled shoulders and very low spires, but such specimens can also be found in Sri Lanka and other localities. The normally seen specimens are somewhat intermediate between *omaria* and *episcopus* in shape with an openly netted pattern with large blotches. The typical variety and *sindon* are uncommon to rare, especially *sindon*.

This is usually a clean species but the early whorls are almost always eroded and there may be some heavy axial growth marks. Although the lip is often thick, the growing edge is fragile and usually chipped. Pinkish specimens are moderately common.

C. magnificus is considered a distinct species although admittedly intermediate specimens are found. *Omaria* is even less distinct with intermediates to variety *episcopus* common.

PERGRANDIS (Iredale, 1937)

1937. *Embrikena pergrandis* Iredale. *Festschrift. . . Prof. Dr. Embrik Strand* (*Riga, Univ. Lettlands*), 3: 407, pl. 18 (Tahli Bay, New Britain).

1972. *Conus fletcheri* Petuch and Mendenhall. *Veliger*, 15(2): 96, figs. 1-2 (SW of Penghu Is. Group, Republic China).

?1973. *Virgiconus tamikoae* Shikama. *Sci. Rpts. Yokohama Nat'l. Univ.*, Sec. II(20): 7, pl. 1, figs. 8-9 (N. of Senkaku Is., Japan).

Description.—Heavy, with a low gloss; biconical, the upper sides convex and tapering to a rather narrow base; body whorl covered with deep spiral grooves, crowded anteriorly and widely spaced posteriorly, separated by flat ribs; some ribs with secondary spiral grooves; whole body whorl covered with heavy and fine spiral and axial threads and growth marks; shell sometimes without grooves on posterior quarter; shoulder broad, roundly angled, concave above; spire tall to moderate, sharply pointed, the sides straight or slightly concave, the whorls weakly stepped; early 4-6 whorls nodulose, later ones finely undulate then straight; tops of whorls concave, with about 3-5 spiral ridges and fine axial threads. Body whorl white to straw, unpatterned or with up to three spiral bands of large and small dark brown blotches; blotches below shoulder often in the form of oblique flammules, sometimes fused to a spiral band of large squarish blotches above midbody, and with a small band above the base; blotches variously developed and fused; spire mostly whitish with some scattered brown blotches. Aperture moderately narrow, nearly uniform in width or slightly wider anteriorly; outer lip fragile, nearly straight; mouth white. Columella heavy, oblique. Length 80-173mm.

Comparison.—In shape this species is quite similar to *C. stupa* or *C. stupella*, but immediately distinguishable by the deep grooving on the body whorl and the color and pattern. It is hard to compare with any other species that comes to mind, but undoubtedly small shells and juveniles would be closer to some familiar species. There may be a superficial resemblance to *C. lynceus*, but that species lacks nodulose early whorls and has a different columella as well as being lighter in weight and smaller. I have seen a very large shell obviously allied to *C. sulcatus* from New Britain which was quite similar to the type of *C. pergrandis* in general appearance, but even this specimen was distinctly coronate (actually strongly undulate) to the shoulder.

Variation.—It is possible that two species are being confounded under this name, but the few specimens so far described and figured have fallen into almost as many shapes and patterns as there were specimens. At one extreme is the type of *C. pergrandis*, which is white, without pattern, has a very high spire, and is heavily grooved to the shoulder. The type of *C. fletcheri* is toward the other extreme with weaker grooving posteriorly, a more convex shape, and very heavy axial brown blotches posteriorly on the dorsal surface. *C. tamikoae* seems to be a smaller specimen with a lower spire, straighter and more concave sides, a straw-colored background, and blotches in distinct bands above and below mid-body plus scattered. Other specimens have had broad spiral bands of brown with rather sharply angled shoulders and lower spires, some with very heavy spiral sculpture and others much smoother. It seems that this large shell is quite variable, as is typical with other deep-water cones.

Distribution.—Probably widely distributed in deep waters of the Western Pacific. Recorded from Japan, Taiwan, New Britain, and Queensland, Australia, but rare.

Synonymy.—The types of the three synonyms are all quite distinct, but it seems that they would fall within the range of variation of a single variable species. See *Variation.*

Notes.—This is not an attractive cone, as it is usually taken dead at this time. The lip breaks easily. There are usually heavy axial growth marks and flaws as well as scars. It is such a rare and seldom offered species at the time of writing, however, that even bad specimens can be sold.

In some ways this is not a very satisfactory species. I know of no really small specimens that can be referred to it, which automatically raises the question of whether it might not be very large specimens of a more common species distorted beyond recognition by the large size, such as senile specimens of other species are sometimes hard to connect with more typical specimens.

PERONIANUS (Iredale, 1931)

1931. *Floraconus peronianus* Iredale. *Rec. Australian Mus.*, 18: 224, pl. 25, fig. 12 (Sydney Harbour, New South Wales, Australia).

Description.—Very thin and light in weight, with a good gloss; low conical to obconical, the upper sides gently convex; body whorl with about a dozen or so weak spiral ridges on the anterior quarter, the rest of the whorl smooth except for numerous fine spiral and axial threads; shoulder broad, roundly angled; spire low or almost flat, the sides concave, the early whorls forming a sharply pointed cone; tops of whorls flat or slightly convex, with about 6-10 narrow spiral ridges crossed by scattered axials. Body whorl white, cream, or fleshy tan, usually with large and small purple, red, or tan blotches above and below a paler midbody band; sometimes with traces of pale brown crosshatching scattered in the pattern or with three narrow bands of color; whole shell often with pinkish tinge; base pale; shoulder and spire about same color as body whorl, with angular brown lines and blotches, early whorls pale brown. Aperture rather wide, especially anteriorly; outer lip thin, fragile, evenly convex; mouth deep pink. Columella long, very narrow, somewhat oblique. Length 40-70mm.

Comparison.—*C. peronianus* is a doubtful species which is obviously closely allied to *C. anemone* and may be just a deep-water ecotype of that species. However, for the moment it is kept as distinct. It is easily distinguished from typical *C. anemone* by having the spiral ridges restricted to the base, by having a more inflated shell with a high gloss and reduced pattern, and by having the mouth often deep pink. This is a very clean-looking cone as compared to most forms of *C. anemone*. From *C. papilliferus* it differs in the more elongated body whorl, very narrow and sometimes indistinct columella that is not free, and the brighter colors.

Variation.—As might be expected with a species so close to *C. anemone*, the pattern varies considerably. In some specimens the shell is fleshy pinkish or tan with remnants of spiral rows of brown spots and blotches especially prominent above midbody; larger blotches vary from dark purple or tan to pale pinkish and may be very obscure. The pattern of the shoulder is often strongly contrasted with the pattern of the body whorl just below the shoulder.

Distribution.—Uncommon in deep water from New South Wales, Victoria, and Tasmania to Western Australia, Australia.

Notes.—This is obviously a large, inflated, smooth *C. anemone* type with bright colors. Even if it should prove to be merely an ecotype, it is still a very distinctive shell.

There are sometimes axial scars and growth marks, and the early spire whorls may be eroded or broken. The lip is very fragile. Specimens often show what is a fairly typical banded *C. anemone* pattern but replaced with bright violets, reds, or purple.

PERTUSUS Hwass, in Bruguiere, 1792

1792. *Conus pertusus* Hwass, in Bruguiere. *Cone*, in *Ency. Method., Hist. Nat. des Vers*, 1: 686 (East Indies). Lectotype figure (here selected): *Tableau Ency. Method.*, Pl. 336, fig. 2 (incorrectly considered the holotype by Kohn, 1968).

1810. *Conus amabilis* Lamarck. *Ann. du Mus. Hist. Nat. (Paris)*, 15: 425 (Mers du grandes Indes). Later reference to Chemnitz, Pl. 182, figs. 1770-1771.

1817. *Conus festivus* Dillwyn. *Descr. Catalogue Recent Shells*, 1: 413 (Moluccas Islands). Based on Chemnitz, Pl. 182, figs. 1770-1771.

?1844. *Conus nitidus* Reeve. *Conchologia Iconica*, 1(*Conus*): Pl. 47, sp. 266 (Locality unknown).

Description.—Moderately heavy, with a high gloss; low conical, the upper sides straight or slightly convex; body whorl with low spiral ridges on the anterior third or so, separated by deeply punctate grooves; more posteriorly the punctate grooves become widely spaced and almost obsolete, while fine spiral threads may still be distinct; sometimes with punctations to shoulder; very fine axial threads sometimes present; shoulder broadly to narrowly rounded, convex above, broad but not very distinct from spire; spire rather low, bluntly pointed if sharp protoconch is not present, the sides usually convex; early 1-2 whorls very weakly nodulose or undulate in some specimens but usually eroded; spire whorls slightly convex to flat above, with about 3-4 low spiral ridges separated by punctate grooves. Body whorl bright orange-red, deep pink, or deep salmon, usually with three spiral bands of large opaque white blotches at shoulder, midbody, and above base, the white blotches apparently formed by fusion of smaller blotches or spots; midbody blotches often fused into a broad continuous spiral band with undulating edges; indistinct curved brownish pink axial flammules sometimes visible; base often tinged with violet; there may be numerous rows of fine brown flecks or dashes alternating with small white spots covering the whorl, these seldom conspicuous other than in white blotches but sometimes almost continuous; opaque white spots sometimes forming short axial flammules above and below midbody; spire white or pink with large and small axial red to salmon blotches and lines; early whorls pink. Aperture rather narrow, slightly wider anteriorly; outer lip thin, sharp, fragile, straight; mouth violet to pink, dark or light. Columella short, narrow, mostly internal. Length 20-41mm.

Comparison.—The bright colors and punctate spiral grooves on the spire whorls are very distinctive. Perhaps small *C. vexillum* are comparable, but they would be yellowish, not reddish, and would have brown blotches on the spire. Some small *C. cernicus* may be very similar in color but have the white spots scattered and not forming large blotches, have a broad spiral band below the shoulder, and have the spire whorls undulate or heavily nodulose.

Variation.—This species seems to have two major color patterns with wide intergradation. In the most brightly colored form the colors are bright reddish with large white blotches, the brown dashes weak or nearly absent; there are no axial pinkish brown flammules. The other pattern is usually found on more slender and less convex-sided shells and has the colors paler, bright pink or salmon, the whitish blotches more pinkish and less contrasted, and commonly with poorly distinguished more brownish axial flammules above and below midbody; the spiral rows of brown and white dashes may be very well developed, the white dots forming axial flammules. Most shells from Hawaii are of the broad, brightly patterned type but are pale salmon instead of reddish, with the white blotches small and the brown dashes often quite conspicuous; more typical shells are taken in Hawaii, however.

Occasional specimens are heavily corded over half or all the body whorl, the cords separated by deep wide grooves with heavy axial ridges; such shells have a very unique appearance because the pattern is very disrupted.

Distribution.—Widespread through the Indo-Pacific from eastern Africa to Hawaii and French Polynesia, but seldom common and often rare. Found in shallow to moderately deep water.

Synonymy.—The type of *C. pertusus* is the slender pale type, while Lamarck's and Dillwyn's names are based on the broader and brighter shell; the type figure of *C. festivus* has an exceptionally large amount of white on the body whorl but seems to belong here. *Conus nitidus* is doubtful but could be referred to the Hawaiian variant.

Notes.—Although locally not uncommon in some areas, this is an expensive shell. Both forms are very colorful and attractive, and dealers do not distinguish between the two. The small Hawaiian shells, however, are drab and fade rapidly, being worth only about a tenth as much as more typical specimens.

The spire is commonly weakly eroded and may have white encrusting deposits; the lip is fragile; and there may be axial growth flaws and occasional inconspicuous scars. Most specimens appear very clean at first glance because of the high gloss.

This species fades rapidly when exposed to light.

PHILIPPII Kiener, 1845

1845. *Conus Philippii* Kiener. *Species gen. et icon. des coqu. viv.*, 2 (*Conus*): Pl. 98, fig. 2. 1848, *Ibid.*, 2: 213 (Coast of Mexico).
1869. *Conus Floridanus* Gabb. *Amer. J. Conch.*, 4: 195, pl. 15, fig. 4 (Tampa Bay, Florida).
1870. *Conus floridensis* Sowerby iii. *Proc. Zool. Soc.* (*London*), 1870: 256, pl. 22, fig. 11 (Florida).
1942. *Conus floridanus burryae* Clench. *Johnsonia*, 1(6): 29, pl. 14, figs. 3-4 (Off Lower Matecumbe Key, Lower Florida Keys).

Description.—Moderately heavy, with a good gloss; biconical, the upper sides straight to slightly convex, the base narrow; body whorl with about a dozen weak spiral ridges anteriorly, separated by shallow grooves; axial threads usually weak or absent; shoulder wide, usually carinate or sharply angled; spire tall, sharply pointed, the sides straight to slightly concave; earliest 2-4 whorls with very fine nodules, the others carinate and often projecting; tops of whorls flat to slightly convex, seldom concave, with numerous fine curved axial threads and sometimes traces of a single spiral ridge. Body whorl white, usually heavily covered with pale yellow to bright reddish brown clouds forming two very broad bands above and below midbody area; usually only base, midbody, and area below shoulder white in adult shells; usually many spiral rows of small reddish brown dots and squares over whorl, these sometimes absent or fused into nearly complete lines; base often tinted waxy salmon; pattern and color very variable; spire and shoulder white, with large and small dark brown to yellowish brown spots and blotches; early whorls often brown. Aperture moderately narrow, slightly widened anteriorly; outer lip sloping below level of shoulder, thin, sharp, straight or slightly convex; mouth white to bluish white. Columella internal. Length 30-52mm.

Comparison.—Pale yellow specimens could be confused with *C. stimpsoni*, but that species has weak but distinct spiral grooves over the body whorl and fine spiral ridges on the spire whorls; also, it is commonly axially plaited. *Conus cingulatus* is similar in shape but is covered with spiral grooves to at least midbody and usually to the shoulder; its pattern is usually made of larger dashes; and the spire is often very dark brown with only small crescentic white spots.

The closest relative of *C. philippii* is *C. delessertii*, which is very similar in shape, sculpture, and general pattern. However, in *C. delessertii* there are usually distinct, rather narrow spiral bands of bright salmon on the body whorl instead of the broad yellowish bands of *C. philippii;* there is a tendency to form oblique axial flammules from fusion of the larger dashes; and the spire whorls are somewhat more concave than in *C. philippii.* Additionally, deep-water *C. philippii* are usually dark orange-brown or reddish brown in tone, are more convex posteriorly, and may have distinctly lower spires than in *C. delessertii.* The early whorls of *C. philippii* are more weakly nodulose than those of *C. delessertii*, which commonly has four nodulose whorls.

Variation.—There is considerable individual and local populational variation in height of spire and convexity of the posterior sides. The pattern falls into three overlapping variants which are presumably ecological in origin.

C. philippii typical: Body whorl covered with many spiral rows of dark brown dashes and spots; color commonly dark reddish brown or orange-brown, the color usually as two broad spiral bands above and below a narrow pale midbody band; base of shell yellowish; usually in moderately deep water.

C. philippii variety *floridanus*: Body whorl pale yellowish to pale tan, sometimes very blotchy and unevenly colored; spiral rows of dashes weakly distinguished, but sometimes visible as nearly complete lines deep under the color pattern; midbody band often indistinct; base usually pale; usually in shallow water.

C. philippii variety *burryae:* Body whorl dark reddish brown, sometimes mottled with paler, the midbody band indistinct or absent; continuous spiral dark brown lines often present; base mostly dark brown; shallow water of the southern tip of Florida and the Keys.

Intergrades between these varieties are common; *burryae* is very poorly distinguishable and a minor variety.

Left: *C. philippii* var. *burryae,* Key Largo, Florida (GG).

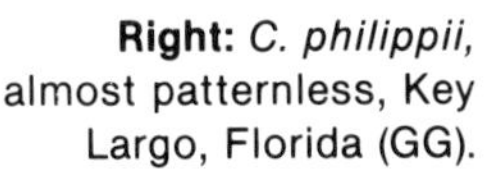

Right: *C. philippii,* almost patternless, Key Largo, Florida (GG).

Distribution.—Common in shallow to moderately deep water from the coast of North Carolina around Florida into the Gulf of Mexico; also recorded from Yucatan, and perhaps more widely distributed in the northern Central America area.

Synonymy.—This species is usually called *C. floridanus*, but I can see no way to distinguish *C. philippii* from the highly patterned deep-water variety. The only difference seems to be that the type of *C. philippii* has a more rounded, not carinate shoulder; however, such a form is sometimes found in specimens of this species which have suffered axial breaks, as apparently the type of *C. philippii* has been injured. The pattern and general shape are easily matched with Florida specimens. *Burryae* is very weakly distinguished and inconstant. *C. floridensis* is the typical deep-water variety, as Sowerby probably suspected.

Notes.—The typical and *floridanus* varieties are easy to obtain, as are intermediates. Each Florida locality seems to have a slightly distinguishable form, some of which are very attractive and interesting. *Floridanus* is usually heavily scarred and broken, while the typical shells are usually very clean except for minor chipped lips. Nearly all-white typical shells have been found, as have very dark reddish brown specimens. *Burryae* is harder to obtain because of the limited range and the inconstancy of the characters; shells without bad breaks can be very attractive if the pattern is uniform, but usually they are mottled and not very attractive. It might be mentioned that juveniles of all three forms are very similar and have dots and dashes.

Specimens from the Mexican coast are seldom available, most specimens coming from both coasts of Florida. This species seems to be restricted to the continental shelf.

PICTUS Reeve, 1843

1843. *Conus pictus* Reeve. *Conchologia Iconica*, 1(*Conus*): Pl. 18, sp. 98 (Locality unknown).

1911. *Conus Beckeri* Sowerby iii. *Proc. Malac. Soc. London*, 9: 352, fig. (St. Francis Bay, South Africa).

Description.—Moderately light in weight with a good gloss; low conical, the sides nearly straight; body whorl with fine crowded spiral ridges basally, otherwise almost smooth; shoulder broad, rather narrowly angled, concave above; spire low to moderate, blunt, the sides slightly convex to straight; whorls distinctly but weakly stepped, concave on top, with many fine axial threads but apparently lacking spiral sculpture; sutures often rude. Body whorl white to pinkish, with three broad spiral bands of bright reddish brown to orange (eroded), a narrow band below the shoulder and wide bands above and below mibody; these bands thus leave a whitish but usually heavily flammulated base, a rather broad midbody white band, and a narrow white band between midbody and the shoulder; these white bands are usually heavily flammulated with short, angular dark brown axial stripes; the flammules are usually restricted to the white bands but may occasionally be longer or even cover most of the shell; occasionally with a few spiral rows of brown dots on whorl; basal oblique flammules usually visible dorsally as well as ventrally; spire and shoulder whitish with many curved brown spots and lines; early whorls whitish. Aperture moderately wide, slightly wider basally; outer lip straight; mouth violet to pale brown. Columella short, narrow. Length 20-50mm.

Comparison.—Although some specimens may vaguely resemble *Conus centurio*, there is no close relationship between these species and they are easily distinguished by size, sculpture, and details of pattern. The closest relative is *C. algoensis scitulus*, and I am not at all sure specimens can be correctly separated. The major difference between the two is the distinctly stepped spire whorls that are concave on top, but this shape is approached by some *scitulus* and by juveniles of *algoensis simplex*. The color pattern is usually distinctive, as *C. pictus* has broad reddish bands separated by narrow white bands and the axial brown flammules are usually restricted to the white bands, but some *C. pictus* have the reddish bands broken, narrow, and heavily infiltrated with flammules.

Variation.—Live-taken specimens seem to be bright reddish brown on white, but beach specimens are usually bright orange on pinkish, the bands not as sharply defined. There is considerable variation in height of the spire. Although the flammules are usually restricted to the white bands, they sometimes are longer, wavy, and may cover much of the body whorl. The base is usually covered with sharply defined oblique brown stripes.

A 13mm juvenile has a nearly flat spire, sharply angled shoulder, and two very indistinct white spiral bands with traces of fine brown axial lines within; the shoulder is pale, as is the unmarked spire. This juvenile is quite distinctive and perhaps the best evidence for the specific status of this taxon.

Distribution.—Restricted to the coast of South Africa, with beach specimens uncommon along the eastern and southeastern areas. Live-taken specimens have been found in rather shallow water.

Synonymy.—*Conus beckeri* is based on two very large specimens, one with a distorted spire. These specimens have the axial flammules very strongly developed and covering much of the body whorl, as well as having spiral rows of brown spots; in these respects they are somewhat intermediate with *C. algoensis scitulus*.

Notes.—Fresh or even good beach specimens of *C. pictus* are very pretty shells, although they are somewhat rough and often have axial flaws and broken lips; the spire may be quite eroded.

Even good beach specimens of *C. pictus* are uncommon, perhaps only slightly less so than *C. bairstowi*. Live-taken specimens are apparently rare at this time, but they could eventually reach the market.

PILKEYI Petuch, 1974

1974. *Conus* (*Phasmoconus*) *pilkeyi* Petuch. *Veliger*, 17(1): 40, figs. 3-6, 9 (Marau Sound, N. of Guadalcanal, British Solomon Islands).

Description.—Moderately heavy, with a high gloss; low biconical, the upper sides slightly but gently convex then tapering to a narrow base; body whorl with about 7-15 deep spiral grooves on anterior half to two-thirds of whorl, the ridges between narrow and rounded anteriorly, broad and flat posteriorly; sometimes shallower grooves partially dividing the ribs; grooves axially threaded within; whole shell covered with fine spiral and axial threads, with a satiny sheen; shoulder roundly or narrowly angled, broad; spire moderate to high, sharply pointed, the sides concave; spire whorls almost flat above or slightly concave, with about 5-9 narrow spiral ridges crossed by many fine axial threads and slightly cancellate; early whorls with somewhat carinate margins. Body whorl bright orange or yellowish orange, usually with traces of three broad spiral bands of pale tan, one below shoulder, the others above and below midbody, or covered with deep mahogany brown obscuring the background except in indistinct paler orange bands at midbody and base;

occasionally with indistinct scattered brown spots visible; spire and shoulder orange or yellow, paler on early whorls and covered with dark brown flammules and spots, tip of spire largely brown. Aperture modev-ately narrow posteriorly, widened anteriorly; outer lip rather thin, sharp, evenly convex or straight; mouth pale to deep orange, darkest in shells with darkest body whorls, sometimes rust-red. Columella long, narrow, indented from inner lip and bounded by a low ridge. Length 40-70mm.

Comparison.—*C. pilkeyi* is very similar in shape and sculpture to *C. ochroleucus* but differs in color and slightly in shape. The dark variety is easily distinguished by the deep mahogany, almost black color or even by the very bright orange that is not found in *C. ochroleucus*, as well as the very dark orange or reddish mouths. Such shells are not a problem in identification. The pale variety, however, is often the same pale orange or yellowish color as typical *C. ochroleucus*, with which it is sympatric. Usually it can be distinguished by the somewhat lower spire and more rounded shoulder, the somewhat more orange mouth, and the common presence of three indistinct tan spiral bands on the body whorl, these often with very indistinct spots. In both color varieties *C. pilkeyi* usually has distinct dark brown axial flammules on the early and middle spire whorls instead of the indistinct light brown striping of *C. ochroleucus;* the margins of the spire whorls appear definitely spotted in *ochroleucus*, but very indistinctly so in *C. pilkeyi*.

Variation.—As indicated, there are two fairly distinct but inter-grading color varieties of this species. In the typical form the body whorl is either very dark mahogany, deep reddish orange, or bright orange; the darkest shells show the bright orange background at midbody and some-times the base and shoulder. Such shells are usually rather narrow, rather sharply angled at the shoulder, and have high spires. The mouth is bright orange to bright rusty red. Juveniles of this form are cylindrical, pale or deep orange, and have two spiral bands of axial dark brown flammules (or deep orange tan) which with growth connect by narrower flammules and eventually fuse spirally to produce very broad spiral bands. The early whorls of the spire are sometimes grayish and heavily flammulated.

Left: Juvenile *C. pilkeyi*, Papua New Guinea, 27mm (B. Parkinson).

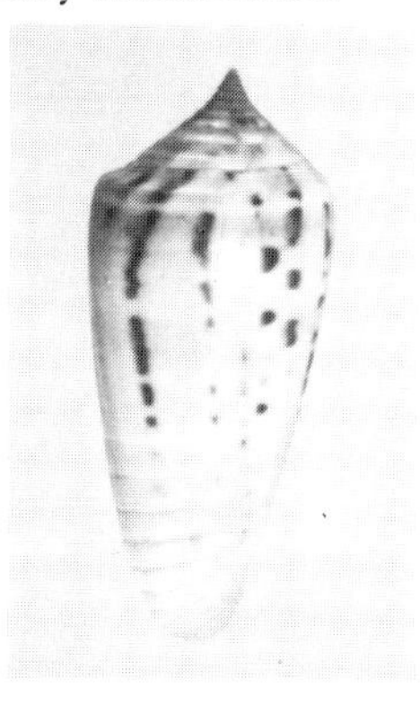

Right: Spotted *C. pilkeyi*, unusual Philip-pine specimen (DM) (photo Jayavad).

The pale variety is sometimes bright orange-yellow but more commonly pale yellowish tan; commonly there are three spiral broad bands of tan, one at the shoulder and two above and below midbody, the bands sometimes containing indistinct darker spots; the mouth is yellowish to pale orange, darker deep within. In this form the shape is somewhat oval, the shoulder roundly angled, and the spire lower; the flammules on the spire are sometimes indistinct. Juveniles of this form have three broad tan bands with indistinct darker spots within at sizes comparable to those of the dark form, the pattern being partially retained in adults.

Distribution.—Known from the Philippines through New Guinea and the Solomons to the New Hebrides and Fiji. The Fiji population is apparently isolated (pale variant). Uncommon in shallow to moderately deep water.

Synonymy.—It is surprising that this species has not been previously named. Collectors use the name *C. daullei* for pale Philippine shells, but that name is a synonym of *C. consors.*

Notes.—This species seems to be fairly distinct from the more common *C. ochroleucus*, although admittedly some specimens are doubtfully placed. Juveniles are apparently quite variable and I cannot be sure of always separating them from juveniles of *C. ochroleucus;* they are heavily grooved like that species.

Dark specimens from the Solomons and New Hebrides are uncommon; dark specimens tending toward the pale variant also occur off New Britain. The pale form is common in the Philippines, uncommon in New Guinea, and apparently rare in Fiji. Possibly these are subspecies, but the distribution seems rather random. Usually this is a clean shell with few flaws.

PLANORBIS Born, 1778

1778. *Conus planorbis* Born. *Index Rerum Naturalium Musei Caesari Vindobonensis*, 1(*Testacea*): 148 (Locality not stated). Lectotype selected and figured by Kohn, 1964.

1791. *Conus polyzonias* Gmelin. *Systema Naturae per Regna Tria Naturae*, ed. 13, 1: 3392 (Locality not stated). Lectotype figure (Kohn, 1966): Born, 1780, Pl. 7, fig. 13 (probably = lectotype of *C. planorbis*).

1792. *Conus vulpinus* Hwass, in Bruguiere. *Cone*, in *Ency. Method., Hist. Nat. des Vers*, 1: 648 (Guinea). Lectotype selected and figured by Kohn, 1968.

1792. *Conus ferrugineus* Hwass, in Bruguiere. *Ibid.*, 1: 649 (Guinea). Lectotype (here selected): specimen figured by Kohn, 1968, as the holotype.

1857. *Conus Chenui* Crosse. *J. Conchyl. (Paris)*, 6: 381, pl. 11, figs. 3-4 (New Caledonia).

?1873. *Conus Loebbeckeanus* Weinkauff. *Syst. Conch. Cab.*, ed. 2, 222 (*Conus*): 221, pl. 36, figs. 3-4 (Locality unknown).

Description.—Heavy, with a good gloss; low conical, the upper sides slightly convex to straight; body whorl with crowded anterior spiral ridges at base, followed by widely spaced heavier ridges with usually heavy granules, the ridges sometimes visible to midbody or higher; many axial threads; shoulder broad, angulate, slightly concave above; spire low to moderate, the sides straight or concave, sharply pointed; spire whorls flat or concave above, with about 3-6 often indistinct spiral ridges crossed by many fine axial threads; grooves between ridges sometimes faintly punctate. Body whorl usually pale yellowish to pale tan, sometimes white, covered with many widely spaced spiral lines of small brown dashes, very thin, sometimes fused into continuous hairlines; base dark brown with reddish tone; two broad spiral bands of darker yellow to tan usually present above and below midbody, the bands sometimes with weak or strong axial flammules and lines within; occasionally dark axial flammules from shoulder to top of posterior band; shoulder and spire pale yellow to white, with a sparse checkering of squarish dark brown spots, heaviest posteriorly; early whorls whitish. Aperture moderately narrow, slightly wider basally; outer lip nearly straight, thin and sharp; mouth creamy white to bluish white, without a violet blotch anteriorly. Columella long and narrow. Length 30-61mm.

Comparison.—*C. planorbis* is part of a complex of four closely related species that are often confused. From *C. vitulinus* it is readily distinguished by the lack of a violet spot anteriorly in the mouth and by sparse spire spots instead of heavy black curved lines. *C. connectens* shares with *C. vitulinus* the pinkish or violet spot in the anterior mouth and is usually toned heavily with bright pink or violet. *Conus striatellus* is very close to *C. planorbis* and similarly lacks the dark spot in the mouth while having reduced spire patterns. However, it is usually covered with many very fine spiral hairlines which are closely spaced and has a very irregular pattern of tan on white with very jagged spiral bands on the body whorl; the sides are somewhat more inflated to give a rather cylindrical, not conical, appearance. I am not sure that all specimens can be separated.

Variation.—Most specimens of *C. planorbis* now available fall into two somewhat distinct forms. The most common type is a yellowish shell with two slightly darker broad bands above and below midbody; there are numerous widely spaced fine brown lines over the body whorl, but axial flammules are rare or absent, although occasionally indicated below the shoulder; the base is reddish to dark violet-brown. The other variant is often called *C. chenui*, but it seems scarcely distinct. In this form the background is very pale yellow or whitish with two irregular spiral tan bands above and below midbody, these often with heavy short axial flammules; the flammules may be restricted to the bands or cover most of the body whorl. The base is reddish to brownish as usual.

According to the literature this is a very variable species, but I suspect some of this variation is due to misidentifications of related species. There is considerable variation in spire height, however, from nearly flat to quite high.

Distribution.—Common in shallow to fairly deep water from eastern Africa to French Polynesia, but not reaching Hawaii. The *chenui* variety seems restricted to the area between the Philippines and New Caledonia but is uncommon or moderately rare.

Synonymy.—The type of *C. planorbis* is an odd specimen that is nearly cylindrical with a flat spire; it is certainly not typical of the species as usually seen. *C. polyzonias* is apparently based indirectly on the same specimen. *C. vulpinus* and *C. ferrugineus* are based on more typical specimens, although with patterns more typical of *C. striatellus* than normal *C. planorbis*. *C. chenui* seems to be an uncommon variant (its pattern fits *C. ferrugineus* more than typical *planorbis*), with *C. loebbeckeanus* apparently a synonym.

Notes.—Except for the *chenui* form, this species has been surprisingly constant in color and pattern. Only one juvenile with a spire pattern similar to that of *C. vitulinus* and apparently lacking the dark spot in the mouth was seen, and it is quite possible that it was a faded *C. vitulinus* not showing the mouth spot clearly.

Some specimens are almost uniformly yellow and very contrasted to the reddish or violet-brown base. Good specimens are easy to obtain if one does not mind a little spire erosion and some limy deposits; the lip is fragile, and there may be fine axial growth marks and occasional unobtrusive scars.

Many misidentified specimens of this species are in circulation, and apparently many dealers cannot distinguish adequately any of the four species in this complex to judge from the mixed series I have seen. Normally this does little harm, but *C. planorbis* has been sold as *C. connectens* to the unwary.

POHLIANUS Sowerby iii, 1887

1887. *Conus Pöhlianus* Sowerby iii. *Thesaurus Conchyliorum*, 5(*Conus* Suppl.): 257, pl. 31(509), figs. 682-683 (New Ireland).

Description.—Moderately light in weight, with a good gloss; rather cylindrical, the upper sides convex, the body whorl elongate; body whorl with narrow low spiral ridges separated by weak grooves over the anterior third of the whorl, rest of body whorl covered with heavy spiral and axial threads in adults, smoother in juveniles; shoulder rounded to roundly angled, not very distinct from spire, rather narrow; spire low, sharply pointed, the sides slightly convex; spire whorls slightly convex on top, the sutures rude, with the first 5-7 whorls with distinct nodules; tops of whorls with about 2-3 distinct spiral ridges on posterior and middle whorls, the sculpture indistinct on later whorls. Body whorl creamy white to very pale straw, with two spiral bands of tan above and below midbody and sometimes another band indicated below the shoulder; bands apparently made from partial fusion of numerous lines of fine brown dots; axial growth marks commonly shaded with pale brown also; spire and shoulder whitish, the early whorls tan, with scattered small light brown dots and streaks. Aperture moderately narrow posteriorly, wider anteriorly; outer lip sharp, nearly straight; mouth white. Columella long and narrow, slightly indented from inner lip. Length 45-70mm.

Comparison.—*C. pohlianus* is obviously very similar to *C. magus* but fairly distinctive in the elongated shape, rounded shoulder, rather convex spire whorls, six nodulose whorls, and the reduced pattern of two pale brown bands above and below midbody on cream. This combination of characters should exclude all *C. magus*. *C. consors*, also an obvious *C. magus* derivative, is much broader and rather bulbous at the shoulder, bright yellow in color, and lacks distinct spire markings in most specimens. If the violet juvenile is truly this species, *C. pohlianus* also vaguely resembles *C. kinoshitai* but is readily distinguished by the very weak sculpture on the body whorl and the lack of large dark brown blotches.

Variation.—I have seen only a few fresh specimens of this species, but the adults are very simply patterned with two pale tan bands on creamy white. The bands are apparently made from fusion of the many small brown dots typical of *C. magus*, as the axial growth lines are also weakly marked with pale brown. It seems likely that some larger shells would be entirely white. Faded shells have only faint traces of the spiral bands.

A 31mm specimen is assigned here although it is basically a very pale violet shell with two very broad darker tan bands and some obvious brown dots in spiral lines; the spire is sculptured and marked like the early whorls of adults. The change in pattern from juvenile to adult is quite distinct and unlike the minor ontogenetic changes that occur in *C. magus*. The importance of juveniles in cone taxonomy is much overlooked, unfortunately.

Distribution.—I have seen shells only from deep water of the New Britain area, but the species probably occurs at least from the Philippines to the Solomons. It is moderately common beach, uncommon live-taken.

Notes.—Strict following of the nomenclatural rules would require spelling the name of this species as *Conus poehlianus*, but I cannot see the logic in changing the original spelling.

Smaller specimens are fairly clean, but larger specimens often have heavy axial breaks and growth marks. The pattern is seldom very dark and is often very inconspicuous. Juveniles are apparently uncommon or misidentified, although I cannot imagine with what.

POLYGRAMMUS Tomlin, 1937

1875. *Conus multilineatus* Sowerby iii. *Proc. Zool. Soc.* (*London*), 1875: 126, pl. 14, fig. 5 (Locality unknown). Non *Conus multilineatus* Pecchioli, 1864, a fossil.

?1913. *Conus filicinctus* Schepman. *Prosobranchia Siboga Exped.*, Monogr. 49(5): 384, pl. 25, fig. 1 (Madura Bay, Indonesia).

1937. *Conus polygrammus* Tomlin. *Proc. Malac. Soc. London*, 22(4): 206. Nomen novum for *Conus multilineatus* Sowerby iii, 1875.

Description.—Light in weight but solid, with a high gloss; low conical, the upper sides nearly straight or slightly convex, the base narrow; body whorl with about 6-12 low spiral ridges on the anterior quarter, the rest of the body whorl with numerous axial threads but otherwise smooth and waxy; shoulder sharply angled, weakly undulate, flat or convex on top; spire low or moderate, sharply pointed, the sides concave; early 1-3 whorls nodulose, the later ones weakly to distinctly undulate or low coronate; tops of whorls flat or slightly concave, with about 3-5 weak spiral threads or low ridges crossed by scattered or crowded curved axial threads. Body whorl glossy whitish, often with wide spiral bands of pale yellow, pink, violet, tan, or even green above and below midbody area; entire shell covered with many spiral rows of dark brown dots or dashes, these often fused into nearly complete lines or sometimes in roughly axial rows; base white, usually with pale tan or yellowish tones; shoulder and spire whitish, with scattered axial lines and spots of dark brown between the undulations. Aperture moderately narrow, uniform in width; outer lip thin, fragile, straight; mouth violet to pink, pale. Columella long, narrow. Length 20-40mm.

Comparison.—This species has a very high gloss and pale, usually pastel, colors. It could be confused only with such species as *C. muriculatus* in some patterns or *C. articulatus*. From both it is usually easily distinguishable by the pale base that is seldom violet and, when violet, is very pale. The spiral rows of dashes are heavier and more striking than in either of the other two species, while the sculpture of the body whorl is weaker and does not extend to midbody. *Conus voluminalis* is more elongate, has subdued dashes, and does not have the later whorls undulate. Once this species is seen it should present no identification problems.

Variation.—The tone of the shell varies from distinctly brown to distinctly yellow, going through pink, violet, and green tones between. In all specimens I have seen the spiral dashes are obvious, although in a few specimens they are somewhat reduced and rather buried in the body color. The shoulder and later spire whorls may be distinctly and strongly undulate, almost coronate, or almost straight; commonly the spiral sculpture is very weak. The base is white, variously tinted with yellow, pinkish, tan or occasionally pale violet.

Larger shells seem to be more elongated than small ones, with a relatively narrower shoulder. There may or may not be heavy black spots between the undulations, pale brown streaks being more common.

If *C. filicinctus* is correctly assigned here, which I doubt, the basal ridges may be granulose.

Distribution.—Apparently a rare or uncommon species dredged in moderately deep water of the western Pacific from the Philippines to the New Hebrides, most specimens coming from the New Guinea-Solomons area.

Synonymy.—It seems likely that the proper name for this species should actually be *C. filicinctus*; however, the type of that taxon is a very dead and eroded specimen with a very low spire, heavy black spots between undulations, and granulose basal ridges. The pattern is similar, consisting of spiral lines on an apparently tan background, but it just does not look quite like the shell described here. There would probably be no harm in using the name *C. filicinctus* if one prefers.

Holotype of *C. filicinctus* Schepman, 27.2mm (Zool. Mus. Amsterdam) (photo Abbott).

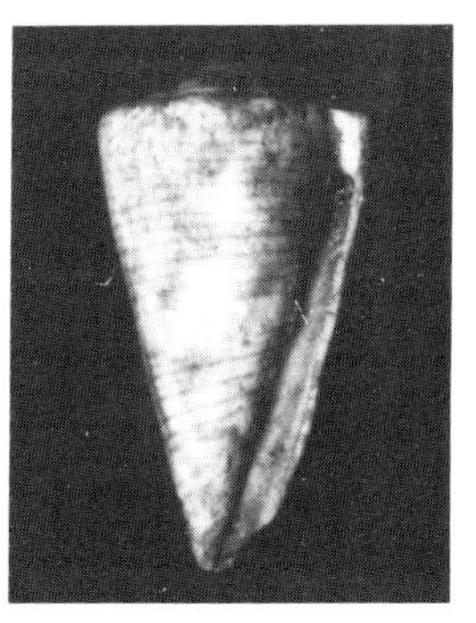

Notes.—The name *C. aureolus* has been used for this shell, but this is almost certainly incorrect. There is a strong resemblance to *C. voluminalis*, and some specimens in collections may be misidentified as that species. The two species are very different in shape, however.

This is a very waxy species with bright pastel colors in young specimens, but it is fragile and usually has minor axial flaws and chips in the lip. The early whorls of the spire may be broken or eroded.

Until recently this was a rare species, but it has now been taken in small numbers by dredging, and further exploration may show it to be only uncommon and more widely distributed.

POORMANI Berry, 1968

1968. *Conus poormani* Berry. *Leaflets in Malacology*, 1(25): 156 (Off Morro Colorado, Sonora, Mexico).

Description.—Moderately heavy, with a good gloss; low biconical, the posterior fifth of the body whorl with parallel sides, then tapering to a narrow base; body whorl with about a dozen or so low widely spaced spiral ridges over the anterior third of the whorl, the ridges weaker and more crowded at the base; rest of whorl with scattered remnants of spiral ridges and many spiral and axial threads; shoulder broad, carinate, concave above; spire moderate, sharply pointed, the sides slightly concave or straight; spire whorls distinctly stepped; first three or so early whorls weakly nodulose, later whorls carinate; tops of whorls concave, with about 3-5 weak spiral ridges (most distinct on early whorls) crossed by faint axial threads, appearing almost smooth on the later whorls. Body whorl white, with broad orange-brown axial flammules which seldom cross the midbody area but commonly have spiral extensions fusing with adjacent flammules; fully developed shells can have a vaguely reticulated appearance; spiral spots absent or very poorly developed; base light; spire and shoulder white, with many curved axial lines and spots of orange-brown; early whorls pale brown. Aperture moderately narrow, slightly wider anteriorly than posteriorly; outer lip thin and sharp, following outline of body whorl, not sloping strongly below shoulder posteriorly; mouth white. Columella very narrow, mostly internal, slightly oblique. Length 30-65mm.

Comparison.—The distinctive shape of the body whorl with the posterior part of the whorl parallel is found only in a few other species, most with similar patterns. The most similar sympatric species is *C. recurvus*, which is a much lighter shell with a darker and more reticulated pattern, about six nodulose spire whorls, and the outer lip strongly sloping below the shoulder level. The shallow-water *C. virgatus* is also similar and probably related, but it does not have the whorls stepped, is often pale salmon in tone and not white, and has the flammules usually extending through the midbody area; the shoulder is more rounded.

In the Caribbean *C. villepinii* and *C. clerii* are both quite similar. *C. villepinii* has the sides less distinctly parallel posteriorly, has more nodulose spire whorls as well as heavier spiral sculpture on the spire, and has a strong tendency to form broken spiral reddish brown bands above and below the midbody area instead of parallel or loosely reticulated flammules. *C. clerii* is very similar but shows several or many spiral rows of squarish brown spots on the body whorl and has the axial flammules much less regular, as well as normally having a less strongly stepped spire.

In the Pacific *C. tribblei* is similar in shape and lacks the undulate spire whorls of allied species, but it differs from *C. poormani* in being covered with many spiral rows of brown dashes, some usually still visible in adults, and in having strong spire sculpture, as well as being thicker and heavier and usually having a lower spire.

Variation.—The spire varies somewhat in height and extent of the stepping. The axial flammules may be straight, distinctly angled near the middle, or wavy and touching the adjacent flammules in several places to produce a loose reticulum. Usually the midbody area is unmarked, but not always; there may occasionally be traces of a few squarish brown spots present. Juveniles tend to have fewer and wider, straighter flammules than adults. Patternless specimens occur.

Distribution.—Uncommon in moderately deep water from the Gulf of California to the Gulf of Panama, possibly more widely distributed in the Panamic area.

Notes.—This recently described species seems quite distinct and it is mystery why it was not described earlier. This is one of several species noted to have long erect hairs on the periostracum of the shoulder, a character of very doubtful systematic value.

This is a rather rough cone with a tendency to heavy growth marks axially, these often causing the flammules to begin a new pattern. The early whorls are often slightly broken or encrusted, and the lip chips easily, as usual.

Some specimens of *C. clerii* of the Brazilian coast closely approach *C. poormani* in shape, sculpture, and pattern, differing mostly in having spots in the pattern.

PRAECELLENS A. Adams, 1854

1833. *Conus bicolor* Sowerby i, in Sowerby ii. *Conchological Illustrations:* Pl. 37, fig. 56 (Locality not stated) (July). Non *Conus bicolor* Sowerby i, in Sowerby ii, 1833 (March).

1833. *Conus Sinensis* Sowerby i, in Sowerby ii. *Conchological Illustrations:* small list. Nomen novum for *Conus bicolor* Sowerby i, in Sowerby ii, July or later, 1833. Non *Conus sinensis* Gmelin, 1791.

1849. *Conus Sowerbii* Reeve. *Conchologia Iconica*, 1(*Conus* Suppl.): 2. Nomen novum for *Conus sinensis* Sowerby i, in Sowerby ii, 1833. Non *Conus sowerbei* Nyst, 1836, a fossil.

1854. *Conus praecellens* A. Adams. *Proc. Zool. Soc.* (*London*), 1853: 119 (China Seas).

1857-1858. *Conus Sowerbyi*, Sowerby ii. *Thesaurus Conchyliorum*, 3 (*Conus*): 12. Emendation of *Conus sowerbii* Reeve, 1849. Non *Conus sowerbyi* Bronn, 1848, a fossil.

1870. *Conus sowerbyi* var. *subaequalis* Sowerby iii. *Proc. Zool. Soc.* (*London*), 1870: 257, pl. 22, fig. 5 (China Sea).

Description.—Light in weight, with a good gloss; tall biconical, the upper sides of the body whorl slightly convex then concave to the narrow base; body whorl covered with narrow rounded to rather flattened spiral ridges of about equal size (narrower at base and shoulder) and separated by axially threaded grooves; entire body whorl often with heavy axial threads; shoulder broad, carinate; spire very tall, over one-third entire length, sharply pointed, the sides straight; spire whorls slightly concave on top, mostly carinate except the early two whorls weakly nodulose; tops of whorls with 6-8 strong spiral ridges and fine axial threads. Body whorl creamy white to bluish white, heavily covered with spiral rows of large dark reddish brown dashes on the ridges; the dashes tend to condense into axial or oblique blotches and flammules to form more-or-less distinct broad spiral bands below the shoulder, above midbody, and below midbody (strongest); base white, usually unmarked; spire same color as body whorl, covered with narrow axial lines and spots of dark brown. Aperture narrow, uniform in width; outer lip thin, straight, sloping strongly below the level of the shoulder; mouth white, the external pattern showing through at the margin. Columella internal. Length 25-50mm.

Comparison.—One of the most elegant of the small cones, the shape alone is usually distinctive. From *C. acutangulus*, *C. orbignyi*, and similar species it is easily distinguished by the non-nodulose middle and later whorls and the exceptionally straight sides of the spire. In some ways it is closest to *C. eugrammatus*, from which it differs in not having stepped spire whorls and having many strong spiral ridges on the whorls, these stronger than the axial threads.

Variation.—*C. praecellens* is a very constant cone, varying little even in minor details of shape and sculpture. The pattern shows little variation other than the degree of condensation of the dashes into blotches and of the blotches into spiral bands. The flammules below the shoulder are commonly very oblique. There may be a yellowish tinge to the shell.

Distribution.—Moderately common in deep water of the western Pacific from southern Japan and Taiwan through the Philippines to the New Guinea-Solomons area and probably Queensland, Australia.

Synonymy.—The nomenclature of this species has been confused since its first introduction into the literature, with preoccupied names galore being applied. It seems that *C. praecellens* is the first available name, with *subaequalis* being a juvenile. It is unfortunate that the name *sowerbyi* (in various spellings) cannot be applied to this familiar cone. The name *C. praecellens* is sometimes incorrectly applied to the Japanese populations of *C. eugrammatus*.

Notes.—Except for minor chips in the lip and sometimes heavy axial growth marks, this is a very clean species. People may quibble over the proper name for the shell, but there is no confusing it with anything else. Most specimens come from the Philippines and Taiwan, but there is no reason to doubt that it is common at least as far south as the Solomons.

PRINCEPS Linnaeus, 1758

1758. *Conus Princeps* Linnaeus. *Systema Naturae per Regna Tria Naturae*, ed. 10, 1: 713 (Locality not stated). Holotype figured by Kohn, 1963.

1792. *Conus regius* Hwass, in Bruguiere. *Cone*, in *Ency. Method.*, Hist.

Nat. des Vers, 1: 617 (Indian Ocean). Lectotype figure (here selected): *Tableau Ency. Method.*, Pl. 318, fig. 3. Non *Conus regius* Gmelin, 1791.

1832. *Conus lineolatus* Valenciennes, in Humbolt and Bonpland. *Voy. inter Amerique—Rec. Observ. Zool. et Anat. Comp.*, 2: 336 (Acapulco).

1910. *Conus princeps* var. *apogrammatus* Dall. *Proc. U.S. Nat'l. Mus.*, 38(1741): 224 (Panama).

Description.—Moderately to very heavy, often cumbersome, with a low gloss; obconical, the sides nearly straight; body whorl with about 6-10 weak spiral ridges above the base, these sometimes visible as fine spiral threads as far as midbody; numerous fine axial threads over whole whorl, these often conspicuous; shoulder roundly angled, with heavy but low coronations, broad; spire low to flat, the early whorls usually badly eroded; spire whorls heavily low coronated, slightly concave above, with numerous fine axial threads and often traces of 2-3 very fine spiral ridges or threads. Body whorl bright deep pink to bright orange, rarely yellowish, with or without narrow to broad and often irregular axial flammules or threads of black or dark brown; spire colored like body whorl, with curved blackish lines; early whorls eroded white. Aperture rather narrow, almost uniform in width; outer lip sharp, straight; mouth dull pink to orange, often with milky white tones deep within. Columella short, narrow. Length 40-130mm.

Comparison.—The combination of bright colors with axial flammules, the obconical and broad shape, and the low but heavy coronations is very distinctive. It could conceivably be confused with *C. regius* var. *citrinus*, but in that species there are usually distinct spiral ridges on the spire whorls, the spire is taller and concave with heavy and erect coronations, and there are often pale brown or orange spiral lines in the pattern; additionally, the anterior ridges are often granulose. Perhaps the closest relative of *C. princeps* is *C. gladiator*, which has a similar shape, low coronations, and reduced body sculpture.

Variation.—The shape and sculpture are quite constant, but the color and pattern vary considerably. The background color varies from bright pink to bright orange, sometimes paler yellowish orange. The major variation of interest to collectors is in the size of the dark axial flammules, where three minor variants can be recognized. In *C. princeps* typical there are heavy, wide, and usually wavy flammules separated by about half their own width; variety *lineolatus* is covered with very fine axial blackish hairlines closely following the axial growth lines; and variety *apogrammatus* totally lacks dark markings. Commonly two or all three patterns are found on the same specimen.

Distribution.—Uncommon to moderately common in shallow water of the Panamic area from the Gulf of California to Ecuador.

Synonymy.—The varieties are obviously of little taxonomic importance as they are very inconstant and possibly reflect variations in growth of individual specimens.

Notes.—This is often a very rough cone, covered with heavy scars and axial growth marks. The spire is often eroded, as may be the body whorl. Usually the colors are bright, however, a necessity for a gem specimen regardless of the condition of the shell.

It is usually presumed that *apogrammatus* is the least common variety, *lineolatus* next so, and the typical form most common. However, at the time of writing the most easily obtainable variety and the cheapest seems to be *lineolatus*, with *apogrammatus* (seldom gem quality) second most easily available. The hardest to obtain has been typical *princeps*, which has drawn a high premium for often very poor shells with only a few poorly developed flammules. Curiously, the typical variety is by far the most attractive, at least to me. Has anyone ever collected juveniles of this species?

PROXIMUS Sowerby ii, 1859

1854. *Conus pulcher* A. Adams. *Proc. Zool. Soc.* (*London*), 1853: 117 (New Caledonia). Non *Conus pulcher* Lightfoot, 1786. Holotype figured by Cernohorsky, 1974, *Rec. Auckland Inst. Mus.*, 11: 139, fig. 35.

1859. *Conus proximus* Sowerby ii. *Ibid.*, 1859: 429, pl. 49, fig. 1 (Locality unknown). Syntype figured by Cernohorsky, *Ibid.*, 11: 139, figs. 36-37.

Description.—Moderately light in weight, with a good gloss; low conical, the sides nearly parallel or slightly convex posteriorly; body whorl covered with usually narrow flat or rounded spiral ridges from base to shoulder, these separated by deep grooves of about equal width; ridges on posterior half of whorl often flat, narrowly separated, or almost obsolete; axial threads reduced or absent in the grooves, but fine axial scratches over whole whorl; shoulder sharply angled, coronate, the coronations erect and sharp or low and rounded; spire low to moderate, sharply pointed, the sides concave; tops of whorls concave, slightly stepped, with strongly coronate margins; whorls with about three spiral ridges and many fine axial threads. Body whorl creamy white to pale straw, covered with fine spiral dashes and lines on the tops of the ridges, these sometimes fused into larger dashes that are interrupted with round white dots; dashes usually brownish, sometimes reddish; often two or three spiral rows of larger squarish dark blotches below shoulder and above and below mid-body area; faint tan axial clouds may be present; base pale; spire whitish, with brownish to reddish streaks and spots between coronations; early whorls white. Aperture moderately wide, wider anteriorly than posteriorly; outer lip nearly straight, thin and sharp; mouth white to violet, variable. Columella long and narrow, bounded by a rounded ridge. Length 20-45mm.

Comparison.—As mentioned earlier, *C. marielae* is probably not distinguishable from *C. proximus*, although it is normally orange-red while most *C. proximus* are brown.

C. proximus is very similar to *C. moluccensis* but usually easily distinguished. In *C. moluccensis* the body whorl is more elongate and usually concave at the middle and at the shoulder coronations; the color pattern usually consists of both spiral dashes and heavy axial flammules of brown or red; the grooves between spiral ridges are axially threaded; and the mouth is white with at most pale violet tones. *C. moluccensis* is usually larger than typical *C. proximus* and has a thickened, pinched-in lip and heavier columella. The basal ridges at least in *C. moluccensis* are heavily granulose, while many *C. proximus* are smooth.

Variation.—Typical specimens from most of the range are fairly constant in shape, although the height of the spire varies considerably. The sculpture is usually of the narrow-ridged type, but sometimes the ridges are broader and flatter, especially posteriorly; the basal ridges are commonly finely granulose, but the granules may be entirely absent or extend to the shoulder. The pattern in most typical specimens is distinctly long-dashed with two or three rows of small dark squarish blotches; there are sometimes distinct small white spots visible in the dark blotches or even between the dashes. Axial threads are very weak or absent in the grooves, and the mouth is usually pale to dark violet.

Philippines specimens are much more variable, although most specimens agree very well with specimens from farther east. However, there seems to be a strong tendency in at least some populations to enlarge the narrow dashes into wide rectangles which are sometimes arranged in distinctly axial lines; the color is commonly brown, but in some specimens the dashes are distinctly bright reddish brown or pale orange-red. Although the basal ridges are narrow and often pustulose, there is a strong tendency for the ridges to widen by midbody and be touching below the shoulder, sometimes with the grooves obsolete. The presence of axial threads in the grooves is not uncommon. The general shape, however, and thickness of the lip still agree completely with *C. proximus* and not with sympatric *C. moluccensis*. The mouth is commonly white or very pale violet in the Philippines specimens.

Distribution.—Found through the western and southern Pacific from the Ryukyus and Philippines south to New Guinea and the Solomons and east to the New Hebrides, New Caledonia, and Fiji. Usually in moderately deep water, but sometimes shallow. Usually rare but sometimes locally not uncommon.

Synonymy.—*C. pulcher* is preoccupied.

Notes.—Although commonly confused with *C. moluccensis*, an admittedly very similar species, *C. proximus* appears to be distinct. It seems that the two species have been too rigidly defined, as some Philippine specimens would not fit either species yet are certainly *C. proximus.*

Except for minor surface erosion and some limy deposits on the spire, *C. proximus* is a clean species. The lip is thin and often filed. In some specimens the shoulder coronations are very low and indistinct. Reddish Philippine specimens are certainly more attractive than the usual brown specimens.

PULCHER Lightfoot, 1786

?1767. *Conus genuanus papilio* Linnaeus. *Systema Naturae per Regna Tria Naturae*, ed. 12, 1: 1168 (Locality not stated). Probably unrecognizable.

1786. *Conus pulcher* Lightfoot. *Catalogue Portland Museum:* 179 (Guinea). Lectotype figure (Kohn, 1964): Lister, Pl. 772, fig. 18.

1791. *Conus leoninus* Gmelin. *Systema Naturae per Regna Tria Naturae*, ed. 13, 1: 3386 (Indian Ocean). Lectotype figure (here selected): Lister, Pl. 772, fig. 18, also lectotype of *C. pulcher.* Non *Conus leoninus* Lightfoot, 1786.

1792. *Conus Siamensis* Hwass, in Bruguiere. *Cone*, in *Ency. Method., Hist. Nat. des Vers*, 1: 662 (Indian Ocean). Lectotype selected and figured by Kohn, 1968.

1792. *Conus papillonaceus* Hwass, in Bruguiere. *Ibid.*, 1: 665 (Guinea). Lectotype selected and figured by Kohn, 1968.

1792. *Conus prometheus* Hwass, in Bruguiere. *Ibid.*, 1: 667 (Africa). Lectotype figure (Kohn, 1968): *Tableau Ency. Method.*, Pl. 331, fig. 5.

1798. *Cucullus trapezium* Roeding. *Museum Boltenianum:* 39 (Locality not stated). Based on Lister, Pl. 773, fig. 19.

1798. *Cucullus archithalassus indiae* Roeding. *Ibid.:* 43 (West Indies). Lectotype figures (Kohn, 1975): Chemnitz, Pl. 138, figs. 1281-1283.

1817. *Conus fluctifer* Dillwyn. *Descr. Catalogue Recent Shells*, 1: 382 (Java, Mozambique, and Zanquebar). Lectotype figure (here selected): *Tableau Ency. Method.*, Pl. 331, fig. 5, also lectotype of *C. prometheus* Hwass.

1831. *Conus Nicolii* Wilson. *Illustrations of Zoology:* Pl. 36 (Locality unknown). Fide Tomlin, 1937.

1833. *Conus bicolor* Sowerby i, in Sowerby ii. *Conchological Illustrations:* Pt. 28, fig. 18 (West Africa) (May). Non *Conus bicolor* Sowerby i, in Sowerby ii, March 1833.

1833. *Conus papillaris* Sowerby i, in Sowerby ii. *Ibid.:* Pt. 37, fig. 54 (Locality not stated).

Description.—Light for its large size, with a good gloss in small specimens; obconical, the upper sides straight or gently convex; body whorl with about 6-12 very low spiral ridges above the base, these becoming nearly obsolete with growth; numerous fine spiral and axial threads over whorl, sometimes forming a very finely cancellate pattern but usually obscure; shoulder roundly angled, broad, concave above; spire flat (young) to low and convex (adults), the early whorls forming an inconspicuous cone; spire whorls slightly concave above, with fine axial threads and traces of 1-2 fine spiral ridges; early whorls carinate. Body whorl white to cream or very pale straw, covered with spiral rows of large and small squarish orange-brown spots and dashes of various sizes, usually in at least 2-3 enlarged rows above midbody and 2 below; sometimes very faint spiral salmon bands present in young; base white, unmarked; pattern changes with growth—see *Variation;* spire and shoulder white or cream, covered with broad curved axial lines of orange-brown converging on the early whorls and as wide as or wider than the white interspaces; tip usually white. Aperture wide, about uniform in width; outer lip thin, sharp, slightly convex; mouth white. Columella long, very narrow, oblique and indistinct. Length 70-222mm.

Comparison.—This very large, thin cone is distinctive in lacking spiral sculpture on the spire and having a subdued orange-brown to dark brown smudged pattern in adults. It has sometimes been confused with *C. spurius*, but only because early illustrations of these two species are very bad. *C. spurius* is a narrower, much thicker and heavier shell with a strongly projecting cone of early whorls, as well as being differently patterned with less smudged and usually darker orange to brown spots and blotches. In hand, the two could never be confused.

I have recognized as distinct the small *Conus byssinus*, which closely resembles juveniles of *C. pulcher*. However, this species, which seldom exceeds 50mm in length, is much thicker and heavier than *C. pulcher* of the same size and has sharply defined bright salmon spiral bands and blackish dashes; the shoulder is relatively more rounded than in juveniles of *C. pulcher* and with a higher spire. In addition, the spire markings of *C. byssinus* are blackish, sharply defined, and narrower than the white interspaces, rather than more orange-brown and wide as in juvenile *C. pulcher*.

Variation.—The shape changes slightly with growth, juveniles having more angulate shoulders and relatively shorter body whorls, while large adults have very rounded shoulders and more elongate body whorls. The pattern varies considerably with growth and goes through three rather distinct patterns.

Small shells (under 60mm) have large dashes in about 10 series, five above and five below the white midbody area, which is usually marked only with small spots; one or two spiral bands of pale brown or salmon are between the dashes above and below midbody; all dashes are dark and well defined. Between about 60 and 100mm the pattern is basically similar but all dashes are relatively lighter in color and more diffuse, showing a tendency to merge into blotches above and below midbody, the midbody area now with several larger dashes and somewhat indistinct; traces of pale brown spiral bands may be still visible; spire still relatively flat as in smaller shells. Adults over about 10cm have the dashes fused into irregular broad brown bands above and below the relatively lightly dashed midbody area; few dashes are distinct; several rows of dashes below the shoulder are also partially fused into a broken band; the spire may be relatively higher and more convex, with the blotching and lines relatively wider and with less white showing. Needless to say, these differences are indistinct and gradual, sometimes with the pattern not necessarily matching the size of the shell.

Distribution.—Moderately common over western Africa south of the Bulge. Shallow water.

Synonymy.—Most early names are based on the same small group of rather bad figures, so there is some doubt about the status of the names. *Conus leoninus* is especially composite, based on at least five different species, but the lectotype restriction places it in *C. pulcher*. *C. papillonaceus* has often been used for the shell here called *C. byssinus*, but it is definitely a *C. pulcher*. *C. prometheus* is commonly used for large specimens of *C. pulcher*.

Notes.—Juveniles are common and have bright patterns, but the shell becomes duller, more flawed, and less common with increasing size, as might be expected. Few specimens over 150mm are ever available, and these are often very badly scarred and broken with very bad patterns. Shells under 100mm but over 60mm usually are clean and have well defined if rather dull patterns and are perhaps most typical of the species.

As in all the other very large cones, gigantic specimens sell for a very high premium when available. Very small juveniles are seldom available.

PULICARIUS Hwass, in Bruguiere, 1792

1792. *Conus pulicarius* Hwass, in Bruguiere. *Cone*, in *Ency. Method., Hist. Nat. des Vers*, 1: 622 (Pacific Ocean Is.). Lectotype (here selected): specimen figured by Kohn, 1968, as holotype.

1792. *Conus fustigatus* Hwass, in Bruguiere. *Ibid.*, 1: 623 (Indian Ocean). Holotype figured by Kohn, 1968.

1798. *Cucullus punctulatus* Roeding. *Museum Boltenianum:* 40 (Locality not stated). Lectotype figure (Kohn, 1975): Martini, Pl. 63, fig. 698.

1845. *Conus Vautieri* Kiener. *Species gen. et icon. des coqu. viv.*, 2 (*Conus*): Pl. 100, fig. 3. 1849-1850, *Ibid.*, 2: 350 (Locality unknown).

Description.—Heavy, thick, with a good or high gloss; low conical, rather ovate, the upper sides convex; body whorl with about 6-12 low, wide spiral ridges at the base, the grooves weakly punctate in some specimens; numerous fine spiral and axial threads often present; shoulder wide to rather narrow, roundly angled, with erect coronations, concave above; spire low, sometimes flat or moderate, sharply pointed, the sides often convex; whorls heavily coronated, concave above, with about four narrow spiral ridges. Body whorl glossy white, sometimes with pale tan axial flammules, covered with large oval black to dark reddish brown spots, these tending to group into rounded clusters above and below mid-body and some narrow axial rows; spots when black often with narrow reddish brown shadows; clusters sometimes covered by pale orange clouds; spire white with large blackish spots or sometimes axial blotches. Aperture narrow posteriorly, wider anteriorly; outer lip thick, sharp, slightly convex posteriorly then nearly straight; mouth white, often with deep pinkish or orange, sometimes brown, tones within; lip sometimes thickened at the middle. Columella long, rather wide, projecting sharply anteriorly, indented from inner lip, and bounded by a heavy raised ridge. Length 30-75mm.

Comparison.—*C. pulicarius* is very distinctive and likely to be confused only with occasional specimens of *C. arenatus*. However, in *C. pulicarius* the majority of the black spots are at least half the width of the base of the shoulder coronations, while they are much smaller in *C. arenatus; C. arenatus* tends to have the spots thickened into wavy axial bands and lacks orange clouds on the body whorl over clusters of spots; *C. arenatus* also usually has many opaque white dashes and blotches under the pattern, usually absent in *C. pulicarius.*

C. zeylanicus is similar in shape to *C. pulicarius* but has very low or obsolete coronations, a distinctly pinkish background color, and a pattern of many small brown spots and axial lines scattered over the body whorl instead of distinct black spots.

Variation.—There is considerable individual variation in height and shape of the spire and width of the shoulder. Some specimens or local populations have the shoulder coronations low and almost obsolete. Occasional specimens have the black spots fused into axial stripes along major growth lines, but this is purely an individual variation as the result of accidents in growth.

Two subspecies can be recognized and seem to be fairly distinctive. The typical subspecies, *C. p. pulicarius*, is found over the entire Indo-Pacific area except the Marquesas and is a white shell with only occasional small blotches of opaque white under the pattern; the spots are very large, blackish or dark reddish brown, and there are no pale orange-tan axial flammules; the spots are usually clustered, sometimes forming spiral bands over and below a weakly patterned midbody area. *C. p. vautieri* appears to be restricted to the Marquesas Islands but shows some influence in the adjacent islands of French Polynesia. In this shell the spots are somewhat smaller, distinctly orange-brown instead of blackish, and often cover the entire body whorl to a greater degree than in typical *C. pulicarius.* In addition, there are heavy axial flammules of opaque white and pale orange-tan usually absent in the typical subspecies. The spire pattern is closer to being axial streaks instead of isolated spots, and the shoulder is probably a bit wider and less strongly coronated on the average.

Distribution.—Found commonly over the entire Indo-Pacific from Africa to Hawaii and French Polynesia. Usually in shallow water, but small juveniles found in deeper water.

Synonymy.—*C. fustigatus* is sometimes used for specimens where the spots are fused into axial black stripes along axial growth marks, but certainly this type of pattern is just an accident of growth and not worth recognition; in addition, the type of *C. fustigatus* is a perfectly normal *C. pulicarius* with widely spaced spots. *C. vautieri* appears very distinctive in color, but its isolated distribution and a tendency for other French Polynesian specimens to be distinctly orange-spotted indicate that it is just a subspecies.

Notes.—A very pretty species in which the minor axial growth marks and eroded early spire whorls are normal and do not detract from the value of a specimen. Although the lip thickens with growth, it occasionally chips easily. The axially striped shells usually sell for a minor premium.

C. p. vautieri is not uncommon in its restricted range, but it is worth a good premium. The shoulder coronations are often very low or absent.

PUNCTICULATUS Hwass, in Bruguiere, 1792

1792. *Conus puncticulatus* Hwass, in Bruguiere. *Cone*, in *Ency. Method.*, *Hist. Nat. des Vers*, 1: 702 (Santo Domingo). Lectotype selected and figured by Kohn, 1968.

1792. *Conus Mauritianus* Hwass, in Bruguiere. *Ibid.*, 1: 703 (Africa). Lectotype (here selected): specimen figured as holotype by Kohn, 1968.

1792. *Conus Columba* Hwass, in Bruguiere. *Ibid.*, 1: 709 (Indian Ocean). Lectotype figure (Clench, 1942, *Johnsonia*, 1(6)): Gualtieri, 1742, Pl. 25, fig. G; illustrated by Kohn, 1968.

1792. *Conus pusio* Hwass, in Bruguiere. *Ibid.*, 1: 710 (Santo Domingo). Lectotype figure (here selected): *Tableau Ency. Method.*, Pl. 334, fig. 4.

1798. *Cucullus minutus* Roeding. *Museum Boltenianum:* 46 (Locality not stated). Lectotype figure (here selected): Martini, Pl. 55, fig. 612a.

1798. *Cucullus mille punctatus* Roeding. *Ibid.:* 47 (Locality not stated). Lectotype selected by Kohn, 1975.

1807. *Conus scaber* Link. *Beschr. Nat.-Samml. Univ. Rostock*, 3: 105 (Locality not stated). Based on Martini, Pl. 55, fig. 612b.

1844. *Conus pygmaeus* Reeve. *Conchologia Iconica*, 1(*Conus*): Pl. 47, sp. 260 (Locality unknown).

1845. *Conus papillosus* Kiener. *Species gen. et icon. des coqu. viv.*, 1

(*Conus*): Pl. 72, fig. 4. 1849, *Ibid.*, 2: 271 (Locality unknown).
1845. *Conus pustulatus* Kiener. *Ibid.*, 2(*Conus*): Pl. 101, fig. 2. 1849, *Ibid.*, 2: 272 (Locality unknown).
1857-1858. *Conus perplexus* Sowerby ii. *Thesaurus Conchyliorum*, 3 (*Conus*): 20, pl. 14 (200), fig. 324 (Gulf of California).
?1857-1858. *Conus Hanleyi* Sowerby ii. *Ibid.*, 3(*Conus*): 20, pl. 17, figs. 399-400 (Mediterranean).
1879. *Conus melvilli* Sowerby iii. *Proc. Zool. Soc.* (*London*), 1878: 795, pl. 48, fig. 1 (Key West).

Description.—Moderately heavy for its size, with a good gloss; low conical, the upper sides weakly to strongly convex; body whorl with broad, flat spiral ridges over the anterior half or so of the whorl, these ridges often granulose; grooves deep, axially threaded; whole shell usually covered with very fine spiral and axial threads; shoulder broad, rounded to narrowly angled, not carinate, slightly concave or flat above; spire rather low to moderate, sharply pointed, the sides slightly to deeply concave; spire whorls nearly flat on top, the margins not projecting carinae, with heavy curved axial threads and sometimes traces of 1-2 weak spiral threads. Body whorl pale to dark bluish gray, sometimes toned with tan, paler with growth; usually with many spiral rows of small reddish brown dashes and squarish dots, at least anteriorly and sometimes extending to shoulder; spotting often weak or absent near midbody and below the shoulder; few to many darker axial flammules and blotches present in most shells, often concentrated in spiral bands at midbody, below the shoulder, and above the base; spire dark to pale bluish gray, with large and small reddish brown spots and streaks; tip brownish. Aperture moderately narrow, wider basally; outer lip thin, sharp, slightly convex to straight; mouth pale violet in two blotches or all dark violet-brown. Columella long, narrow, sharp anteriorly, bounded by a low to heavy ridge. Length 15-40mm.

Comparison.—*C. puncticulatus* in the Caribbean has often been confused with *C. jaspideus* and *C. mindanus* but appears quite distinct. From *C. mindanus* it differs in the lower and more concave spire which lacks stepped whorls and seldom has distinct small dark dots on the margins; the spire whorls in *C. puncticulatus* merge smoothly and are not separated by narrow ridges; the color in *C. mindanus* is white with orange or tan, not bluish gray as in *C. puncticulatus*.

C. jaspideus is sympatric with *C. puncticulatus* in the southern Caribbean and thus the two are not subspecies; they appear to be full species. *C. jaspideus* is usually white to deep reddish brown, the shoulder distinctly carinate and usually with a flattened rim-like margin as opposed to the rounded or sometimes rather narrowly angled shoulder in *C. puncticulatus*, which is usually distinctly bluish gray or, in whitish forms, distinctly violet in tone. The patterns overlap between the two species, but generally the spots in *C. puncticulatus* are much more uniform and not mottled as in *C. jaspideus.* There are also apparently differences in shape and size of the protoconch, but this is not always a good character.

The Panamic *C. p. perplexus* has sometimes been confused with *C. ximenes*, but it has a deep anal notch, is broader with a lower spire, and lacks distinct small spots at the margins of the whorls in most specimens; there are many differences between typical patterns of the two.

Variation.—The differences between Caribbean and Panamic populations of this species are so slight that specimens from some areas would be very difficult to identify if accidentally placed in the same tray. Although the Panamic subspecies is larger and distinct in pattern, its juveniles are virtually identicial to some fairly typical Caribbean specimens. For this reason I feel that two subspecies rather than two full species should be recognized.

Granulose color variant of *C. puncticulatus,* off Magdalena River, Colombia, 21mm (EP).

C. puncticulatus puncticulatus: Smaller, seldom more than 25mm in adults, distinctly bluish gray, with a rounded shoulder and relatively tall spire; sides of body whorl usually more distinctly convex below the shoulder; spiral rows of spots often absent in midbody area or poorly developed; color in adults bluish gray with only very faint tan tones; usually without large squarish dark blotches at the shoulder. Caribbean, from the Lesser Antilles and the Central American coast to northern Brazil.

Of this subspecies three major color pattern types can be recognized. In the typical form the axial flammules are usually few in number and the spiral rows of dots are conspicuous; the form commonly called *pygmaeus* is slightly more slender than normal with a perhaps less concave-sided spire, the axial flammules very conspicuous, while the dots are reduced and often absent over most of the body whorl. In some localities of the Lesser Antilles a white form, *columba*, occurs along with intermediates to more typical *puncticulatus*; this form is rather broad across the shoulder, solid white with a distinctly violet tone, and usually has the basal ridges heavily granulose; intermediates are pale bluish gray, almost white, with small axial flammules near the midbody area and scattered spots.

The major individual variation here, other than color pattern, is in the degree of granulosity of the spiral ridges. Usually the anterior ridges are smooth or at most have a few small granules, but not uncommonly the anterior ridges are very heavily granulose; occasional specimens are heavily ridged to the shoulder and covered with granules.

C. puncticulatus perplexus: Adults larger, commonly 30mm or more, very pale bluish gray with distinct tan tones, the shoulder more narrowly angled and broader, with the spire lower and flatter; the sides of the body whorl in adults are relatively straight, but more convex posteriorly in juveniles; spiral rows of distinct dots showing few dashes from base to shoulder, thickened into larger squarish spots in series above the base and at midbody; usually large brownish spots at shoulder and many tan axial flammules and clouds. Panamic, from the Gulf of California and Baja California to Ecuador.

Juveniles often have more rounded shoulders and are uniformly covered with dots from base to shoulder with little blotching other than at the shoulder. There is a strong tendency in some populations for juveniles to be covered from base to shoulder with large spiral rows of granules between the dots. Axial flammules may be absent or heavy; the tan tones may be very dark to almost absent, especially in juveniles.

Distribution.—In the Caribbean: common in shallow to moderately deep water from the Lesser Antilles and the Central American coast to northern Brazil; usually sold from Venezuela, the Netherlands Antilles, or Martinique at this time. In the Panamic region: from the Gulf of California to Ecuador, common intertidally to moderate depths.

Synonymy.—The early figures, as usual, are confused and difficult to distinguish; for some reason this species was repeatedly confused with the Mediterranean cone. *Conus columba* is certainly just a local variety, as intermediates occur; the name has been incorrectly used for *C. parius* as well. *C. pygmaeus* seems to be a barely differentiated form from some bays in northern Venezuela and perhaps other localities; it is barely deserving of varietal status.

Conus papillosus and *C. pustulatus* of Kiener surely are pustulose individuals of *C. p. perplexus* because of the shape and pattern as well as the occurrence of similar shells in several known populations. Of course, they could also be pustulose individuals of the Caribbean subspecies, which is often very similar. Since these two names are much older than the very familiar *perplexus*, I prefer to consider them synonyms of the Caribbean subspecies to preserve common usage. The Kiener names do not seem to apply to the species here called *C. axelrodi*, as has commonly been stated.

C. hanleyi seems to be this species, as does *C. melvilli*. In *C. melvilli* the axial flammules are poorly developed but the spiral rows of dots have fused into nearly continuous spiral hairlines; I have seen specimens very similar to the type figure in series of *C. puncticulatus* from Venezuela, but the Key West locality is almost certainly wrong.

Notes.—Both subspecies are common small shells within their ranges. Although often marred by axial growth flaws and chipped lips, their small size makes such flaws inconspicuous.

Granulose individuals of both subspecies are worth a premium as they are not common. The *columba* variety currently sells for a high premium, but it appears to be relatively common once the proper habitat and localities are known.

I have not seen any specimens that I would consider intermediate between *C. puncticulatus* and *C. jaspideus*, but surely these must occur. However, *C. puncticulatus* is the common southern Caribbean species while *C. jaspideus* is more typical of the northern Caribbean. Although broadly sympatric, *C. jaspideus* seems to be uncommon in the southern Caribbean and perhaps is found in deeper water than *C. puncticulatus*.

PURPURASCENS Sowerby i, in Sowerby ii, 1833

1833. *Conus purpurascens* Sowerby i, in Sowerby ii. *Conchological Illustrations:* Pt. 25, figs. 13-13* (Panama).
1834. *Conus regalitatis* Sowerby i, in Sowerby ii. *Ibid.:* Pt. 56/57, fig. 87 (Real Llejos).
1853. *Conus comptus* A.A. Gould. *Boston J. Nat. Hist.*, 6: 387, pl. 14, fig. 23 (Santa Barbara, California).
1910. *Conus purpurascens* var. *rejectus* Dall. *Proc. U.S. Nat'l. Mus.*, 38 (1741): 219 (Gulf of California).

Description.—Moderately heavy, with a good or low gloss; low conical, rather ovate at times, the upper sides distinctly or slightly convex then concave and tapering to the base; body whorl with low spiral ridges at the base, these usually soon obsolete but occasionally continuing to midbody and sometimes weakly granulose; numerous spiral and axial threads, sometimes conspicuous; shoulder sharply angulate to roundly angled, flat or concave above; spire rather low, bluntly pointed or sharp, the sides straight or somewhat concave; spire whorls usually concave above, weakly but distinctly stepped, the margins rude; first 2-4 whorls weakly undulate; tops of whorls with about 3-5 low spiral ridges and weak, scattered axial threads. Body whorl generally purplish to bluish gray, sometimes whitish or tan with violet tones, covered with numerous spiral rows of brown dashes and small opaque white spots, the dashes often very long and almost continuous; large irregular brownish to reddish or blackish blotches above and below midbody, these sometimes fused into broad bands or reduced in size and indistinct; a row of small blotches just below the shoulder in some specimens; midbody area usually pale and lightly marked; spire and shoulder bluish to whitish, with large or small dark brown axial lines and blotches, margins of whorls often brown; early whorls usually eroded, pinkish. Aperture moderately wide, almost uniform in width; outer lip thin, sharp, straight to weakly convex; mouth deep bluish white to violet. Columella long, narrow, mostly fused. Length 30-84mm.

Comparison.—*C. purpurascens* seems to be the Panamic cognate of the Atlantic *C. ermineus*, from which it is often almost inseparable. As a rule, however, it is wider and more strongly angled at the shoulder, with the spire whorls stepped and more irregular at the margins. The color is usually some tone of bluish or purple, moreso than in typical *C. ermineus*, and with the dashes longer and more conspicuous than in *C. ermineus*, often forming almost continuous lines. Typical *C. purpurascens* have many small spots and blotches of opaque white in the pattern, these usually absent in *C. ermineus*.

Some *C. monachus* may be vaguely similar in pattern and color to *C. purpurascens*, but that species is usually narrower and more oval in shape with heavier sculpture and heavy axial bluish clouds.

Variation.—A variable species, *C. purpurascens* is almost always bluish or purple in tone, although some almost unmarked specimens may be mostly whitish with pale violet tones. The dark blotches are very variable in development, sometimes almost completely covering the body whorl but sometimes nearly absent; the margins are often very irregular and may be margined with opaque white. The dashes and opaque white spots are usually very well developed and may constitute most of the pattern in some shells; occasional shells are encircled by many almost complete brown hairlines.

The shape is very variable also, with some shells very round-shouldered like many *C. ermineus*, but more typically with angulate, sometimes narrowly angulate, shoulders. The spire is usually low, with distinctly but weakly stepped whorls, the margins often very rude and irregular. The sides of the shell may be almost straight or strongly convex posteriorly then deeply concave near the middle.

Distribution.—Moderately common in shallow water from the Gulf of California to Peru.

Synonymy.—The synonyms are based on poorly developed shells with the purplish color weak or absent (probably faded) and the spiral lines often exaggerated. They certainly fall easily within the range of variation of this species and have long been considered synonyms.

Notes.—Other than minor spire erosion and some minor axial flaws, this is a relatively clean species. Commonly the axial growth lines are deep and cause major changes of pattern and color on the new growth side.

Specimens from the Galapagos Islands seem to be especially variable, with all extremes from nearly solid dark brown to unmarked white sometimes found at the same locality.

QUERCINUS Solander, in Lightfoot, 1786

1786. *Conus quercinus* Solander, in Lightfoot. *Catalogue Portland Museum:* 67 (Locality not stated). Lectotype figure (Kohn, 1964): Martini, Pl. 59, fig. 657.
1791. *Conus cingulum* Gmelin. *Systema Naturae per Regnum Tria Naturae,* ed. 13, 1: 3378 (Tonga). Holotype, a corded specimen, figure (Kohn, 1966): Martyn, Pl. 39.
1807. *Conus buxeus* Link. *Beschr. Nat.-Samml. Univ. Rostock,* 3: 99 (Locality not stated). Based on Martini, Pl. 59, fig. 657. Non *Conus buxeus* (Roeding, 1798).
?1887. *Conus Akabensis* Sowerby iii. *Thesaurus Conchyliorum,* 5(*Conus* Suppl.): 273, pl. 36(512*), figs. 512-513 (Akaba (Red Sea)).
1915. *Conus quercinus* var. *albus* Shaw. *Proc. Malac. Soc. London,* 11: 210 (Aden). Non *Conus albus* Sowerby iii, 1887.
1942. *Conus fulvostriatus* Fenaux. *Bull. l'Inst. Oceanogr.* (*Monaco*), No. 814: 2, fig. 4 (I. Bourbon).
1966. *Cleobula albonerosa* Garrard. *J. Malac. Soc. Australia,* No. 10: 11, figs. (35 fms. off Wide Bay, southern Queensland, Australia).

Description.—Very heavy and cumbersome, with a low gloss; obconical, the sides nearly straight, slightly convex posteriorly; body whorl with a few low spiral ridges above the base, these usually continued as spiral threads to about midbody, especially in smaller specimens; fine axial scratches sometimes prominent; shoulder broad, rounded in adults, more angulate in juveniles; spire very low or flat in adults, the early whorls forming a small pointed cone; spire whorls flat above, covered with about 8-12 fine spiral ridges, the early whorls with fairly conspicuous axial threads. Body whorl usually deep yellow to pale yellow or occasionally white, covered with many fine, closely spaced continuous brown hairlines, these sometimes partially or entirely absent; often a paler midbody band visible; spire uniformly pale yellow to whitish; early whorls darker brownish. Aperture wide, nearly uniform in width; outer lip straight or nearly so, thick in adults; mouth white. Columella long, narrow, not very distinct. Length 50-144mm.

Comparison.—No other cone looks quite like *C. quercinus.* The pale form of *C. concolor* may be vaguely similar, but it has a higher spire with finely cancellate spire whorls, often has more convex sides, and usually shows at least a few dark brown spiral dashes near midbody. *C. virgo* and *C. coelinae* are much more elongated and have very weak spiral sculpture on the spire. *C. fergusoni* are somewhat different in shape and tone, with fine nodules on the early whorls and indistinct spiral spire sculpture.

Variation.—Juveniles of *C. quercinus* are very slender relative to adults, with a sharply angled shoulder; the body whorl is waxy yellow with widely spaced brown lines. With growth the shell becomes much broader, the spire lower, and the lines more closely spaced. There is some variation in the yellow tone of adults and the distinctness of the brown lines, but only two variants are worthy of mention.

Some specimens from the Red Sea seem to retain the narrow body whorl and relatively high spire of juveniles but are completely white without pattern; such shells, which are apparently rare freaks, were named *akabensis*. From the Solomons and New Guinea to Queensland there is a tendency for adult shells to lose the lines and yellow color, becoming completely white; intermediates to the normal form are found. Such shells were named *albonerosa*; they are not uncommon.

Distribution.—*C. quercinus* is a very common species in shallow to moderately deep water from eastern Africa to Hawaii and French Polynesia.

Synonymy.—*C. cingulum* is based on a corded specimen, a rare occurrence in cones. *Buxeus* and *fulvostriatus* seem to be rather normal specimens of *C. quercinus*. *C. akabensis* was discussed under *Variation*, as was *C. albonerosa*; both variants seem to be well within the range of *C. quercinus*, although *akabensis* is unusual in combining a white color and juvenile shape. *Albus* was used for a white form from the Red Sea, probably not the same variant as *akabensis*; white shells are occasional throughout the range of the species, but usually only as isolated individuals.

Notes.—If you've seen one *quercinus*, you've seen them all—or almost, white shells and juveniles being rather different at first glance. This is a clean species except for minor spire erosion and some axial growth marks; of course very large specimens are usually eroded and scarred, as well as being extremely heavy.

White shells from the Queensland area are easily available at a moderate premium, but the *akabensis* form is apparently rare and seldom offered. Juveniles are uncommonly seen but well worth obtaining.

RADIATUS Gmelin, 1791

1791. *Conus radiatus* Gmelin. *Systema Naturae per Regna Tria Naturae*, ed. 13, 1: 3386 (Locality not stated). Lectotype figure (Kohn, 1966): Martini, Pl. 53, fig. 584.

1844. *Conus Martinianus* Reeve. *Conchologia Iconica*, 1(*Conus*): Pl. 40, sp. 217 (Putao, Province of Albay, Luzon, Philippines).

Description.—Rather light in weight, with a low gloss; cylindrical, the upper sides gently convex then straight or slightly concave to the base; body whorl with many narrow spiral ridges at the base, these rapidly becoming broad flat ribs separated by deep, punctate grooves; ribs extend to about midbody and then disappear; rest of body whorl with fine spiral and heavier axial threads; shoulder rounded to roundly angled, usually concave above; spire low, the sides concave to slightly convex, sharply pointed; spire whorls concave above, sunken at the margins, with about 3-6 heavy spiral ridges and some fine axial threads. Body whorl yellowish tan, fleshy gray, or dark brown; base, a narrow spiral band below mid-body (often absent), and a narrow to wide band at the shoulder white or bluish gray, as may be the grooves on the body whorl; spire with later whorls whitish or grayish, the margins often dark brown, earlier whorls same color as body whorl; tip dark brown. Aperture moderately wide, uniform in width; outer lip thin, sharp, slightly convex; mouth white (pale external colors) to bluish white or grayish (dark external colors). Columella narrow, very long, slightly indented from inner lip and with a low ridge posteriorly. Length 40-77mm.

Comparison.—This species is obviously related to *C. pilkeyi* and *C. ochroleucus* as well as *C. cinereus* and *C. parius*. From the latter two species it is readily distinguished by the presence of heavy spiral ridges on the spire whorls; *C. cinereus* is more slender and usually has at least traces of a dashed pattern, while *C. parius* is smaller, has a deep violet mouth, and is often more uniformly colored without a distinct pale shoulder band. *C. pilkeyi* and *C. ochroleucus* have higher spires without sunken margins, have yellowish to bright orange mouths, and have varicolored spire whorls; they usually lack a distinct pale shoulder band and seldom duplicate the body colors of *C. radiatus*.

Variation.—This species is constant in shape although the spire may be variable from almost flat with strongly concave or sunken sides to rather moderate with slightly convex sides. A whitish to pale gray band is almost always present at the shoulder and sharply divides the spire color from the body whorl color. The range of colors of the body whorl varies from pale yellowish tan or darker brown to bluish gray, pale or dark, and almost black. The spiral grooves of the body whorl usually reach to mid-body but may be reduced or occasionally reach almost to the shoulder.

Distribution.—Common but apparently restricted to the western and southern Pacific from the Philippines through the Solomons area to Fiji. Most specimens come from moderately deep water in the Philippines.

Synonymy.—*C. martinianus* is certainly this species.

Notes.—Although seldom damaged, this is a rough cone in having heavy axial growth marks showing lighter colors and disturbing the solid coloration. Some specimens give the impression of being covered with white axial hairlines on the otherwise dark body whorl. The lip is fragile.

Dark specimens are sometimes incorrectly sold as *C. gubba*, a dark variety of *C. cinereus* that has weaker grooving and a slender body.

RAOULENSIS Powell, 1958

1958. *Conus* (*Kermasprella*) *raoulensis* Powell. *Rec. Auckland Inst. Mus.*, 5(1/2): 83, pl. 9, fig. 1 (Off Raoul Island, Kermadecs). See also Cernohorsky, 1976, *Hawaiian Shell News*, 24(9): 3-4.

Description.—Light in weight, without a gloss; biconical, the sides nearly straight or slightly convex; body whorl covered with about 20-30 closely spaced spiral ridges, about every third or fourth ridge enlarged and usually bearing widely spaced large granules; grooves between ridges with fine axial threads; shoulder wide, carinate, nodulose, concave above; spire tall, pointed, the sides straight or nearly so; spire whorls stepped, the margins projecting and nodulose; tops of whorls concave, with many fine curved axial threads and traces of 1-2 fine spiral ridges. Body whorl pale orange-tan or salmon, with three or four spiral bands of indistinct whitish blotches, the whole pattern very indistinct in available specimens; shoulder and spire white, with faint tan blotches. Aperture narrow, uniform in width; outer lip thin, fragile, nearly straight; mouth pale to deep pink within. Columella internal. Length 16-20mm.

Comparison.—Although this species closely resembles a small *C. acutangulus* in general shape and sculpture, it seems to be quite distinct from other described species. From *C. acutangulus* and *C. eugrammatus* of similar size it differs in being granulose over the body whorl on every few narrow spiral ridges and in having a pinkish mouth; the spire sculpture is weaker than in either of the two species. *Conus insculptus* and similar species are more elongated and have different sculpture on the spire and body whorl. The 19.6mm paratype examined appeared to be quite adult, and I cannot imagine what better known species it might be the juvenile of.

Variation.—Only a half-dozen or so specimens are known, all in rather poor condition with the patterns indistinct. At first glance this appears to be a pale tan shell with the spiral ridges slightly darker tan, but when wet a pattern of indistinct broad spiral bands of orange-tan or pale salmon can be seen, separated by indistinct squarish blotches of white in about three or four spiral bands. The spire may have faint color blotches or be all whitish.

About 20 to 26 spiral ridges have been counted on the known specimens, with every third or fourth (varies) ridge slightly enlarged and usually granulose; the granules appear to be very regular and are very large for such a small shell. The shoulder is heavily granulose, as are the spire margins.

Distribution.—Known only from moderately deep water off the Kermadec Islands; it should be looked for off Queensland, as two other Kermadec cones (of a total of four) occur there.

Notes.—Not available on the market as yet or perhaps in the near future. The known specimens are very dull and possibly do not give a good picture of what fresher specimens would look like. The relationships of this species are very uncertain, but it seems to be adult and with the granules the normal condition.

RATTUS Hwass, in Bruguiere, 1792

1792. *Conus rattus* Hwass, in Bruguiere. *Cone*, in *Ency. Method., Hist. Nat. des Vers*, 1: 700 (America). Lectotype figure (Kohn, 1968): *Tableau Ency. Method.*, Pl. 338, fig. 7.

1792. *Conus Taitensis* Hwass, in Bruguiere. *Ibid.*, 1: 713 (Tahiti). Holotype figure: *Tableau Ency. Method.*, Pl. 336, fig. 9. This name has been emended two times: *taheitensis* Reeve, 1843, and *tahitiensis* Dautzenberg, 1933 (*J. Conchyl.* (*Paris*), 77: 89).

1817. *Conus Chemnitzii* Dillwyn. *Descr. Catalogue Recent Shells*, 1: 363 (Ceylon). Holotype figures: Chemnitz, Pl. 182, figs. 1764-1765.

1857-1858. *Conus viridis* Sowerby ii. *Thesaurus Conchyliorum*, 3(*Conus*): 20, pl. 5(191), fig. 102 (Locality unknown).

1882. *Conus semivelatus* Sowerby iii. *Proc. Zool. Soc.* (*London*), 1882: 118, pl. 5, fig. 3 (Red Sea).

Description.—Moderately heavy, with a good gloss; low conical, the upper sides gently convex or straight and sometimes concave to the base; body whorl covered with heavy spiral ridges on the anterior third, these rapidly becoming obsolete before midbody in most specimens; grooves deeply punctate, the spiral rows of punctations sometimes continuing to shoulder but widely spaced posteriorly; numerous axial threads over entire shell; shoulder broad, roundly angled to sharply angled, flat or convex above; spire low to moderate, sometimes nearly flat, bluntly pointed but with sharp protoconch, the sides straight or slightly concave to convex; earliest 1-2 whorls with very weak nodules, occasionally later whorls very weakly undulate or irregular; tops of whorls flat or convex, with about 3-5 spiral ridges and heavy or weak axial threads. Body whorl pale bluish white to opaque white, heavily covered with two broad irregular spiral bands of yellowish tan, olive, or dark brown blotches that are usually fused and cover most of the background except irregular blotches of white at base, midbody, and a broad scalloped band at the shoulder; many small flecks of white scattered over whorl, concentrated on basal half, sometimes fused into small blotches; very fine brown flecks numerous on base in most specimens, sometimes obscured by dark brown and violet color of base, the flecks vaguely spiral in alignment; spire and shoulder white with many to few large and small brownish blotches; early whorls sulfur yellow. Aperture rather narrow, uniform in width or slightly wider anteriorly; outer lip rather thin, sharp, straight or evenly convex; mouth purple, often with lighter bands at middle and below shoulder. Columella very indistinct, mostly internal. Length 25-64mm.

Comparison.—The numerous small brown flecks at the base indicate a relationship with *C. vexillum* and *C. namocanus* and allies. From these species it differs in the heavy basal ridges and deeply punctate grooves in adults as well as the brown and white blotching with scattered white flecks in many specimens. There is a strong resemblance to *C. balteatus* and *C. cernicus* in pattern, shape, and to some extent sculpture, but these two species are distinctly coronated and usually have the spiral ridges extending to near the shoulder.

Variation.—Typical adults are highly variable in tone of the dark blotching and its fusion into spiral bands; some specimens are mostly light, some mostly dark with greatly reduced blotching of white. Small white flecks are usually obvious anteriorly but may be absent or incorporated into larger white blotches. The basal ridges are usually heavy and separated by deep grooves, but they may be reduced and weak; the rows of punctations above midbody may be distinct or absent. Often the base is strongly tinged with dark violet, but it may also be brown.

Juveniles are somewhat different in color, having the upper half or so of the body whorl mostly white, either in an even-edged band or as a deeply scalloped band; the dark brown color is restricted to the anterior half of the shell, and the entire shell may have a distinctly yellowish tinge.

The above variation applies to specimens from Africa to French Polynesia and Hawaii, but in the Marquesas there is a rather distinctive shell which is obviously related to *C. rattus* but seems distinctive enough to recognize as a separate subspecies, *C. rattus taitensis.* This Marquesan subspecies differs from typical *C. rattus rattus* in being broadly conical with nearly straight sides, the shoulders wide and sharply angled; the spire is very low, almost flat. The body color is a nearly uniform dark brown, with the white restricted to a narrow band of small and rather regular blotches at midbody and equally small blotches at the edge of the shoulder; the anterior ridges are relatively weak, while the axial threads of the spire whorls are relatively heavy. Juveniles are similar to adults in shape and reduced amount of white, contrasting with *C. rattus rattus* juveniles from Tahiti which have a broad scalloped band of white at the shoulder.

Distribution.—Widely distributed throughout the Indo-Pacific from eastern Africa to Hawaii and French Polynesia. Usually in shallow water, common.

Synonymy.—*C. taitensis* (the emendations are not valid) seems to be too similar to typical *rattus* and too restricted in distribution to recognize as a full species; like *C. pulicarius vautieri*, its influence can be seen in specimens from nearby islands. *C. chemnitzii* seems to fit fairly well in *C. rattus*, although it has most often been referred to *C. capitaneus*. *C. viridis* and *C. semivelatus* seem to be based on juveniles, the former with a scalloped shoulder band, the latter with an even-edged band; both types of juveniles have been seen in collections.

Notes.—Although a common species, it is sometimes difficult to obtain specimens in good condition. There is a strong tendency to strongly erode and calcify the spire whorls, as well as to have large axial breaks and scars on the body whorl. Specimens from some populations have very dull patterns that are poorly contrasted. *C. rattus taitensis* seems to be not uncommon in its restricted range, but it too is often chipped, scarred, and eroded.

RECLUZIANUS Bernardi, 1853

1853. *Conus Recluzianus* Bernardi. *J. Conchyl.* (*Paris*), 4: 148, pl. 6, fig. 6 (China Seas).

1961. *Conus* (*Rhizoconus*) *gloriakiiensis* Kuroda and Ito. *Venus*, 21(3): 248, pl. 17, figs. 6-7 (Nada Coast, Gobo, Kii Peninsula, Japan).

Description.—Moderately light in weight, with a low gloss; elongate obconical, the upper sides nearly parallel for a short distance below the shoulder, then straight or concave to a narrow base; body whorl with a few low spiral ridges above the base, these rapidly becoming widely spaced and then obsolete; many spiral and axial threads over whorl, sometimes heavy; shoulder carinate, distinctly undulate or low coronate, nearly flat above; spire low to flat, the early whorls forming a small pointed cone; early 1-2 whorls weakly nodulose; later whorls flat or concave above, the margins strongly undulate or low coronate, with about 5-8 narrow spiral ridges crossed by heavy axial threads so distinctly cancellate. Body whorl usually pale cream or pale straw, often with strong violet or yellow tones; young and some adult shells with two broad spiral bands of darker tan above and below a light midbody area; the brown bands in juveniles contain fine spiral dashes or lines of darker brown, these fading with growth; there are commonly small dark brown squares at the posterior edge of the midbody band, these also disappearing with growth; some fully adult shells dirty cream without distinct markings; base usually pale; spire about same color as body whorl or slightly paler, sparsely covered with narrow pale brown axial lines or blotches. Aperture narrow, about uniform in width; outer lip thin, fragile, straight or concave at the middle; mouth white to very pale violet, sometimes darker in very dark shells. Columella not visible externally. Length 40-80mm.

Comparison.—*C. recluzianus* belongs to a closely knit group of western Pacific deep-water cones. The closest species is *C. urashimanus*, which is perhaps not distinct and just an ecotype. *Urashimanus* differs in usually being darker brown, having the spire whorls heavily covered with dark and light brown lines and blotches, and usually having the mouth deep violet; it may also have very heavy shoulder coronations. *C. recluzianus* and *C. urashimanus* both have the spire whorls and shoulder distinctly undulate or low coronate, while the similar *C. tribblei* and *C. sugimotonis* have the shoulder and whorl margins straight and not regularly undulate. *C. tribblei* may be similar in appearance to some *C. recluzianus*, but it is a white shell, seldom violet, with the orange-tan pattern tending to form axial flammules grouped into two broad and indistinct spiral bands above and below midbody; the juveniles are heavily covered with long spiral brown dashes, at least some still visible in adults on occasion. *C. sugimotonis* is uniformly white without pattern.

Conus voluminalis has been widely confused with *C. recluzianus*, especially when the latter is brightly colored, but it has a higher and more pointed spire with more nodulose whorls; the spire sculpture is very weak, and the margins of the whorls are not undulate. *C. voluminalis* is thinner in texture and waxy.

Variation.—As indicated above, the coloration of this species is very variable. Small shells are more contrastingly patterned with spiral dashes or lines and dark spots above the midbody band. The contrasts are usually lost with growth and the general coloration fades to uniform tan or cream, sometimes completely white. The mouth is pale, except in dark violet shells, where it is darker. With few exceptions, the spire is lightly pigmented and has only a few dark spots and lines.

Distribution.—Moderately common in deep water from southern Japan and Taiwan to the Philippines; probably more widely distributed, but confused with *C. tribblei.*

Synonymy.—There are no obvious differences between the holotype of *C. gloriakiiensis* and the type of *C. recluzianus* that are not very minor and easily matched in even small series. Certainly they are the same species. Collectors usually use the name *gloriakiiensis* for either white shells or those with bright colors, and even occasionally for very large specimens.

Notes.—This is a fragile species, but except for minor axial growth flaws and scars plus a filed lip, satisfactory specimens are easy to obtain. Juveniles are better patterned than adults and sometimes have bright violet or yellowish colors lacking in adults. Bright specimens are worth a small premium, as are solid white specimens.

This species has been badly confused with *C. tribblei*, although the species are quite distinct in sculpture and pattern. The description by Bernardi is very clear and the figure of the type is very characteristic and easily matched among rather small specimens. *Conus urashimanus* is only doubtfully distinct, however, and may be just dark, rather waxy, and heavily undulate specimens with dark violet mouths.

RECURVUS Broderip, 1833

1832. *Conus cinctus* Valenciennes, in Humboldt and Bonpland, *Voy. Inter. Amer. -Recueil Obs. Zool. et Anat. Comp.*, 2: 337 (Acapulco). Non *Conus cinctus* Bosc, 1801.

1833. *Conus recurvus* Broderip. *Proc. Zool. Soc.* (*London*), 1833: 54 (Monte Christi).

1833. *Conus incurvus* Sowerby i, in Sowerby ii. *Conchological Illustrations:* Pt. 33, fig. 36 (Monte Christe).

1839. *Conus arcuatus* Gray. *Zool. Beechey's Voyage:* 119, pl. 36, fig. 22 (Pacific Ocean). Non *Conus arcuatus* Broderip and Sowerby i, 1829.

1844. *Conus emarginatus* Reeve. *Conchologia Iconica*, 1(*Conus*): Pl. 43, sp. 232 (Pacific Ocean). Nomen novum for *Conus arcuatus* Gray, 1839.

1910. *Conus scariphus* Dall. *Proc. U.S. Nat'l. Mus.*, 38(1741): 225 (Off Cocos Island, Gulf of Panama).

Description.—Light in weight, with a high gloss; low biconical, the upper sides slightly convex then tapering to a narrow base; body whorl with a dozen or so low narrow spiral ridges above the base, these separated by narrow grooves with very weak axial threads; body whorl smooth or nearly so above anterior third; shoulder broad, carinate, concave above; spire moderately high, sharply pointed, the sides concave; whorls slightly stepped, concave above, with numerous fine curved axial threads and traces of a few very faint spiral threads; first six or so spire whorls with strong and distinct nodules. Body whorl white, covered with numerous wavy deep reddish brown axial flammules which may be continuous from above base to shoulder or broken into three roughly spiral bands below shoulder, at midbody, and above the base; often the flammules are connected to each other by narrower flammules or fused where the angled portions of the flammules touch, producing in extreme cases an irregularly reticulated pattern over at least part of the shell; base white; spire and shoulder white, with few to many curved reddish brown lines or blotches, continuations of the body flammules; early whorls pale tan. Aperture moderately wide, nearly uniform in width; outer lip thin, strongly sloping below level of shoulder, nearly straight; mouth white. Columella not visible externally. Length 40-92mm.

Comparison.—*C. recurvus* has often been confused with the *scalaris* variant of *C. gradatus*, but the two are not closely related. In *C. gradatus* the shell is usually thicker, slightly different in shape, has the outer lip less sloping, usually has spiral rows of brownish spots in the pattern, and has only 1-3 weakly nodulose spire whorls. *C. poormani* is more similar in shape but has the posterior sides distinctly parallel, the lip not strongly sloping, and the flammules pale orange-tan instead of dark reddish brown. *C. villepinii* is more slender than *C. recurvus*, usually has a higher spire with distinct spiral sculpture, and a pattern of flammules radiating from two broad spiral bands of reddish brown above and below midbody. *C. clerii* is somewhat thicker, has less concave spire whorls which are barely nodulose, and has distinct spiral rows of spots in the pattern.

Variation.—The shape is constant, although the height of the spire and the concavity of the whorls vary somewhat. The axial flammules may be nearly continuous from the base to the shoulder or weak and often broken, forming three irregular spiral bands. Some individuals have a strong tendency to connect the flammules into a loose reticulum giving the appearance of large oval or rhomboidal white spots on reddish brown; such shells were called *C. scariphus*, but the variation is purely individual and apparently the result of the flammules following axial growth lines; specimens with typical patterns on one side and *scariphus* patterns on the other are found.

Distribution.—Moderately common in deep to moderately deep water from the west coast of Baja California and the Gulf of California south in the Panamic area to Ecuador.

Synonymy.—It seems likely that the type of *C. recurvus* is actually a *C. gradatus* var. *scalaris*, but at the moment it seems best not to look too closely at the problem; early figures of species of this general shape all look very much alike and cannot be safely used to determine just what species an author had. *C. incurvus* and *C. emarginatus* probably can be referred to this species, especially the latter. *C. scariphus* is an individual variant.

Notes.—This fragile species commonly has heavily filed lips, spire whorls missing, and heavy axial flaws and scars. The axial growth marks affect the pattern to some extent, so unusual patterns may develop in injured specimens, including mostly white specimens and specimens apparently covered with spots—actually broken flammules and not the same as the spots of *C. gradatus*.

REGIUS Gmelin, 1791

1791. *Conus Ammiralis regius* Gmelin. *Systema Naturae per Regna Tria Naturae*, ed. 13, 1: 3379 (Jaimanitas, Cuba). Lectotype figure (Clench, 1942): Martini, Pl. 62, fig. 684.

1791. *Conus leucostictus* Gmelin. *Ibid.*, 1: 3388 (American Ocean). Lectotype figure (Kohn, 1966): Knorr, Pl. 13, fig. 5.

1791. *Conus solidus* Gmelin. *Ibid.*, 1: 3389 (Locality not stated). Holotype figure: Chemnitz, Pl. 141, fig. 1310.

1791. *Conus citrinus* Gmelin. *Ibid.*, 1: 3389 (Curacas). Holotype figure: Martini, Pl. 61, fig. 681.

1792. *Conus nebulosus* Hwass, in Bruguiere. *Cone*, in *Ency. Method., Hist. Nat. des Vers*, 1: 606 (Santo Domingo, as restricted by Clench, 1942). Lectotype selected and figured by Kohn, 1968. Non *Conus nebulosus* Gmelin, 1791.

Description.—Moderately light in weight, with a low gloss; ovately low conical, the upper sides convex then straight or slightly concave to the base; body whorl usually with about a dozen or so widely spaced narrow spiral ridges from the base to midbody, sometimes higher; ridges usually bear fine to large granules, and the whole body whorl may be rough to the touch; shoulder broad, roundly angulate, with heavy, rounded coronations; spire moderately low, sometimes high, blunt, deeply concave, the early whorls forming a thick rounded cone; earlier spire whorls often fused, the sutures not visible; all whorls coronate, usually concave above, the later whorls with about 6-12 narrow and sometimes indistinct spiral ridges crossed by fine axial threads. Body whorl white to straw, usually heavily covered with brown to olive, sometimes yellowish or reddish, blotches usually arranged above and below a clearer midbody area;

blotches commonly partially or entirely fused to leave only the base, mid-body area, and some whitish blotches below the shoulder pale on a dark shell; fine spiral lines of dark brown, interrupted by small white dots encircled by brown, are commonly well developed and obvious somewhere in the pattern; pale areas often crosshatched with irregular dark lines; spire usually heavily mottled with brown or other base color on later whorls, the early whorls mostly white with a few small flecks or darker. Aperture moderately narrow, nearly uniform in width; outer lip often thickened, straight or slightly concave at the middle; mouth white to bluish white (juveniles sometimes with a faint violet patch at the middle). Columella internal. Length 40-71mm.

Comparison.—Although several large cones have the same general shape and appearance as *C. regius*, few are actually close. *C. imperialis*, *C. zonatus*, and *C. brunneus* are obviously distinct in shape, pattern, and spire. *C. bartschi* is similar but lacks the spiral sculpture on the spire whorls, has the spire lower and less concave, and usually lacks the strongly granulose basal sculpture; in addition, the pattern of *C. bartschi* is comprised largely of dark dashes, not the blotches of *C. regius*. *C. aurantius* and *C. cedonulli* are also close to *C. regius* but are more elongated in the body whorl and narrower at the shoulder; the spire of these two species is usually less concave than in *C. regius*, lacks distinct spiral sculpture, and has fine axial brown lines at the margins of the whorls. The patterns of the two elongated species are usually quite different from those of *C. regius*. *C. cardinalis* is usually a red or orange shell with a reddish spire; when it isn't red, it is axially flammulated with brown or olive; the mouth is commonly pinkish, and the spire sculpture commonly lacks spirals or has them very reduced. *C. cabritii* and *C. varius* are smaller and differ in color pattern and body sculpture.

Variation.—Young specimens have the spire relatively lower with straighter sides, the body whorl narrower and straighter-sided, and the pattern often of broad axial brownish flammules going through the mid-body area. With growth the spire becomes more concave with the early whorls forming a thick cone with largely fused whorls. The pattern of adult shells, as indicated earlier, is quite variable but usually consists of

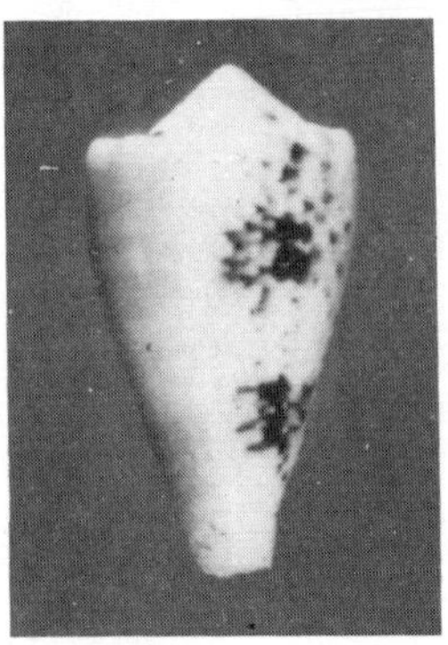

C. regius, half normal, half *citrinus,* Key Largo, Florida (GG).

two spiral rows of large blotches that may be variably fused. Occasional specimens completely lack distinct blotching and appear mottled.

Only one variant is dignified with a name, variety *citrinus*. This shell is a totally yellow or orange shell, the spire white but sometimes with some pattern. Shells are commonly found that are partially normal and partially *citrinus*, so the value of the variety is minimal.

One other possible variety is *C. caledonicus*, here treated as a full—but very doubtful—species; it could easily be a *citrinus* variety showing the usual dark brown spiral lines as brighter orange or pale brown spiral hairlines.

Distribution.—Common in shallow to moderately deep water from off Georgia (specimen from Mr. Ted Yocius), around Florida, and south through the Antilles and Central American coast to central Brazil.

Synonymy.—The early names probably belong here, but the figures are subject to interpretation. This is especially true of *C. citrinus*, which has also been used as a name for *Conus mus*; however, the figure is almost certainly a rather typical *C. lividus* and not Caribbean; if treated this way, however, *citrinus* would have one year priority over *lividus* and the name would have to change.

Notes.—Most *C. regius* are at best unexciting and dull cones with smudgy and indistinct color patterns. The eroded condition of the early whorls is the usual condition in this species and cannot be used for grading. There are commonly heavy axial growth marks and scars, all these affecting the pattern to some extent. The lip is often filed. Bright yellow or orange specimens are more attractive, and the occasional shell with both a normal and a *citrinus* pattern is especially interesting. Juveniles are moderately common and often more attractive than adults. Specimens in the *citrinus* variety and with unusual colors or patterns usually get a small premium.

RETIFER Menke, 1829

1829. *Conus retifer* Menke. *Verzeichn. Conch.-Samml. Malsburg:* 68 (Locality not stated).

1834. *Conus textile* var. *sulcata* Sowerby i, in Sowerby ii. *Conchological Illustrations:* Pt. 56/57, fig. 76 (Locality not stated). Non *Conus sulcatus* Hwass, 1792.

1834. *Conus solidus* Sowerby i, in Sowerby ii. *Conchological Illustrations:* large list. Nomen novum for *Conus textile* var. *sulcata* Sowerby i, in Sowerby ii, 1834. Non *Conus solidus* Gmelin, 1791.

Description.—Heavy, thick, with a high gloss; ovate, the sides strongly convex, almost bulbous posteriorly; body whorl covered with low, broad spiral ridges, closely spaced, these heavier anteriorly, weaker above midbody, sometimes obsolete below the shoulder; axial growth marks and scratches often prominent; shoulder broad, rounded, not distinct from spire; spire moderately high, pointed, the sides straight or slightly convex; early 2-3 whorls strongly nodulose if not eroded; spire whorls flat above, with 3-5 low but usually distinct spiral ridges and fine axial threads. Body whorl pinkish white, white, or pale straw, covered with narrow zigzag axial chestnut or orange-brown lines which overlap to produce large tents in about three spiral bands, below shoulder, at midbody, and above base; two broad spiral bands of dark orange above and below midbody, these usually with a few tents and many wide, deep reddish to black 'textile lines', rather evenly spaced and occasionally extending above and below the margins of the band; much of body whorl covered with exaggerated 'textile lines' on occasion; spire with continuation of body pattern, mostly pinkish or whitish with a few dark spots and lines over the orange-brown netting; early whorls pink or eroded white. Aperture moderately narrow, wider anteriorly; outer lip thick, slightly and evenly convex; mouth deep violet to pinkish, faded white. Columella long but narrower than usual for this group. Length 30-70mm.

Comparison.—Although obviously a textile cone, *C. retifer* seems more closely related to *C. aureus* than to *C. textile*, *C. canonicus*, or *C. legatus*. This is shown by the rather heavy spiral ridges on the body whorl and the strong spire nodules, as well as the tendency to exaggerate the 'textile lines' to cover much of the body whorl. The thick shell with a very inflated body whorl is very distinctive, as is the rather orange coloration.

Variation.—Other than the size of the tents and the intensity of the coloration, the main variation is in whether the 'textile lines' are widely or closely spaced and stay within the spiral bands or extend beyond them. At one extreme are shells with only a single band of short lines (usually above midbody) that are widely spaced; at the other extreme are shells which are mostly covered with wide 'textile lines' of various widths and orange to black in color. After major breaks in the body whorl the lines tend to be elongated and often wide, usually accompanied by a few very large white tents, but this is usually soon replaced by a more normal pattern. Large shells tend to be smoother than small shells posteriorly. The width of the body whorl varies individually, but all are distinctly convex posteriorly and thicker than any *C. textile*. Pink or white mouths are probably faded.

Distribution.—*C. retifer* is an uncommon species of the Pacific Ocean from the Ryukyus and Philippines to French Polynesia and Hawaii. I have seen only a few Indian Ocean specimens. Usually shallow water.

Synonymy.—This species is still sometimes called *C. solidus* in the collector literature, a fitting name but unfortunately one that is both preoccupied and second in priority to the obscure name of Menke.

Notes.—This is often a beautiful shell tinted pink and orange with strongly contrasting blackish 'textile lines'. Unfortunately it is rather rough for a textile cone, often with small spots of erosion on the body whorl and spire and even a few small burrowing sponge holes. The axial growth lines are often heavy and marked by drastic pattern changes, and there may be bad scars. The lip is thick but still chips easily on the new growth.

Juveniles are very lightly patterned and heavily ridged to the shoulder.

RUFIMACULOSUS Macpherson, 1959

1959. *Conus rufimaculosus* Macpherson. *Mem. Nat'l. Mus. Victoria*, No. 24: 54, fig. 5 (Tweed Head, New South Wales, Australia).

Description.—Moderately light in weight, with a high gloss; rather cylindrical, the upper sides slightly convex, widest below shoulder, then straight or slightly concave; body whorl with a few weak and low spiral ridges at the base, the grooves between with closely set punctations; numerous fine axial scratches over the rest of the whorl and a few spiral threads; shoulder narrow, with a raised carina, canaliculate, the edge sometimes weakly undulate or irregular; spire very low, flat, or actually depressed, with slightly convex sides; spire whorls slightly concave (early) to canaliculate (late), earliest whorls flat; tops of whorls with numerous fine axial threads and faint traces of spiral threads. Body whorl pinkish or white, irregularly covered with large and small clouds and blotches of brownish or orange-red, these often rather regularly distributed over the whorl or arranged in concentrations of indistinct axial blotches and flammules below the shoulder, at midbody, and above the base; spire white, with a few axial blotches about the same as those of the body whorl. Aperture rather narrow, slightly widened anteriorly; outer lip thin, sharp, slightly convex or concave at middle; mouth strongly toned with pink, darkest deep within. Columella narrow, very slightly free posteriorly. Length 30-50mm.

Comparison.—The mottled pattern and deeply canalicuate whorls make this species difficult to compare with any others. Presumably it is related to *C. sculletti*, but it differs in not having the pattern distinctly banded, having canaliculate whorls and shoulder, and not having the sides deeply concave. *C. nielsenae*, especially the southern WAust. form, has some resemblance, but that species has the whorls much flatter and the pattern not mottled. *C. nobilis* is somewhat similar in body shape but differs in spire form, pattern, and presence of nodules on the early whorls. *C. angasi* and allies have the columella free over a longer distance and differ radically in shape and pattern.

Variation.—The distinctive shape and canaliculate spire vary little, although the spire varies from low to actually depressed. The major variation is in the arrangement of the color blotches. In some specimens the color is rather uniformly spread over the shell as rather triangular or ovate spots, with few concentrations or blotches. In other shells there are distinct but nebulous blotches below the shoulder and above the base, with smaller spots of color scattered over the intervening area. Still other shells have large axial blotches below the shoulder, similar but squarer and smaller blotches at midbody and the base, and nearly continuous axial flammules connecting the three series of blotches. The color varies from distinctly brownish to distinctly reddish; often the entire shell is suffused with bright pink, darkest in the mouth.

Distribution.—Moderately common in deep water off southern Queensland and northern New South Wales.

Notes.—This is a clean species except for sometimes heavy axial growth lines. The lip is sometimes filed.

The mottled pattern on a pale ground is unusual in cones, as most other species have at least traces of spotting or tenting somewhat developed in the pattern.

RUTILUS Menke, 1843

1843. *Conus rutilus* Menke. *Moll. Nov. Holl.:* Spec. 27 (South Seas).

?1870. *Conus semisulcatus* Sowerby iii. *Proc. Zool. Soc.* (*London*), 1870: 257, pl. 22, fig. 13 (Locality unknown). Non *Conus semiculcatus* Bronn, 1831, a fossil.

1876. *Conus Tasmanicus* Tenison-Woods. *Pap. Pr. Rpt. Roy. Soc. Tasmania for 1875:* 139 (Tasmania). Non *Conus tasmaniae* Sowerby ii, 1866 (not truly a homonym).

1877. *Conus Macleayana* Tenison-Woods. *Ibid., for 1876*: 134. Nomen novum for *Conus Tasmanicus* Tenison-Woods, 1876, thought to be preoccupied.

1877. *Conus* (*Stephanoconus*) *smithi* Angas. *Proc. Zool. Soc.* (*London*), 1877: 36, pl. 5, fig. 8 (Cape Solander, Botany Bay, New South Wales, Australia).

Description.—Fragile, very thin and light in weight, with a good gloss; low conical to obconical, the upper sides slightly to strongly convex then tapered to the base; body whorl with a few low spiral ridges at the base and a few faint spiral and axial scratches; shoulder broad, roundly to sharply angled, concave above, the margin undulate to heavily coronate, sometimes smooth; spire low to very low, the protoconch very large; sides of spire nearly straight to concave; tops of spire whorls concave above, the margins usually coronate, tops with heavy axial threads. Body whorl very variable in color from bright red or orange, straw, yellow, or violet, variably covered with fine spiral rows of brown spots, dashes, and blotches or immaculate; spire often paler than body whorl, with a few or many large brownish blotches; early whorls pale. Aperture moderately wide, slightly widened anteriorly; outer lip very thin and fragile, straight or slightly convex; mouth white, pink, violet, etc., reflecting external color. Columella narrow, short. Length 8-15mm.

Comparison.—The small size, coronated whorls, and variable pattern prevent comparison with other cones. I am not really sure why this species is almost universally considered to be an adult shell, as the specimens I have seen surely look like the first 3-5 whorls of a much larger cone or perhaps several species of cones. For instance, there is really nothing that would separate it conchologically from *C. papilliferus* early whorls, although of course that species has relatively distinct spiral sculpture even on the early whorls and is only partially sympatric with *C. rutilus*. For the moment *C. rutilus* must be considered the adult of one of the smallest species of cones with a very large protoconch in the adult.

Variation.—The color range is extreme but probably much of this is due to fading and erosion after death. Patterns fall into two general groups that seem to be generally related to shape as well. In the first pattern the body whorl is almost obconical with sharply angled shoulder; the body whorl is often uniformly reddish or orange without obvious pattern. The second pattern has the shoulder more rounded, the spire higher; the body whorl is often pale, yellowish or straw, with spiral dashes and small spots. The presence or absence of coronations seems variable in this species but may possibly be related to minor local populations.

Distribution.—Moderately common as beach shells, uncommon live-taken, from New South Wales to Freemantle, Western Australia, including Tasmania. Shallow water.

Synonymy.—*C. semisulcatus* is usually assigned here but certainly seems to belong elsewhere; the shape and sculpture are wrong for this species but I cannot place it anywhere else at the moment.* *C. tasmanicus* is certainly this species, while *C. smithi* probably belongs here. The description does not mention coronations and the figure does not show them, while the body pattern belongs to pattern two, being straw with dashes and spots and a rounded shoulder. The presence or absence of coronations may be determined by local populations or perhaps individual variation. Of course, if *C. rutilus* were the juvenile of another species, then *C. smithi* could be the juvenile of a second species.

Notes.—This is an almost impossible species to obtain in gem condition, as might be expected. The juvenile characters go with a juvenile fragility; beach specimens often are completely eroded and show many cracks. Live-taken specimens are few and far between in reaching the market, and often they are not in much better condition than beach specimens.

*See *commodus* in doubtful species; *semisulcatus* is perhaps actually *sennottorum*.

SANGUINOLENTUS Quoy and Gaimard, 1834

1834. *Conus sanguinolentus* Quoy and Gaimard. *Voy. Astrolabe, Zool.*, 3: 99, pl. 53, fig. 18 (Carteret, New Guinea).

Description.—Moderately heavy, with a low gloss; low conical, the upper sides slightly to distinctly convex, then sometimes slightly concave; body whorl with about a dozen low spiral ridges on the anterior half to quarter, some or all the ridges bearing large granules; more posterior ridges widely spaced, sometimes extending well above midbody; numerous spiral and axial threads over whorl; shoulder broad, roundly angled, with heavy but rather low and rounded coronations; spire low, bluntly pointed, the sides straight to concave; spire whorls with heavy rounded coronations, slightly concave above, with about 3-5 spiral ridges and very fine scattered axial threads. Body whorl uniform pale yellowish tan to dark olive or pale brown, the base strongly toned with brownish violet dorsally and ventrally; granules pale; sometimes with indistinct pale area at midbody, but no distinct band present; narrow pale band just below the shoulder angle; spire and shoulder yellowish or pale olive, the whorls sometimes with faint darker spots between coronations; early whorls white from erosion, the tip pinkish. Aperture moderately narrow, slightly widened anteriorly; outer lip thin, straight or very slightly convex, sometimes slightly concave at middle; mouth purple to pale violet, without well defined paler bands. Columella very narrow. Length 25-45mm.

Comparison.—This somewhat doubtful species is very similar to *C. lividus* but appears to be distinguishable. It differs from that species in having the basal violet tone visible both dorsally and ventrally, lacking a midbody white band (although the midbody area may be somewhat paler than the rest of the shell), having the shoulder pale band very narrow and often not sharply defined, and by having the spire distinctly tinged with pale olive, sometimes with indistinct spots on the later whorls. On the average it also seems somewhat more convex posteriorly than in *C. lividus*.

Variation.—The color of the shell varies from a pale but distinctly yellowish tan to dark brown, although usually it is dark olive. The midbody band is not present, although sometimes a narrow and very indistinct paler band can be seen under the darker coloring. The shoulder band is narrow and has indistinct anterior edges. Although the spire is very pale, at least the later whorls and the shoulder are usually toned with pale olive, rather reddish or greenish, sometimes with indistinct spots between the coronations. The mouth is usually almost uniformly violet, sometimes lighter deep within or with a darker blotch posteriorly. Of course the granulose ridges are variable, sometimes absent, but usually reach to midbody and sometimes almost to the shoulder.

Distribution.—Widely distributed in the Indo-Pacific, but often confused with *C. lividus*. Known from the Natal coast and Red Sea east to

French Polynesia; no records for Hawaii, but probably confused with *C. lividus* there and overlooked. Common in shallow water.

Notes.—*C. sanguinolentus* is usually a clean species for a small reef-dweller. Except for sometimes heavy axial growth marks and occasional small growth scars on the spire, the condition is good. In some specimens there is a persistent periostracum which is hard to remove completely.

I believe that *C. sanguinolentus* is a full species although closely related to *C. lividus*. I have seen no real intermediates, but such probably exist.

SAZANKA Shikama, 1970

?1850. *Conus dilectus* A.A. Gould. *Proc. Boston Soc. Nat. Hist.*, 3: 172 (Feejee Islands).

1970. *Conus* (*Rhizoconus*) *sazanka* Shikama. *Sci. Rpts. Yokohama Nat'l. Univ.*, Sec. II(16): 25, pl. 1, figs. 24-25 (SW Kochi Pref., Japan).

1974. *Conus kurzi* Petuch. *Veliger*, 17(1): 41, figs. 1-2 (32 km. SE of Tosa Shimizu, Shikoku I., Japan).

Description.—Moderately heavy, with a high gloss; low conical, the upper sides nearly straight or slightly convex; body whorl with a few low spiral ridges at base, otherwise with only spiral and axial threads and scratches; occasional specimens with few to many spiral rows of granules over body whorl; shoulder roundly angled, broad, with low, rounded coronations, sometimes almost straight; spire low to moderate, sharply pointed, the sides straight to slightly concave; whorls slightly stepped, slightly concave or slightly convex above, with about 2-5 low spiral ridges and fine indistinct axial threads; margins of whorls low coronate. Body whorl bright yellow to bright orange, sometimes with tan tones; base white, usually contrasted; squarish white blotches may develop at midbody in a narrow band, the blotches sometimes shadowed with brown; a few brown flecks may be present at the base; shoulder and spire white to deep pink, with small pale to deep orange or tan spots between the coronations; early whorls bright pink. Aperture moderately narrow, slightly widened anteriorly; outer lip nearly straight; mouth pale violet to pinkish. Columella internal. Length 23-40mm.

Comparison.—*C. sazanka* has been confused with *C. cumingii*, but the two species are quite distinct. In *C. cumingii* the whorls have smooth margins and are not stepped; the body whorl is dark, with a dark base, and is covered with many spiral rows of strong brown dots and dashes. Some shells assigned to *C. sazanka* have distinct brown flecks at the base and some brown in the midbody band, but nothing to compare with *C. cumingii*. There seems to be a general resemblance between *C. sazanka* and *C. amphiurgus*, but this is probably purely superficial due to the bright colors; both *C. amphiurgus* and *C. daucus* have at most weakly nodulose early whorls, the later whorls smooth. Large specimens, if properly assigned here, somewhat resemble *C. capitaneus* in having axial brown marks in the midbody band.

Variation.—Based on the descriptions and photos of two specimens described by Shikama and a paratype of *C. kurzi*, this is a rather constant species. The spire is usually distinctly pinkish with orange spots, the body whorl varying from unmarked bright yellow to bright orange with tan tones. The spire varies from almost flat to quite high and stepped, but it is at least undulate on the margins. The largest specimen has traces of pinkish squares in a midbody band on bright orange. All have contrasting white bases.

Kohn and Weaver (1962, *Pacific Science*, 16(4): 354-356) reported a large series of what seems to be *C. sazanka* under the name of *C.* sp. cf. *cumingii* from Hawaii. Although most of the shells illustrated could easily fall into *C. sazanka*, there are some differences. Several of the specimens have a few to many spiral rows of large granules on the body whorl; the spots on the spire seem to be darker and larger than in Japanese specimens. The largest specimens are somewhat elongated, have distinct squares of whitish at midbody bordered by brown, and have brown spots at the base; the spire is higher and heavily marked. To judge from the description, the spiral ridges of the spire whorls are more heavily sculptured with heavier axial threads. The base is not mentioned.

Distribution.—So far known only from moderately deep water off southern Japan and Hawaii, about the same distribution as *C. smirna* and *C. eugrammatus*. It is rare.

Synonymy.—*C. dilectus* is certainly a juvenile of another species, possibly *C. sazanka*. I have examined specimens from the Philippines, and it is also reported from Fiji. The shell is very small and delicate, covered with broad spiral ridges and axial threads, has distinctly coronated or heavily nodulose early and later whorls, and is pinkish and red with white flecks clustered at the base and midbody. It is rare.

C. kurzi is apparently a synonym of *C. sazanka*, agreeing well with Shikama's material. The types were obtained from trawlers and could just as easily have come from off Hawaii as off Japan (pers. comm. Ed Petuch).

Notes.—The few specimens at the moment in private collections have mostly come from Hawaii, where the species can be taken in moderately deep water in good numbers. Presumably the series described by Kohn and Weaver comprise a single species, although the larger specimens certainly look different from smaller ones or the large Japanese holotype. The actual relationships of this species are unknown, although it is probably related to the *C. vexillum* group.

SCABRIUSCULUS Dillwyn, 1817

1817. *Conus scabriusculus* Dillwyn. *Descr. Catalogue Recent Shells*, 1: 406 (Guinea). Based on Chemnitz, Pl. 182, figs. 1768-1769.
1833. *Conus fabula* Sowerby i, in Sowerby ii. *Conchological Illustrations:* Pt. 24, figs. 5-5* (Locality not stated).
?1844. *Conus plumbeus* Reeve. *Conchologia Iconica*, 1(*Conus*): Pl. 46, sp. 253 (Locality unknown).

Description.—Moderately light in weight but thick, with a good gloss; ovately conical, the upper sides distinctly convex and somewhat bulbous, then tapering to the base; body whorl covered with rounded spiral ridges which are most prominent on the anterior half of the whorl and may be obsolete or nearly so below the shoulder; ridges usually with low, uniformly spaced granules; grooves covered with fine axial threads; shoulder rounded, angled, somewhat distinct, convex above; spire low, the sides straight to convex; first three or so whorls with nodules; spire whorls sunken at suture, convex above, with about 4-6 weak spiral ridges which rarely show granules; last spire whorl above edge of upper lip with only weak ridges; axial threads inconspicuous. Body whorl white with violet tones, the base deep violet; anterior third of whorl mostly deep brown from fusion of large blotches below midbody; a band of large to small dark brown blotches below the shoulder, these often connected to the anterior blotches by narrow and wavy axial flammules; spire mostly white with a few small brown blotches on the later whorls; tip pink. Aperture rather narrow, wider anteriorly than posteriorly; outer lip sharp, thick, straight or slightly concave at the middle; mouth pale to dark violet, with a short white band at the middle. Columella short and narrow. Length 20-40mm.

Comparison.—*C. scabriusculus* is closely related to *C. glans* and *C. tenuistriatus*. It is the most inflated of the three species, with the upper sides distinctly convex and the shoulder weak but fairly distinct; the body whorl is relatively smooth, the granules usually weak, and the spire ridges not usually granulose. The other two species have the shoulder almost indistinguishable from the spire, the granulose ridges distinct to the shoulder, and the body whorl mostly brown or brownish violet with a few whitish bands or blotches, not the large white areas with relatively small brown areas of *C. scabriusculus*. The species is easily recognized.

C. scabriusculus has been badly confused with *C. tenuistriatus* but seems very distinctive in shape, pattern, and sculpture. Some *C. tenuistriatus* have large white blotches at the midbody area and shoulder but are still mostly brownish and violet shells, greatly elongated compared to *C. scabriusculus*.

Variation.—The relatively short and very wide body whorl is constant. The spiral ridges at the shoulder may be weak or obsolete but are seldom distinctly granulose like the basal ridges. The mostly brown anterior area seems constant, but the band of brown blotches above mid-body is variable from mostly dark to almost absent, with or without axial flammules to the base. If the blotches are all large and the axial flammules are strongly developed, the white areas between them are still large and give the shell a mostly white appearance. Violet toning may be weak or strong.

Distribution.—Known with certainty from the western Pacific from the Ryukyus and Philippines to Fiji and French Polynesia, but perhaps more widely distributed and confused with *C. tenuistriatus.* Common in shallow water.

Synonymy.—*Conus fabula* is certainly this species and should not be used. The status of *C. plumbeus* is doubtful, but it looks like a small specimen of either *C. scabriusculus* or *C. glans.*

Notes.—A very attractive species, *C. scabriusculus* is usually brightly patterned with dark brown and white with pale violet tones. There may be some heavy axial growth marks, but this is an otherwise clean shell.

This species probably occurs in the Indian Ocean, but the few specimens from that area I have seen proved to be *C. tenuistriatus.*

SCALPTUS Reeve, 1843

1843. *Conus scalptus* Reeve. *Conchologia Iconica,* 1(*Conus*): Pl. 37, sp. 203 (Locality unknown).

Description.—Moderately light in weight, with a high gloss in living specimens; rather narrowly low conical, the upper sides gently convex then tapering to a narrow base; body whorl covered with about 10-20 flat heavy spiral ribs separated by shallow grooves often containing secondary ridges; ribs often absent or greatly reduced on posterior third of whorls; grooves with heavy axial threads crossing the secondary ridges, producing a distinctly square cancellation; shoulder roundly to narrowly angulate, narrow, fairly distinct from spire; spire moderately tall, sharply pointed, deeply concave; spire whorls flat on top, with 4-6 heavy spiral ridges crossed by heavy axial threads so usually strongly cancellate; no nodulose whorls. Body whorl creamy white, sometimes with pale violet tones in living specimens, covered with about 20-25 broad deep reddish or orangish brown spiral bands of very regular arrangement and width; sometimes with minor secondary brown lines, and occasionally with scattered indistinct short axial brownish blotches; area below shoulder often with the bands less distinct and partially fused or covered with short oblique flammules, sometimes producing a single broad band; spire and shoulder whitish, with few to many narrow axial reddish brown streaks; early whorls white; base of shell white, very contrasting, unmarked. Aperture moderately narrow, slightly widened anteriorly; outer lip sharp, slightly convex; mouth white with pale violet tones. Columella internal. Length 30-50mm.

Comparison.—*C. scalptus* is usually a very distinctive species because of the rather slender shape, broad spiral ribs at least anteriorly, and the broad bright reddish spiral bands over the entire whorl. Dead specimens with reduced pattern might be confused with *C. mucronatus* but *scalptus* is more slender, has a more rounded shoulder on the average, and usually has at least remnants of the banded pattern.

This species is closely related to *C. tegulatus* and in some patterns that species distinctly approaches *C. scalptus*. However, almost any pattern of *C. tegulatus* has a much less regularly banded pattern with blotches and spots visible over much of the shell, usually in axial flammules; the shoulder tends to be wider and more angled, giving the shell a different appearance; and the spiral sculpture is narrower and less regular.

Such striped species as *C. hirasei*, *C. caillaudii*, and *C. kimioi* may have similar patterns on the body whorl but lack the distinctive sculpture. *C. grangeri* and *C. sulcatus* sometimes have similar body sculpture but have coronated whorls and different color patterns.

Variation.—The stripes are constant, although in beach specimens they fade and may appear to break into dashes. There are sometimes short axial lines or clouds in juveniles, but these are not conspicuous and are apparently absent in adults. The secondary ridges may be white or reddish. Below the shoulder the bands are almost always obscured and partially fused with indistinct short oblique reddish flammules that may be visible at the very edge of the shoulder.

The sculpture is easily visible and feelable anteriorly but is commonly reduced or obsolete above midbody; occasionally the ribs are partially or entirely divided into two narrower ridges. The shoulder is usually rounded, although it may be slightly angulate. The spire pattern may be heavy or sparse.

Distribution.—*Conus scalptus* seems to be restricted to the Philippines, New Guinea, Solomons, and probably Indonesia; it is uncommon to rare in live condition but moderately common as beach shells.

Synonymy.—For some reason this brightly colored species has been called *C. mucronatus* for many years in the collector literature. The description and figure of *C. scalptus* are both good and certainly apply here. The figure of *C. mucronatus* occurs on the same plate, so perhaps someone carelessly confused the two figures and names at some time? It seems highly unlikely that *C. scalptus* falls within the range of variation of *C. mucronatus*.

Notes.—Even beach specimens of this species are in clean condition except for minor chips in the lip and sometimes the earliest spire whorls missing. However, beach specimens change colors from pale creamy with violet tones and bright reddish bands to pale straw or dirty white with narrow and faded brownish bands that may be broken into dashes. Such shells could easily be confused with some patterns of *C. tegulatus* even though spots and blotches are absent, but they are not truly representative of the species. Live-taken shells are very uncommon, however.

Juveniles are uncommon and very similar to adults except in having some axial markings; the type of *C. scalptus* is apparently a large juvenile. The range of this species is uncertain, but it seems to be restricted to the western Pacific.

SCHEPMANI Fulton, 1936

?1898. *Conus (Leptoconus) saecularis* Melvill. *Mem. Proc. Manchester Lit. Phil. Soc.*, 42(4): 10, pl. 1, fig. 23 (Malcolm Inlet, Persian Gulf).

1913. *Conus elegans* Schepman. *Prosobranchia Siboga Exped., Monogr.* 49(5): 393, pl. 25, fig. 4 (Bougainville Straits). Non *Conus elegans* Sowerby iii, 1895.

1936. *Conus schepmani* Fulton. *Proc. Malac. Soc. London*, 22: 7. Nomen novum for *Conus elegans* Schepman, 1913.

Description.—Light in weight, with a good gloss; elongate biconical, the sides convex posteriorly then deeply concave and tapering to a narrow base; body whorl covered from base to shoulder with flat ribs, narrowest anteriorly and sometimes weak posteriorly; grooves between ribs narrow, deep, with weak or strong axial threading; whole whorl covered with fine axial scratches; shoulder broad, carinate, concave above; spire tall, sharply pointed, the sides concave; whorls stepped, concave above, with about 3-4 heavy spiral ridges crossed by numerous heavy axial curved threads so deeply cancellate; first six whorls with strong nodules, next two or so whorls weakly undulate. Body whorl pale straw or yellowish, heavily covered with small brownish to reddish dashes that tend to form oblique blotches in two broad bands above and below midbody and a weaker band below shoulder; midbody area with a few dashes and many large spots or blotches of opaque white, strongly contrasting; pale bands above base and below shoulder with a few smaller opaque white dashes; shoulder and spire pale straw, heavily or weakly covered with reddish brown or orange-tan streaks and dots; early whorls pale. Aperture narrow, uniform; outer lip thin, strongly concave at middle; mouth white to pale violet. Columella not visible externally. Length 18-30mm, perhaps larger.

Comparison.—*C. schepmani* is allied to the slender form of *C. insculptus* but differs in having more numerous strongly nodulose early whorls, a stronger brownish pattern, and the constant presence of large opaque white spots at least in the midbody area. The distinct but normal spire sculpture will readily separate it from all forms of *C. aculeiformis*, while the pattern is different from that of *C. comatosa*, which also has distinctly carinate and rim-like shoulder margins. *C. memiae* is similar in general shape and in having heavy white spotting at midbody, but in that species only the first 1-3 or fewer whorls are nodulose, there is a distinctly orange spiral band below midbody but not above, and there is a pattern of enlarged spots above and below midbody; *C. memiae* is less elongate, more weakly sculptured, and is often deep pinkish in tone.

Variation.—The few specimens I have seen were quite constant in shape and general pattern as well as sculpture. The spots may form almost solid bands above and below midbody or may remain distinct and cover much of the body whorl in vaguely axial rows; the midbody white spots are constant, but other white spots may be present or absent in the other clear bands.

Distribution.—I have examined specimens from moderately deep water in the Ryukyus, Philippines, Indonesia, New Guinea, and the Solomons. Presumably the species is widespread and not uncommon in the western Pacific, but confused with *C. aculeiformis* and *C. insculptus*. If *C. saecularis* truly belongs here, as seems likely, the species also occurs in the Indian Ocean.

Synonymy.—*Conus elegans* was preoccupied and renamed by Fulton. Although Tomlin (1937) assigned *C. saecularis* to the synonymy of *C. dictator* by saying that the type was a broken and distorted juvenile of that species (itself doubtful), this seems very unlikely. The type of *C. saecularis*, one of several specimens described, was 36mm in length, appears adult from the figure, and is very different in shape, sculpture, and pattern from *C. dictator*, which apparently is related to *C. inscriptus*. Melvill mentioned the white midbody spots in his species, and the shape and sculpture agree well with *C. schepmani*. Unfortunately the description of *C. saecularis* is based on dead and badly broken shells so for the moment *C. schepmani* is retained for this species; I am not aware of any other Indian Ocean record nor any western Pacific specimens over 30mm in length.

Notes.—Dead-dredged specimens are badly faded with broken lips and worm holes, but they still show the distinctive white blotching at midbody. Live-taken specimens are pale tan or straw with sometimes rather bright orange-tan spotting and very contrasting opaque white markings. The lip is fragile.

At the time of writing the species does not appear to be available on the market, but it is apparently not uncommon (dead) in the Solomons.

SCOPULORUM Van Mol, Tursch, and Kempf, 1971

1971. *Conus scopulorum* Van Mol, Tursch, and Kempf. *Zoologische Mededelingen*, 45(15): 162, figs. 3-4; pl. 1, figs. 5-6 (Brazil, off Fernando de Noronha).

Description.—Moderately light in weight, with a low gloss; biconical, the upper sides convex; body whorl with a few low spiral ridges at the base and numerous axial and spiral threads, appearing rough; shoulder roundly to sharply angled, with low to high, rounded coronations, concave above; spire tall, sharply pointed, the sides concave; whorls weakly stepped; tops of whorls slightly concave, with a single low median spiral ridge and numerous fine axial threads; early whorls not nodulose, later whorls with rounded coronations. Body whorl white or pale straw, covered with many spiral rows of dark brown dashes and two spiral bands of large orange-brown to dark brown blotches above and below midbody; where rows of dashes go through dark blotches they alternate with small white dots; spire and shoulder whitish, with some scattered brown spots and lines, especially fine brown hairlines between coronations; early whorls white. Aperture narrow, slightly widened anteriorly; outer lip thin, slightly sloping below level of shoulder, straight; mouth whitish. Columella very narrow or internal. Length 20-26mm.

Comparison.—*C. scopulorum* is obviously very closely related to *C. aurantius*, which seems not to reach northern Brazil with certainty. It is distinguished mainly by its smaller size, supposedly more numerous nuclear whorls (at least 3, but *C. aurantius* has at least 2, both species usually broken or eroded), and the lower and sometimes indistinct or absent coronations on the early whorls. In addition, the spire is more slender and concave-sided than in *C. aurantius* and the color pattern less complicated. Probably the two are just subspecies.

Variation.—The shoulder may be rounded with low coronations or sharply angled with heavy pointed coronations. At least the later spire whorls have distinct coronations, but occasionally they may also be seen on the early whorls. The pattern is of two spiral bands of large squarish orangish to dark brown blotches above and below midbody, these usually fused spirally at least in spots and containing dark and light dots and dashes. Fine axial lines connecting the blotches across the midbody area may be present or absent. The maximum size seems to be under 30mm.

Distribution.—Known only from moderately deep water around the offshore islands and continental shelf of northern Brazil. Moderately common but seldom reaching the market.

Notes.—This is a very minor and doubtful species that probably cannot be separated with confidence from *C. aurantius* of similar size. The differences in radula and nuclear whorls used in the original description seem at best doubtful and certainly of no use to the collector.

Although it very rarely reaches the market, this species is not rare in its range, occurring in 14 of 16 dredge hauls on offshore reefs, for a total of 77 specimens as reported by the original authors.

SCULLETTI Marsh, 1962

1962. *Conus (Leptoconus) sculletti* Marsh. *J. Malac. Soc. Australia*, No. 6: 42, pl. 4, figs. 3-4 (Off Cape Moreton, South Queensland, Australia).

Description.—Moderately light in weight, with a good gloss; elongate low conical, the sides distinctly concave; body whorl with faint traces of spiral ridges at the base and numerous axial threads and scratches, otherwise smooth and polished; shoulder projecting, carinate, concave or flat above; spire very low, slightly domed, the whorls slightly to deeply concave above; tops of whorls with numerous fine axial threads, the margins often rim-like and erect.

Body whorl creamy to pinkish or straw white, with numerous large and small irregular blotches of deep orange-brown; these blotches tend to form continuous spiral bands above and below midbody, the anterior band widest; other blotches may form an indistinct band at the shoulder; axial flammules or bands of small blotches often present, irregular, sometimes forming heavy axial flames above midbody; anterior band usually with small whitish spots in 2-3 spiral rows within; spire about same color as body whorl, with radiating blotches of orange-brown; early whorls white. Aperture rather wide, uniform in width; outer lip very thin, straight or slightly concave; mouth creamy white with large pink blotch deep within. Columella indistinct. Length 30-60mm.

Comparison.—*C. sculletti* is most similar to *C. angasi* and *C. rufimaculosus* in sculpture and pattern, though very different from either species. *C. angasi* and *C. advertex* are shorter, broader, have a free columella, and have a pattern of large spiral dashes. *C. rufimaculosus* has more nearly straight sides, has canaliculate tops to the shoulder and late spire whorls, and has a very mottled pattern that does not form broad and distinct bands.

C. monile has sometimes been confused with *C. sculletti*, but the two are not related. The early whorls of *C. monile* are nodulose and form a projecting cone, the shell is thicker, the pattern is completely different in nature, and the sides are straight; more-or-less the same differences apply between *C. bayani* and *C. sculletti*. *C. nielsenae* is broader, has at least a single spiral ridge on the spire whorls, and is usually covered with fine brown hairlines, never large blotches or bands.

Variation.—The shape is constant, although the spire varies from low to nearly flat and the sides may be barely concave to strongly concave. The color varies from white or pinkish to distinctly pale straw or tan, with the blotching orange-tan to dark brown. Development of the blotches varies considerably, although the band below midbody is usually well developed and there is usually at least a narrow band (sometimes two) above midbody. There may be few or many blotches at midbody, while the area below the shoulder may be almost clear or covered with heavy squarish or axial blotches.

Distribution.—Moderately common in deep water off southern Queensland and northern New South Wales, Australia.

Notes.—This shell is often very waxy in texture, but it is fragile and often has rather bad axial growth scars on the body whorl. The lip is usually heavily filed. In some patterns the blotches are very indistinct and smeared, while in others they are very sharply defined. The shoulder is often strongly projecting.

SELENAE Van Mol, Tursch, and Kempf, 1967

1967. ***Conus selenae* Van Mol, Tursch, and Kempf.** *Ann. l'Inst. Oceanogr.*, **45(16): 250, fig. 13; pl. 8, fig. 2 (Fortaleza, Ceara, Brasil).**

1967. *Conus yemanjae* Van Mol, Tursch, and Kempf. *Ibid.*, 45(16): 251, fig. 15; pl. 8, fig. 1 (Off mouth of Rio San Francisco, Brasil).

Description.—Light in weight, with a high gloss; ovately low conical, the sides very convex; body whorl with about 4-6 low spiral ridges anteriorly, these becoming nearly flat ridges more posteriorly and crossed by axial threads and plicae; where the spiral and axial sculpture cross large granules usually form, about 10-20 spiral rows over the body whorl maximum; granules may be entirely absent, restricted to anterior part of whorl, or continued to shoulder; shoulder rounded, sometimes appearing sharply angulate when large granules occur at its margin; spire moderately high, the sides concave or nearly straight, blunt; whorls slightly concave above, with 2-3 heavy spiral ridges crossed by heavy axial grooves; the anterior ridge is often granulose, the posterior ridge sometimes obsolete; occasionally the margins of the whorls are also granulose, resulting in three or four spiral rows of granules per whorl. Body whorl white, pink, or pale salmon, with small to large blotches of orange brown to violet or darker pink; blotches usually arranged in two roughly spiral bands above and below midbody area, but often very indistinct or completely absent; numerous spiral rows of short dark brown dashes over entire body whorl, usually some visible even in whitish specimens; shoulder and spire pinkish or whitish, with few to many orange-brown blotches and spots; early whorls white to orange. Aperture moderately narrow posteriorly, widened anteriorly; outer lip thickened, convex; mouth white, pink, or violet. Columella short, rather wide, thick, with low ridges above. Length 12-20mm.

Comparison.—Although it could be confused at first glance with *C. jaspideus*, *C. puncticulatus*, or *C. mindanus*, the presence of distinct heavy spiral ridges on the spire whorls will readily separate these species.

Conus selenae is very closely related to *C. hieroglyphus*, and the two would perhaps be better treated as subspecies. However, typical *C. hieroglyphus* is a dark shell with distinct spiral ridges over the whorl, a contrasted white pattern, and about four narrow spiral ridges on the spire whorls. Occasional specimens, however, are pale reddish tan or even white with reduced patterns that look very much like *C. selenae*. As far as I know, *C. hieroglyphus* does not have granules on the spire whorls, but it is cancellate. Specimens very much like *C. selenae* have been seen from southern Florida and the Central American coast. Much more work is needed with these species.

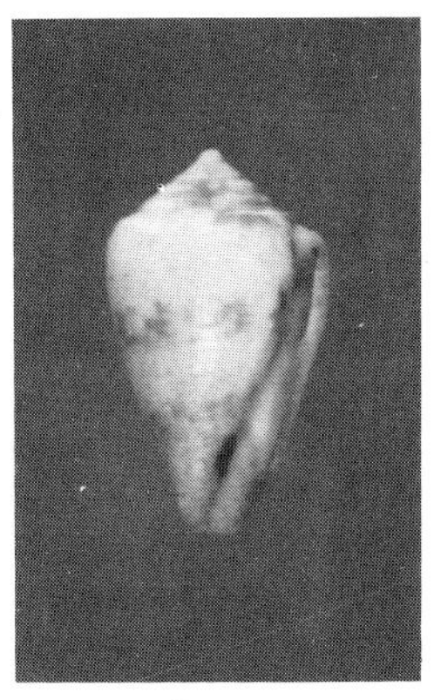

C. aff. *selenae,* Pompano Beach, Florida, 22mm (EP).

Variation.—In shape this species is rather constant, appearing very inflated with a moderately low spire. The shell may be entirely white or pale violet, lightly mottled with reddish or lilac, pale or dark pink, or any combination of these colors. There are usually distinct spiral rows of dashes visible between the granules even in mostly white shells.

The sculpture is variable, the body whorl being almost smooth except for spiral and axial threads or low ridges or heavily granulose over the entire whorl; the granules may be very widely spaced and sparse or form a cobblestone texture over the whorl. The spire whorls are often indistinctly ridged, but often the anterior ridge is heavily granulose where axial grooves cut across the spiral ridge; sometimes the entire whorl surface is covered with axial rows of large granules, not only the ridge(s) being granulose but the margins as well.

Distribution.—Rather common along the continental shelf of northern and northeastern Brazil.

Synonymy.—Typical *C. selenae* had distinct spiral spire ridges with low granules, *C. yemanjae* had indistinct ridges but large granules at the anterior margin; it was later discovered that these characters were individually variable and the two species synonymized.

Notes.—Although this little species is reported to be common all along the continental shelf of northern and northeastern Brazil, it seldom reaches the American market and then at a high price. The few specimens I have seen were clean enough with a tendency to have small scars on the spire whorls. The color patterns are very variable as is the sculpture, but at the moment it seems all are about equally common.

SENNOTTORUM Rehder and Abbott, 1951

1951. *Conus sennottorum* Rehder and Abbott. *Revista de la Sociedad Malac. Carlos de la Torre*, 8: 63, pl. 9, figs. 1-2 (50 mi. off Campeche, Yucatan, Mexico).

Description.—Moderately light in weight, with a good gloss; ovately biconical, the upper sides strongly convex then concave and shortly tapered to a narrow base; body whorl with about 5-8 low flat spiral ridges above the base, otherwise smooth except for sometimes conspicuous axial threads; shoulder broad, carinate, nearly flat above; spire tall, sharply pointed, deeply concave; earliest 2-3 whorls weakly nodulose, later whorls with carinate margins, the earlier ones projecting; tops of whorls nearly flat, covered with numerous crowded curved axial threads. Body whorl white to pale cream, with few to many spots and dashes of pale brown in vaguely spiral and axial rows, these sometimes restricted to the midbody area and below the shoulder, where sometimes fused into short axial flames; spire and shoulder whitish, with few to many spots and streaks of pale brown. Aperture narrow, uniform in width; outer lip thin and fragile, convex above then concave; mouth white. Columella internal. Length 25-35mm.

Comparison.—*C. sennottorum* has been badly confused with *C. lorenzianus* juveniles, although the two are distinct. In *C. lorenzianus* the shoulder is rounded, the early whorls are not or weakly carinate and not projecting, the narrow spiral ridges extend to midbody or higher, and the dark brown or reddish brown spots are large and regular, covering most of the body whorl and spire. In *C. spurius* juveniles the shoulder is rounded, the body whorl is narrower, and the spire has nearly straight sides and appears rather sunken into the shoulder.

C. kerstitchi from the Panamic region is very similar in shape and general appearance to *C. sennottorum* although they are probably not really related. *C. kerstitchi* often shows a single spiral ridge on the spire whorls and traces of 2-3 others, has salmon toning or banding at the base and shoulder and on the spire, and is seldom creamy white in color, the spots reddish brown instead of pale orange-tan. *C. stimpsoni* is also very similar and probably closely related; its sides are straighter, the sculpture of the body whorl is weak but often extends to the shoulder, there are weak but distinct spiral ridges on the spire whorls, and color pattern consists of spiral bands of pale yellow or orange.

Variation.—The major differences between specimens are in the number of spots and their color. Typically the spots are rather few, mostly in the midbody and shoulder area, and pale orange-tan in color, quite inconspicuous. However, in some specimens the spots are darker brown and scattered in vaguely spiral or axial rows over much of the body whorl, weak only at the base. The spots may be rounded, dash-like, or irregular, sometimes covered by clouds of tan. The concavity of the spire varies a bit, as does the convexity of the body whorl.

Distribution.—*C. sennottorum* is found in deep water of the Gulf of Mexico from south rn Florida to Yucatan, Mexico. It is uncommon or rare.

Notes.—This is a very difficult species to obtain today, many of the specimens being sold under this name being young *C. lorenzianus* or even young *C. spurius*. The fragile nature of the shell means that the lip is often broken, there are bad growth scars and breaks on the body whorl, and the pattern may be very indistinct and irregular.

SIBOGAE Schepman, 1913

1913. *Conus mucronatus* var. *sibogae* Schepman. *Prosobranchia Siboga Exped., Monogr.* 49(5): 390, pl. 25, fig. 2 (Aru Islands).

Description.—Moderately light in weight, with a good gloss; low conical, almost obconical, the sides nearly straight or very gently convex; body whorl covered from base to shoulder with narrow flat spiral ridges, slightly narrower anteriorly and wider below shoulder, separated by narrow to moderately wide grooves; entire whorl covered by distinct straight axial grooves giving the whole whorl a faintly cancellate or beaded appearance; shoulder broad, sharply angled to carinate, sometimes irregular because of the axial grooves interrupting its growth; spire low to very low, the sides straight to slightly concave; early whorls forming a low, pointed cone; tops of whorls flat, with about four narrow but strong spiral ridges, these crossed by heavy axial threads so the ridges appear malleated or faintly beaded. Body whorl white to cream, covered with two broad spiral bands of pale brown to dark brown squarish blotches above and below midbody, these often fused into solid bands; smaller and squarish spots below the shoulder; usually with wavy axial flammules of the same brown running from shoulder to above base, sometimes broken into small dashes; base white; spire and shoulder white, with a few to many radiating streaks and spots of brown. Aperture narrow posteriorly, somewhat widened anteriorly; outer lip sharp, straight; mouth white. Columella long, narrow. Length 25-40mm.

Comparison.—Although described as a variety of *C. mucronatus*, which has a higher spire and much rougher sculpture, it seems that *C. sibogae* is very closely related to *C. tegulatus*, that complex and very variable species or species complex that is very poorly understood. From most *C. tegulatus* it is easily distinguished by the sharply carinate or angled shoulder, the low spire with nearly straight sides, the nearly straight sides of the body whorl, the narrow and very regular spiral ridges over the body whorl, and the pattern with most of the color concentrated into two broad spiral bands with the axial flammules reduced. The axial grooves covering the body whorl are very regular and give the entire shell a minutely beaded appearance, with a similar appearance on the spire whorls. Some *C. tegulatus* approach *C. sibogae* very closely in pattern, but these shells usually have less regular and often reduced spiral ridging, a higher spire, broader and more convex body whorl, and a less carinate shoulder; the mouth is often bright violet. I believe the two can be separated with some certainty and hesitate to add another form to the synonmy of the already badly confused *C. tegulatus*.

Conus illawarra is very similar to *C. sibogae* in general appearance but has a more rounded shoulder, reduced spiral sculpture on the body and spire whorls, and a less fused pattern. *C. minnamurra* has heavier sculpture, nodulose early whorls, and usually dark spots at the shoulder.

Variation.—The shape and sculpture are very constant, although there is a weak tendency to reduce the grooves posteriorly on the body whorl. The pattern sometimes consists of just two broad spiral bands of pale tan to dark tan without axial flammules or with a few dashes; in other patterns the large blotches are separated and connected by axial wavy flammules that are broken and often indistinct. The shoulder is very narrowly angled to strongly carinate, the spire very low or almost flat with nearly straight or very slightly concave sides (ignoring the early whorls, which apparently break easily). The mouth may have pale violet tones.

Distribution.—Known from eastern Indonesia to the Solomons and south to Queensland, Australia. Uncommon in moderately deep water.

Notes.—The relationship of this species and some patterns of *C. tegulatus* is very close, but for the moment it does no harm to keep this species distinct as most or all specimens can be placed in one species or the other. The type specimen of *C. sibogae* seems to be this species rather than *C. tegulatus* in having the sculpture even, the blotches large and with reduced flammules, and the shoulder sharply angled though not carinate; the spire is higher than usual, but much of this appears to be due to a complete set of protoconch whorls.

The lip is often weakly filed and there may be heavy axial growth marks distorting the pattern. Most specimens have very pale patterns, sometimes with little contrast.

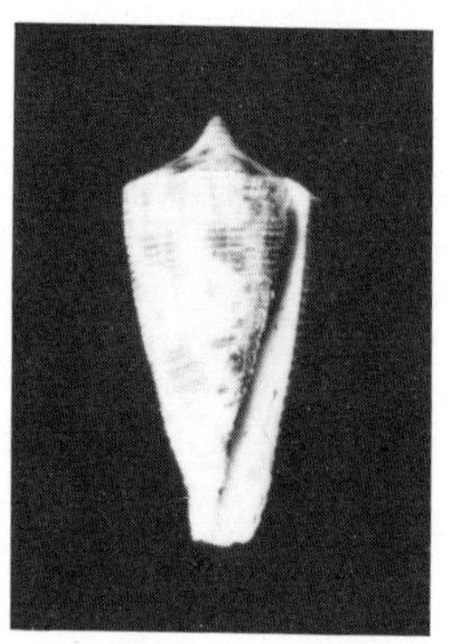

Type of *C. mucronatus* var. *sibogae*, 26mm (Zool. Mus. Amsterdam) (photo Abbott).

SIEBOLDII Reeve, 1848

1848. *Conus Sieboldii* Reeve. *Conchologia Iconica,* 1(*Conus* Suppl.): Pl. 1, sp. 269 (Japan).
1870. *Conus rarimaculatus* Sowerby iii. *Proc. Zool. Soc.* (*London*), 1870: 257, pl. 22, fig. 4 (China Seas).

Description.—Moderately heavy, with a low gloss; elongate biconical, the upper sides slightly convex then nearly straight to a narrow base; body whorl with a few low spiral ridges near the base, these at first rather narrow then becoming broad flat ribs separated by narrow, axially threaded grooves that are often undulate; ribs extend to about midbody, sometimes higher; numerous heavy axial threads and growth marks and fine spiral threads over rest of whorl; shoulder moderately broad, concave above, with a raised margin that forms a ridge; spire very tall to moderate, sharply pointed, the sides nearly straight or slightly concave; spire whorls strongly stepped, deeply concave above, with erect rim-like margins; first 3-4 whorls faintly nodulose, the later ones straight or very weakly irregular; tops of whorls with heavy (early) to weak (late) curved axial threads. Entire shell white to creamy white, usually with about three spiral rows of small to large reddish brown rather axial spots, the rows usually visible above and below midbody and below the shoulder; the two posterior bands may be fused into long flammules, while another band of smaller spots may be present above the base; spire white, with a few scattered reddish brown spots and lines, heaviest on earlier whorls; early whorls tan. Aperture moderately wide, uniform in width; outer lip thin, strongly sloping below level of shoulder, often concave at middle. Columella very narrow, long. Length 60-100mm.

Comparison.—*C. sieboldii* can be confused with only three vaguely similar species, *C. teramachii*, *C. ione*, and *C. altispiratus.* From *C. ione*, with which it is sometimes very similar at first glance, it differs in being dull white instead of pale violet, having very irregular larger spots of reddish brown instead of regular spiral rows of small dots as well as larger spots, and being generally more elongate and thicker as well as more deeply grooved anteriorly. *C. teramachii* is wider at the shoulder, often has the posterior sides almost parallel, has about half the spire whorls distinctly nodulose, is often pale flesh in color, and lacks the deep anterior grooves. *C. altispiratus* in some forms can be very similar in shape and even pattern, but it lacks the strong rims on the spire whorls and the deep anterior grooving as well as being lighter in weight.

Variation.—The reddish brown pattern varies considerably and is often almost entirely absent. The three spiral rows of rather small spots are commonly present, although any one is often absent or the two posterior bands may be fused into axial flammules. A row of smaller spots is commonly present above the base. Some specimens seem to have a few scattered very long and irregular axial flammules only instead of spots.

The spire pattern is often almost absent.

The spire varies from very tall to quite low, apparently individual variation. The anterior grooving usually reaches midbody and often continues more posteriorly; sometimes the grooving is weak and indistinct above the base. The columella is exceptionally variable, from long and quite wide, sometimes oblique or bounded by a heavy ridge, to very narrow and indistinct.

Distribution.—Apparently restricted to the northern China Sea from Japan to Taiwan and perhaps the Philippines. Common in deep water.

Synonymy.—*Conus rarimaculatus* seems to be the juvenile of *C. sieboldii*, with the spots large and regular. I have never seen a specimen that matches Sowerby's description, but see no reason to doubt this synonymy.

Notes.—This is usually a chalky white shell with heavy axial growth flaws and scars, broken spire whorls, and a chipped lip. The pattern is seldom well developed and is commonly almost absent or very irregular. Although a distinctive species, it is not an attractive one.

It seems curious that this shell should be quite commonly dredged off southern Japan and Taiwan but not recorded from anywhere else in the China Sea or the western Pacific. Almost certainly it is found much further south in deep water.

SMIRNA Bartsch and Rehder, 1943

1943. *Conus smirna* Bartsch and Rehder. *Proc. Biol. Soc. Washington*, 56: 87 (Near Kauai, Hawaiian Islands).

1956. *Chelyconus* (?) (*Profundiconus*) *profundorum* Kuroda. *Venus*, 19 (1): 5, text figs. 8-9 (SW part of off Tosa, Japan).

1964. *Profundiconus soyomaruae* Okutani. *J. Fac. Sci., Univ. Tokyo*, Sec. II(15-3): 436, pl. 4, fig. 1 (6 mi. off Hachiji Island, Japan).

?1972. *Profundiconus scopulicola* Okutani. *Bull. Tokai Reg. Fish. Res. Lab.*, 72: 98, pl. 2, fig. 12 (Hyotanse Bank, nr. Izu-Shichito Is., Japan).

Description.—Light in weight, thin, with a low gloss; biconical, the posterior sides distinctly convex, greatest width below shoulder; body whorl with 5-15 low but usually distinct spiral ridges anteriorly, often undulating, rapidly becoming obsolete; many spiral and axial threads over rest of whorl, the axial threads often coarse and irregular; shoulder rather narrow, indistinct from spire, slightly concave above; spire tall, pointed, the sides straight or slightly concave, the whorls sometimes weakly stepped; early whorls sometimes nodulose, the nodules extending to the third to sixth whorls before becoming obsolete; tops of whorls flat or slightly concave, the margins sometimes weakly carinate and projecting; tops of whorls with traces of 1-2 weak spiral threads on early whorls, none on later ones, and many curved axial threads. Body whorl creamy white to pale straw, with broad spiral bands of pale to dark brown above and below midbody area, sometimes with secondary brown bands weakly set off near midbody; base whitish; shoulder and spire whitish to pale straw, sometimes with tan on some whorls. Aperture wide, especially anteriorly; outer lip thin, fragile, slightly convex, its posterior edge sloping well below level of shoulder; mouth creamy white to pale tan. Columella long and narrow, sometimes indistinct. Length 60-84mm.

Comparison.—The shape and texture of this shell, combined with the brown-banded pattern, are very distinctive. It is probably related to *C. sieboldii* but differs greatly in texture and pattern.

Variation.—The number of nodulose whorls varies greatly, from none to at least three in Japanese specimens and with about six in Hawaiian specimens. The body whorl may be relatively narrow to rather broad, the spire very tall to low with the whorls flat or distinctly concave with projecting margins. The pattern is very indistinct but consists of a broad band above midbody and a narrower band below, sometimes with a narrow band at midbody. If *C. scopulicola* is truly the juvenile of this species, then very small shells are axially flammulated in three bands, at shoulder, above and below midbody, with a few rows of smaller spots between. It would take little for this pattern to fuse into the broad bands of the adult.

It is perhaps possible to recognize two very similar subspecies of *C. smirna*, although larger series will probably show that they vary too greatly to be distinguished.

C. smirna smirna: Rather narrow, the spire a bit taller, the nodules heavier and extending to about the sixth whorl. Hawaiian Islands.

C. smirna profundorum: Somewhat broader, the spire a bit lower, nodules on from 0-3 whorls and weak. Southern Japan.

Distribution.—Uncommon to rare in deep water off the Hawaiian Islands and southern Japan.

Synonymy.—*C. profundorum* is extremely similar to Hawaiian *smirna* and not separable at the specific level—nor perhaps even the subspecific. *C. soyomaruae* falls easily within the range of variation of Japanese *smirna*, and *C. scopulicola* certainly appears to be a juvenile of *C. smirna*, differing only in minor pattern and sculptural features that might be expected in a juvenile.

Notes.—This is one of the deepest-living cones, having been taken in more than 250 fathoms. It is very thin in texture and as such is often damaged or shows large and heavy healed and unhealed breaks and flaws. The early whorls may be partially nacred-over, possibly as an aid in healing or preventing breaks.

As might be expected of such a deep-water shell, gems are practically unavailable, and even quite bad shells sell for good prices. Probably it is not uncommon, just seldom collected. It would be very surprising if the range was restricted to southern Japan and Hawaii, although *C. sazanka* and *C. eugrammatus* are almost as restricted.

SPECTRUM Linnaeus, 1758

1758. *Conus spectrum* Linnaeus. *Systema Naturae per Regna Tria Naturae*, ed. 10, 1: 717 (Asia). Lectotype figure (Tomlin, 1937): Rumphius, Pl. 32, fig. S.
1798. *Cucullus carota* Roeding. *Museum Boltenianum:* 47 (Locality not stated). Based on Martini, Pl. 52, fig. 581. See Kohn, 1975.
1798. *Cucullus Chinensis* Roeding. *Ibid.:* 47 (Locality not stated). Lectotype figures (here selected): Chemnitz, Pl. 144A, figs. g-h.
1807. *Conus felinus* Link. *Beschr. Nat.-Samml. Univ. Rostock*, 3: 104 (Locality not stated).
?1833. *Conus inflatus* Sowerby i, in Sowerby ii. *Conchological Illustrations:* Pt. 33, fig. 41 (Locality not stated).
1844. *Conus Broderipii* Reeve. *Conchologia Iconica*, 1(*Conus*): Pl. 46, sp. 254 (Locality unknown).
1848. *Conus Pica* A. Adams and Reeve. *Zool. Voy. Samarang, Moll.*, 1: 18, pl. 15, figs. 10a-10d (Balambougan Is., Philippines).
1849. *Conus stillatus* Reeve. *Conchologia Iconica*, 1(*Conus* Suppl.): Pl. 5, sp. 247 (Moluccas).
1849. *Conus filamentosus* Reeve. *Ibid.*, 1(*Conus* Suppl.): Pl. 6, sp. 260 (Locality unknown).
?1854. *Conus stigmaticus* A. Adams. *Proc. Zool. Soc.* (*London*), 1853: 119 (New Caledonia).
1865. *Conus straturatus* Sowerby ii. *Proc. Zool. Soc.* (*London*), 1865: 518, pl. 32, fig. 14 (Borneo).
1867. *Conus Blanfordianus* Crosse. *J. Conchyl.* (*Paris*), 15: 66, pl. 11, fig. 4 (Locality unknown).

Description.—Light in weight, with a good gloss; ovate, the sides convex; body whorl with widely spaced shallow spiral grooves extending from above base to about midbody, leaving broad flat ribs between; a few crowded low ridges at the base; axial threads may be conspicuous; shoulder roundly angled, flat above; spire low, the early whorls forming a small sharp cone; spire whorls flat to slightly convex above, with about 3-5 spiral ridges crossed by weak, scattered axial threads; early whorls weakly carinate. Body whorl white, cream, or very pale straw, with a very variable pattern of usually axial irregular blotches of black, orange, bright yellow, tan, or reddish brown, the blotches tending to concentrate into three vague bands below shoulder and above and below midbody; spiral rows of fine brown dots usually present, these often inconspicuous or even fused into continuous hairlines; base white or cream, sometimes pale straw; spire whitish, lightly or heavily marked with streaks and spots of about the same color as body whorl pattern; early whorls white. Aperture very wide, much wider anteriorly than posteriorly but wide throughout; outer lip sharp, thin, gently to greatly convex; mouth white, occasionally pale yellow to bluish white. Columella long, narrow, seldom strongly indented from inner lip, but bounded posteriorly by a low ridge. Length 30-50mm.

Comparison.—This is a very variable shell with rather indefinite specific limits, the many patterns capable of being compared with or confused with many species. From *C. cinereus* it is easily distinguished by the distinct spire sculpture and the wide, ovate shape in most varieties. The shape usually will also distinguish it from *C. subulatus*, which in comparable forms has a violet mouth and often a strongly indented columella. *C. tegulatus* has a more strongly angled shoulder, heavier spiral sculpture, a higher spire, and a pattern of flexuous axial stripes; some patterns are quite similar, but this species has a narrower mouth, often violet, than in *C. spectrum*. *C. wittigi* may be vaguely similar in shape but has a higher spire and a distinctly tented pattern. *C. conspersus* is covered with large dashes (often inconspicuous) and has a bright pinkish orange mouth.

Variation.—The form is usually very wide and ovate, with the spire very low and the aperture very wide. Beyond this the different major and minor varieties vary greatly in sculpture of the body whorl and the pattern and color.

Major varieties include: *spectrum* typical: large Y- or H-shaped blotches of orange-brown to reddish brown in two very irregular rows above and below midbody, the rows often connected; *filamentosus:* more elongated, with a rather typical pattern covered with numerous spiral hairlines, whole shell sometimes with distinct spiral ridges; *stillatus:* fine axial lines extending from small dark brown blotches, with large brown areas above the base, whole shell covered with fine spiral ridges; *pica:* blotching deep blackish brown, often reduced, resulting in solid white shells. One of the interesting minor varieties is a rather slender shell of pale straw heavily covered with indistinct brownish dashes of large size in rather spiral and axial rows, the whole shell then divided by fine white or tan spiral lines; such shells can be called *C. spectrum* var. *broderipii* if desired.

There is much individual variation in color and pattern within each variety, even in shells from the same locality. The major varieties all over-

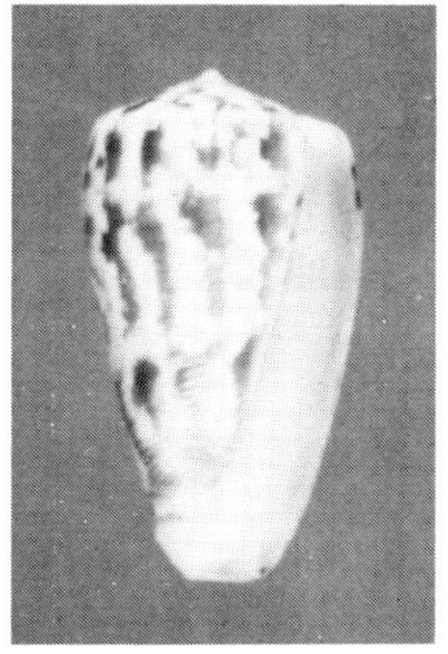

C. spectrum var. *stillatus*, Exmouth Gulf, Western Australia (JB).

lap broadly, and all are of about the same degree of rarity.

Distribution.—Supposedly widely distributed across the Indo-Pacific, but I have seen specimens only from the western Pacific and from Western Australia. Most specimens come from the Philippines, where it is moderately common and all varieties except *stillatus* can be found; *C. spectrum* var. *stillatus* seems to be restricted to the area between Western Australia and Queensland and might be a valid subspecies.

Synonymy.—Until the types, where still existent, can all be examined, the synonymy of this and related species such as *C. subulatus*, *C. cinereus*, and even *C. tegulatus* will be in doubt. The listed names seem to belong here, with most of them recognizable varieties. *C. straturatus* could be a form of *C. subulatus*, but seems to belong here instead. *C. inflatus* is known from one miserable figure and has been assigned to *C. spectrum* in the past (among other species), but it is anyone's guess what it really represents. *C. blanfordianus* seems to be a *broderipii*-like shell, while *C. stigmaticus* could be a slender specimen of the typical form with the flammules in rather even rows. . .but *stigmaticus* could also be a form of the species here called *C. tegulatus*.

Notes.—Few specimens from other than the Philippines and Australia are ever available at the time of writing. As mentioned, all the major varieties except *stillatus* come from the same localities in the Philippines, along with shells intermediate between the varieties. When the spiral rows of brownish dots are visible, they are a good hallmark of this species.

Specimens are usually quite clean, and even the thin lip is stronger than it looks.

SPONSALIS Hwass, in Bruguiere, 1792

1792. *Conus sponsalis* Hwass, in Bruguiere. *Cone*, in *Ency. Method., Hist. Nat. des Vers*, 1: 635 (St. George Is.). Holotype figure: *Tableau Ency. Method.*, Pl. 322, fig. 1.

1792. *Conus puncturatus* Hwass, in Bruguiere. *Ibid.*, 1: 635 (Botany Bay, Australia). Holotype figure: *Tableau Ency. Method.*, Pl. 322, fig. 9.

1801. *Conus maculatus* Bosc. *Hist. Nat. Coq.*, 5: 124, pl. 40, fig. 5 (Locality unknown).

1833. *Conus nux* Broderip. *Proc. Zool. Soc.* (*London*), 1833: 54 (Insulas Gallapagos).

1833. *Conus nanus* Sowerby i, in Sowerby ii. *Conchological Illustrations:* Pt. 24, fig. 6 (Lord Hood's Is.).

1844. *Conus liratus* Reeve. *Conchologia Iconica*, 1(*Conus*): Pl. 47, sp. 268 (Locality unknown).

1853. *Conus pusillus* A.A. Gould. *Boston J. Nat. Hist.*, 6: 388, pl. 14, fig. 22 (Mazatlan, Mexico). Non *Conus pusillus* Lamarck, 1810.

Description.—Moderately heavy for its size, with a low gloss; low conical to obconical, narrow to wide, the upper sides gently to strongly convex; body whorl with heavy, broadly spaced spiral ridges over the anterior half or so, the ridges nearest the base often strongly granulose; axial growth marks and threads often prominent; shoulder angulate to rounded, usually with small but heavy coronations, these often obsolete or nearly so; spire flat to low, blunt, the sides straight or slightly concave; spire whorls heavily coronate (sometimes obsolete), with a weak spiral ridge and fine axial threads. Body whorl white to pale straw, the base bright purple; pattern may be absent except for broad bluish zone anteriorly, or with a spiral band of short reddish brown flammules at midbody, or with two or more rows of wavy reddish brown axial flammules, or with broad spiral bands of fused brownish flammules covering most of body; spire and shoulder pale, usually with dark spots between the coronations. Aperture moderately narrow, somewhat narrower anteriorly; outer lip slightly convex; mouth mostly white at the margin, deep purple within, with a purple band extending to the margin at middle. Columella short, narrow. Length 15-32mm.

Comparison.—Although some specimens of *C. sponsalis* bear a close resemblance to the related *C. ebraeus* and *C. chaldaeus*, those species have black markings, are heavier and thicker, and have larger though often lower coronations. *C. musicus* is a distinct species distinguished by the narrow body whorl with straighter sides, the regular spiral rows of blackish dashes or dots, the general presence of fine pustules over the body whorl in the Pacific form, and the often clean, distinctive spire pattern of large black or brown dots. There is no apparent overlap between the two species, with the patterns quite distinct.

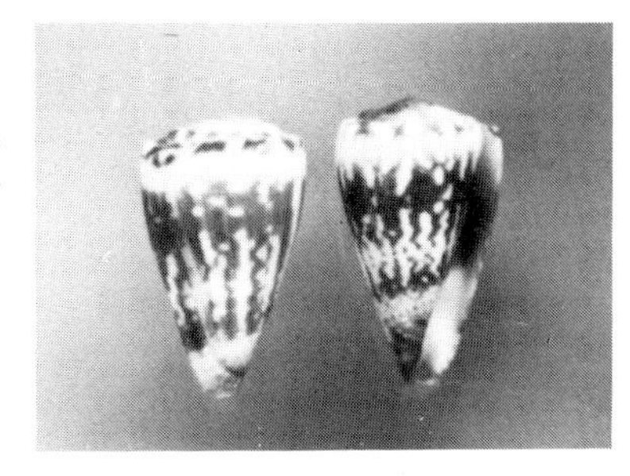

C. sponsalis nux, Baja California, Mexico, 22mm (photo Kerstitch).

Variation.—As treated here, *C. sponsalis* is a variable species with two weakly defined subspecies and two major patterns in the western subspecies along with intermediates. The subspecies may be defined as follows.

C. sponsalis sponsalis: Sides of body whorl in adults usually strongly convex, the coronations erect; reddish brown flammules may touch and may extend from above base to below shoulder but do not form nearly solid spiral bands. Indo-Pacific to Hawaii and French Polynesia.

C. sponsalis nux: Sides of body whorl in adults usually slightly convex to nearly straight; axial flammules brownish, usually extended over much of whorl and heavily connected spirally to produce a dense reticulation or solid spiral bands. Eastern Pacific, Mexico to Ecuador.

Within *C. sponsalis nux* there is considerable variation in the extent of fusion of the flammules, from nearly free flammules to solid bands with few breaks or pale spots within.

Although *Conus nux* is generally considered a full species, it has usually been treated as very closely related to *Conus sponsalis.* Several lots of shells from the Indian Ocean and the southern Pacific are virtually indistinguishable from Panamic shells, although admittedly typical *Conus nux* is usually quite distinct. For the moment I consider the taxa to be subspecies.

C. sponsalis sponsalis is more variable. Individual variation in adults includes smooth and completely granulose individuals, high and flat spires, a single median band of reddish spots or broad wavy reddish brown flammules from below the shoulder to above the base or in up to three spiral series; the coronations may be high and triangular or almost obsolete.

The more important problem in this subspecies is the status of the *nanus* form. In at least some populations there is a definite ontogenetic development of patterns: the young are white with a broad pale bluish band anteriorly, the sides nearly straight; subadults are very pale straw with a midbody band of indistinct reddish flammules that are very even and give the band a jagged appearance; and adults have convex sides and from one to three spiral series of rounded spots or wavy flammules. *C. nanus* was a white shell with a bluish anterior band, corresponding very closely with the juveniles in this series. This is a common pattern transition in some areas, for instance Tahiti and perhaps Hawaii and the Philippines.

In some other areas, for instance some Solomons and Philippines populations, there are two distinctive juvenile types, one of the white *nanus* type, the other looking like a dwarf flammulated adult. In such areas fully adult normal flammulated *sponsalis* occur alongside fully adult whitish *nanus* types with no apparent intergrades. Clearly the status of the shell called *nanus* is complex and perhaps does not fit our current species concepts. For the moment I follow a conservative path and recognize *C. nanus* as a variety of *C. sponsalis* instead of a full species.

Distribution.—Widespread and common in shallow water from the eastern African coast to Hawaii, French Polynesia, and the Panamic region.

Synonymy.—Typical *sponsalis* is granulose basally, very convex-sided, and has a central row of irregular reddish blotches. *Puncturatus* probably represents the intermediates between typical *sponsalis* and *nanus* described earlier. *Maculatus* is a granulose *sponsalis* with two rows of smallish rounded spots. *Nanus* has been discussed above. *Liratus* seems to be a specimen with an abnormally high spire, but it could also be a specimen of *C. boeticus*. *C. pusillus* is a faded *nux*.

Notes.—This small shell is often rough, with heavy axial flaws, eroded spires, and limy encrustations. As with other very common small shells, however, grading is usually not followed closely because the specimens are individually nearly worthless to the collectors. *C. s. nux* is often very poorly patterned with very eroded spires; very *nux*-like shells are often found outside the Panamic region.

Very interesting growth series from the *nanus* form to typical flammulated adults can be formed in some areas; dealers often call the intermediate shells *'ceylanensis'*, here considered a synonym of *C. musicus*.

SPURIUS Gmelin, 1791

1791. *Conus spurius* Gmelin. *Systema Naturae per Regna Tria Naturae*, ed. 13, 1: 3396 (Locality not stated). Lectotype figure (Clench, 1942): Gualtieri, Pl. 21, fig. D.

?1791. *Conus insularis* Gmelin. *Ibid.*, ed. 13, 1: 3389 (Locality not stated). Based on Martini, Pl. 61, fig. 683.

1792. *Conus proteus* Hwass, in Bruguiere. *Cone*, in *Ency. Method., Hist. Nat. des Vers*, 1: 682 (Santo Domingo and Guadeloupe). Lectotype figure (Kohn, 1968): *Tableau Ency. Method.*, Pl. 334, fig. 1.

1792. *Conus leoninus* Hwass, in Bruguiere. *Ibid.*, 1: 683 (Mexico to Brazil). Lectotype selected and figured by Kohn, 1968. Non *Conus leoninus* Lightfoot, 1786.

1798. *Cucullus ferugineus* Roeding. *Museum Boltenianum:* 41 (Locality not stated). Lectotype figure (Kohn, 1975): Martini, Pl. 56, fig. 626.

1798. *Cucullus quadratus* Roeding. *Ibid.:* 41 (Locality not stated). Lectotype figure (Kohn, 1975): Chemnitz, Pl. 140, fig. 1300. This shell looks much more like *Conus inscriptus*.

1798. *Cucullus Syriacus* Roeding. *Ibid.:* 41 (Locality not stated). Lectotype figure (Kohn, 1975): Martini, Pl. 56, fig. 627.

1798. *Cucullus Gualterianus* Roeding. *Ibid.:* 42 (Locality not stated). Lectotype figure (Kohn, 1975): Gualtieri, 1742, Pl. 21, fig. F.

1798. *Cucullus leoninus* Roeding. *Ibid.:* 44 (Locality not stated). Lectotype figure (Kohn, 1975): Martini, Pl. 57, fig. 640. Not *Conus leoninus* Lightfoot, 1786.

1810. *Conus ochraceus* Lamarck. *Ann. du Mus. Hist. Nat. (Paris)*, 15: 275 (Locality unknown). Holotype figured by Kiener, Pl. 37, fig. 2.

1823. *Conus grandis* Sowerby i. *Genera Recent and Fossil Shells*, 2(16): Pl. 266, fig. 2 (Locality not stated).

1827. *Conus reticularis* Bory. *Tableau Ency. Method.*, 10: 160, pl. 334, fig. 7 (Locality not stated).
1833. *Conus breviculus* Sowerby i, in Sowerby ii. *Conchological Illustrations:* Pt. 37, fig. 55 (Locality not stated). Spelled *Conus breviusculus* by Weinkauff, 1874, *Jahrb. Deutsch. Malak. Ges.*, 1: 245.
1882. *Conus Weinkauffii* Loebbecke. *Jahrb. Deutsch. Malak. Ges.*, 9: 90 (? New Caledonia). 1882, *Ibid.*, 9: 188, pl. 4, figs. 1-3.
1942. *Conus spurius atlanticus* Clench. *Johnsonia*, 1(6): 20, pl. 10, figs. 1-3 (Bonita Springs, Florida).
1951. *Conus spurius aureofasciatus* Rehder and Abbott. *Revista de la Sociedad Malac. Carlos de la Torre*, 8: 64, pl. 9, figs. 3-4 (Off Tortugas, Florida).
1968. *Conus spurius arubaensis* Usticke. *Caribbean Cones from St. Croix and the Lesser Antilles:* 12, pl. 1, sp. 995 (Aruba, dredged in 25 ft.).

Description.—Heavy, with a good gloss; broadly conical, the upper sides straight or slightly convex; body whorl with a few to many spiral ridges on the anterior third, these either weak and inconspicuous or heavy and distinct; rest of body whorl without sculpture or with axial threads; shoulder roundly angled, broad; spire moderate to tall, the sides concave, leaving the earlier whorls as a large cone on a flat or depressed plate; spire whorls with rounded margins, slightly stepped, often concave above, with weak axial scratches and no spiral sculpture. Body whorl shining white, covered with about 15-20 or more spiral rows of small to moderately large usually squarish orange-brown to dark brown spots, these often tending to fuse to produce larger squarish blotches above and below midbody, these sometimes fused into broad spiral bands or axial flammules; spire white, with continuation of spotted pattern of body whorl. Aperture moderate, wider anteriorly than posteriorly; outer lip thin, sharp, straight or nearly so; mouth white. Columella broad, heavy, with heavy ridges above. Length 40-99mm.

Comparison.—*C. spurius* is very similar at first glance to *C. pulcher* and *C. tessulatus*, both of which are similarly spotted. *C. pulcher* is lighter in weight than *C. spurius* and has a very low spire, wider aperture, reduced columella, and very smeared or blotchy pattern. *C. tessulatus* has the spots reddish, the base violet, and a pair of heavy spiral ridges on the spire whorls. *C. sennottorum* has the shoulder carinate, the posterior sides convex then concave anteriorly, the early whorls carinate, and the pattern often pale and irregular. *C. lorenzianus* is very similar and possibly intergrades with *C. spurius* in Central America; however, it has much heavier spiral ridges extending above midbody in most specimens, a different spire shape at similar sizes, is thinner and lighter, and has many more smaller spots crowded onto the body whorl and very heavy on the base.

Variation.—Although relatively constant in shape of the body whorl and spire (occasional specimens have nearly straight-sided spires) as well as sculpture, *C. spurius* varies considerably in color and degree of fusion of the spotted pattern. In many specimens the spotting is distinctly reddish or orange-brown, the orange tones being most common in beach specimens. Other shells have the spots deep brown, even black. Although three subspecies have been recognized by the degree of fusion of the spots, this is a highly variable character, and specimens with all spots distinct are sometimes found with specimens having extensive fusion of the spots axially or spirally. Specimens with the spots fused into broad spiral bands with even edges have been called *C. s. aureofasciatus*, but this is certainly just a freak pattern that occurs rarely among normal *C. spurius*; intermediates are found. Specimens from the Bahamas and Antilles tend to have the spots fused into vaguely zigzag axial bands over part of the shell, while those from continental waters have the spotting distinct or fused into squarish blotches; the latter specimens have been called *C. s. atlanticus*, but the differences are very minor and inconstant and not worthy of recognition at a formal level, especially since specimens from the southern Caribbean occur with both types of patterns at some localities.

Distribution.—Common in the western Atlantic from the Carolinas (rare) through Florida, the southwestern Gulf of Mexico, the Antilles, and south to Venezuela; the status of Central American specimens is uncertain, and the species appears to be uncommon in most of the Lesser Antilles. Shallow to deep water.

Synonymy.—Interpretation of early figures of this species is largely guess-work regardless of what anyone says; the figures just are not trustworthy. Of the early names, *C. spurius* has now been used for this species for many years, but in actuality it is possibly unrecognizable. *C. proteus*, long used for this species, seems to truly be the Caribbean shell. *Conus insularis* has often been assigned to the *cedonulli*-type shells, but the figure lacks coronations, is shaped like *C. spurius*, and has a pattern possible for this species. *C. ochraceus* seems to be a very dark specimen of *C. spurius*. *C. weinkauffii* was described as probably from New Caledonia, but it appears to be a fairly normal specimen of *C. spurius*; I am not aware of any shell similar to the type being taken in New Caledonia. *Atlanticus*, *aureofasciatus*, and *arubaensis* are all minor varieties.

Notes.—*Aureofasciatus* is a very pretty banded shell and is certainly rarely taken or available. Specimens from Yucatan and vicinity often have unusual spires and odd sculpture on the body whorl, with very variable patterns, sometimes being axially flammulated; unfortunately these shells are usually trawled and in bad condition. Some southern Caribbean specimens have very dark and contrasting spots or blotches.

Axial growth flaws are common in the species, as are eroded spire whorls. Juveniles are sometimes sold as *C. sennottorum* but are easily separable by shape. This species has caused relatively minor stings in humans but should be considered possibly dangerous.

STERCUSMUSCARUM Linnaeus, 1758

1758. *Conus stercus muscarum* Linnaeus. *Systema Naturae per Regna Tria Naturae*, ed. 10, 1: 715 (Asia). Lectotype selected and figured by Kohn, 1963.
1798. *Cucullus arenatus* Roeding. *Museum Boltenianum:* 49 (Locality not stated). Lectotype figures (Kohn, 1975): Martini, Pl. 64, figs. 711-712. Non *Conus arenatus* Hwass, 1792.
1798. *Cucullus sabella* Roeding. *Ibid.:* 49 (Locality not stated). Lectotype figure (Kohn, 1975): Martini, Pl. 64, fig. 713.

Description.—Moderately heavy, with a high gloss; cylindrical, the sides slightly convex or straight; body whorl with about a dozen rather low spiral ridges anteriorly, rest of whorl smooth except for often heavy spiral and axial ridges and growth marks; shoulder sharply angled, deeply concave above, not very distinct from shoulder; spire low, rather domed, the sides straight or slightly convex; spire whorls slightly stepped, concave on top, with traces of 2-3 low spiral ridges, obsolete or nearly so on later whorls, and minute axial scratches; protoconch sharp. Body whorl cream, heavily covered with closely spaced spiral rows of blackish or deep brown, often triangular, spots and opaque white spots; spots show a strong tendency to partially fuse into larger dark areas which are vaguely axial and may form two vague spiral bands above and below the heavily marked midbody area; often with many opaque white axial flammules; spire cream, sparsely or heavily covered with axially elongated black spots in the form of small groups alternating with the light ground color; early whorls with fine radiating brownish streaks; protoconch pink. Aperture broad posteriorly, even broader anteriorly; outer lip sharp, nearly straight; mouth white, with large oval blotch of deep reddish orange within. Columella long, rather wide, twisted, bounded by a heavy rounded ridge. Length 40-60mm.

Comparison.—*C. stercusmuscarum* is one of the more attractive common cones. It is somewhat similar to *C. bullatus* and allies in shape, but the pattern could only be confused with *C. arenatus*, *C. pulicarius*, and perhaps *C. zeylanicus*, which are coronated, different in texture, and lack the bright orange mouth.

Variation.—Minor variations in width of the body whorl and its shape exist, but they are insignificant. The pattern may be very open with all the spots sharp and well defined, or very dark and smudgy with a great deal of fusion of the spots into large blotches. The opaque white may be restricted to small dots or may be formed into many large flammules that can be very conspicuous. Basal area is often tinted dark brown.

Distribution.—A common shallow-water species in the western Pacific from the Ryukyus to at least the Solomons. Supposedly Indo-Pacific, but I have not seen Indian Ocean specimens or specimens from the central Pacific.

Synonymy.—*Cucullus arenatus* of Roeding is probably just one of Roeding's usual misidentifications. His *sabella* is this species.

Notes.—A very pleasing shell with interesting pattern, high gloss, and a bright mouth; unmistakable. Except for sometimes heavy axial growth marks and very minor spire erosion, this is a very clean species. Specimens with very dark patterns are not uncommon but are less attractive than typical specimens.

STIMPSONI Dall, 1902

1902. *Conus stimpsoni* Dall. *Proc. U.S. Nat'l. Mus.*, 24: 503, pl. 29, fig. 7 (Key West, Florida).

Description.—Moderately heavy, with a good gloss; biconical, the upper sides straight or very slightly convex then tapered to a very narrow base; body whorl covered with shallow spiral grooves, close-set above the base, more widely spaced posteriorly, sometimes almost obsolete below the shoulder; entire whorl covered with fine axial threads, most distinct in the grooves, and scattered axial plicae; shoulder carinate, broad, nearly flat above; spire tall, sharply pointed, the sides slightly concave or straight; early 3-4 whorls weakly nodulose, later whorls carinate, the margins sometimes projecting; tops of whorls flat or slightly concave, with about four narrow spiral threads or low ridges, these often indistinct on late whorls and shoulder, crossed by fine curved axial threads to produce a weakly cancellate appearance. Body whorl creamy white to glossy white, with two or three spiral series of yellowish tan or bright yellow closely spaced oblique flammules, strongest above midbody, weaker below midbody and below shoulder; flammules usually partially or entirely fused into broad spiral bands that are often indistinct from the background; base often solid pale yellow; spire pale yellow, sometimes with indistinct yellowish spots and blotches; tip tan. Aperture narrow, uniform in width; outer lip thin, straight; mouth white or pale bluish white. Columella extremely narrow or internal. Length 25-52mm.

Comparison.—The sharply biconical shape and carinate shoulder eliminate from consideration several similar species, such as *C. patae,* which is smaller, has a more rounded shoulder and lower spire, and has the body whorl distinctly cancellate in appearance. *C. cancellatus* and *C. arcuatus* differ in shape and color pattern, as does *C. armiger*, which has much heavier sculpture. *C. sennottorum* is more distinctly concave posteriorly with a distinctly turnip-like shape and a deeply concave-sided spire as well as a spotted pattern. The closest relative of *C. stimpsoni* appears to be *C. atractus*, especially subspecies *atractus*, which is very similar in shape but has heavier spire sculpture, is relatively longer in the body whorl, has finer and more cancellate body whorl sculpture, and is

C. cf. *stimpsoni,* sold as 'demarcoi,' Campeche, Mexico, 40mm (photo Kerstitch).

not banded with yellow; the two are very close, however.

Variation.—In shape *C. stimpsoni* is quite constant, varying slightly in concavity of the spire and whether or not the whorls are stepped. The early whorls are distinctly spirally and axially sculptured, but the later whorls and shoulder are often only faintly threaded or smooth. Axial plicae on the body whorl may be heavy or absent, while the spiral grooving may be restricted to the anterior half of the body whorl or continued as deep grooves to the shoulder. Juveniles and adults in some populations have distinct short flammules in three spiral bands, while other specimens are brightly and smoothly banded. Occasional specimens are white, without pattern on body whorl or spire.

Distribution.—Uncommon in deep water off the southeastern United States, Florida, the western Gulf of Mexico, and south to about the Yucatan area. Specimens from off Honduras and Campeche are apparently *C. stimpsoni* but have a higher and more distorted spire and deeper grooving with an indistinct color pattern.

Notes.—The white shells banded with yellow are sometimes very pale and blurred, while other specimens may be sharply contrasted yellow and white and make better specimens. The lip is fragile and chips easily, but other than this and occasional small scars, it is a clean species of one that lives in deep water. Almost all specimens usually seen on the market come from off Florida.

STRAMINEUS Lamarck, 1810

1810. *Conus stramineus* Lamarck. *Ann. du Mus. Hist. Nat.* (*Paris*), 15: 273 (Asiatic Ocean).

1833. *Conus alveolus* Sowerby i, in Sowerby ii. *Conchological Illustrations:* Pl. 25, fig. 11 (Locality not stated).

1845. *Conus nisus* Kiener. *Species gen. et icon. des coqu. viv.*, 2(*Conus*): Pl. 59, fig. 4. 1848, *Ibid.*, 2: 217 (Madagascar). Also reference to Chemnitz, Pl. 183, figs. 1786-1787. Non *Conus nisus* Dillwyn, 1817.

1845. *Conus roseus* Kiener. *Ibid.*, 2(*Conus*): Pl. 107, fig. 4. 1849, *Ibid.*, 2: 265 (Locality unknown). Spelled *rosaceus* on the plate, in error. Non *Conus roseus* Fischer, 1807, or *C. rosaceus* Dillwyn, 1817.

1849. *Conus Kieneri* Reeve. *Conchologia Iconica*, 1(*Conus* Suppl.): 5, pl. 9, sp. 282 (Locality unknown).

1877. *Conus fuscomaculatus* E.A. Smith. *Ann. Mag. Nat. Hist.*, (Ser. 4), 19: 224 (Locality unknown).

?1913. *Conus hedgesi* Sowerby iii. *Ibid.*, (Ser. 8), 11: 558, pl. 9, fig. 4 (Locality unknown).

Description.—Moderately heavy, with a good gloss; low conical, somewhat cylindrical, the upper slides gently convex to nearly straight; body whorl with shallow to deep grooves separating broad flat ribs that may continue to or above midbody; basal ridges rarely granulose; numerous fine axial scratches over whorl; shoulder roundly angled, moderately broad and distinct from the spire, slightly convex above; spire low to moderate, the sides deeply concave, sharply pointed; spire whorls slightly convex on top, with about 3-4 low spiral ridges on posterior part, not very distinct when viewed from the side, covered by minute axial threads; early whorls sometimes weakly carinate. Body whorl white, cream, pale straw, or bluish, covered with many spiral rows of reddish to reddish brown spots, these usually grouped into larger blotches in three or four spiral bands, below shoulder, above and below midbody, and sometimes above base; spots in intervening areas usually distinct regardless of the size and shape of the larger spots, but sometimes fused into vaguely axial wavy flammules; base white to pale straw; shoulder and spire whitish with curved reddish brown lines and blotches, continuations of spots at shoulder; early whorls white to pale brown. Aperture moderately narrow above, much wider anteriorly; outer lip sharp, straight or slightly convex; mouth with the edge whitish, the rest deep reddish violet to chocolate brown, dark and never clear violet. Columella long, narrow, sometimes set off posteriorly by a low ridge. Length 30-60mm.

Comparison.—*C. stramineus* is the principle species of a group of species including *C. inscriptus*, *C. lynceus*, *C. subulatus*, and *C. janus*, among others. From *C. inscriptus* and *C. lynceus* it differs in having fewer and less distinct spiral ridges and axial threads on the spire whorls, having the mouth dirty off-violet instead of white to clear violet, and usually differs in minor aspects of shape and pattern. *C. janus* is more elongated, weakly sculptured, and has a white to yellowish mouth; it often has a very axial pattern that is heavily smeared. *Conus subulatus* in the Pacific is often very similar in shape, pattern, and sculpture; however, on the average it is a more slender and thinner shell with a more rounded shoulder; the basal sculpture is weaker, with flatter ribs than in many *C. stramineus*; the spotting is often greatly reduced or fused into axial blotches; the simplest character, however, is the pale violet to whitish mouth of *C. subulatus*, often (in the Pacific) with a large pale brown spot at the anterior edge of the mouth.

Variation.—This species is quite variable in shape, color, and pattern, sometimes also in sculpture. Although the grooves are usually rather deep and the ribs distinct but flat, sometimes the grooves are very shallow and the ribs almost obsolete. At the other extreme are heavily corded shells with the anterior ridges granulose at the margins. The body whorl may be narrowly cylindrical or broadly ovate, the shoulder rounded or rather sharply angled. Typical Philippines specimens are glossy white with the spotting distinct and rectangular, only occasionally fused into blotches; more often in other localities the spots form large blotches especially above midbody and may be very irregular; occasionally there are axial flammules, but with the spotting usually still visible through them. Deep-water specimens from the Philippines have blackish spotting on white, with a high contrast. The mouth color in all these forms is rather constant, dirty off-violet to chocolate brown, usually with a reddish tinge, never clear or pale violet.

Distribution.—*C. stramineus* as here treated ranges from the New Guinea-Solomons area through the Philippines into the Bay of Bengal and across the Indian Ocean to the Natal coast. It is usually moderately common in shallow to rather deep water.

Synonymy.—Once again the early figures are very subject to interpretation. Poor figures of *C. stramineus*, *C. subulatus*, and *C. inscriptus* all look very much alike and specimens would have been available to the early iconographers. There was much confusion over the non-binomial names of Chemnitz, which resulted in the names *nisus* Kiener, *roseus* Kiener, and *kieneri*. *Conus alveolus* is almost certainly this species, while *C. fuscomaculatus* seems to represent a heavily corded and granulose specimen. *C. hedgesi* is very doubtful but looks like *C. stramineus* to some extent. It is possible that some of the names listed under *C. subulatus* also belong here.

Notes.—The color of the mouth should be the first character noted when identifying shells of this complex, as it often is possible to separate some of the forms by mouth color. *C. stramineus* is the most distinctive species of this group, well separated from the taxa clustered around *C. inscriptus* and *C. subulatus*.

This is usually a clean species although there seems to be a tendency in some populations to heavy axial breaks that heal fairly evenly. Specimens with whitish backgrounds and clear spots are easier on the eye than those with bluish backgrounds and smudged reddish brown blotches.

Many misidentified specimens of this species are in collections, usually under the names *C. collisus*, *C. zebra*, or even *C. janus*.

STRIATELLUS Link, 1807

1792. *Conus lineatus* Hwass, in Bruguiere. *Cone*, in *Ency. Method., Hist. Nat. des Vers*, 1: 645 (Indian Ocean). Lectotype (here selected): specimen figured by Kohn, 1968, as the holotype. Non *Conus lineatus* Solander, 1766, a fossil.

1807. *Conus striatellus* Link. *Beschr. Nat.-Samml. Univ. Rostock*, 3: 103 (Locality not stated). Based on Chemnitz, Pl. 138, fig. 1285.

?1865. *Conus mirmillo* Crosse. *J. Conchyl.* (*Paris*), 13: 300, pl. 9, fig. 2 (Locality unknown).

1921. *Conus pulchrelineatus* Hopwood. *J. Conchol.* (*London*), 16: 151. Nomen novum for *Conus lineatus* Hwass, 1792.

1933. *Conus lineatus granulosus* Barros e Cunha. *Mem. Estud. Mus. Zool. Univ. Coimbra*, (Ser. 1), 71: 118 (Locality unknown). Non *Conus granulosus* (Roeding, 1798).

Description.—Moderately heavy, with a low gloss; low conical, the upper sides gently convex; body whorl with about a dozen or so narrow spiral ridges near the base, these crowded at first then more widely spaced; some basal ridges often granulose; rest of body whorl with often heavy spiral threads and heavy axial growth marks and threads; shoulder broad, concave or flat above, sharply angled or even carinate; spire low to flat, the tip blunt, with the sides straight or slightly concave; spire whorls usually concave, with about 3-4 spiral ridges crossed by fine axial threads, only vaguely punctate in the grooves. Body whorl mostly white, usually with two broad spiral bands of pale to dark brown above and below midbody, these sometimes blotchy and connected by heavy or weak axial wavy flammules from the spire to the base, the midbody area usually weakly marked; entire whorl covered with many (50 or more) fine spiral dark brown hairlines which are usually easily visible and closely spaced, sometimes broken into weaker dashes in the white areas; juvenile shells are mostly brown with squarish white blotches at midbody and below the shoulder, the spiral lines still very visible; base of shell pale brown to dark blackish violet (Indian and Pacific Oceans respectively); spire white with rather closely spaced curved blackish blotches, about equal to white areas or smaller; tip white, eroded. Aperture moderately wide posteriorly, wider anteriorly; outer lip thin, sharp, often slightly concave at the middle; mouth white to pale bluish white. Columella long, narrow, often indented from inner lip. Length 30-55mm.

Comparison.—*C. striatellus* is closely related to *C. connectens*, *C. vitulinus*, and especially *C. planorbis*. Some specimens are very similar to *C. connectens* in shape and pattern, even to being faintly tinged with pink or violet. However, in *C. connectens* the base of the shell is distinctly violet or reddish and there is a large blotch of a similar tone in the anterior end of the mouth, absent in all *C. striatellus*. *C. vitulinus* may also be very similar in shape and have numerous spiral hairlines, but it has heavy curved black markings on the spire, is usually axially flammulated, and has a large blackish violet blotch at the anterior end of the mouth. *C. planorbis* agrees with *C. striatellus* in lacking a dark blotch anteriorly in the mouth, but differs in having the hairlines weaker and much more widely spaced over the body whorl, having the spire pattern typically reduced to squarish spots on the earlier whorls, and being more regularly conical in typical specimens; the pattern is usually less axially flammulated and on a yellowish background. This group is closely related and some specimens may be impossible to place satisfactorily.

Variation.—As treated here this species is fairly constant in shape, sculpture, and pattern. The change from mostly brown juveniles to mostly white adults is the opposite of at least some populations of *C. vitulinus* and *C. planorbis*, where the juveniles are light and gain pattern with growth. The extent of the spiral bands and flammules in adult shells is quite variable, with a general reduction first of the flammules then of the bands; very large shells may be mostly whitish with little dark pattern. The hairlines are distinct in all specimens examined, although they may break into closely spaced dashes.

There seems to be a fairly constant variation in the color of the base in at least some populations from the Indian and Pacific Oceans. At least some Indian Ocean populations have the base pale brown, about the same color as the bands or paler; the Pacific populations examined have the base dark blackish violet, distinctly darker than the bands. This does not seem to vary greatly with size, but it may not be really constant over the entire range of the species.

The extent of concavity of the outer lip and occasionally the left margin varies with growth, larger shells tending to be more concave. This is seldom a distinctly conical species, the shape often being distinctly cylindrical with gently convex or nearly parallel sides.

Distribution.—Widely distributed over the Indo-Pacific from the eastern African coast to at least Fiji. It does not seem to be recorded from Hawaii or French Polynesia. Usually moderately common in shallow water.

Synonymy.—The familiar name *lineatus* is preoccupied and the next available name is *C. striatellus*, which must be used as it has over a century of priority over *pulchrelineatus*. *C. mirmillo* looks like a *C. striatellus* but might be a *C. planorbis*. I do not see what *granulosus* was based on other than a more extensive granulation than usual.

Notes.—This is a duller and rougher species than the others in the group, often with heavy axial flaws, badly eroded spires, and faded patterns. The outer lip is commonly distorted. Occasional shells are pleasingly tinted with pink, however. Juveniles are moderately common and often very glossy brown.

The status of the base color from the two oceans is interesting, and from the limited samples I have seen the difference is constant. However, much larger samples from many more populations throughout the range must be studied; if constant, the difference would be worthy of subspecific recognition but it seems that a new name would have to be coined for the Pacific subspecies.

STRIATUS Linnaeus, 1758

1758. *Conus striatus* Linnaeus. *Systema Naturae per Regna Tria Naturae*, ed. 10, 1: 716 (Hitoe). Lectotype (here designated): the specimen figured by Kohn, 1963, as holotype.
1786. *Conus leoninus* Lightfoot. *Cat. Portland Museum:* 72 (Locality not stated). Lectotype figure (Kohn, 1964): Knorr, Pl. 12, fig. 5.
?1830. *Conus doreyanus* Blainville. *Dict. Sci. Nat.*, 60: 119 (New Guinea). See Tomlin, 1937.
1857-1858. *Conus floridus* Sowerby ii. *Thesaurus Conchyliorum*, 3 (*Conus*): 47, frontispiece fig. 558 (Locality unknown).

Description.—Moderately heavy, with a good gloss; cylindrical, the greatest width well below the shoulder so rather ovate; body whorl with about 10-20 distinct spiral ridges above the base, then the whole whorl covered with distinct spiral threads separated by shallow grooves and easily feelable; shoulder broad, carinate, deeply canaliculate; spire low to moderate, the sides straight to concave, sharply pointed; tops of spire whorls nearly flat for first 4-6 whorls, then becoming deeply concave; early whorls with 3-5 distinct spiral ridges and axial threads, later whorls with indistinct sculpture. Body whorl white, pink, bluish, brown, or even reddish, covered with very variable darker mottlings of purplish, reddish brown, bluish brown, or blackish, the blotches often forming two broad spiral bands above and below the midbody area; under magnification the blotches are seen to be comprised of long and short, closely set dark spiral lines on top of the spiral threads, these very variable and producing the irregular blotches by changes in their length and width, pale areas with at most a few fine dots or unmarked; base pale; spire whitish, with revolving blotchy or reticulated bands of bright reddish brown, often highly contrasted to body color; tip pinkish. Aperture wide, especially anteriorly; outer lip angled outward posteriorly then nearly straight, thick but sharp; mouth white, sometimes with pinkish or bluish tones. Columella long, wide, pointed, bounded by a heavy rounded ridge. Length 60-130mm.

Comparison.—This large cone is very distinctive by its shape and the mottled pattern formed from numerous rows of narrow dark lines. *Conus terminus* is the most closely related species in shape and the canaliculate spire, but it is readily separated by being almost smooth compared to *C. striatus* and lacking the strongly developed hairlines, having only rows of small spots instead. *C. gubernator* is conical in shape instead of cylindrical and has the small spots very reduced in typical shells. *C. barthelemyi* is conical, thicker, bright reddish with black rounded spots in most specimens, and has a very contrasting black spire pattern; occasional specimens show some spiral dashes as individual variations.

Variation.—The shape is very constant although the height of the spire varies from low to actually depressed. The colors of the background and mottling are highly variable as indicated in the description, often being reduced in large specimens to just two broad irregular spiral bands of dark on almost unmarked pale backgrounds. The later spire whorls are deeply canaliculate. Some specimens are almost smooth and sculptureless.

Specimens identified as juveniles are very different from adults. They are smooth except for a few basal ridges, narrowly conical with the tops of the whorls slightly concave and with distinct sculpture, and the shoulder rounded; the spire pattern is as in adults, but the body whorl is pale tan, orange, or violet with fine brown dots alternating with opaque white dots in widely spaced spiral rows.

Distribution.—Widespread and common across the Indo-Pacific from eastern Africa to Hawaii and French Polynesia; shallow water.

Synonymy.—*Conus doreyanus* is very doubtful and may even be just an emendation of *Conus dorreensis*. *C. floridus* seems to be an exceptionally ovate specimen with reduced pattern and bright pink background that is exceptionally smooth; it is usually assigned here, but actually it looks much more like an aberrant *C. terminus*.

Notes.—This is one of the most dangerous cones, second only to *C. geographus* but fortunately less aggressive.

Hawaiian specimens have exceptionally low spires that seem constant for the islands, but this spire shape is also seen in other areas on occasion. There do not seem to be any other major differences between Hawaiian and other Indo-Pacific specimens.

Small adults are usually very clean and brightly colored, but large adults fade, have reduced patterns, and are commonly covered with major axial flaws and breaks.

STUPA (Kuroda, 1956)

1956. *Embrikena stupa* Kuroda. *Venus*, 19(1): 1, pl. 1, fig. 1 (Off Tosa, Shikoku Island, Japan).

Description.—Very heavy, with a low gloss; biconical, the upper sides strongly convex, then straight or slightly concave and tapering to the base; body whorl with a few low spiral ridges above the base, otherwise with only weak spiral and axial threads and flaws; shoulder broad, roundly angled, slightly concave above; spire tall, sharply pointed, the whorls stepped; sides nearly straight to concave, the whole spire appearing relatively slender; early whorls weakly to strongly nodulose, later whorls smooth, concave above, with 1-3 low spiral ridges, sometimes nodules and ridges very weak. Body whorl dirty white with pale violet tones, with spiral rows of small reddish brown spots most prominent on the posterior half of the whorl; usually small to large reddish brown to pale brown irregular blotches above midbody; shoulder usually white, with traces of scattered short narrow oblique reddish brown lines; spire whorls whitish, the margins with short oblique reddish brown lines, these disappearing on earlier whorls. Aperture moderately narrow, uniform in width; outer lip thick, straight or slightly concave at the middle; mouth white. Columella long, narrow, often oblique. Length 75-100mm.

Comparison.—The large size, biconical shape, weakly sculptured whorls, and lack of grooves on the body whorl allow comparison only with the very similar *C. stupella*, with which it is sympatric. In *C. stupella* the spots are larger, squarish, in fewer rows but covering almost the entire body whorl; the shoulder and spire whorls are marked with large squarish spots of dark reddish brown, almost black; and the shell is commonly toned with deep violet. In addition, for some reason the spire seems to be shorter (it isn't) and broader, with more convex sides (illusion). The two appear quite distinct on direct comparison and overlap in size, so they are not juvenile and adult.

Variation.—Some specimens are broader at the shoulder than others, with the shoulder roundly to more narrowly angled. The spots are always small and often indistinct, but they do not cover the whole body whorl uniformly and are often completely absent over much of the shell. The spire and shoulder may be very weakly marked.

Distribution.—Rarely dredged in deep water from southern Japan to Taiwan.

Notes.—The few specimens of this species I have seen were very coarse and badly scarred, with injuries on the spire whorl often prominent. Worm holes and burrowing sponges were common, as were bad mud stains. Amazingly, one of these specimens, just as decrepit as the others, was collected alive. Although rare, the ugly appearance does not allow this shell to be sold at the price its rarity probably deserves.

STUPELLA (Kuroda, 1956)

1956. *Embrikena stupella* Kuroda. *Venus*, 19(1): 3, pl. 1, fig. 3 (Off Tosa, Japan).

Description.—Heavy, with a good gloss; biconical, the upper sides convex, then concave to straight and tapered to the base; body whorl with a few low spiral ridges at the base, otherwise smooth except for spiral and axial threads; shoulder broad, roundly to somewhat narrowly angulate, slightly concave above; spire tall, sharply pointed, the sides straight or slightly concave but appearing broad and convex; spire whorls stepped, concave above; early whorls with distinct nodules, these usually disappearing by the middle whorls; tops of whorls with 1-3 rather distinct spiral ridges. Body whorl with deep violet tones in some specimens, covered with about 7-12 spiral rows of rather large roughly squarish spots of purple-brown, the spots not very regular in size or shape and sometimes more axially than spirally aligned; shoulder with enlarged and rather oblique spots similar to those of the body whorl; spire whorls with similar large squarish spots at the margins; early whorls unmarked. Aperture rather narrow, about uniform in width; outer lip thickish, nearly straight or slightly concave at middle; mouth white, often with violet or pinkish tones. Columella long and narrow, straight or oblique. Length 60-78mm.

Comparison.—*C. stupella* at first glance is obviously very close to *C. stupa* and might appear to be just a clean juvenile form. However, the two overlap in size so this is not the case. *C. stupella* has larger and more regular spots than *C. stupa*, these covering most of the body whorl and enlarged on the shoulder, with the spire margins similarly marked with large spots; the spots are usually darker, more purple than the pale reddish brown spots of *C. stupa*. *C. stupella* seems to have a lower, thicker, more convex spire than *C. stupa*, although this is probably an illusion; it is a quite obvious difference in shape between the two species, however.

Variation.—Most fresh specimens are distinctly toned with violet, sometimes quite darkly so. The spots vary considerably in size, shape, and placement, often being absent over small portions of the body whorl. There may be small blotches present between the rows of spots. The shoulder is usually rounded, but the sides of the body whorl vary from distinctly convex posteriorly then concave to nearly straight.

Distribution.—Moderately rare in deep water off southern Japan and Taiwan.

Notes.—This species has been considered synonymous with *C. stupa*, but it appears quite distinct when the two are directly compared. This is usually a clean, virtually flawless shell with strong violet tones compared to the dull off-white and rough *C. stupa*. Although apparently collected more often than *C. stupa*, it commonly sells at a higher price. The lip chips easily and there may be erosion on the early spire whorls, as well as sometimes heavy axial growth marks.

It would be interesting to know what the juveniles of *C. stupa* and *C. stupella* look like, as the adults are quite different from most other cones, except perhaps *C. pergrandis*, which is deeply grooved.

SUBULATUS Kiener, 1845

1845. *Conus subulatus* Kiener. *Species gen. et icon. des coqu. viv.*, 2 (*Conus*): Pl. 70, fig. 2. 1849, *Ibid.*, 2: 243 (Locality unknown).

1849. *Conus collisus* Reeve. *Conchologia Iconica*, 1(*Conus* Suppl.): Pl. 8, sp. 273 (Locality unknown) Lectotype figure (here selected): Reeve, Pl. 8, sp. 273. Tomlin, 1937, states that the British Museum 'types' do not agree with the figure.

1878. *Conus andamanensis* E.A. Smith. *Proc. Zool. Soc. (London)*, 1878: 804, pl. 50, figs. 1-1a (Port Blair, Andaman Islands).

1936. *Conus mulderi* Fulton. *Proc. Malac. Soc. London*, 22: 9, pl. 2, fig. 1 (Mindoro Is., Philippines).

Description.—Moderately light in weight, with a good gloss; low conical, the sides slightly convex; body whorl with widely spaced spiral grooves basally, these demarking wide flat ribs that extend to midbody or higher, the grooves sometimes axially threaded within; shoulder roundly angled, not very distinct from spire; spire moderate, sharply pointed, the sides deeply concave; spire whorls with about 2-4 low and indistinct spiral ridges on the posterior portion, crossed by fine scattered axial threads, all indistinct; whorls convex above. Body whorl white to bluish white, sometimes tan, covered with axial series of pale brown to blackish spots that are heavier in bands below the shoulder and above and below midbody; usually the spots are fused into either axial flammules or spiral broken bands, usually with indistinct and smeared margins; pattern sometimes very blotchy; spots between larger fused spots usually small and irregular; base white or pale straw; spire and shoulder colored like body whorl, the spots or blotches moderately wide and heavy, radiating from the pale early whorls. Aperture moderately narrow posteriorly, much wider anteriorly; outer lip thin, sharp, straight or gently convex; mouth dark violet, pale violet, or white, without reddish or brownish tones, often with a small pale brown spot at anterior margin. Columella long, narrow, slightly indented from inner lip. Length 20-46mm.

Comparison.—From *C. stramineus*, with which it is often confused, *C. subulatus* differs in being somewhat narrower and lighter in weight, often distinctly bluish, having the spots usually fused and irregular or smudged, and having the mouth white to dark violet without reddish or brown tones. Although the two species sometimes are very similar in appearance, the differences in spotting patterns and mouth color usually allow them to be separated easily. *C. inscriptus* usually has somewhat different colors and patterns than *C. subulatus*, but the main difference is in the heavier and more numerous spiral ridges on the spire whorls, the ridges covering the entire whorl instead of being restricted to the posterior portion.

C. tegulatus has presented a problem because it is often very similar in pattern to *C. subulatus andamanensis* in the Bay of Bengal. However, in most *C. tegulatus* of this type the shoulder is more angled, the spire ridges heavier, covering most of the whorl top and covered with axial threads, and the anterior sculpture narrow ridges instead of broad ribs. Admittedly some specimens are very similar; in most *C. tegulatus* the pattern is obviously derived from broad flexuous axial flammules thickened into irregular bands above and below midbody.

Variation.—This species can be divided into two rather distinct subspecies which seem to intergrade around both sides of the Malaysian peninsula.

C. subulatus subulatus: Often smaller, under 35mm, the mouth pale to deep violet but not white; spots tend to fuse into axial blotches and sometimes flammules, usually dark bluish brown, almost black at times. Western Pacific.

C. subulatus andamanensis: Often larger, over 30mm, the mouth white to very pale violet; spots tend to fuse into spiral bands and blotches, seldom axial in alignment, the spots yellowish brown to reddish brown. Northeastern Indian Ocean.

The pattern in both subspecies is very variable. In some *C. s. subulatus* the spots are rather distinct and not fused except above the base, but in others three spiral series of dark bluish brown or blackish spots may be the only pattern. Occasionally the fused spots are connected into axial flammules. In some areas, especially the Solomons, the blotches are shadowed with tan that may eventually cover the entire shell, almost obscuring the blotches. The mouth is usually dark violet with a small brown spot anteriorly.

Indian Ocean *C. s. andamanensis* are much paler on the average, the spots often yellowish brown. There are usually three bands (often doubled) of spirally smeared spots or blotches and rather scattered small spots axially arranged over the rest of the shell. The mouth is white or very pale violet, more violet and with more distinct spotting the closer one gets to Singapore. The region of intergradation with the typical is rather broad, with influence of both subspecies easily visible in populations on both sides of Malaysia.

C. s. andamanensis may be transitional to *C. janus*, but specimens from the area between Madagascar and Sri Lanka are not available to me.

Distribution.—Rather locally distributed from southern India and Sri Lanka to the Solomons, moderately common in deep to shallow water. Several local populations seem to exist within this range as well as two subspecies.

Synonymy.—Although usually called *C. collisus*, there seems little doubt that Kiener's *subulatus* is also this species; as the proper date of Kiener is 1845, his name has four years of priority over Reeve's more familiar name. *Andamanensis* is certainly at best a subspecies of Pacific *subulatus*, as intergrades are common; this taxon was described from a juvenile but seems to be the taxon here described. *C. mulderi* seems to be a specimen with the spots smeared into a tan background like some Solomons shells. *C. straturatus* may belong here rather than in *C. spectrum*; the two species are easily distinguished by shape and mouth color, but sometimes the patterns do overlap.

Notes.—This is usually a clean species but it often has very poorly developed and indistinct patterns. The outer lip is thin but does not chip easily except at the anterior margin. There may be axial growth flaws.

Locality data on shells coming out of Thailand are apparently very doubtful. Lately *C. tegulatus* has appeared on the market from this area mixed with *C. s. andamanensis;* admittedly the two are sometimes extremely difficult to separate.

SUGIMOTONIS Kuroda, 1928

1928. *Conus sugimotonis* Kuroda. *Venus,* 1: 81, pl. 1, fig. 6 (Tosa, Japan).

Description.—Moderately heavy, with a low gloss; obconical, the posterior sides nearly parallel for a short distance then straight or concave and tapering to a narrow base; body whorl with a few low spiral ridges above the base and often coarse axial growth lines, otherwise smooth; shoulder broad, carinate or sharply angled, nearly flat on top; spire flat, the early whorls forming a slightly projecting cone; spire whorls nearly flat, with about 3-6 narrow spiral ridges and weak axial threads; first 3-4 whorls weakly nodulose, margins of later whorls straight. Body whorl solid white, without pattern, sometimes with a small brown spot at base and tan shading along major growth flaws. Aperture moderately narrow, of uniform width; outer lip straight to slightly concave, thin and sharp; mouth white. Columella short and narrow. Length 60-95mm.

Comparison.—*C. sugimotonis* is closely related to *C. tribblei* in general shape and sculpture of the spire, but that species differs in having a pattern of spiral dashes and brownish spiral bands; it commonly has a more stepped spire than *C. sugimotonis,* with distinct brown spotting on the spire. Solid white specimens are narrower at the shoulder than *C. sugimotonis* and heavier, usually with higher spiral ridges anteriorly. *C. sieboldii* is deeply grooved anteriorly and has a high spire.

Variation.—All specimens are white, but some have a small brownish spot at the base and brown streaks along the often numerous axial flaws. The spire may be completely flat except for a small cone formed from the early whorls, or the middle whorls may also be incorporated into the projecting cone; the whorls are apparently never stepped.

Distribution.—Uncommonly dredged in deep water off southern Japan and Taiwan; also reported from off Queensland, but this record may represent white *C. tribblei.*

Notes.—An ugly white shell with many breaks and flaws, often with heavy scars and unhealed breaks. Rarely available on the market in any condition, though not really more than uncommon.

SULCATUS Hwass, in Bruguiere, 1792

1792. *Conus sulcatus* Hwass, in Bruguiere. *Cone*, in *Ency. Method., Hist. Nat. des Vers*, 1: 618 (East Indies). See Kohn, 1968.

(1802. *Conus costatus* Holten. *Enumeratio Systematica Conchyl. beat Chemnitzii:* 39. Based on Chemnitz, Pl. 181, figs. 1745-1747.)

1810. *Conus asper* Lamarck. *Ann. du Mus. Hist. Nat.* (*Paris*), 15: 39 (China).

1817. *Conus costatus* Dillwyn. *Descr. Catalogue Recent Shells*, 1: 399 (South Seas and coasts of China). Based on Chemnitz, Pl. 181, figs. 1745-1747. Non *Conus costatus* Gmelin, 1791.

?1857-1858. *Conus crebisulcatus* Sowerby ii. *Thesaurus Conchyliorum*, 3 (*Conus*): 21, pl. 14(200), fig. 321 (Locality unknown). Misspelled '*crebisulcus*' on page 21, but corrected on plate and in index.

1857-1858. *Conus undulatus* Sowerby ii. *Ibid.*, 3(*Conus*): 34, pl. 4 (190), fig. 63 (Locality unknown). Non *Conus undulatus* Solander, 1786.

1881. *Conus bocki* Sowerby iii. *Proc. Zool. Soc.* (*London*), 1881: 636, pl. 56, fig. 7 (Amboyna).

Description.—Moderately heavy, with a low gloss; low conical, the upper sides distinctly convex to nearly straight, then tapering to a narrow base; body whorl covered with closely spaced spiral ridges, the grooves between usually with axial threads; often every second or third ridge is enlarged and usually heavily granulose; sculpture often reduced or absent on posterior third to half of whorl; shoulder broad, sharply angulate, with flat triangular horizontal coronations or undulate; spire low to nearly flat, the early and middle whorls often raised into a large, straight-sided or concave cone, sharply pointed; early whorls finely nodulose, becoming undulate to coronate; shoulder sometimes undulate; tops of whorls flat or slightly concave, with about 5-7 strong spiral ridges, crossed by numerous and heavy axial threads so strongly and finely cancellate. Body whorl yellowish tan to deep reddish brown, sometimes white; base white; occasionally the spiral ridges are slightly darker than the grooves, so shell appears weakly spirally lineated; spire white to pale tan, the margins often with a narrow dark brown line; occasionally with scattered indistinct short stripes. Aperture moderately wide, especially anteriorly; outer lip sharp, solid, slightly to distinctly convex; mouth white. Columella long, narrow. Length 50-90mm.

Comparison.—*C. mucronatus* may be vaguely similar but is smaller with straight spire margins, the sculpture of the body whorl is usually not so heavily beaded, and brown mottling is often present or else the shell is distinctly white. *C. grangeri* is very similar but is more inflated and has distinct spots or flammules of dark brown on a whitish body. The juveniles could be confused with several small deep-water cones, but they have projecting triangular horizontal coronations.

Variation.—The ridging is quite variable, apparently individually.

Some specimens are heavily ridged, almost every spiral ridge from base to shoulder being beaded, while others have very weak anterior sculpture that is not beaded (*undulatus* Sowerby). Some specimens have the posterior third to quarter of the body whorl smooth, without erect ridges. The coronations on the shoulder may be very large or reduced and undulate. The coloration varies from pale to dark tan, sometimes white or reddish brown. Two somewhat distinct subspecies may be recognized.

C. sulcatus sulcatus: Body whorl relatively narrow at the shoulder; spire relatively high; posterior half of body whorl in adults not normally smooth; color pale to dark tan in both juveniles and adults. Southern Japan to the Philippines.

C. sulcatus bocki: Body whorl relatively wide at the shoulder; spire relatively low, sometimes with brown axial lines; posterior half of body whorl in adults usually smooth; color tan in juveniles, becoming dark reddish brown on new growth in adults. Bay of Bengal through the Indonesian area to the Solomons.

The subspecies are only weakly defined and individual specimens may not be assignable with certainty.

Distribution.—Commonly dredged in moderately deep water from southern Japan, Taiwan, and the Philippines; less common in the Bay of Bengal, Indonesia, New Guinea, and Solomons.

Synonymy.—This species was long known as either *C. costatus* or *C. asper*, but the earliest name under our rules is *C. sulcatus. C. crebisulcatus* appears to be a juvenile, but this is not certain. *Undulatus* is a minor individual variant and not worthy of recognition.

Juveniles of *C. sulcatus*, Marinduque, Philippines, 18-28mm (AK).

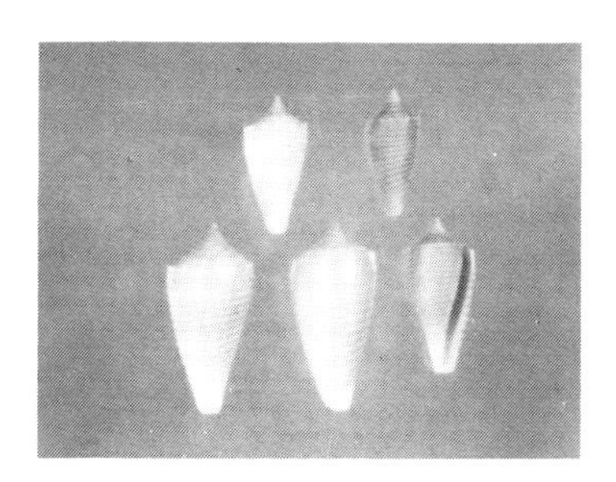

Notes.—Although a deep-water shell, this sturdy species is in good condition when seen. The major flaws are axial growth marks and occasionally small scars on the spire; there may be asphalt on the spire whorls. *Bocki* tends to have heavy axial growth marks on new growth in adults, with dark reddish-brown pigment accumulating here and producing the distinctive color of adults of this subspecies.

Juveniles are real gems of little shells, with very heavy coronations; they are seldom available under their own name, however, usually being sold under various names of unrelated species.

SURATENSIS Hwass, in Bruguiere, 1792

1792. *Conus suratensis* Hwass, in Bruguiere. *Cone*, in *Ency. Method., Hist. Nat. des Vers*, 1: 669 (East Indies). Holotype figured by Kohn, 1968.

?1850. *Conus* (*Dendroconus*) *agrestis* Moerch. *Cat. Conch. Kierulf:* 16, 31 (Nicobars). Fide Tomlin, 1937.

Description.—Very heavy, with a good gloss; obconical, the upper sides convex then straight to the base; body whorl with about a dozen fairly weak spiral ridges above the base, the rest of the whorl smooth except for fine spiral and axial threads and scratches; shoulder broad, rounded, flat or slightly concave above; spire flat or nearly so, the early whorls forming a small pointed cone at the middle; tops of whorls concave to flat, with only fine spiral and axial threads, no distinct sculpture; early whorls commonly eroded. Body whorl white or cream, the base and sometimes the shoulder broadly stained bright orange or yellow; numerous small dark brown dashes over most of the whorl, these in spiral rows but also aligned into axial bands; little fusion occurs between dashes, but sometimes they form short axial flammules at the shoulder; spire white, often with orange tones, with curved blackish axial lines and spots often narrower than the light areas between them; early whorls unmarked. Aperture moderately wide, about uniform in width; outer lip thick, sharp, straight or with a slight concavity at the middle; mouth white, cream, or stained with pale pink. Columella long, narrow, with a heavy sharp ridge posteriorly. Length 60-150mm.

Comparison.—This species can be confused only with *C. betulinus*, with which it has sometimes been considered conspecific. *C. suratensis* is narrower and more tapered than *C. betulinus* of the same size, with the columellar area more elongated; in *C. betulinus* of any form the spots are larger, squarish or rounded, often alternating with white areas, and more widely spaced and fewer, in spiral, not axial, rows; *C. betulinus* is usually deep yellow to orange, *C. suratensis* white to cream with orange extremities. Certainly the two species are close, but they seem fully distinct.

Also close is *C. glaucus*, which has a pattern much like that of *C. suratensis* but is dark or light bluish gray instead of white or cream; *C. glaucus* is also much shorter and broader in relation to size than *C. suratensis*.

Variation.—*C. suratensis* is very constant in shape and pattern, varying only in minor details from specimen to specimen. Juveniles have the base and shoulder bright orange, this fading in adults but often still present as a tint. In juveniles the dashes are in distinctly spiral rows, the axial alignment becoming more apparent with growth. The anteriormost dashes in adults may be more spot-like than dash-like.

Distribution.—Apparently widely distributed in the Indo-West Pacific, but the limits of the range poorly known because of confusion with *C. betulinus*. From the eastern African coast to southern India and

Sri Lanka, then to the Philippines and east to Solomons; perhaps also from further east. Common in shallow water.

Synonymy.—*C. agrestis* was considered a hybrid between *C. glaucus* and *C. betulinus*, so it probably belongs here.

Notes.—This is a very pretty species when young. Adults are seldom as cumbersome as *C. betulinus* adults. There is a tendency for strong axial growth flaws and weakly eroded early whorls.

SUTURATUS Reeve, 1844

?1844. *Conus incarnatus* Reeve. *Conchologia Iconica*, 1(*Conus*): Pl. 41, sp. 221 (Malacca) (January).

1844. *Conus suturatus* Reeve. *Ibid.*, 1(*Conus*): Pl. 45, sp. 250 (Locality unknown) (February).

1857-1858. *Conus turbinatus* Sowerby ii. *Thesaurus Conchyliorum*, 3 (*Conus*): 25, pl. 10(196), fig. 227 (Locality unknown).

Description.—Moderately heavy, with a good gloss; low conical to obconical, sometimes ovate, the sides usually convex posteriorly but occasionally straight; body whorl with shallow spiral punctate grooves, usually ending below midbody in adults and separating broad flat ribs, but often continuing to shoulder in juveniles and sometimes granulose; shoulder broad to narrow, rounded to sharply angulate, occasionally undulate, concave above; spire flat with the early whorls projecting or relatively high and with straight or concave sides, sharply pointed; spire whorls with thickened rude margins, the middle of each whorl with a large ridge separated by deep grooves from the margins; occasionally a second small ridge is present; early whorls nodulose. Body whorl white or pinkish, usually with the base strongly tinted with deep violet or violet brown; two broad bands of deeper pink, pinkish tan, or pale brown usually present above and below midbody, sometimes with another band at the shoulder; in specimens from some populations, especially Hawaii, this is overlaid by a pattern of pale reddish or brownish spots or dashes in spiral bands or covering the entire whorl, often with other blotches or spots of opaque white; spire pinkish to whitish, unmarked or with small brownish red spots or streaks. Aperture relatively wide, almost uniform in width; outer lip thin, sharp, convex to straight; mouth white to pinkish, usually with a violet blotch at the anterior end. Columella short and narrow but thick. Length 25-46mm.

Comparison.—This species is very similar to *C. tessulatus* in color and in the sculpture of the spire. Generally *tessulatus* can be separated from typical *suturatus* by the presence of many reddish or orangish dashes and rectangles over the body whorl, the absence of spiral pinkish or pale tan bands above and below midbody, and the absence of punctate grooves to the shoulder in juveniles. These differences will not always work in Hawaii or occasionally in Queensland, but at these localities the shells which have reddish rectangles without pinkish or tan spiral bands are *C. tessulatus*, those with the combination of rectangles and bands, plus usually a higher spire, *C. suturatus. C. kashiwajimensis* is somewhat similar also, but it is much more elongated, has the later whorls with normal spiral sculpture, and lacks the violet base.

Variation.—It has long been recognized that Hawaiian specimens assigned to *C. suturatus* are quite distinct from those occurring elsewhere in the Indo-Pacific, but such shells are apparently without a name. One problem has been the presence of shells somewhat intermediate between *C. suturatus* and *C. tessulatus* in Queensland and Japan, these shells combining indications of *tessulatus* pattern with *suturatus* shape and pattern. Since such shells are usually uncommon in otherwise normal populations of *C. suturatus*, I see no reason not to consider them just interesting specimens that may show that *C. suturatus* and *C. tessulatus* occasionally hybridize. They are certainly not the same as the Hawaiian shell. For this reason two subspecies, one of them new, are recognized.

C. suturatus suturatus: Body whorl usually ovate or ovately conical, the sides convex; spire low or flat; pattern of two or three spiral bands of pale tan to darker pinkish tan on white or pink, lacking reddish or orange rectangles on top of the bands; opaque white spots seldom developed. Indo-West Pacific.

C. suturatus **SANDWICHENSIS** Walls, **new subspecies:** Body whorl low biconical, narrow, the sides nearly straight; spire tall to moderately tall, with straight or concave sides; pattern of two broad spiral bands of pale tan or pinkish tan on white, juveniles and many adults with numerous rectangles of brown or orange over the bands or over the entire shell; small and large spots of opaque white developed at least in juveniles; spiral ridges commonly weakly granulose in juveniles. Hawaiian Islands (Holotype: Delaware Museum of Natural History; 14.4 x 7.2mm granulose juvenile from Pokai Bay, Oahu, Hawaii; paratypes in the author's collection).

Distribution.—Range uncertain; relatively common in the western Pacific from southern Japan through the Philippines and the New Guinea-Solomons area south to Queensland, Australia. Apparently also in the eastern or northeastern Indian Ocean. Hawaii. Common to uncommon in shallow to deep water.

Synonymy.—*C. incarnatus* seems to be this species, but the colors seem rather dull and the bands broad; it has one month priority over the more familiar *C. suturatus*; this name has also been referred to *C. fumigatus*, which seems unlikely. *C. turbinatus* is fairly typical. *Sandwichensis* seems to be a well defined subspecies endemic to the Hawaiian Islands.

Notes.—Typical specimens of *C. s. suturatus* are fairly common in the western Pacific, but I have not seen any Indian Ocean specimens, although *C. tessulatus* is common in the Indian Ocean. This is usually a quite clean shell with bright patterns that fade when exposed to light; the lip is thin and often filed. Specimens from Queensland and southern Japan with the rectangles of *C. tessulatus* are uncommon and usually worth a premium.

C. s. sandwichensis is uncommon in deep water of the Hawaiian Islands and shows a greater variability than in the typical subspecies, although the shape is quite constant and the juveniles are also usually very similar in having both bands and rectangles. The colors are often more subdued and tan instead of pinkish. In Hawaii this is a rather rough shell often with heavy axial breaks. Salisbury, 1977, *Hawaiian Shell News*, 25 (5): 13-14, figs. 6-9, illustrated and discussed some interesting variants; granulose adults are uncommon or possibly rare.

TAENIATUS Hwass, in Bruguiere, 1792

1792. *Conus taeniatus* Hwass, in Bruguiere. *Cone*, in *Ency. Method., Hist. Nat. des Vers*, 1: 628 (China). Lectotype (here selected): specimen figured by Kohn, 1968, as the holotype.

1798. *Cucullus Fernambucinus* Roeding. *Museum Boltenianum:* 39 (Locality not stated). Lectotype figure (Kohn, 1975): Martini, Pl. 57, fig. 632.

1798. *Cucullus genuinus* Roeding. *Ibid.:* 42 (Locality not stated). Lectotype figure (Kohn, 1975): Chemnitz, Pl. 144A, fig. n.

Description.—Moderately heavy but not thick, with a good gloss; low conical, the upper sides slightly convex then tapering to the base; body whorl with a few crowded spiral ridges above the base, then 4-5 broader flat ribs more posteriorly, usually extending to about anterior third of the body whorl; rest of whorl with fine axial and spiral scratches; shoulder broad, roundly angled, undulate or with weak rounded coronations; spire low, the sides convex, the tip usually eroded; tops of whorls slightly concave, the margins undulate, weakly coronate, or just rude; whorls with numerous fine axial and spiral threads and scratches, sometimes with traces of 2-3 low spiral ridges on the early whorls, which are not more strongly coronate than the later whorls. Body whorl white, usually with strong bluish gray, bluish tan, or bright blue tones; base pale bluish white; whorl covered with spiral rows of large blackish dashes alternating with white dashes in a generally very precise pattern; usually the midbody area and the area below the shoulder are more weakly marked, the subshoulder area often plain pale bluish; shoulder and spire bluish white, with lighter margins, covered with regularly spaced, axially elongated black blotches between the whitish coronations or undulations, pattern sometimes reduced or absent; early whorls yellowish. Aperture wide, uniform in width; outer lip sharp, rather thick, sometimes thickened at the middle, evenly convex; mouth brownish violet, with a pale band at the middle and below the shoulder. Columella long, very narrow. Length 25-43mm.

Comparison.—This beautiful little bluish cone is very similar to *C. miliaris* and *C. coronatus* in general appearance but is much more distinctly patterned with alternating black and white dashes; the coronations are also much weaker, often reduced to mere undulations. This species is also very similar to *C. tiaratus* in sharing a distinctly dashed pattern. However, *C. tiaratus* is more parallel-sided than *C. taeniatus*, is narrower at similar body lengths, and is heavier and thicker with a generally rougher texture often heavily marked with growth flaws; few *C. tiaratus* are distinctly bluish, usually being some tone of brown. *C. genuanus* is larger, has a completely smooth shoulder, a white mouth, and lacks the blue tones.

C. taeniatus is perhaps most closely related to *C. musicus* as the juvenile looks much like a bluish *C. musicus* Indian Ocean form with dashes instead of spots; the shoulder in the juvenile I have seen is white, unmarked, and smooth.

Variation.—*Conus taeniatus* is very constant in pattern and shape, the major variations being the height of the spire, from nearly flat to moderately high, and the density and intensity of the black dashes. The coronations are seldom very distinct, usually just low undulations most easily seen by their slightly whiter shade than the rest of the whorl margin. Color is distinctly blue, usually a pale blue-gray or bluish tan, sometimes bluish white or deep blue; the base, midbody, and shoulder are generally paler than the rest of the body whorl.

Distribution.—*C. taeniatus* seems to be restricted to the northwestern Indian Ocean, including the Red Sea through the Gulf of Iran.

Synonymy.—The Roeding names are meaningless, as usual.

Notes.—The specimens I have seen had the coronations extremely low, like weak undulations. There are occasional minor axial flaws and erosion on the spire whorls, as well as chipping of the lip. A very pretty species.

TASLEI Kiener, 1845

1845. *Conus Taslei* Kiener. *Species gen. et icon. des coqu. viv.*, 2(*Conus*): Pl. 110, fig. 3. 1849-1850, *Ibid.*, 2: 360 (Locality unknown).

1845. *Conus olivaceus* Kiener. *Ibid.*, 2(*Conus*): Pl. 111, fig. 3. 1849-1850, *Ibid.*, 2: 359 (Locality unknown). Non *Conus olivaceus* Salis, 1793.

1854. *Conus luridus* A. Adams. *Proc. Zool. Soc.* (*London*), 1853: 118 (Australia).

1854. *Conus desidiosus* A. Adams. *Ibid.*, 1853: 119 (West Africa).

?1857-1858. *Conus irregularis* Sowerby ii. *Thesaurus Conchyliorum*, 3 (*Conus*): 29, pl. 118(201), fig. 418-419 (West Africa).

Description.—Moderately heavy, with a low gloss; low conical, the upper sides convex then tapering to the base; body whorl with a few low and irregular spiral ridges at the base; rest of whorl with numerous fine spiral and axial threads; shoulder rather broad, roundly angled, slightly convex on top; spire low to moderate, often turbinate, the earlier whorls forming an erect cone; later whorls nearly flat; early whorls with traces of a few indistinct nodules, later whorls with 1-3 indistinct low spiral ridges and scattered axial threads; spire usually eroded. Body whorl pale yellowish white, covered with very fine axial brown lines that are very close to each other and nearly parallel, seldom easily visible except under a lens; lines thickened into three rather wide straight-edged bands above and below midbody and below the shoulder, appearing darker brown on casual examination; pale bands sometimes with traces of irregular white blotches within; base pale; spire and shoulder dirty whitish with traces of small brown spots and streaks usually heaviest on shoulder, sometimes eroded white. Aperture moderately wide, uniform or slightly widened anteriorly; outer lip thin, nearly straight; mouth whitish to dark violet with two light bands at midbody and below shoulder. Columella short, narrow. Length 20-51mm.

Comparison.—As with all other West African cones related to *C. mercator* and *C. ermineus*, this is a very indistinct species of doubtful nomenclatural history. The name is used more for convenience to identify certain color patterns than out of certainty that: 1) this species is valid, or 2) this is the proper name for it.

Obviously *C. taslei* is very close to *C. mercator* and there are obvious intermediates; at the same time, it is allied to *C. ermineus* and there are intermediates in that direction also. From typical forms of both these species it is conveniently separated by the rather broader shoulder with a more turbinate spire and the presence of a greatly increased number of nearly parallel axial hairlines in three dark and obvious bands. From *C. balteus* it differs in usually lacking crosshatching and being paler in color and larger in size.

Conus cloveri, another species here described more out of convenience than out of certainty of its validity, is also very closely related though I have seen no intermediates. In *C. cloveri* the axial lines are larger, almost uniform in thickness over the shell except at midbody, so at most two vague darker spiral bands are present, and they tend to be obviously wavy, trapping circular or triangular white spots between the lines; the mouth is always white or very pale violet; and the shoulder is patternless or nearly so and sharply separated from the body pattern.

Variation.—As treated here, this shell varies in shape and pattern. The shoulder is often broad and forms a plate on which the high earlier whorls stand erect, but the shoulder may be roundly angled and the spire with convex sides. The banded pattern is usually distinct, with the axial lines thickened to form the three bands although the same number of lines is present in the pale bands. Irregular white spots may be present or absent. The mouth varies from white to pale violet to dark violet in very dark specimens. The overall color of the shell may be pale yellow and tan, yellowish and dark brown, or dark brown and almost black; the spire is pale and sparsely marked.

Distribution.—Shallow to moderately deep water of western Africa below the Bulge and the offshore islands. Uncommon.

Synonymy.—*C. taslei* and *C. olivaceus* seem to be this species, although the color patterns illustrated by Kiener are rather pale to have dark violet mouths as he indicated. *C. luridus* was said by Sowerby ii to be a synonym of *C. taslei*, but it has not been illustrated. *C. irregularis* is seemingly this species, at least in part, although it has a very flammulated pattern, moreso than any specimen I have seen. *C. desidiosus* seems to be this species also, although the pattern seems to be rather strongly cross-hatched; the spire and shoulder shape resemble some variants of *C. taslei* very well. As mentioned, *C. mercator*, *C. ermineus*, *C. cloveri*, and even *C. balteus* all seem to vary in the direction of *C. taslei* but are kept distinct until western African cones are better understood.

Notes.—Most specimens are badly eroded on the spire and often the body whorl, with irregular axial growth marks and breaks. This is seldom an attractive species, being poorly contrasted in color.

TEGULATUS Sowerby iii, 1870

1870. *Conus tegulatus* Sowerby iii. *Proc. Zool. Soc.* (*London*), 1870: 256, pl. 22, fig. 12 (China Seas).

1904. *Conus planiliratus* var. *batheon* Sturany. *Denkschr. K. Akad. Wiss., Wien, 74, Exp. "Pola" Rothe Meer:* 227, pl. 4, figs. 6a-6c, 7a-7b (Stations 127, 128, 143, 145, Red Sea).

Description.—Moderately light in weight, with a good gloss; low conical, the upper sides slightly convex; body whorl covered with narrow spiral ridges, often in pairs, divided by wide grooves containing fine axial threads; ridges often weak posteriorly, but usually still present as heavy spiral threads and axial threads so conspicuous under a lens; shoulder broad, sharply angled, flat or slightly concave on top; spire low or moderate, sharply pointed, the sides deeply concave, the early whorls forming a pointed cone; tops of whorls flat or slightly concave, with 4-6 heavy spiral ridges usually crossed by distinct axial threads so very faintly cancellate; entire surface of whorl with ridges in most shells. Body whorl white, covered with a variable pattern usually consisting of two spiral bands of rather large, squarish but irregular light to dark brown blotches above and below midbody, sometimes with a smaller row at the shoulder; the blotches are usually connected by wavy axial flammules or rows of small brown spots and dashes; occasionally most of pattern reduced, the row below the shoulder disappearing first, then the one below midbody; sometimes blotches partially fused into broad broken bands; base light or with a continuation of the axial flammules; shoulder white, with small and large brown spots and blotches; early whorls white. Aperture moderately wide, slightly wider anteriorly; outer lip gently convex or nearly straight, thin and sharp; mouth white to pale violet. Columella long but very narrow. Length 20-50mm.

Comparison.—This is a highly variable species with a close similarity to several species, although it is uncertain what its actual relationships are. In typical patterns it is close to *C. sibogae*, which is a straighter-sided shell with very regular spiral ridges and grooves crossed by fine but distinct axial threads so the whole shell usually appears beaded; the shoulder is very sharp and often distinctly carinate, with the spire low. The two seem quite distinguishable. There is also a distinct similarity to *C. subulatus andamanensis* in the northeastern Indian Ocean, and it is often difficult to tell the two apart. However, most *C. tegulatus* have strong but narrow spiral ridges at least anteriorly, with the pattern distinctly axial instead of being nearly smooth with a pair or three smeared spiral bands. In *C. subulatus* the spiral ridges on the spire whorls are weak and usually limited to the posterior portion of the whorl, the entire structure seldom nearly flat; in *C. tegulatus* the ridges are sharp and distinct, covering most or all the surface of the nearly flat or slightly concave whorls.

Specimens of *C. tegulatus* var. *batheon* from India have been circulated as *C. alabaster* (here *C. mucronatus* var. *alabaster*), but the two are only superficially similar, though admittedly whitish shells of both look very much alike. In whitish *C. mucronatus* there are seldom distinct dark brown spots, usually just a posterior zone of pale brown; the spire appears thick compared to *C. tegulatus* and less sharply pointed; the spiral ridges on the body whorl are heavier and separated by narrower ridges with 0-1 small spiral ridge within and heavy axial threads, while most *C. tegulatus* even when fully sculptured have the spiral ridges narrow, often paired, separated by broader spiral grooves usually containing two weaker spiral ridges and finer axial threads; *C. tegulatus* is often more inflated posteriorly than in *C. mucronatus*, with the base relatively short and not elongated as in *C. mucronatus*; the spiral ridges of *C. tegulatus* are rarely or never granulose, while they are often granulose in *C. mucronatus*. Very inflated *C. t.* var. *batheon* may resemble *C. grangeri* in shape and pattern, but that species is coronated.

The resemblance of some patterns of *C. spectrum* to *C. tegulatus* is certainly superficial. The two are separated by the rounded shoulder and much wider mouth occurring in *C. spectrum* and the longer and wider columella, as well as the much weaker sculpture of *C. spectrum*.

Variation.—Although this is a variable species, it falls readily into only a few groups. In the most common patterns the shell is relatively elongate with a sharp shoulder and deeply concave but not flat spire, covered with 2-3 rows of light brown blotches only partially connected by broken axial flammules; the spiral sculpture usually extends to the shoulder with little reduction, and the mouth may be pale violet. A second type, most common off Western Australia, is relatively shorter and broader with the spire nearly flat except for the projecting early whorls; the pattern is similar to the first type, but dark chocolate brown and with heavy and often continuous flammules from shoulder to base; the sculpture is often reduced posteriorly. The third pattern is most common in the Indonesian area and is like a small, weakly patterned and nearly smooth type one, the shell relatively narrower and the spire taller; such shells often have the pattern very weak and the posterior half of the body whorl with only fine spiral threads. The fourth type, seemingly restricted to the northern Indian Ocean where it occurs with type one, is heavily sculptured, more inflated posteriorly and sometimes actually oval, with the pattern usually reduced to scattered spots and blotches of dark and pale brown. A fifth type, of uncertain assignment to this species, is found from the Philippines to the Solomons and resembles type four in being heavily sculptured, except the brown tends to accumulate on top of the ridges and forms long broken spiral dashes or lines somewhat like those of eroded *C. scalptus*, with a greater suffusion of pigment over the posterior half of the shell. Intermediates between the five patterns are common.

Distribution.—As treated here, *C. tegulatus* ranges from the northern Indian Ocean (patterns 1 and 4) to Western Australia (pattern 2) and the Indonesian area (pattern 3); it also occurs from the Philippines to the Solomons (pattern 5). Uncommon in moderately deep water.

Synonymy.—*C. batheon* appears to be pattern 4 above; it is often quite distinct in the reduced pattern and inflated shape with heavy sculpture and can be recognized as *C. tegulatus* variety *batheon* if desired; it occurs with type 1 shells. Type 3 shells have been called *C. stigmaticus*, a name here assigned to *C. spectrum* on the basis of the original description and the figure in *Thesaurus Conchyliorum*. Several names placed in the synonymy of other species could possibly apply here; they would be older than *C. tegulatus*, so the name of the species may change when someone examines types. The holotype of *C. tegulatus* is a juvenile shell with the pattern only partially formed (probably type 1).

Notes.—*C. alabaster* from the Indian Ocean is variety *batheon;* shells called *wallangra* from that area are *C. tegulatus* type 1. *C. stigmaticus* from the Banda Sea is type 3. This species is usually clean except for minor axial growth marks and small chips in the lip.

TELATUS Reeve, 1848

1848. *Conus telatus* Reeve. *Conchologia Iconica*, 1(*Conus* Suppl.): Pl. 1, sp. 270 (Locality unknown).

Description.—Moderately light in weight, with a low or good gloss; weakly cylindrical or biconical, the upper sides weakly convex or almost parallel then tapering to a rather long base; body whorl with low spiral ridges at the base, rest of whorl covered with low spiral ridges that usually bear small but feelable granules; axial threads numerous but weak; shoulder broad, sharply angled, usually undulate, concave above; margins sometimes with projecting rounded coronations instead of undulations; spire tall, sharply pointed, the sides concave to nearly straight; early 3-4 whorls with fine nodules, later whorls with raised margins bearing low or heavy rounded coronations; whorls concave on top with traces of 3-5 low and often indistinct spiral ridges. Body whorl white, yellowish, to pale straw, covered with fine axial orange lines that are zigzag and tend to overlap to produce large and small white tents over the body whorl, especially in broad spiral bands at base, midbody, and below the shoulder; two broad spiral bands of darker orange-brown to dark brown above and below midbody, the bands sometimes broken into irregular blotches by axial bands of tents; brown bands contain long and heavy dark brown 'textile lines' that are usually longest in the posterior band; base pale; shoulder and spire orange-brown with large darker spots between the coronations and fine axial dark brown lines; coronations often white, forming semicircles so pattern very regular in some specimens; scattered white tents also present on the spire; early whorls white to tan. Aperture moderately wide, wider anteriorly; outer lip thin, sharp, evenly convex to nearly straight; mouth white, sometimes with pale violet tone. Columella long, very narrow. Length 35-70mm.

Comparison.—*C. telatus* is very similar in general appearance to *C. textile*, but even in smooth specimens it is usually narrower, more elongated, with a taller spire and sharper whorl margins; the 'textile lines' are longer than in most *C. textile.* There is a marked similarity to *C. aureus*, which is also slender, has a high spire, is covered with fine spiral ridges, and has long 'textile lines'. However, *C. telatus* has the spiral ridges usually granulose, the spire whorls concave above with raised and coronate margins, and the 'textile lines' restricted to two spiral bands rather than continuing through the midbody area as in many *C. aureus. C. victoriae* is also similar.

The regular spire pattern of white semicircles or crescents brings to mind *C. ammiralis*, which however is a broad-shouldered shell with spiral brown lines rather than axial 'textile lines'. These differences also apply to granulose and coronate admirals.

Variation.—Some specimens are very pale, others very dark, but the pattern of two broad spiral bands of brown above and below a lightly marked midbody band is usually constant. The posterior band is usually wider and more continuous than the anterior band, which is often broken into blotches. The 'textile lines' are distinct and long, but they seldom penetrate the midbody area. The tenting may be very fine or rather coarse, but very large tents are rare. The dark blackish brown spots between shoulder and spire coronations are usually distinct even in specimens without coronations.

Typically specimens are covered from above the base to the shoulder with fine low spiral ridges covered with weak granules; the shoulder and spire whorls have raised margins with low coronations or undulations; the spiral ridges of the spire whorls are heavier and more distinct than normal in textile cones. However, in some specimens, including live-taken specimens which could not be eroded, the spiral ridges are absent or are present but not granulose and the shoulder and spire margins are not coronate or even undulate. However, such specimens usually still have the slightly raised margins of the spire whorls and the same shape and pattern as typical specimens.

Distribution.—As far as known, *C. telatus* is restricted to the islands of the southern Philippines. Moderately common in moderately deep water.

Synonymy.—See Old and D'Attilio for a history of this species and its rediscovery (*Veliger*, 1963, 6(2): 60-61). It might be mentioned that the name *telatus* has been used for a form of *C. textile* from the Red Sea, apparently in error.

Notes.—In the last decade this species has become readily available from the Philippines, and dealers now commonly offer both the granulose form and the smooth form. Although some smooth shells are obviously eroded or acid-etched from decomposition products, others appear to be typically live-collected shells with full gloss and with the full pattern of granulose shells. Intermediates in spire sculpture between coronate and smooth shells occur.

TENUISTRIATUS Sowerby ii, 1857-1858

1857-1858. *Conus tenuistriatus* Sowerby ii. *Thesaurus Conchyliorum*, 3 (*Conus*): 46, pl. 22(208), figs. 532-533 (Philippines).

Description.—Moderately heavy, thick, with a low or good gloss; cylindrical, the upper sides convex then tapering to a long base; body whorl covered from base to shoulder with narrow, rather low spiral ridges separated by narrow grooves that may be axially threaded; ridges often bear fine granules, seldom conspicuous; area below shoulder with the ridges often weaker or obsolete compared to those at midbody; shoulder rounded, narrow, not distinct from spire, the spiral ridges usually very weak or obsolete, not granulose; spire low, convex-sided, domed; early 3-4 whorls with weak nodules and 3-5 distinct but low spiral ridges; later whorls convex, the sutures rude and overlapping, with traces of spiral threads and axial threads, none sometimes distinct or 1-3 visible, not granulose although may appear finely cancellate. Body whorl pale brown to dark brown, tinted with violet especially strong at the base; young shells entirely dark, then developing a midbody and subshoulder band of small whitish squarish blotches; with growth the blotches often fuse into an indistinct midbody band, sometimes with another below the shoulder; large shells assigned to this species may be broadly whitish at midbody and below the shoulder, leaving two broad brown spiral bands above and below midbody, the entire basal area brown; shoulder and spire whitish, darker in small shells, with large or small axial brown squarish spots separated by wide white or violet areas; early whorls pinkish. Aperture narrow, slightly wider anteriorly; outer lip thickish, sharp, nearly straight; mouth pale to dark violet, often with traces of a paler band at midbody and below the shoulder. Columella very narrow or internal. Length 25-40mm, perhaps to 55mm.

Comparison.—*C. tenuistriatus* is usually considered a form or synonym of *C. scabriusculus* or *C. glans*, and admittedly it is somewhat intermediate between these two species. When removed and treated as a full species, all three taxa become easier to recognize and more clearly defined.

Like *C. scabriusculus*, the sides of the body whorl are not parallel posteriorly but are convex; however, the body whorl is narrow and elongated rather than bulbous as in *C. scabriusculus*. The spiral sculpture is weaker than in *C. glans*, seldom being distinct on the spire whorls or even shoulder and not granulose; when the body ridges are granulose, the granules are small and fine, barely feelable. The ridges are, however, somewhat stronger than those of *C. scabriusculus* and more commonly granulose than in that species.

The body whorl shape alone will usually distinguish most specimens of the three species, but when combined with pattern there are very few shells that require more than a good look for positive identification. In *C. glans* the shell is usually deep violet but is always distinctly brownish-banded with even bands. In *C. scabriusculus* the tendency is to form large brown blotches above and below midbody, leaving very large white patches. In *C. tenuistriatus* the development of the pattern is from an entirely dark shell to one with squarish pale blotches at midbody and sometimes below the shoulder, these occasionally fusing to produce rather indefinite pale spiral bands broken by pale tan axial stripes. The very large specimen assigned here is mostly brown with a narrow violet band at midbody and a broader one below the shoulder containing squarish brown spots. This type of pattern is not found in either *C. glans* or *C. scabriusculus* to my knowledge. I do not think there will be much problem separating the three species.

Variation.—The body whorl is usually distinctly elongate but occasionally is rather short; the posterior sides are gently convex, neither bulbous nor parallel. The body whorl ridges are narrow and usually weakly granulose at least anteriorly, though sometimes smooth; the ridges are commonly weak or obsolete just below the shoulder and do not continue onto the shoulder. Spire sculpture is weak or indistinct. The body whorl

Giant (52.6mm) Zanzibar specimen assigned to *C. tenuistriatus* (MF).

varies from dark brown to pale brown, usually with a pale violet tone. The midbody band is usually well developed but broken in typical adults, with the subshoulder band weak or absent and pale axial stripes absent. The spire pattern seems constant.

A 55mm specimen from Zanzibar seems to be this species, as it agrees well in shape and pattern as well as sculpture. It bears a rather striking resemblance to *C. violaceus*, but that species is more evenly cylindrical with a pattern of spiral and axial broad brown bands leaving white rectangles and a whitish base.

Distribution.—*C. tenuistriatus* seems to be widely distributed and common from the eastern African coast to at least the Marshall Islands and French Polynesia. Usually found in shallow water.

Notes.—I think this is a distinct and recognizable species that is perhaps more common than either of its allies and very wide-ranging. Bright dark brown specimens are less common than pale brown specimens. Usually this is a very clean species with at most a little limy encrustation on the early whorls and some axial growth lines. Some specimens are more violet than others. Juveniles are dark violet brown with virtually no pattern and have bright pink early whorls.

TERAMACHII (Kuroda, 1956)

1901. *Conus torquatus* von Martens. *Sitzber. Ges. Naturf. Freunde Berlin*, 1901: 15 (East Africa). Type figured in *Tief.-See Exped.*, 7 (1): 75, pl. 1, fig. 1. Non *Conus torquatus* (Roeding, 1798).

1956. *Asprella* (*Endemoconus ?*) *teramachii* Kuroda. *Venus*, 19(1): 8, pl. 1, fig. 4 (Off Tosa, Japan).

Description.—Moderately light in weight with a good gloss; biconical, the upper sides broadly convex or nearly parallel then concave and tapering to a long base; body whorl with about 10-12 narrow oblique and spiral ridges closely spaced above the base; rest of whorl with many spiral and axial threads that are conspicuous and give a satiny sheen to the shell; shoulder broad, with a rim-like carina, concave above, with fine spiral and axial threads; spire tall to moderate, sharply pointed, the sides straight to concave; spire whorls strongly stepped, deeply concave above, with raised, rim-like margins; early 5-8 whorls with strong marginal nodules, the later whorls often with smaller nodules visible under the marginal rim; early whorls with about 4-8 very narrow and indistinct spiral ridges, these increasing in number and decreasing in sharpness on the later whorls, all covered with many very fine axial threads or scratches; at first glance the whorls appear smooth above. Body whorl and spire uniformly creamy white to pale flesh, without any pattern; early whorls and base sometimes stained pale tan. Aperture wide, nearly uniform in width; outer lip thin, sharp, fragile, following outline of body whorl or nearly straight and concave at the middle; mouth creamy white near the margin then deep fleshy pink within. Columella long, narrow, twisted and partially free posteriorly. Length 50-110mm.

Comparison.—This is one of the most distinctive cones, only being approached closely by some forms of *C. howelli*. However, that species is very small when apparently adult and has more-or-less distinct brown blotches and axial opaque white lines. It should be noted, however, that I have never seen specimens of *C. teramachii* under 50mm in length, but these 50mm specimens had no traces of pattern.

There is a superficial similarity between *C. teramachii* and some other large white deep-water cones, but probably no close relationship. *C. sieboldii* is thicker, narrower, and has deep spiral grooves on the anterior half of the body whorl, usually with reddish spots. *C. ione* is the closest approach, but it is smaller, pale violet with fine reddish dots and blotches, and usually has a lower spire and different texture. *C. altispiratus* in some forms is superficially similar, but it lacks the rim-like carinate margins of the whorls, the heavy nodules on early whorls, the tall and deeply stepped spire whorls, and has a different texture as well as being clear white in color with small reddish spots or blotches often present. *C. sugimotonis* is thicker, has a flat or nearly flat spire with distinct spiral sculpture, and has a very different texture.

Variation.—The color is best described as pale to dark flesh, with the mouth similarly variable; some specimens are distinctly pinkish, while a few are almost white; I have seen no traces of pattern on any specimen regardless of size (juveniles unknown).

The African populations are slightly differentiated, but I doubt if individual specimens could be told with certainty from Japanese specimens. African specimens are generally somewhat narrower at the shoulder and taper to a narrower, longer base. The nodules on the spire whorls seem to be weaker on the later whorls with the carinae slightly less distinct. The mouth may be darker than typical for China Sea specimens, while the color of the shell may be slightly darker as well.

Individual variation in shade of color, height of spire, strength of nodules and carinae, and width of body whorl is shown by small samples from the China Sea. The smallest specimen seen (from South Africa) was much like the adults.

I guess it must be mentioned that the operculum is very large in this species and has a series of distinct teeth on the margin; the value of such a character is unknown.

Distribution.—As yet known from deep water off southern Japan and Taiwan as well as populations in deep water off the coast of South Africa, Mozambique, and Madagascar (on the west coast). Probably the species is much more widely distributed as this pattern somewhat resembles the range of *C. orbignyi*. Rare in good condition.

Synonymy.—*C. torquatus* is unfortunately preoccupied, and the African specimens are too weakly distinguished to recognize a subspecies or even variant. The relationship of this species to *C. howelli* is certainly close, and I would certainly like to see a 25mm specimen of *C. teramachii.*

Notes.—A fragile species often with the lip badly broken, the spire broken and chipped, and heavy axial scars distorting the body whorl; worm holes are common in dead-dredged specimens. This is a rare and distinctive species collectable in almost any grade.

Be careful of *Strombus* opercula placed with dead-dredged specimens by unscrupulous dealers.

TEREBRA Born, 1778

1778. *Conus terebra* Born. *Index Rerum Naturalium Musei Caesari Vindobonensis*, 1(*Testacea*): 146 (Locality not stated, restricted to Fiji by Cernohorsky, 1964). Holotype figured by Kohn, 1964.

1791. *Conus fusus* Gmelin. *Systema Naturae per Regna Tria Naturae*, ed. 13, 1: 3390 (Locality not stated). Based on Martini, Pl. 52, fig. 576.

1798. *Cucullus albeolus* Roeding. *Museum Boltenianum:* 47 (Locality not stated). Lectotype figure (Kohn, 1975): Martini, Pl. 52, fig. 577.

1807. *Conus fusiformis* Fischer. *Museum Demidoff*, 3: 144 (Locality not stated).

1843. *Conus coelebs* Hinds. *Ann. Mag. Nat. Hist.*, 11: 256 (Ambow, Feejee Islands). Also spelled *'caelebs'*.

1881. *Conus thomasi* Sowerby iii. *Proc. Zool. Soc.* (*London*), 1881: 635, pl. 56, fig. 4 (Red Sea).

Description.—Heavy, with a low gloss or dull; low conical, sometimes rather cylindrical, the upper sides convex then straight or slightly concave; body whorl covered from base to shoulder by numerous evenly spaced narrow spiral ridges of almost equal strength throughout; grooves with fine axial threads, sometimes the threads widely spaced; shoulder rounded to roundly angled, not distinct from spire, convex to slightly concave above; spire low, blunt, the sides straight to slightly convex; spire whorls slightly convex to slightly concave on top, with rather rude and overlapping sutures, sculptured with very fine and often indistinct spiral threads (about 3-10) and finer axial threads, whole surface sometimes appearing smooth. Body whorl white to pale yellow or pale olive, sometimes strongly tinted with pale violet; base pale to dark violet or violet-brown; usually two broad spiral bands of light or dark yellow, pale olive, or violet-tan above and below midbody; spire whitish to yellowish or olive. Aperture narrow, slightly wider anteriorly; outer lip straight or slightly convex; mouth white, yellowish, pale or dark violet, sometimes with two broad paler bands within in dark specimens. Columella long and narrow. Length 40-98mm.

Comparison.—*Conus emaciatus* is somewhat similar but with a narrower body whorl and often concave sides; the spiral ridges are weaker, do not extend to the shoulder, and are sometimes granulose; shoulder is broader, the spire often lower; and the color is often bright yellow or even orange, with a darkly stained violet base.

C. terebra is related to *C. tenuistriatus* and similar shells but easily distinguished by the very uniform spiral ridges on the body whorl, the pale color pattern without spots, and the often white mouth. *Conus virgo* and *C. coelinae* are smooth and have lower spires, usually more sharply angled shoulders, and a deeply contrasting base (in *C. virgo*) or little brownish staining (*C. coelinae*).

Variation.—Juveniles are narrower, usually darker olive or lavender in color, and often heavily stained violet. There is considerable variation in tone of the adults, from distinctly white to yellow or pale olive, with the bands sharply contrasting or very poorly defined; the posterior band may be absent, and some shells have no obvious bands. The base usually is pale violet, but not uncommonly it is not stained. The body whorl is straight-sided, slightly convex, or slightly concave. A poorly defined variant or possibly a subspecies has been recognized and seems to be distinctive although I am not sure if it is constant. This variant, *C. terebra* variety *thomasi*, seems to be restricted to northwestern Indian Ocean localities and is distinguishable by the broader and more sharply angled shoulder, straighter-sided spire, and weaker posterior spiral ridges; the mouth is violet, but violet mouths also occur in Sri Lankan specimens of the typical variety.

A curious specimen from Zanzibar could be interpreted as a hybrid between *C. terebra* and *C. virgo*; certainly it does not appear to fall within the range of variation of either species. The body whorl is deep yellow with very poorly contrasted darker spiral bands, the base being broadly and darkly stained with violet continued dorsally and into the anterior edge of the aperture; spiral and axial threads are numerous but very weak; the shoulder is bounded by a sharp ridge, is moderately low with slightly convex sides, rude and overlapping sutures, and fine but distinct spiral and axial threads on the earlier whorls.

Distribution.—Widely distributed and moderately common in shallow waters from the eastern African coast to the Red Sea and east to French Polynesia; apparently absent from Hawaii.

Synonymy.—The names listed here all belong with *C. terebra* with considerable certainty. *C. coelebs* is based on a juvenile. *C. thomasi* is either a very weak subspecies or a fairly distinct variety. In the older literature this species is usually called *C. clavus* Linnaeus, a taxon probably composite between *C. terebra* and *C. auricomus* but now usually considered unrecognizable.

Notes.—Often a very undistinguished or even ugly shell with poorly defined patterns, axial flaws and breaks, chipped lips, and persistent periostracum. *C. t.* var. *thomasi* is sometimes brighter but more expensive.

TERMINUS Lamarck, 1810

1810. *Conus terminus* Lamarck. *Ann. du Mus. Hist. Nat.* (*Paris*), 15: 426 (Asian Ocean).

Description.—Moderately heavy, with a high gloss; cylindrical, rather elongate, the upper sides convex below shoulder then straight; body whorl with about a dozen low and narrow spiral ridges clustered above the base, the rest of the whorl covered with fine spiral and axial threads and scratches, appearing satiny; shoulder relatively narrow, sharply angled to carinate, concave to deeply canaliculate above; spire low to moderate, concave, the early whorls forming a sharp, raised cone; early 3-5 whorls with small nodules, the later whorls with straight margins but sometimes rude; early whorls with about 3-5 very fine spiral ridges and axial threads, nearly flat above; later whorls progressively more concave, the spiral ridges becoming less and less distinct until by the shoulder they are mere threads or scratches. Body whorl pale violet on pinkish white, the violet tones usually as cloudy narrow to broad wavy axial flammules; numerous spiral rows of very small brownish dots over entire shell, these occasionally fusing to form short inconspicuous brown undulating dashes; usually the whorl is covered with large very irregular blotches or mottlings of pale brown to deep purple, these in the form of irregular spiral bands above and below midbody and connected by very irregular narrow and broad axial flammules; blotches commonly invading midbody area; pattern may be reduced to axial pale brown blotches above and below midbody, sometimes with another series at the shoulder; blotches under magnification commonly show short axial undulate flammules of brown within; base white; spire pinkish white with violet tones, covered with radiating lines and blotches of bright to pale reddish tan; early whorls pinkish. Aperture rather wide posteriorly, wider anteriorly; outer lip sharp, fragile, straight or very slightly convex;

mouth glossy white with pale violet tones. Columella long, narrow, twisted, bounded with a heavy rounded ridge posteriorly. Length 40-85mm.

Comparison.—*C. terminus* is very similar to *C. striatus* in shape and pattern, but it differs in lacking the distinct spiral ridges over the body whorl and the long and short dark hairlines which form the pattern of *C. striatus*; it is usually narrower and more elongated, and the spire whorls are less concave on top. It has been very badly confused with *C. gubernator* as described earlier, but it seems to be quite distinct in being cylindrical rather than conical, having the greatest width below the shoulder rather than at it, being covered with strong violet tones usually in axial flammules, having the pattern much more mottled and usually not restricted to dark blotches in or near two broad spiral salmon or rusty bands, having the spire markings bright reddish tan and usually contrasted with the body patterns, and having numerous rows of fine brown dots over the entire body whorl rather than just anteriorly. There are admittedly shells which appear intermediate, but even these are elongated and mottled with different patterns than in *C. gubernator*, as well as being covered with small dots. In addition, typical *C. gubernator* has the columella set off by only a low and often indistinct ridge, while in *C. terminus* there is an extremely heavy ridge that causes the columella to appear indented.

Variation.—Three types are recognizable, although they intergrade completely. In the most commonly seen form (from the African coast and Madagascar), the sides of the body whorl are nearly parallel or even concave, appearing very long; the shell is heavily covered from shoulder to above the base by dark purplish to reddish brown mottlings in three vague bands; and there are heavy flammules of violet obvious over much of the shell. The second type, from Madagascar and the other southern Indian Ocean islands, is somewhat broader and more shouldered although still elongate; the body whorl commonly has two ragged bands of dark purplish or blackish brown mottlings on an otherwise almost unmarked shell; and the violet flammules are weak and often spiral. The third type is typical of Sri Lanka and the neighboring islands; it is rather

Left: *C. terminus*, northern Indian Ocean form (DM) (photo Jayavad).

Right: Deep-water form of *C. terminus* from Sri Lanka, 75mm (T.C. Lan).

ovately cylindrical, usually small (although it reaches at least 75mm long), pale violet in broad spiral bands, and often with a reduced pattern of pale orange-tan and opaque white.

The spire varies considerably from individual to individual and so does the concavity of the spire whorls.

Distribution.—Apparently restricted to the Indian Ocean, typically found on the islands but also on the mainland of eastern Africa. Exact range unknown. Moderately common in shallow water.

Notes.—As mentioned earlier, I feel this species is distinct from the species here called *C. gubernator*. They look different and feel different.

Eastern African specimens may be among the most unusually colored cones, with dark purple and violet over white. Unfortunately many specimens of this species are far from gem, having heavy axial flaws, often bad spires, and sometimes indistinct and muddled patterns.

An unusual deep-water form with a constantly distorted spire occurs off Sri Lanka; it is very thin at times and very pale.

TESSULATUS Born, 1778

1778. *Conus tessulatus* Born. *Index Rerum Naturalium Musei Caesari Vindobonensis*, 1(*Testacea*): 131 (Africa). Lectotype figure (Kohn, 1964): Martini, Pl. 59, fig. 653. Commonly misspelled 'tessellatus' or 'tesselatus'.

1798. *Cucullus parvimentum* Roeding. *Museum Boltenianum:* 41 (Locality not stated). Lectotype figure (Kohn, 1975): Martini, Pl. 59, fig. 653.

1910. *Conus edaphus* Dall. *Proc. U.S. Nat'l. Mus.*, 38(1741): 223 (Off Clarion Island).

Description.—Heavy, with a high gloss; low conical, the upper sides straight to slightly convex; body whorl with about 6-8 low spiral ridges at the base, otherwise smooth; shoulder broad, roundly to narrowly angled, concave above; spire very low to moderate, sharply pointed, the sides slightly to deeply concave; early 2-3 whorls sometimes with very faint nodules; later whorls concave above, with broad, rounded margins and usually a single heavy spiral ridge at the middle separated from the margins by deep grooves; sometimes there are 1-2 weaker ridges present on some whorls; axial threads very faint. Body whorl glossy white, the base brightly stained with deep violet; entire whorl covered with large rectangular blocks of bright orange to pale red, these heaviest and often partially fused in two bands above and below midbody, sometimes forming vaguely axial flammules on the rest of the shell; in very large specimens the blocks become more irregular and can be of several different sizes and tones on the same specimen; small shells have the blocks more reddish tan or brown than in adults, but they lack spiral bands of pink above and below midbody or blotches of opaque white; shoulder and spire checkered with large and small squarish spots of orange or red, these sometimes nearly axial in alignment; early whorls white. Aperture moderately narrow, slightly widened anteriorly; outer lip sharp, straight or slightly convex; mouth white, usually with pink tones, the anterior edge stained violet. Columella narrow, rather short, sometimes with a ridge posteriorly. Length 30-65mm.

Comparison.—The conical shell with nearly straight sides and heavy covering of reddish to orange blocks of color is very distinctive. It is possible to confuse it only with *C. suturatus sandwichensis* and occasional specimens of *C. s. suturatus*, which share the unusual spire sculpture of a large median ridge. Such shells as *C. spurius*, which occasionally has orange rectangular spots, lack such sculpture on the spire.

C. tessulatus reaches a larger size than *C. s. sandwichensis*, so only juveniles are likely to be confused. In juvenile and small *C. tessulatus* there are no spiral bands of pink under the blocks, nor are there obvious spots and blotches of opaque white. While the spiral grooving in small *C. s. sandwichensis* is deep and commonly goes to the shoulder, in small *C. tessulatus* the grooving is shallow and usually obsolete after the anterior third (specimens down to 15mm). The early whorls of *C. s. sandwichensis* are usually distinctly nodulose, and even the body ridges may be granulose, while in *C. tessulatus* the early whorls are commonly smooth or at best very weakly nodulose, with the body whorl smooth. The occasional *C. s. suturatus* with traces of a *tessulatus* pattern have spiral pinkish bands visible and have convex sides.

Variation.—Juveniles are more slender than adults, with higher spires; commonly in juveniles the spots are brownish instead of reddish, while in large shells they may be pale orange. The arrangement of the color blocks is very variable even on different parts of the shell, but usually there are broad spiral bands of free or partially fused blocks above and below midbody and often axial rows of blocks as well; another partially fused band may occur below the shoulder. The spire may be heavily or sparsely checkered, or the pattern may be strong axial reddish lines in small shells. Although the base is usually heavily stained bright violet, it is occasionally almost white in large specimens.

Distribution.—This is one of four cones (*C. chaldaeus*, *C. ebraeus*, *C. orbignyi*, and *C. tessulatus*) that range in typical form from the eastern African coast to the Panamic area. It is usually common in shallow water, but is rare on the offshore islands and mainland of Mexico to Costa Rica.

Synonymy.—*C. edaphus* was compared to *C. suturatus*, and the differences given are those separating *C. suturatus* and *C. tessulatus*. I have seen *C. tessulatus* from the Panamic area (courtesy Mr. Alex Kerstitch) and see no differences from other Indo-Pacific specimens.

Notes.—This is a very brightly colored shell of sturdy construction. Smaller specimens have brighter patterns than larger ones and are usually clean, while large specimens commonly have heavy axial scars or growth marks. This is a rare or at least uncommon species in the Panamic area, but it is common elsewhere. Its absence from the Hawaiian islands is possibly explained by postulating earlier intergradation there between *C. tessulatus* and *C. suturatus*, resulting in *sandwichensis*.

TEXTILE Linnaeus, 1758

1758. *Conus textile* Linnaeus. *Systema Naturae per Regna Tria Naturae*, ed. 10, 1: 717 (Banda). Lectotype selected and figured by Kohn, 1963.

1786. *Conus undulatus* Solander, in Lightfoot. *Cat. Portland Museum:* 180 (Locality not stated). Lectotype figure (Kohn, 1964): Gualtieri, Pl. 25, fig. I.

1792. *Conus archiepiscopus* Hwass, in Bruguiere. *Cone*, in *Ency. Method., Hist. Nat. des Vers*, 1: 747 (East Indies). Lectotype selected and figured by Kohn, 1968.

1798. *Cucullus auriger* Roeding. *Museum Boltenianum:* 49 (Locality not stated). Lectotype figure (Kohn, 1975): Martini, Pl. 54, fig. 599.

1798. *Cucullus gloria maris* Roeding. *Ibid.:* 49 (Locality not stated). Lectotype figure (Kohn, 1975): Martini, Pl. 54, fig. 598. Non *Conus gloriamaris* Chemnitz, 1777.

1798. *Cucullus auratus* Roeding. *Ibid.:* 50 (Locality not stated). Lectotype figure (Kohn, 1975): Knorr, Pl. 8, fig. 3. Non *Conus auratus* Hwass, 1792.

1810. *Conus panniculus* Lamarck. *Ann. du Mus. Hist. Nat. (Paris)*, 15: 435 (Mers des grandes Indes). See *Tableau Ency. Method.*, Pl. 347, fig. 1.

1811. *Conus gloria maris* Perry. *Conchology:* Pl. 25, no. 1 (South Seas). Non *Conus gloriamaris* Chemnitz, 1777.

1811. *Conus rete-aureum* Perry. *Ibid.:* Pl. 25, no. 5 (African Seas).

1843. *Conus verriculum* Reeve. *Conchologia Iconica*, 1(*Conus*): Pl. 38, sp. 208 (Ceylon, etc.).

1845. *Conus concatenatus* Kiener. *Species gen. et icon. des coqu. viv.*, 2 (*Conus*): Pl. 110, fig. 1. 1848, *Ibid.*, 2: 362 (Locality not known).

?1881. *Conus prevosti* Sowerby iii. *Proc. Zool. Soc. (London)*, 1881: 636, pl. 56, fig. 3 (New Caledonia).

1900. *Conus Cholmondeleyi* Melvill. *J. Conch. (London)*, 9: 308, fig. (Locality unknown).

1926. *Conus eumitus* Tomlin. *Ann. Natal Mus.*, 5(3): 288, pl. 16, fig. 3 (Scottburgh and Umtwalumi, South Africa).

1931. *Darioconus textilis osullivani* Iredale. *Rec. Australian Mus.*, 18: 224, pl. 25, fig. 13 (Black Rock, Richmond River, New South Wales, Australia). 'Textilis' is a lapsus.

1932. *Conus textile* var. *ponderosa* Dautzenberg. *J. Conchyl. (Paris)*,

76: 16 (Ste. Marie, Madagascar). Non *Conus ponderosus* Brocchi, 1814, a fossil.

1937. *Conus textile* var. *abbreviata* Dautzenberg. *Mem. Mus. Roy. d'Hist. Nat. Belg.*, 2(18): 255. Given as a nomen novum for *Conus vicarius* Lamarck, a misidentification, so this is first description. Refers to *Tableau Ency. Method.*, Pl. 346, figs. 2 and 5. Non *Conus abbreviatus* Reeve, 1843.

1937. *Conus textile* var. *loman* Dautzenberg. *Ibid.*, 2(18): 257 (Senegal). Holotype figure: Adanson, 1757, Pl. 6, fig. 7.

1943. *Conus sirventi* Fenaux. *Bull. l'Inst. Oceanogr.* (*Monaco*), No. 834: 4, fig. 10 (Madagascar).

Description.—Moderately light in weight to moderately heavy, with a high gloss; ovately low conical, the upper sides gently to strongly convex; body whorl with about 5-10 low and indistinct spiral ridges at the base, these very soon obsolete; rest of whorl smooth except for very fine spiral threads and slightly heavier axial growth marks; shoulder broad, occasionally rather narrow, usually rounded but sometimes more sharply angled; shoulder usually weakly concave above; spire moderate, the sides nearly straight to slightly convex, occasionally slightly concave, sharply pointed; early 2-5 whorls with very fine nodules, these sometimes eroded, the protoconch continuous with early whorls; tops of whorls flat or slightly concave, with traces of 2-5 very weak spiral ridges and very fine curved axial threads. Body whorl white to pale bluish white, covered with rather heavy axial orange-brown to blackish or yellowish wavy to zigzag lines that usually overlap repeatedly to produce a tented pattern of usually medium and large tents; two well developed spiral bands of large rather squarish or spirally elongated orange-brown blotches above and below midbody, often with a series of smaller blotches below the shoulder; blotches contain numerous axial darker brown to blackish rather straight lines called 'textile lines'; patterns often differ from this basic type, but 'textile lines' are usually present; pattern may be very large tents or mostly small ones, but does not give the effect of snake scales; in some forms the axial lines do not overlap and a lineate pattern is produced with the blotches indistinct; shoulder and spire whitish with many axial brown lines continued from body whorl and some orange-brown spots, with a few tents; early whorls pinkish. Aperture wide, especially anteriorly; outer lip sharp, slightly convex to nearly straight; mouth glossy white to pale bluish white, rarely violet or pinkish. Columella long or very long, narrow, slightly indented from inner lip. Length 40-129mm, probably larger.

Comparison.—*C. textile* is such a familiar cone that it is seldom misidentified, but several other species are commonly confused with it. First, eliminate the *pennaceus* cones, which also have a tented pattern, by checking for the sharp, protuberant protoconch of that group and the presence of fine spiral rows of brown dots and dashes; this group lacks spire nodules. Second, eliminate those species with tented patterns that are blotched or that lack spire nodules or that have different columellas. Then eliminate species with deep spiral grooving on the anterior half of the body whorl, which is lacking in *C. textile.* The remaining species should be only *C. textile* and allies.

Conus retifer, *C. telatus*, *C. aureus*, *C. auricomus*, *C. gloriamaris*, and allies all differ greatly from *C. textile* in either general shape of the body whorl, height of spire, presence of distinct spiral ridges over the body whorl, or pattern. This should only leave a few species, of which *C. legatus* can be separated by the narrow, small shell that is bright pink with large white tents and black axial bands. *C. victoriae* has all the whorls nodulose and has secondary series of dash-like blotches. *C. dalli* has a very finely tented body whorl often with a dark mouth, broadly continuous spiral bands, and often a bright dark blue tone. *C. abbas* has the spire concave-sided, the tenting very small and regular, the spiral banding often spirally continuous, and the mouth usually distinctly violet. *C. canonicus* is usually narrower, thicker, has a pink mouth, and has the blotches reduced, often with the 'textile lines' indistinct or absent. For more details, check the individual discussions.

Variation.—This is a variable shell with many patterns that at times seem quite constant and have been given specific or varietal rank. However, all the variants can be traced back through series to the typical fully tented form with two or three rows of blotches as described above. Even the typical pattern is very variable in color, size of the blotches, length and darkness of 'textile lines', and especially the size of the tents, from rather small and crowded to very large and open. The shape varies from distinctly widely ovate to narrow with nearly straight sides, probably as an ecological factor. Typical pattern specimens are distinctly *C. textile* and have seldom been given other names.

Four groups of rather ill-defined variants can still be recognized, although all intergrade fully with typical patterns and intermediates are usually more common than the extremes. The first pattern is the one called variety *verriculum*; this is simply a shell that is mostly white, with the tents so large they are barely recognizable and look more like a weak reticulum; the blotches are very small or completely absent. Such shells are moderately common in the Indian Ocean but occur sporadically in other populations as well.

The second pattern is very minor but many people find it attractive. In this type two, the pattern is rather finely tented and the shell is distinctly toned with blue; the blotches may be small or large. Rather common in the Indian Ocean on the African coast and sometimes sold as 'scriptus', a synonym of *C. canonicus*; also sold under the name *C.* 'magnificus' on occasion.

Pattern three is uncommon but not very attractive. In this type of shell the axial lines break into short lengths of usually pale orange, isolating the reduced blotches on a mostly white background; with further reduction only the 'textile lines' may still be visible as straight axial orange or pale brown lines. Such shells have been called 'vicarius', but that name is a synonym of a similar pattern of C. *ammiralis* and should not be used; if a name must be given to this reduced pattern, variety *sirventi* seems available.

The fourth major pattern is one that is usually considered rare, but to judge from the number of specimens I have seen, it is actually moderately uncommon along the Natal-Mozambique-Zanzibar coast area of Africa, although presumably it occurs only as small populations among larger numbers of typical patterns. This pattern is commonly called *cholmondeleyi* or *eumitus*, but the first available name seems to be variety *archiepiscopus* (usually applied to *abbas* by collectors). This is a very variable pattern, but generally it consists of the axial lines being parallel or nearly so instead of overlapping to produce tents; this results in a lineate pattern which may or may not be broken by small blotches; in more extreme shells the blotches are absent, the 'textile lines' have disappeared into the other axial lines, and the lines themselves have become narrower and more numerous. Extreme shells are rare, but in such patterns the shell may seem to be uniformly orange with two broader orange spiral bands above and below midbody; under a lens the shell is seen to be finely covered with very narrow and closely spaced lines (a similar variety occurs in C. *pennaceus*, but that shell can be distinguished by the protoconch). Fully developed patterns without blotches are rare, but intermediate shells with partially lineate patterns and small blotches are not uncommon. This pattern, like the typical one, occurs in both narrow and ovate shells.

Distribution.—Widespread across the Indo-Pacific from eastern Africa to Hawaii and French Polynesia. Common in shallow water.

Synonymy.—Most of the names are minor variants or are based on early figures. Several have been described under *Variation* above. *Osullivani* is just a rather normal *textile* named without description strictly because it came from New South Wales. C. *prevosti* looks like a fairly typical C. *textile*, but the mouth is said to be yellow, which is outside the normal range of variation of this species; if the color is not a lapsus, it certainly deserves further investigation.

Notes.—Typical patterns are commonly available, but the variants are less commonly offered, sometimes at very high premiums which they probably don't deserve. Very large shells have recently been coming from the Philippines, averaging over 75mm in length. Except for often conspicuous axial growth marks that cause major pattern changes, this is usually a clean shell with a very pleasing appearance and high gloss; the lip is delicate in some populations.

It might be mentioned that the variant patterns occur most commonly in the southern Indian Ocean, at the edge of the range of the species. Presumably there is stress on the species in this area, resulting in abnormal development of patterns with growth; however, juvenile shells of the variants can be found, so this is not just a simple case of growth malformations.

This common species has been linked to several human deaths, so it must be considered very dangerous although it is a molluscivore.

THALASSIARCHUS Sowerby i, in Sowerby ii, 1834

1834. *Conus thalassiarchus* Sowerby i, in Sowerby ii. *Conchological Illustrations:* Pt. 56/57, figs. 80, 85 (Locality not stated).
1842. *Conus castrensis* A.A. Gould. *Boston J. Nat. Hist.*, 4(1): back cover (Locality unknown). 1843, *Proc. Boston Soc. Nat. Hist.*, 1: 138. Lectotype figured by Johnson, 1964, *Bull. U.S. Nat'l. Mus.*, 239.

Description.—Moderately light in weight, with a good gloss; low conical, the sides nearly straight, the body whorl rather elongate; body whorl with a few low, crowded spiral ridges at the base, otherwise smooth except for axial growth marks; shoulder broad, carinate, usually concave on top; spire nearly flat to low, the sides concave, sharply pointed; early whorls usually eroded, but sometimes as many as six nodulose whorls; margins usually straight; tops of whorls slightly to deeply concave, the margins often slightly projecting, with numerous crowded axial threads and occasional traces of 1-2 spiral threads. Body whorl white to cream or yellowish, heavily covered with very close set spiral rows of small pale brown dots, the dots connected axially by fine brown lines, giving the shell the appearance of being covered with fine, rather wavy brown lines that may form small tents; usually there are several large triangular tents scattered over the whorl, commonly with nearly clear whitish bands at midbody and below the shoulder, the bands sometimes absent or covering most of the shell and leaving only a few rows of brown dots and some axial threads; base black, the color extending over about the anterior fourth of the whorl and visible both dorsally and ventrally, apparently never reduced or absent; shoulder and spire heavily or sparsely covered with a dark brown reticulum; early whorls whitish, eroded. Aperture moderately wide, slightly widened anteriorly; outer lip sharp, straight or nearly so; mouth white to pale yellow, the anterior edge blackish. Columella very long, narrow, twisted, with a low ridge posteriorly. Length 40-100mm.

Comparison.—*C. thalassiarchus* is presumably related to *C. amadis* or *C. ammiralis*, but it really bears no resemblance to these species other than the occasional presence of a few tents. The greatest similarity is perhaps to *C. furvus*, with which it shares the dark base (much smaller in *C. furvus*), the almost unsculptured spire, and the numerous spiral rows of small brown dots. However, this is a much more profusely patterned shell than any *C. furvus*, the accent being on axial threads of brown as well as the spiral dots, giving the shell a minutely tented appearance; there should be no real confusion. Specimens with open black reticulations look quite different from the normal form but can be readily separated by the blackish base and the carinate shoulder from any *pennaceus* or *textile* cone. Some patterns of *C. thomae* may apparently look vaguely similar, but that species is covered with spiral rows of orange spots and has a pale or orange base and a higher spire.

Variation.—The tendency is apparently to reduce the pattern from uniformly lined to one with broad white bands and scattered threads and reticulations. At one extreme are completely lineate shells with two slightly darker bands above and below the base. Typical shells are heavily lineate but with scattered large clear triangles or tents and white bands at midbody and below shoulder. Extreme patterns are creamy white with widely spaced spiral rows of brown dots and a few scattered areas containing short brown threads. A rather different pattern occurs on some islands and at first glance seems quite different. This black form is heavily or loosely reticulated with dark brown or black thickened into two slightly defined bands above and below a clearer midbody area. In all patterns the spire markings may be heavy or sparse, but the base is always dark. The mouth is commonly white, but in several shells it was distinctly yellowish. The height of the spire is very variable, but commonly it is almost flat.

Distribution.—Apparently restricted to islands in the southern Philippines, where common.

Synonymy.—*C. castrensis* is a rather reticulated specimen of this species.

Notes.—This is now a common species, but it was once considered rare. Specimens over 60mm are usually rather rough and faded in appearance, with heavy axial flaws, badly eroded spires, and dull patterns; the lip is fragile. Smaller specimens, however, are also commonly eroded and have axial growth marks that are sometimes heavy. The black variety for some reason seems to be a cleaner shell; perhaps it inhabits a somewhat different habitat.

It has been hinted that different islands have distinct patterns, but commercial shells are almost always labelled with a commercial shipping point such as Sulu, Mindanao, Zamboango, etc., so it is hard to tell what patterns come from where.

THOMAE Gmelin, 1791

1791. *Conus Thomae* Gmelin. *Systema Naturae per Regna Tria Naturae*, ed. 13, 1: 3394 (Indian Ocean). Lectotype figure (Kohn, 1966): Chemnitz, Pl. 138, fig. 1331(2).

1792. *Conus Omaicus* Hwass, in Bruguiere. *Cone*, in *Ency. Method.*, *Hist. Nat. des Vers*, 1: 714 (Oma Island). Lectotype (here selected):

specimen figured by Kohn, 1968, as the holotype.
1891. *Conus Jousseaumei* Couturier. *J. Conchyl.* (*Paris*), 39: 212, pl. 4, fig. 1 (Island of Oma).

Description.—Moderately light in weight, with a high gloss; low conical, rather cylindrical, the sides straight; body whorl smooth except for a few weak spiral ridges above the base and fine axial scratches; shoulder broad, narrowly angled, concave above; spire moderate, sharply pointed, the sides straight or slightly concave; whorls weakly stepped, concave above on the later and middle whorls, nearly flat on the early whorls; tops of whorls with only traces of spiral and axial threads, no distinct sculpture; whorls sometimes weakly undulate or rude. Body whorl glossy white, the base heavily mottled with bright orange or mostly pale; body whorl covered from base to shoulder with many spiral rows of small bright orange squares that are commonly fused into long or short dashes; lines closely spaced, sometimes partially fused into rather squarish blotches; usually the lines are clustered into two or three broad spiral bands of darker orange, the bands widest below midbody, then above midbody, usually poorly developed below the shoulder; shoulder and spire white, the margins with irregular short oblique axial orange lines; few orange lines and spots on top of whorls; early whorls pink. Aperture narrow posteriorly, distinctly wider anteriorly; outer lip straight or nearly so; mouth white. Columella long, narrow, slightly oblique. Length 65-85mm.

Comparison.—*C. thomae* is closely related to *C. amadis*, but seems to differ from all variants of that species in having the pattern consisting only of spiral rows of orange spots. Although some *C. amadis* have many rows of spiral orange dashes or dots, they also have large and small brownish or orange blotches and often indications of a tented pattern. *C. amadis* is also usually wider and more carinate at the shoulder than *C. thomae*, with the spire higher and less concave on the whorls; the dark spots on the margins of the whorls of *C. amadis* are apparently absent in *C. thomae*, replaced by the short irregular lines.

Variation.—This species is apparently known from only a handfull of good specimens, but there is much variation in the placement of the rows of orange spots, whether almost uniformly spaced over the whorl from base to shoulder or clustered and partially fused into broad bands above and below midbody and below the shoulder. The height of the spire varies individually, as does the concavity of the whorls. The color in fresh specimens can apparently vary from glossy white to bright yellow. Some shells are so completely covered with orange lines that the entire shell appears orange except for a paler midbody area and the spire.

Distribution.—Known with certainty only from Misool Island, Moluccas, but apparently restricted to the Indonesian area. Very rare, apparently not taken live this century, most specimens taken by the Dutch in the nineteenth century and almost never present in private collections. Hopefully this situation will change soon.

Synonymy.—*C. thomae* and *C. omaicus* are based on a specimen and a figure of the same specimen. *C. jousseaumei* certainly seems to be this species, but it was lower-spired, yellow, the orange lines crowded and covering the entire shell heavily except more widely spaced at midbody; the whorls were said to be weakly nodulose (undulate).

Notes.—A large, rare, bright orange shell not likely to be confused with anything. The relationship with *C. amadis* is certainly close based on the shape and pattern, but so far no one has reported anything like an intermediate. Like other Indonesian species, it is still waiting to be rediscovered and be made available to collectors.

Specimens seldom change hands as they are largely in museum collections and have been there since the turn of the century or earlier. The distribution is very poorly known, of course, and collectors in the central Indo-West Pacific should keep an eye out for specimens resembling this species—perhaps it is much more widely distributed than suspected, as has been shown lately for several other rare and supposedly restricted species.

TIARATUS Sowerby i, in Sowerby ii, 1833

1833. *Conus tiaratus* Sowerby i, in Sowerby ii. *Conchological Illustrations:* Pt. 25, fig. 10 (Gallapagos).
1939. *Conus roosevelti* Bartsch and Rehder. *Smithsonian Misc. Coll.*, 98 (10): 3, pl. 1, figs. 4, 7 (Clipperton Island).

Description.—Heavy, thick, with a good gloss; low conical, the upper sides slightly convex then concave at about middle; body whorl with narrow, widely spaced spiral ridges on the anterior third of the body whorl, these sometimes continued as far as the shoulder, the ridges sometimes granulose at the base; numerous fine axial threads in the grooves and over the rest of the whorl, but spiral rows of punctations absent or greatly reduced in at least some shells; shoulder broad, sharply angled, with heavy erect coronations, concave above; spire low to moderately tall and domed, the sides usually convex or straight, blunt; all whorls heavily coronated, although these can be eroded and inconspicuous; tops of whorls slightly concave, with traces of about 2-3 weak spiral ridges and sparse axial threads. Body whorl variable from pale or dark brown to bluish, grayish, pinkish, or whitish in tone, but constantly covered with spiral rows of rather large brownish to reddish dashes alternating with white dots and dashes, the rows rather widely spaced; midbody area usually weakly marked, as is band below the shoulder; there may be solid bands of color around the whorl above and below midbody, accentuating the color of the dots and dashes; base pale; spire whitish, sparsely marked with dark spots between the coronations. Aperture rather narrow, slightly wider anteriorly; outer lip thick, slightly convex; mouth dirty brown to dirty violet, with pale bands at middle and below shoulder. Columella short, narrow, thick, with low ridge posteriorly. Length 15-31mm.

Comparison.—*C. tiaratus* is obviously closely related to *C. miliaris* and *C. abbreviatus*, with all three probably best treated as subspecies with allopatric distribution. From *C. abbreviatus* it is easily distinguished by having the dark dashes large, not small square spots, and by having a distinct midbody band in most specimens. *C. miliaris* is much harder to separate, but as a rule it has much smaller spots and dashes in more rows, is not as brightly colored nor as dark, is relatively broader at the shoulder, lacks defined spiral bands of dark color accentuating the dot and dash bands, and has the spiral rows of punctations usually well defined to the shoulder instead of apparently absent or reduced (this character is uncertain). The two could generally be distinguished with little trouble if mixed, but extremes of the two are very similar.

Sympatric coronated species usually are larger, lack a strongly dashed pattern, and seldom have a dark mouth. The shapes are usually also different, as are the textures, and there should be no identification problems.

Variation.—The spire may be flat or high and domed through distortion, with the coronations very badly eroded or clean and erect. The constriction near midbody may be deep or shallow, slightly changing the shape of the body whorl. The dashes in the pattern may be relatively small or quite large, but usually they are at least as large as the white dots. The color, as mentioned, is very variable, but colors such as brown and bluish are not uncommon. The mouth is usually distinctly dark with two pale bands, but is very variable in color and may be uniformly suffused with pale brown instead of blotched. Usually the ridges are rather restricted to at most the anterior half of the body whorl, but occasionally they reach the shoulder and are heavily granulose.

Distribution.—Moderately common in shallow water of the Panamic region from Mexico to Ecuador, usually on the offshore islands, more uncommon on the mainland.

Synonymy.—*C. roosevelti* is a small specimen with heavy ridges and small spots.

Notes.—Most of the specimens I have seen were badly flawed, with heavy axial growth marks, scars, erosion on the body whorl, and nearly formless spires. True gems are few in this species, although even real dogs have interesting patterns. Most specimens come from the Galapagos, but the species is not really that common in collections.

It would probably be a wise move to make *C. abbreviatus* and *C. tiaratus* subspecies of *C. miliaris* because of their variability and allopatric distribution. It has often been mentioned that *C. tiaratus* was very similar to *C. diadema* and *C. brunneus*, but other than they are all coronated I cannot see any similarity.

TIMORENSIS Hwass, in Bruguiere, 1792

1792. *Conus Timorensis* Hwass, in Bruguiere. *Cone*, in *Ency. Method., Hist. Nat. des Vers*, 1: 731 (East Indies). Holotype figured by Kohn, 1968.
1825. *Conus vespertinus* Sowerby i. *Cat. Tankerville:* 91, no. 2446, plate figs. 3-4 (Locality not stated).
1828. *Conus gracilis* Wood. *Index Test., Suppl.:* 8, pl. 3, fig. 3 (Timor). Non *Conus gracilis* Sowerby i, 1823.
1937. *Conus euschemon* Tomlin. *Proc. Malac. Soc. London*, 22(4): 206. Nomen novum for *Conus gracilis* Wood, 1828.

Description.—Light in weight, with a high gloss; cylindrical, the sides slightly convex posteriorly, sometimes slightly concave at middle; body whorl smooth except for a few spiral ridges at the base; shoulder narrow, roundly angled, concave above; spire low or moderate, rather blunt, the sides straight or nearly so; spire whorls nearly flat on the early whorls, becoming deeply concave anteriorly; tops of whorls with about 3-4 low spiral ridges. Body whorl glossy white with a strong pinkish tint, the base and midbody area pale; about 10-15 spiral rows of widely spaced large reddish or pinkish orange dashes alternating with opaque white spots over body whorl, plus scattered and variable opaque white axial blotches and streaks; two vague spiral bands of large and very irregular bright pink to reddish or bluish orange blotches above and below midbody, these sometimes with axial extensions crossing the midbody area and extending to the base and shoulder; blotches sometimes outlined with narrow line of darker reddish or with opaque white; spire whitish, heavily mottled and streaked with bright reddish pink; early whorls white. Aperture moderately wide posteriorly, much wider anteriorly; outer lip thin, straight or slightly concave at middle; mouth glossy white to deep pinkish with violet tones. Columella long, narrow, with a low ridge posteriorly. Length 30-45mm.

Comparison.—I can think of no other species which combines the deep pinkish to reddish orange coloration with a cylindrical, lightweight shell and large spiral dashes. This species is probably related to *C. magus* and *C. bullatus*, but it is very distinct.

Variation.—Concavity of the spire whorls and shoulder varies from just barely concave to almost canaliculate. The blotches are very variable in shape and sharpness, sometimes appearing more like clouds than actual blotches. Opaque white may be prominent in the pattern or virtually absent, while the dashes may be relatively larger and bright or smaller and less distinct. The sculpture of the base varies somewhat in heaviness, but seldom extends more than one-third up the whorl.

Distribution.—Uncommon to rare in moderately deep water of the Indonesian area west at least to Mauritius. The limits of the range are poorly known.

Synonymy.—The figures of *C. vespertinus* and *C. gracilis* leave no doubt they are this species.

Notes.—This is a gorgeous pastel species which is seldom available in live-taken condition. Dead specimens are much faded, dull, with the blotches almost absent, the dashes very pale and indistinct, and much of the pink and reddish tones gone with the gloss. The lip and for that matter the entire shell are very fragile and easily broken. The few specimens I have seen came from Mauritius, but it occurs at least east to Papua.

TINIANUS Hwass, in Bruguiere, 1792

1792. *Conus Tinianus* Hwass, in Bruguiere. *Cone*, in *Ency. Method., Hist. Nat. des Vers*, 1: 713 (Tinian). Holotype figured by Kohn, 1968.
1810. *Conus aurora* Lamarck. *Ann. du Mus. Hist. Nat. (Paris)*, 15: 423 (Locality unknown). Holotype figured by Kiener, Pl. 45, fig. 1.
1817. *Conus rosaceus* Dillwyn. *Descr. Catalogue Recent Shells*, 1: 433 (East Indies). Based on Chemnitz, Pl. 181, figs. 1756-1757.
1833. *Conus tenuis* Sowerby i, in Sowerby ii. *Conchological Illustrations:* Pt. 24, fig. 7 (Locality not stated).
1844. *Conus lautus* Reeve. *Conchologia Iconica*, 1(*Conus*): Pl. 46, sp. 255 (Locality unknown).
1848. *Conus Loveni* Krauss. *Sudaf. Moll.:* 131, pl. 6, fig. 25 (Cape and Natal).
1865. *Conus secutor* Crosse. *J. Conchyl. (Paris)*, 13: 303, pl. 9, fig. 3 (Locality unknown). Misspelled 'scutor' by Hedley, 1913, *Proc. Linn. Soc. NSW*, 38: 308.
1889. *Conus fulvus* Sowerby iii. *J. Conch. (London)*, 6: 10, pl. 1, fig. 1 (South Africa). Non *Conus fulvus* Schroeter, 1803.
1915. *Conus lavendulus* Bartsch. *Bull. U.S. Nat'l. Mus.*, 91: 12, pl. 1, fig. 12 (Port Alfred, South Africa).
1915. *Conus alfredensis* Bartsch. *Ibid.*, 91: 13, pl. 1, fig. 12 (Port Alfred, South Africa).
1932. *Conus kraussi* Turton. *Mar. Shells Pt. Alfred:* 13, pl. 3, no. 108 (Pt. Alfred, South Africa).
1932. *Conus lavendulus* var. *approximata* Turton. *Ibid.:* 13 (Pt. Alfred, South Africa).

Description.—Moderately light in weight, with a low gloss or dull finish; low conical, the sides gently convex, rarely almost straight; body whorl with weak, narrow spiral ridges clustered at the base, rest of whorl smooth except for often heavy spiral and axial threads; shoulder roundly angled, often convex on top, with traces of 3-5 weak undulating spiral ridges and some axial threads; spire low, blunt, the sides straight or slightly convex; whorls with rude sutures, usually slightly convex on top, with 3-5 low narrow spiral undulating ridges and axial threads, still visible in beach specimens; whorls often weakly stepped. Body whorl pale straw, covered from base to shoulder with narrow and indistinct spiral lines of brown and white dashes and dots; usually three spiral rows of large irregular dark brown blotches, below shoulder and above and below midbody area; base pale; with erosion the blotches smear into broad spiral bands and then merge to produce a unicolored shell with a pale midbody band spotted with white; colors of eroded specimens often pale brown, deep red, pink, orange, yellow, etc.; spire straw, with scattered large and small brown blotches; early whorls usually pinkish. Aperture moderately wide, especially anteriorly; outer lip thin, convex to nearly straight; mouth dark violet posteriorly, pale violet anteriorly, becoming dark brown to dark red in eroded specimens. Columella long and narrow, partially free and projecting into the mouth posteriorly. Length 30-50mm.

Comparison.—The blotched and indistinct pattern lacking regular dark and light spots on the shoulder should eliminate most *C. algoensis* that are still in decent condition, while the pattern lacking strong dashes or heavy axial mottling should eliminate *C. guineensis*. The best character seems to be the partially projecting talon of the columella, which seems to be restricted in this group to *C. tinianus*, although it may not be present in all specimens. Any uniformly colored bright cone of this general shape and texture from South Africa is likely to be this species. Certainly it is almost impossible to tell for sure that beach cones from this area are *not C. tinianus*.

Variation.—As indicated in the description, the color and pattern change drastically with erosion, the brightest specimens apparently being the most eroded. Live-taken specimens are dull straw or tan with three spiral rows of dark brown blotches and indistinct spiral dot and dash rows. Solid yellowish brown shells with the hint of a blotchy midbody band probably belong with *C. tinianus*, but apparently fresh specimens with this pattern have been seen.

Some specimens are rather narrow with nearly straight sides, others very inflated; the spire may be flat to quite high with the whorls stepped and distinctly convex. The sculpture of spiral ridges is usually visible on the spire of all but the most eroded specimens, but it is sometimes just as fine as in *C. algoensis*. The status of all South African shells of this general type is doubtful, although the three species (*tinianus*, *algoensis*, *guineensis*) usually found seem distinct when live-taken.

Distribution.—Restricted to the southern coasts of South Africa, but found much further northeast as beach shells. Live-taken in shallow water. Common.

Synonymy.—Most of the synonymy is probably correct, but as most of the names are based on beach shells and separated by details of pattern and shape or by color, they couldn't be certainly placed with the right species even if the types were available. *C. aurora* and *C. rosaceus* are more familiar names than *C. tinianus*, but all names seem to refer to various eroded colors of this species. *C. lautus* seems to be better placed here than in *C. guineensis*. Bartsch's names are certainly this species. I have not seen the original descriptions of the names by Krauss and Turton.

Notes.—This is one species where very bad beach specimens are much more attractive than live-taken specimens and usually sell for a better price. Live-taken specimens are about as ugly as a cone can get, with persistent periostracum, bad axial growth marks, scars, broken lip, and often a distorted spire. Terminal beach specimens (sold as *rosaceus*), on the other hand, are bright red or reddish orange with small brown spots in the midbody area. Gem specimens probably do not exist in this species, or they are certainly rare.

I have seen several *C. algoensis* with bad patterns mixed with *C. tinianus*, as well as some very nice *C. guineensis*. Not all specimens even in selected lots of beach specimens can be identified.

TORNATUS Sowerby i, in Sowerby ii, 1833

1833. *Conus tornatus* Sowerby i, in Sowerby ii. *Conchological Illustrations:* Pt. 29, fig. 25 (Locality not stated).

1879. *Conus catenatus* Sowerby iii. *Proc. Zool. Soc.* (*London*), 1878: 796, fig. 3 (?Panama). Non *Conus catenatus* Sowerby i, 1850, a fossil.

1887. *Conus concatenatus* Sowerby iii. *Thesaurus Conchyliorum*, 5 (*Conus* Suppl.): 249. Misspelling of *Conus catenatus* Sowerby iii. Non *Conus concatenatus* Kiener, 1845.

1937. *Conus desmotus* Tomlin. *Proc. Malac. Soc. London*, 22(4): 206. Nomen novum for *Conus catenatus* Sowerby iii, 1879.

Description.—Moderately light in weight, with a good gloss; biconical, the sides nearly straight; body whorl with broad flat ribs anteriorly, separated by rather narrow spiral grooves containing fine axial threads; rest of body whorl with widely spaced shallow or deep spiral grooves or rounded spiral ridges, commonly with widely spaced spiral rows of large granules to shoulder; shoulder broad, carinate, sometimes projecting, nearly flat to concave above; spire tall, sharply pointed, the sides straight; whorls commonly stepped, often with projecting margins, the margins carinate; whorls nearly flat to deeply concave, with fine curved axial threads and sometimes traces of a single spiral thread. Body whorl pale to dark straw, often with bluish tones, and usually with heavy irregular axial flammules of deep brown to black; entire body whorl covered with spiral rows of squarish deep brown to blackish spots, the spots seldom fused or irregular; shoulder and spire with large oval or axially elongated blackish or dark brown blotches regularly spaced over surface; earlier whorls often with small regular dark brown spots at the margins; early whorls whitish. Aperture moderately narrow, somewhat widened anteriorly; outer lip thin, straight; mouth white to bluish white or grayish. Columella short and very narrow. Length 20-45mm.

Comparison.—*C. tornatus* need be compared only with *C. ximenes*, which is practically identical in shape and pattern. Although the radular teeth are said to be very different, conchologically the two species are extremely close and separated only by details. In *C. ximenes* the shoulder is rounded, not distinctly carinate even in juveniles; the spire whorls are never distinctly concave with projecting and carinate margins as in many *C. tornatus;* and the body whorl is seldom covered with either grooves to the shoulder or granulose spiral ridges in adult shells. Adult *C. ximenes* tend to have more and smaller spots than *C. tornatus* of equal size, but this is variable. Both species have a very shallow anal notch. *C. ximenes* usually is found intertidally, while *C. tornatus* is often found in moderately deep water.

Variation.—First, the shape of the spire is variable. Typically the whorls are concave above, with projecting carinate margins; the shoulder is also concave and is strongly carinate; however, commonly the spire whorls are almost flat, the carinate margins slightly or not projecting. The shoulder is constantly carinate and usually separated by a groove from the body whorl. The basal ridges are broad and raised on both sides, apparently made from fused narrower ridges; when the grooves extend more posteriorly than the anterior third of the whorl, they are usually widely separated by angular ridges. The grooves may abruptly stop and be replaced by narrow spiral ridges. Any of this sculpture may have heavy granules developed or absent.

The color varies from white to bluish gray, with or without weak or heavy axial dark flammules that may form bands on each side of the mid-body area or be continued from shoulder to base. The spots are large or relatively small, but apparently always present in distinct rows. The spire is usually very dark brown with a regular pattern.

Distribution.—Moderately common in moderately deep water of the Panamic region from the Gulf of California to Ecuador, including some offshore islands.

Synonymy.—*C. catenatus* is this species and fairly typical although lightly patterned; the other names are replacements.

Notes.—Nybbaken (1970, *Amer. Mus. Novitates*, No. 2414: 1-29) indicates that the radular teeth of *C. tornatus* and *C. ximenes* are extremely different. Conchologically, however, there are no differences between *C. ximenes* and *C. tornatus* that would not be allowable between intertidal and offshore ecotypes of a single species. I think that it surely cannot be coincidence that both these species have very shallow anal notches and virtually identical color patterns; juveniles of *C. ximenes* are spirally grooved to the shoulder on occasion. This situation is a good example of the flaws in using radular teeth to show relationships. I prefer conchological characters and believe they are more reliable and useful in almost any situation.

TRIBBLEI Walls, 1977

1977. *Conus tribblei* Walls. *The Pariah*, No. 1: 2, figure (China Sea off Taiwan).

Holotype: Delaware Mus. Nat. Hist.; 62.3mm; China Sea off Taiwan.
Paratypes: Collections of George Gundaker, Alex Kerstitch, and the author.
Diagnosis: Similar to *C. sugimotonis* in having smooth spire margins, but colored with brown dashes and bands; differs from *C. recluzianus* in lacking undulate shoulder and spire whorls and in details of pattern. 60-120mm length.

Description.—Heavy, with a good gloss; obconical to low conical, elongate, sides of body whorl parallel for a short distance anterior to shoulder then straight or slightly concave to a narrow base; body whorl with about 12 heavy rounded spiral ridges separated by deeply punctate grooves of variable width; ribs soon become obsolete, but the punctate grooves continue to midbody in small adults and often reappear just anterior to shoulder as 2-3 isolated spiral rows; posterior part of body whorl often with spiral ridges and grooves still visible, but widely scattered; spiral rows of small granules sometimes developed in three to many rows at base, sometimes extending to shoulder; axial scratches and threads dense; shoulder strongly angled, not undulate, carinate; spire low, usually the later whorls forming a nearly flat plate with early whorls projecting as a small cone; protoconch 2 whorls, glossy brown; first 3-5 whorls weakly nodulose, next 2-3 undulate, remaining whorls with carinate, non-undulating margins; tops of whorls weakly concave or flat, with about 4-5 (sometimes 3) strong narrow spiral ridges crossed by widely spaced curved axial threads. Body whorl white to cream, with a basic pattern of two spiral bands above and below midbody area of tan to reddish brown blotches which may be: 1) fused into nearly continuous broad spiral bands; 2) extended axially with heavy axial flammules sometimes covering most of body whorl; or 3) greatly reduced, especially the anterior band; combinations of these conditions may occur on a single shell, with the pattern usually better developed and more distinct in smaller specimens; visible in brown blotches and sometimes in white background are a few to many short brown dashes arranged in spiral and sometimes axial rows, the dashes most conspicuous in specimens under 65mm in length and often obsolete in shells over 75mm; spire whitish, with sparse pale brown to reddish brown squarish spots sometimes in indistinct axial rows; early whorls pale brown. Aperture narrow, nearly uniform in width or slightly expanded anteriorly; outer lip thin, slightly concave at middle, following outline of body whorl; mouth white to pale bluish white. Columella very narrow, indistinct, sometimes with a low posterior ridge. Length 60-120mm.

Comparison.—*C. tribblei* is very closely related to *C. sugimotonis*, which is a pure white shell without pattern at any size, somewhat wider at the shoulder than *C. tribblei*, giving it a different appearance, and with weaker anterior sculpture. *C. recluzianus* has been badly confused with *C. tribblei*, but it has the spire whorls and shoulder with distinctly undulate margins; the anterior spiral grooves are not as deeply punctate and are generally absent above the basal area; the typical pattern is of two broad spiral brown bands with small square spots often visible above the midbody area; the anterior band is usually strong; and *C. recluzianus* often has bright colors, violet predominating. *C. urashimanus* is darker brown overall with undulating shoulder and spire whorls. Other species with similar shapes usually differ greatly in pattern and spire sculpture.

Conus bayani and *C. voluminalis* are sometimes confused with *C. tribblei*, but these species have tall early whorls, weak or absent spiral sculpture on the spire whorls, very weak body sculpture, and a much lighter weight and smoother texture.

Variation.—Although constant in shape, *C. tribblei* is very variable in color pattern. As a general rule, the pattern becomes sparser and less distinct with growth, the spiral dashes and the anterior spiral brown band usually being the first elements to be lost. With few exceptions the posterior spiral band is developed as rather dark axial blotches often with extensions which may reach the shoulder and the base. Small shells may be entirely covered with rows of spiral brown dashes, while shells over 75mm in length may have no dashes visible. Basal sculpture is constant, although becoming weaker with growth; punctate grooves and traces of spiral ridges above midbody are more likely to be found in small specimens and may be obsolete in large shells. Although not usually granulose, occasional specimens may have 3 or more spiral rows of small granules in the basal area, and rare specimens may have strong granulations developed to the shoulder.

Sympatric with *C. tribblei* off Taiwan is a very similar cone differing mostly by being broader in width relative to length, somewhat heavier,

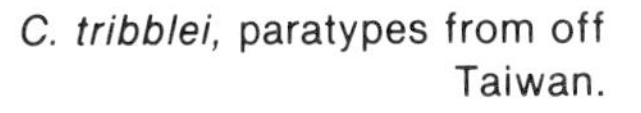

C. tribblei, paratypes from off Taiwan.

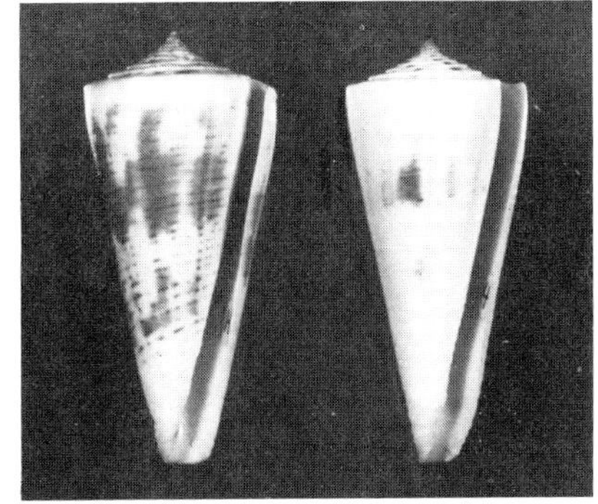

and having a reduced number of spiral dashes in young shells. This cone is for the moment considered a variant of *C. tribblei* but may be distinct. See Lopez, 1970, *Hawaiian Shell News*, 19(5): 8 (his *recluzianus* = *tribblei*, *bayani* = *tribblei* variety).

Distribution.—Moderately common in deep water from the China Sea off Taiwan and in the Philippines; I have recently seen a specimen from the Solomons, and shells resembling *C. tribblei* have been reported from Queensland, Australia.

Notes.—There should be no problem distinguishing *C. tribblei* from *C. recluzianus*, as collectors and dealers have long been certain that the two were distinct. The absence of undulate shoulders and spire margins and the different patterns are quite distinct.

All-white specimens are collected occasionally, but these would not be confused with *C. sugimotonis* because of the narrower shell and heavier sculpture. This species is often badly flawed, with heavy axial growth marks and scars; the lip is usually filed. Many specimens have an unusual 'twist' at the extreme anterior end of the body that stands out at almost a right angle to the axis of the shell.

Named in honor of my favorite cat, Tribble, who in turn was named after the featureless balls of fur made famous in the science fiction play "The Trouble with Tribbles" by David Gerrold, from the TV show *Star Trek*.

TRIGONUS Reeve, 1848

1848. *Conus trigonus* Reeve. *Conchologia Iconica*, 1(*Conus* Suppl.): Pl. 3, sp. 286 (Locality unknown).

Description.—Light in weight, with a good gloss; low conical or obconical, the upper sides slightly convex to nearly straight; body whorl with about a dozen low and closely spaced spiral ridges anteriorly, these separated by shallow grooves with fine axial threads; rest of body whorl with sometimes distinct spiral threads and fine axial threads and scratches, almost smooth posteriorly; shoulder roundly angled to sharp, flat or slightly concave on top; spire depressed, flat, or low convex, blunt; sutures between whorls deep and broad; tops of whorls slightly concave, with about five strong spiral ridges crossed by fine, scattered axial threads. Body whorl white to cream, with two broad spiral bands of bright orange to dark brown above and below midbody area, the bands containing numerous spiral rows of dark and light long dashes, these usually contrasted; base white; midbody area white, with spiral rows of small brown dots and dashes, sometimes arranged into irregular axial flammules that are seldom complete; a narrow white band below shoulder, also containing indistinct and broken short brown flammules; spire white, heavily covered with radiating curved brown or orange lines and blotches; early whorls white. Aperture moderately wide, uniform in width; outer lip very thin, straight or slightly convex; mouth white, sometimes with pale violet tones. Columella long, narrow, with a large free talon posteriorly, very obvious. Length 35-70mm.

Comparison.—The thin shell with a distinct talon on the columella is very similar in shape to *C. angasi* and *C. advertex*, with a pattern easily derived from theirs. *C. trigonus* differs in having two broad spiral bands of orange or brown with light and dark dashes instead of being covered with large and small squares and rectangles on a white ground; *C. angasi* and *C. advertex* lack flammules of any type in the midbody area, which is not contrasted. The sculpture is also weaker and less punctate. From unrelated species it differs in the very low spire, the partially free columella, and the three distinct white bands.

Variation.—In orange specimens the dashes are brown hairlines, sometimes nearly continuous and seldom with distinct white dots between. Dark brown specimens seem to have broader and more smeared dashes, often with distinct white dots between; there may appear to be spiral white dashes in the pattern as well. The flammules in the midbody and shoulder bands are poorly developed from partially fused or consolidated small dots and dashes; commonly there are fairly distinct squarish darker brown spots at the posterior margin of the midbody band. In some specimens there are distinct heavy spiral threads to the shoulder, but most specimens are nearly smooth above midbody. Smaller specimens seem to have nearly straight sides and flat or depressed spires, while large specimens are more convex both on the sides and spire; the supposedly specific differences between *C. angasi* and *C. advertex* occur as normal variation in this species.

Distribution.—Apparently restricted to Western Australia and Northern Territory, Australia, where moderately uncommon in moderately deep water. I have seen specimens labelled Cape Moreton, Queensland, but presumably this data is incorrect.

Notes.—Although reasonably common in nice dead-dredged condition, live-taken specimens are very uncommon. The lip is very fragile and usually badly filed in dead or live specimens.

The free columella is very distinct in the specimens I have seen and is not likely to be missed. Orange specimens are more attractive than brown specimens because the dashes are more distinct. Large specimens may have many partially healed axial breaks and growth flaws.

TULIPA Linnaeus, 1758

1758. *Conus tulipa* Linnaeus. *Systema Naturae per Regna Tria Naturae*, ed. 10, 1: 717 (Locality not stated). Lectotype selected and figured by Kohn, 1963.

1798. *Cucullus purpureus* Roeding. *Museum Boltenianum:* 47 (Locality

not stated). See Kohn, 1975.
1868. *Conus* (*Chelyconus*) *borbonicus* H. Adams. *Proc. Zool. Soc.* (*London*), 1868: 288, pl. 28, fig. 1 (Isle of Bourbon).

Description.—Very light in weight, with a low gloss, not fragile; ovately cylindrical, the sides smoothly convex; body whorl with a few indistinct spiral ridges at the base, otherwise smooth, although axial threads may be common; shoulder narrow, rounded, weakly undulate, slightly concave on top with fine spiral and axial threads; spire low, bluntly pointed, the sides nearly straight; early whorls nearly flat to slightly concave above, later whorls more concave and progressively wider at each whorl; whorls with distinct nodules at first, then becoming undulate, later whorls almost smooth; tops of whorls with about 5-10 narrow spiral ridges and widely spaced axial threads, the sculpture becoming finer on later whorls. Body whorl reddish violet to whitish violet, with two or three spiral rows of large and irregular brownish blotches often connected axially and spirally by narrower flammules; whorl covered from base to shoulder with many closely spaced spiral lines of large and small dark brown and white dots and dashes, these sometimes forming short undulating hairlines near major growth flaws; spire and shoulder pale violet, the margins with small and rather regular dark brown spots, only a few brown streaks on top; early whorls pinkish. Aperture wide posteriorly, much wider anteriorly, greatest width more than half greatest width of body whorl proper; outer lip thin, sharp, evenly convex; mouth pale violet. Columella very long, indented, forming a very long and narrow basal area. Length 30-84mm.

Comparison.—*C. tulipa* is closely related to *C. geographus* and *C. obscurus* because of its shape and sculpture as well as texture. There is a similarity to *C. cuvieri* also, but that species is readily distinguished by being more angled at the shoulder and not having undulate or low coronate later spire whorls as well as being brownish instead of violet.

C. geographus of sizes similar to *C. tulipa* are narrower with straighter sides and a pattern including remnants of a tented or at least crosshatched pattern; the spire whorls are distinctly coronated with erect brownish coronations, the tops of the whorls pinkish to pale tan. *C. obscurus* is somewhat more similar in shape and texture to *C. tulipa*, but it is also narrower, as well as being smaller when adult; the early whorls are nodulose, the later whorls smooth or nearly so; the tops of the spire whorls are mostly dark brown with small pale areas, not mostly pale violet; juveniles are bright orange with dark spire whorls. The juveniles of *C. tulipa* are orange-violet with violet spire whorls having distinct dark dots at the margins.

Variation.—Adult *C. tulipa* differ mostly in the convexity of the sides, the strength of undulation of the shoulder (very weakly undulate to distinctly low coronate), and the development of the large brownish blotches, which may be almost absent or very large and fused spirally and axially to cover most of the body whorl. The many small dots and dashes in the background are usually distinct, though in some patterns they may be visible over only part of the shell, while in other patterns they may be fused to form rather long undulating brownish hairlines, especially near growth flaws. Juveniles are orange-violet, the blotches sometimes absent, and are covered with widely spaced rows of relatively large dark dashes which are very regular; the margins of the spire whorls are very regularly marked with dark spots still visible in adult shells.

Distribution.—Widely distributed through the Indo-Pacific from eastern Africa to French Polynesia but apparently absent from Hawaii. Moderately common in shallow water.

Synonymy.—Juveniles are seldom seen and look very different from adults, although their shape and pattern are easily visible in the early whorls of good adult shells. They were named *C. borbonicus*, a name which for some reason refuses to die completely, sometimes still being used for any *C. tulipa* of small size and bright pattern, especially those from the Indian Ocean. There is little doubt that *borbonicus* is just a juvenile of *tulipa*; the name should no longer be used.

Notes.—Like *C. obscurus*, this species is subject to bad axial breaks and growth flaws which result in major pattern changes. The tone varies from distinctly reddish violet to distinctly whitish with the violet reduced. Juveniles seem to be uncommon.

This is a venomous cone that could easily cause human fatalities (several cases of human death have already been recorded from the bite of this species), so it should be handled very carefully or not at all.

TYPHON Kilburn, 1975

1975. *Conus* (*Leptoconus*) *typhon* Kilburn. *Durban Museum Novitates*, 10(15): 213, fig. 1; pl. 1, figs. a-d (Off Ilha Chiloane, Mocambique).

Description.—Moderately light in weight, with a high gloss; low conical, the upper sides slightly convex, almost straight; body whorl with a few low spiral ridges at the base, otherwise almost smooth; shoulder broad, sharply angled, slightly concave on top; spire almost flat to low, sharply pointed, the sides concave; early 1-2 whorls very weakly nodulose, later whorls slightly concave above with 2-4 weak spiral ridges and scattered usually weak curved axial threads, the margins sometimes appearing slightly irregular. Body whorl white to cream, sometimes tinted with pale yellow or violet tones, covered from above base to shoulder with numerous fine brown spiral hairlines, these usually continuous or only occasionally broken; base white, unmarked; midbody area usually light; sometimes with two broad spiral bands of yellowish to pale tan above and below midbody area; spiral hairlines sometimes absent; spire and shoulder whitish with heavy curved axial blotches of pale and dark brown; early whorls whitish. Aperture moderately wide, slightly wider anteriorly; outer lip thin, straight or slightly concave at middle; mouth white, with violet tones sometimes darker posteriorly. Columella narrow. Length 30-49mm.

Comparison.—The low spire and striped pattern resemble only two species at all closely. *C. caillaudii* is narrower and more elongate in the body whorl, with the spire strongly undulate; the color is bright yellow with the spiral lines more widely spaced. Some Natal specimens of *C. typhon* bear a superficial resemblance to *C. caillaudii*, but this could be accidental. The closest relative seems to be the Australian *C. nielsenae*, which is very similar in general appearance. However, many *C. nielsenae* have the base tan, not white, and have the spiral ridges of the spire whorls very weak or absent; there are never two broad slightly contrasted spiral bands above and below midbody; and the axial threads of the spire in *C. nielsenae* are more crowded in at least some populations. The two species are very closely related.

Variation.—This shell varies in tone from white or pale golden brown to distinctly yellowish, pinkish, or violet. The two broad spiral bands above and below midbody are seldom well defined, but they may be the only pattern present in the shell. The hairlines are usually strong and heavy although narrow, and are seldom broken except near axial flaws; they are occasionally entirely absent, however. The base is usually unmarked white but in beach specimens may be pale yellowish.

The spiral ridges on the spire whorls are usually distinct though weak, but the axial threads vary from weak and sparse to numerous and heavy enough to produce an irregular appearance at the margins. The spire is always low, sometimes flat. The nodules on the early whorls may be absent through even minor erosion.

Distribution.—Uncommonly dredged in moderately deep water off the Natal and Mozambique coasts of South Africa; probably more widely distributed.

Notes.—Specimens are often heavily axially flawed, with major disruptions of the pattern. Specimens without hairlines are seldom collected or sold. *C. typhon* appears to be not uncommon in some areas of the known range, at least to judge from the number of specimens on the market.

URASHIMANUS Kuroda and Ito, 1961

1961. *Conus* (*Rhizoconus*) *urashimanus* Kuroda and Ito. *Venus*, 21(3): 259, pl. 17, figs. 13-14 (Nada, Gobo, Kii Peninsula, Japan).

Description.—Moderately heavy, with a low gloss; obconical, the upper sides nearly parallel for a short distance below the shoulder then straight or concave and tapered to a narrow base; body whorl with about 6-10 low rounded spiral ridges separated by narrow grooves with fine axial threads; rest of whorl with a rather waxy texture, covered with many spiral and axial threads, some of the spiral threads appearing as narrow broken ridges posteriorly; shoulder sharply angled, strongly undulate or low coronate, slightly concave above; spire flat or nearly so, the early whorls forming a small sharp cone; tops of whorls slightly concave, the margins strongly undulate, with about 4-5 narrow but strong spiral ridges covered by heavy axial curved threads so distinctly cancellate; first 1-3 whorls weakly nodulose. Body whorl pale tan with a violet tone, with two broad bands of dark tan above and below midbody; there are commonly other spiral bands of slightly different tan or brown tones covering the rest of the whorl and overlapping, producing a mostly brown shell; base paler; often a pale midbody band is visible, but it is seldom well defined; there may be traces of many spiral rows of brown dots or dashes over much of the shell in small specimens; spire white to pale straw, heavily covered with curved blotches and lines of dark brown so spire looks mostly dark and not with squarish spots. Aperture moderately narrow, slightly wider anteriorly; outer lip sharp, straight or slightly concave at the middle; mouth dark violet or violet gray, fading to paler violet after death. Columella long, very narrow. Length 40-75mm.

Comparison.—*C. urashimanus* is very similar to the sympatric *C. recluzianus* but seems sufficiently distinct to recognize as a full species. The spire is usually flatter than in *C. recluzianus*, with much heavier cancellate sculpture and heavier dark brown curved markings than in *C. recluzianus*. The body whorl is more uniformly tan or brown, with the bands above and below midbody often lost in the other brown bands that are commonly present; small squarish spots above the midbody area seem to be normally absent, while the spiral rows of small dots or dashes are very inconspicuous. The mouth is usually distinctly dark violet to pale violet, not white with a violet tone. *C. urashimanus* has a distinctly waxy texture usually absent from *C. recluzianus* taken at the same locality.

From *C. tribblei*, *C. urashimanus* is easily distinguished by the heavily undulate or low coronate whorls and shoulder, as well as the more uniform brown color, the spiral rather than axial brown pattern, the heavily marked spire, and the more distinctly cancellate spire sculpture.

Variation.—As indicated, there is considerable variation in the tones of brown or tan and their placement on the shell. Normally this species appears mostly dark and pale brown in even spiral bands that overlap occasionally, the spiral rows of brown dots and dashes barely apparent through the darker brown bands. This species seems to be more constant in shape, sculpture, and pattern than its relatives, but some specimens from the Philippines assigned to this species are rather slender, have bright violet tones under the brown, and have very large rounded coronations instead of the usual undulations or low coronations; they are still waxy shells, however.

Distribution.—*C. urashimanus* is fairly commonly dredged in deep water of southern Japan; I have not seen it from Taiwan, but have seen a few specimens from deep water in the Philippines.

Notes.—The relation of this species to *C. recluzianus* is very close, with occasional specimens hard to place, but it seems best to retain it as a full species for the moment. It is a cleaner shell than its allies and seemingly less subject to bad patterns and heavy flaws. The violet tones fade badly after collecting.

VARIEGATUS Kiener, 1845

1845. *Conus variegatus* Kiener. *Species gen. et icon. des coqu. viv.*, 2 (*Conus*): Pl. 106, figs. 1-1a. 1849, *Ibid.*, 2: 261 (Locality unknown).
1845. *Conus africanus* Kiener. *Ibid.*, 2(*Conus*): Pl. 104, fig. 2. 1849, *Ibid.*, 2: 260 (Coast of Guinea).
1845. *Conus guttatus* Kiener. *Ibid.*, 2(*Conus*): Pl. 105, fig. 4. 1849, *Ibid.*, 2: 259 (Locality unknown). Non *Conus guttatus* (Roeding, 1798).
1883. *Conus figulinus* var. *chytreus* Melvill, in Tryon. *Manual of Conchology*, 6(1): 17, pl. 27, fig. 1 (Locality not stated).
1905. *Conus fuscolineatus* Sowerby iii. *Proc. Malac. Soc. London*, 6: 282, fig. 6 (Sierra Leone).
1957. *Conus lucirensis* Paes da France. *Trab. Miss. Biol. Marit.*, No. 13: 79, pl. 1, figs. 5-6; pl. 2 (Baia de Lucira, Angola).
1975. *Conus cepasi* Trovao. *Centro Port. Activ. Subaq.*, 4(1): 3 of separate, pl. 1, fig. 1 (Angola, 14°27'S, 12°20'E).

Description.—Heavy, thick, with a low gloss or dull surface; pyriform, the upper sides strongly convex; body whorl with about 4-8 narrow and weak spiral ridges, widely spaced, above the base, with many fine spiral and axial threads and flaws over the rest of the whorl; shoulder broad, rounded, not distinct from spire, with fine spiral threads over the convex surface; spire low, rounded, convex, usually heavily eroded; sutures rude; tops of whorls slightly convex, with 3-5 indistinct spiral ridges or heavy threads and scattered axial threads. Body whorl white or cream, covered with numerous spiral rows of small dots or brownish hairlines, narrowly to widely spaced, often broken into long or short dashes; over this may develop broad brown spiral bands that obscure some of the lines and result in irregular patterns of narrow and wide spiral lines; occasional patterns are mostly brown except for small irregular white blotches near midbody; shoulder usually mostly dark brown or reddish brown, sometimes with just some axial brown lines or narrow blotches; spire whitish, with few or many brownish blotches, usually eroded white. Aperture moderately narrow, a bit wider anteriorly; outer lip evenly convex, thin; mouth pure white, white with small violet blotches, or white with large brownish blotches. Columella short, narrow, often slightly indented and with a heavy ridge posteriorly. Length 20-40mm.

Comparison.—*C. variegatus* is very similar to *C. bulbus*, which is about the only species which is likely to be confused with it. The two species are very similar and probably hybridize, but for the moment they are kept separate as the variations and tendencies of the patterns are quite different. In *C. bulbus* the patterns are basically axial, sometimes vaguely tented, while in *C. variegatus* the patterns are all modifications of spiral rows of dashes or narrow brown hairlines, sometimes heavily obscured by brown areas, not axial or tented. *C. bulbus* is also somewhat thinner and narrower than *C. variegatus*, which is a thick, strongly convex shell.

C. variegatus with strongly spirally lineate patterns were twice confused with *C. figulinus*, but it is doubtful if anyone would do this today.

Variation.—Although constant in shape and usually in color (white or cream and brown or reddish brown), the pattern of this species is very variable. However, even the most extreme patterns can be easily connected with more typical patterns with little effort, and there is no reason to dignify any minor variants with names. The basic pattern is probably one of many rows of dashes of various sizes; these commonly unite into many continuous spiral hairlines that cover the entire whorl, but the dashes may also be broken into smaller dots of very regular size and arrangement; with every other line becoming obsolete, a pattern of widely spaced hairlines is developed. In some patterns the area between lines becomes suffused with brown, producing various patterns combining hairlines and broad bands. With further addition of brown the entire shell may become brown except for squarish white areas near midbody; further development of this pattern results in brown shells with irregular white blotches usually heaviest near midbody.

Left: *C. cepasi* Trovao = *C. variegatus* (photo Clover).

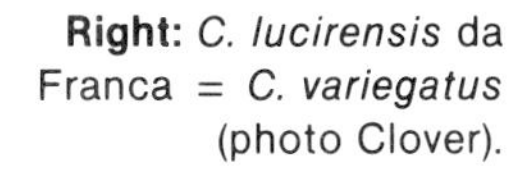

Right: *C. lucirensis* da Franca = *C. variegatus* (photo Clover).

Distribution.—Western Africa below the Bulge to southern Angola, but most common south of the Congo. Moderately common in shallow water.

Synonymy.—Based on development of patterns, the names listed can all easily fit within *C. variegatus* and most can be found in small series from various localities. Because *C. africanus* has been used for a variety of shells, usually *C. aemulus* or *C. bulbus* patterns, I prefer to use the less familiar name *C. variegatus* so this species can be given a 'fresh start' with collectors; the name is also very appropriate. Tryon once listed *C. variegatus* as a variety of *C. africanus*, which might classify him as first revisor, but his concept of the species was very different from mine and his choice of names was apparently based on page priority.

Notes.—Like *C. bulbus*, this species appears to have a slightly different variety of pattern in every population, and it looks as though certain workers have determined to name each variety as a full species, with numerous manuscript names already in circulation among dealers and collectors. Beware of specimens of this species sold at high prices under various new names or as newly described species. Also remember that a name without an author is of very doubtful standing, and that museums cannot be listed as authors of taxa.

Like other western African shells, *C. variegatus* usually is a real dog, with heavily eroded and distorted spires, often areas of erosion on the body whorl, heavy axial flaws and scars, and badly broken lips. The pattern may be indistinct with a very dull surface.

VARIUS Linnaeus, 1758

1758. *Conus varius* Linnaeus. *Systema Naturae per Regna Tria Naturae*, ed. 10, 1: 715 (Locality not stated). Supposed holotype figured by Kohn, 1963.

1798. *Cucullus annularis* Roeding. *Museum Boltenianum:* 40 (Locality not stated). See Kohn, 1975.

1798. *Cucullus granulosus* Roeding. *Ibid.:* 40 (Locality not stated). Lectotype figure (Kohn, 1975): Chemnitz, Pl. 138, fig. 1284.

1798. *Cucullus radula* Roeding. *Ibid.:* 40 (Locality not stated). See Kohn, 1975.

?1807. *Conus laevigatus* Link. *Beschr. Nat.-Samml. Univ. Rostock*, 3: 106 (Locality not stated).

1828. *Conus interruptus* Wood. *Index Test., Suppl.:* 8, pl. 3, fig. 2 (Locality unknown).

1834. *Conus pulchellus* Sowerby i, in Sowerby ii. *Conchological Illustrations:* Pt. 54, fig. 61 (Locality not stated). Non *Conus pulchellus* (Roeding, 1798).

1854. *Conus Hevassii* A. Adams. *Proc. Zool. Soc.* (*London*), 1853: 118. Nomen novum for *Conus pulchellus* Sowerby i, in Sowerby ii, 1834.

1874. *Conus Hwassi* Weinkauff. *Syst. Conch. Cab.*, ed. 2, 227(*Conus*): 252. Emendation of *C. hevassii* Adams, 1854.

Description.—Moderately heavy, with a good gloss; low biconical, the sides convex posteriorly or nearly straight; body whorl with heavy widely spaced rounded spiral ridges on at least anterior third, these often continued to shoulder or reduced to spiral threads posterior to midbody; basal ridges usually granulose, and the entire body whorl is commonly covered with widely spaced granules; axial threads often heavy; shoulder with heavy, triangular, rather horizontal coronations, concave above; spire moderate to tall, sharply pointed, the sides straight or nearly so; early whorls apparently only weakly nodulose, but later whorls heavily coronate with very sharp coronations; tops of whorls flat or slightly concave with 4-6 narrow spiral ridges separated by faintly punctate grooves and with scattered weak axial threads. Body whorl white, creamy, often with pale yellow or violet tones; entire whorl covered with spiral rows of small and very small dark brown to reddish brown spots on the spiral ridges or threads, and with a spiral band of brown above the base; this band is usually pale brown and continuous, containing larger darker brown blotches; usually another spiral band of large and small reddish to blackish brown blotches above midbody and extending to shoulder; these blotches may be connected into a solid or nearly solid band like the anterior one; shoulder and spire white, usually without pattern or with a few small brown spots. Aperture narrow posteriorly, wider anteriorly; outer lip rather thick, slightly convex; mouth white, usually with a large yellow blotch deep within. Columella short, narrow. Length 25-61mm.

Comparison.—*C. varius* is related to such species as *C. brunneus*, *C. bartschi*, and *C. cabritii* in sculpture, coronated whorls, and general pattern. From the first two species it is readily distinguished by the generally smaller adult size, the higher spire, slender shape with narrow shoulder, the heavily granulose spiral ridges at least basally, and the pattern of a heavy anterior brown band and a blotchy posterior band. It is very similar in size to *C. cabritii*, but that species has deep punctations between the spiral ridges at least anteriorly, lacks the small dots heavily developed over the body whorl, and has a dark mouth. *C. aurantius* may be superficially similar in shape and pattern, but that species lacks distinct spiral ridges on the spire whorls and has a usually well developed spire pattern and less horizontal coronations; it does not have a yellow mouth; the same differences more-or-less apply to *C. scopulorum*.

Variation.—There is considerable variation in height of the spire and convexity of the sides, but the shape is generally quite distinct and rather elongated. Spire pattern is seldom present. The spiral ridges in adults commonly extend to midbody and often are present to the shoulder; only the basal ridges may be granulose, but more commonly at least the anterior half of the body whorl is distinctly granulose, and the entire whorl may be rough.

The pattern apparently changes with growth. In small shells the color may be pinkish or yellowish, rather waxy, with the small dots and dashes very obscure. The anterior band is usually visible but may be very pale and seldom contains large darker blotches; the posterior band is usually absent. With growth dark blotches develop within the anterior band and small blotches develop above midbody; a third row of blotches may develop below the shoulder. The bands become darker brown with age, and also tend to become larger, the two bands above midbody becoming single large blotches often with spiral flammules connecting them to adjacent blotches. Very large specimens may have two broad spiral bands of dark blackish brown above and below midbody plus numerous rather large squarish spots in regular spiral rows over the entire body whorl. The mouth is commonly bright yellow in adults.

Distribution.—Widespread and common in shallow water throughout the Indo-Pacific from eastern Africa to Fiji; apparently not recorded from Hawaii and French Polynesia.

Synonymy.—The Roeding names are rather doubtful, but *C. interruptus* is certainly this species, as is *C. pulchellus* and its replacements.

Notes.—Adults are commonly heavily covered with chalky deposits which are very difficult to remove and which may closely coat not only the spire but the entire body whorl. There may be erosion of the early whorls and some axial growth marks, but these are seldom detracting. Juveniles with faint patterns and waxy texture are not common and presumably occur in deeper water than adults; they may look very different from adults.

VENTRICOSUS Gmelin, 1791

Note: The following 75 names have been extracted from the literature with no real effort to see the original descriptions. Most of the names apply to strictly individual variants and, as is common in this literature, are preoccupied by earlier names; this has not always been indicated.

1791. *Conus ventricosus* Gmelin. *Systema Naturae per Regna Tria Naturae*, ed. 13, 1: 3397 (Locality not stated). Lectotype figure (Kohn, 1966): Kaemmerer, Pl. 6, fig. 3.

1792. *Conus Franciscanus* Hwass, in Bruguiere. *Cone*, in *Ency. Method., Hist. Nat. des Vers*, 1: 698 (Mer d'Afrique). Lectotype selected and figured by Kohn, 1968.

1792. *Conus Jamaicensis* Hwass, in Bruguiere. *Ibid.*, 1: 700 (Jamaica). Lectotype selected and figured by Kohn, 1968.

1792. *Conus Mediterraneus* Hwass, in Bruguiere. *Ibid.*, 1: 701 (Mediterranean). Lectotype figure (Kohn, 1968): *Tableau Ency. Method.*, Pl. 330, fig. 4.

1792. *Conus Madurensis* Hwass, in Bruguiere. *Ibid.*, 1: 709 (Indian Ocean). Lectotype figure (here selected): *Tableau Ency. Method.*, Pl. 333, fig. 3.

1792. *Conus ignobilis* Olivi. *Zool. Adriat.:* 133 (Istria). See Kohn, 1968.

1793. *Conus jaspis* Salis. *Reisen Koenigreichs Neapel:* 363 (Mediterranean). See Kohn, 1975.

1793. *Conus olivaceus* Salis. *Ibid.:* 363 (Taranto). See Kohn, 1975.
1793. *Conus humilis* Salis. *Ibid.:* 364 (Taranto and Naples). See Kohn, 1975.
(1804. *Conus Listeri* Renier. *Tav. Alfab.:* 8 (Locality not stated). Based on Lister, Pl. 765, fig. 14.)
1810. *Conus Adansonii* Lamarck. *Ann. du Mus. Hist. Nat. (Paris)*, 15: 424 (Mers du Senegal). Holotype figured by Kiener, Pl. 61, fig. 4; also a secondary reference to Adanson, Pl. 6, fig. 6.
1816. *Conus ziczac* Megerle. *Ges. Naturf. Freunde Berlin*, 8(1): 4, pl. 1, fig. 3 (Mediterranean).
1828. *Conus siculus* Chiaje. *Mem. Storia e Notomia An. s. Vert. Napoli*, 3: 205, 219, 228, pl. 49, figs. 1-3 (Caiola).
1834. *Conus glaucescens* Sowerby i, in Sowerby ii. *Conchological Illustrations:* Pt. 54, fig. 60 (Locality not stated).
1836. *Conus ater* Philippi. *Enum. Moll. Sicil.*, 1: 238, pl. 12, figs. 20-21 (Lacuna, Pantano del Faro prope Messinam).
1836. *Conus mediterraneus* var. *marmoratus* Philippi. *Ibid.*, 1: 238, pl. 12, figs. 17, 19, 22 (Sicily). Non *C. marmoratus* Schroeter, 1803.
1836. *Conus ignobilis* var. *rufa* Scacchi. *Cat. Conch. Regn. Neapol.:* 9 (Italy).
1845. *Conus Bruguiersi* Kiener. *Species gen. et icon. des coqu. viv.*, 2 (*Conus*): Pl. 56, fig. 2. 1848, *Ibid.*, 2: 221 (Locality unknown). Emended in text to *C. bruguieri.* Non *Conus bruguierii* Nyst, 1843.
1846. *Conus Cailliaudi* Jay. *Ann. Lyc. Nat. Hist. New York*, 4: 169, pl. 10, figs. 8a-8b (Locality unknown).
1847. *Conus adriaticus* Nardo. *Sinon. Moderna*, pt. ii: 39-40 (Adriatic).
1847. *Conus amazonicus* Nardo. *Ibid.:* 41-42 (Adriatic).
1847. *Conus chersoideus* Nardo. *Ibid.:* 41-42 (Isle of Cherso).
1847. *Conus clodianus* Nardo. *Ibid.:* 41-42 (Chioggia).
1847. *Conus cretheus* Nardo. *Ibid.:* 41-42 (Adriatic).
1847. *Conus epaphus* Nardo. *Ibid.:* 41-42 (Adriatic).
1847. *Conus herillus* Nardo. *Ibid.:* 41-42 (Isle of Cherso).
1847. *Conus istriensis* Nardo. *Ibid.:* 41-42 (Istria).
1847. *Conus pallans* Nardo. *Ibid.:* 39-40 (Adriatic).
1847. *Conus phegeus* Nardo. *Ibid.*: 41-42 (Adriatic).
1847. *Conus stercutius* Nardo. *Ibid.:* 41-42 (Locality not stated).
1847. *Conus thuscus* Nardo. *Ibid.:* 39-40 (Adriatic).
1848. *Conus mediterraneus* var. *fasciata* Requien. *Cat. Coq. Corse*: 86 (Corse). Non *C. fasciatus* Schroeter, 1803.
1848. *Conus mediterraneus* var. *obtusa* Requien. *Ibid.:* 86 (Corse). Non *C. obtusus* Kiener, 1845.
1853. *Conus grossii* Maravigna. *Atti Accad. Gioenia Sci. Nat. Catania*, (Ser. 2), 8: 137 (Sicily).
1860. *Conus Guestieri* Lorois. *J. Conchyl. (Paris)*, 8: 329, pl. 12, fig. 5 (Locality unknown).
1880. *Conus mediterraneus* var. *minor* Monterosato. *Bull. Soc. Mal.*

Ital., 5: 230 (Barbary).

1882. *Conus mediterraneus* var. *albina* Bucquoy, Dautzenberg, and Dollfus. *Moll. Mar. Roussillon*, 1: 84 (Barbary).

1882. *Conus m.* var. *caerulescens* B.D.D. *Ibid.*: 83, pl. 13, fig. 5 (Barbary). Non *C. coerulescens* Schroeter, 1803.

1882. *Conus m.* var. *carinata* B.D.D. *Ibid.*: 82, pl. 13, figs. 16-17 (Barbary). Non *C. carinatus* Swainson, 1832.

1882. *Conus m.* var. *elongata* B.D.D. *Ibid.*: 82, pl. 13, figs. 14-15 (Barbary). Non *C. elongatus* Borson, 1820.

1882. *Conus m.* var. *flammulata* B.D.D. *Ibid.*: 83, pl. 13, fig. 16 (Barbary).

1882. *Conus m.* var. *fusca* B.D.D. *Ibid.*: 83, pl. 13, fig. 14 (Barbary).

1882. *Conus m.* var. *lutea* B.D.D. *Ibid.*: 83 (Barbary). Non *C. luteus* Sowerby i, 1883.

1882. *Conus m.* var. *major* B.D.D. *Ibid.*: 82 (Locality not stated).

1882. *Conus m.* var. *oblonga* B.D.D. *Ibid.*: 82, pl. 13, figs. 12-13 (Locality not stated).

1882. *Conus m.* var. *pallida* B.D.D. *Ibid.*: 83, pl. 13, fig. 20 (Locality not stated).

1882. *Conus m.* var. *rubens* B.D.D. *Ibid.*: 83, pl. 13, figs. 21-22 (Barbary).

1885. *Conus mediterraneus* var. *alalmus* Gregorio. *Bull. Soc. Mal. Ital.*, 11: 112 (Barbary).

1885. *Conus m.* var. *amigus* Gregorio. *Ibid.*: 113 (Mediterranean).

1885. *Conus m.* var. *emisus* Gregorio. *Ibid.*: 112 (Mediterranean).

1885. *Conus m.* var. *endorus* Gregorio. *Ibid.*: 96 (Mediterranean).

1885. *Conus m.* var. *imelus* Gregorio. *Ibid.*: 95 (Mediterranean).

1885. *Conus franciscanus* var. *Pereirae* Gregorio. *Ibid.*: 105 (Mediterranean).

1885. *Conus m.* var. *rufatra* Gregorio. *Ibid.*: 112 (Barbary).

1885. *Conus m.* var. *steppus* Gregorio. *Ibid.*: 96 (Barbary).

1885. *Conus m.* var. *subviridis* Gregorio. *Ibid.*: 112 (Barbary).

1886. *Conus galloprovincialis* Locard. *Ann. Soc. Agric. Hist. Nat. Arts Utiles Lyon*, (Ser. 5), 8: 105 (Mediterranean).

1886. *Conus submediterraneus* Locard. *Ibid.*: 105 (Mediterranean).

1899. *Conus galloprovincialis* var. *lineolata* Locard. *Ann. Soc. Lyon*, 46: 227 (Corsica).

1899. *Conus g.* var. *minor* Locard. *Ibid.*: 227 (Corsica). Non *C. m.* var. *minor* Monterosato, 1880.

1899. *Conus trunculus* Monterosato. *J. Conchyl. (Paris)*, 47: 401 (Kerinia, Cyprus).

1904. *Conus mediterraneus* var. *alticonica* Pallary. *J. Conchyl. (Paris)*, 52: 217 (Golfe de Gabes).

1906. *Conus mediterraneus* var. *persistens* Kobelt. *Icon. Schalentr. Europ. Meeres-Conch.*, 4: 6, pl. 100, figs. 15-16 (?North Africa).

1906. *Conus provincialis* Kobelt. *Ibid.*: 4. Misspelling of *C. galloprovin-*

cialis Locard, 1886.

1906. *Conus m.* var. *Vayssieri* Kobelt. *Ibid.:* 6, pl. 100, figs. 13, 14, 17, 18 (Golfe de Gabes).

1911. *Conus m.* var. *scalare* Dautzenberg. *J. Conchyl.* (*Paris*), 58: 209, pl. 10, fig. 9 (St. Raphael). Non *C. scalaris* Valenciennes, 1832.

1917. *Conus m.* var. *arenaria* Monterosato. *Bull. Soc. Zool. Ital.*, (Ser. 3), 4: 23 (Tripoli).

1917. *Conus m.* var. *debilis* Monterosato. *Ibid.:* 24 (Tripoli).

1917. *Conus m.* var. *ossea* Monterosato. *Ibid.:* 23 (Tripoli and Tunis).

1917. *Conus m.* var. *pretunculus* Monterosato. *Ibid.:* 23 (Tripoli).

1923. *Lautoconus m.* var. *noeformis* Monterosato. *R. Comit. Talassogr. Ital.*, Mem. 107: 11 (Coste Cirenaiche).

1933. *Conus m.* var. *castanea* Coen. *Ibid.*, Mem. 192: 70 (Adriatic). Non *C. castaneus* Kiener, 1845.

1933. *Conus* (*Chelyconus*) *m.* var. *interrupta* Coen. *Ibid.*, Mem. 192: 70, 175 (Adriatic). Non *C. interruptus* Wood, 1828.

1933. *Conus* (*Chelyconus*) *m.* var. (anomalous) *producta* Coen. *Ibid.*, Mem. 192: 175 (Adriatic).

1933. *Conus m.* var. *galloprovincialis* anom. *turrita* Coen. *Ibid.*, Mem. 192: 175 (Adriatic). Non *Conus turritus* (Roeding, 1798).

Description.—Moderately light in weight, with a good gloss; low conical, the upper sides slightly convex to almost straight; body whorl with a few low spiral ridges above the base and many axial and spiral threads and flaws, some heavy; shoulder rounded, narrow, not very distinct from spire; spire low to moderate, blunt, the sides straight to slightly convex; spire whorls with rude sutures, the tops flat to slightly convex, with 1-2 low spiral ridges and weak curved axial threads. Body whorl usually tan, olive, grayish, yellowish brown, or bluish gray, covered from base to shoulder with closely spaced spiral rows of fine dashes and dots of brown, these sometimes alternating with opaque white; pattern very variable, usually with two broad spiral bands of large brownish blotches above and below midbody, these mostly fused into nearly continuous bands; base and midbody area pale, whitish to grayish, sometimes with a spiral band of variable whitish blotches between the shoulder and midbody, these rounded or flammulated; spire whitish, with large or small brownish blotches rather regularly arranged; early whorls usually badly eroded. Aperture moderately wide, nearly uniform in width; outer lip thin, evenly convex; mouth white with large brownish to dirty violet blotches separated by pale stripes at middle and below shoulder. Columella narrow, short. Length 20-65mm.

Comparison.—The general shape, rows of dashes, dark mouth, and generally bad condition call to mind only *C. aemulus* of the West African fauna, a closely related and possibly conspecific species. *C. aemulus* in most of its forms is readily separable by being distinctly bluish or bluish gray with darker blotches in two rows above the midbody area, the blotches forming either axial flammules or spiral bands. *C. ventricosus* is commonly darker brown or olive, broadly banded or mottled with dark. I am not sure that specimens of *C. aemulus* mixed into a large series of *C. ventricosus* would always be separable as both species are very variable in color and pattern as well as shape, but typical forms of each are no problem.

Variation.—Legion. I will make no attempt to even touch on the variation of *C. ventricosus;* an interpretation of the synonymy will indicate the broad variety of colors, sizes, and shapes that have been dignified with names. One variant, however, is of importance to collectors because it is sold as a rare species from northwestern Africa. This is the shell called *C. bruguieri* or *C. adansonii* by most collectors, a narrow, high-spired nearly black or deep purplish brown shell with faint traces of a spiral pattern of dashes and commonly a sharply contrasted white band at midbody and another at the shoulder. The proper name of this variant should be *C. ventricosus* var. *franciscanus* if any name is used, and it definitely intergrades fully with typical African *C. ventricosus.*

C. ventricosus varies from narrow to ovate or inflated, obconical to tall biconical with stepped whorls, has a rounded to carinate shoulder, may vary in color from white to pink to blue to green to brown to black and any shade between, may be spirally banded or have axial stripes, and may be adult at any size between 15mm and 65mm. The pattern described earlier is the most common type and the only one likely to be seen unless the collector is deeply absorbed by the Mediterranean fauna.

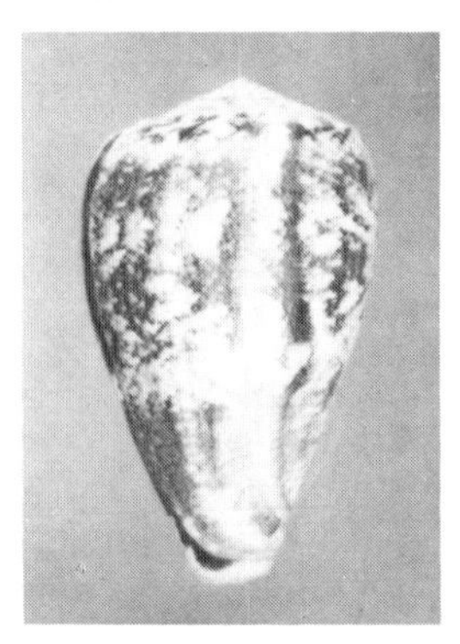

Left: *C. ventricosus,* Tunisia, 47mm (GG).

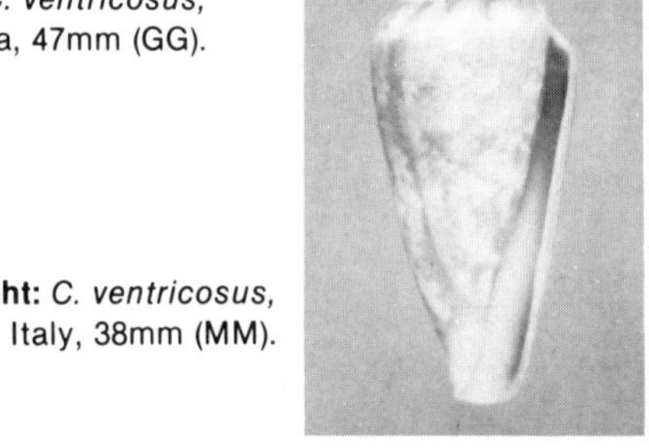

Right: *C. ventricosus,* Italy, 38mm (MM).

Distribution.—Entire coasts of the Mediterranean Sea and Adriatic, north to the southern coast of Spain and south perhaps to Angola (the *franciscanus* variety in various forms seems to be the typical western African form). Abundant to rare, shallow to deep water.

Synonymy.—Probably one of the most laborious synonymies of any animal is that accumulated by *C. ventricosus*, not even the common rat and mouse having been named as many times. It has long been decided, with little possible doubt, that there is just one very variable species of cone in the Mediterranean, 75 names not withstanding. The listed names, except for most of those before 1847, were applied before it was realized what was going on in that sea. After that date the names are worthless and just an exercise in the total 'stamp collecting' philosophy of conchology: if it's different, name it. Most of the names require no comment, but it should be noted that *C. adansonii* (pale) and *C. bruguieri* (dark) are extremes of the same type of variant, which should be called *C. ventricosus* var. *franciscanus*.

C. ventricosus is the same species as that called *C. mediterraneus* for over two centuries. Both type figures are very bad and probably not recognizable by themselves, so it is a shame that steps were not taken to preserve the name *C. mediterraneus*.

Notes.—This is usually an ugly, rough shell with little to appeal to the collector who does not wish to specialize in this single species. Although it is highly variable, the shells sold by most dealers in the U.S., even though from different localities and obviously different populations, look very much alike and no one would have trouble realizing they represent a single species. The extreme variants are uncommon to rare and often based on single shells that are unique.

This is one of the few species of cones reported to have sinistral specimens, some of which appear on lists every few years.

Bandel and Wils (*Basteria*, 1977, 41(1/4): 33-45) have recently argued for retention of the name *C. mediterraneus* and recognize *C. guianicus* as a valid sympatric species. As mentioned earlier, the type figures of both *C. ventricosus* and *C. mediterraneus* are probably almost unrecognizable, and *C. mediterraneus* is certainly the more familiar name, but *C. ventricosus* does seem to be the same species. Bandel and Wils' *C. guianicus* is apparently a small northern form of *C. ermineus*, but they extend the name to include *C. aemulus* as well.

For the moment I prefer to retain *C. ventricosus*, refer *C. guianicus* to *C. ermineus*, and retain *C. aemulus* as valid. As in several other cases, it would be easy to refer all these names to *C. ermineus*.

VENULATUS Hwass, in Bruguiere, 1792

1792. *Conus venulatus* Hwass, in Bruguiere. *Cone*, in *Ency. Method., Hist. Nat. des Vers*, 1: 695 (America). Lectotype figure (here selected): *Tableau Ency. Method.*, Pl. 337, fig. 9.

1803. *Conus marmoratus* Schroeter. *Arch. Zool. Zootom.*, 3(2): 67 (Locality not stated).

1810. *Conus nivosus* Lamarck. *Ann. du Mus. Hist. Nat. (Paris)*, 15: 278 (?American seas).

1833. *Conus nivifer* Sowerby i, in Sowerby ii. *Conchological Illustrations:* Pt. 25, fig. 14; pt. 36, fig. 47; pt. 56/57, fig. 84 (Cape Verde).

1844. *Conus trochulus* Reeve. *Conchologia Iconica*, 1(*Conus*): Pl. 45, sp. 246 (Locality unknown).

1845. *Conus ateralbus* Kiener. *Species gen. et icon. des coqu. viv.*, 2 (*Conus*): Pl. 108, fig. 4. 1849-1850, *Ibid.*, 2: 313 (Locality unknown).
1849. *Conus Crotchii* Reeve. *Conchologia Iconica*, 1(*Conus* Suppl.): Pl. 6, sp. 254 (Saldanha Bay, South Africa).

Description.—Moderately heavy, thick, with a good gloss; low conical, the sides slightly and gently convex to nearly straight; body whorl with about 5-10 low spiral ridges above the base and numerous spiral and axial threads over the rest of the whorl; shoulder broad, roundly to sharply angled, concave on top; spire low to moderate, rather sharply pointed, the sides straight to concave; early whorls eroded; later whorls with rude sutures, slightly concave above, with 0-3 low and weak spiral ridges and scattered weak axial threads. Body whorl white to pinkish or pale straw, with a very variable pattern; usually the body whorl is covered with pale brown spiral lines of variable width and strength as well as spacing, the lines often broken into dashes; overlying this pattern are many to few large or small white spots and blotches, often rather triangular or squarish and evenly spaced over the whorl or tending to cluster at the base, midbody, and below the shoulder; large or small brownish blotches and clouds may be present, eventually obscuring most of the pattern except for a few white spots at midbody and the shoulder; some specimens are deep black with sparse white rather ragged triangular spots at midbody; spire colored much as body whorl, sparsely spotted with darker. Aperture moderately narrow above, wider anteriorly; outer lip sharp, evenly convex to nearly straight; mouth white, pinkish, pale violet, or dark brownish, depending on color of body whorl, sometimes with pale bands at middle and below shoulder. Columella narrow, rather short. Length 20-50mm.

Comparison.—*C. venulatus* is rather distinctive among the western Africa cones as several patterns from light to dark are present in most series and are a sure sign of *C. venulatus*. Very pale pinkish or white shells possibly cannot be identified with certainty without connecting specimens, although they are thicker and commonly have a higher spire than most other sympatric cones. *C. balteus* is the most similar species in size and color, but it differs in pattern, tending to have heavy axial blackish or dark brown flammules and spiral rows of white broken into distinct bands and often crosshatched; they do not show the fine spiral brown lines of *C. venulatus* nor white spots rather randomly distributed over the shell. Very dark specimens are easily recognized by the blackish color with triangular white spots clustered at midbody and below the shoulder.

Variation.—As indicated in the description, this is a variable species. The sides are nearly straight or gently convex, but the spire is often rather mucronate if not badly eroded. The spiral ridges on the spire whorls may be absent or represented by up to three low ridges, often with different counts on specimens from the same collection. As indicated, typical shells have at least traces of spiral brown lines or dashes in the pattern on a whitish background, with white spots of variable size and number scattered over the whorl or clustered in roughly three spiral bands. As the shell becomes darker the lines become less distinct; as the pattern becomes lighter the lines also tend to disappear. Solid white or pinkish shells are found. Of described varieties, four names could be used if desired, but they intergrade broadly with four varieties common in any sizable collection: 1) all white or pink (*trochulus*); 2) white spots priminent (*venulatus*); 3) brown lines prominent (*nivifer*); 4) very dark, blackish (*ateralbus*). Very light or very dark shells are least common.

Distribution.—Apparently restricted to shallow water in the Cape Verde Islands and presumably the adjacent mainland. Uncommon.

Synonymy.—*C. marmoratus* is possibly unrecognizable. *C. crotchii* is rather heavily banded like *C. variegatus*, but it has a more angled shoulder and higher spire as in *C. venulatus*. *C. nivosus* seems to be this species, with a pattern close to typical, while the other names represent forms found in series from the same localities.

Notes.—Like other western African cones, the spire and body whorl are often eroded and have heavy axial flaws. This is the oldest name for a Cape Verde cone, and it is likely that the other species of similar form from the same locality (*C. anthonyi*, *C. balteus*) are just ecotypes. This is the least common of the three and the largest.

VEXILLUM Gmelin, 1792

1792. *Conus vexillum* Gmelin. *Systema Naturae per Regna Tria Naturae*, ed. 13, 1: 3397 (Locality not stated). Lectotype figure (Kohn, 1966): Rumphius, Pl. 31, fig. 5.

1792. *Conus Sumatrensis* Hwass, in Bruguiere. *Cone*, in *Ency. Method.*, *Hist. Nat. des Vers*, 1: 655 (Sumatra). Lectotype figures (here selected): Chemnitz, Pl. 144A, figs. a-b.

1798. *Cucullus canonicus* Roeding. *Museum Boltenianum:* 43 (Locality not stated). Lectotype figure (Kohn, 1975): Martini, Pl. 57, fig. 629. Non *Conus canonicus* Hwass, in Bruguiere, 1792.

1817. *Conus leopardus* Dillwyn. *Descr. Catalogue Recent Shells*, 1: 364 (East Indian Seas). Lectotype figures (here selected): Chemnitz, Pl. 144A, figs. a-b. Non *Conus leopardus* (Roeding, 1798).

1845. *Conus sulphuratus* Kiener. *Species gen. et icon. des coqu. viv.*, 2 (*Conus*): Pl. 66, fig. 3; pl. 78, fig. 4. 1847, *Ibid.*, 2: 130 (Locality unknown).

1858. *Conus Robillardi* Bernardi. *J. Conchyl. (Paris)*, 7: 182, pl. 7, figs. 2-3 (Locality unknown).

Description.—Heavy, with a good gloss; low conical, the upper sides convex then almost straight or distinctly concave at the middle; body whorl with a few low spiral ridges at the base, the rest of the whorl covered with fine to heavy spiral threads, the grooves between sometimes punctate, and heavy, closely spaced axial threads and growth marks; shoulder broad or very broad, roundly angled, flat or convex on top; spire low to moderate, blunt, the sides nearly straight; spire whorls flat on top, with about 5-10 narrow spiral ridges separated by punctate grooves and crossed by fine axial threads so spire appears malleated. Body whorl white, broadly covered with two spiral bands of pale or dark brown above and below midbody, leaving midbody and shoulder bands of large white spots; axial extensions from the white spots may be large and nearly continuous from shoulder to above base, producing a nearly white shell; along the major growth lines are wavy darker brown axial flammules, often dense and numerous, tending to cluster into wide bands above and below midbody; occasionally the axial flammules are nearly solid and very heavy; base and area above it dark brown; spire and shoulder white, with light to dark brown squarish blotches about as wide as white areas or smaller; early whorls bright yellow. Aperture wide, uniform in width; outer lip thin, sharp, straight or concave at middle; mouth white. Columella mostly internal, slightly oblique. Length 50-156mm.

Comparison.—The usually heavy axial flammules on a white ground with two broad brown bands and a dark base are very characteristic. *C. namocanus* has spiral brown lines in the pattern and is usually lighter in weight. *C. trigonus* usually has a very low spire, a partially free columella, and a whitish base. *C. miles* has the dark base strongly contrasted with the rest of the whorl and a brown spiral band at midbody, with large dark blotches in the mouth. *C. capitaneus* is somewhat similar but often has a dark mouth, has the anterior basal ridges separated by punctate grooves even in adults and commonly visible to midbody, and has distinct blackish spots on each side of a white midbody area; *C. mustelinus* is similar.

Variation.—Adult shells have two extreme patterns that are commonly held as distinct species even though intermediates are common. In the typical pattern the shell is mostly light to dark brown with rather distinct white spots at midbody and the shoulder, the axial flammules narrow but often crowded and fairly distinct from the background brown. At the other extreme are the shells often called *sumatrensis*, here considered just a variety. This is a mostly white shell with the spiral bands of brown absent or very weak and the axial flammules dark brown to almost black and forming very irregular wavy bands over the shell; the base color and spire are often very light in this form. In both variants the spire, width at shoulder, and density of the pattern vary from shell to shell.

Considerable ontogenetic changes occur from juvenile to adult. Shells up to about 30mm in length are bright to pale yellow, often with a greenish tone, marked only by a row of widely spaced round black spots above midbody and sometimes smaller spots below midbody and above the base; the spire has only a few squarish blotches (seldom more than four); and the body whorl is covered with shallow spiral grooves, widely spaced, that appear weakly punctate. The mouth may have two brown spots within. At about 40mm or so the body whorl begins to develop dark brown flammules along the growth lines, these eventually producing the adult pattern on the body whorl and spire. Yellow juveniles were called *sulphuratus*.

Distribution.—*C. vexillum* is widely distributed and rather common throughout the Indo-Pacific from the eastern African coast to Hawaii and French Polynesia, usually in shallow water. Juveniles seem to occur in deeper water and are uncommon. Var. *sumatrensis* is typically most common in the Indian Ocean and East Indies, but it occurs sporadically to at least French Polynesia.

Synonymy.—Since *C. vexillum* and *C. sumatrensis* intergrade broadly and specimens combining both patterns are not uncommon, *sumatrensis* cannot be recognized at more than the varietal level. The juvenile was described as *C. sulphuratus*, with *C. robillardi* apparently also a juvenile. *C. canonicus* and *C. leopardus*, both preoccupied, are *C. vexillum*.

Notes.—Average specimens of *C. vexillum* are not gem quality, having rather dull and uncontrasted patterns, eroded spires, often heavy axial flaws, and filed lips. However, good gems with bright colors and contrasted patterns are not uncommon. Large specimens often become very heavy and cumbersome, though a 100mm specimen I have seen is relatively light in weight for the species and has a good gloss with bright pattern and is very clean. Juveniles and subadults with both brown and yellow patterns are uncommon to rare. The *sumatrensis* pattern is not any more uncommon than is the typical form.

VICTORIAE Reeve, 1843

1843. *Conus Victoriae* Reeve. *Conchologia Iconica*, 1(*Conus*): Pl. 37, sp. 202 (Mouth of the Victoria River, New Holland).

1864. *Conus nodulosus* Sowerby ii. *Descr. Three New Shells:* unnumbered page and fig. (Swan River). 1866, *Thesaurus Conchyliorum*, 3 (*Conus* Suppl.): 328, pl. 27(288), fig. 635.

1866. *Conus complanatus* Sowerby ii. *Thesaurus Conchyliorum*, 3(*Conus* Suppl.): 330, pl. 28 (289), figs. 650-651 (Australia). Also called *C. victoriae* variety, *T. C.*, 1857-1858, Pl. 23(209), fig. 576. Misspelled 'compactus' by Weinkauff, *Syst. Conch. Cab.*, ed. 2, 230(*Conus*): 277.

Description.—Light in weight, with a good gloss; low conical, the sides convex; body whorl with very weak spiral ridges on the anterior third, these replaced by fine spiral and axial threads more posteriorly;

shoulder rather broad, sharply to roundly angled, usually concave above; spire moderate or low, sharply pointed, the sides straight to slightly concave; spire whorls weakly stepped, concave above, with about 2-4 low and indistinct spiral ridges and axial threads, not very distinct; first 3-5 whorls rather heavily nodulose, next 2-3 distinctly undulate. Body whorl cream to yellowish or pinkish, covered with fine orange to blackish axial lines that overlap to produce a finely to coarsely tented pattern; usually three spiral rows of small orange blotches (below shoulder, above and below midbody) containing indistinct 'textile lines', the blotches sometimes connected into bands or almost absent; narrower dash-like blotches often present above the main series; often with distinctly bluish axial flammules visible over parts of the shell; shoulder and spire whitish or yellowish with orange spots and white tents; margins of shoulder and spire whorls sometimes with distinctly blackish dashed line where blotches become very dark; early whorls usually eroded. Aperture moderately wide, wider anteriorly; outer lip sharp, straight to slightly convex; mouth dark bluish white, white, or bright pink. Columella long and narrow, somewhat indented from columella. Length 35-70mm.

Comparison.—*C. victoriae* seems to be very close to *C. textile* in pattern and appearance, and it goes through some of the same patterns. However, it is structurally quite distinct and perhaps actually most closely related to *C. telatus*. Thus the early spire whorls are heavily nodulose and later ones undulate, as well as being distinctly stepped or tabled with concave tops, characters virtually never found constantly in *C. textile* populations. In addition, many *C. victoriae* have distinctly bluish or bluish gray axial flammules that are absent in *C. textile*. The dash-like blotches above the major series are also quite distinct when present. From *C. telatus*, *C. victoriae* is quite distinct in being less elongate, lacking heavy spiral ridges and granules, and having the orange blotches small with rather indistinct 'textile lines', plus the presence of bluish flammules.

Variation.—*C. victoriae* can be divided into two distinct but allopatric subspecies that seem too closely related to retain as full species. Each of these varies considerably in height of spire, convexity of the sides, and the sharpness of the shoulder.

C. victoriae victoriae: Often with strong bluish axial flammules; dash-like blotches usually present above major series; 'textile lines' usually obvious and the orange blotches generally well developed; mouth bluish to white. North-West Cape, Western Australia to Northern Territory, Australia.

Typical shells are heavily covered with small tents, have bluish flammules, two or three major series of blotches, and two minor series. Very pale shells, variety *complanatus*, differ in lacking the bluish flammules and having little contrast between the blotches and the tents. Very dark shells correspond to the variety *archiepiscopus* of *C. textile* in having the fine lines almost parallel and not producing tents, but additionally have the bluish flammules which give the shell a very dark gray color. Occasional specimens are white with blue flammules but without the axial brown lines.

C. victoriae nodulosus: Bluish axial flammules absent; orange blotches reduced or absent, only the major series usually present; mouth bright pink. Western Australia between Freemantle and Shark Bay, to the south of the range of the typical subspecies.

The pattern is often very loose, with the orange blotches completely absent in some specimens and the shell appearing off-pink with little pattern. The mouth may be bright pink or only pinkish white.

Distribution.—Restricted to shallow waters of Western Australia and Northern Territory, Australia, where moderately common.

Synonymy.—*Complanatus* appears to be an individual variant that falls easily within the range of variation of the typical subspecies. *Nodulosus* apparently is absent immediately south of the range of the typical subspecies, so intergradation between the two is not known; however, the differences between them do not seem to be of the specific level in tented cones, being minor pattern and color differences. The lineate pattern is sometimes called 'cholmondeleyi', but that name is a synonym of the same pattern in *C. textile.*

The type figure of *C. nodulosus* does not really bear a close resemblance to the taxon now called by that name; however, it does show the pink mouth, the open tenting, the nodules are mentioned, and the locality given is Australia (though doubtful).

Notes.—This species is rougher than most tented cones, often having heavy axial growth flaws and scars. The early and often middle whorls are usually heavily eroded. Most specimens of *C. v. nodulosus*, which sells for a much higher price than the typical subspecies, have very weak patterns and are rather ugly shells of indifferent pinkish tones, although only the nicest specimens are usually figured in the literature. The *complanatus* and lineate varieties are not uncommon but are seldom sold.

VICWEEI Old, 1973

1973. *Conus (Textilia) vicweei* Old. *Veliger*, 16(1): 58, figs. 1-3, 5 (Off northwest coast of Sumatra, Indonesia).

Description.—Light in weight but solid, with a high gloss; ovately cylindrical, the greatest width below the shoulder; body whorl with about 6-12 low wide spiral ridges basally, otherwise smooth except for spiral and axial threads and sometimes strong axial growth lines; shoulder narrow, sharply angled or carinate, flat or slightly concave above; spire very low to moderately low, sharply pointed; early two whorls nodulose, the later whorls with carinate and sometimes slightly projecting margins; whorls flat or concave on top with about 5-7 narrow spiral ridges crossed by distinct curved axial threads so finely cancellate. Body whorl pale tan to brown or reddish brown, covered with many axial wavy lines of white or cream, the waves exaggerated and tightly packed to produce 2-4 spiral bands, above and below midbody and sometimes below the shoulder and above base, which appear white until examined closely; pattern often disappears near new growth of lip, becoming white with broad tan bands; spire and shoulder whitish with many large brownish blotches. Aperture wide, especially anteriorly; outer lip rather thick but sharp, following the outline of the body whorl proper; mouth white, sometimes with a pale yellow tint. Columella long, narrow. Length 60-81mm.

Comparison.—*C. vicweei* is a recently described large cone of very distinctive pattern; the texture allies it to *C. bullatus* and *C. cervus*, with which it shares the shape of the body whorl and the outer lip. In size it is intermediate between the two, but its pattern of wavy axial whitish lines is unique in this group of shells and very different from the mottled or weakly dashed pattern of *C. bullatus* or the spirally dashed pattern of *C. cervus*. There should be no confusion.

Variation.—The color varies from pale golden brown to dark brown or occasionally reddish brown, but the pattern is quite uniform within the limits to be expected of a lineate pattern. There is a tendency for the broad bands above and below midbody to be evenly divided into two or three narrower bands. The spire is sometimes nearly flat but more commonly moderately low with nearly straight sides. Like related cones, the pattern disappears near the outer lip in many shells.

Distribution.—*C. vicweei* is considered moderately rare in deep water of the eastern Indian Ocean from off Sumatra to Thailand.

Notes.—Although this species sells for a high price, it is apparently dredged quite regularly off Thailand and many specimens are on the market. The specimens offered are usually far from gem, however, with badly broken lips, dulled patterns, and axial growth flaws on many specimens.

The absence of a full pattern near the lip should not be considered a flaw as it seems to be an adult character. Small specimens of this species have apparently not been reported.

VILLEPINII Fischer and Bernardi, 1857

1857. *Conus Villepinii* Fischer and Bernardi. *J. Conchyl.* (*Paris*), 5: 292, pl. 9, fig. 12 (Marie Galante).

1942. *Conus fosteri* Clench and Aguayo. *Johnsonia*, 1(6): 34, pl. 12, fig. 5 (Off Sagua la Grande, Santa Clara Prov., Cuba).

Description.—Moderately light in weight, with a good gloss; low biconical, the upper sides convex then concave or nearly straight and tapering to a narrow base; body whorl with about a dozen weak spiral flat ridges above the base, these usually obsolete more posteriorly but sometimes continued in very weak fashion to the shoulder; numerous weak spiral and axial threads over whorl; shoulder wide, carinate or sharply angled, flat or slightly concave above; spire moderate to high, sharply pointed, the whorls slightly stepped, the sides nearly straight or concave; first 4-6 whorls distinctly nodulose; later whorls with carinate or sharply angled margins that often project slightly; tops of whorls flat or slightly concave, with about 4-6 weak spiral ridges crossed by weak axial threads so appear faintly cancellate. Body whorl white, with two rather even-edged spiral bands of reddish brown axial flammules on pale reddish brown; the flammules sometimes extend irregularly outside the bands, often reaching to the shoulder but sometimes passing through the mid-body area and reaching the base as well; anterior band often weak and reduced to small axial spots; occasional traces of squarish reddish brown spots below and above the posterior band may be present, probably actually reduced flammules rather than true spots; shoulder and spire white, with curved reddish or orange-brown blotches; early whorls whitish or tan. Aperture moderately narrow, nearly of uniform width; outer lip thin, sharp, straight or slightly concave at the middle; mouth white to bluish white, the external pattern showing through. Columella internal. Length 40-72mm.

Comparison.—Several American deep-water cones resemble *C. villepinii* fairly closely. In the Pacific are *C. recurvus* and *C. poormani*, which are similarly elongated. In *C. recurvus* the posterior portion of the body whorl is wider, the spire is higher with the whorls deeply concave on top, and the spiral ridges on the spire whorls are very weak; the body pattern is of axial flammules that are either continuous, reticulated, or form three indistinct spiral bands, but not two strong spiral bands. *C. poormani* is perhaps a bit thicker, with the posterior portion of the body whorl distinctly parallel below the shoulder then tapering to the base; the spire tends to be lower with weaker nodules on the early whorls; and the flammules are usually paler orange-brown instead of reddish brown. *C. clerii* is usually less elongated with very weak nodules on only the first 1-2 or no

C. villepinii with reduced pattern, St. Augustine, Florida, 44mm (TY).

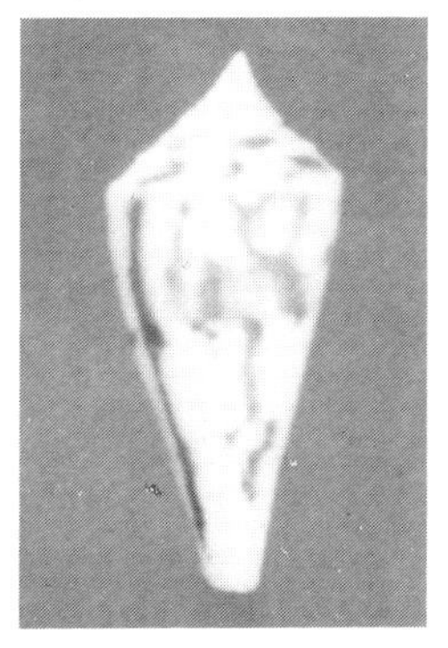

whorls, the pattern of the body whorl having many squarish spots. *C. centurio* may be similar in pattern, but it is relatively much wider and shorter in the body whorl and has many other differences.

Variation.—*C. villepinii* varies in height of the spire from moderately low to distinctly high, with straight to distinctly concave sides. The body whorl sculpture is seldom well developed above midbody but weak remnants can sometimes be seen to the shoulder. The spiral reddish brown bands are usually distinct, but they may be very irregular and blotchy; when extensions of the axial flammules are present, they are usually widely spaced and seldom penetrate the midbody area. There may be traces of brown dashes buried in the spiral bands in some specimens. Occasional specimens are white with only a few scraggly pale brown flammules.

Juveniles are less elongated, almost evenly biconical, with a heavy spire pattern and a wide pale brown band above midbody and a weaker band anteriorly. With growth the body whorl elongates, the sculpture becomes more restricted to the anterior end, and the bands consolidate by developing darker blotches within them.

Distribution.—Moderately common in deep water from off northern Florida south to southern Brazil. It seems to be most common near the continental shelf.

Synonymy.—*C. fosteri* was described as having a low spire and concave sides, both of which characters are very variable in this species. There is still some doubt as to whether the holotype of *C. villepinii* (32mm long) is actually the species here discussed. Although it agrees well in shape, there is no mention of nodules on the spire whorls, although they are usually quite distinct in juveniles of the present species. Also, the pattern consists of three irregular rows of large reddish brown blotches, which to the best of my knowledge is certainly not a typical pattern at any size in this species.

Notes.—Fine, but not gem, *C. villepinii* are easy to obtain, but they are dead-dredged and as such have the usual minor and major flaws. Live-taken specimens in true gem condition are relatively rare. Juveniles are also very uncommon.

VIMINEUS Reeve, 1849

1849. *Conus vimineus* Reeve. *Conchologia Iconica*, 1(*Conus* Suppl.): Pl. 7, sp. 269 (Cagayan, Is. of Mindanao, Philippines).

Description.—Light in weight but fragile, with a low gloss; elongate biconical, the upper sides convex then concave and tapering to a very long and narrow base; body whorl covered with rather narrow spiral ridges, rounded or nearly flat, separated by deep grooves containing heavy axial threads and often 1-2 weak secondary spiral ridges; sculpture usually distinct to shoulder; shoulder broadly rounded, not distinct from spire, slightly concave on top and with narrow raised margins and 1-2 spiral ridges and heavy axial threads; spire moderately high, sharply pointed, the sides straight or convex; early whorls apparently not nodulose; whorls with narrow but thickened raised margins and a concave or nearly flat area between; this area contains one heavy spiral ridge or one heavy and one narrow ridge, crossed by heavy axial threads that are widely spaced. Body whorl pale tan, sometimes with very indistinct traces of reddish brown pattern in three spiral bands below shoulder, at midbody, and above the base, spotting never distinct; spire also pale tan, sometimes with very faint pale brown spots; early whorls tan. Aperture narrow, of about equal width; outer lip very thin, strongly sloping below level of shoulder, straight; mouth whitish with large brownish blotch sometimes visible deep within. Columella internal, not visible. Length 25-50mm.

Comparison.—*Conus vimineus* is a doubtful species very closely related to *C. aculeiformis* var. *longurionis* in shape and sculpture. It seems to be constantly separable by combining a pale brownish shell with no obvious spotting or blotching with wide spiral grooving containing very heavy axial threads so the body whorl looks very strongly punctate, much more so than typical *C. aculeiformis*. The sculpture of the spire is slightly different also, with the margins of the whorls narrower though still thickened and having a wider, flatter area between, this area containing 1-2 strong spiral ridges crossed by heavy axial threads instead of a single often indistinct ridge in a deep groove between very wide margins. The body whorl also seems more inflated posteriorly, but this is perhaps an illusion due to lack of distinct pattern.

These differences seem to be constantly associated in Philippine shells and different in sympatric *C. aculeiformis* var. *longurionis*, but probably large series would prove overlapping or random distribution of the supposed distinguishing characters, in which case *C. vimineus* would become a synonym of *C. aculeiformis*.

Distribution.—Apparently restricted to moderately deep waters of the Philippines. Rather common.

Notes.—This species is even more fragile than normal *C. aculeiformis*, and the lip is never intact. The protoconch whorls are commonly broken off, but since there are at least four of them this is to be expected.

VIOLACEUS Gmelin, 1791

1791. *Conus violaceus* Gmelin. *Systema Naturae per Regna Tria Naturae*, ed. 13, 1: 3391 (Locality not stated). Holotype figures: Martini, 1774, *Allg. Gesch. Nat.*, 1: Pl. 2, figs. 18-19.

1792. *Conus tendineus* Hwass, in Bruguiere. *Cone*, in *Ency. Method.*, *Hist. Nat. des Vers*, 1: 733 (Africa). Lectotype (here selected): specimen figured by Kohn, 1968, as holotype.

1834. *Conus tendineus* var. *granulosus* Sowerby ii. *Proc. Zool. Soc. (London)*, 1834(II): 18 (Insulam Anna). Non *Conus granulosus* (Roeding, 1798).

Description.—Moderately light, with a low gloss; elongate cylindrical, the sides slightly convex below the shoulder then straight; body whorl covered from base to shoulder with narrow spiral ridges separated by deep grooves, sometimes weaker or obsolete at the shoulder; grooves with fine axial threads and sometimes strong spiral threads within; shoulder narrow, not distinct from spire; spire low or moderate, convex, the whorls overlapping, the tip rounded with a sharp protoconch; early 4-5 whorls strongly nodulose; later whorls with fine spiral and axial threads, none very distinct or enlarged, the surface slightly convex. Body whorl bright white, with strong violet tints at the base; this is covered by three broad dark brown to purplish brown spiral bands at shoulder and above and below midbody, these crossed by 4-7 broad straight axial bands of the same color to cut off two spiral rows of large axial white rectangles above and at midbody, the basal area largely white; occasionally there may be extra small axial flammules or some of the broad bands may be broken or absent; shoulder and spire dark brown or purplish brown, with or without large white blotches; early five whorls contrasting white in small specimens, tan in older specimens. Aperture narrow posteriorly, wider anteriorly; outer lip rather thick, nearly straight, often strongly sloping below level of shoulder. Columella narrow, not very distinct. Length 40-80mm.

Comparison.—*C. violaceus* is similar to *C. tenuistriatus* and *C. mitratus* in shape, but has the regular spiral body sculpture of *C. terebra*. From all these species it is readily distinguished by the regular axial and spiral bands of dark brownish, with the basal area largely white. Large *C. tenuistriatus* may have a vaguely similar pattern, but the background is not bright white, the banding is not as sharp, and the base is largely brownish and violet.

Variation.—This is a very uniform species in shape and sculpture as well as pattern. The color of the bands varies a bit from rather pale brown to deep purplish brown, almost black. Occasionally part of the pattern, usually the axials, is missing or reduced, but the spiral bands are not uncommonly broken. Occasionally smaller secondary flammules are present alongside the major bands, but these seldom change the pattern drastically. The base may be weakly or strongly tinted with violet. The shoulder and late whorls of the spire are usually covered with a broad dark brown band, but sometimes this is broken by white blotches. Rarely the basal ridges are weakly granulose.

Distribution.—Although usually stated to be Indo-Pacific in range, I have seen specimens and records only from the southern Indian Ocean, from Mozambique to Zanzibar and from the Seychelles to Mauritius and

Reunion. It is uncommonly collected in shallow water.

Synonymy.—This shell has long been known as *C. tendineus*, but it now seems certain that Gmelin's *violaceus* is the same species and has a year's priority. The change must be made.

Notes.—Fresh specimens have good color and few flaws other than chips or breaks in the lip, with the early spire whorls slightly eroded. Live-taken specimens are uncommon but not rare. Specimens are usually very regularly patterned, but sometimes the pattern deposited near the lip is irregular.

VIRGATUS Reeve, 1849

1849. *Conus virgatus* Reeve. *Conchologia Iconica*, 1(*Conus* Suppl.): 1. New name for a shell previously misidentified as *Conus zebra* Lamarck. Based on *C.I.*, 1843-44, Pl. 16, fig. 87 (Salango, Central America).

1849. *Conus Cumingii* Reeve. *Ibid.*, 1(*Conus* Suppl.): Pl. 8, sp. 277 (Salango, West Colombia). Non *Conus cumingii* Reeve, 1848.

1937. *Conus signae* Bartsch. *Nautilus*, 51(1): 3, pl. 2, fig. 8 (Guaymas, Mexico).

Description.—Moderately light in weight, with a good gloss; low conical, the upper sides slightly convex then tapering to a narrow base, body whorl rather elongated; body whorl with weak and usually widely spaced spiral ridges anteriorly, these often continued to about midbody, but weak; spire threads above midbody sometimes heavy, with many fine axial threads; shoulder sharply angled, usually flat or slightly concave above, broad, distinct from spire; spire low to moderate, the sides concave, sharply pointed; early whorls apparently not nodulose, somewhat carinate and often projecting; later whorls slightly concave or flat, covered with many curved axial threads and fine spiral threads, sometimes 1-2 spirals slightly heavier and almost ridge-like; whorls with slightly projecting margins except the last 2-3. Body whorl whitish or cream, heavily suffused with pale orange or salmon, occasionally yellowish; base white or pale; usually numerous wavy and often irregular axial flammules or pale brown from below shoulder to above base, with a strong tendency to break into large staggered dashes in two weakly distinguishable spiral bands at midbody and below the shoulder; fine spiral dashes sometimes present in the flammules, seldom distinct; pattern may be mostly or completely absent, resulting in a pale orangish or salmon shell; spire and shoulder heavily or lightly covered with curved brownish blotches and lines like those of body, sometimes forming slightly darker dashes at the margins; early whorls tan. Aperture moderately narrow, slightly widened anteriorly; outer lip sharp, straight or slightly concave at the middle; mouth white. Columella indistinct. Length 40-58mm.

Comparison.—*C. virgatus* is a very distinctive if simply patterned shell. It has often been confused, more nomenclaturally than in specimens, with *Conus lorenzianus* or *C. cingulatus*, but it is easily distinguished from both; *C. lorenzianus* has a spotted pattern that is heavy on the anterior end dorsally, while *C. cingulatus* has large spiral dashes under the flammules (if present) and is usually heavily grooved to at least midbody. *Conus daucus* is perhaps its closest ally, but that species is much less elongated in the body whorl, has spiral dashes and not flammules, and is usually yellowish to reddish, not pale salmon or orange on white. *Conus poormani* has the posterior sides parallel below the shoulder, is white, and has the flammules seldom going through the midbody area. *Conus recurvus* is different in shape, has the spire whorls strongly concave and heavily nodulose, and has a different pattern of flammules. *Conus clerii* has distinct squarish spots in the pattern. *Conus kerstitchi* is much less elongated and spotted, not flammulated.

Variation.—The major difference from shell to shell is the development of the flammules. Occasional shells are pure white with just a suffusion of salmon or tan over them; others have widely spaced rows of indistinct dashes forming faint axial lines; most have at least part of the body whorl covered with heavy brown axial flammules that run from below the shoulder to above the base with little interruption although wavy and often irregular in the midbody area. In some heavily patterned specimens there are indistinct spiral white bands at midbody and below the shoulder in which the flammules have broken into rather triangular spots. The base is apparently always unmarked, with the spire nearly white to solid brown.

Distribution.—Moderately common in shallow water from Baja California and the Gulf of California south to Ecuador.

Synonymy.—The first figure of this species was misidentified as *C. zebra* Lamarck, possibly a synonym of *C. ximenes*, and later named as new by Reeve. In the same supplement he described a poorly marked specimen under the name *C. cumingii*, preoccupied. The pale form with greatly reduced markings was later again redescribed as *C. signae*. They are all the same species, sharing a very distinctive shape and general appearance regardless of whether flammules are present or absent.

Notes.—This species is very clean under normal circumstances, the only major flaws being minor chips or thins in the lip. Specimens with weak and heavy patterns are not distinguished.

VIRGO Linnaeus, 1758

1758. *Conus virgo* Linnaeus. *Systema Naturae per Regna Tria Naturae*, ed. 10, 1: 713 (Locality not stated). Lectotype figure (Kohn, 1963): Rumphius, Pl. 31, fig. E.

1807. *Conus flavocinctus* Link. *Beschr. Nat.-Samml. Univ. Rostock*, 3: 103 (Locality not stated). Based on Martini, Pl. 53, fig. 586.

Description.—Heavy, with a low gloss; obconical, the sides nearly straight, slightly convex at the shoulder; body whorl with narrow widely spaced low spiral ridges on the anterior third or less of whorl, remainder covered with very fine spiral and axial threads, apparently smooth in large specimens; shoulder broad, sharply angled, flat to concave above; spire low to flat, the sides slightly concave to slightly convex, the early whorls rounded but not usually forming a distinctly set off cone; tops of whorls with very fine spiral and axial threads, none obviously enlarged. Body whorl pale yellow to white, the base broadly stained with deep purple visible both dorsally and ventrally; occasionally with traces of paler or darker spiral bands, never distinct; spire same color as body whorl, usually paler, the early whorls pale or eroded white. Aperture moderately narrow posteriorly, somewhat wider anteriorly; outer lip sharp, straight or nearly so; mouth glossy white with a large dark purple blotch anteriorly. Columella long and narrow, slightly oblique. Length 50-152mm.

Comparison.—*Conus coelinae* is often held to be a synonym of *C. virgo*, but it seems sufficiently distinct to be held as a full species. In all its forms the base is not stained with violet, being either white or pale tan; the shoulder is often very sharp and concave above, with the early whorls forming a small but distinct cone that is often slightly darker in color than the remaining whorls. Additionally, the spiral ridges on the body whorl are usually heavier than in *C. virgo* and extend further up the whorl, sometimes still easily visible at the shoulder. There is also a stronger tendency in *C. coelinae* to be indistinctly banded with yellow, tan, violet, or pink on a white ground, instead of very poorly contrasted white to pale yellow.

Conus emaciatus has often been considered just a juvenile or dwarf form of *C. virgo*, but it also appears quite distinct. In typical *C. emaciatus* there are heavy spiral ridges, usually granulose, to at least midbody and sometimes to the shoulder. Small specimens are yellowish with two paler yellow bands that are often quite distinct. The simplest difference is that in even small (35mm) *emaciatus* the spire whorls are heavily eroded and encrusted, moreso than in typical *C. virgo* over 80mm in length.

Variation.—Juveniles are said to be orange in color with distinct spiral ridges on the body anteriorly. Adults vary somewhat in tone from distinctly yellow to distinctly white, but the contrasting purple base is constant. The posterior half or more of the body whorl is usually nearly smooth to the eye in specimens over 70mm.

Distribution.—Widely distributed and common throughout the Indo-Pacific from the eastern African coast to French Polynesia, absent from Hawaii (where it is replaced by *C. coelinae spiceri*). Shallow water.

Notes.—Large specimens are usually cumbersome and eroded, with bad lips, but normal adults are very clean. Juveniles are rare, and I am not at all sure I could distinguish them from juveniles of *C. emaciatus.* Few adults have any distinct colors.

VITTATUS Hwass, in Bruguiere, 1792

1792. *Conus vittatus* Hwass, in Bruguiere. *Cone*, in *Ency. Method., Hist. Nat. des Vers*, 1: 704 (Indian Ocean). Lectotype selected and figured by Kohn, 1968.

Description.—Moderately light in weight, with a good gloss; low conical, rather cylindrical; sides gently and evenly convex; body whorl with a few crowded spiral ridges at the base, these soon replaced by narrow, broadly spaced regular spiral ridges usually extending to shoulder; fine axial threads and scratches often obvious; shoulder roundly angled, flat or slightly concave above; spire low to moderate, sharp, the sides flat to convex; early 1-2 whorls weakly nodulose; remaining whorls with straight margins and flat or concave on top, with 3-4 weak spiral ridges sometimes visible and numerous curved axial threads, sculpture weak. Body whorl tan, reddish or pinkish, even yellow, usually with a broad white midbody band and occasionally weaker white bands or series of white blotches above the base and below the shoulder; midbody band with dark brown indistinct squarish spots posteriorly, these usually with irregular axial flammules extending to the anterior edge of the band; entire whorl covered with large and small brown dots and dashes on top of the spiral ridges, these sometimes partially continuous; base yellowish or, more commonly, deep salmon; spire white, heavily covered with curved dark brown to blackish or occasionally reddish blotches; early whorls pale. Aperture moderately narrow, wider basally; outer lip thin, sharp, slightly convex to straight; mouth white with pinkish tones or pale violet, often with a narrow salmon blotch anteriorly. Columella inconspicuous. Length 25-45mm.

Comparison.—*C. vittatus* is an often brightly colored cone with a strong resemblance to *C. coccineus*. It shares with that species the white midbody band and the spiral dashes, but differs in being regularly spirally ridged, having the spire whorls with straight and not distinctly undulate margins, and having a strongly contrasting dark and light spire pattern instead of a mostly dark spire. From *C. cardinalis*, which is often very similar in color and pattern, it is distinguishable by the thinner shell, the evenly spaced spiral ridges over the whorl with few or no distinct granules, and the straight spire margins rather than undulate or low coronate spire whorls. Some specimens of *Conus amphiurgus* are very similar in shape and pattern, although usually the shoulder is broader and the spire has more concave sides than in *C. vittatus; C. amphiurgus* has the distinct spiral ridges restricted to the anterior third of the whorl, with only scattered and irregular ridges and threads above this; the colors are usually paler in *C. amphiurgus*, with the base pale, not bright salmon.

Although *C. vittatus* has often been confused with *C. orion*, the two species have little in common except the salmon base. *C. orion* has a sharp shoulder, heavy axial flammules, is usually brownish and not reddish, and has the spiral dashes weak or absent; in addition, the body whorl is smooth and polished, not covered with strong spiral ridges.

Variation.—The shape is rather constant, though the extent of convexity of the sides and width at the shoulder vary a bit. The spiral ridges are strong and seldom obsolete below the shoulder. Although a white midbody band with brown dashes and spots is normally present, the two bands at the base and below the shoulder may be absent, indicated only by whitish blotches, or as fully developed as the midbody band. In occasional specimens pale brown axial clouds may be developed along axial growth marks, but these are seldom distinct. The color varies from light or dark brown to pale pink or bright red; occasional specimens are yellow; the spire is always heavily patterned with dark and light. The salmon blotch at the base is usually present but may sometimes be yellowish or tan.

Distribution.—Uncommon in the Panamic region from the Gulf of California south to Ecuador; seldom intertidal.

Synonymy.—The type figure of *C. coccineus* seems to be *C. vittatus* instead of the western Pacific species, but this is preferably ignored for the moment. *Conus henoquei* is often considered a synonym of *C. vittatus*, but for the moment it is considered as a dubious species and discussed later.

Notes.—*C. vittatus* is an uncommon species but an often brightly colored one. Specimens are usually some shade of reddish tan or dark pinkish, with yellow specimens apparently uncommon. Normally only the midbody band is well developed, but specimens with three distinct bands are not very uncommon. The early spire whorls are commonly eroded, and there may be major axial growth marks, but *C. vittatus* is otherwise quite clean.

Other than the bright salmon base, I can see no reason why this species has been confused with *C. orion*.

VITULINUS Hwass, in Bruguiere, 1792

1792. *Conus vitulinus* Hwass, in Bruguiere. *Cone*, in *Ency. Method.*, *Hist. Nat. des Vers*, 1: 648 (Indian Ocean). Lectotype selected and figured by Kohn, 1968.

1807. *Conus cinctus* Link. *Beschr. Nat.-Samml. Univ. Rostock*, 3: 102 (Locality not stated). Non *Conus cinctus* Bosc, 1801.

1829. *Conus vulpinus* Schubert and Wagner. *Syst. Conch. Cab.*, 12: 56, pl. 222, fig. 3073 (Locality not stated). Non *Conus vulpinus* Hwass,

in Bruguiere, 1792.
1942. *Conus praeclarus* Fenaux. *Bull. l'Inst. Oceanogr.* (*Monaco*), No. 814: 2, fig. 3 (Bermudes).

Description.—Moderately heavy, with a good gloss; low conical or obconical, the sides nearly straight or slightly convex; body whorl with about 10-20 narrow spiral ridges at the base, the more posterior ones widely spaced and usually heavily granulose; rest of whorl with fine spiral threads, some widely spaced ones sometimes granulose to shoulder; fine axial scratches and threads sometimes present; shoulder broad, sharply angled, slightly concave on top; spire low to flat, blunt, the sides straight or nearly so; tops of whorls slightly concave to flat, with about 4-7 narrow spiral ridges crossed by strong axial threads, appearing cancellate. Body whorl white, the base dark violet brown; two broad spiral bands of tan to dark brown above and below midbody, the midbody band usually white or very pale yellowish; fine widely spaced colorless spiral lines or dashes are often visible within the spiral bands; usually a narrow to moderately wide white band below the shoulder is present; many specimens are heavily covered with narrow and rather straight dark brown axial flammules which may be indistinct and confined to the spiral bands or heavier and longer, penetrating the midbody and shoulder bands; spire white, with usually heavy covering of curved blackish brown blotches and lines, more dark than light in typical specimens; early whorls eroded. Aperture moderately narrow, distinctly wider anteriorly; outer lip thin, sharp, evenly convex or straight; mouth white, the margin often brown, with a large violet-brown blotch at the anterior end. Columella long and narrow, sometimes indented. Length 35-77mm.

Comparison.—Although some patterns closely resemble *C. g. generalis*, that species has a tall turbinate spire, weak anterior sculpture that is not granulose, and weak or obsolete spiral sculpture on the spire whorls. This species is closely related to *C. planorbis*, *C. striatellus*, and *C. connectens*. From *C. planorbis*, which it sometimes closely resembles in pattern, it differs in having heavier black markings on the spire and a large violet-brown blotch anteriorly in the mouth. From *C. striatellus* it differs in the same ways as from *C. planorbis*, plus lacking the numerous closely spaced brown hairlines of *C. striatellus*, usually having instead fine widely spaced colorless lines. From *C. connectens* it differs in the darker flammules, the violet-brown base darker, the usual absence of spiral brown hairlines, and the absence of bright pinkish to pale violet tones; the spire markings are usually heavier than in *C. connectens*.

Variation.—There are two general groups of patterns in this species, banded and flammulated, with the two sometimes occurring on the same shell. Small specimens commonly have a broad tan spiral band posteriorly, the anterior band broken or indistinct. Many adults retain this juvenile pattern, having only weak flammules visible in the shoulder band; the midbody band is often yellowish in such shells, with faint colorless or yellowish hairlines visible in the spiral bands. In the other pattern the shell is often darker brown, the axial flammules heavy and often extending from the shoulder to the base with little interruption; the clear hairlines in the spiral bands appear as distinct dashes because of the contrasting colors of the spiral band, and there may be large white areas breaking the spiral bands into large blotches along the sides of the flammules. Granules may be restricted to the base or extend as widely spaced rows to the shoulder in occasional specimens. The base is always very dark, not tan or pinkish tan.

Distribution.—Widespread and common in the Indo-Pacific from eastern Africa to Hawaii and French Polynesia in shallow water.

Synonymy.—Although described from Bermuda, *C. praeclarus* seems to be just a specimen of *C. vitulinus;* like Fenaux's other species, it is very poorly described. The earlier names are both preoccupied.

Notes.—This variable shell is the most distinctive of its group because of the often heavy pattern and indistinct or absent hairlines replaced by clear lines. Juveniles bear some resemblance to *Conus tessulatus* in having distinct blocks of tan anteriorly before the spiral band is formed.

Except for minor erosion of the spire and some axial growth marks, this is usually an easy species to obtain in gem. The various patterns are about equally common, although very small specimens are seldom seen.

VOLUMINALIS Reeve, 1843

1843. *Conus voluminalis* Reeve. *Conchologia Iconica*, 1(*Conus*): Pl. 37, sp. 206 (Straits of Malacca).

1857. *Conus Macarae* Bernardi. *J. Conchyl.* (*Paris*), 6: 56, pl. 2, fig. 2 (Locality unknown).

Description.—Light in weight, waxy, with a high gloss; elongate low conical, the sides straight to slightly concave and strongly tapering to a very narrow base; body whorl with about 4-6 weak spiral ridges at the base, otherwise smooth except for fine axial growth lines; shoulder very broad, carinate and sometimes projecting; spire moderate, the later whorls forming a flat plate from which projects a tall, sharply pointed cone made from the earlier whorls; first 5-7 whorls distinctly nodulose, the later ones with straight or rude margins; tops of later whorls flat or slightly concave, with traces of about two spiral ridges (often threads) and numerous fine axial threads; earlier whorls with the spiral ridges more distinct but usually eroded. Body whorl white to pale lavender, covered from above base to shoulder with fine brown spiral hairlines; two broad spiral bands of yellow to red, usually tan, above and below a weakly marked midbody band; the midbody band with squarish dark brown spots posteriorly; base white to pale tan; usually with a narrow white band below the shoulder, this band with short dark brown axial flammules; spire white, heavily checkered with pale to dark brown blotches; early whorls whitish. Aperture narrow, slightly wider anteriorly; outer lip fragile, straight or concave at middle; mouth white to pale yellow or pale violet, the external pattern showing through. Columella internal, inconspicuous. Length 30-50mm.

Comparison.—The extremely slender and tapered body whorl will distinguish this species from almost anything else. It is very close to the species here called *C. polygrammus*, but that shell is relatively broader and less attenuated, has the shoulder narrower and less prominent, seldom has the spire as flat on the later whorls, usually has only the first three or so whorls nodulose, the later whorls undulate or low coronate, and lacks the strongly contrasted midbody band with large squares posteriorly as well as the narrow flammulated band below the shoulder. *C. recluzianus* is larger, heavier, has strongly undulate spire whorls with only the first 2-3 whorls weakly nodulose, and has strong spiral ridges on the spire whorls; the colors are seldom as bright as those of *C. voluminalis* or the body whorl so attenuated anteriorly.

C. eximius has a much more mottled and irregular pattern, deep grooves anteriorly on the body whorl, and strong spiral ridges on the spire whorls; it is also relatively broader. *C. lemniscatus* is grooved at least anteriorly, has distinctive spire sculpture, and has a spirally dashed pattern.

Variation.—The color varies considerably, from bright yellow to bright red, but is usually a pale reddish brown. The large square spots at the posterior edge of the midbody band are usually distinct, but they are sometimes weak or absent, while the hairlines of brown may be heavy and continuous, broken into dashes, or almost absent. The band below the shoulder is usually fairly distinct, though it is not uncommonly absent. The shape is constant, very slender and elongated with the shoulder broad and sharp, the spire very deeply concave.

Distribution.—*C. voluminalis* appears to be moderately uncommon in the eastern Indian Ocean from the Bay of Bengal into the Indonesian area and in the western Pacific from the Indonesian area north to the Philippines and Taiwan. Its more eastern range is uncertain, though there seems no reason why it should not occur in New Guinea and the Solomons. Dredged in deep water.

Synonymy.—Use of the name *voluminalis* for this species seems very uncertain as Reeve's figure looks more like *C. planorbis* var. *chenui* than the present species, but his description emphasizes the texture, broad sharp shoulder, and fragile feel, which do fit the species. Probably the figure is just poor. *C. macarae* is indeed this species; that name is often used by dealers for poorly patterned specimens.

Notes.—This fragile shell usually has heavily filed lips and heavy axial growth marks which disrupt the pattern. In specimens with many such growth marks the spiral bands may appear to break into large axial flammules. Poorly marked specimens with large colorless areas are common, perhaps more common than well marked specimens.

This species has lately been sold under the name *C. bayani*, but that species is a thicker shell with a streaky and spotted pattern, not spiral hairlines; the textures of the two species are very different, *C. bayani* not nearly as waxy as *C. voluminalis*; the spire in *C. bayani* is sparsely marked.

WALLANGRA (Garrard, 1961)

1961. *Floraconus wallangra* Garrard. *J. Malac. Soc. Australia*, 5: 29, pl. 1, fig. 3 (East of Stanwell Park, New South Wales, Australia).

Description.—Light in weight, with a good gloss; low conical, the sides nearly straight, only slightly convex posteriorly; body whorl with about 5-10 low spiral ridges at the base and scattered low spiral threads and punctate grooves sometimes visible above midbody; numerous fine axial growth marks usually present; shoulder sharply angled, flat or slightly concave above; spire low, bluntly pointed, the sides straight to slightly concave or convex; spire whorls slightly stepped, the margins often rude; tops of whorls flat or very slightly concave, with about 5-7 distinct narrow spiral ridges crossed by axial threads so faintly cancellate. Body whorl cream to pale straw with violet tones, usually with two broad spiral bands of irregular large pale brown blotches above and below mid-body, these sometimes fused into continuous bands; base pale; rest of whorl usually heavily mottled with small triangular or rounded spots and streaks of brown, these sometimes reduced and in more distinct spiral rows; shoulder usually with slightly larger spots and streaks; spire colored like body whorl, streaked and blotched with pale brown. Aperture moderately wide, slightly wider anteriorly; outer lip nearly straight; mouth pale violet to deep pink within. Columella long, narrow, posteriorly free and projecting. Length 30-40mm.

Comparison.—*Conus wallangra* is very similar to *C. papilliferus* in shape, sculpture, pattern, and free columella. It is only doubtfully distinct at the specific level, perhaps being just a deeper water ecotype of the common intertidal *C. papilliferus*. However, it does seem to have a slightly higher spire that is sometimes weakly stepped, a sharper shoulder angle on the average, slightly heavier spiral sculpture on the spire whorls, and a moderately distinctive pattern of two spiral brown bands and small mottled spots instead of distinct axial flammules and spiral rows of dots. The mouth is perhaps brighter, but this seems very variable.

Variation.—Considerable variation exists in the height and shape of the spire, from low and distinctly concave to higher and convex. The margins of the whorls often appear rude or weakly undulate. The original description mentions punctate rows over the body whorl, but I have seen only the merest traces of punctations in the few specimens I have examined.

The pattern is fairly constant, although the spiral bands are more broken and made of smaller blotches in N.S.W. specimens than in those from Queensland. In southern specimens the mottling is stronger and more definite, while in northern specimens the spotting is reduced and the spots may appear to again be in spiral rows and rather regular. The ground color varies from distinctly violet with pink tones in southern specimens to straw with pale violet tones in northern specimens. Mouth color is quite variable.

Distribution.—Known only from rather deep water off southern Queensland and New South Wales, Australia. Uncommon.

Notes.—The lip is commonly broken and filed, and the axial growth marks may be heavy. Southern specimens are more attractively colored than Queensland specimens. This shell is almost certainly not a full species, which should be remembered if high prices are asked for specimens.

WITTIGI Walls, 1977

1977. *Conus wittigi* Walls. *The Pariah,* No. 1: 1, figures (Lesser Sunda Islands north of Timor).

Holotype: Delaware Mus. Nat. Hist.; 31.8mm; Lesser Sunda Islands north of Timor.

Paratypes: Collections of Mr. and Mrs. Renate W. Skinner, George Gundaker, and the author.

Diagnosis: A rather small tented cone related to *C. spectrum* in lacking spire nodules and having distinct spire sculpture, but differing in the tented pattern and juveniles with angulate shoulders and deep spiral grooves over the body whorl. 24-41mm length.

Description.—Moderately light in weight, with a high gloss; low conical, the sides slightly convex, widest just anterior to shoulder; body whorl with usual crowded oblique ridges, these changing to 2-4 broad flat ribs separated by narrow, indistinctly punctate spiral grooves below midbody; grooves and ribs obsolete above midbody in adults but continuing to shoulder in juveniles; numerous fine axial scratches over entire whorl; shoulder angulate in juveniles, weakly rounded in adults; spire low, sharply pointed, the sides straight or nearly so, the earliest three whorls and occasionally 1-2 later whorls weakly stepped; protoconch pale tan, glossy, bulbous, about two whorls, broken in adults to produce a hard, cylindrical projecting tip; first three whorls rounded, with 1-2 deep spiral grooves; later whorls flat above (rarely weakly concave or convex) with about four weak spiral flattened ridges separated by shallow grooves; whorls with indistinct axial scratches. Body whorl glossy white or pale cream, covered with thin wavy axial lines of pale to dark yellowish brown; the lines extend from the shoulder to above the base, tending to overlap and merge to produce fine to coarse tents; lines vary from nearly straight with little tenting to obsolete in a broad band near midbody; there is often a tendency to form squarish spots above and below the mid-body area through filling in of some tents with nebulous lines and clouds of brown; basal area largely white; spire and shoulder with continuations of axial body pattern to produce irregular spots or jagged markings on a whitish ground; early whorls dirty white to pale tan, unmarked. Aperture moderately wide posteriorly, widened anteriorly; outer lip sloping below level of shoulder in adults, thin, slightly convex; mouth pale pink, fading to white or bluish white, with traces of a small pale tan spot at anterior margin. Columella long, narrow, sometimes weakly set off by an oblique ridge posteriorly but more often in line with inner lip. Length 24-41mm.

Comparison.—Although usually with a distinctly tented pattern, *C. wittigi* lacks the spire nodules of the textile cones, *C. lienardi*, and *C. neptunus*, while lacking the small spiral brown and white dots and dashes of the pennaceus cones. Juvenile specimens resemble small *C. amadis* in the angulate shoulder, widely spaced spiral grooves over the body whorl, and the pattern, but that species is distinctly nodulose on the early whorls; adult *C. amadis* are not similar to adult *C. wittigi.*

The closest relative of *C. wittigi* seems to be *Conus spectrum*, which similarly lacks nodules on the early whorls and has distinct spiral ridges on the spire whorls, though with a somewhat wider mouth. The spire of *C. wittigi* is higher than in most forms of *C. spectrum*, with a more slender shape to the body whorl; the columella in *C. spectrum* is longer than in *C. wittigi* and is more strongly set off posteriorly and indented from the inner lip. I know of no color variety or individual variant of *C. spectrum* which has a distinctly tented pattern as in *C. wittigi* or which even approaches it. Also, juvenile *C. spectrum* lack the sharply angulate shoulder and deep spiral grooves over the body whorl of juvenile *C. wittigi*. The small spiral brown dots of most *C. spectrum* patterns are apparently absent or very greatly reduced in *C. wittigi.*

Variation.—Adults of *C. wittigi* are constant in shape and sculpture, differing mostly in the intensity of the color pattern as indicated in the

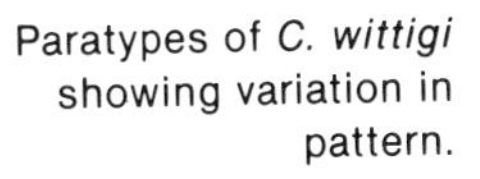

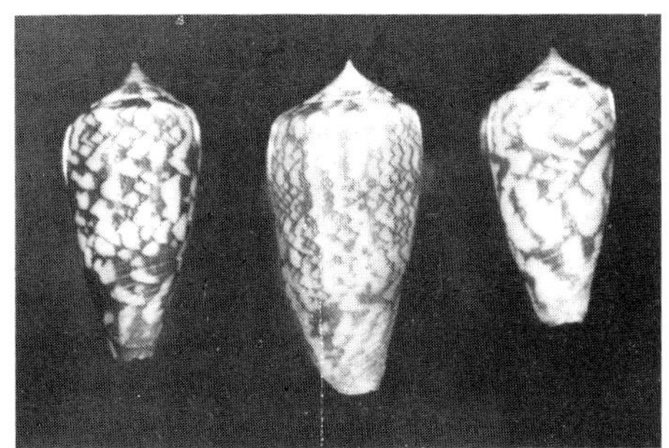

Paratypes of *C. wittigi* showing variation in pattern.

description; a nearly all-white shell has been reported. Juveniles are distinguished from adults by the more strongly angulate shoulder and the widely spaced distinctly punctate spiral grooves extending from the basal area to the shoulder. The posteriormost grooves begin to disappear at about 15mm in length and are restricted to the anterior half of the body whorl by a length of at least 24.6mm.

Distribution.—So far known only from one island in the Lesser Sundas, Indonesia, north of Timor. Uncommon in rather shallow water.

Notes.—*C. wittigi* has been marketed under the name *C. convolutus*, but that name seems applicable to a slender form of *C. pennaceus* with a tall and very concave spire. Other than minor chips in the lip, this is a very clean species.

Almost all specimens have been collected by Renate Skinner and her husband Ed, who loaned me a good series for description.

The species is named after Mr. Fritz Hermann Paul Wittig at the request of his daughter, Renate.

XANTHICUS Dall, 1910

1910. *Conus xanthicus* Dall. *Proc. U.S. Nat'l. Mus.*, 38(1741): 225 (Off Guaymas, Mexico).

1968. *Conus chrysocestus* Berry. *Leaflets in Malacology*, 1(25): 157 (Off Morro Colorado, Sonora, Mexico).

Description.—Moderately light in weight, with a good gloss and waxy texture; low conical, the upper sides nearly parallel just below the shoulder then straight or slightly concave to base; body whorl with fine spiral ridges at the base, these soon obsolete and replaced by many fine spiral threads and scratches; axial growth marks sometimes heavy; shoulder broad, carinate, concave above, with only axial threads or faint spirals on occasion; spire very low to moderately high, sharply pointed, the sides concave; first three or so whorls weakly nodulose, then next three or so undulate or irregular; early whorls stepped slightly; whorls concave above, with about three weak spiral ridges and axial threads, the spirals becoming weaker on later whorls. Body whorl deep orange-tan to bright yellow, sometimes with a distinctly greenish tone; two more-or-less distinct spiral bands of white blotches at midbody and below the shoulder, the one at midbody usually wide and distinct, sometimes continuous or with only narrow axial flammules or a narrow spiral line through the middle; band below shoulder usually weaker and indistinct; base colored or white; margin of shoulder narrowly white and with axial flammules in some specimens; spire white, heavily checkered with orange-brown to yellowish; early whorls pale. Aperture moderately narrow, nearly uniform in width; outer lip thin, straight or slightly concave at the middle, the posterior margin level with shoulder or projecting slightly above it; mouth white. Columella internal, not distinct. Length 30-60mm.

Comparison.—*C. xanthicus* has long been considered a juvenile of *C. fergusoni*, but it seems certain now that it is quite distinct and perhaps not actually related. *C. fergusoni* is a thicker, heavier shell with a rounded shoulder, weaker sculpture on the spire whorls, a paler color at all sizes, lacks the distinct spiral banding of *C. xanthicus*, and has a whitish spire without strong markings. *C. fergusoni* has the posterior edge of the outer lip below the level of the shoulder, while it is even with or above the level of the shoulder in *C. xanthicus*.

Perhaps more closely related to *C. xanthicus* is *C. excavatus*, a species which is very similar in shape, thin, has weak sculpture much as in *C. xanthicus*, and often has a similar pattern of spiral white bands at midbody, below the shoulder, and at the margin of the shoulder. However, it is seldom as waxy or as bright in color, usually shows at least some brownish dashes somewhere in the pattern, has the early whorls straight or undulate, not nodulose, and has the outer lip slightly sloping.

Variation.—The color varies from yellow with greenish tones to bright dark orange or pale orange-tan; the spire is heavily checkered in the specimens seen. The spiral bands are variably developed but the midbody band at least is present and well developed; the other two bands may be absent or strongly developed, clear or blotched. The spire may be almost flat or quite tall and distinctly stepped, while the outer lip seems to be constantly at least level with the shoulder.

C. xanthicus, Guaymas, Mexico, 58.5mm (photo Kerstitch).

Distribution.—Presently known only from deep water off Mexico in the Panamic region, but probably more widely distributed in the eastern Pacific. Very uncommon or rare.

Synonymy.—The holotype of *C. xanthicus* was illustrated in color by Hanna, 1963, *Occas. Pap. California Acad. Sci.*, No. 35: Pl. 5, fig. 5. *Conus chrysocestus* has not been figured, but it is certainly this species.

Notes.—*Conus xanthicus* is a rare and colorful cone with an unusual pattern, but it unfortunately is rather fragile. All specimens apparently have heavy growth marks and broken lips, but this does not detract greatly from the appearance.

The confusion of this species with *Conus fergusoni* does not seem reasonable, as the two have quite different color patterns and textures, as well as one species being thick and heavy, the other thin and fragile.

XIMENES Gray, 1839

?1810. *Conus zebra* Lamarck. *Ann. du Mus. Hist. Nat. (Paris)*, 15: 273 (?Asian Ocean). See Kiener, Pl. 76, fig. 2.

1829. *Conus interruptus* Broderip and Sowerby i. *Zool. J.*, 4: 379 (Near Mazatlan). Non *Conus interruptus* Wood, 1828.

1839. *Conus ximenes* Gray. *Zool. Beechey's Voy., Moll.:* 119 (Panama).

1843. *Conus mahogani* Reeve. *Conchologia Iconica*, 1(*Conus*): Pl. 22, sp. 126 (Salango, West Colombia).

1844. *Conus exaratus* Reeve. *Ibid.*, 1(*Conus*): Pl. 44, sp. 238 (Locality unknown). See Emerson and Old, 1962, *Amer. Mus. Novitates*, No. 2112: 36.

1945. *Hermes triggi* Cotton. *Rec. South Australian Mus.*, 80(2): 267, pl. 4, fig. 11 (? New Hebrides).

Description.—Moderately heavy, with a good gloss; low conical, the upper sides slightly convex, almost straight; body whorl with a few spiral ridges above the base and widely spaced punctate shallow grooves on the anterior third or so of whorl in adults, the grooves sometimes reaching shoulder in juveniles; numerous axial growth marks; shoulder broad or moderately narrow, roundly to narrowly angled, not carinate; spire moderate, sharply pointed, the sides concave to deeply concave; tops of whorls slightly concave to flat, the margins rounded to weakly carinate and not distinctly projecting; tops of whorls with fine axial threads and fine spiral threads, sometimes 1-2 threads a bit heavier than the others. Body whorl whitish or creamy to pale straw, covered from base to shoulder with many spiral rows of small squarish brown spots; tendency in some populations to develop large and small blotches above and below midbody, sometimes also light or heavy axial dark brown to purplish brown axial flammules over the entire whorl; shoulder with large rounded brown spots in most specimens, occasionally with only small blotches; rest of spire may be covered with similar large brown oval spots or may be patternless except for a row of small brown dots at the margins; early whorls commonly eroded white. Aperture moderately narrow, slightly wider anteriorly; outer lip straight or nearly so, rather thick; the anal notch is very shallow or obsolete; mouth white to deep violet, variable. Columella narrow, not very distinct. Length 25-59mm.

Comparison.—The finely spotted pattern is similar only to that of *C. puncticulatus perplexus* and *C. tornatus*. *C. p. perplexus* is relatively broader and stubbier, with a lower spire; there are usually large squarish blotches on the body whorl just under the margin of the shoulder; the columella is wide and distinct, sharp and projecting; and the anal notch is normal and deep. *C. ximenes* and *C. p. perplexus* are not closely related, the former perhaps related to *C. ermineus*, the latter to *C. puncticulatus* and allies.

The relationship of *C. ximenes* and *C. tornatus*, as mentioned earlier under the latter species, is very close and they are probably not separable conchologically in all cases. However, usually in *C. tornatus* the shoulder is distinctly carinate rather than rounded, the spire whorls are often deeply concave with projecting margins, and the spiral grooving of the body whorl is commonly deep and extends above midbody in adults; *C. tornatus* is usually granulose, while *C. ximenes* rarely has more than a few granules on the basal ridges. Juveniles of *C. ximenes* are sometimes grooved to the shoulder, but they have rounded shoulders.

Variation.—The shape varies from rather fusiform with a slender body whorl and narrow shoulder to broader, the body whorl relatively broader at the shoulder. The spire may have almost straight sides or be deeply concave, but it is always moderately tall or taller. The brown spots are somewhat more numerous in some populations than in others, and are sparse in small specimens. Usually the spire is heavily spotted with large rounded or oval spots, but in some populations the spots are absent and replaced by a row of small dots at the margins and a few scattered blotches.

Extremely flammulated *C. ximenes*, Nicaragua, 46mm (GG).

Two varieties or full species are commonly recognized in *Conus ximenes*, the typical variety with limited or no dark blotching and flammules and with a violet mouth, and the *mahogani* variety with heavy axial blotching sometimes obscuring much of the body whorl and a white mouth. However, I cannot see how these two can be separated as clear specimens commonly have white mouths, heavily flammulated specimens may have deep violet mouths, half a specimen may be clear with no blotching and the other half heavily axially flammulate, and both types can be taken in the same population. The variation in color seems to be partially ecological, with accumulations of pigments on axial growth lines.

Distribution.—Common intertidally and in moderately deep water from the Gulf of California to Peru.

Synonymy.—*Conus zebra* has been used as a form of *C. subulatus* or *C. stramineus* by various authors, but the specimen figured by Kiener would seem to be an unusual specimen of *C. ximenes*; the status of this name is very uncertain. *C. interruptus* was of course preoccupied, and *C. ximenes* is the first available name. *C. mahogani* is considered a minor variety, while it seems that *C. exaratus* is a juvenile. The status of *C. triggi* is uncertain; it was described from a single specimen of uncertain locality found with some shells from New Hebrides. Since its description no shell agreeing with the figure and description of the type has been found, although a very dissimilar but unidentified shell has been incorrectly identified with *C. triggi.* To me the figure and description appear to be a rather normal specimen of *C. ximenes* belonging to a population with very little blotching on the body whorl and fine dots at the margins of the spire whorls. A comparison of the type figure from Cotton's paper with the specimen figured as 517 (bottom) of Keen's *Sea Shells of Tropical West America*, 2nd edition, p. 668, will show the great similarity.

Notes.—This is a very common species that is easily collected in series to show the variation in blotching and body shape. Juveniles are uncommon, as are specimens over about 50mm. Specimens seldom have granules developed as in the similar *C. p. perplexus.* It commonly has heavy axial growth flaws and heavily eroded spire whorls.

ZEYLANICUS Gmelin, 1791

1791. *Conus Zeylanicus* Gmelin. *Systema Naturae per Regna Tria Naturae*, ed. 13, 1: 3389 (Locality not stated). Lectotype figure (Kohn, 1966): Martini, 1777, *Neuste Mannigfaltigkeiten*, 1: Pl. 2, fig. 20.

1792. *Conus obesus* Hwass, in Bruguiere. *Cone*, in *Ency. Method., Hist. Nat. des Vers*, 1: 623 (East Indies). Lectotype selected and figured by Kohn, 1968.

1798. *Cucullus theobroma* Roeding. *Museum Boltenianum:* 43 (Locality not stated). Lectotype figure (Kohn, 1975): Chemnitz, Pl. 142, fig. 1318.

1798. *Cucullus meningeus* Roeding. *Ibid.:* 49 (Locality not stated). Lectotype figure (Kohn, 1975): Chemnitz, Pl. 142, fig. 1318.

Description.—Heavy, thick, with a good gloss; low conical, the sides convex posteriorly and inflated; body whorl with about 10-15 shallow grooves anteriorly forming low spiral ridges, smoother posteriorly with many low spiral and axial threads and growth marks; shoulder wide, rather convex, the margin with low coronations or undulations; spire low to moderate, sharply pointed, the sides slightly convex; early whorls weakly nodulose, later whorls low coronate to undulate; tops of whorls flat to concave, with traces of 1-2 low spiral ridges and fine axial threads. Body whorl white, with strong pinkish tones, covered with a very variable pattern of small brown dashes, opaque white blotches and dashes, larger blackish triangles, and very fine axial wavy pale brown hairlines; combinations of these features result in patterns usually comprising two broad and indistinct spiral bands of blackish triangles and blotches, numerous brown dashes, and the opaque spots, the bands usually above and below midbody; along axial flaws are often developed concentrations of short hairlines and blackish triangles to produce indistinct and irregular axial dark flammules; the whole pattern is very inconstant and hard to describe; shoulder whitish with large dark blotches formed from concentrations of hairlines; spire whorls whitish with smaller blotches; early whorls white; base of shell usually strongly tinted pale violet-pink. Aperture narrow posteriorly, much wider below; outer lip rather thick, straight or nearly so; mouth white, sometimes with a deep pink blotch within. Columella long, wide, thick, twisted, bounded by a very heavy ridge posteriorly, the anterior tip partially projecting and pointed. Length 35-60mm.

Comparison.—The general shape and thickness of the shell resemble only *C. pulicarius* or *C. arenatus* among other cones. From both these species it is easily distinguished by the variety of dots, dashes, blotches, and hairlines in the pattern instead of the small or large dark dots which comprise the patterns of the other two species, especially the small hairlines which are never present in any form in *C. pulicarius* or *C. arenatus*. Other differences are the very low or obsolete shoulder coronations and the pale violet-pink base, as well as the generally pinkish tinge of the entire shell.

Variation.—Specimens vary in shape from rather cylindrical with almost straight to slightly concave sides to extremely inflated or obese, almost as wide across the shoulders as it is high. The spire varies from flat to quite tall, apparently as normal individual variation. The pinkish tones may be heavy or reduced, but they are apparently always present. In some shells the short hairlines are heavy and obvious, forming large clusters along axial flaws, but in others they may be absent to the eye until examined closely with a lens. As indicated in the description, combinations of the pigment elements result in spiral bands or axial flammules that may be very indistinct or dark and very obvious; the midbody area is usually lightly marked. The mouth is white but commonly has at least a distinct pinkish tone, often with a large dark pink blotch deep within.

Distribution.—*C. zeylanicus* is moderately uncommon and appears to be restricted to the Indian Ocean, where it occurs from the Kenya coast north to India and south to the Reunion-Mauritius area and presumably the Malaysian coast. Records are incomplete, even though this is a shallow-water species.

Synonymy.—This species was long known as *C. obesus*, a very fitting name for some specimens, but Gmelin's name of *zeylanicus* has priority. The Roeding names appear to be this species.

Notes.—Although often brightly colored with pink and opaque white as well as fine brown points, this species is usually rather rough. The spire is often badly eroded, with spots of erosion on the body whorl as well, and there are almost always heavy axial growth marks which seem to be normal in this species. Specimens with open patterns are less attractive than those with heavy patterns concentrated near the midbody area.

Although not hard to obtain, this species seems to be uncommon in most of its range, with definite records hard to find.

ZONATUS Hwass, in Bruguiere, 1792

1792. *Conus zonatus* Hwass, in Bruguiere. *Cone*, in *Ency. Method., Hist. Nat. des Vers*, 1:613 (Indian Ocean). Lectotype figure (here selected): *Tableau Ency. Method.*, Pl. 318, fig. 4.
1798. *Cucullus turritus* Roeding. *Museum Boltenianum*: 38 (Locality not stated). Lectotype figure (Kohn, 1975): Chemnitz, Pl. 139, fig. 1286.
1908. *Conus edwardi* Preston. *Rec. Indian Museum*, 2: 190 (Andaman Is.).

Description.—Heavy, with a high gloss; low conical, the sides straight; body whorl with 4-6 low widely spaced spiral ridges at the base, otherwise smooth except for fine axial growth marks; shoulder broad, sharply angled, with large erect coronations; spire low, deeply concave, the earlier whorls forming a large rounded cone at the middle; early whorls badly eroded; later whorls concave on top, with heavy erect coronations and scattered axial threads. Body whorl dark bluish gray, covered from the blackish base to the shoulder with regularly spaced reddish spiral hairlines; the whorl is heavily covered with many oval white blotches about the size of a coronation, these forming dense spiral bands above the base, at midbody, and below the shoulder, usually leaving only three broad spiral bands of bluish gray, above and below midbody and at the shoulder; the white spots may further break the grayish zones into smaller blotches, or they may be reduced in number and allow more bluish gray to show through; shoulder and spire white, the coronations white, with bluish gray blotches between; most of spire usually eroded and encrusted with no visible pattern. Aperture very narrow, only slightly wider anteriorly, often with a heavy ridge at the middle; outer lip rather thick, straight; mouth bluish white, the external pattern showing through, with a large black blotch at the anterior margin. Columella very narrow, indistinct. Length 50-78mm.

Comparison.—*Conus zonatus* is obviously related to *C. imperialis* in shape and sculpture, differing greatly, however, in the high gloss, the numerous reddish spiral lines, and the pattern of bluish gray with many oval white spots. Although the pattern of *C. zonatus* is quite variable, there is usually little approach to the patterns of *C. imperialis* even in the occasional specimens of *C. imperialis* that have bluish gray flammules.

There is also some resemblance in shape and general appearance to *C. araneosus nicobaricus*, but that species is distinctly tented with axial pale brown lines, usually has a light base, and has a large yellow spot in the mouth.

Variation.—In shape this species varies from the usual slender body whorl with straight sides to inflated specimens that are broad and have slightly convex to concave sides. The spire may be flat or low, occasionally distorted into a tall, stepped structure.

The presence of the fine reddish lines is constant, although sometimes the lines are broken or incomplete or at least wavy. There are usually three spiral bands of bluish gray visible, but the proportion of white to gray is highly variable from specimen to specimen, some being mostly bluish gray with only a few scattered flecks and blotches of white, others mostly white with just blotches of bluish gray. There is never any doubt about the identification of this species, however.

Distribution.—*Conus zonatus* seems to be restricted to the Indian Ocean from the Sumatra coast to India and to the Seychelles. It is uncommon on shallow reefs.

Synonymy.—There is not much doubt about the names listed here.

Notes.—Once considered quite rare, *C. zonatus* is now on the market in good numbers. Most specimens come from the northern Indian Ocean where there is now much commerical collecting. This is a rather rough cone although it has a high gloss. The spire is usually badly eroded and encrusted almost beyond recognition. The lip is often badly broken, and there are commonly large scars near the aperture. The color and pattern are generally good, however, with only a little erosion at the shoulder.

Extremely light or extremely dark specimens are uncommon but draw little premium.

Doubtful and Unrecognizable Species

The following partially annotated lists are of: 1) doubtful taxa which for one reason or the other I have been unable to place, though they are almost certainly synonyms of recognized species; 2) unrecognizable names too poorly described to ever be placed and usually without figures or type material; and 3) taxa described as cones but now believed to belong to other families.

1) DOUBTFUL TAXA

ALBICANS Sowerby ii, 1857-1858. *Conus albicans* Sowerby ii. *Thesaurus Conchyliorum*, 3(*Conus*): 3, pl. 5(191), fig. 98 (Locality not stated).
Somewhat similar in shape to *C. recluzianus* or a young *C. imperialis*, but white with the basal area brown; shoulder obsoletely coronate.

ALBOSPIRA E.A. Smith, 1880. *Conus albospira* E.A. Smith. *Proc. Zool. Soc.* (*London*), 1880: 480, pl. 48, fig. 4 (Locality unknown).
This is perhaps a small, very poorly colored *C. tegulatus;* mostly white, with axial olive flammules, striated.

ANABATHRUM Crosse, 1865. *Conus anabathrum* Crosse. *J. Conchyl.* (*Paris*), 13: 304, pl. 9, fig. 4 (Locality unknown).
Normally this would be considered a synonym of *C. articulatus*, but the shape and general appearance closely resemble a small cone from Mexico usually called *C. 'dispar'* by collectors. The species is small,

C. cf. *anabathrum,* off Los Frailes, Baja Calif. del Sur, Mexico, 22.5mm (EP).

waxy, slender with a high spire, and has a subdued midbody band and spiral brown dashes on tan. Crosse compares it with *C. gradatus*.

BAYLEI Jousseaume, 1872. *Conus Baylei* Jousseaume. *Rev. Mag. Zool.*, (Ser. 2), 23: 198, pl. 18, fig. 2 (Locality unknown).

A whitish shell with sparse spots and no strong sculpture. Perhaps a bad *C. bairstowi?*

CAELATUS A. Adams, 1854. *Conus caelatus* A. Adams. *Proc. Zool. Soc. (London)*, 1853: 117 (Chinese Seas). See *Thes. Conch.*, Pl. 5, fig. 107.

Very small, striated, minutely coronated, with a marbled pattern; probably an as yet unrecognizable juvenile.

COMMODUS A. Adams, 1854. *Conus commodus* A. Adams. *Proc. Zool. Soc. (London)*, 1853: 117 (Locality unknown). See *Thes. Conch.*, 2nd Suppl., fig. 690.

Small, fusiform with very high spire and deep grooves anteriorly; yellowish, without pattern. Probably the same as *C. semiculcatus* Sowerby iii, here very tentatively and probably incorrectly referred to *C. rutilus*. Perhaps *C. sennottorum* or *C. kerstitchi?*

DICTATOR Melvill, 1898. *Conus (Leptoconus) dictator* Melvill. *Mem. Proc. Manchester Lit. Phil. Soc.*, 42(4): 9, pl. 1, fig. 10 (Sheikh Shuaib I., Persian Gulf).

This seems to be a strongly striated *C. inscriptus* with an exceptionally tall spire.

GLOYNEI Sowerby iii, 1881. *Conus gloynei* Sowerby iii. *Proc. Zool. Soc. (London)*, 1881: 637, pl. 56, fig. 5 (Locality unknown).

Vaguely similar to *C. sanguinolentus*, brown with two darker brown bands, coronated, 3 spiral ridges per whorl, but the mouth said to be white.

GRATACAPII Pilsbry, 1904. *Conus gratacapii* Pilsbry. *Proc. Acad. Nat. Soc. (Phila.)*, 1904: 7, pl. 1, figs. 10-10a (Kikai, Osumi, Japan; ?Pliocene).

A Japanese fossil name variously applied in Recent cones; very like *C. praecellens* in shape, size, and sculpture, but seems narrower. I have seen no live-collected specimens referable to this name.

HENOQUEI Bernardi, 1860. *Conus Henoquei* Bernardi. *J. Conchyl. (Paris)*, 8: 380, pl. 13, fig. 4 (Locality unknown).

Usually referred to *C. vittatus*, this taxon has a stepped spire, a distinct median groove on the whorls, only basal ridges strong, is reddish brown with an interrupted white midbody band and a narrower one at shoulder, the spire pattern strong and continued to the shoulder band; 38mm. It does not look like *C. vittatus*, *C. coccineus*, or *C. cumingii*, nor even large *C. sazanka*, in shape or sculpture. See NADAENSIS below.

JAPONICUS Hwass, in Bruguiere, 1792. *Conus Japonicus* Hwass, in Bruguiere. *Cone*, in *Ency. Method., Hist. Nat. des Vers*, 1: 710

(Japan). Lectotype figure: *Tableau Ency. Method.*, Pl. 330, fig. 3 (here designated).

Variously used for *C. articulatus*, *C. eugrammatus*, *C. praecellens*, etc., this taxon appears to apply to a spotted shell of low conical form, not coronated, with the anterior fourth deeply grooved. In pattern and shape it closely resembles the species here called *C. polygrammus*.

LAMELLOSUS Hwass, in Bruguiere, 1792. *Conus lamellosus* Hwass, in Bruguiere. *Cone*, in *Ency. Method.*, *Hist. Nat. des Vers*, 1: 636 (Ceylon). Holotype figure: *Tableau Ency. Method.*, Pl. 322, fig. 5.

A small cone, poorly figured, perhaps related to *C. chiangi*, where it is discussed.

LATIFASCIATUS Sowerby ii, 1857-1858. *Conus latifasciatus* Sowerby ii. *Thesaurus Conchyliorum*, 3(*Conus*): 35, pl. 20(206), fig. 485 (Locality unknown).

Unrecognizable to me. Described near *C. inscriptus*, so perhaps a very dark *C. inscriptus* or an ally covered with dark brown. Rather shouldered, with a concave, pointed spire.

LISCHKEANUS Weinkauff, 1875. *Conus Lischkeanus* Weinkauff. *Syst. Conch. Cab.*, ed. 2, 233(*Conus*): 311, pl. 56, figs. 2-3 (Japan).

A large (52mm) cone with slightly stepped spire, no spiral ridges on the spire, yellowish color, and rough appearance. Perhaps the Japanese *C. kermadecensis* form, but seems a bit broad.

MACULIFERUS Sowerby i, in Sowerby ii, 1833. *Conus maculiferus* Sowerby i, in Sowerby ii. *Conchological Illustrations:* Pt. 29, fig. 23 (Locality not stated).

A small cone that looks much like a distorted *C. boeticus* or *C. sponsalis*. However, very bad specimens of a somewhat similar shell have been seen from the Caribbean, so perhaps this is a juvenile of *C. cardinalis?*

MARIEI Jousseaume, 1899. *Conus Mariei* Jousseaume. *Le Naturaliste*, Annee 21 (Ser. 2, Annee 13): 8, figure (Locality unknown or not stated).

The description is brief and the photo poor, but this looks like a partially melanistic *C. araneosus nicobaricus;* the base is dark violet and granulose, the photo shows low coronations, and what can be seen of the pattern is tented. The photo is miscaptioned as *Petraeus Schoukraensis*, a previous species.

MAURUS Gray, 1826. *Conus maurus* Gray, in King, P.P. *Narrative Surv. Coasts Australia, Appendix*, 2: 486 (presumably Australia).

"Shell very plain, top-shaped, crowned, and whitish, with two brown bands; spire rather depressed; crowned, blunt; the epidermis pale greenish-brown; the inside white, with two broad blue bands, in the front of which is enclosed the canal; axis one and a half, diameter one inch." Perhaps *C. lividus?*

MELANCHOLICUS Lamarck, 1810. *Conus melancholicus* Lamarck. *Ann. du Mus. Hist. Nat.* (*Paris*), 15: 430 (Locality unknown). Holotype figured by Kiener, Pl. 45, fig. 3 and Reeve, Pl. 21, sp. 117 (Kiener's plate is misnumbered, 3 and 4 (*nimbosus*) being reversed).

A slender dark brown *magus*-type shell with a blotchy midbody band and a low spire; the shoulder is rather sharp and somewhat separated from the whorl. Similar specimens have been seen from the Philippines; this may be a synonym of *C. magus*.

* NADAENSIS (Azuma and Toki, 1970). *Endemnoconus nadaensis* Azuma and Toki. *Venus*, 29(3): 77, figs. 2-3, 4, 4a (Off Nada, Wakayama Pref., Kii Peninsula, Japan).

Very similar in shape, sculpture, and color pattern to *C. henoquei*, discussed earlier. I cannot help feeling the two are the same, but even the status of *C. nadaensis* is uncertain to me. It seems overly close to *C. otohimeae*.

Left: Holotype of *C. nadaensis*, 25.5mm (orig. descr.).

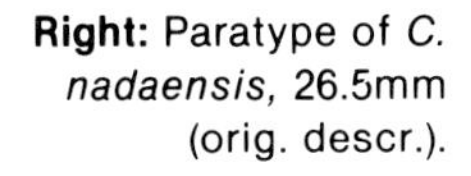

Right: Paratype of *C. nadaensis*, 26.5mm (orig. descr.).

NEBULOSUS (Azuma, 1973). *Rhizoconus nebulosus* Azuma. *Venus*, 32(2): 34, fig. 1 (Off Nada, Kii, Japan). Non *Conus nebulosus* Gmelin, 1791.

Very much like *C. spectrum* in shape, spiral grooving on anterior half, and pattern, but the early 3 whorls apparently nodulose. *Nebulosus* is preoccupied in *Conus*.

PETRICOSUS Azuma, 1961. *Conus* (*Asprella*) *petricosus* Azuma. *Venus*, 21(3): 302, figs. 8, 11 (Off Tosa, Japan).

Very much like *C. sieboldii* in shape, white, without pattern, the spiral grooving weak. Radula said to be different from that of *C. sieboldii*. The shell would fall within the normal range of variation of *C. sieboldii*.

PHOEBEUS Jousseaume, 1894. *Conus phoebeus* Jousseaume. *Bull. Soc. Philom. Paris*, (Ser. 8), 6: 99 (Aden).

A solid shell, oblong-oval, turbinate, the spire prominent, conical, with granules anteriorly; yellow and white, the body whorl with two mottled spiral bands of yellow; spire margins carinate, with spiral ridges; aperture oblong, narrow, anteriorly dilated. 27mm long,

* A recently examined specimen indicates this may be just a form of *C. articulatus*.

12.5mm wide, spire 6mm. I cannot place this unfigured species with anything from the meager description. There is no comparison.

PLANAXIS Deshayes, 1863. *Conus planaxis* Deshayes. *Cat. Moll. de l'Isle de la Reunion:* 134, pl. 13, figs. 11-12 (Reunion).

A juvenile shell with a nearly flat spire and narrow brownish spiral hairlines like those of *C. caillaudii.* Perhaps it is that species?

SOPHIAE Brazier, 1875. *Conus* (*Rhizoconus*) *Sophiae* Brazier. *Proc. Linn. Soc. New South Wales*, 1: 7 (Hammond's or Bannietta Island, Solomon Arch.).

A small, heavily granulose shell, the type of which has apparently been lost and never figured. It could be a granulose *C. catus*, but there is insufficient information in the original description.

SPECIOSUS Sowerby ii, 1857-1858. *Conus speciosus* Sowerby ii. *Thesaurus Conchyliorum*, 3(*Conus*): 7, pl. 6(192), fig. 138 (Locality unknown).

A small cone with weak coronations and a very jagged flammulated pattern of reddish brown on white. Could be almost anything.

SPHACELATUS Sowerby i, in Sowerby ii, 1833. *Conus sphacelatus* Sowerby i, in Sowerby ii. *Conchological Illustrations:* Pt. 37, fig. 51 (Locality unknown).

A small granulose shell with a violet base and large blotches. Perhaps related to *C. sponsalis?*

SUPERSTES Hedley, 1911. *Conus superstes* Hedley. *Commonw. Australia, Dept. Trade and Customs, Fisheries: Zool. Res. Endeavour* (*1909-1910*), 1: 111, pl. 20, figs. 35-36 (Off Cape Wiles, South Australia, Australia).

Based on very small juveniles with grooved body whorls and a twisted protoconch (as found in several other cones); unrecognizable until the life histories of South Australian cones are better known; the adult need not be a deep-water cone, as the juveniles of several cones are known to live in much deeper water than the adults.

SUPERSTRIATUS Sowerby ii, 1857-1858. *Conus superstriatus* Sowerby ii. *Thesaurus Conchyliorum*, 3(*Conus*): 37, pl. 13(199), fig. 282 (Locality unknown).

Described between *C. nimbosus* and *C. rhododendron* and close to *C. aurisiacus* and *C. stercusmuscarum.* Oblong, obscurely clouded with brown, sulcate above and below (on the body whorl) but smooth at the middle; white midbody band spotted with white; spire depressed, white, striated, spotted with brown. I cannot place this.

TRISTIS Reeve, 1844. *Conus tristis* Reeve. *Conchologia Iconica*, 1 (*Conus*): Pl. 45, sp. 252 (Locality unknown).

Small, whitish, low conical, grooved anteriorly; spire whorls striated, nodulose, the shoulder without nodules. A very plain little shell with no recognizable characters.

TUBERCULOSUS Tomlin, 1937. *Conus tuberculosus* Tomlin. *Proc.*

Malac. Soc. London, 22(4): 206. Nomen novum for the preoccupied *Conus tuberculatus* Yokoyama, 1920, *J. Coll. Sci., Imp. Univ. Tokyo*, 39(6): 34, pl. 1, figs. 15-16 (Miyata zone, Koshiba zone, Japan).

This is a small fossil shell with fine tubercules on the spire and shoulder, but with a smooth body whorl. Possibly close to *C. otohimeae*, but I have seen nothing Recent that matches the original description and figures. Juveniles of several Japanese species are sold under this name at high prices.

UNICOLOR Sowerby i, in Sowerby ii, 1833. *Conus unicolor* Sowerby i, in Sowerby ii. *Conchological Illustrations:* Pt. 28, fig. 20 (Locality unknown).

Elongate low conical, brownish with purple tones, spire convex, shoulder and whorls nodulose; mouth apparently pinkish. No one has recognized this species since the description of the type, only repeating the same descriptions. Perhaps related to *C. moreleti*?

2) UNRECOGNIZABLE NAMES

ALATUS Fischer, 1807. *Museum Demidoff*, 3: 143 (Locality unknown).

ALBIDUS Schroeter, 1803. *Arch. Zool. Zootom.*, 3(2): 70 (Locality unknown).

AMERICANUS Gmelin, 1791 (*C. ammiralis americanus*). *Systema Naturae per Regna Tria Naturae*, ed. 13, 1: 3378 (Locality not stated).

AMETHYSTEUS Link, 1807. *Beschr. Nat.-Samml. Univ. Rostock*, 3: 105 (Locality not stated). Based on Martini, Pl. 63, fig. 708.

AUSTRALIS Schroeter, 1803. *Arch. Zool. Zootom.*, 3(2): 71 (Southern countries). Non *C. australis* Holten, 1802.

BANDATUS Perry, 1811. *Conchology:* Pl. 25, no. 4 (Locality not stated).

BICINCTUS Donovan, 1826. *Naturalist's Repository*, 5: Pl. 170 (Indian Seas).

CAFFER (Roeding, 1798) (*Cucullus*). *Museum Boltenianum:* 48 (Locality not stated). Lectotype figure (here selected): Martini, Pl. 56, fig. 618, also lectotype of *C. coffeae*.

CINCTUS Bosc, 1801. *Hist. Nat. Coqu.*, 5: 140, pl. 40, fig. 2 (Locality not stated). Perhaps *C. miles*?

CINEREUS Schroeter, 1803. *Arch. Zool. Zootom.*, 3(2): 72 (Locality not stated). Non *C. cinereus* Hwass, 1792.

CLAVATUS (Roeding, 1798) (*Cucullus*). *Museum Boltenianum:* 46 (Locality not stated). Based on Martini, Pl. 52, fig. 578. Possibly *C. magus*?

CLAVUS Linnaeus, 1758. *Systema Naturae per Regna Tria Naturae*, ed. 10, 1: 716. Composite, *C. terebra* and *C. auricomus.*

COERULESCENS Schroeter, 1803. *Arch. Zool. Zootom.*, 3(2): 67 (Locality not stated).

COFFEAE Gmelin, 1791. *Systema Naturae per Regna Tria Naturae*, ed. 13, 1: 3388 (Locality not stated). Holotype figure: Martini, Pl. 56, fig. 618. See Kohn, 1966.

COSTATUS Gmelin, 1791. *Ibid.*, ed. 13, 1: 3388 (Locality not stated). Based on Gualtieri, Pl. 20, fig. 0. See Kohn, 1966.

EGREGIUS Sowerby iii, 1914. *Ann. Mag. Nat. Hist.*, (Ser. 8), 14: 475, pl. 19, fig. 9 (New Caledonia). A 4mm juvenile!!

FASCIATUS Schroeter, 1803. *Arch. Zool. Zootom.*, 3(2): 69 (Locality not stated).

FULGURANS Hwass, in Bruguiere, 1792. *Cone*, in *Ency. Method.*, *Hist. Nat. des Vers*, 1: 687 (Grandes Indes). Lectotype figure (Kohn, 1968): Martini, Pl. 58, fig. 644, same as *C. fulmineus.*

FULMINEUS Gmelin, 1791. *Systema Naturae per Regna Tria Naturae*, ed. 13, 1: 3388 (Locality not stated). Holotype figure: Martini, Pl. 58, fig. 644. Sometimes used for *C. kermadecensis*, but the figure is exaggerated and unrecognizable. Could be *C. monachus?*

FULVUS Schroeter, 1803. *Arch. Zool. Zootom.*, 3(2): 73 (Locality not stated).

GIGAS Fischer, 1807. *Museum Demidoff*, 3: 140 (Locality not stated).

GUIENENSIS Schroeter, 1803. *Arch. Zool. Zootom.*, 3(2): 71 (Guinea coast).

LAPIDEUS Holten, 1802. *Enum. Syst. Conchyl. beat Chemnitzii:* 37.

LOCUMTENENS Blumenback, 1791. *Handb. Naturgeschichte*, ed. 4: 448 (Red Sea). Used for *C. acuminatus*, but no valid references and description very poor.

MARMORATUS Holten, 1802. *Enum. Syst. Conchyl. beat Chemnitzii:* 33.

MINUTUS Schroeter, 1803. *Arch. Zool. Zootom.*, 3(2): 74 (Possibly Guinea). Non *C. minutus* (Roeding, 1798).

NIVEUS Gmelin, 1791. *Systema Naturae per Regna Tria Naturae*, ed. 13, 1: 3392 (Locality not stated). Holotype figure same as *C. candidus* Born, so possibly *C. marmoreus?*

NOVEMSTRIATUS (Roeding, 1798) (*Cucullus*). *Museum Boltenianum:* 48 (Locality not stated). Lectotype figure (Kohn, 1975): Martini, Pl. 58, fig. 644.

OLEA Schroeter, 1803. *Arch. Zool. Zootom.*, 3(2): 76 (Locality not stated).

PORCELLANEUS Fischer, 1807. *Museum Demidoff*, 3: 134 (Locality not stated).

PULCHELLUS (Roeding, 1798) (*Cucullus*). *Museum Boltenianum:* 45 (Locality not stated). Based on Favanne and Favanne, Pl. 79, fig. L.

PUNICEUS Schroeter, 1803. *Arch. Zool. Zootom.*, 3(2): 75 (Locality not stated).

RUBESCENS Schroeter, 1803. *Ibid.:* 73 (Locality not stated).

SENATOR Linnaeus, 1758. *Systema Naturae per Regna Tria Naturae*, ed. 10, 1: 714 (Locality not stated). Used for *C. planorbis*, but unrecognizable.

SEURATI Fenaux, 1942. *Bull. l'Inst. Oceanogr. (Monaco)*, No. 814: 4, fig. 13 (I. Paumotou). Juvenile.

SULCATUS Link, 1807. *Beschr. Nat.-Samml. Univ. Rostock*, 3: 104 (Locality not stated). Non *C. sulcatus* Hwass, 1792.

TRAILLII A. Adams, 1855. *Proc. Zool. Soc. (London)*, 1855: 121 (Malacca). Juvenile.

VENUSTUS (Roeding, 1798) (*Cucullus*). *Museum Boltenianum:* 47 (Locality not stated).

VITIFERA Perry, 1811. *Conchology:* Pl. 25, no. 2 (Locality not stated).

VIVIDUS Sowerby iii, 1914. *Ann. Mag. Nat. Hist.*, (Ser. 8), 14: 476, pl. 19, fig. 8 (New Caledonia). Juvenile.

Undoubtedly several other names tentatively included in synonymy with other species should properly be placed here.

3) NON-CONES

ATRAMENTOSUS Reeve, 1849. *Conchologia Iconica*, 1(*Conus* Suppl.): Pl. 7, sp. 265 (Island of Mindoro, Philippines). A turrid, *Mitromorpha (Lovellona)*. Recently figured in Salvat and Rives, 1975, *Coquillages de Polynesie*, p. 343. Uncommon, Philippines to French Polynesia.

CONCINNULUS Crosse, 1858. *Rev. Mag. Zool.*, (Ser. 2), 10: 200. Nomen novum for *Conus concinnus* Broderip, 1833. A columbellid, *Parametaria dupontii* Kiener, Panamic.

CONCINNUS Broderip, 1833. *Proc. Zool. Soc. (London)*, 1833: 53 (Gulf of California). Non *Conus concinnus* J. de C. Sowerby, 1821, a fossil. A columbellid, *Parametaria dupontii* Kiener, Panamic.

COROMANDELICUS E.A. Smith, 1894. *Ann. Mag. Nat. Hist.*, (Ser. 6), 14: 159, pl. 4, figs. 1-2 (Off Coromandel Coast). Now considered a turrid, *Conorbis coromandelicus*, but very much like a cone. 30-40mm, biconical, slender, whitish, heavy spiral ridges from base onto spire, the shoulder with a flat rim that winds onto the spire, spire whorls with about 3 heavy ridges, often nodulose on all ridges of body and spire. Northern Indian Ocean, deep water.

CORONATUS Reeve, 1849. *Conchologia Iconica*, 1(*Conus* Suppl.): Pl. 7, sp. 263 (Island of Ticao, Philippines). Non *Conus coronatus* Gmelin, 1791. See *papalis* Weinkauff, below.

DENTATUS Schroeter, 1803. *Arch. Zool. Zootom.*, 3(2): 74 (Locality not stated). Very narrow mouth with teeth, certainly not a cone.

DUPONTII Kiener, 1845. *Species gen. et icon. des coqu. viv.*, 2(*Conus*): Pl. 61, fig. 2. 1849-1850, *Ibid.*, 2: 273 (Locality unknown). A columbellid, *Parametaria.*

EDENTULUS Reeve, 1844. *Conchologia Iconica*, 2(*Mitra*): Pl. 11, fig. 80 (Locality unknown). A miter, *Dibaphus.*

FUSIFORMIS Pease, 1861. *Proc. Zool. Soc.* (*London*), 1860: 398 (Sandwich Islands). Non *C. fusiformis* Fischer, 1807. A turrid, *Mitromorpha* (*Lovellona*) *peaseana* Finlay.

LUTEUS Quoy and Gaimard, 1834. *Voy. Astrolabe, Zool.*, 3: 103, pl. 53, figs. 23-24 (Port du Roi, Georges, Australia). Non *Conus luteus* Sowerby i, 1833. Oval, whitish with large black spots; looks like a marginellid or columbellid.

MICARIUS Hedley, 1912. *Rec. Australian Mus.*, 8: 147, pl. 43, fig. 32 (15 mi. SW of Cape York, Queensland, Australia). A turrid, *Mitromorpha* (*Lovellona*).

PAPALIS Weinkauff, 1875. *Syst. Conch. Cab.*, ed. 2, 237(*Conus*): 359, pl. 66, fig. 10. Nomen novum for *Conus coronatus* Reeve, 1849. Body whorl greenish brown with heavy basal ridges, biconical, shoulder and spire whitish with few dark blotches; shoulder coronations very large and rounded, projecting. This certainly does not appear to be a cone, but I cannot place it anywhere else. Specimens are taken rarely in the Philippines.

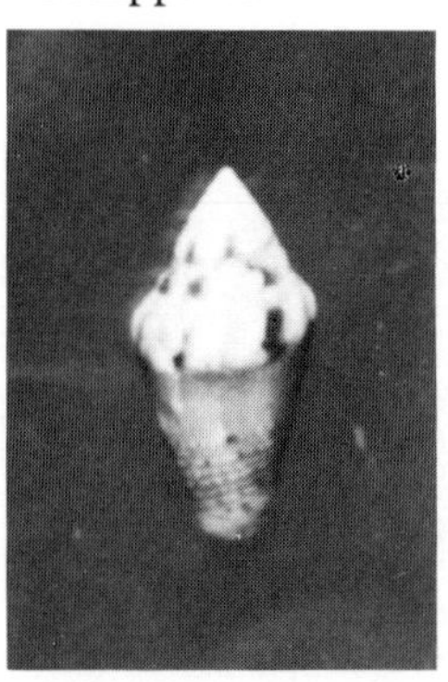

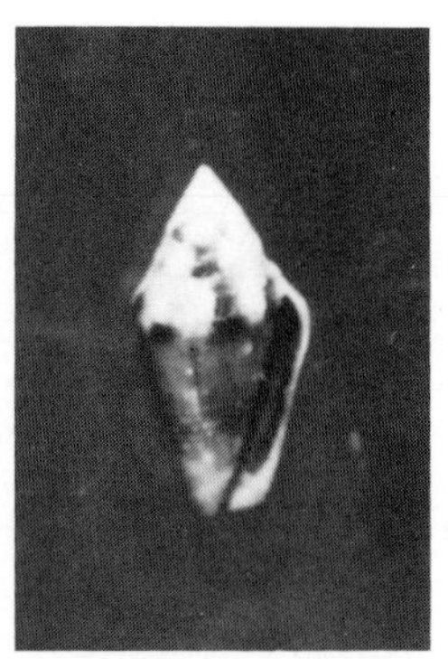

"Conus" papalis Weinkauff, Philippines, 20mm (photo Burgess).

PARVUS Pease, 1868. *Amer. J. Conch.*, 4: 126. Nomen novum for *Conus fusiformis* Pease, 1861; non *C. parvus* (Roeding, 1798). A turrid, *Mitromorpha* (*Lovellona*) *peaseana* Finlay.

SINENSIS Gmelin, 1791. *Systema Naturae per Regna Tria Naturae*, ed. 13, 1: 3394 (Locality not stated). Lectotype figure (here selected): Martini, Pl. 56, fig. 615. Appears to be a columbellid?

SPINOSUS Linnaeus, 1758. *Systema Naturae per Regna Tria Naturae*, ed. 10, 1: 715 (Locality not stated). A turrid, *Volutospina spinosus*.

TEREBELLUM Linnaeus, 1758. *Ibid.*, ed. 10, 1: 718 (Asia). A strombid, *Terebellum terebellum*.

VIADERI Fenaux, 1942. *Bull. l'Inst. Oceanogr.* (*Monaco*), No. 814: 3, fig. 9 (I. Maurice). A columbellid, possibly *Conella?*

Selected Bibliography

Adanson, M. 1757. *Histoire Naturelle de Senegal. Coquillages.* Paris.
Barnard, K.H. 1958. *Ann. S. Afr. Mus.*, 44: 73-163. (South Africa.)
Cernohorsky, W.O. 1964. *Veliger*, 7(2): 61-94. (Fiji.)
Chemnitz, J.H. 1788-1795. *Neues Systematisches Conchylien-Cabinet*, vol. 10, 11. Nuernberg.
Clench, W.J. 1942. *Johnsonia*, 1(6): 1-40. (Caribbean; later suppl.)
Cotton, B.C. 1945. *Rec. South Australian Mus.*, 8: 229-280. (Catalog; Brazier types.)
Dance, S.P. 1969. *Rare Shells.* Faber and Faber, London.
D'Asaro, C.N. 1970. *Bull. Mar. Sci.*, 20(2): 414-440. (Spawning *C. spurius.*)
Dautzenberg, P. 1937. *Mem. Mus. Roy. d'Hist. Nat. Belg.*, 2(18): 1-184. (Indonesian area, literature only.)
Dillwyn, L.W. 1817. *Descriptive Catalogue Recent Shells*, vol. 1. London.
Emerson, W.K. and W.E. Old, Jr. 1962. *Amer. Mus. Novitates*, No. 2112: 1-44. (Panamic cones; list of generic names.)
Gualtieri, N. 1742. *Index Testarum Conchyliorum.* Florence.
Hanna, G.D. 1963. *Occas. Pap. California Acad. Sci.*, No. 35: 1-103. (Panamic species.)
Hinton, A.G. 1972. *Shells of New Guinea and the Central Indo-Pacific.* Jacaranda Press, Melbourne.
Keen, A.M. 1971. *Sea Shells of Tropical West America*, ed. 2. Stanford Univ. Press, Calif.
Kiener, L.C. 1845-1850. *Species General et Iconographie des Coquilles Vivantes*, vol. 2. Paris.
Kilburn, R.N. 1971. *Ann. Natal Mus.*, 21(1): 37-54. (Cape, South Africa).
Knorr, G.W. 1757-1772. *Vergnuegen der Augen und des Gemueths*, vols. 1-6. Nuernberg.
Kohn, A.J. 1959. *Ecol. Monogr.*, 29: 47-90. (Ecology, Hawaii.)
________. 1959. *Pacific Sci.*, 13(4): 368-401. (Hawaiian list.)
________. 1959. *Ann. Mag. Nat. Hist.*, (Ser. 13), 2: 309-320. (Sri Lanka.)
________. 1961. *Bull. Bingham Oceanogr. Coll.*, 17(4): 1-51. (Spawning and development.)
________. 1963. *J. Linn. Soc. London*, 44(302): 740-768. (Linnaean types.)
________. 1964. *Ibid.*, 45(304): 151-167. (Born types.)
________. 1966. *Ibid.*, 46(308): 73-102. (Gmelin types.)
________. 1968. *Ecology*, 49(6): 1046-1062. (Ecology, Maldives.)
________. 1968. *J. Linn. Soc. (Zool.)*, 47(313): 431-503. (Hwass types.)
________. 1975. *Zool. J. Linn. Soc.*, 57: 185-227. (Roeding types.)
________, J.W. Nybakken, and J.J. Van Mol. 1972. *Science*, 176: 49-51. 7 April. (Structure of radular tooth.)

________ and A. Riggs. 1977. (Untitled). Privately printed, Seattle. (Additions to Tomlin catalog.)
________ and R. Robertson. 1968. *J. Mar. Biol. Assoc. India*, 8(2): 273-277. (List for Maldives and Chagos.)
________ and P.R. Saunders. 1960. *Ann. N.Y. Acad. Sci.*, 90(3): 706-725. (Venoms.)
________ and C.S. Weaver. 1961-1962. *Hawaiian Marine Mollusks*, 1(10-19): 38-74. (Hawaii.)
________ and ________. 1962. *Pacific Sci.*, 16(4): 349-355. (Hawaii.)
Lamarck, J.B.P. 1822. *Histoire Naturelle des Animaux sans Vertebres*, vol. 7. Paris.
Lim, C.F. 1970. *Veliger*, 19(2): 160-164. (Radular types.)
Lister, M. 1688. *Historiae sive Synopsis Methodicae Conchyliorum*, vol. 4. London.
Martini, F.H.W. 1773. *Neues Systematisches Conchylien-Cabinet*, vol. 2, Nuernberg.
Melvill, J.C. 1900. *J. Conch. (London)*, 9(10): 303-311. (Tented cones.)
Mermod, G. 1947. *Revue Suisse Zool.*, 54: 155-217. (Lamarck types.)
Nybakken, J. 1970. *Veliger*, 12(3): 316-318. (Food *C. imperialis*, etc.)
________. 1970. *Amer. Mus. Novitates*, No. 2414: 1-29. (Panamic radulae.)
________. 1974. *Pacific Discov.*, 27(5): 23-27. (General feeding.)
Peile, A.J. 1939. *Proc. Malac. Soc. London*, 23(6): 24-35. (Radulae.)
Reeve, L.A. 1843-1849. *Conchologia Iconica*, vol. 1 and suppl. London.
Salvat, B. and C. Rives. 1975. *Coquillages de Polynesie.* Edit. Pacifique, Tahiti. (French Polynesia, several errors.)
Songdahl, J.H. 1973. *Bull. Mar. Sci.*, 23(3): 600-612. (Venom *C. spurius.*)
Sowerby, G.B. i, in Sowerby, G.B. ii. 1832-1841. *Conchological Illustrations.* London.
Sowerby, G.B. ii. 1857-1887. *Thesaurus Conchyliorum*, vol. 3 and suppl., vol. 5 suppl. London.
Swainson, W. 1840. *Treatise on Malacology.* London.
Thiele, J. 1929. *Handbuch der Systematischen Weichtierkunde*, vol. 1. Jena. (Generic diagnoses.)
Tomlin, J.R.L. 1937. *Proc. Malac. Soc. London*, 22: 205-330. (*The* catalog; an essential item of literature.)
Tryon, G.W. 1883-1884. *Manual of Conchology*, vol. 6. Philadelphia.
Usticke, G.W. Nowell. 1968. *Caribbean Cones from St. Croix and the Lesser Antilles.* Privately printed, St. Croix.
Van Mol, J.J., B. Tursch, and M. Kempf. 1967. *Ann. l'Inst. Oceanogr.*, 45(16): 233-254. (Brazil.)
Warmke, G.L. 1960. *Nautilus*, 73(4): 119-124. (Puerto Rican radulae.)
Wenz, W. 1943. *Handbuch der Palaeozoologie.* 6(6): 1400-1505. Geneva. (Recent and fossil genera.)
Wilson, B.R. and K. Gillett. 1971. *Australian Shells.* Tuttle, Rutland, Vt.

Checklist and Pricing Guide to the Cones

The following prices reflect market values in the U.S. at the time of writing (1977). Unless otherwise indicated, the price range is for average patterns in *gem* or *very fine* condition *for the species*. Many things affect pricing in cones, so it is very difficult to establish a definite value for a species. Values at the lower end of the range are generally for smaller or slightly immature shells or those with weakly developed patterns; higher prices in the range are for larger or exceptionally perfect specimens or for uncommon and interesting patterns. Prices for some major varieties and subspecies are also listed.

Species and variants with prices over $500 are seldom sold on the open market, so values are very uncertain in several cases. Often a rare specimen will sell immediately for whatever the market will bear, but just as often it sits in a dealer's cabinet for months before finding a buyer. Species with especially unstable markets are indicated by an asterisk(*).

Prices of cones vary considerably from dealer to dealer and often from month to month. Locality often has an effect; Philippine specimens of some species, for example, are often cheaper than equivalent shells of the same species from the Red Sea or Polynesia. Many minor but desirable variants (such as granulose or high-spired individuals) fetch higher prices than indicated for normal specimens, while unattractively patterned but still *gem* shells may sell for lower prices than usual. Conditions below *very fine* often result in large discounts for more common species, but remember that some species almost never occur in better than *fine* condition (*C. tinianus*, *C. balteatus*, etc.).

It should also be noticed that several species of cones (such as *C. balteatus*, *C. cernicus*, *C. aplustre*, *C. excavatus*, etc.) listed below are almost never available on the open market, although they are listed here at relatively low prices. These are usually small or nondescript species which are difficult to identify, although they are not really uncommon in nature.

The price lists of reputable dealers are always your best guides to current cone prices. Find a couple of good dealers and stick with them and you will seldom go wrong. Finally, I must offer a bit of advice which should prove helpful: never attempt to invest in "rare" cones. All too often a species said to be very rare and represented by only a handful of specimens may sell for $1000 this month, while two or three months later the same "rare" species may be on the market by the hundreds and retailing for $15 or less. This has happened many times in the shell market and continues to happen. Before you purchase a rare shell or a newly described one, check with several dealers and advanced collectors for their opinions —often the "unique" shell is one of dozens sitting in an unscrupulous collector's basement waiting for the right moment to be dumped on the market. Be cautious and always know your dealer.

Neither the author nor the publisher offers any shells for sale; this listing is purely a guide to relative values and not an offer to buy or sell.

(**L**): live condition (indicated for only a few species); (**b**): good beach or good dredged dead (often the only available condition for certain species); **NP**: not priced (no sales known to me). Subspecies and variants are listed after the nominate subspecies; (**varieties**): indicates the more uncommon variants discussed in the text.

____ abbas—$8.-20.
____ abbreviatus—$.50-3.
____ aculeiformis—$5.-15.
____ acuminatus—$2.-15.
____ acutangulus—$3.-15.
____ adamsoni—(L)$200.-600.
(b)$50.-150.
____ advertex—$5.-8.
____ aemulus—$2.-5.
____ algoensis—(b)$3.-10.
(L)$15.-40.
____ altispiratus—$60.-150.
____ amadis—$2.-7.
(dark)$5.-15.
____ ambiguus—$8.-25.
____ ammiralis—$7.-15.
(granulose)$50.
(coronate)$150.-300.
____ amphiurgus—$10.-35.
____ anemone—$.50-2.
(varieties)$2.-15.
____ angasi—$20-40.
____ anthonyi—$15.-25.
____ aplustre—$1.-3.
____ araneosus—$3.-7.
____ archon—$10.-20.
____ arcuatus—$1.-3.
____ arenatus—$.50-1.
____ argillaceus—$20.-40.
____ armiger—$10.-25.
(bajanensis)$50.-100.*
____ articulatus—(L)$15.-25.
(b)$7.-12.
____ artoptus—$8.-25.
____ atractus—$50.-80.*
(austini)$4.-10.
____ attenuatus—$50.-100.
____ augur—$4.-8.
____ aulicus—$5.-15.
____ aurantius—$15.-35.
____ auratus—$200.-300.*
____ aureus—$15.-60.
____ auricomus—$10.-50.
____ aurisiacus—(L)$75.-150.
(b)$20.-50.
____ australis—$1.-5.
____ axelrodi—$3.-10.

____ bairstowi—(b)$6.-10.
____ balteatus—$3.-5.
____ balteus—$10.-25.
____ barthelemyi—$70.-120.
____ bartschi—$15.-30.
____ bayani—$3.-8.
____ beddomei—$10.-15.
____ bengalensis—$160.-250.
____ betulinus—$2.-6.
____ biliosus—$2.-5.
____ boeticus—$1.-5.
____ boivini—NP
____ boschi—$6.-15.
____ brunneus—$1.-3.
____ bruuni—NP
____ bulbus—$3.-15.
____ bullatus—$40.-90.
____ byssinus—$3.-10.

____ cabritii—$10.-20.
____ caillaudii—(b)$100.-200.
____ caledonicus—NP
____ californicus—$1.-3.
____ cancellatus—$1.-2.
____ canonicus—$1.-5.
____ capitanellus—$15.-40.
____ capitaneus—$1.-2.
____ caracteristicus—$2.-5.
____ cardinalis—(b)$10.-20.
____ catus—$1.-2.
____ cedonulli—$50.-100.
____ centurio—$20.-75.
____ cernicus—$2.-5.
____ cervus—(b)$1500.*
____ chaldaeus—$.50-1.
____ chiangi—$10.-25.
____ cinereus—$1.-3.
(gubba)$10.-20.*

____ cingulatus—(L)$20.-50.
(b)$5.-10.
____ circumcisus—$30.-80.
____ clarus—$10.-20.
____ clerii—$8.-25.
(clenchi)$40.-75.*
____ cloveri-$15.-25.
____ cocceus—$5.-15.
____ coccineus—$5.-15.
____ coelinae—$10.-15.
(spiceri)$80.-120.
(berdulinus)$150.-300.*
____ colubrinus—$60.-150.
____ comatosa—$25.-75.
____ concolor—$15.-50.
____ connectens—$30.-60.
____ consors—$3.-8.
____ conspersus—$8.-15.
____ coronatus—$.50-2.
____ cumingii—$15.-30.
____ cuvieri—$30.-50.
____ cyanostoma—$4.-8.
____ cylindraceus—$5.-15.

____ dalli—$5.-15.
____ daucus—$5.-15.
____ delessertii—$5.-30.
____ diadema—$2.-5.
____ distans—$1.-5.
____ dorreensis—$1.-3.
____ duplicatus—$50.-100.
____ dusaveli—$900.-1500.*

____ ebraeus—$.50-1.
____ eburneus—$.50-2.
(crassus)$20.-60.
____ elegans—$10.-30.
____ emaciatus—$1.-3.
____ encaustus—$3.-10.
____ ermineus—$1.-5.
____ erythraeensis—$3.-6.
____ eucoronatus—$50.-100.
____ eugrammatus—$8.-20.
(Hawaiian)$75.-200.*
____ excavatus—$10.-30.
____ excelsus—$1500.-2500.
____ eximius—$4.-10.

____ fergusoni—$2.-5.
____ figulinus—$1.-5.
____ flavidus—$.50-2.
____ floccatus—$30.-60.
____ frigidus—$2.-5.
____ fulgetrum—$2.-5.
____ fulmen—$8.-15.
____ fumigatus—$3.-5.
____ furvus—$1.-5.

____ generalis—$1.-5.
____ genuanus—$8.-15.
____ geographus—$1.-5.
____ gilvus—$5.-15.
____ gladiator—$1.-3.
____ glans—$1.-3.
____ glaucus—$5.-10.
____ gloriamaris—$300.-800.*
____ gradatus—$1.-5.
____ grangeri—$10.-20.
____ granulatus—(L)$75.-100.
(b)$5.-15.
____ gubernator—$50.-100*
____ guineensis—(b)$5.-10.

____ havanensis—NP
____ hieroglyphus—$5.-15.
____ hirasei—$50.-100.
____ howelli—NP
____ hyaena—$3.-10.
____ hypochlorus—$10.-20.*

____ ichinoseana—$25.-50.
____ illawarra—$10.-20.
____ imperialis—$1.-5.
(varieties)$5.-10.
____ infrenatus—(b)$3.-5.

____ inscriptus—$3.-7.
____ insculptus—$7.-15.
____ iodostoma—$7.-15.
____ ione—$10.-15.*

____ janus-$10.20.
____ jaspideus—$1.-3.
____ jickelii—$30.-50.
____ julii—$100.-200.

____ kashiwajimensis—$40.-60.
____ kermadecensis—$5.-20.
____ kerstitchi—$50.-75.
____ kimioi—$100.-200.
____ kinoshitai—$20.-50.
____ klemae—$15.-30.

____ legatus—$10.-15.
____ lemiscatus—$50.-75.*
____ lentiginosus—$5.-15.*
____ leopardus—$2.-5.
____ lienardi—$10.-20.
____ litoglyphus—$1.-5.
____ litteratus—$2.-5.
____ lividus—$.50-2.
____ lohri—NP
____ lorenzianus—$20.-50.
____ lucidus—$3.-10.
____ luteus—$10.-30.
____ lynceus—$10.-20.

____ magellanicus—$5.-15.
____ magnificus—$1.-5.
____ magus—$1.-10.
____ malacanus—$5.-15.
____ marielae—NP
____ marmoreus—$.50-3.
(varieties)$10.-25.
(nocturnus)$75.-150.
____ mazei—$30.-50.
(juvenile)$10.-15.
____ memiae—$30.-60.

____ mercator—$5.-15.
(varieties)$15.-25.
____ miles—$1.-3.
____ miliaris—$.50-2.
____ milneedwardsi—$500.-1000.*
____ mindanus—$15.-30.
(Caribbean)$25.-50.*
____ minnamurra—$10.-20.
____ mitratus—$5.-15.
____ moluccensis—$100.-200.
____ monachus—$1.-5.
(varieties)$10.-20.
____ monile—$3.-5.
____ moreleti—$2.-5.
____ mucronatus—$1.-5.
____ muriculatus—$1.-3.
(granulose)$25.-50.
(high-spired)$30.-50.
____ mus—$1.-3.
____ musicus—$.50-2.
____ mustelinus—$1.-5.

____ namocanus—$2.-5.
____ natalis—(b)$3.-5.
(gilchristi)(b)$5.-10.
____ neptunus—$100.-200.
____ nielsenae—$10.-20.
____ nigropunctatus—$1.-3.
____ nimbosus—$15.-25.*
(dark)$30.-50.
____ nobilis—$10.-20.
(victor)$80.-150.
(marchionatus)$6.-15.
____ nussatella—$2.-10.

____ obscurus—$2.-5.
____ ochroleucus—$2.-3.
____ orbignyi—$1.-5.
____ orion—$10.-25.
____ otohimeae—$15.-40.

____ papilliferus—$1.-5.
____ parius—$3.-7.

____ parvulus—$1.-5.
____ patae—(L)$60.-100.*
(b)$20.-30.
____ patricius—$2.-10.
____ pennaceus—$1.-10.
(elisae)$20.-40.*
____ pergrandis—$200.-500.
____ peronianus—$20.-50.
____ pertusus—$10.-50.
(Hawaiian)$3.-5.
____ philippii—$1.-5.
____ pictus—(b)$5.-15.
____ pilkeyi—$5.-15.
(Philippines)$3.-5.
____ planorbis—$.50-3.
(chenui)$25.-50.
____ pohlianus—$10.-25.
____ polygrammus—$30.-50.
____ poormani-$5.-10.
____ praecellens—$2.-5.
____ princeps—$2.-6.
____ proximus—$40.-60.
____ pulcher—$3.-10.
(over 120mm)$15.-30.
____ pulicarius—$.50-2.
(vauteri)$3.-10.
____ puncticulatus—$1.-3.
(white)$5.-10.
____ purpurescens—$1.-3.

____ quercinus—$1.-5.
(white)$20.-30.

____ radiatus—$1.-3.
____ raoulensis—NP
____ rattus—$.50-2.
(taitensis)$3.-7.
____ recluzianus—$6.-15.
____ recurvus—$2.-10.
____ regius—$2.-10.
____ retifer—$6.-20.
____ rufimaculosus—$3.-6.
____ rutilus—(b)$2.-5.

____ sanguinolentus—$1.-3.
____ sazanka—$50.-100.*
____ scabriusculus—$1.-3.
____ scalptus—(L)$25.-50.
(b)$5.-10.
____ schepmani—$25.-50.
____ scopulorum—$50.-75.*
____ sculletti—$6.-12.
____ selenae—$25.-50.*
____ sennottorum—$15.-30.
____ sibogae—$10.-20.
____ sieboldii—$1.-5.
____ smirna—$75.-150.
____ spectrum—$4.-15.
____ sponsalis—$.50-1.
____ spurius—$1.-5.
____ stercusmuscarum—$1.-3.
____ stimpsoni—$6.-15.
____ stramineus—$3.-6.
____ striatellus—$3.-6.
____ striatus—$2.-5.
____ stupa—(b)$75.-150.
____ stupella—$100.-200.
____ subulatus—$3.-8.
____ sugimotonis—$20.-50.
____ sulcatus—$1.-5.
(bocki)$5.-15.
____ suratensis—$3.-8.
____ suturatus—$2.-5.
(Hawaiian)$6.-20.

____ taeniatus—$2.-6.
____ taslei—$10.-25.
____ tegulatus—$10.-25.*
____ telatus—$10.-15.
____ tenuistriatus—$1.-3.
____ teramachii—$60.-150.
____ terebra—$1.-5.
(thomasi)$4.-10.
____ terminus—$3.-6.
____ tessulatus—$.50-3.
____ textile—$1.-6.
(lineate)$50.-100.*

____ thalassiarchus—$2.-6.
____ thomae—$2000.*
____ tiaratus—$3.-6.
____ timorensis—(L)$50.-80.
(b)$20.-50.
____ tinianus—(b)$1.-5.
____ tornatus—$2.-4.
____ tribblei—$6.-15.
____ trigonus—$8.-25.
____ tulipa—$2.-4.
____ typhon—$50.-100.

____ urashimanus—$6.-$12.

____ variegatus—$10.-25.
____ varius—$1.-3.
____ ventricosus—$.50-2.
(varieties)$5.-25.
____ venulatus—$15.-25.
____ vexillum—$2.-5.
____ victoriae—$3.-6.
(nodulosus)$8.-15.
____ vicweei—$1500.-2500.*
____ villepinii—$5.-12.
____ vimineus—$2.-7.
____ violaceus—$4.-10.
____ virgatus—$1.-3.
____ virgo—$1.-3.
____ vittatus—$6.-15.
____ vitulinus—$1.-3.
____ voluminalis—$6.-15.

____ wallangra—$10.-20.
____ wittigi—$60.-150.

____ xanthicus—$75.-150.*
____ ximenes—$.50-3.

____ zeylanicus—$2.-6.
____ zonatus—$7.-25.

INDEX I

Partial List of Dealer-names and Their Equivalents

Names commonly used by dealers and collectors will often be found to refer to very different shells than the same names as used in this book. In some cases this is because of a simple misidentification which has become firmly entrenched in the collector literature (such as use of *mucronatus* for *scalptus*) while in other cases it is the result of a name change in this book (such as *clarki* for *armiger*). The following list presents a few of the more familiar dealer-names and equivalents as used here. Remember, though, that the name is not the final proof of a specimen's identity. If you bought a shell sold as *C. lizardensis*, it is probably the one called *C. minnamurra* in this book, as most specimens sold under *lizardensis* have been this species, **but** occasionally *lizardensis* is used for *C. sibogae* or even *C. arcuatus*. Dealer-names are commonly synonyms of the species being sold, **but not always.** Always check your purchase against the descriptions and illustrations. Hopefully in a few years dealer-names will mostly come to agree with usage in this book and the following list will become unnecessary.

Dealer-name (what you thought you bought)	**Name used in this book** (what you got)
adansonii	*ventricosus* variety
adenensis	*inscriptus*
africanus	*aemulus* or *variegatus*
agassizii	*villepinii* or *mindanus*
alabaster	*mucronatus alabaster*
anabathrum	*articulatus*
anaglypticus	*axelrodi*
andamanensis	*subulatus andamanensis*
archithalassus	coronate *admiralis*
aristophanes	*coronatus*
armadillo	*duplicatus*
aurora	*tinianus*
austini	*atractus austini*
bandanus	*marmoreus*
barbara	*monachus*
bayani	often *tribblei*
beddomei	*mindanus* or *jaspideus*
bermudensis	*mindanus*
brasiliensis	*beddomei*
brugieri	*ventricosus* variety
caledonicus	*marmoreus* var. *suffusus*
castaneofasciatus	*amadis*
ceylanensis	*sponsalis*
chenui	*planorbis* var.

circumactus	*connectens*
clarki	*armiger*
clarus	*nielsenae*
classiarius	*fumigatus* or *excavatus*
clenchi	*clerii*
collisus	*subulatus subulatus*
columba	*puncticulatus* var.
convolutus	*wittigi*
conspersus	*spectrum* var. *filamentosus*
crassus	*eburneus* var.
crocatus	*colubrinus*
cuneolus	*balteus*
daphne	*conspersus*
daullei	*pilkeyi* var.
decurtata	*nigropunctatus*
desidiosus	*anthonyi*
dispar	*gradatus*
dominicanus	*aurantius*
elisae	*pennaceus* var. *sindon*
episcopus	*magnificus*
flavescens	*magellanicus*
floridanus	*philippii*
floridulus	*muriculatus* var.
fulmineus	*kermadecensis*
fuscolineatus	*variegatus*
gilvus	*nielsenae*
gloriakiiensis	*urashimanus, recluzianus,* or *tribblei*
gradatulus	*altispiratus*
grayi	*aemulus* or *ermineus*
gubba	*cinereus* var. or *radiatus*
gubernator	*terminus*
inflatus	*tinianus* or *guineensis*
insularis	*aurantius*
japonicus	*eugrammatus*
juliae	*amphiurgus*
keati	*inscriptus*
kenyonae	*kermadecensis*
'kintoki'	*coelinae berdulinus*
lamarckii	*mercator* var.
lamberti	*colubrinus*
lignarius	*furvus*
lizardensis	*minnamurra*
locumtenens	*acuminatus*
longurionis	*aculeiformis* var.
loroisii	*figulinus* var.
macarae	*voluminalis*
magnificus	*pennaceus* var. or *textile* var.
maldivus	*generalis maldivus*
mappa	*aurantius* or *cedonulli*
mcgintyi	*mazei*
mediterraneus	*ventricosus*
mozambicus	*guineensis guineensis*
mucronatus	*scalptus*
mutabilis	*hyaena*
nanus	*sponsalis*
nicobaricus	*araneosus nicobaricus*

nocturnus	*marmoreus* var. *vidua*
nux	*sponsalis nux*
obtusus	*aemulus*
okamotoi	*kermadecensis*
olivaceus	*ermineus* or *taslei*
papilionaceus	*byssinus* or juv. *pulcher*
papillosus	*axelrodi*
patens	*altispiratus*
pauperculus	*gilvus*
perplexus	*puncticulatus perplexus*
piperatus	*biliosus*
praecellens	*eugrammatus*
'pramparti'	*gubernator*
propinquus	*cernicus*
prytanis	*lividus* var.
pseudosulcatus	*mucronatus*
pulchrelineatus	*striatellus*
puncturatus	*axelrodi*
pygmaeus	*puncticulatus*
rainesae	*mazei*
ranunculus	*nigropunctatus* or *ermineus*
recluzianus	often *tribblei*
regularis	*gradatus*
scalaris	*gradatus*
scriptus	*textile* var.
segravei	*clarus*
semisulcatus	*articulatus*
semivelatus	*rattus*
sowerbyi	*praecellens*
sozoni	*delessertii*
sphacelatus	*boeticus*
spirogloxus	*generalis maldivus* juv.
stainforthii	*moluccensis*
stigmaticus	*tegulatus*
submarginatus	*furvus*
sugillatus	*muriculatus*
sumatrensis	*vexillum* var.
surinamensis	*aurantius*
tendineus	*violaceus*
testudinarius	*ermineus*
tigrinus	*canonicus*
torquatus	*teramachii*
traversianus	*lemniscatus*
urashimanus	usually *recluzianus*
verrucosus	*jaspideus*
vidua	*marmoreus* var. *vidua*
wakayamaensis	*eugrammatus*
zebra	*stramineus* or *subulatus*
zebroides	*bulbus*

INDEX II
List of Fossil Names

The following alphabetical listing of specific, subspecific, and varietal names applied to cones is a strict compilation from the catalog of Tomlin (1937) and the recent supplement of Kohn and Riggs (1976). I have made no effort to check original descriptions or to interweave the fossil literature with that on Recent species, and this list is presented only as a convenience to aid in preventing use of preoccupied names. Certainly many of the fossil names are preoccupied by earlier names of Recent species and a possibly equal number refer to sub-Recent occurrences of living species. In my opinion, use of a name based on a fossil should be avoided for living species unless there is no other choice—fossils are simply not comparable with living specimens at the present time.

In the following list all taxa are treated as of the same rank whether described as species, subspecies, or varieties, and parentheses are not used around authors' names even if the taxon was originally described as other than *Conus*. Some of these names are surely just misidentifications. Spellings follow the catalogs mentioned above, with no effort to correct any but obvious errors; dates also follow the catalogs.

anomalosulcata Sacco, 1894
antediluvianoides Sacco, 1894
antediluvianus Deshayes, 1835
antediluvianus Bell, 1918
antidiluvianus Bruguiere, 1792
antidiluvianus Grateloup, 1847
antiquus Lamarck, 1810
apenninensis Orbigny, 1852
apenninicus Bronn, 1838
apiceperlonga Sacco, 1893
apium Woodring, 1928
approximata Deshayes, 1865
aquensis Orbigny, 1852
aquitanicus Mayer, 1858
aquoreus Hoerle, 1976
aratispira Pilsbry, 1905
aratus Gabb, 1873
argillicola Eichwald, 1830
aristos Jung, 1965
armoricus Suter, 1917
arntzenii Martin, 1916
ascalarata Sacco, 1893
ascalaratospira Sacco, 1894
ascalaris Sacco, 1893
asdensis Gregorio, 1885
asparagiformis Sacco, 1893
asperula Sacco, 1893
asphaltodes Beets, 1953
astensinflata Sacco, 1893
astensis Sacco, 1894
astensiscalaris Sacco, 1893
asterope Tomlin, 1937
astriolata Sacco, 1894
asyli Gregorio, 1880
atractoides Tate, 1890
auingeri Gregorio, 1885
aulacophorus Cossmann, 1900
austriaconoe Sacco, 1893
austriacus Hoernes & Auinger, 1879
avaensis Noetling, 1901
avellana Lamarck, 1810
avellana Hoernes, 1851
avellana da Costa, 1866
avus Pilsbry, 1905

baldichierensis Borson, 1830
baldichieri Borson, 1820
baldus Orbigny, 1852
bantamensis Oostingh, 1938
barbadensis Trechmann, 1925
bareti Cossmann, 1896
bartonensis Brown, 1838
basteroti Mayer-Eymar, 1890
bathis Orbigny, 1852
baudoni Cossmann, 1889
beali Carson, 1926
belus Orbigny, 1852
benoisti Mayer-Eymar, 1890
berghausi Michelotti, 1847
berghausi Hoernes, 1851
berghausopsis Gregorio, 1885
berryi Spieker, 1922
berwethi Hoernes & Auinger, 1879
betulinoides Lamarck, 1810
betulinoides Grateloup, 1847
beyrichii Koenen, 1865
bhagothorensis Vredenburg, 1925
biapproximatus Gregorio, 1880
biconolonga Sacco, 1894
bicoronatus Melleville, 1843
bifasciolata Sacco, 1894
bimutata Finlay, 1924
birmanica Vredenburg, 1921
bistriata Deshayes, 1834
bisulcatus Bellardi & Michelotti, 1840
bitorosus Fontannes, 1880
bittneri Hoernes & Auinger, 1879
blagravei Vredenburg, 1925
bloetei Koperberg, 1931
blountensis Mansfield, 1935
bocapanensis Spieker, 1922
bocasensis Olsson, 1922
boggsi Olsson, 1964
bonaczyi Gabb, 1873
bonneti Cossmann, 1900
bonus Nomura, 1935
borsoni Mayer, 1864
boussaci Cossmann, 1913
brachys Pilsbry & Johnson, 1917
bramkampi Hanna & Strong, 1949
bravoi Spieker, 1922
bredai Michelotti, 1847
brevicaudata Sacco, 1894
brevidepressula Sacco, 1894
brevipupoides Sacco, 1893
brevis J. de C. Sowerby, 1840
brevispira Newton, 1891
brevispiratus Noszky, 19--
brezinae Hoernes & Auinger, 1879
briarti Geinitz, 1875
britannus Cossmann, 1896
brocchii Bronn, 1828
brocchii Philippi, 1844
brongniartii Orgibny, 1850
bronnii Michelotti, 1847
broteri da Costa, 1866
bruguierii Nyst, 1843
burdigalensis Mayer, 1858
burckhardti Boese, 1906
busuegoi Shuto, 1969

caboblanquensis Weisbord, 1962
cacellensis da Costa, 1866
cacuminatus Spieker, 1922

gratacapii Pilsbry, 1904
gratteloupi Orbigny, 1852
gravis Peyrot, 1938
grinzingensis Sacco, 1893
grolpus Gregario, 1885
grotriani Koenen, 1865
gugurensis Altena & Beets, 1944
gulemani Eruenal-Erentoez, 1958
guppyi Woodring, 1928
gurabensis Maury, 1917
gymnogyra Moerch, 1874
gypsus Olsson, 1964
gyratus Morton, 1834

haighti Gardner, 1945
hamiltonensis Tate, 1890
hanza Noetling, 1901
hardi Martin, 1879
harrisi Olsson, 1922
haueri Hoernes, 1851
haughtoni King, 1947
hayesi Arnold, 1909
haytensis Sowerby i, 1850
helveticus Mayer, 1877
hemilissa Edwards, 1856
hendersoni Marwick, 1931
herklotsi Martin, 1879
heterospira Tate, 1890
highgatensis Brown, 1838
hochstetteri Martin, 1879
hochstetteri Hoernes & Auinger, 1879
hornesi Doderlein, 1863
hornii Gabb, 1864
hortensis Regny, 1897
hulshofi Martin, 1906
humerosus Pilsbry, 1921
humilispirata Sacco, 1893
hungaricus Hoernes & Auinger, 1879
huttoni Tate, 1890
hypermeces Cossmann, 1900

ickei Martin, 1906
ighinai Michelotti, 1861
iheringi Frenguelli, 1946
illiolus Dall, 1915
imitator Brown & Pilsbry, 1911
improvidus Gregorio, 1890
inaequistriata Deshayes, 1865
incomptus Deshayes, 1865
indefatigabilis Dall & Ochsner, 1928
ineditus Michelotti, 1861
infirmus Gregorio, 1896
inflatulospira Sacco, 1893
inflatulospira Sacco, 1893(2)
inflatus Eruenal-Erentoez, 1958
infragradatus Cossmann, 1889
infulatus Hoerle, 1976
injucundus Hanna, 1923
insculptus Koenen, 1890
intermedius Lamarck, 1810
intermedius Grateloup, 1847
interposita Deshayes, 1865
interstinctus Guppy, 1866
invalidus Link, 1808
irregularilineata Sacco, 1893
isgolpus Gregorio, 1885
isomitratus Dall, 1896
ixion Orbigny, 1852

jacksonensis Meyer, 1885
javanus Martin, 1879
jenkinsi Martin, 1879
jocus Finlay, 1926
johannae Hoernes & Auinger, 1879
juanensis Wiedey, 1928
junghuhni Martin, 1879
jungi Boettger, 1887
junior Grateloup, 1847
juttingae Pannekoek, 1936
juvenoasulcata Sacco, 1893
juvenodepressa Sacco, 1894
juventula Sacco, 1893

kanayai Shuto, 1969
karamanensis Eruenal-Erentoez, 1958
karikalensis Cossmann, 1900
karlschmidti Maury, 1917
karreri Hoernes & Auinger, 1879
kikaiensis Pilsbry, 1904
kitteredgei Maury, 1917
kressenbergensis Schlossen, 1925
kueneni Koperberg, 1931
kueneni Koperberg, 1931(2)
kyudawonensis Vredenburg, 1921

labiata Deshayes, 1834
laevigatus Defrance, 1818
laevigata Melleville, 1843
laevimutinensis Sacco, 1893
laeviponderosus Sacco, 1893
laevis Crosse, 1858
laevispira Sacco, 1893
laevissima Sacco, 1893
lamarckii Edwards, 1849
lapugyensis Hoernes & Auinger, 1879
laqua Mansfield, 1935
larraldei Mayer-Eymar, 1890
larvatus Pilsbry & Johnson, 1917
lavacillensis Gregorio, 1895
lavillei Cossmann, 1913
lebruni Deshayes, 1865
leunieri Cossmann & Pissarro, 1901
levigatus Defrance, 1818
ligans Gregorio, 1880

ligatus Tate, 1890
ligeriana Peyrot, 1938
ligusticoconica Sacco, 1893
ligusticofusulata Sacco, 1893
ligusticomamilla Sacco, 1893
ligusticovulata Sacco, 1893
limonensis Olsson, 1922
lineatus Solander, 1766
lineatus Borson, 1820
lineatus Cristofori & Jan, 1832
lineata Grateloup, 1847
lineoclavata Sacco, 1893
lineofasciata Sacco, 1893
lineolata Lamarck, 1804
lineolatus Cocconi, 1873
lissitzensis Sacco, 1893
lius Woodring, 1928
loczyi Eruenal-Erentoez, 1958
longanfractus Sacco, 1893
longispirata Sacoo, 1894
longitudinalis Pilsbry & Johnson, 1917
longoastensis Sacco, 1894
longobiconica Sacco, 1893
longogracilis Sacco, 1893
longogracilis Sacco, 1893(2)
longopyrulata Sacco, 1893
longoturbinata Sacco, 1893
longovulata Sacco, 1893
longovuloides Sacco, 1893
loochooensis MacNeil, 1960
loomisi Dall   Ochsner, 1928
losariensis Martin, 1895
lyratus Marshall, 1918

maaboensis Icke   Martin, 1907
macrocentrus Bayan, 1873
maculosus Grateloup, 1835
maduranus Tomlin, 1937
madurensis Martin, 1906
maga Vokes, 1938
magnoconica Sacco, 1893
magnolapugyensis Sacco, 1893
magnomamillata Sacco, 1893
magnovata Sacco, 1893
mamillaris Green, 1830
mamillatocrasso Sacco, 1893
mamillatoides Sacco, 1893
mamillatospira Sacco, 1893
mamillospira Sacco, 1893
mantovani Seguenza, 1880
manzanillaensis Mansfield, 1925
marginata Lamarck, 1804
marginatus J. de C. Sowerby, 1840
marginatus Sowerby i, 1850
marginatus Cossmann, 1913
mariae Hoernes & Auinger, 1879
marii Sacco, 1893
marksi Olsson, 1964
marshalli Finlay, 1926
marticensis Matheron, 1843
martini Wanner & Hahn, 1935
martini Shuto, 1969
martinicensis Cossmann, 1913
marylandicus Green, 1830
mauryi Finlay, 1926
maussieri Cossmann, 1923
maynardensis McWilliams, 1971
medialira Vokes, 1938
mediosulcata Sacco, 1893
medioventrosa Sacco, 1893
melficus Desio, 1929
melitensis Gregorio, 1882
melitosiculus Gregorio, 1885
menengtenganus Martin, 1895
menkrawitensis Beets, 1941
mercati Brocchi, 1814
mercati Grateloup, 1835
militaris J. de C. Sowerby, 1840
minbuensis Vredenburg, 1921
minimespirata Sacco, 1893
minimum Conti, 1864
mioanomalospira Sacco, 1893
mioantiqua Sacco, 1893
mioatra Sacco, 1893
mioblita Sacco, 1894
miocenica Sacco, 1894
miofusuloides Sacco, 1893
miofusuluvoides Sacco, 1893
miopermamillata Sacco, 1893
mioperovata Sacco, 1893
mioplenispira Sacco, 1893
miopraecedens Sacco, 1893
miosubagranosa Sacco, 1894
miosubmamillata Sacco, 1893
miosubscalarata Sacco, 1893
miosubtypica Sacco, 1893
miosubuloides Sacco, 1893
mioventrosa Sacco, 1893
miovulea Sacco, 1893
mitus Gregorio, 1885
mojsvari Hoernes & Auinger, 1879
molis Brown & Pilsbry, 1911
mompichiensis Olsson, 1964
monairyensis Wade, 1917
moniwaensis Nomura, 1940
monstruosa Sacco, 1893
montisclavus Sacco, 1893
moravicoides Sacco, 1894
moravicus Hoernes & Auinger, 1879
mucronata Eruenal-Erentoez, 1958
mucronatina Sacco, 1893
mucronatolaevis Sacco, 1893
mucronatula Sacco, 1893
multilineatus Pecchioli, 1864

vasseuri Mayer-Eymar, 1890
vaughanensis Mansfield, 1935
vaughani Dall, 1916
veatchi Olsson, 1922
velatus J. de C. Sowerby, 1841
ventricosus Schlotheim, 1820
ventricosus Bronn, 1831
vergrandis Hoerle, 1976
veridicus Gregorio, 1880
verneuilli Vilanova, 1859
vetustus Quenstedt, 1836
vicinus Schlotheim, 1820
vindobonensis Hoernes, 1851
virginalis Brocchi, 1814
vitius Hoerle, 1976
vlerki Cox, 1948
voeslauensis Hoernes & Auinger, 1879
vouensis Ladd, 1945
vredenburgi Nath & Chiplonker, 1938

waccamawensis Smith, 1930
wakullensis Mansfield, 1937
walli Mansfield, 1925
waltonensis Aldrich, 1903
warreni Hendon, in Turner, 1938
washingtonensis Winkle, 1918
weaveri Dickerson, 1915
wheatleyi Michelotti, 1847
wiedenmayeri Jung, 1965
williamgabbi Maury, 1917

xenicus Pilsbry & Johnson, 1917

yabei Nomura, 1935
yanuyanuensis Ladd, 1945
yaquensis Gabb, 1873
ynezanus Loel & Corey, 1932
yoshidensis Kanno, 1958
yuleianus Noetling, 1901

zalleigrus Gregorio, 1885
zebrinus Foresti, 1887
zonarius Grateloup, 1835

INDEX III

Names Applied to Living Cones

The following is a list of validly proposed names applied to living cones. An attempt has been made to bring the list up-to-date through 1976, but it is to be expected that some names have been omitted through either accident or ignorance. An attempt has also been made to evaluate the status of each name in a conservative taxonomic system, although I am sure many will disagree with my conclusions. References and dates are given in the appropriate synonymy within the text.

Although I have tried to make all spellings as correct as possible, using the spelling given in the original description except for reducing capital letters and dropping accents, the usual typographical errors should be watched for. Obvious misspellings (except when very commonly used in the literature) and invalidly described names are generally omitted, as are non-binomial names. Page numbers in bold refer to color photos.

CONVERSION

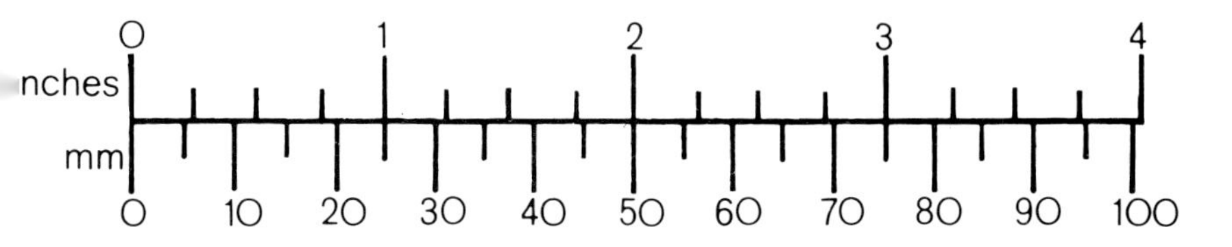

0
1
2
3
4
nches
mm
0
10
20
30
40
50
60
70
80
90
100

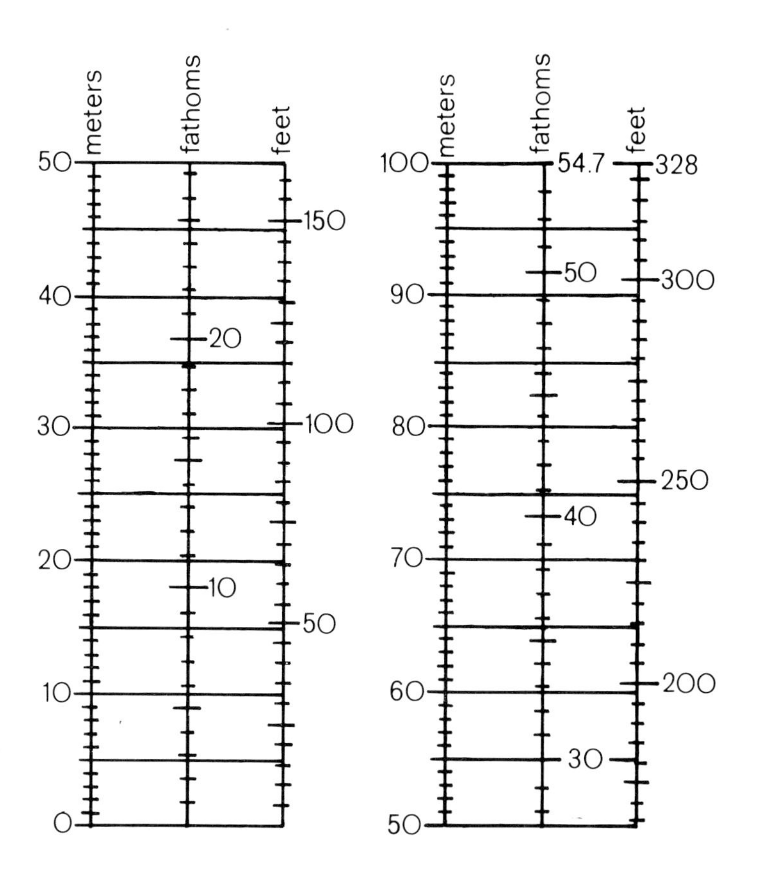

meters
fathoms
feet
50
40
30
20
10
0
150
20
100
10
50
meters
fathoms
feet
100
54.7
328
50
300
90
80
250
40
70
60
200
30
50

About the Author

Born in 1946 in Bunkie, Louisiana, Jerry G. (for Glen) Walls has long been interested in invertebrates and lower vertebrates. The interest has usually been taxonomic in nature, which has led to a wide acquaintance with the classification of reptiles and amphibians, crustaceans, beetles and dragonflies, fishes, and some families of gastropod molluscs. After obtaining a Master of Science degree in zoology from McNeese State University in Lake Charles, Louisiana, he worked briefly as a marine biologist before becoming an editor for T.F.H. Publications.

Cone Shells is Jerry's fourth book on marine invertebrates and fishes, and he is also the author of many technical and popular articles on diverse animal groups from crawfishes (including the descriptions of several new species and subspecies) to reptiles. For the last four years shells have occupied much of his time, with research for this cone book taking almost two years. A book on conchs and tibias is in the works, along with continuing research on herps, insects, and fishes.

Jerry now lives in central New Jersey with his wife, Maleta, four cats (one of which served as the namesake for *Conus tribblei*), several thousand cones, cowries, olives, and conchs, and over 300 editions and printings of the Sherlock Holmes and Nero Wolfe mystery stories.

Checklist and Pricing Guide to the Cones

The following list and pricing guide is taken from the one beginning on page 975 of *Cone Shells: A Synopsis of the Living Conidae,* by Jerry G. Walls (T.F.H. Publications, Inc., Neptune, N.J.). It is reproduced here to simplify trading and selling among cone enthusiasts by standardizing the names and providing reasonable price ranges for all valid species and many of their varieties. For full details and illustrations of all species, a copy of *Cone Shells* should be at hand. Feel free to duplicate this checklist so all your trading friends will know just what you have or need.

____ abbas—$8.-20.
____ abbreviatus—$.50-3.
____ aculeiformis—$5.-15.
____ acuminatus—$2.-15.
____ acutangulus—$3.-15.
____ adamsoni—(L)$200.-600.
(b)$50.-150.
____ advertex—$5.-8.
____ aemulus—$2.-5.
____ algoensis—(b)$3.-10.
(L)$15.-40.
____ altispiratus—$60.-150.
____ amadis—$2.-7.
(dark)$5.-15.
____ ambiguus—$8.-25.
____ ammiralis—$7.-15.
(granulose)$50.
(coronate)$150.-300.
____ amphiurgus—$10.-35.
____ anemone—$.50-2.
(varieties)$2.-15.
____ angasi—$20-40.
____ anthonyi—$15.-25.
____ aplustre—$1.-3.
____ araneosus—$3.-7.
____ archon—$10.-20.
____ arcuatus—$1.-3.
____ arenatus—$.50-1.
____ argillaceus—$20.-40.
____ armiger—$10.-25.
(bajanensis)$50.-100.*
____ articulatus—(L)$15.-25.
(b)$7.-12.
____ artoptus—$8.-25.
____ atractus—$50.-80.*
(austini)$4.-10.
____ attenuatus—$50.-100.
____ augur—$4.-8.
____ aulicus—$5.-15.
____ aurantius—$15.-35.
____ auratus—$200.-300.*
____ aureus—$15.-60.
____ auricomus—$10.-50.
____ aurisiacus—(L)$75.-150.
(b)$20.-50.
____ australis—$1.-5.
____ axelrodi—$3.-10.

____ bairstowi—(b)$6.-10.
____ balteatus—$3.-5.
____ balteus—$10.-25.
____ barthelemyi—$70.-120.
____ bartschi—$15.-30.
____ bayani—$3.-8.
____ beddomei—$10.-15.
____ bengalensis—$160.-250.
____ betulinus—$2.-6.
____ biliosus—$2.-5.
____ boeticus—$1.-5.
____ boivini—NP
____ boschi—$6.-15.
____ brunneus—$1.-3.
____ bruuni—NP
____ bulbus—$3.-15.
____ bullatus—$40.-90.
____ byssinus—$3.-10.

____ cabritii—$10.-20.
____ caillaudii—(b)$100.-200.
____ caledonicus—NP
____ californicus—$1.-3.
____ cancellatus—$1.-2.
____ canonicus—$1.-5.
____ capitanellus—$15.-40.
____ capitaneus—$1.-2.
____ caracteristicus—$2.-5.
____ cardinalis—(b)$10.-20.
____ catus—$1.-2.
____ cedonulli—$50.-100.
____ centurio—$20.-75.
____ cernicus—$2.-5.
____ cervus—(b)$1500.*
____ chaldaeus—$.50-1.
____ chiangi—$10.-25.
____ cinereus—$1.-3.
(gubba)$10.-20.*

____ cingulatus—(L)$20.-50.
(b)$5.-10.
____ circumcisus—$30.-80.
____ clarus—$10.-20.
____ clerii—$8.-25.
(clenchi)$40.-75.*
____ cloveri-$15.-25.
____ cocceus—$5.-15.
____ coccineus—$5.-15.
____ coelinae—$10.-15.
(spiceri)$80.-120.
(berdulinus)$150.-300.*
____ colubrinus—$60.-150.
____ comatosa—$25.-75.
____ concolor—$15.-50.
____ connectens—$30.-60.
____ consors—$3.-8.
____ conspersus—$8.-15.
____ coronatus—$.50-2.
____ cumingii—$15.-30.
____ cuvieri—$30.-50.
____ cyanostoma—$4.-8.
____ cylindraceus—$5.-15.

____ dalli—$5.-15.
____ daucus—$5.-15.
____ delessertii—$5.-30.
____ diadema—$2.-5.
____ distans—$1.-5.
____ dorreensis—$1.-3.
____ duplicatus—$50.-100.
____ dusaveli—$900.-1500.*

____ ebraeus—$.50-1.
____ eburneus—$.50-2.
(crassus)$20.-60.
____ elegans—$10.-30.
____ emaciatus—$1.-3.
____ encaustus—$3.-10.
____ ermineus—$1.-5.
____ erythraeensis—$3.-6.
____ eucoronatus—$50.-100.
____ eugrammatus—$8.-20.
(Hawaiian)$75.-200.*
____ excavatus—$10.-30.
____ excelsus—$1500.-2500.
____ eximius—$4.-10.

____ fergusoni—$2.-5.
____ figulinus—$1.-5.
____ flavidus—$.50-2.
____ floccatus—$30.-60.
____ frigidus—$2.-5.
____ fulgetrum—$2.-5.
____ fulmen—$8.-15.
____ fumigatus—$3.-5.
____ furvus—$1.-5.

____ generalis—$1.-5.
____ genuanus—$8.-15.
____ geographus—$1.-5.
____ gilvus—$5.-15.
____ gladiator—$1.-3.
____ glans—$1.-3.
____ glaucus—$5.-10.
____ gloriamaris—$300.-800.*
____ gradatus—$1.-5.
____ grangeri—$10.-20.
____ granulatus—(L)$75.-100.
(b)$5.-15.
____ gubernator—$50.-100*
____ guineensis—(b)$5.-10.

____ havanensis—NP
____ hieroglyphus—$5.-15.
____ hirasei—$50.-100.
____ howelli—NP
____ hyaena—$3.-10.
____ hypochlorus—$10.-20.*

____ ichinoseana—$25.-50.
____ illawarra—$10.-20.
____ imperialis—$1.-5.
(varieties)$5.-10.
____ infrenatus—(b)$3.-5.

____ inscriptus—$3.-7.
____ insculptus—$7.-15.
____ iodostoma—$7.-15.
____ ione—$10.-15.*

____ janus-$10.20.
____ jaspideus—$1.-3.
____ jickelii—$30.-50.
____ julii—$100.-200.

____ kashiwajimensis—$40.-60.
____ kermadecensis—$5.-20.
____ kerstitchi—$50.-75.
____ kimioi—$100.-200.
____ kinoshitai—$20.-50.
____ klemae—$15.-30.

____ legatus—$10.-15.
____ lemiscatus—$50.-75.*
____ lentiginosus—$5.-15.*
____ leopardus—$2.-5.
____ lienardi—$10.-20.
____ litoglyphus—$1.-5.
____ litteratus—$2.-5.
____ lividus—$.50-2.
____ lohri—NP
____ lorenzianus—$20.-50.
____ lucidus—$3.-10.
____ luteus—$10.-30.
____ lynceus—$10.-20.

____ magellanicus—$5.-15.
____ magnificus—$1.-5.
____ magus—$1.-10.
____ malacanus—$5.-15.
____ marielae—NP
____ marmoreus—$.50-3.
(varieties)$10.-25.
(nocturnus)$75.-150.
____ mazei—$30.-50.
(juvenile)$10.-15.
____ memiae—$30.-60.
____ mercator—$5.-15.
(varieties)$15.-25.
____ miles—$1.-3.
____ miliaris—$.50-2.
____ milneedwardsi—$500.-1000.*
____ mindanus—$15.-30.
(Caribbean)$25.-50.*
____ minnamurra—$10.-20.
____ mitratus—$5.-15.
____ moluccensis—$100.-200.
____ monachus—$1.-5.
(varieties)$10.-20.
____ monile—$3.-5.
____ moreleti—$2.-5.
____ mucronatus—$1.-5.
____ muriculatus—$1.-3.
(granulose)$25.-50.
(high-spired)$30.-50.
____ mus—$1.-3.
____ musicus—$.50-2.
____ mustelinus—$1.-5.

____ namocanus—$2.-5.
____ natalis—(b)$3.-5.
(gilchristi)(b)$5.-10.
____ neptunus—$100.-200.
____ nielsenae—$10.-20.
____ nigropunctatus—$1.-3.
____ nimbosus—$15.-25.*
(dark)$30.-50.
____ nobilis—$10.-20.
(victor)$80.-150.
(marchionatus)$6.-15.
____ nussatella—$2.-10.

____ obscurus—$2.-5.
____ ochroleucus—$2.-3.
____ orbignyi—$1.-5.
____ orion—$10.-25.
____ otohimeae—$15.-40.

____ papilliferus—$1.-5.
____ parius—$3.-7.

____ parvulus—$1.-5.
____ patae—(L)$60.-100.*
(b)$20.-30.
____ patricius—$2.-10.
____ pennaceus—$1.-10.
(elisae)$20.-40.*
____ pergrandis—$200.-500.
____ peronianus—$20.-50.
____ pertusus—$10.-50.
(Hawaiian)$3.-5.
____ philippii—$1.-5.
____ pictus—(b)$5.-15.
____ pilkeyi—$5.-15.
(Philippines)$3.-5.
____ planorbis—$.50-3.
(chenui)$25.-50.
____ pohlianus—$10.-25.
____ polygrammus—$30.-50.
____ poormani-$5.-10.
____ praecellens—$2.-5.
____ princeps—$2.-6.
____ proximus—$40.-60.
____ pulcher—$3.-10.
(over 120mm)$15.-30.
____ pulicarius—$.50-2.
(vauteri)$3.-10.
____ puncticulatus—$1.-3.
(white)$5.-10.
____ purpurescens—$1.-3.

____ quercinus—$1.-5.
(white)$20.-30.

____ radiatus—$1.-3.
____ raoulensis—NP
____ rattus—$.50-2.
(taitensis)$3.-7.
____ recluzianus—$6.-15.
____ recurvus—$2.-10.
____ regius—$2.-10.
____ retifer—$6.-20.
____ rufimaculosus—$3.-6.
____ rutilus—(b)$2.-5.

____ sanguinolentus—$1.-3.
____ sazanka—$50.-100.*
____ scabriusculus—$1.-3.
____ scalptus—(L)$25.-50.
(b)$5.-10.
____ schepmani—$25.-50.
____ scopulorum—$50.-75.*
____ sculletti—$6.-12.
____ selenae—$25.-50.*
____ sennottorum—$15.-30.
____ sibogae—$10.-20.
____ sieboldii—$1.-5.
____ smirna—$75.-150.
____ spectrum—$4.-15.
____ sponsalis—$.50-1.
____ spurius—$1.-5.
____ stercusmuscarum—$1.-3.
____ stimpsoni—$6.-15.
____ stramineus—$3.-6.
____ striatellus—$3.-6.
____ striatus—$2.-5.
____ stupa—(b)$75.-150.
____ stupella—$100.-200.
____ subulatus—$3.-8.
____ sugimotonis—$20.-50.
____ sulcatus—$1.-5.
(bocki)$5.-15.
____ suratensis—$3.-8.
____ suturatus—$2.-5.
(Hawaiian)$6.-20.

____ taeniatus—$2.-6.
____ taslei—$10.-25.
____ tegulatus—$10.-25.*
____ telatus—$10.-15.
____ tenuistriatus—$1.-3.
____ teramachii—$60.-150.
____ terebra—$1.-5.
(thomasi)$4.-10.
____ terminus—$3.-6.
____ tessulatus—$.50-3.
____ textile—$1.-6.
(lineate)$50.-100.*

____ thalassiarchus—$2.-6.
____ thomae—$2000.*
____ tiaratus—$3.-6.
____ timorensis—(L)$50.-80.
(b)$20.-50.
____ tinianus—(b)$1.-5.
____ tornatus—$2.-4.
____ tribblei—$6.-15.
____ trigonus—$8.-25.
____ tulipa—$2.-4.
____ typhon—$50.-100.

____ urashimanus—$6.-$12.

____ variegatus—$10.-25.
____ varius—$1.-3.
____ ventricosus—$.50-2.
(varieties)$5.-25.
____ venulatus—$15.-25.
____ vexillum—$2.-5.
____ victoriae—$3.-6.
(nodulosus)$8.-15.
____ vicweei—$1500.-2500.*
____ villepinii—$5.-12.
____ vimineus—$2.-7.
____ violaceus—$4.-10.
____ virgatus—$1.-3.
____ virgo—$1.-3.
____ vittatus—$6.-15.
____ vitulinus—$1.-3.
____ voluminalis—$6.-15.

____ wallangra—$10.-20.
____ wittigi—$60.-150.

____ xanthicus—$75.-150.*
____ ximenes—$.50-3.

____ zeylanicus—$2.-6.
____ zonatus—$7.-25.